Advances in Experimental Medicine and Biology

Volume 1191

More information about this series at http://www.springer.com/series/5584

Yong-Ku Kim
Editor

Anxiety Disorders

Rethinking and Understanding Recent Discoveries

Springer

Editor
Yong-Ku Kim
College of Medicine
Korea University
Gyeonggido, South Korea

ISSN 0065-2598 ISSN 2214-8019 (electronic)
Advances in Experimental Medicine and Biology
ISBN 978-981-32-9704-3 ISBN 978-981-32-9705-0 (eBook)
https://doi.org/10.1007/978-981-32-9705-0

This Springer imprint is published by the registered company Springer Nature Singapore Pte Ltd.
The registered company address is: 152 Beach Road, #21-01/04 Gateway East, Singapore 189721, Singapore

Contents

Part I Brain Network Concepts

1 **Task MRI-Based Functional Brain Network of Anxiety** 3
Chien-Han Lai

2 **Phenotype Network and Brain Structural Covariance Network of Anxiety** 21
Je-Yeon Yun and Yong-Ku Kim

3 **Linear and Nonlinear EEG-Based Functional Networks in Anxiety Disorders** 35
Poppy L. A. Schoenberg

4 **White Matter-Based Structural Brain Network of Anxiety** 61
Kang Soo Lee and Sang Hyuk Lee

5 **Anxiety Disorders and the Brain's Resting State Networks: From Altered Spatiotemporal Synchronization to Psychopathological Symptoms** 71
Georg Northoff

Part II Neurobiological Aspects

6 **Gene-Environment Interactions and Role of Epigenetics in Anxiety Disorders** 93
Eugene Lin and Shih-Jen Tsai

7 **The Role of the Oxytocin System in Anxiety Disorders.** 103
Seoyoung Yoon and Yong-Ku Kim

8 **Translational Studies in the Complex Role of Neurotransmitter Systems in Anxiety and Anxiety Disorders** 121
Jocelien D. A. Olivier and Berend Olivier

9 **The Role of Early Life Stress in HPA Axis and Anxiety** 141
Mario F. Juruena, Filip Eror, Anthony J. Cleare, and Allan H. Young

10 Immune-Kynurenine Pathways and the Gut Microbiota-Brain Axis in Anxiety Disorders 155
Alper Evrensel, Barış Önen Ünsalver, and Mehmet Emin Ceylan

11 Experimental Anxiety Model for Anxiety Disorders: Relevance to Drug Discovery 169
Michel Bourin

Part III Diagnostic and Clinical Issues of Anxiety Disorders

12 Anxiety Disorders in the DSM-5: Changes, Controversies, and Future Directions 187
Seon-Cheol Park and Yong-Ku Kim

13 Biological and Clinical Markers to Differentiate the Type of Anxiety Disorders 197
Fiammetta Cosci and Giovanni Mansueto

14 Comorbid Anxiety and Depression: Clinical and Conceptual Consideration and Transdiagnostic Treatment 219
Kwan Woo Choi, Yong-Ku Kim, and Hong Jin Jeon

15 Anxiety Disorders and Medical Comorbidity: Treatment Implications 237
Alicia E. Meuret, Natalie Tunnell, and Andres Roque

Part IV Therapeutic Issues

16 Biofeedback and Neurofeedback for Anxiety Disorders: A Quantitative and Qualitative Systematic Review 265
David F. Tolin, Carolyn D. Davies, Danielle M. Moskow, and Stefan G. Hofmann

17 Cognitive Behavioral Therapy, Mindfulness-Based Cognitive Therapy and Acceptance Commitment Therapy for Anxiety Disorders: Integrating Traditional with Digital Treatment Approaches 291
Jennifer Apolinário-Hagen, Marie Drüge, and Lara Fritsche

18 Neurostimulation in Anxiety Disorders, Post-traumatic Stress Disorder, and Obsessive-Compulsive Disorder 331
Rafael Christophe Freire, Casimiro Cabrera-Abreu, and Roumen Milev

19 Current and Novel Psychopharmacological Drugs for Anxiety Disorders 347
Borwin Bandelow

20 Role of Benzodiazepines in Anxiety Disorders 367
Richard Balon and Vladan Starcevic

21 Virtual Reality for Anxiety Disorders: Rethinking a Field in Expansion 389
Javier Fernández-Álvarez, Daniele Di Lernia, and Giuseppe Riva

22 Current Research on Complementary and Alternative Medicine (CAM) in the Treatment of Anxiety Disorders: An Evidence-Based Review 415
Vladimir Trkulja and Hrvoje Barić

23 Contemporary Psychodynamic Approaches to Treating Anxiety: Theory, Research, and Practice 451
Seth R. Pitman and Daniel P. C. Knauss

24 Well-Being Therapy in Anxiety Disorders 465
Fiammetta Cosci

Part V Anxiety and Precision Psychiatry

25 Personalized Clinical Approaches to Anxiety Disorders 489
Giampaolo Perna, Alessandra Alciati, Erika Sangiorgio, Daniela Caldirola, and Charles B. Nemeroff

26 The Role of Hormonal and Reproductive Status in the Treatment of Anxiety Disorders in Women 523
Samantha Tang and Bronwyn Margaret Graham

27 Risk Factors and Prevention Strategies for Anxiety Disorders in Childhood and Adolescence 543
Maria Demma Cabral and Dilip R. Patel

28 Anxiety Disorders in the Elderly 561
Carmen Andreescu and Soyoung Lee

Part I

Brain Network Concepts

Task MRI-Based Functional Brain Network of Anxiety

1

Chien-Han Lai

Anxiety, Fear, and Task MRI

The anxiety is tended to be associated with fear. The origin of anxiety and fear might be the functional and structural alterations in the amygdala of anxious subjects, which would be more specific to the rating and response of anxiety [1]. A heightened degree of amygdala-anterior cingulate cortex (ACC) connectivity would predispose the anxious subjects to focus on the attention related to the environmental threat [2]. The specific risk genotype of anxiety would be also associated with increased reactivity of the amygdala and hippocampus to threat stimuli [3]. A study of clinically anxious individuals showed that the frontal region would evaluate the meaning of stimuli and exhibit the inhibitory action toward the heightened responses of the amygdala, which suggested a significant role of prefrontal-limbic control circuit in the pathophysiology of anxiety [4]. The explicit threat memory and threat appraisal condition would also provoke alterations in the amygdala-frontal circuit [5]. The cannabinoid administration-related anxiety in healthy subjects during the fear-processing task would also be associated with the cannabinoid receptors of the amygdala [6]. The threat-related anxiety due to fear was related to the inflammatory biomarker of human body [7]. The neuroticism personality, which would be predisposed to anxiety, has been mentioned to be altered in the anterior cingulate cortex while processing the fearful facial expressions in the lower serotonin status [8]. The pathological fear network activations might be related to the attention arousal while feeling fearful subjectively [9]. The fear model has been mentioned in certain kinds of anxiety disorders, such as social anxiety disorder (SAD) [10–12], panic disorder (PD) [13–15], specific phobia [16–18], and generalized anxiety disorder (GAD) [9,

C.-H. Lai (✉)
Psychiatry & Neuroscience Clinic, Taoyuan, Taiwan

Institute of Biophotonics, National Yang-Ming University, Taipei, Taiwan

Department of Psychiatry, Yeezen General Hospital, Taoyuan, Taiwan

Y.-K. Kim (ed.), *Anxiety Disorders*, Advances in Experimental Medicine and Biology 1191, https://doi.org/10.1007/978-981-32-9705-0_1

19, 20]. For the concept of fear-related network in the brain, the typical example would be the hypothesis of "fear network" [14], which included many regions of frontal and limbic areas. The frontal part of this network included the medial frontal gyrus and anterior cingulate. The limbic part of fear network included the amygdala, thalamus, and hippocampus. According to the update study results, the potential extension of original fear network model to other regions, such as the limbic and cortical regions, has been proposed in the modified version of fear network model [21]. The modified version of fear network model also insisted that amygdala hyper-responses would be source of anxiety symptoms and would be aberrantly linked with frontal and cortical regions in the impairments of cognitive functions. The modified fear network was derived from imaging modalities, such as the magnetic resonance imaging (MRI) and positron emission tomography. In spite of the current study, mentioning about fear network would be mostly related to PD; however, other types of anxiety disorders were also significantly associated with alterations in the fear-related areas, such as limbic and frontal regions [9–20]. Therefore the fronto-limbic network seemed to play the major role in the brain pathophysiology of anxiety disorders.

To understand the functional and structural perspectives of the brain, the neuroimaging would be a great tool for researchers. Among the neuroimaging modalities, the characteristics of relatively sufficient spatial and temporal resolutions of MRI would contribute to the substantial and significant roles in the translational medicine, which can help the researchers explore the puzzle of mechanisms in the brain pathophysiology of anxiety disorders. The MRI also includes two major parts, such as functional and structural MRI, to survey the brain network. In the perspective of functional MRI, the task MRI and resting-state MRI would be the state-of-the-art modalities to realize the process of explicit mode and default mode in cognitive function of neurosciences. The task MRI can also be applied as the tool to discover the putative biomarkers of neurological and psychiatric illnesses, which can be utilized as the target option for the medication treatment, psychotherapy, and the further innovative treatment strategies. In this chapter, the content will be focused on the perspective of task MRI-based functional network in DSM-5 anxiety disorders, such as SAD, PD, specific phobia, GAD, separation anxiety disorder (SeAD), and selective mutism (SM). In this chapter, the phobia section will not include SAD. The common and distinct pathways of task MRI-based functional network between these anxiety disorders would also be discussed in the following sections.

Task MRI-Based Functional Network in SAD

The SAD is characterized by aberrant socio-emotional processing [22], which might be linked with alterations in the frontal and limbic systems. The brain areas responsible for complex social cognitions include the prefrontal regions, such as the medial and superior frontal cortex. The face encoding during initial phase of memory formation might also play a role in the symptoms of social anxiety [23]. The task MRI study of scrutiny perception showed significant alterations in the thalamocortical and fronto-striatal circuits. The increased functional connectivity between the ACC and limbic-related areas was observed in SAD during the heightened scrutiny perception due to the possibility of enhanced inhibitory ability for behavioral control [24]. The fearful face vs happy face task in SAD demonstrated insula hyperactivity and reduced functional connectivity between the insula and ACC for the cognitive control and emotional regulation [25], which also confirmed the important role of the ACC. In the study of emotional face task, greater responses of the ACC would be needed to shift the attention from social signals in controls when compared to SAD. In addition, a negative relationship between ACC and insula activities occurred. Therefore the ACC might play a role in regulating the attention for social signals in SAD [26]. The GAD and SAD in the explicit emotional regulation task showed reduced capacity for emotion regulation brain regions, such as the ACC, amygdala, and medial prefrontal cortex, which implied about the crucial role of fronto-limbic network in the task MRI-based pathophysiology of SAD [27]. The limbic and paralimbic regions might be linked with the attentional bias toward the internal threat and external threat in SAD. An internal threat task might activate the responses of the ACC, insula, and orbitofrontal cortex. On the contrary, the external threat task might be associated with hyperactivity in the posterior cingulate cortex and middle temporal gyrus [28]. This study suggested that the different threat perceptions would activate different parts of brain regions and the ACC might be more specific to internal threat perception.

A study of Internet-based cognitive behavioral therapy in SAD showed that the therapy would attenuate the amygdala responses and increase frontal activities in the affective face processing task [29], which also corresponded to the concept of fronto-limbic network in the task MRI-based brain network for SAD. Another study of unattended neutral face task in SAD also demonstrated a possible differentiation from healthy controls in the functional connectivity between the temporal pole and hippocampus [30], which suggested an intra-limbic network alteration in SAD. The whole brain analysis of amygdala-based functional connectivity in the emotion identification task showed that the network between the amygdala and frontal areas, such as the dorsomedial prefrontal cortex/ACC, would have greater circuit coupling during fearful faces sessions of the task. The maladaptive responses and chronic engagements of this fronto-limbic network would predispose the SAD patients to have anxiety symptoms [31]. The SAD patients also had dysfunctional amygdala regulation during viewing the emotional faces, such as the harsh face [32]. The working memory task containing the face of self-referential negative comments would stimulate the responses in the frontal lobe, and the task-related regions would have significant associations with the amygdala during resting-state analysis [33]. The GAD and SAD patients both had significant reduced activations in the dorsolateral prefrontal cortex during working memory task, which suggested that clinical anxiety might be associated with cognitive deficits, rather than just a consequence of threat [34]. The task MRI and resting-state MRI also showed a modest distinction in the brain network representations for SAD. The fearful face task would reduce a functional connectivity between the amygdala, dorsolateral prefrontal cortex, and rostral ACC. The resting-state MRI showed a reduced connectivity between the amygdala and rostral ACC. The amygdala-dorsolateral prefrontal cortex coupling might be a phasic abnormality, and the amygdala-rostral ACC might represent both phasic and tonic abnormalities [12]. A meta-analysis of gray matter study results in SAD, GAD, PD, and specific phobia also confirmed the significant structural pathophysiology of frontal areas, such as the ACC and medial prefrontal cortex, in the anxiety symptoms [35]. In spite of the structural research of the meta-analysis, the results can support the crucial role of frontal lobe in the fronto-limbic network alterations in task MRI studies of DSM-5 anxiety disorders. The social anxiety symptoms can be relieved by oxytocin using the mechanism to dampen the amygdala over-response and modulate the functional connectivity between the amygdala and ACC during the processing of fearful faces [36]. A similar fronto-limbic network alteration in SAD was also found in the amygdala-orbitofrontal cortex/dorsolateral prefrontal cortex during the facial emotion identification task, which can confirm the importance of fronto-limbic network in the task MRI-based network for SAD [37]. The network-based statistics for resting-state MRI based on the seeds of previous task MRI study results also confirmed the importance of fronto-limbic circuit (fear circuit) in SAD [38]. The GAD and SAD patients also had the common alterations of less engagement of the ACC and less functional connectivity between the amygdala and ventrolateral prefrontal cortex during the reappraisal-based emotion regulation task [39], which implied about the significance of fronto-limbic network in anxiety disorders. Another meta-analysis of task neuroimaging studies in SAD confirmed the hyper-activation of fear network, such as the amygdala, insula, ACC, and prefrontal cortex. However, the study revealed additional regions beyond the fronto-limbic circuit, such as the hyper-activation in medial parietal and occipital regions and reduced functional connectivity of parieto-fronto-limbic regions [40]. The role of insula might be related to the dense connections with fronto-limbic network for the processing and integration of sensory inputs from the somatosensory cortex and visual cortex [41]. In addition to the insula, another region regulating the complex sensory inputs to integrate frontal and limbic systems, the pulvinar nucleus, had heightened influences on the high-order visual areas and frontal lobe in the effective connectivity analysis under the emotional face processing task [22].

In summary, the core components of task MRI-based network in SAD should include the fronto-limbic network. The frontal region, such as the ACC, might be responsible for emotional face, scrutiny perception, reappraisal-based emotion regulation, and internal threat perception tasks. Another frontal region, the dorsolateral prefrontal cortex, might be crucial for working memory task. The limbic region, such as the amygdala, might be responsible for emotional face, scrutiny perception, reappraisal-based emotion regulation, and unattended neutral face tasks. The fronto-limbic network should be the crucial regions for task MRI-related pathophysiology. However, the significance of additional components, such as insula, pulvinar, parietal, and occipital regions, should not be ignored and would be discussed in the eighth section of this chapter.

Task MRI-Based Functional Network in PD

The anticipatory anxiety about hyperventilation and potentially threatening bodily symptoms would be the core symptoms of PD. The anticipation of hyperventilation in guided hyperventilation task would activate the ACC, insula, and other frontal regions. The activation of the ACC was significantly correlated with reported fear of body somatic symptoms, which can support the role of the ACC in the anxious apprehension and pathophysiology of PD [42]. The amygdala heightened response has been discovered in PD patients when they were processing fearful faces, which might suggest the importance of the amygdala in the maladaptive responses to threatening visual stimuli in PD [43]. The study of masked affective face processing task in PD and phobia also found the fronto-limbic network alterations with increased activity in the amygdala and decreased activity in the ventromedial prefrontal cortex [44]. The fronto-limbic network still plays a role in the task MRI-based pathophysiology of PD. A study of fearful facial affect recognition task showed that PD patients had less activations in the ACC. The chronic PD patients would have less amygdala activity in response to fearful face. The activities of the ACC were negatively correlated with anxiety level in PD. The study results suggested that chronic hyperarousal might diminish attention and emotional response via the atypical findings in amygdala-ACC network [45]. The study about discrimination between SAD and PD during emotional face perception task showed that reductions in hippocampus-temporal pole functional connectivity would be more specific for SAD, while unattended angry face perception [30], which might suggest the alterations in the intra-limbic network, would not occur in PD.

In the study of emotion regulation task and threat processing task in PD, a better predictor for the treatment response of cognitive behavioral therapy would be greater activations in the dorsolateral prefrontal cortex and insula during threat processing [46]. It also corresponded that the cognitive behavioral therapy could reduce the hyper-activation in several limbic regions and frontal regions, which were responsible for the fear memory, threat monitor, and maladaptive emotion regulation [47]. The study also emphasized the significance of frontal area in the task MRI-based pathophysiology of PD. Another study of emotion regulation task found that PD patients would have increased limbic-prefrontal activations while viewing negative images naturally and cognitive reappraisal with intentional regulation might attenuate the limbic-prefrontal activations during the perception of negative images. The maladaptive regulation strategy in PD might be related to the abnormal hyper-activation in frontal areas [48]. However, a study of emotion regulation task in GAD and PD suggested that prefrontal cortex hypofunction and the subsequent impaired top-down regulation would predispose PD patients to have emotion dysregulation [49]. The emotional Stroop task also increased the activity of the frontal regions, which suggested the altered processing of emotion-related stimuli in PD [50]. The structural study focusing on the common structural traits of anxiety pathophysiology also showed that the dorsolateral prefrontal cortex might be linked with worry and fear. In addition, the ACC neuroanatomical model might be more specific for anxiety [35], which can support the importance of frontal part in the fronto-limbic network [51]. The role of additional area, such as the insula, might be related to the dense connections with fronto-limbic network for integrating sensory inputs from the somatosensory cortex and visual cortex [41]. The insula is crucial for the integration of the filtered sensory information via the thalamus from the visuospatial and other sensory inputs related to the occipital, parietal, and temporal lobes [51]. Another study of fear conditioning task demonstrated the inhibitory coupling of ACC-amygdala network would predict the treatment response toward the exposure therapy in PD patients with long allele polymorphism of serotonin genes [52]. The pure PD without comorbid depression would have reduced activation of fronto-insula network after cognitive behavioral therapy [53]. The fear conditioning task would activate fear circuit during early acquisition, especially in the insula and amygdala. Then extinction recall task would have significant fronto-insula activations in PD patients [54]. The panic-related scene task also provoked the activations of amygdala-ACC network accompanied with the insula, thalamus, and brain stem. The fear of cardiovascular symptoms and the fear of respiratory symptoms were associated with insula and brain stem activations, respectively [55]. In a study to find the common and distinct abnormalities in PD, GAD, and SAD during the facial emotion matching task, the activation of the insula seemed more specific to the pathophysiology of PD [56]. A different task, attention network task, showed an alerting network consisted of fronto-parietal regions in the task MRI-based pathophysiology in PD [57]. The complex motor paradigm task of non-dominant hand showed decreased activation in bilateral putamen of PD patients [58].

From the above references, I summarized that different kinds of tasks would activate different regions and networks in PD. In the fearful face recognition task, panic-related scene task, hyperventilation task, and emotion regulation task, the task MRI-based functional connectivity would be altered in the fronto-limbic network, especially for the ACC and amygdala. The frontal hyper-activation or hypo-activation would influence the ability to control limbic-related fear and anxiety response. The insula's role would be more emphasized in the panic-related scene task and hyperventilation task. The inconsistent phenomenon of frontal activity might be related to the course of PD, such as acute and chronic phase. In the cognition-related task, such as the fear conditioning task, emotional Stroop task, and attention network task, the role of frontal regions would be more significant and combined with the insula to form fronto-insula network. However, the role of limbic system can't be ignored in the cognition-related task. At last, the motor task would predispose to activate the motor-related regions, such as the putamen. The fronto-limbic network is still crucial for the task MRI-based pathophysiology in PD. However, the role of the insula in the pathophysiology and treatment seemed to be more significant in PD than other kinds of anxiety disorders, except SAD.

Task MRI-Based Functional Network in Specific Phobia

The identification task for phobic stimuli in specific phobia patients showed a pattern of alterations in the fronto-limbic regions, including the amygdala, ACC, and dorsomedial prefrontal cortex. The amygdala was responsible for the automatic processing of stimuli. The ACC and dorsomedial prefrontal cortex were associated with direct threat evaluation under sufficient attentional resources [59]. In the study of masked affective processing task, the significant alterations in fronto-limbic network were also found in ventromedial prefrontal cortex and amygdala areas of specific phobia patients [44]. The blood-injection-injury phobia and animal phobia patients significantly had lack of ability to inhibit the exaggerated amygdala responses by the ACC, which meant that certain alterations in the ACC-amygdala network would disturb the emotion regulation process within fear circuitry networks [60]. The gray matter alterations in the ACC and prefrontal cortex can also provide the structural basis to prove the importance of frontal areas in the pathophysiology of specific phobia [35]. The fear processing task in spider phobia would induce amygdala activity which was originated from a fast subcortical pathway mediated by the pulvinar nucleus. The glucocorticoid administration would suppress the aberrant amygdala-occipital network interactions [61]. The patients with animal phobia had no greater activation in the striatal regions during implicit sequential learning paradigm, which was different from the results of obsessive compulsive disorder [62]. The emotional counting Stroop task in the patients with animal phobia also showed alterations in the fronto-limbic network, such as the ACC, inferior frontal cortex, dorsomedial prefrontal cortex, amygdala, and thalamus, which was associated with attentional abnormalities [63]. The anticipation and immediate perception phase of phobia picture stimuli caused the significant activations in the defensive behavior-related regions, such as the amygdala and hippocampus, of animal phobia patients. However, such an activation pattern was not observed in dental phobia patients [64], which might suggest that greater impacts on the limbic system of patients with animal phobia. The introspection task of dental images would elicit the activations of pain modulatory areas in the frontal lobe, such as the dorsomedial prefrontal cortex [65], which emphasized the important role of frontal regions in the task MRI pathophysiology of dental phobia. However, another study of dental phobia using audiovisual phobic stimuli in dental phobia showed ACC-amygdala-insula network alterations, which might be alleviated by the brief hypnosis therapy [66]. The study about the exposure therapy augmented with D-cycloserine in the treatment of snake phobia showed a relieving effect on the frontal areas, including the ACC and medial orbitofrontal cortex [67]. Another study using cue-conditioning task in spider phobia showed that early fear acquisition was associated with activations in occipital regions and extinction recall was associated with deactivations in the precuneus, which suggested enhanced differential fear retention and altered brain activation patterns from other kinds of anxiety disorders [68]. However, no further studies replicated the results mainly focused on the parieto-occipital network.

In summary, different tasks can determine the findings in the brain network of specific phobia. The major components of brain network involving in the task MRI-based functional connectivity model for specific phobia still relied on fronto-limbic network, especially for the ACC and amygdala. In the fear processing task, emotional regulation task, and emotional counting Stroop task, the crucial role of fronto-limbic network seemed established in specific phobia. The visual cue condition in the specific phobia might also activate parieto-occipital regions. However, the importance of visual-related network might need further study to be proved in specific phobia.

Task MRI-Based Functional Network in GAD

Excessive and uncontrollable worry is the core symptom of GAD, which might be associated with the worry perseveration and "as many as can" worry. The pathological worry pattern might be linked with alterations of fronto-limbic regions, such as the amygdala and prefrontal cortex, to handle the external threat via the heightened arousal and distress state [69]. The GAD patients had greater response to fearful face in the face emotion rating task. The activations would be distributed in the fronto-limbic network, including the amygdala, ACC, and ventral prefrontal cortex. The functional connectivity analysis showed positive correlations among these regions [9]. The masked angry face task in pediatric GAD also showed a negative coupling between the amygdala and ventrolateral prefrontal cortex, which suggested that frontal region might be unable to suppress the limbic hyperactivity in GAD [70] and the fronto-limbic network alterations might be endophenotype for GAD due to such consistent findings in pediatric and adult GAD. A study of emotional face task using region-of-interest strategy of hippocampus found that a decreased anterior hippocampus connectivity network with the ACC would be observed in GAD [71]. Another study of facial emotion processing task showed that cognitive behavioral therapy would attenuate the amygdala-ACC activation to fear/angry faces [72]. Another kind of task, the fear generalization task, showed that GAD patients would have less discriminating activities in the meso-cortical and mesolimbic networks, which might suggest the importance of ventral tegmental area-related network in the maladaptive threat processing and response in GAD [73]. The same study team used generalized conditioning of fear task to survey the endophenotype of GAD. They found that GAD patients were impaired in identifying the fear and safety cues due to the functional connectivity alterations between the ventromedial prefrontal cortex and other cortico-limbic structures, which might support the crucial role of fronto-limbic network in the task MRI-based pathophysiology of clinical anxiety [20]. The later-life GAD in the face shape task showed a possible association between the anxiety and cerebrovascular burden, which might be mediated by emotion reactivity in the fronto-limbic network [74]. A dot-probe task study on neutral and fearful faces showed that youths with GAD, SeAD, and SAD had similar alterations in the fronto-limbic network, such as the ACC and hippocampus. The abnormal disengagement of limbic regions still existed even the attention was directed from threat of fearful faces [75]. It suggested common fronto-limbic network alterations in youths with GAD, SeAD, and SAD, which also confirmed the core structures of fronto-limbic network in task MRI-based pathophysiology of anxiety disorders. The later study of emotion identification task in GAD and SAD patients also found a pattern of alterations of fronto-limbic network while identifying the fearful faces [31].

In the emotional regulation task, the GAD patients were unable to engage the ACC-related ability to dampen the elevated activity of the amygdala [76], which implied about the impairment in the spontaneous regulation of emotional processing. The emotional conflict task also found that GAD patients failed to implicitly regulate emotion conflict, which was associated with the deficits in the activation and connectivity between the amygdala and ACC [77]. In a later study using the emotion regulation task, GAD and SAD patients were unable to increase the fronto-limbic network while viewing emotion pictures. Therefore GAD and SAD might have a reduced capacity to engage the emotion regulation network, such as the amygdala and medial prefrontal cortex [27]. In another study of emotion regulation task, GAD and PD had significantly lower activity in the dorsolateral and dormedial prefrontal cortex during reappraisal and maintenance [49]. The reappraisal-based emotion regulation task also replicated that less engagement of ACC activities and less functional connectivity of the amygdala-ventrolateral prefrontal cortex during anxiety [39]. The common deficits of gray matter over the ACC and prefrontal cortex also could provide another kind of support to insist the importance of frontal regions in the pathophysiology of anxiety [35].

In addition to the emotion regulation and face-related tasks, some cognition-related tasks have been applied in the GAD. A kind of cognitive task, the continuous performance task, also demonstrated decreased functional connectivity between the ventrolateral prefrontal cortex and medial prefrontal cortex. In addition, the altered functional connectivity between the amygdala, posterior cingulate, and precuneus was observed during the continuous performance task. The continuous performance task would recruit additional mentalization-related regions, such as the precuneus and posterior cingulate [78]. The uncertainty condition, such as the gambling task, showed a decreased activity in the amygdala and might represent the disengagement of limbic regions while receiving the uncertainty challenge, which might be associated with the core symptoms of excessive worry [79]. Another study of gambling task showed that GAD patients had impairments in the fronto-limbic network, such as the amygdala and prefrontal cortex, which suggested that fronto-limbic network was also involved in the decision-making process of cognitive function. In addition, the repetitive transcranial magnetic resonance stimulation could normalize the functional connectivity within the ACC regions of GAD patients [80]. A study of passive avoidance task in GAD also demonstrated the impaired function of reinforcement-based decision-making. The prediction error was significantly correlated with the reduced activities in the fronto-limbic regions, such as the ventromedial prefrontal cortex, ventral striatum, and other limbic areas. The impairments in prediction error and decision-making might suggest that GAD patients had alterations in the correct evaluation of reward and punishment for the threat reality check [81]. The study of explicit verbal memory task in GAD patients showed that anxiety-inducing words would be associated with significantly increases in activities of ventrolateral prefrontal cortex [82]. GAD patients also would have an impaired performance in the working memory task, which was associated with altered activities in the fronto-limbic network, such as the dorsolateral prefrontal cortex, ventrolateral prefrontal cortex, amygdala, and hippocampus [83]. The cognition deficits in the frontal regions might be more responsible for the anxiety symptoms, rather than a threat-related consequence [34]. The cognitive inflexibility and autonomic rigidity might be the core issues of the cognitive deficits in GAD [84].

In summary, the emotion-related and fearful face tasks, such as the face emotion recognition task, emotion regulation task, emotion identification task, masked angry face task, and dot-probe task, might show the consistent findings in the fronto-limbic network of GAD patients. The fear generalization task showed the alterations in additional ventral tegmental area, which is combined with the fronto-limbic regions to form a hypothesis in the meso-cortico-limbic network in GAD. The cognitive task, such as gambling decision-making task, working memory task, and continuous performance task, mostly showed a consistent pattern of alterations in the fronto-limbic network, except for the additional findings in the posterior cingulate and precuneus during the continuous performance task in GAD. Therefore the meso-cortico-limbic and fronto-limbic network might still play a crucial role in the task MRI-based functional connectivity model of GAD.

Task MRI-Based Functional Network in SeAD and SM

The task MRI studies in SeAD and SM were relatively few compared to SAD, PD, GAD, and specific phobia. The reduced activation in the frontal region, such as the dorsolateral prefrontal cortex, has been found in SeAD and other anxiety disorders, which suggested that hypoactive prefrontal responses to the error processing of multisource interference task in anxiety disorders [85]. The emotional face matching task in SeAD, GAD, and SAD showed that anxious youths would have alterations in the activity of frontal regions. In addition, greater activities in the frontal regions would predict a better response to cognitive behavioral therapy and antidepressant treatment [86]. The mindfulness behavioral therapy could also relieve the anxiety symptoms of SeAD, GAD, and SAD. The improvements in anxiety symptoms were correlated with the changes in activation of the ACC and insula [87]. The dot-probe task of fearful faces in SeAD showed that reduced functional connectivity between ACC and hippocampus was correlated with greater anxiety, which also supported the importance of fronto-limbic network in the pathophysiology of SeAD [75]. Strawn et al. mentioned that fear-based anxiety disorders, such as SeAD, GAD, and SAD, would have neuroanatomical and functional alterations in the fronto-limbic network, such as the ACC and amygdala. The altered functional connectivity among the anterior limbic network regions and alterations of neurochemistry in the ACC also contributed to the core symptoms of the fear-based anxiety disorders. The anxious youths with behavioral inhibition temperament also were prone to have the dysregulation status in the frontal regions, such as the medial prefrontal cortex. Their review article also emphasized the crucial role of fronto-limbic network [88].

For the SM, there was not enough study to explore the possible pattern of task MRI-based functional connectivity in this kind of anxiety disorder. However, several studies about psychological mutism might shed a light on the pathophysiology of SM. The hysterical mutism would have alterations in the frontal regions, such as the inferior frontal, middle frontal, and sensorimotor cortex, during the vocalization task. During the speech recovery process, the activity of the inferior frontal cortex was positively correlated with that of the ACC and negatively correlated with that of the amygdala [89]. The connectivity to the ACC was also associated with the mutism status [90].

From the above limited literatures, the multisource interference task, emotional face matching task, and the dot-probe task during fearful faces in SeAD showed a possible pattern of alterations in the fronto-limbic network, especially for the ACC and amygdala. For the SM, the role of frontal regions seemed more important than limbic regions within the fronto-limbic architecture. However, more studies of task MRI would be needed for SeAD and SM to confirm the possible role of fronto-limbic network in the brain network model for the pathophysiology of these two anxiety disorders.

Fronto-limbic: Major Components of Task MRI-Based Functional Network

From the above literature review, the major components of task MRI-based functional network might be the fronto-limbic network, including the ACC, dorsolateral prefrontal cortex, ventrolateral prefrontal cortex, amygdala, and hippocampus. The cognition-related tasks, such as gambling decision-making task, working memory task, continuous performance task, and attention network task, seemed to be consistent to detect the alterations of frontal regions in anxiety disorders, especially for the prefrontal regions and ACC. The limbic regions would also be altered in these cognition-related tasks. However, the role of limbic regions would be not superior to the frontal regions in the cognition domain of patients with anxiety disorders.

The emotion-related and fear-related tasks, such as face emotion recognition task, emotion regulation task, emotion identification task, masked angry face task, dot-probe task, fearful face recognition task, panic-related scene task, hyperventilation task, fear processing task, scrutiny perception task, reappraisal-based emotion regulation task, and emotional counting Stroop task, would detect the alterations of limbic regions in anxiety disorders, such as the amygdala and hippocampus. The role of frontal regions, such as the ACC, might be not such significant as that of the limbic regions in emotion-related and fear-related tasks. However, the regulation role of frontal regions toward such altered emotion and fear originating from the limbic regions would be the major factor for the formation of clinic anxiety symptoms in anxiety disorders.

The dorsolateral prefrontal cortex, ventrolateral prefrontal cortex, and ACC could regulate the excessive fear through the top-down mechanism to enhance the executive function processing and resolve the conflicts, which can improve the social understanding and play a major interface between cognition and emotion [45, 91–96]. The amygdala would show greater responses toward the environmental threat and fear, which would subsequently process the elevated noradrenergic input from the locus nucleus [2, 6, 97]. The avoidance behaviors would also be associated with the phasic pattern of network associated with the amygdala [98]. The rostral-ventral amygdala pathway might be responsible for the preverbal processing of fear without attention and conscious monitoring. The fronto-limbic pathway might be responsible for high-order processing of signals attended by conscious cognition [99]. The alterations in fronto-limbic network would also predispose the patients to clinical anxiety symptoms [100]. The early-life anxiety model of primate species also indicated that reduced function connectivity between frontal regions and central nucleus of the amygdala would be associated with elevated anxiety due to loss of control over the cognition and emotion [101]. Therefore the frontal regions can use the top-down mechanism, such as the cognition and executive function, to control the emotional responses and fears originating from the limbic regions in anxiety disorders. The alterations in fronto-limbic network of patients with anxiety disorders would be crucial for the pathophysiology and might provide a potential target for future therapy and pharmacological agent.

Figure 1.1 would summarize the role of fronto-limbic regions in the task MRI-based functional network of anxiety disorders. The regions of fronto-limbic network might be major components in such functional network model of task MRI field. However, the additional components beyond the fronto-limbic network should not be ignored, especially for the visual and sensory-related tasks. The detailed information of additional components would be addressed in the following section.

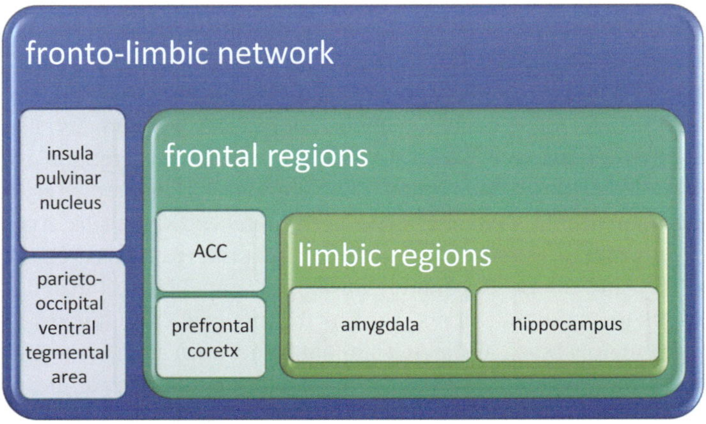

Fig. 1.1 The fronto-limbic network model and the additional components in the task MRI-based functional connectivity of anxiety disorders. (*ACC* anterior cingulate cortex)

Additional Components

The additional components of task MRI-based functional connectivity in anxiety disorders might be related to different kinds of tasks, especially for the visual- and sensory-based tasks. The neuroimaging meta-analysis study of SAD showed hyper-activation of fear circuit, which included insula and fronto-limbic network. In addition, the hyper-activation of parieto-occipital regions, such as the posterior cingulate cortex, cuneus, and precuneus, would also be observed in SAD [40]. The parieto-occipital regions would be more easily activated by the visual- and sensory-based tasks. The insula and pulvinar nucleus receive sensory inputs from somatosensory and visual areas. Therefore the alterations in the insula and pulvinar nucleus might be related to the sensory input characteristics of the cognition and emotion tasks. However, the dense connection of the insula and pulvinar nucleus with the limbic regions and frontal regions might play a possible role for task MRI-based functional connectivity in anxiety disorders. The alterations of parieto-occipital regions have been observed in SAD, GAD, and specific phobia. The alterations in the pulvinar nucleus were also observed in spider phobia. The alterations of the insula seemed more significant in PD and SAD than other anxiety disorders. The ventral tegmental area, which is also a part of limbic region, might be altered in GAD. The parieto-occipital regions might receive the sensory and visual inputs from these functional MRI tasks, such as face emotion recognition task, emotion regulation task, emotion identification task, masked angry face task, dot-probe task, fearful face recognition task, panic-related scene task, fear processing task, scrutiny perception task, reappraisal-based emotion regulation task, and emotional counting Stroop task. Then the pulvinar nucleus of thalamus would filter the sensory inputs from parieto-occipital regions. The filtered inputs, such as the visual, auditory, and tactile inputs, would be integrated in the insula due to its dense connection with the ACC, amygdala, hippocampus, parieto-occipital regions, and motor cortex. The integrated and filtered sensory inputs would be processed by the fronto-limbic network to send the response to the motor cortex to generate the motor and behavioral responses while receiving those cognition, emotion, and fear tasks. Therefore the insula, pulvinar nucleus, parieto-occipital regions (posterior cingulate cortex, precuneus, and cuneus), and ventral tegmental area might be the additional components of fronto-limbic model for task MRI-based functional connectivity network in anxiety disorders, especially for SAD, PD, GAD, and specific phobia (Fig. 1.1). The role of additional components might not be so significant in the SeAD and SM. However, due to the limited study in SeAD and SM, the role of additional components might be underestimated. The supportive role of additional components under the architecture of fronto-limbic network model would help us explain the task MRI-based pathophysiology of anxiety disorders in a more comprehensive viewpoint. The cooperation of fronto-limbic network with additional components should be crucial for researchers to understand the underlying theory of anxiety disorders and possible aids for the future treatment development.

References

1. Thomas KM, Drevets WC, Dahl RE, et al. Amygdala response to fearful faces in anxious and depressed children. Arch Gen Psychiatry. 2001;58:1057–63.
2. Carlson JM, Cha J, Mujica-Parodi LR. Functional and structural amygdala - anterior cingulate connectivity correlates with attentional bias to masked fearful faces. Cortex. 2013;49:2595–600.
3. Stevens JS, Almli LM, Fani N, et al. PACAP receptor gene polymorphism impacts fear responses in the amygdala and hippocampus. Proc Natl Acad Sci USA. 2014;111:3158–63.
4. Cha J, DeDora D, Nedic S, et al. Clinically anxious individuals show disrupted feedback between inferior frontal gyrus and prefrontal-limbic control circuit. J Neurosci. 2016;36:4708–18.
5. Gold AL, Shechner T, Farber MJ, et al. Amygdala-cortical connectivity: associations with anxiety, development, and threat. Depress Anxiety. 2016;33:917–26.
6. Bhattacharyya S, Egerton A, Kim E, et al. Acute induction of anxiety in humans by delta-9-tetrahydrocannabinol related to amygdalar cannabinoid-1 (CB1) receptors. Sci Rep. 2017;7:15025.
7. Swartz JR, Prather AA, Hariri AR. Threat-related amygdala activity is associated with peripheral CRP concentrations in men but not women. Psychoneuroendocrinology. 2017;78:93–6.
8. Hornboll B, Macoveanu J, Nejad A, et al. Neuroticism predicts the impact of serotonin challenges on fear processing in subgenual anterior cingulate cortex. Sci Rep. 2018;8:17889.
9. McClure EB, Monk CS, Nelson EE, et al. Abnormal attention modulation of fear circuit function in pediatric generalized anxiety disorder. Arch Gen Psychiatry. 2007;64:97–106.
10. Warwick JM, Carey P, Jordaan GP, Dupont P, Stein DJ. Resting brain perfusion in social anxiety disorder: a voxel-wise whole brain comparison with healthy control subjects. Prog Neuropsychopharmacol Biol Psychiatry. 2008;32:1251–6.
11. Schneier FR, Pomplun M, Sy M, Hirsch J. Neural response to eye contact and paroxetine treatment in generalized social anxiety disorder. Psychiatry Res. 2011;194:271–8.
12. Prater KE, Hosanagar A, Klumpp H, Angstadt M, Phan KL. Aberrant amygdala-frontal cortex connectivity during perception of fearful faces and at rest in generalized social anxiety disorder. Depress Anxiety. 2013;30:234–41.
13. Windmann S. Panic disorder from a monistic perspective: integrating neurobiological and psychological approaches. J Anxiety Disord. 1998;12:485–507.
14. Gorman JM, Kent JM, Sullivan GM, Coplan JD. Neuroanatomical hypothesis of panic disorder, revised. Am J Psychiatry. 2000;157:493–505.
15. Goddard AW, Mason GF, Appel M, et al. Impaired GABA neuronal response to acute benzodiazepine administration in panic disorder. Am J Psychiatry. 2004;161:2186–93.
16. Lueken U, Kruschwitz JD, Muehlhan M, Siegert J, Hoyer J, Wittchen HU. How specific is specific phobia? Different neural response patterns in two subtypes of specific phobia. Neuroimage. 2011;56:363–72.
17. Lueken U, Hilbert K, Stolyar V, Maslowski NI, Beesdo-Baum K, Wittchen HU. Neural substrates of defensive reactivity in two subtypes of specific phobia. Soc Cogn Affect Neurosci. 2014;9:1668–75.
18. Schienle A, Scharmuller W, Leutgeb V, Schafer A, Stark R. Sex differences in the functional and structural neuroanatomy of dental phobia. Brain Struct Funct. 2013;218:779–87.
19. Hettema JM, Kettenmann B, Ahluwalia V, et al. Pilot multimodal twin imaging study of generalized anxiety disorder. Depress Anxiety. 2012;29:202–9.
20. Cha J, Greenberg T, Carlson JM, Dedora DJ, Hajcak G, Mujica-Parodi LR. Circuit-wide structural and functional measures predict ventromedial prefrontal cortex fear generalization: implications for generalized anxiety disorder. J Neurosci. 2014;34:4043–53.
21. Dresler T, Guhn A, Tupak SV, et al. Revise the revised? New dimensions of the neuroanatomical hypothesis of panic disorder. J Neural Transm. 2013;120:3–29.
22. Tadayonnejad R, Klumpp H, Ajilore O, Leow A, Phan KL. Aberrant pulvinar effective connectivity in generalized social anxiety disorder. Medicine (Baltimore). 2016;95:e5358.

23. Holsen LM, Dalton KM, Johnstone T, Davidson RJ. Prefrontal social cognition network dysfunction underlying face encoding and social anxiety in fragile X syndrome. Neuroimage. 2008;43:592–604.
24. Gimenez M, Pujol J, Ortiz H, et al. Altered brain functional connectivity in relation to perception of scrutiny in social anxiety disorder. Psychiatry Res. 2012;202:214–23.
25. Klumpp H, Angstadt M, Phan KL. Insula reactivity and connectivity to anterior cingulate cortex when processing threat in generalized social anxiety disorder. Biol Psychol. 2012;89:273–6.
26. Klumpp H, Post D, Angstadt M, Fitzgerald DA, Phan KL. Anterior cingulate cortex and insula response during indirect and direct processing of emotional faces in generalized social anxiety disorder. Biol Mood Anxiety Disord. 2013;3:7.
27. Blair KS, Geraci M, Smith BW, et al. Reduced dorsal anterior cingulate cortical activity during emotional regulation and top-down attentional control in generalized social phobia, generalized anxiety disorder, and comorbid generalized social phobia/generalized anxiety disorder. Biol Psychiatry. 2012;72:476–82.
28. Choi SH, Shin JE, Ku J, Kim JJ. Looking at the self in front of others: Neural correlates of attentional bias in social anxiety. J Psychiatr Res. 2016;75:31–40.
29. Mansson KN, Carlbring P, Frick A, et al. Altered neural correlates of affective processing after internet-delivered cognitive behavior therapy for social anxiety disorder. Psychiatry Res. 2013;214:229–37.
30. Pantazatos SP, Talati A, Schneier FR, Hirsch J. Reduced anterior temporal and hippocampal functional connectivity during face processing discriminates individuals with social anxiety disorder from healthy controls and panic disorder, and increases following treatment. Neuropsychopharmacology. 2014;39:425–34.
31. Robinson OJ, Krimsky M, Lieberman L, Allen P, Vytal K, Grillon C. Towards a mechanistic understanding of pathological anxiety: the dorsal medial prefrontal-amygdala 'aversive amplification' circuit in unmedicated generalized and social anxiety disorders. Lancet Psychiatry. 2014;1:294–302.
32. Minkova L, Sladky R, Kranz GS, et al. Task-dependent modulation of amygdala connectivity in social anxiety disorder. Psychiatry Res Neuroimaging. 2017;262:39–46.
33. Yoon HJ, Kim JS, Shin YB, Choi SH, Lee SK, Kim JJ. Neural activity during self-referential working memory and the underlying role of the amygdala in social anxiety disorder. Neurosci Lett. 2016;627:139–47.
34. Balderston NL, Vytal KE, O'Connell K, et al. Anxiety patients show reduced working memory related dlPFC activation during safety and threat. Depress Anxiety. 2017;34:25–36.
35. Shang J, Fu Y, Ren Z, et al. The common traits of the ACC and PFC in anxiety disorders in the DSM-5: meta-analysis of voxel-based morphometry studies. PLoS One. 2014;9:e93432.
36. Gorka SM, Fitzgerald DA, Labuschagne I, et al. Oxytocin modulation of amygdala functional connectivity to fearful faces in generalized social anxiety disorder. Neuropsychopharmacology. 2015;40:278–86.
37. Sladky R, Hoflich A, Kublbock M, et al. Disrupted effective connectivity between the amygdala and orbitofrontal cortex in social anxiety disorder during emotion discrimination revealed by dynamic causal modeling for FMRI. Cereb Cortex. 2015;25:895–903.
38. Yang X, Liu J, Meng Y, et al. Network analysis reveals disrupted functional brain circuitry in drug-naive social anxiety disorder. Neuroimage. 2017.
39. Fitzgerald JM, Klumpp H, Langenecker S, Phan KL. Transdiagnostic neural correlates of volitional emotion regulation in anxiety and depression. Depress Anxiety. 2018;
40. Bruhl AB, Delsignore A, Komossa K, Weidt S. Neuroimaging in social anxiety disorder-a meta-analytic review resulting in a new neurofunctional model. Neurosci Biobehav Rev. 2014;47:260–80.
41. Nagai M, Kishi K, Kato S. Insular cortex and neuropsychiatric disorders: a review of recent literature. Eur Psychiatry. 2007;22:387–94.
42. Holtz K, Pane-Farre CA, Wendt J, Lotze M, Hamm AO. Brain activation during anticipation of interoceptive threat. Neuroimage. 2012;61:857–65.

43. Ottaviani C, Cevolani D, Nucifora V, et al. Amygdala responses to masked and low spatial frequency fearful faces: a preliminary fMRI study in panic disorder. Psychiatry Res. 2012;203:159–65.
44. Killgore WD, Britton JC, Schwab ZJ, et al. Cortico-limbic responses to masked affective faces across ptsd, panic disorder, and specific phobia. Depress Anxiety. 2014;31:150–9.
45. Pillay SS, Gruber SA, Rogowska J, Simpson N, Yurgelun-Todd DA. fMRI of fearful facial affect recognition in panic disorder: the cingulate gyrus-amygdala connection. J Affect Disord. 2006;94:173–81.
46. Reinecke A, Thilo K, Filippini N, Croft A, Harmer CJ. Predicting rapid response to cognitive-behavioural treatment for panic disorder: the role of hippocampus, insula, and dorsolateral prefrontal cortex. Behav Res Ther. 2014;62:120–8.
47. Reinecke A, Thilo KV, Croft A, Harmer CJ. Early effects of exposure-based cognitive behaviour therapy on the neural correlates of anxiety. Transl Psychiatry. 2018;8:225.
48. Reinecke A, Filippini N, Berna C, et al. Effective emotion regulation strategies improve fMRI and ECG markers of psychopathology in panic disorder: implications for psychological treatment action. Transl Psychiatry. 2015;5:e673.
49. Ball TM, Ramsawh HJ, Campbell-Sills L, Paulus MP, Stein MB. Prefrontal dysfunction during emotion regulation in generalized anxiety and panic disorders. Psychol Med. 2013;43:1475–86.
50. Dresler T, Hindi Attar C, Spitzer C, et al. Neural correlates of the emotional Stroop task in panic disorder patients: an event-related fMRI study. J Psychiatr Res. 2012;46:1627–34.
51. Lai CH. Fear network model in panic disorder: the past and the future. Psychiatry Investig. 2018;
52. Lueken U, Straube B, Wittchen HU, et al. Therapygenetics: anterior cingulate cortex-amygdala coupling is associated with 5-HTTLPR and treatment response in panic disorder with agoraphobia. J Neural Transm (Vienna). 2015;122:135–44.
53. Kunas SL, Yang Y, Straube B, et al. The impact of depressive comorbidity on neural plasticity following cognitive-behavioral therapy in panic disorder with agoraphobia. J Affect Disord. 2019;245:451–60.
54. Schwarzmeier H, Kleint NI, Wittchen HU, Strohle A, Hamm AO, Lueken U. Characterizing the nature of emotional-associative learning deficits in panic disorder: An fMRI study on fear conditioning, extinction training and recall. Eur Neuropsychopharmacol. 2019;29:306–18.
55. Feldker K, Heitmann CY, Neumeister P, et al. Cardiorespiratory concerns shape brain responses during automatic panic-related scene processing in patients with panic disorder. J Psychiatry Neurosci. 2018;43:26–36.
56. Fonzo GA, Ramsawh HJ, Flagan TM, et al. Common and disorder-specific neural responses to emotional faces in generalised anxiety, social anxiety and panic disorders. Br J Psychiatry. 2015;206:206–15.
57. Neufang S, Geiger MJ, Homola GA, et al. Cognitive-behavioral therapy effects on alerting network activity and effective connectivity in panic disorder. Eur Arch Psychiatry Clin Neurosci. 2018;
58. Marchand WR, Lee JN, Healy L, et al. An fMRI motor activation paradigm demonstrates abnormalities of putamen activation in females with panic disorder. J Affect Disord. 2009;116:121–5.
59. Straube T, Mentzel HJ, Miltner WH. Neural mechanisms of automatic and direct processing of phobogenic stimuli in specific phobia. Biol Psychiatry. 2006;59:162–70.
60. Stefanescu MR, Endres RJ, Hilbert K, Wittchen HU, Lueken U. Networks of phobic fear: functional connectivity shifts in two subtypes of specific phobia. Neurosci Lett. 2018;662:167–72.
61. Nakataki M, Soravia LM, Schwab S, et al. Glucocorticoid Administration Improves Aberrant Fear-Processing Networks in Spider Phobia. Neuropsychopharmacology. 2017;42:485–94.
62. Martis B, Wright CI, McMullin KG, Shin LM, Rauch SL. Functional magnetic resonance imaging evidence for a lack of striatal dysfunction during implicit sequence learning in individuals with animal phobia. Am J Psychiatry. 2004;161:67–71.

63. Britton JC, Gold AL, Deckersbach T, Rauch SL. Functional MRI study of specific animal phobia using an event-related emotional counting stroop paradigm. Depress Anxiety. 2009;26:796–805.
64. Lueken U, Hilbert K, Stolyar V, Maslowski NI, Beesdo-Baum K, Wittchen HU. Neural substrates of defensive reactivity in two subtypes of specific phobia. Soc Cogn Affect Neurosci. 2014;9:1668–75.
65. Schienle A, Wabnegger A, Schoengassner F, Scharmuller W. Neuronal correlates of three attentional strategies during affective picture processing: an fMRI study. Cogn Affect Behav Neurosci. 2014;14:1320–6.
66. Halsband U, Wolf TG. Functional changes in brain activity after hypnosis in patients with dental phobia. J Physiol Paris. 2015;109:131–42.
67. Nave AM, Tolin DF, Stevens MC. Exposure therapy, D-cycloserine, and functional magnetic resonance imaging in patients with snake phobia: a randomized pilot study. J Clin Psychiatry. 2012;73:1179–86.
68. Lange I, Goossens L, Bakker J, et al. Functional neuroimaging of associative learning and generalization in specific phobia. Prog Neuropsychopharmacol Biol Psychiatry. 2019;89:275–85.
69. Meeten F, Davey GC, Makovac E, et al. Goal directed worry rules are associated with distinct patterns of amygdala functional connectivity and vagal modulation during perseverative cognition. Front Hum Neurosci. 2016;10:553.
70. Monk CS, Telzer EH, Mogg K, et al. Amygdala and ventrolateral prefrontal cortex activation to masked angry faces in children and adolescents with generalized anxiety disorder. Arch Gen Psychiatry. 2008;65:568–76.
71. Chen AC, Etkin A. Hippocampal network connectivity and activation differentiates post-traumatic stress disorder from generalized anxiety disorder. Neuropsychopharmacology. 2013;38:1889–98.
72. Fonzo GA, Ramsawh HJ, Flagan TM, et al. Cognitive-behavioral therapy for generalized anxiety disorder is associated with attenuation of limbic activation to threat-related facial emotions. J Affect Disord. 2014;169:76–85.
73. Cha J, Carlson JM, Dedora DJ, Greenberg T, Proudfit GH, Mujica-Parodi LR. Hyper-reactive human ventral tegmental area and aberrant mesocorticolimbic connectivity in overgeneralization of fear in generalized anxiety disorder. J Neurosci. 2014;34:5855–60.
74. Karim H, Tudorascu DL, Aizenstein H, Walker S, Good R, Andreescu C. Emotion reactivity and cerebrovascular burden in late-life GAD: a neuroimaging study. Am J Geriatr Psychiatry. 2016;24:1040–50.
75. Price RB, Siegle GJ, Silk JS, et al. Looking under the hood of the dot-probe task: an fMRI study in anxious youth. Depress Anxiety. 2014;31:178–87.
76. Etkin A, Prater KE, Hoeft F, Menon V, Schatzberg AF. Failure of anterior cingulate activation and connectivity with the amygdala during implicit regulation of emotional processing in generalized anxiety disorder. Am J Psychiatry. 2010;167:545–54.
77. Etkin A, Schatzberg AF. Common abnormalities and disorder-specific compensation during implicit regulation of emotional processing in generalized anxiety and major depressive disorders. Am J Psychiatry. 2011;168:968–78.
78. Strawn JR, Bitter SM, Weber WA, et al. Neurocircuitry of generalized anxiety disorder in adolescents: a pilot functional neuroimaging and functional connectivity study. Depress Anxiety. 2012;29:939–47.
79. Yassa MA, Hazlett RL, Stark CE, Hoehn-Saric R. Functional MRI of the amygdala and bed nucleus of the stria terminalis during conditions of uncertainty in generalized anxiety disorder. J Psychiatr Res. 2012;46:1045–52.
80. Assaf M, Rabany L, Zertuche L, et al. Neural functional architecture and modulation during decision making under uncertainty in individuals with generalized anxiety disorder. Brain Behav. 2018;8:e01015.

81. White SF, Geraci M, Lewis E, et al. Prediction Error Representation in Individuals With Generalized Anxiety Disorder During Passive Avoidance. Am J Psychiatry. 2017;174:110–7.
82. Moon CM, Yang JC, Jeong GW. Explicit verbal memory impairments associated with brain functional deficits and morphological alterations in patients with generalized anxiety disorder. J Affect Disord. 2015;186:328–36.
83. Park JI, Kim GW, Jeong GW, Chung GH, Yang JC. Brain Activation Patterns Associated with the Effects of Emotional Distracters during Working Memory Maintenance in Patients with Generalized Anxiety Disorder. Psychiatry Investig. 2016;13:152–6.
84. Ottaviani C, Watson DR, Meeten F, Makovac E, Garfinkel SN, Critchley HD. Neurobiological substrates of cognitive rigidity and autonomic inflexibility in generalized anxiety disorder. Biol Psychol. 2016;119:31–41.
85. Fitzgerald KD, Liu Y, Stern ER, et al. Reduced error-related activation of dorsolateral prefrontal cortex across pediatric anxiety disorders. J Am Acad Child Adolesc Psychiatry. 2013;52:1183–91.. e1
86. Kujawa A, Swain JE, Hanna GL, et al. Prefrontal reactivity to social signals of threat as a predictor of treatment response in anxious youth. Neuropsychopharmacology. 2016;41:1983–90.
87. Strawn JR, Cotton S, Luberto CM, et al. Neural function before and after mindfulness-based cognitive therapy in anxious adolescents at risk for developing bipolar disorder. J Child Adolesc Psychopharmacol. 2016;26:372–9.
88. Strawn JR, Dominick KC, Patino LR, Doyle CD, Picard LS, Phan KL. Neurobiology of pediatric anxiety disorders. Curr Behav Neurosci Rep. 2014;1:154–60.
89. Bryant RA, Das P. The neural circuitry of conversion disorder and its recovery. J Abnorm Psychol. 2012;121:289–96.
90. Darby RR, Joutsa J, Burke MJ, Fox MD. Lesion network localization of free will. Proc Natl Acad Sci USA. 2018;115:10792–7.
91. Allen M, Dietz M, Blair KS, et al. Cognitive-affective neural plasticity following active-controlled mindfulness intervention. J Neurosci. 2012;32:15601–10.
92. Jasinska AJ, Yasuda M, Rhodes RE, Wang C, Polk TA. Task difficulty modulates the impact of emotional stimuli on neural response in cognitive-control regions. Front Psychol. 2012;3:345.
93. Atzil S, Hendler T, Feldman R. The brain basis of social synchrony. Soc Cogn Affect Neurosci. 2014;9:1193–202.
94. Lai CH. The neural markers of MRI to differentiate depression and panic disorder. Prog Neuropsychopharmacol Biol Psychiatry. 2018;
95. Matthews SC, Paulus MP, Simmons AN, Nelesen RA, Dimsdale JE. Functional subdivisions within anterior cingulate cortex and their relationship to autonomic nervous system function. Neuroimage. 2004;22:1151–6.
96. Palm ME, Elliott R, McKie S, Deakin JF, Anderson IM. Attenuated responses to emotional expressions in women with generalized anxiety disorder. Psychol Med. 2011;41:1009–18.
97. Bremner JD, Krystal JH, Southwick SM, Charney DS. Noradrenergic mechanisms in stress and anxiety: I. Preclinical studies. Synapse. 1996;23:28–38.
98. Brinkmann L, Buff C, Feldker K, et al. Distinct phasic and sustained brain responses and connectivity of amygdala and bed nucleus of the stria terminalis during threat anticipation in panic disorder. Psychol Med. 2017;47:2675–88.
99. Williams LM, Liddell BJ, Kemp AH, et al. Amygdala-prefrontal dissociation of subliminal and supraliminal fear. Hum Brain Mapp. 2006;27:652–61.
100. Herringa RJ, Birn RM, Ruttle PL, et al. Childhood maltreatment is associated with altered fear circuitry and increased internalizing symptoms by late adolescence. Proc Natl Acad Sci USA. 2013;110:19119–24.
101. Birn RM, Shackman AJ, Oler JA, et al. Evolutionarily conserved prefrontal-amygdalar dysfunction in early-life anxiety. Mol Psychiatry. 2014;19:915–22.

Phenotype Network and Brain Structural Covariance Network of Anxiety

2

Je-Yeon Yun and Yong-Ku Kim

Anxiety Phenomena Interacting with Other Clinical-Neurocognitive Phenotypes

What Is the Phenotype Network?

Alike the Research Domain Criteria (RDoC) [1, 2] raised from the National Institute of Mental Health, network-based approach for psychological phenotypes does not presume the latent features or disorders from which several psychiatric symptoms, psychological characteristics, and neurocognitive performances arise, but assume the dynamical interactions among these phenotypes as they coexist, propagate, and inhibit other components within the network of mental phenomena [3, 4]. For differential types of dataset from which the phenotype network is to be estimated, a Gaussian graphical model (regularized partial correlation network in which undirected groupwise associations among the ordinal or continuous variables are estimated) [5], an Ising model (a co-activation network to reconstruct the groupwise undirected interactions among the dichotomous variables) [6], a directed acyclic graph (a graphical presentation of Bayesian network, comprised of the probabilistic dependencies among the parental nodes (affecting) and sibling nodes (affected) and meeting the Markov property as the conditional probability distribution of sibling nodes within this network depends only upon the parental nodes) [7], or intraindividual covariance network (to derive the undirected intraindividual variation of z-transformed values for several psychological features per individual) [8] could be applied. For patients primarily diagnosed with anxiety disorder as well as who

J.-Y. Yun (✉)
Seoul National University Hospital, Seoul, South Korea

Yeongeon Student Support Center, Seoul National University College of Medicine, Seoul, South Korea

Y.-K. Kim
Department of Psychiatry, College of Medicine, Korea University, Seoul, South Korea

Y.-K. Kim (ed.), *Anxiety Disorders*, Advances in Experimental Medicine and Biology 1191, https://doi.org/10.1007/978-981-32-9705-0_2

suffer from comorbid anxiety symptoms, this framework of phenotype network might further uncover the precipitating factors of anxiety symptoms in diverse clinical population, mediating the role of anxiety symptoms in propagating the aftermaths of stressors to other psychopathology, and also will be able to demonstrate the possible target of more effective therapeutic intervention for these patients.

Gaussian Graphical Model

In the community-living children and adolescents, symptoms of panic disorder as well as social anxiety are well-demarcated as distinctive clinical syndrome that could influence other psychiatric symptoms of depressive mood and anxiety [9]. In the regularized partial correlation network (Gaussian graphical network) estimated from the diverse items of internalizing symptoms measured using a self-reporting questionnaire of "Revised Children's Anxiety and Depression Scale" [10], clinical symptoms of panic disorder as well as social phobia are more selectively interconnected as communities in the middle of other clinical symptoms such as major depression, separation anxiety, generalized anxiety, and obsessive-compulsive disorder for community-dwelling children and adolescents ($N = 37{,}162$); of note, individual clinical symptom items related to panic, fear of making a fool of oneself in public, and worry are more highly ranked for their level of interconnectedness to several clinical symptoms within the network [9]. Moreover, the Gaussian graphical model applied to the childhood self-reporting measures of the Social Behavior Questionnaire [11] (acquired at 6, 8, and 10 years of age by parental reports) combined with follow-up data regarding the prevalence of anxiety disorder at adolescence (15 years of age) or early adulthood (23 years of age) could demonstrate the possible predictors for future prevalence of anxiety disorder from the childhood clinical symptoms of disruptive symptom (not liked by other children) in girls; as sustained suffering of "not liked by others" as observed by parents might be predictive of future prevalence for anxiety disorders at adolescence or early adulthood, more attentive detection for as well as more targeted management of such clinical symptom would be required [12] (Fig. 2.1).

Moreover, trans-diagnostic interactions among the several coexisting clinical symptoms might be estimated using the phenotype network approach. For example, in patients diagnosed with eating disorder and having been recently cared for by way of the inpatient setting or partial hospital setting, anxiety symptoms (reported using the State-Trait Anxiety Inventory-Trait subscale [13]) were bridged to eating-related clinical symptoms (measured using the self-reporting questionnaire of Eating Disorder Examination Questionnaire version 4 [14]) including the "fear of losing control over eating" by way of "feeling overwhelmed" [15]; as the key player symptom [16] of "feeling overwhelmed" (in terms of the closeness and betweenness centralities; reflects the degree of direct information flow through the given node as well as the degree of mediating several nodes simultaneously, respectively) connects several anxiety symptoms to the community of eating-related symptoms in the

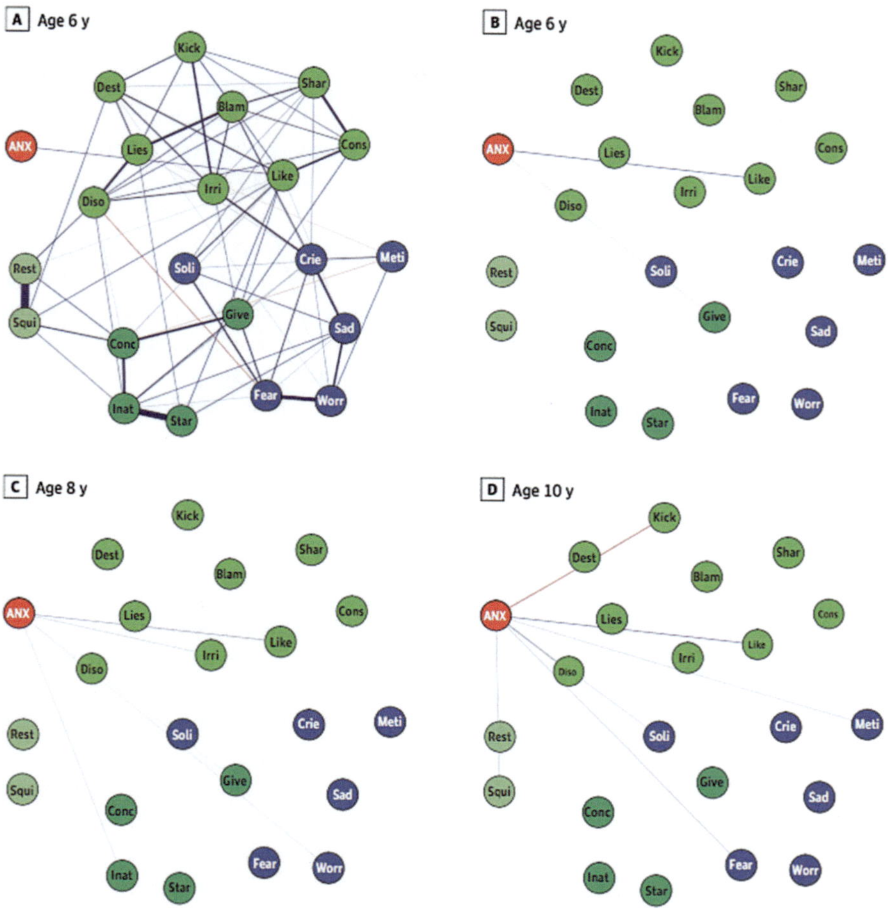

Fig. 2.1 Links between the Social Behavioral Questionnaire (SBQ) item network during childhood and occurrence of anxiety disorders at follow-up. (**a**) Network of the 21 items of the hyperactivity (light green), attention (sea green), disruptive (green), and internalizing (blue) communities in the SBQ at age 6 years. The ANX node (red) represents the occurrence of anxiety disorders at follow-up. Each edge corresponds to a partial correlation (positive in blue, negative in red, with an absolute magnitude >0.03) between two items, and its thickness corresponds to the absolute magnitude of the correlation. For instance, an edge between ANX and the like implies that indication of the like item at age 6 years was indicative of anxiety disorder diagnosis at follow-up (age 15 or 22 years). (**b–d**) All edges are removed except relevant edges involving the ANX node in the network at ages 6, 8, and 10 years, with no minimum absolute weight magnitude appearing in the graph (Figure 4 in [12])

Gaussian graphical model of eating symptom-depressive mood-anxiety phenomena for patients diagnosed with eating disorder, at least we hypothetically presume that more targeted therapeutic intervention for anxiety symptom of "feeling overwhelmed" could result in more efficient reduction of eating-related symptoms by way of the tight association between "feeling overwhelmed" and "fear of losing control over eating" [15]. Further, by applying the modularity-based community detection analysis to the Gaussian graphical model estimated from the dataset of State-Trait Anxiety Inventory-Trait form gathered from adult community population, current conceptualization of the trait anxiety (tendency of experiencing higher level of state anxiety in response to stimuli related to threat) [17] as single coherent construct in which two items of "presence of intrusive thoughts" and "being unable to get disappointments out of one's mind" are tightly connected to other items of anxiety [as calculated using the centrality index of node strength (sum of the edge weights connecting a node to other nodes within a given network; reflects the summated strength of correlation between a component and other components)] is shown [18]. In short, network approach using the Gaussian graphical model could be successfully applied to uncover the patterns of interconnectedness as well as segregation among various psychiatric symptoms and psychological constructs.

Ising Network

Patterns of interconnectedness among the clinical or psychological features converted into the binary form (presence or absence of specific symptom) could be shown by way of the Ising network model [6]. The Ising model is originally benchmarked from the physics and assumes the activation of a node within a given network depends on the activation of its neighboring nodes, estimated using the combination of logistic regression with model selection based on a goodness-of-fit measures [6]. Of note, applying this Ising network framework into several datasets gathered from community population, coexisting patterns of comorbid anxiety with depressive mood [6] in addition to the effects of suffering from anxiety regarding the suicidal ideation [19] as well as empathic communication capacity [20] have been illustrated. In other words, specific anxiety symptom of "feeling anxious" shows exceptionally tighter connectedness [as evidenced from centrality measures (of node strength and betweenness centrality)] with other symptoms of cognitive depressive symptoms or somatic symptoms, measured using the self-reporting tool named Inventory of Depressive Symptomatology [21] from healthy adults and patients diagnosed with depressive or anxiety disorders [6]. Further, for school-attending adolescents, social anxiety symptom of "I feel I cannot get close to people" could mediate the psychotic experiences (such as bizarre experiences, positive symptoms, and negative psychotic experiences) and suicidal ideation; this specific social anxiety component was highly scored for the betweenness centrality in the Ising network estimated from the binary dataset of the Community Assessment of Psychotic Experiences-Positive scale [22], the Brief Self-Report Questionnaire for Screening Putative Pre-psychotic States [23], and Columbia-Suicide Severity Rating Scale [24] [19]. From the perspective of social cognition and interpersonal communication, with self-efficacy for predicting others' feelings, emotional concealment, active emotional expression, and emotional leakage, an anxiety feature of "agitation" does a key role for combining the diverse components of empathic communication network for medical students, as shown by higher scores of node strength as well as betweenness and closeness centralities [20] (Fig. 2.2). Collectively, recent network studies using Ising network demonstrate more detailed profile of symptom interaction among the cognitive and somatic facets of depressive symptoms or among the psychotic experiences and suicidal ideations, as mediated by general anxiety or social anxiety, respectively. Moreover, importance of controlling somatic anxiety (agitation) for better performance in empathic social communication is also revealed.

Directed Acyclic Network

The Bayesian networks model the overall dependence structure of these multiple variables as visualized by way of directed acyclic graphs; the directed acyclic graph is comprised of the joint probability distribution of directed associations – from the parental nodes (upstream of arrow) to sibling nodes (downstream of arrow) – that can be decomposed as a product of conditional distribution of parental node(s). Of note, directional associations among the various clinical features including anxiety and worry as well as diverse environmental factors including stress-related features have been recently explored using the framework of Bayesian networks. In community-living adults, "worry" directly mediates the effect of stressors such as bullying toward several psychological discomforts including insomnia, depressive mood, and generalized anxiety [25, 26]. On the other hand, another study for community-living college students raised the possible effect of educational or therapeutic efforts to relieve the "fear of assertiveness in social situations," a form of social anxiety, in enhancing the students' immune against the peer-related verbal abuse (Fig. 2.3) [27]. In addition, by applying the dynamic Bayesian network [28] framework to the longitudinal dataset, all these psychological features of generalized anxiety, worry, and insomnia are strongly self-predictive for the status of 18-month follow-up time point; on the contrary, other types of anxiety such as social anxiety and situational anxiety tend to fluctuate over the 18-month time passage [26]. In turn, as estimated from the community-dwelling female college students by way of the directed acyclic graph, intensity of social anxiety could directly influence the severity of pathologic subjective halitosis (a preoccupation with unpleasant mouth odor without objective confirmation) and olfactory reference syndrome (a preoccupation with the false belief that one emits a foul or offensive body odor) [29]. Collectively, these recent Bayesian network-based studies demonstrated the mediating role of "worry" as transmitting the influence of environmental stressors to be manifested as diverse forms of psychological discomforts, the enduring nature of generalized anxiety within an individual across time passage, and the cognitive bias for self-image and social interaction affected from the severity of social anxiety.

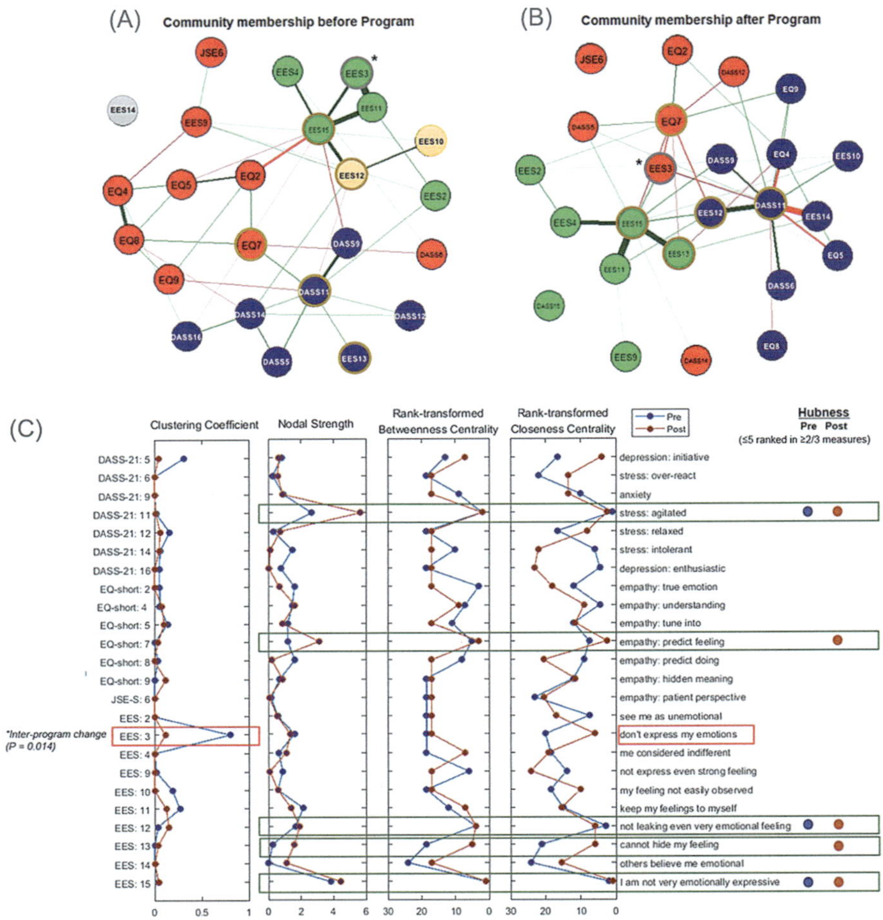

Fig. 2.2 Changed community profiles in the emotion-empathy network (**a**) before and (**b**) after five modules of the Empathy-Enhancement Program for Medical Students. The emotion-empathy networks were estimated using the Ising model; community memberships were detected from the transformed weighted, undirected network using the InfoMap algorithm after the negative connections (red-colored edges; cf. positive connections tagged with green) were converted into absolute values. Spheres of a given color identify each distinctive community; among these spheres, a total of five hubs – stress, agitated (DASS-21: 11); empathy, predict feelings (EQ-short: 7); not showing even very intense feelings (EES: 12); cannot hide feelings (EES: 13); and not very emotionally expressive (EES: 15) – are indicated with tan-colored circles. The node identified by a gray-colored circle ("I do not express my emotions to other people" (EES: 3)) demonstrated a significant change in the clustering coefficient value ($*p < 0.015$, based on the distribution of values calculated from the graph theory analyses for 5000 pseudo-networks, produced using random permutations for 80 participant time points into 2 subgroups). (**C**) Regional network characteristics of the clustering coefficient, nodal strength, betweenness centrality, and closeness centrality values before (blue-colored dots) and after (brown-colored dots) the Empathy-Enhancement Program for Medical Students. The global and regional network characteristics were calculated using the Brain Connectivity Toolbox and MATLAB R2016b software after the negative connections were converted into absolute values. As a result, five nodes ranked ≤5 for two of the three centrality measures

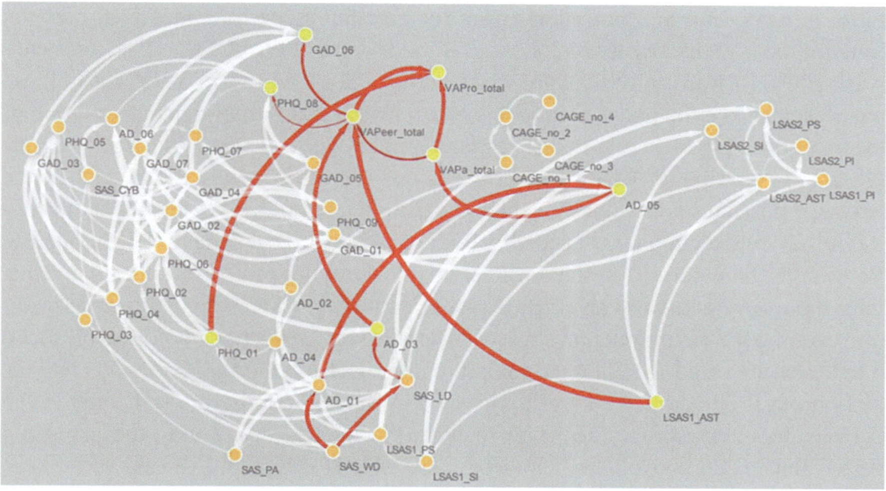

Fig. 2.3 Directed acyclic network comprised of perceived, verbal abuse severity; psychopathology; and social interaction patterns. Perceived, verbal abuse severity components of parents (VAPa_total), peers (VAPeer_total), and supervisors (VAPro_total) are shown, as are six further components directly connected to these components (red arrows) including (1) fidgeting when sitting for a long time (AD_05); (2) problems remembering appointments or obligations (AD_03); (3) fear of assertiveness (LSAS1_AST); (4) low levels of interest and/or pleasure (PHQ_01); (5) psychomotor change (PHQ_08); and (6) irritability (GAD_06); all are rimmed with yellow circles (Figure 2 in [27])

Intraindividual Covariance Network

The majority of phenotype networks mentioned above were only able to estimate the groupwise interacting patterns of psychological features and environmental factors including stressors. On the contrary, the intraindividual covariance network could uncover the covarying profile of several psychological features within an individual that might be further used to predict the individual-level outcome. For instance, interacting patterns of anxiety, depressive mood, personality traits including the sociality, and subdomains of neurocognitive performance in community-dwelling college students were uncovered by way of the graph theory approach for the intraindividual covariance network, calculated using the Beck Depression Inventory total score, the Beck Anxiety Inventory total score, t-score of nine clinical scales of the Minnesota Multiphasic Personality Inventory-2, and performance scores of neurocognitive

Fig. 2.2 (continued) (node strength, betweenness centrality, and closeness centrality) were selected as hubs (right-hand side of the figure). Moreover, the statistical significance of the changes in the clustering coefficient values was estimated from the distribution of values retrieved from network analyses for 5000 pseudo-networks (produced by way of random permutations for 80 participant time points into 2 subgroups) ($*p < 0.015$). Abbreviations: DASS, Depression Anxiety Stress Scale-21; EES, Emotional Expressivity Scale; EQ, Empathy Quotient-Short Form; JSE, Jefferson Scale of Empathy-S version (Figures 1 (for (**a**) and (**b**)) and 2 (for (**c**)) in [20])

tasks (commission and omission errors for continuous performance test, reaction time for the trail making test parts A and B, perseverative error of the Wisconsin card sorting test, forward and backward spans of the digit span test); in other words, these variables became the nodes comprising the network, and intraindividual variation of the two different network nodes (z-transformed) was encoded as edge weights. Groupwise distribution of the node betweenness centrality values as well as the edge betweenness centrality values calculated per individual networks uncovers the central influencing component of sociality (clinical scale 0(Si) of Minnesota Multiphasic Personality Inventory-2) in the intraindividual covarying profile of college students with more severe anxiety level [8]. Another exploration for the patterns of associations among the environmental stressors (verbal abuse), psychopathology, and social interaction revealed the roles of anxiety components including "nervous, anxious, and on edge" in addition to "restlessness" as principal connectors [reflected in the betweenness centrality values] that mediate the interactions among the perceived verbal abuse, depressive mood, anxiety, adult ADHD symptoms, diverse patterns of social interaction, and avoidance (social anxiety and avoidance, alcohol abuse, and smartphone overuse) with other key connectors of depressive mood components ("low interest or pleasure" and "poor appetite or overeating") and verbal abuse components (self-esteem damage and unjust blame) [27]. These exemplary studies [8, 27] illustrate how to find the central influential component(s) to become a potential target of clinical follow-up as well as of therapeutic intervention among the several interacting psychological symptoms including anxiety and environmental factors, by applying the graph theory approach to the intraindividual covariance networks and deriving the centrality measures (Fig. 2.4 [27]).

Conclusion

Network-based approach for anxiety-related psychological phenomena has been helpful in quantitative and pictorial understanding of qualitative dynamics among the diverse psychological phenomena as well as mind-environment interactions. Further trans-diagnostic approaches using larger dataset acquired with more sophisticated measure of psychopathology-neurocognition-social cognition-environmental factors could yield more robust and informative study finding, to be applied for focused clinical follow-up and targeted therapeutic intervention of anxiety disorder patients.

Neural Correlates of Anxiety Reflected in the Brain Structural Covariance Networks

What Is the Brain Structural Covariance Network?

Brain structural covariance refers to the correlative patterns of diverse brain morphological features such as cortical thickness [30], cortical surface area [31],

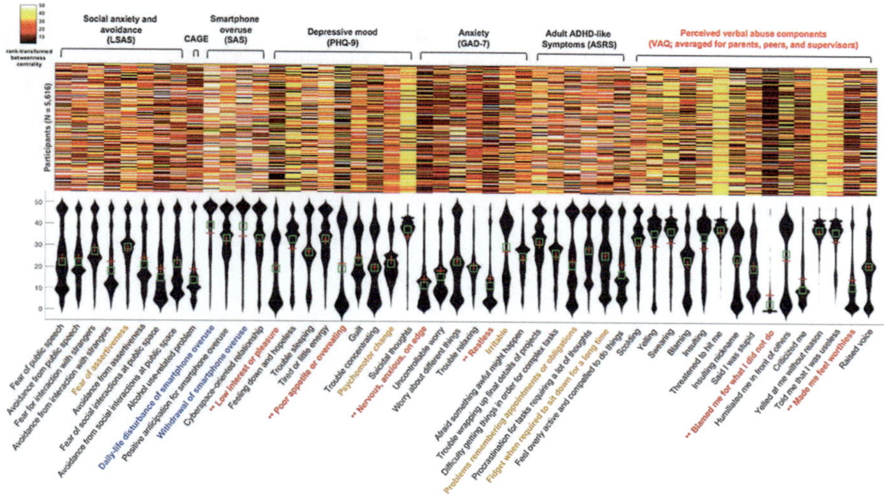

Fig. 2.4 Heatmap (upper) and violin plot (lower) of rank-transformed betweenness centrality values calculated from the intraindividual covariance network ($N = 5616$) featuring perceived verbal abuse components (averaged over parents, peers, and supervisors), psychopathology, and social interaction patterns. In the x-axis of the violin plot, the six most influential components (hubs; the top 12% nodes for rank-transformed betweenness centrality in >25% of participants at a network sparsity level of $K = 0.1$) are (1) low-level interest or pleasure; (2) poor appetite or overeating; (3) nervousness; (4) restlessness; (5) blaming oneself for what one has not done; and (6) feeling worthless, which are colored brown and marked with asterisks (Figure 3 in [27])

cortical and subcortical volumes [32], regional gray matter density [33], gyrification pattern [34], and myelination patterns [35], among differential brain regions comprising the brain, as calculated per participant (intraindividual brain structural covariance) [36] or across the participants (inter-individual brain structural covariance) [37]. These covarying patterns of brain morphology partly overlap with longitudinal patterns of brain cortical maturation from childhood to early adulthood [38–41] (Fig. 2.5) as well as propagating pattern of brain morphological changes such as cortical thinning and brain volume reduction in patients diagnosed with neurologic or psychiatric disorders along the trajectory of disease progression [42–44]. Further, graph theory approach for these brain structural covariance networks could decipher the hierarchical distribution of brain morphological changes (including small-worldness and modularity), in addition to the potential influencing brain regions that show stronger associations with several other brain regions (hubs) [45–47]. Collectively, using the frame of brain structural covariance network, researchers could estimate the dynamics of brain morphological changes among several brain regions from the cross-sectional data and also might be able to find the potential target(s) of therapeutic intervention among the hub regions (as uncovered from the graph theory approach).

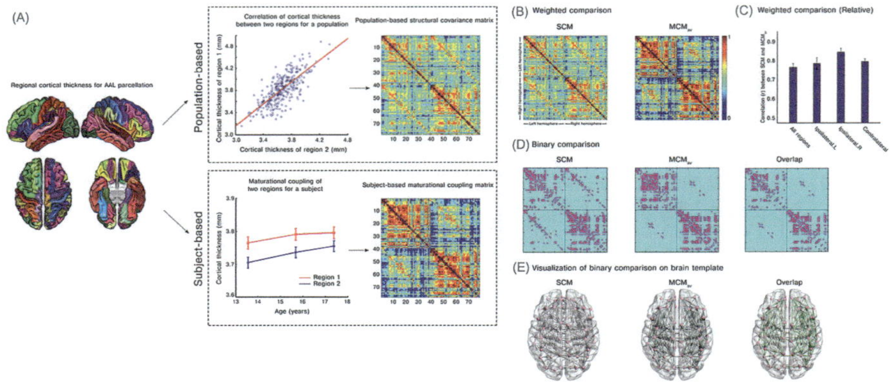

Fig. 2.5 Comparison of the population-based structural covariance and subject-based maturational coupling approach. (**a**) Regional cortical thickness based on autonomic anatomical labeling (AAL) template is first computed. Correlation of cortical thickness between two regions across a population is computed; doing so for all regions leads to the population-based structural covariance matrix (first row). In the proposed method, maturational coupling (defined as the similarity in the change of cortical thickness across the three time points; see Methods for details) between two regions of a subject is computed; doing so for all regions leads to the subject-based maturational coupling matrix (second row). (**b**) Whole-brain and (**c**) hemispheric-level comparison of structural covariance matrix (SCM) and averaged maturational coupling matrix (MCM_{av}). Note the strong correlation ($r > 0.8$) between the two matrices at whole-brain and hemispheric comparisons. For better interpretation, the binarized matrices (at sparsity 13%) were (**d**) compared and (**e**) visualized on a brain template (see Methods for details). Note the strong overlap between the two matrices (71%) localized in homologous and near-neighbor cortical regions (Figures 3 (for (**a**)) and 4 (for (**b**)–(**e**)) in [41])

Neural Underpinning of Anxiety Disorder Explored Using the Brain Structural Covariance Network

Till now, only a few studies applied the brain structural covariance network to the brain of patients diagnosed with anxiety disorder(s). Among others, possibly reflecting the abnormal profile of cortical development in panic disorder regarding the cortical surface area expansion as well as cortical folding, a previous study for the brain structural covariance network of local gyrification index diagnosed with panic disorder at adulthood demonstrated attenuated interconnectedness between the posterior cingulate cortex and precuneus versus other brain regions of default mode network (including the medial prefrontal, inferior parietal, and medial temporal cortices) in patients who suffer from severe panic symptoms (defined as the total score of the Panic Disorder Severity Scale >9) in terms of the covariance of local gyrification index, compared to not only healthy controls but also to patients with milder panic symptoms [48].

Anxiety in Other Psychiatric Disorders and Nonclinical Population

Brain structural covariance network that reflects the groupwise covarying patterns of regional cortical-subcortical volumes for children and adolescents who suffer from epilepsy and comorbid anxiety revealed higher harmonic mean values (a measure of network integration; higher values indicate lower level of network integration) and transitivity (an index of network segregation; reflects degrees of triplet connections among the network nodes) compared to both healthy controls and other non-anxious epilepsy patients [49]. On the other hand, in adult patients diagnosed with Parkinson's disease, severity of anxiety measured using the Beck Anxiety Inventory was inversely correlated with the strengths of brain structural covariance [of regional gray matter densities among different brain regions] between the left dorsal caudate nucleus-putamen and right caudate nucleus, between the right dorsal caudate nucleus and the right inferior frontal gyrus, and between the left nucleus accumbens and right caudate nucleus as well as the left dorsolateral prefrontal cortex; in other words, reduced covarying tendency of regional gray matter density among the contralateral striatal regions as well as in the prefrontal-striatal regions was associated with more severe level of anxiety for patients diagnosed with Parkinson's disease [43]. Further, exposure to severe stressors such as intimate partner violence, even without prevalence of psychiatric disorder(s), distorts the covarying patterns of brain morphology (named cortical thickness and subcortical volume) so that cortical thickness of the caudal anterior cingulate and precuneus and the thalamic volume become the most densely associated brain regions within the victims' brain compared to controls without prior exposure to intimate partner violence [50].

Conclusion

Previous studies that used the brain structural covariance network with or without graph theory approach could show neural correlates of specific anxiety disorder such as panic disorder and also elucidate the neural underpinning of anxiety symptom severity in diverse psychiatric and neurologic disorder patients. Further, recent suffering from interpersonal stressors or trauma was also reflected in the altered profile of brain structural covariance network. More rigorous cohort-based studies for neural correlates of diverse anxiety symptoms such as social anxiety and specific phobia, in addition to the more active exploration for neural imprints of diverse stressors, from the perspective of altered brain structural covariance profile might be helpful for understanding the whole-brain dynamics of brain morphological changes in association with suffering from stress and anxiety.

Acknowledgments This research was supported by the Basic Science Research Program through the National Research Foundation of Korea (NRF) funded by the Ministry of Education (NRF-2017R1D1A1B03028464).

References

1. Hyett MP, McEvoy PM. Social anxiety disorder: looking back and moving forward. Psychol Med. 2018;48(12):1937–44.
2. Frank B, Jacobson NC, Hurley L, McKay D. A theoretical and empirical modeling of anxiety integrated with RDoC and temporal dynamics. J Anxiety Disord. 2017;51:39–46.
3. Borsboom D. A network theory of mental disorders. World Psychiatry. 2017;16(1):5–13.
4. Fried EI, van Borkulo CD, Cramer AO, Boschloo L, Schoevers RA, Borsboom D. Mental disorders as networks of problems: a review of recent insights. Soc Psychiatry Psychiatr Epidemiol. 2017;52(1):1–10.
5. Epskamp S, Fried EI. A tutorial on regularized partial correlation networks. Psychol Methods. 2018;23(4):617–34.
6. van Borkulo CD, Borsboom D, Epskamp S, Blanken TF, Boschloo L, Schoevers RA, et al. A new method for constructing networks from binary data. Sci Rep. 2014;4:5918.
7. McNally RJ, Mair P, Mugno BL, Riemann BC. Co-morbid obsessive-compulsive disorder and depression: a Bayesian network approach. Psychol Med. 2017;47(7):1204–14.
8. Yun JY, Choi Y, Kwon Y, Lee HY, Choi SH, Jang JH. Hubness of strategic planning and sociality influences depressive mood and anxiety in College Population. Sci Rep. 2017;7(1):17856.
9. McElroy E, Patalay P. In search of disorders: internalizing symptom networks in a large clinical sample. J Child Psychol Psychiatry. 2019;60(8):897–906.
10. Chorpita BF, Yim L, Moffitt C, Umemoto LA, Francis SE. Assessment of symptoms of DSM-IV anxiety and depression in children: a revised child anxiety and depression scale. Behav Res Ther. 2000;38(8):835–55.
11. Tremblay RE, Vitaro F, Gagnon C, Piché C, Royer N. A prosocial scale for the preschool behaviour questionnaire: concurrent and predictive correlates. Int J Behav Dev. 1992;15(2):227–45.
12. Rouquette A, Pingault JB, Fried EI, Orri M, Falissard B, Kossakowski JJ, et al. Emotional and behavioral symptom network structure in elementary school girls and association with anxiety disorders and depression in adolescence and early adulthood: a network analysis. JAMA Psychiat. 2018;75(11):1173–81.
13. Spielberger DC. State-trait anxiety inventory. In: Weiner IB, Craighead WE, editors. The Corsini Encyclopedia of Psychology. Hoboken: Wiley; 2010.
14. Fairburn CG, Beglin SJ. Assessment of eating disorders: interview or self-report questionnaire? Int J Eat Disord. 1994;16(4):363–70.
15. Smith KE, Mason TB, Crosby RD, Cao L, Leonard RC, Wetterneck CT, et al. A comparative network analysis of eating disorder psychopathology and co-occurring depression and anxiety symptoms before and after treatment. Psychol Med. 2019;49(2):314–24.
16. Borgatti SP. Identifying sets of key players in a social network. Comput Math Organ Theory. 2006;12(1):21–34.
17. Eysenck HJ. Cicero and the state-trait theory of anxiety: another case of delayed recognition. Am Psychol. 1983;38(1):114–5.
18. Heeren A, Bernstein EE, McNally RJ. Deconstructing trait anxiety: a network perspective. Anxiety Stress Coping. 2018;31(3):262–76.
19. Nunez D, Fresno A, van Borkulo CD, Courtet P, Arias V, Garrido V, et al. Examining relationships between psychotic experiences and suicidal ideation in adolescents using a network approach. Schizophr Res. 2018;201:54–61.
20. Yun JY, Kim KH, Joo GJ, Kim BN, Roh MS, Shin MS. Changing characteristics of the empathic communication network after empathy-enhancement program for medical students. Sci Rep. 2018;8(1):15092.
21. Rush AJ, Gullion CM, Basco MR, Jarrett RB, Trivedi MH. The inventory of depressive symptomatology (IDS): psychometric properties. Psychol Med. 1996;26(3):477–86.
22. Capra C, Kavanagh DJ, Hides L, Scott J. Brief screening for psychosis-like experiences. Schizophr Res. 2013;149(1–3):104–7.

23. Liu CC, Tien YJ, Chen CH, Chiu YN, Chien YL, Hsieh MH, et al. Development of a brief self-report questionnaire for screening putative pre-psychotic states. Schizophr Res. 2013;143(1):32–7.
24. Posner K, Brown GK, Stanley B, Brent DA, Yershova KV, Oquendo MA, et al. The Columbia-Suicide Severity Rating Scale: initial validity and internal consistency findings from three multisite studies with adolescents and adults. Am J Psychiatry. 2011;168(12):1266–77.
25. Moffa G, Catone G, Kuipers J, Kuipers E, Freeman D, Marwaha S, et al. Using directed acyclic graphs in epidemiological research in psychosis: an analysis of the role of bullying in psychosis. Schizophr Bull. 2017;43(6):1273–9.
26. Kuipers J, Moffa G, Kuipers E, Freeman D, Bebbington P. Links between psychotic and neurotic symptoms in the general population: an analysis of longitudinal British National Survey data using Directed Acyclic Graphs. Psychol Med. 2019;49(3):388–95.
27. Yun JY, Shim G, Jeong B. Verbal abuse related to self-esteem damage and unjust blame harms mental health and social interaction in college population. Sci Rep. 2019;9(1):5655.
28. Friedman N, Murphy K, Russell S. Learning the structure of dynamic probabilistic networks. Proceedings of the fourteenth conference on uncertainty in artificial intelligence; Madison, Wisconsin 2074111: Morgan Kaufmann Publishers Inc; 1998. p. 139–47.
29. Tsuruta M, Takahashi T, Tokunaga M, Iwasaki M, Kataoka S, Kakuta S, et al. Relationships between pathologic subjective halitosis, olfactory reference syndrome, and social anxiety in young Japanese women. BMC Psychol. 2017;5(1):7.
30. Khundrakpam BS, Lewis JD, Reid A, Karama S, Zhao L, Chouinard-Decorte F, et al. Imaging structural covariance in the development of intelligence. NeuroImage. 2017;144(Pt A):227–40.
31. Sharda M, Foster NEV, Tryfon A, Doyle-Thomas KAR, Ouimet T, Anagnostou E, et al. Language ability predicts cortical structure and covariance in boys with autism spectrum disorder. Cereb Cortex. 2017;27(3):1849–62.
32. Chou KH, Lin WC, Lee PL, Tsai NW, Huang YC, Chen HL, et al. Structural covariance networks of striatum subdivision in patients with Parkinson's disease. Hum Brain Mapp. 2015;36(4):1567–84.
33. Nosarti C, Mechelli A, Herrera A, Walshe M, Shergill SS, Murray RM, et al. Structural covariance in the cortex of very preterm adolescents: a voxel-based morphometry study. Hum Brain Mapp. 2011;32(10):1615–25.
34. Das T, Borgwardt S, Hauke DJ, Harrisberger F, Lang UE, Riecher-Rossler A, et al. Disorganized gyrification network properties during the transition to psychosis. JAMA Psychiat. 2018;75(6):613–22.
35. Ma Z, Zhang N. Cross-population myelination covariance of human cerebral cortex. Hum Brain Mapp. 2017;38(9):4730–43.
36. Yun JY, Kim SN, Lee TY, Chon MW, Kwon JS. Individualized covariance profile of cortical morphology for auditory hallucinations in first-episode psychosis. Hum Brain Mapp. 2016;37(3):1051–65.
37. Alexander-Bloch A, Raznahan A, Bullmore E, Giedd J. The convergence of maturational change and structural covariance in human cortical networks. J Neurosci. 2013;33(7):2889–99.
38. Geng X, Li G, Lu Z, Gao W, Wang L, Shen D, et al. Structural and maturational covariance in early childhood brain development. Cereb Cortex. 2017;27(3):1795–807.
39. Shaw DJ, Marecek R, Grosbras MH, Leonard G, Pike GB, Paus T. Co-ordinated structural and functional covariance in the adolescent brain underlies face processing performance. Soc Cogn Affect Neurosci. 2016;11(4):556–68.
40. Sandini C, Zoller D, Scariati E, Padula MC, Schneider M, Schaer M, et al. Development of structural covariance from childhood to adolescence: a longitudinal study in 22q11.2DS. Front Neurosci. 2018;12:327.
41. Khundrakpam BS, Lewis JD, Jeon S, Kostopoulos P, Itturia Medina Y, Chouinard-Decorte F, et al. Exploring individual brain variability during development based on patterns of maturational coupling of cortical thickness: a longitudinal MRI study. Cereb Cortex. 2017;29(1):178–88.

42. Voevodskaya O, Pereira JB, Volpe G, Lindberg O, Stomrud E, van Westen D, et al. Altered structural network organization in cognitively normal individuals with amyloid pathology. Neurobiol Aging. 2018;64:15–24.
43. Oosterwijk CS, Vriend C, Berendse HW, van der Werf YD, van den Heuvel OA. Anxiety in Parkinson's disease is associated with reduced structural covariance of the striatum. J Affect Disord. 2018;240:113–20.
44. Guo S, Palaniyappan L, Liddle PF, Feng J. Dynamic cerebral reorganization in the pathophysiology of schizophrenia: a MRI-derived cortical thickness study. Psychol Med. 2016;46(10):2201–14.
45. Bethlehem RAI, Romero-Garcia R, Mak E, Bullmore ET, Baron-Cohen S. Structural covariance networks in children with autism or ADHD. Cereb Cortex. 2017;27(8):4267–76.
46. Mak E, Colloby SJ, Thomas A, O'Brien JT. The segregated connectome of late-life depression: a combined cortical thickness and structural covariance analysis. Neurobiol Aging. 2016;48:212–21.
47. Palaniyappan L, Marques TR, Taylor H, Mondelli V, Reinders A, Bonaccorso S, et al. Globally efficient brain organization and treatment response in psychosis: a connectomic study of Gyrification. Schizophr Bull. 2016;42(6):1446–56.
48. Yoon S, Jun CS, Jeong HS, Lee S, Lim SM, Ma J, et al. Altered cortical gyrification patterns in panic disorder: deficits and potential compensation. J Psychiatr Res. 2013;47(10):1446–54.
49. Garcia-Ramos C, Lin JJ, Bonilha L, Jones JE, Jackson DC, Prabhakaran V, et al. Disruptions in cortico-subcortical covariance networks associated with anxiety in new-onset childhood epilepsy. Neuroimage Clin. 2016;12:815–24.
50. Roos A, Fouche JP, Stein DJ. Brain network connectivity in women exposed to intimate partner violence: a graph theory analysis study. Brain Imaging Behav. 2017;11(6):1629–39.

Linear and Nonlinear EEG-Based Functional Networks in Anxiety Disorders

3

Poppy L. A. Schoenberg

Concepts: Overview and Some Definitions

The Electrical Structure of the Brain

The neuroelectrical architecture of the brain is intricately complex, with the average system consisting of an estimated 2.0–5.0 × 10^{11} cortical neurons arranged in multileveled neuronal assemblies that transfer information via poly-frequency electrical pulses transmitted by a conjectured 10^{14} synapses, approximately 2000–5000 "tagged" to each neuron. Neuronal cells, and their network dynamics, are fundamental to higher-order brain function in terms of coding and computation, such as calculation, strategy, decision-making, and consequent directed output or behaviors [1]. Electroencephalography (EEG) provides a noninvasive and economical technique to record postsynaptic activity from pyramidal cells in the brain constituting action potentials "locked" with millisecond resolution to default and other "states," in addition to specific responses to stimuli and processes during tasks/stimuli. Macroscopically, firing of densely numbered "nets" of neurons synchronizes and/or desynchronizes at frequency patterns extracted from the EEG signal. The utility of equating discrete oscillatory patterns with particular microcircuitry or distinct functional processes may be questionable in absolute terms since the EEG signal and its morphology reflect a larger scope of microcircuit configurations [2, 3]. The EEG state-space multidimensionality of neurophysiological activity is vast wherein neural microcircuitry gives rise to dynamics at cellular, cortical, and cortical sheet scales producing the time/frequency/spatial dimensions of the EEG signal. Physiologically, however, populations of neurons are evident in relatively small subsets of possible configurations across all these scales instantaneously [4]. This

P. L. A. Schoenberg (✉)
Department of Physical Medicine and Rehabilitation, Vanderbilt University Medical Center, Osher Center for Integrative Medicine, Nashville, TN, USA
e-mail: poppy.schoenberg@vanderbilt.edu

Y.-K. Kim (ed.), *Anxiety Disorders*, Advances in Experimental Medicine and Biology 1191, https://doi.org/10.1007/978-981-32-9705-0_3

added level of complexity has yet to be fully translated into clinical biomarker research, since the majority of EEG clinical studies focus on the resulting oscillations that are both linearly and nonlinearly related to these dynamics. From this clinical perspective, however, relative differences in oscillatory dynamics in anxiety disorders (vs. nonclinical controls) do have merit in terms of elucidating disorder etiology, course trajectory, and predictors of treatment response. For example, oscillatory patterns characterizing disorders provide electrocortical targets for subsequent alteration of electrical activity and brain organization via active (e.g., neurofeedback, brain training, cognitive restructuring) and passive (e.g., neuromodulation via pharmacology, transcranial magnetic simulation, optogenetics) precision medicine modalities.

EEG Functional Network Connectivity

Neural oscillations are generated from multiple biophysical mechanisms, across various spatial/temporal scales and microcircuitry comprising intrinsic/extrinsic properties, making oscillation-to-symptom correlations challenging. Nevertheless, oscillations are the most ubiquitous property of EEG and observed across the nervous system at multiple scales, thus providing an informative window into the complexity of the neurophysiological system. Globally, neural oscillations appear to have functional significance in terms of human consciousness, pertaining to affective, cognitive, motor, and perceptual domains [2] (albeit via multidimensional mechanisms across scales of cortical organization and function). This appears to be modulated by context-dependent temporal synchronization of oscillatory architectures that facilitate network organization and binding dynamics towards specialized functionality [6]. Neuronal oscillations appear necessary for (1) specific information coding, (2) orchestrating brain attentional states, and (3) facilitating communication between neuronal assemblies towards the creation of specific dynamic workspaces, particularly pertinent for executive functions [5].

Therefore, stratifying patterns of oscillatory network connectivity across space-time-frequency domains so to grant mechanistic value to anxiety disorders potentially offer more accessible clinical insights. Neural connectivity, in broad terms, pertains to an array of techniques defining cortical function as connectivity "networks" via spatially discrete neurophysiological events [7]. Theoretically, networks are described in terms of "nodes" and "edges," wherein edges connect nodes within the network topology. Additional geometric structure pertains to the nodes and edges embedded within the state-space of the brain. Graphical representation of the network topology visualizes germane connections by a vertex-edge matrix, assigned with real or complex values depending upon the connectivity analysis conducted [7]. Categorically, neural connectivity can be defined as (1) the arrangement of the fiber tracts that physically connect different brain regions (*structural connectivity*) [since structural connectivity is primarily visualized through neuroimaging such as diffusion tensor methods, then this chapter will focus on functional and effective EEG connectivity patterns in anxiety]; (2) the characterization of functional interactions of neuronal networks, with no assumptions regarding the underlying biology (*functional connectivity*); and (3) the influence that one constellation within the neuronal system, or network cluster, exerts over another (causality) towards integrated cortical action and output (*effective connectivity*) [7, 8] .

The function of any brain component, whether it be the cortical-, subcortical-, population- (e.g., neuronal assemblies), or cellular- (e.g., neuron) level, is defined primarily by its connection within and between the entire brain system. While this concept largely represents a theoretical abstraction based on mathematics/statistics (outlined later), it is also rooted in an embodied physiological reference via the neuroanatomy and physiology of intrinsic and extrinsic connections [9]. The latter couple discrete cortical areas, whereas the former are confined to the cortical sheets (layers in the perpendicular dimension defined by the density/morphology of neurons). Cortico-cortical connections have distinct properties that dictate functional applicability/suitability within a hierarchical structure comprising forward and backward connections. Whether a connection serves a forward/backward function is based on its cortical trajectories or layers of origin and termination, i.e., connectivity routes/pathways [7, 9]. Albeit, one cannot assume any linearity regarding such pathways, since the brain as a neuronal system is inherently nonlinear representing a highly open and complex dynamic system, and as such brain symptom dynamics and associated clinical pathways require the assumption of multidimensionality [10, 11]. This has implications at the level of functional and effective EEG connectivity applications and ensuing interpretation of data for anxiety disorders.

Methods: EEG Brain Connectivity and Network Measures

This section provides a (rather coarse-grain) primer of the main methods, and related theory, to extract functional/effective connectivity network information from the EEG signal, particularly for the clinician or researcher without a background in neurophysiology/electrophysiology, computational neuroscience, and/or modeling. Neural connectivity is often synonymous with EEG "synchrony" or "coherence," although these terms can be referred to, and interchanged, rather loosely, since their calculation may be more nuanced. Mathematically, both functional and effective connectivity are ascertained via statistical principles and mathematical algorithms so to characterize linear and nonlinear connectivity between and within cortical regions. Since functional connectivity is derived from temporal correlations between remote neurophysiological events [10, 12], such as correlational and coherence analyses in the frequency-spatial and temporal-frequency-spatial EEG domains, then mathematically the edge matrices (between nodes within the network) are undirected, thus symmetric (unless time-lagged correlations) although not necessarily binary [7]. Alternatively, effective connectivity is derived from estimations as to the causal influence one neural component exerts on another [13] (at whichever system level), and consequently the edge matrices are directed, possibly asymmetric and non-binary [7]. For comprehensive technical particulars and mathematical formulae, if of interest, see here [8, 13]. There are many techniques to characterize (bivariate, partial, undirected, and directed) connectivity in the brain, with one study benchmarking 42 methods with ~10^4 simulated datasets generated from five types of models characterized by differing connectivity structures [14]. Some theoretical overview as to the most viable applications for EEG clinical data, most salient to the extant anxiety evidence base, and perhaps to also spark interest in the multiplicity of applications for readers' own anxiety patient datasets/research programs is as follows:

Functional Connectivity Measures

Bivariate Connectivity (1) *Coherence* is presently widely utilized, largely due to its simplicity, and essentially pertains to a covariance statistical application to the processed EEG data. This may include linear cross-correlation between the time series from two separate electrodes of a specific frequency band. The cross-correlation function ascertains unidirectional interactions that exert the largest influence within the data at a specific time delay; alternatively, Pearson correlation coefficient is a nondirectional measure of connectivity that does not factor the temporal structure of the EEG data [13]. (2) *Phase synchronization* is another bivariate measure in multichannel EEG providing a preferable alternative to coherence [15] since it is independent from the spectra amplitude that can be contaminated by artifact, although it has shown comparable statistical outcomes applied to real-world data [16]. Phase synchronization is rooted in the concept of chaotic systems and pertains to the difference of instantaneous phases of two synchronized oscillators that remain constant during a "phase-locked" state, while the amplitudes of phase-synchronized systems can be uncorrelated. Complete synchronization is when both phase and amplitudes of the two oscillators are highly correlated [17]. There are various measures of phase synchronization, dependent upon hypotheses, experimental design, and preprocessing steps of the EEG signal. Some examples include phase slope index (PSI) (assumes time-lagged linear interactions), phase locking value (PLV) (the application of coherence equation to amplitude normalized FFT), phase lag index (PLI) (estimates distribution of phase differences across observations so to determine phase coupling that are invariant against common sources, such as the issue of volume conduction and/or active reference electrodes in EEG applications [18]), and pairwise phase consistency (quantifies distribution of phase pairwise differences across observations) [13].

Multivariate Decomposition such as functional principle component analysis (PCA) and independent component analysis (ICA) methods. PCA orthogonally transforms the connectivity of each patient into subject-specific vectors or functional principal component (FPC) scores. FPC values lend the visualization of variability in connectivity between patients, such as examining joint and marginal distributions of FPC values so to detect subpopulations, in addition to using regression models to quantify predictors of response [19]. ICA defines spatial maps and associated temporal courses (or "independent components") as representations of network nodes. Moreover, if a low number of components are estimated (low ICA dimensionality), then each ICA component could be considered as a discrete network, albeit representing a rather gross-level complex functional system [20]. The value of decompositions from PCA and ICA is to provide independent variables for variance and regression models so to detect relationships between neural connectivity and dependent clinical/outcome variables, for example, in anxiety populations. Partial least squares (PLS) regression also aims to explain as much as possible the covariance between two sets of explanatory/independent and dependent variables via uncorrelated variables that represent components or "latent (intuitive or hidden) vectors" [21].

Effective Connectivity Measures

Structural Equation Modeling (SEM) (a) Specifies the causal connectivity among EEG variables (i.e., source localization, current density) as an anatomical structure, (b) describes the causal effects, and (c) assigns explained and unexplained interregional variance within the structure [22, 23]. This is achieved by modeling a set of linear structural equations comprising observed variables and parameters of the EEG signal. Variables in the equation may be endogenous (dependent) or exogenous (independent). The ensuing model, or path analysis, represents the influence of regions on each other via the putative anatomical connections.

Dynamic Causal Modeling (DCM) Specifies EEG spatiotemporal connectivity (in a bilinear state-space) derived from differential equation modeling, where inferences regarding the temporal dimension are formulated in terms of architecture and underlying neuronal dynamics. This is based on the assumption that the EEG data may be modeled as the response to experimental perturbations within a dynamic input-output system. Various components are estimated from forward and inverse modeling so to define the model parameters, and thus a certain degree of a priori knowledge about the network is required. The forward process involves ascertaining differential responses to discrete experimental factors giving rise to changes in connection strength within or between sources, constituting a final spatiotemporal model representing nonlinear state-space with unobserved (or hidden) neuronal states. The inverse (backward) process estimates the model parameters that best fit how the observed data have been generated [24].

Granger Causality (GC) Assumes that a causal influence is present if the statistical information about the first (e.g., temporal frequency) series improves prediction of the second series [25]. Partial directed coherence (PDC) and directed transfer function (DTF) describe the frequency domain of GC [26, 27]. Linear vector autoregressive (VAR) modeling is often utilized for GC and estimates the stochastic time series data in which the value of a variable at one time point is modeled as a (linear) weighted sum from its own preceding discrete time points and the preceding time points of another/set of other variables [28]. Univariate (VAR) and multivariate vector autoregressive (MVAR) modeling ascertain "granger causality" in EEG data, albeit due to the nonlinearity of the brain state space, it is not without its limitations [29]. For example, prerequisites for GC application involves requiring that the interaction between variables is well approximated in linear terms and that low noise and signal cross-talk data is entered into the model.

Graph Theoretical Analysis Based on network science, applications of graph theory permit the topological properties of complex networks to be characterized in terms of modularity, hierarchy, centrality, and distribution of network hubs in graphical format. Specifically, applications aim to describe the network segregation (system segregation, modularity, local efficiencies, number of subpopulation hubs) and network integration (global efficiency, number of connector hub nodes) of the EEG brain system [12, 30]. While various approaches are possible, the overarching technique comprises (a) defining network nodes (i.e., EEG high-density/multielectrode array points), (b) estimating the edges as a continuous measure of association between nodes (i.e., via GC or spectral coherence), (c) generating an association/connectivity matrix with all pairwise associations between nodes to produce an undirected graph, and (d) calculating the network parameters of interest in the resulting brain network graphical model and overlaying to the equivalent parameters of a population of random networks, providing a directed graphical network [31]. Point (d) represents a normalization process that can also be achieved by setting an average degree or connectivity threshold or modeling a weighted graph using minimum spanning tree (MST) [32]. The latter method ensures all nodes are connected to one another without forming circles/loops creating subgraphs that represent different populations/experimental conditions that can then be compared as quantitative indices. Graph theory metrics, pertinent to extant anxiety disorder applications (outlined later [32–34]), wherein each EEG electrode represents a node in the graph network, include (i) *characteristic path length (CPL)* that quantifies global information integration, (ii) *clustering coefficient (CC)* that measures the network's local segregation, (iii) *minimum spanning tree (MST)* as above, and (iv) *small-world index (SWI)* describing the overall topology of the network where most nodes are not directly connected but peripherally connected via a small number of edges mapped out by network paths (node-to-node points).

Transfer Entropy (TE) Based on information theory allows the detection of effective (causal) connectivity in nonlinear systems/dynamics. Unlike modeling techniques TE is not determined upon a model of pathways, and thus no a priori definitions/knowledge of networks/pathways is required. Furthermore, it detects nonlinear interactions (and causality) even with wide distribution of interaction delays between signals within the system and is immune against linear signal cross talk (that may be a particular issue with EEG data) [35].

Applications: EEG Connectivity Networks in Anxiety Disorders

Here, we map the extant evidence base pertaining to EEG-based functional networks in anxiety disorders as defined by the *Diagnostic and Statistical Manual for Psychiatric Disorders 5 (DSM-5)* [36]. The DSM-5 stratifies distinct classifications, with some overlap in anxiety symptomatology, yet the diagnosis of anxiety disorder proper pertains exclusively to point (1) as follows: (1) *fear circuitry-based anxiety disorders*, e.g., phobias, panic, generalized anxiety, and social anxiety; (2) *anxiety related to obsessions and compulsions*, e.g., obsessive-compulsive disorder (OCD) (previously subsumed within the previous category in DSM-IV); (3) *trauma- and stressor-related disorders*, e.g., posttraumatic stress disorder (previously subsumed within category (1) in DSM-IV); and (4) *those characterized by dissociation*, i.e., dissociative disorders, depersonalization disorder, and dissociative amnesia [37]. Clinically, anxiety disorders include disorders that share features of excessive fear (emotional response to real or perceived imminent threat) and anxiety-related behavioral disturbances (in the anticipation of future threat), often in the absence of true danger. Logically, fear-processing networks, comprising the amygdala (a central node), hippocampus, insular cortex, dorsal anterior cingulate cortex (dACC), and medial prefrontal cortex (mPFC), are salient considerations for understanding anxiety [38, 39]. However, these clinical conjectures regarding fear neurocircuitry were primarily based on animal models and clinical studies prior to the DSM5, and thus patient samples also included trauma-/stress-based disorders (i.e., PTSD) [38, 39] that may have skewed findings and reduced specificity of outcomes. Here, we move towards rethinking the underlying neural architecture of anxiety disorders from a premise that incorporates previous models of fear neurocircuitry and integrates the emerging interest in, and new findings from, innovative connectivity modeling techniques for examining EEG-based functional networks. A summary of the extant studies is presented in Table 3.1 and further discussed henceforth.

Panic Disorder (PD)

Morphometric and functional neuroimaging investigations into panic disorder (PD) strongly align with the fear neurocircuitry hypothesis, suggesting input may be processed via two prime (distally discrete) circuitry networks in the disorder: (1) via a short-loop network wherein stimuli gated by the sensory thalamus are first transmitted to the lateral amygdala and then its central nucleus and (2) via a long-loop network wherein sensory input projects to the thalamus that in turn sends this information to a sensory cortex, insula, and prefrontal cortex (PFC) relay for higher-level analysis/processing, the outcome of which is transmitted to the amygdala. In both preprocessing loops, once input transfers to the amygdala, signals project to

Table 3.1 Overview of EEG connectivity studies in anxiety disorders

First author (date)	Clinical population ($\bar{x}$ age/range)	N	EEG parameter	Connectivity measure	Results	Significance (according to authors)
Xing (2019)	Social anxiety disorder (18–55 range)	20 patients vs. 20 HC	θ-freq. only EEG collected during ERT	Weighted phase lag index Graph theory analysis (MST)	(1) MST showed overall ↑ θ connectivity in patients. (2) SAD = longer phase-space trajectories during "neural" and "maintain" epochs of ERT; trajectory of "maintain" correlated with anxious state. (3) Patients ↑ use of reappraisal outside the laboratory as an emotion regulation strategy = ↓ trajectory lengths for the "reappraisal" epochs during the ERT	Results suggest MST manifold trajectory length may be a proxy measure of cognitive load of emotion processing / regulation in anxiety
Imperatori (2019)	Trait anxiety (22.6 mean)	47 HC (high STAI-T scorers)	EEG-DMN Band-pass filtering eLORETA	Lagged phase synchronization	(1) ↑ TA = ↓ θ right intra-hemispheric connectivity between mPFC and PCC (2) ↑ TA = ↓ β right intra-hemispheric connectivity in between mPFC and ACC	↑ TA = ↓ synchronization in DMN regions that reflect intrinsic brain activity (hypervigilance) and a top-down cognitive control deficit
Pattyn (2018)	Panic disorder (comorbidity) (47.93 mean)	16 tinnitus patients with comorbid PD vs. 16 tinnitus patients without PD	EEG-DMN spectra power (δ, θ, α-1, α-2, β-1, β-2, β-3, γ) sLORETA	Lagged phase coherence	(1) ↓ θ-activity in precuneus (BA7) and extending into retrosplenial part of PCC (BA31) in comorbid PD patients vs. controls (2) ↓ connectivity between dACC (BA24/32), insula (BA13), amygdala (BA34/35), and subgenual ACC (BA25) specific to α-1 freq, in comorbid PD patients	Brain networks related to tinnitus and anxiety overlap. However, co-morbid PD showed ↓ θ + α activity and connectivity in limbic system involved in emotional distress

Xing (2017)	Generalized social anxiety disorder (18–55 range)	32 gSAD patients vs. 32 HC	FFT (δ, θ, α, β)	WPLI Graph theory analysis (CPL, CC)	(1) WPLI = θ in frontal midline (Fz, Cz) electrodes showed ↑ connections to most electrodes in gSAD patients (vs. HC). α highest connectivity across all Ss. (2) gSAD Ss = state anxiety ↑ correlated with θ WPLI, trend for trait anxiety (3) Graph theory measures = ↓ global segregation + ↑ local integration of θ network in gSAD vs. HC	θ-dependent interconnectivity associated with state anxiety + ↑ information processing efficiency in gSAD (vs. HC). This may represent ↑ baseline self-focused attention
Klados (2017)	Generalized social anxiety disorder: test anxiety (22.2 mean)	16 high math test anxiety vs. 16 low scorers		Magnitude-squared coherence Graph theory analysis (CPL, CC, SWN)	(1) HMA ↑ CC (reflecting ↑ local segregation) across all freq. vs. LMA (2) Main effect of anxiety in anticipation of upcoming math task evident in SWN connectivity, i.e. ↑ CC + ↓ CPL	Differences in segregation + integration network connectivity measures across freq. in anxiety, indicative of anticipatory threat
Saunders (2016)	Comorbid generalized anxiety in autism spectrum disorder (25.5/16–47)	N = 46 (10 anxiety)	EEG-DMN – FFT (θ, α)	Coherence (linear cross-correlation)	(1) Generalized anxiety symptoms = ↑ θ intra-hemispheric frontal-occipital coherence, in addition to ↑ α interhemispheric coherence.	↑ coherence may reflect ↓ flexibility processing visual stimuli + sensorimotor integration. ↑ coherence limited to visual sensations in anxiety, i.e. threat sensitivity

(continued)

Table 3.1 (continued)

First author (date)	Clinical population ($\bar{x}$ age/range)	N	EEG parameter	Connectivity measure	Results	Significance (according to authors)
Knyazev (2016)	Trait anxiety (18–39 range)	44 HC	ERSP sLORETA	Graph theory analysis	(1) ↑ TA = smaller (i.e., more negative) difference in connectivity estimates for between stimuli (angry vs. neutral social interactions) and the baseline, exclusive to α-freq. (2) principle nodes in the above connectivity dynamics pertained to the PCC	Anxiety appears to be associated with a hyper-active attentional system, exclusive to α-freq
Xing (2016)	Primary anxiety disorder diagnosis (27.6 mean)	20 patients vs. 20 HC	EEG-DMN - CSD	WPLI Graph theory analysis (CPL, CC)	(1) Network properties of CPL + CC in θ-freq. show distinct order of (1) resting, (2) neutral, (3) maintain, (4) reappraise related to ERT (2) The (1) resting epoch for θ-freq. shows the largest differentiation between anxiety patients vs. HC	↑ θ-freq. network integration, via CPL, as cognitive load ↑ during emotion regulation. EEG-based connectivity promising marker of anxiety
Lackner (2014)	Trait anxiety composite measure (12–14 range)	84 HC	Complex demodulation (α-1, α-2, full α)	Phase locking Phase shifting Coherence (linear cross-correlation)	(1) ↑ anxiety levels correlated with full α-freq. phase locking at F3-C3, F3-P3, C3-O1, F7-T5, and F4-F8, F4-C4, F4-P4, C4-O2, and P4-O2	Anxiety levels were associated with full α-connectivity, whereas aggression levels related to α-1 freq. connectivity
Kikuchi (2011)	Panic disorder (30.2/16–52)	18 drug-naïve patients vs. 18 HC	EEG-DMN	Microstate analysis: global field power	(1) PD = ↑ duration/stability of microstate class with right-anterior to left-posterior orientation (vs. HC) (2) PD = ↑ duration/stability of microstate class with symmetric anterior-posterior orientation (vs. HC)	EEG-microstates may represent a promising marker of brain network dynamics in PD at the millisecond scale

Knyazev (2011)	State anxiety (20.2/19–30)	39 HC	ICA sLORETA	Cross-frequency SW-FW coupling Phase-amplitude coupling (δ-β corr.)	(1) ↑ δ-β correlation for anxiogenic stimuli, associated w/ sLORETA OFC/PFC + ACC. (2) ↑ δ-power and connectivity in same regions. (3) Within subject δ-β showed ↑ state anxiety scores associated with ↑ amplitude-to-amplitude + phase-amplitude coupling in OFC + ACC	Anxiogenic situations ↑δ-β coupling + δ-power, in regions relevant to δ-global network as part of the motivational brain circuitry that coordinates behavioral response
Lopes (2010)	Panic disorder (30.7 mean)	15 patients (within-Ss design)	FFT (δ, θ, α, β)	Coherence (linear cross-correlation) + Fisher's Z normalization	(1) During 35% CO_2 inhalation (relative to rest) = ↓ coherence in δ-freq. for electrode pairs O1-F7 + O2-F8 (2) Following 35% CO_2 inhalation = interhemispheric asymmetry in FP1-FP2 for β-freq, ↓ RH activation (3) Following 35% CO_2 inhalation = interhemispheric asymmetry in O1-O2 for δ-freq, ↓ LH activation	The data suggests PD patients may have disturbed frontal cortical processing + ↓ coherence between frontal + occipital networks + ↑ right posterior activity during high arousal states such as panic
Hanaoka (2005)	Panic disorder (31.5/16–52)	18 patients vs. 18 HC	FFT (δ, θ, α-1, α-2, β)	Coherence (linear cross-correlation) + Fisher's Z normalization	(1) PD = ↓ coherence in electrode pairs: F3-F4, C3-C4, and P3-P4, across freq. (2) PD = ↓ coherence in electrode pairs: F7-T5 and F8-T6, across freq. (3) Spearman +correlations between δ and α-2 + electrode pairs C3-C4 + F7-F8 (-corr for δ) and T5-T6 for α-2 only, with duration of disorder (4) Spearman +correlations between α-2 pairs P3-P4 + T5-T6, F4-C4 + F8-T6, and β C4-O2 with severity of panic attacks	PD patients have ↓ inter-hemispheric connectivity in frontal regions, and ↓ intra-hemispheric connectivity in bilateral temporal regions. ↓ α-coherence related to both length of disorder and severity of symptoms

(continued)

Table 3.1 (continued)

First author (date)	Clinical population ($\bar{x}$ age/range)	N	EEG parameter	Connectivity measure	Results	Significance (according to authors)
Knyazev (2005)	State anxiety (21.2/18–25)	30 HC (all male)	FFT (α-peak freq.)	Magnitude-squared coherence	(1) Active α-oscillators linearly ↓ in Ss with ↓ anxiety levels and linearly ↑ in Ss with ↑ state anxiety levels (2) Large part of the relationship between anxiety and α-power is accounted for by α-synchrony	Alpha synchronization allows dynamic regulation of α system preparedness for processing external stressors/stimuli
Wiedemann (1998)	Panic disorder (35.1 mean)	27 patients vs. 28 HC	EEG-DMN + neu/neg image viewing	Microstate analysis: global field power	(1) Microstate durations shorter in PD patients for all conditions (rest, neg, neu). (2) PD vs. HC = opposite microstate fields for neg/anxiety image viewing	PD is characterized by increased cortical activation, and different neuronal arrays when viewing anxiety stimuli

Notation: δ = delta; θ = theta; α = alpha; β = beta; γ = gamma

Acronym: *ACC* anterior cingulate cortex, *CC* clustering coefficient, *CPL* characteristic path length, *CSD* cross spectral density, *DMN* default-mode network (EEG resting state), *ERT* emotion regulation task, *ERSP* event-related spectral perturbations, *freq.* frequency, *gSAD* generalized social anxiety disorder, *HMA* high math anxiety, *LMA* low math anxiety, *MST* minimum spanning tree, *OFC/PFC* orbitofrontal cortex/prefrontal cortex, *PCC* posterior cingulate cortex, *Ss* subject/participant, *SWN* small-world index, *TA* trait anxiety, *WPLI* weighted phase lag index

the parabrachial nucleus (increases respiration), locus coeruleus (norepinephrine release), lateral and paraventricular nucleus of the hypothalamus (sympathetic autonomic arousal and adrenocorticoid release), and the periaqueductal gray (associated behavioral responses), in the brain stem and hypothalamus so to activate acute fear response in the autonomic, endocrine, and behavioral systems [40–42].

Linear coherence methods of EEG-based functional connectivity in PD support the downregulation of frontal electrode coherence, across frequencies at the trait level (vs. healthy controls) [43], and are specific to fronto-occipital coherence in the delta-band at the state level (i.e., during a panic attack) [44]. Moreover, following recovery from carbon dioxide-induced panic attack (35% CO_2 inhalation), attenuation in frontal coherence has been observed in the beta-band (this frequency also correlating to severity of panic attacks), and occipital coherence in the delta-band [44]. These findings indicate decoupling of the frontal cortico-occipital loops suggesting reduced neuroelectrical signaling in prefrontal and anterior-posterior circuitry during stimuli that would be more optimally processed with top-down executive control. Thus, as with the neuroimaging evidence base, information processing bypasses higher-level cortical response and remains "short-circuited" in the limbic system, signaling inappropriate hyperarousal response. Interestingly, a study investigating transcranial alternating current stimulation (tACS) intervention yielded enhanced posterior-frontal connectivity (ascertained by granger causality) in alpha-frequency accompanied by ameliorated anxious arousal and negative perception of sensory stimuli in controls administered with repeated alpha-tACS over 4 consecutive days [45].

Nonlinear EEG connectivity approaches enrich the above conjecture, since dysregulation between anterior and posterior neural system networks was also apparent in a study examining comorbid panic disorder in tinnitus patients implementing lagged phase coherence on EEG-DMN (rest) current density data ascertained by standardized low-resolution brain electromagnetic tomography (LORETA). Reduced connectivity was evident in the precuneus and retrosplenial plane of the posterior cingulate cortex (PCC) for theta-frequency in patients with comorbid PD compared to those without. Furthermore, attenuated connectivity in a specific network comprising the dorsal anterior cingulate cortex (dACC), insular cortex, amygdala, and subgenual ACC (sgACC) within the alpha-1 frequency (8–10 Hz) was observed. Presence of the dACC and sgACC in this network is pertinent, since the dACC is associated with cognitive processing as part of the frontoparietal attention networks, compared to the ventral stratification of the ACC associated with emotion processing networks that are cytoarchitecturally and neurochemically disparate [46]. The sgACC (Brodmann's area 25) forms part of the ventral ACC and is highly specialized and possibly a separate region, since its cell structure and receptor mapping are significantly distinct from the remaining ACC, comprising dense distribution of serotonin receptors and the most reciprocal connections with the amygdala [47]. It also appears to be part of a wider "visceromotor network" that modulates autonomic/neuroendocrine and neurotransmitter responses during the neural processing of reward, fear, and stress [48]. Ergo, nonlinear EEG functional connectivity measures would suggest the reduced connectivity of ACC in the fear circuitry network might be more central than previously considered and that the attenuation of the ACC as an executive neuro-regulator in fear circuitry networks in PD consequently means less top-down control and short-circuit looping straight to autonomic and brain stem systems thus ensuing the inappropriate response to non-threatening stimuli.

The presence of the insula as a node within the identified network exclusive to EEG-DMN in tinnitus patients with comorbid PD is interesting, as other LORETA-based connectivity studies examining state and trait anxiety did not find insula activity to be significant in associated connectivity networks, while ACC and PCC (reported above) were [49, 50] (discussed later). The insula, in conjunction with the ACC, forms an integral hub of the cortical "saliency network" [51]. Within this system the insula serves heterogeneous functions: (1) predominantly in bottom-up saliency detection, (2) co-facilitation with a wider network in the administration of attention and memory-related neural resources once saliency has been detected, (3) interaction of the anterior and posterior striations of the insular cortex towards the modulation of autonomic reactivity, and (4) co-facilitation with the ACC towards the administration of rapid motor system activity [52]. Hence, the insula (bottom-up pathway) in collaboration with the ACC (top-down control) is instrumental in assessing incoming saliency and mediating germane autonomic response towards optimal emotion regulation [53], whereby when connectivity is diminished, processing is less discriminate, in turn, behaviorally less appropriate (i.e., panic outcomes).

Network stability may be another functional connectivity property of significance in PD. EEG "microstates" (dynamic field networks that change configuration at the millisecond temporal scale) show increased duration and stability of symmetric anterior-posterior and right-anterior/left-posterior microstate orientation in panic disorder patients compared to healthy controls, during EEG-DMN resting state [54]. However, when viewing anxiety-arousing stimuli, PD patients show hypostability compared to controls [55]. This may suggest resting state is more "sticky" and rigid, in neuronal dynamics terms, in PD reflected by overall less neural flexibility (i.e., hyperstability) in default mode, whereas microstate properties in response to anxiogenic stressors are characterized by hyperactive neuronal arrays with diminished configuration stability during threat processing.

Generalized Anxiety Disorder (GAD)

Scant EEG functional connectivity investigations have been conducted in GAD to date. Regarding EEG nonlinear dynamic correlates comparing functional changes in the cerebral cortex and impact of symptom severity, higher D2 (correlation dimension) values indexing complexity of information processing and cerebral cortical dynamics have been observed in a sample of 64 clinically diagnosed GAD patients [56]. Conversely, *lower* D2 values characterize other disorders, including Alzheimer's disease, schizophrenia, seizure epilepsy, and depression. Of significance, lower D2 values have been found in posttraumatic stress disorder (PTSD) [57], representing reduced signal complexity suggesting that discrete electrocortical dynamics and CNS information processing underpin PTSD vs. GAD during rest (interesting since prior to the DSM5, these disorders were subsumed under converging diagnostic classification). As an aside, tasks involving divergent versus convergent thinking also show correlation of higher D2 values with the former [58]. One interpretation regarding increased EEG network complexity of GAD patients at rest may allude to increased worry/internal cognitive processing in the form of divergent negatively biased mind wandering (i.e., considering all possible outcomes leading to "catastrophic thinking") during non-specific information processing.

From a network perspective, nonlinear EEG-based functional connectivity approaches, examining weighted phase lag index (WPLI) and graph theory metrics so to visualize spatiotemporal "small-world network" (SWN) dynamics (the degree of global integration and local segregation within networks), highlight differences in the theta-frequency in GAD vs. controls, both at rest and during an emotion regulation task (ERT) [59]. Briefly, the ERT requires participants to view neutral and negative (i.e., anxiogenic) images with instructions to simply "look" (to neutral images, *neutral*), or *maintain* (to view negative images without any strategy) or *reappraise* (to view negative images and attempt to reduce emotional arousal by reinterpreting meaning of the images). Increased characteristic path length (CPL) indexing lower network integration unfolded as cognitive load progressed through *neutral*, *maintain*, and *reappraise* ERT conditions, in patients vs. controls. This was not the case for the rest condition, however, as patients yielded significantly attenuated CPL, indexing higher network integration, and higher cluster coefficient (CC) values indexing local network segregation. This would suggest enhanced SWN efficiency in the theta-frequency during EEG-DMN (resting state) in GAD indicative of neuronal hyperexcitability, whereas SWN properties were comparable during task-related emotion regulation between patients and controls. In line with this, the presence of comorbid generalized anxiety in patients primarily diagnosed with autism spectrum disorders (ASD) yield increased theta- and alpha-frequency coherence during EEG-DMN compared to ASD patients without clinical levels of generalized anxiety [60]. The specificity of differences in EEG-DMN network connectivity in GAD is interesting, viably representing basal/tonic aberrations opposed to state-based dysregulation in electrocortical connectivity dynamics.

Social Anxiety Disorder (SAD)

Network connectivity dynamics ascertained via WPLI and SWN metrics highlight dominant local and infrequent long-range theta-based connectivity pertaining to generalized social anxiety disorder (gSAD). The most prominent neural differentiation between gSAD patients versus controls is at rest/EEG-DMN. Patients yield increased clustering coefficient (CC), the degree to which nodes in the network cluster locally or network segregation, and attenuated characteristic path length (CPL), the average of shortest connections between nodes in the network, collectively reflecting greater efficiency in theta-frequency connectivity. Moreover, phase synchronization analyses point to increased phase frontal midline theta phase synchronization exclusively in gSAD, implicating the ACC and anterior attentional system that may be hyperactive at rest in such patients, possibly reflecting heightened basal self-focused negative internal processing [32]. These findings were replicated by another study utilizing the same graph theory and SWN metrics in gSAD, in this case test anxiety in anticipation of an upcoming mathematics task. In terms of graphical network parameters, increased CC and decreased CPL were evident across all frequencies (delta to gamma), representing heightened global connectivity efficiency, that is, increased nodal activity and more efficient remote connections [34]. Additionally, those with high test anxiety yielded greater network density across all frequencies, suggesting overall greater neuronal network activity in anticipation of a socially defined evaluation (test performance).

Task-processing connectivity network dynamics in SAD, measured with WPLI and nonlinear graph theory, also suggest discrepancies. The phase-space connectome (or map) of EEG connectivity (opposed to sensor-space connectivity) can be visualized using minimum spanning tree (MST) analysis that represents a symmetric geodesic distance matrix (GDM) from dynamic phase-space connectomes (a concept akin to phase-space "microstates" or "snapshots") using a moving window technique. Each connectome comprising the matrix is consequently mapped as a point on a manifold embedded in a high-dimensional state-space wherein the intrinsic geodesic topology of the manifold is thus encoded and visualized as MST (subset of edges of a connected undirected graph that connects all vertices without any overlays/loops) with independent "branches" or "trajectories" representing experimental conditions. Using this technique on ERT data (task described above), SAD patients yield significantly longer trajectory lengths to *neutral* and *maintain* conditions, versus controls. Moreover, *maintain* trajectory lengths positively correlated with severity of negative effect. Interestingly, those patients who used *reappraise* as an emotion self-regulation strategy in everyday life (outside the experimental setting) yielded shorter trajectory lengths during *reappraise* conditions of the experiment [33]. To note, the experimenters reported MST analysis for the theta-frequency only. One interpretation is that the length of the trajectory as mapped by the MST technique represents the temporal transitions through the snapshots comprising it (much like the reconfigurations of microstates), wherein longer trajectory lengths reflect more rapidly transitioning dynamical cognitive processing. Phenomenologically, this may be experienced clinically as "racing thoughts" or "overthinking" apparent even in low-complexity cognitive load associated with *neutral* and *maintain* conditions of the ERT in patients [33]. This may be differentiated from a simple stress response, since psychosocial stress reactivity has been shown to correlate with attenuated global brain network efficiency and amygdala centrality using graph theory measures [61].

In sum, the accumulating evidence supports a theory of neural overstimulation/activation in SAD at rest and during specific processing likely leading to misattribution in socially salient contexts. Its behavioral manifestation however (i.e., avoidance) is distinct from psychosocial "shutdown" in the face of overwhelming stress, since it appears to be underlined by higher neural efficiency and more complex neural mechanisms.

Trait Anxiety (TA)/State Anxiety (SA) via Psychometric Composites

A social interaction model, as mediated by degrees of anxiety and depression in a nonclinical sample, has emerged via connectivity analysis applications to sLORETA EEG data [62]. Concomitant to EEG, an event-related design comprising valenced facial stimuli (e.g., angry, happy, neutral) alongside imagining virtual social scenarios ("attack," "avoid," "make friends") with the character/face onscreen was administered. High-scoring trait anxiety (TA) was associated with a larger drop in alpha-based connectivity (desynchronization) to angry and neutral faces, relative to baseline. This was not apparent in the high depression scorers, thus a distinct anxiety effect. Interestingly, the clinical manifestation of anxiety and depression often co-occurs, yet this EEG-based functional connectivity study identified distinct brain topographical regions that differentiated anxiety from depression. Namely, while processing negative (angry) faces, the precentral gyrus, inferior temporal gyrus, medial prefrontal cortex (mPFC), and PCC were pertinent anxiety-network nodes. Additionally, distinct nodes including the inferior temporal gyrus and inferior parietal sulcus were evident for TA networks during neutral face processing, not evident in depression networks. Interestingly, insula and frontal regions comprised depression EEG functional networks during neutral socially salient stimuli processing yet were absent in TA networks (see Fig. 3.1).

Alpha connectivity aberrations have also been observed in the functional networks of high TA at rest/EEG-DMN. Adolescent externalizing (aggression) vs. internalizing (anxiety) appears to be underlined by distinct alpha-phase reset dynamics that gauge cortical activation and network communication between electrode sites. Longer phase locking in the full alpha range of (inter- and intra-hemispheric) electrode pairs was evident in TA [63]. Additionally, exact (e)LORETA in adults highlights decreased theta connectivity between mPFC and PCC/retrosplenial cortex and decreased beta connectivity in mPFC and ACC [50] during EEG-DMN. Interestingly, the neurobiology underlying the mechanism of attentional "set-shifting" that precedes "inflexible decision-making," or cognitive rigidity, points to a mPFC-ACC relay [64]. Ergo, reduced EEG-based connectivity of this circuit in people with high TA would suggest diminished flexible decision-making and ensuing adaptive behavior as a dispositional trait. Moreover, connectivity between the mPFC and PCC is linked to post-choice shift of attitude towards chosen options [65]. Thus, reduced EEG-based connectivity in this relay would suggest a level of incongruency in high TA scorers between (anxiety-driven) chosen outcomes and actual assessment, i.e., the experience of sustained/non-specific anxiety devoid of an obvious stimulus/stressor. Related peripheral evidence supports decreased EEG prefrontal-posterior coherence is linked to diminished PFC control in social-emotional information entailing heightened emotional affectivity/absorption during emotional/mood contagion [66]. Thus, high TA scorers may be unable to adapt their mood appropriately/effectively due less top-down regulation via the mPFC-PCC circuit.

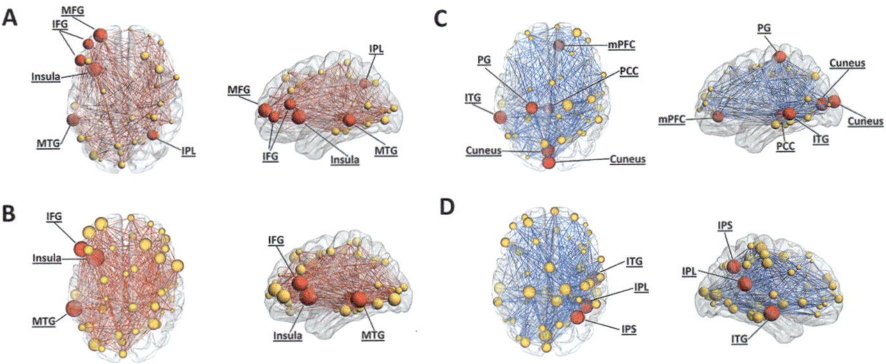

Fig. 3.1 EEG-based functional connectivity maps from Knyazev et al. (2016) [62] sLORETA study disentangling connectivity networks associated with depression (**a**+**b** theta networks) vs. anxiety (C+D alpha networks) during virtual social interactions. Networks represent regression of BDI (**a**), BDI-r (**b**), and TA (**c** and **d**) scores on the difference in connectivity estimates between test (**a** and **b** = neutral faces; **c** = angry faces; **d** = neutral faces) and baseline. Node size is proportional to its degree (i.e., larger nodes have more connections with other nodes). Red circles show network hubs. Edges of warm color show positive association of *psychometric* variable with the test *minus* baseline connectivity; edges of cool color show the opposite effect. *BDI* beck depression Inventory, *TA* State Trait Anxiety Inventory (STAI-T), *IFG* inferior frontal gyrus, *ITG* inferior temporal gyrus, *IPL* inferior parietal lobule, *IPS* inferior parietal sulcus, *MFG* middle frontal gyrus, *MTG* middle temporal gyrus, *PG* precentral gyrus [Reprinted with Permission]

Turning to state anxiety (SA), a linear relationship between SA levels and alpha synchrony has been found, supporting a concept of the "alpha system" in attentional preparedness for stressors in the environment (to note, only alpha-frequency was examined) [67]. Using topographical methods (sLORETA), increased coupling between delta and beta-frequencies in PFC and ACC regions alongside increased delta power overall has been shown in response to anxiogenic stimuli in high SA scorers [49]. The pertinence of delta and beta connectivity networks, also related to PCC and ACC regions, is in line with the PD findings. State anxiety may be recruiting similar networks as fear circuitry in PD, suggesting that state anxiety is distinct from trait and generalized anxiety neurobiologically, opposed to a similar state that is either acute or chronic.

Connectivity in Oscillatory Networks: Significance for Anxiety Disorders

Disentangling frequency-related connectivity patterns, (1) delta-, theta-, and beta-frequencies appear pertinent to PD, (2) theta connectivity aberrations appear germane to GAD and SAD, (3) alpha-frequency seems to connect with TA, and (4) SA, as with PD, is primarily differentiated by delta-beta coupling in connectivity networks. As mentioned, in absolute terms microcircuit-to-oscillation and thus oscillation-to-symptom mapping is not plausible due to the inherent complexity of the electrocortical system [2, 3]. However, oscillatory activity allows a macroscopic analysis of changes/differentiation in frequency dynamics observed in relation to behavioral outputs (i.e., performance on sensorimotor, cognitive, affective, perceptual tasks). From such inductive reasoning, distinct oscillation frequencies appear to have some functional significance (or correlation). Knyazev has synthesized the data into two functionally meaningful stratifications: (a) global integrative oscillations (related to synchronizing coherent activity across spatially distributed neural assemblies) reflected by lower-frequency delta, theta, and alpha ranges and (b) local processing oscillations (distributed in segregated assemblies) reflected by higher-frequency beta and gamma [68]. Albeit, gamma synchronization is also conjectured to serve an integrative mechanism in higher-order cortical computation via the segmentation and selection of converging neuronal inputs (or "binding") [69]. Thus, the complexity and multileveled functionality of EEG oscillations are beyond the scope of this commentary.

To simplify, delta and theta appear correlated with saliency detection, motivation, emotional appraisal and learning [67], and thus networks activating the neural "saliency system." Interestingly, delta oscillations are linked with the phylogenetically oldest neural subsystems related to information transmission (evidenced by the EEG signals of reptiles predominantly emanating the 0.5–4 Hz range [70]) and associated with behavioral inhibition (i.e., "freeze" response in panic [71], regulated by the phylogenetically ancient "social" polyvagal response [72]). Such neural activity represents a mechanism of descending inhibition, in which higher systems inhibit lower systems [73, 74]. Viability of this postulation has been observed during tasks requiring considerable internal concentration/mentation concomitant to increased delta interpreted to inhibit external interference via cortical deafferentation, i.e., dampening of sensory/external afferences [74]. This would fit the delta and theta EEG-based connectivity findings differentiating PD and SA from other anxiety subtypes, in alignment with the "long-loop" (refer above) fear circuitry. That is, higher-order ACC, insula, and prefrontal cortex (PFC) relays serve to inhibit new viscerosensory information from informing an appropriate behavioral response to an otherwise non-threatening stimuli (as the case with panic/SA), extending well beyond an initial reactionary period (where the assessment of genuine threat may be required).

Despite theta-frequency being correlated with an array of cognitive, affective, and behavioral outcomes, robust evidence suggests theta is principally implicated in memory and emotion regulation. Neurons in the amygdala produce theta activity during heightened emotional arousal, further to theta generation and synchronization in memory encoding and consolidation [65, 75, 76]. This has considerable functional significance for GAD and SAD based on increased efficiency in connectivity of the theta networks characterizing both disorders. In that, it may suggest overactivation of electrocortical networks implicated in the synthesis of emotional learning. Social and/or more generalized contexts are encoded as emotionally aversive and threatening via a maladaptive mechanism wherein neurocircuitry conditioned to specific threatening stimuli couples and overextends to nonconditioned/generalized stimuli.

Contrary to the traditional conceptualization of alpha enacting an inhibitory mechanism upon cortical processing systems [77], alpha synchrony appears to have multiple functional significance. For example, accruing evidence links alpha synchrony with increased cortical excitability specific to the attentional system [78]. Furthermore, alpha synchrony has been implicated in top-down functioning related to attention akin to a binding property within the frontoparietal "global neuronal workspace" [79], i.e., intricate parallel distributed assemblies of specialized neuronal networks. This also links to the LORETA findings in terms of increased alpha connectivity in mPFC, ACC, and PCC relays, suggesting that TA (and partially SA since it also shares functional neurocircuitry with PD) is primarily underlined by an overactivated attention system wherein specialized/germane attentional networks for threatening stimuli become dysregulated and activated during inappropriate stimuli exposure and/or the absence of stimulation altogether. This simplification is only part of the puzzle, however, since (as discussed previously) functional connectivity between mPFC-PCC and mPFC-ACC relays appear to have specific mechanistic involvement with cognitive set-shifting, decision-making, and subsequent post-choice attitude. An emerging premise could be that this reduced integrative connectivity among multileveled networks associated with set-shifting and congruency of behavioral response suggest that on one level high TA scorers are experiencing incongruency related to their dispositional anxiety, albeit the anxiety-like experience ensues due to continued hypervigilance or over-reactive attention system networks.

Future Trajectories

Linear and nonlinear connectivity analyses enrich examinations of associated mechanisms propagating anxiety disorders. Such approaches afford the disentanglement of functional and effective/causal mechanisms necessary for deciphering bioelectrical dynamics within complex neuroanatomical architecture. New vistas may entail further utilization of mathematical frameworks/modeling applications so to understand network complexity and structural stability of transient multidimensional properties with more clarity [80]. Non-model-based conceptualizations, such as transfer entropy, also provide apt nonlinear facilitation towards extracting information transfer in multileveled EEG topologies [81, 82]. Furthermore, multimodal imaging designs paired with connectivity techniques offer scope to examine electrocortical network integration and effectivity with other bio-regulatory/neuroanatomical systems pertinent to trait/state functioning in anxiety disorders, i.e., electrocortical brain dynamics in relation to vagal flexibility/heart rate variability (HRV) spectra or alternate neuroimaging parameters [83, 84]. Micro- and meta-state stability of electrocortical dynamics may also contribute to identifying anxiety-specific neurobehavioral interactions [85]. Elucidating aberrations in underlying network activity may inform biomarker targets for neuromodulation, particularly in the development of interventions in place of psychopharmacology [86, 87], for the treatment of anxiety disorders. The innovative application of data-driven applications of graphical networks (i.e., dissimilarity embedding and nonlinear dimensionality reduction) will allow the illumination of hidden properties of electrocortical activity tagged to clinically salient functionality (i.e., emotion regulation) within the framework of high-dimensional manifolds in topological state-space [88]. Replicated examinations into EEG network dynamics in anxiety disorders are warranted, albeit the extant purview would suggest that the current DSM-5 classification is viably stratified by distinct electrocortical network subtypes that may aid progress for intervention advancement.

Author Declarations None.

References

1. Eliasmith C, Anderson CH. Neural engineering: computation, representation, and dynamics in neurobiological systems, Computational Neuroscience Series. Cambridge-London: MIT Press; 2004.
2. Cohen MX. Where does EEG come from and what does it mean? Trends Neurosci. 2017;40(4):208–18.
3. Schoenberg PLA, Vago DR. Mapping meditative states and stages with electrophysiology: concepts, classifications, and methods. Curr Opin Psychol. 2019;28:211–7.
4. Moran A, Bar-Gad I. Revealing neuronal functional organization through the relation between multi-scale oscillatory extracellular signals. J Neurosci Methods. 2010;186(1):116–29.
5. Lopes da Silva F. EEG and MEG: relevance to neuroscience. Neuron. 2013;80(5):1112–28.
6. Uhlhaas PJ, Pipa G, Lima B, Melloni L, Neuenschwander S, Nikolić D, Singer W. Neural synchrony in cortical networks: history, concept and current status. Front Integr Neurosci. 2009;3:17. https://doi.org/10.3389/neuro.07.017.2009.
7. Friston KJ, Büchel C. Functional connectivity: eigenimages and multivariate analyses. In: Statistical parametric mapping: the analysis of functional brain images. 2007. p. 492–507.
8. Greenblatt RE, Pflieger ME, Ossadtchi AE. Connectivity measures applied to human brain electrophysiological data. J Neurosci Methods. 2012;207(1):1–16.
9. Chen CC, Henson RN, Stephan KE, Kilner JM, Friston KJ. Forward and backward connections in the brain: a DCM study of functional asymmetries. NeuroImage. 2009;45:453–62.
10. Friston K. Non-linear coupling and kernels. In: Statistical parametric mapping: the analysis of functional brain images. 2007. p. 522–33.
11. Schoenberg PLA, Speckens AEM. Multi-dimensional modulations of α and γ cortical dynamics following mindfulness-based cognitive therapy in Major Depressive Disorder. Cogn Neurodyn. 2015;9:13–29.
12. Friston K. Functional integration. In: Statistical parametric mapping: the analysis of functional brain images. 2007. p. 471–91.
13. Bastos AM, Schoffelen J-M. A tutorial review of functional connectivity analysis methods and their interpretational pitfalls. Front Syst Neurosci. 2016;9:175. https://doi.org/10.3389/fnsys.2015.00175.
14. Wang HE, Bénar CG, Quilichini PP, Friston KJ, Jirsa VK, Bernard C. A systematic framework for functional connectivity measures. Front Neurosci. 2014;8:405. https://doi.org/10.3389/fnins.2014.00405.
15. Lachaux J-P, Rodriguez E, Martinerie J, Varela FJ. Measuring phase synchrony in brain signals. Hum Brain Mapp. 1999;8(4):194–208.

16. Mezeiová K, Paluš M. Comparison of coherence and phase synchronization of the human sleep electroencephalogram. Clin Neurophysiol. 2012;123(9):1821–30.
17. Rosenblum MG, Pikovsky AS, Kurths J. Phase synchronization of chaotic oscillators. Phys Rev Lett. 1996;76(11):1804–7.
18. Stam CJ, Nolte G, Daffertshofer A. Phase lag index: assessment of functional connectivity from multi channel EEG and MEG with diminished bias from common sources. Hum Brain Mapp. 2007;28(11):1178–93.
19. Petersen A, Zhao J, Carmichael O, Müller H-G. Quantifying individual brain connectivity with functional principal component analysis for networks. Brain Connect. 2016;6(7):540–7.
20. Sockeel S, Schwartz D, Pélégrini-Issac M, Benali H. Large-scale functional networks identified from resting-state EEG using spatial ICA. PLoS One. 2016;11(1):e0146845. https://doi.org/10.1371/journal.pone.0146845.
21. Alin A, Kurt S, McIntosh AR, Öniz A, Özgören M. Partial least squares analysis in electrical brain activity. J Data Sci. 2009;7:99–110.
22. Astolfi L, Cincotti F, Babiloni C, Carducci F, Basilisco A, Rossini PM, Salinari S, Mattia D, Cerutti S, Ben Dayan D, Ding L, Ni Y, He B, Babiloni F. High-resolution EEG and structural equation modelling: simulations and application to finger tapping data. IEEE Trans Biomed Eng. 2005;52(5):757–68.
23. Babiloni F, Cincotti F, Basilisco A, Maso E, Bufano M, Babiloni C, Carducci F, Rossini P, Cerutti S, Ben Dayan Rubin D. Frontoparietal cortical networks revealed by structural equation modelling and high resolution EEG during a short term memory task. In: Proceedings of the 1st International IEEE EMBS, Conference on Neural Engineering. 2003, March 20–22. p. 79–82.
24. Kiebel SJ, Garrido MI, Moran R, Chen C-C, Friston KJ. Dynamic causal modeling for EEG and MEG. Hum Brain Mapp. 2009;30:1866–76.
25. Granger CWJ. Investigating causal relations by econometric models and cross-spectral methods. Econometrica. 1969;37(3):424–38.
26. Huang D, Ren A, Shang J, Lei Q, Zhang Y, Yin Z, Li J, von Deneen KM, Huang L. Combining partial directed coherence and graph theory to analyse effective brain newtorks of different mental tasks. Front Hum Neurosci. 2016;10:235. https://doi.org/10.3389/fnhum.2016.00235.
27. De Vico Fallani F, Richiardi J, Chavez M, Achard S. Graph analysis of functional brain networks: practical issues in translational neuroscience. Philoso Trans R Soc B. 2014;369:20130521.. doi.org/10.1098/rstb.2013.0521
28. Seth AK, Barrett AB, Barnett L. Granger causality analysis in neuroscience and neuroimaging. J Neurosci. 2015;35(8):3293–7.
29. Maziarz M. A review of the granger-causality fallacy. J Philos Econ. 2014;8(2):86–105.
30. Cohen JR, D'Esposito M. The segregation and integration of distinct brain networks and their relationship to cognition. J Neurosci. 2016;36(48):12083–94.
31. Bullmore E, Sporns O. Complex brain networks: graph theoretical analysis of structural and functional systems. Nat Rev. 2009;10:186–98.
32. Xing M, Tadayonnejad R, MacNamara A, Ajilore O, DiGangi J, Phan KL, Leow A, Klumpp H. Resting-state theta band connectivity and graph analysis in generalized social anxiety disorder. Neuroimage Clin. 2017;13:24–32.
33. Xing M, Lee H, Morrissey Z, Chung MK, Phan KL, Klumpp H, Leow A, Ajilore O. Altered dynamic electroencephalography connectome phase-space features of emotion regulation in social anxiety. NeuroImage. 2019;186:338–49.
34. Klados MA, Pandria N, Micheloyannis S, Margulies D, Bamidis PD. Math anxiety: brain cortical network changes in anticipation of doing mathematics. Int J Psychophysiol. 2017;122:24–31.
35. Vicente R, Wibral M, Lindner M, Pipa G. Transfer entropy – a model-free measure of effective connectivity for the neurosciences. J Comput Neurosci. 2011;30(1):45–67.
36. American Psychiatric Association. *Diagnostic and statistical manual of mental disorders*. 5th ed. American Psychiatric Association (APA): Arlington; 2013.

37. Regier DA, Kuhl EA, Kupfer DJ. The DSM-5: classification and criteria changes. World Psychiatry. 2013;12(*2*):92–8.
38. Kent JM, Rauch SL. Neurocircuitry of anxiety disorders. Curr Psychiatry Rep. 2003;5:266–73.
39. Shin LM, Liberzon I. The neurocircuitry of fear, stress, and anxiety. Neuropsychopharmacology. 2010;35:169–91.
40. de Carvalho MR, Dias GP, Cosci F, de Melo Neto VL, Bevilaqua MCDN, Gardino PF, Nardi AE. Current findings of fMRI in panic disorder: contributions for the fear neurocircuitry and CBT effects. Expert Rev Neurother. 2014;10(2):291–303.
41. Gorman JM, Kent JM, Sullivan GM, Coplan JD. Neuroanatomical hypothesis of panic disorder, revised. Am J Psychiatr. 2000;157(4):493–505.
42. Coplan JD, Lydiard RB. Brain circuits in panic disorder. Biol Psychiatry. 1998;44(12):1264–76.
43. Hanaoka A, Kikuchi M, Komuro R, Oka H, Kidani T, Ichikawa S. EEG coherence analysis in never-medicated patients with panic disorder. Clin EEG Neurosci. 2005;36(1):42–8.
44. Lopes FL, Oliveira MM, Freire RC, Caldirola D, Perna G, Bellodi L, Valença AM, Nascimento I, Piedade RA, Ribeiro P, Zin WA, Nardi AE. Carbon dioxide-induced panic attacks and quantitative electroencephalogram in panic disorder patients. World J Biol Psychiatry. 2009;11(2–2):357–63.
45. Clancy KJ, Baisley SK, Albizu A, Kartvelishvili N, Ding M, Li W. Lasting connectivity increase and anxiety reduction via transcranial alternating current stimulation. Soc Cogn Affect Neurosci. 2018;13(12):1305–16.
46. Cersosimo MG, Benarroch EE. Chapter 5: Central control of autonomic function and involvement in neurodegenerative disorders. In: Buijs RM, Swaab DF, editors. Handbook of clinical neurology: autonomic nervous system, Vol. 117(3). 2013. p. 45–57.
47. Stevens FL, Hurley RA, Taber KH. Anterior Cingulate Cortex: unique role in cognition and emotion. J Neuropsychiatr Clin Neurosci. 2011;23(2):120–5.
48. Drevets WC, Savitz J, Trimble M. The subgenual anterior cingulate cortex in mood disorders. CNS Spectr. 2009;13(8):663–81.
49. Knyazev GK. Cross-frequency coupling of brain oscillations: an impact of state anxiety. Int J Psychophysiol. 2011;80:236–45.
50. Imperatori C, Farina B, Adenzato M, Valenti EM, Murgia C, Della Marca G, Brunetti R, Fontana E, Ardito RB. Default mode network alterations in individuals with high-trait-anxiety: an EEG functional connectivity study. J Affect Disord. 2019;246:611–8.
51. Seeley WW, Menon V, Schatzberg AF, Keller J, Glover GH, Kenna H, Reiss AL, Greicius MD. Dissociable intrinsic connectivity networks for salience processing and executive control. J Neurosci. 2007;27(9):2349–56.
52. Menon V, Uddin LQ. Saliency, switching, attention, and control: a network model of insula function. Brain Struct Funct. 2010;214:655–67.
53. Terasawa Y, Fukushima H, Umeda S. How does interoceptive awareness interact with the subjective experience of emotion? An fMRI study. Hum Brain Mapp. 2013;34(3):598–612.
54. Kikuchi M, Koenig T, Munesue T, Hanaoka A, Strik W, Dierks T, Koshino Y, Minabe Y. EEG microstate analysis in drug-naïve patients with panic disorder. PLoS One. 2011;6(7):e22912. https://doi.org/10.1371/journal.pone.0022912.
55. Wiedemann G, Stevens A, Pauli P, Dengler W. Decreased duration and altered topography of electroencephalographic microstates in patients with panic disorder. Psychiatry Res Neuroimaging. 1998;84:37–48.
56. Wang Y, Chai F, Zhang H, Liu X, Xie P, Zheng L, Yang L, Li L, Fang D. Cortical functional activity in patients with generalized anxiety disorder. BMC Psychiatry. 2016;16:217. https://doi.org/10.1186/s12888-016-0917-3.
57. Chae JH, Jeong J, Peterson BS, Kim DJ, Bahk WM, Jun TY, Kim SY, Kim KS. Dimensional complexity of the EEG in patients with posttraumatic stress disorder. Psychiatry Res. 2004;131(*1*):79–89.
58. Molle M, Marshall L, Wolf B, Fehm HL, Born J. EEG complexity and performance measures of creative thinking. Psychophysiology. 1999;36:95–104.

59. Xing M, Tadayonnejad R, MacNamara A, Ajilore O, Luan Phan K, Klumpp H, Leow A. EEG based functional connectivity reflects cognitive load during emotion regulation. In: IEEE 13th International Symposium on Biomedical Imaging (ISBI). 2016. p. 771–4. https://doi.org/10.1109/ISBI.2016.7493380.
60. Saunders A, Kirk IJ, Waldie KE. Hemispheric coherence in ASD with and without comorbid ADHD and anxiety. Biomed Res Int. 2016. 4267842.
61. Wheelock MD, Rangaprakash D, Harnett NG, Wood KH, Orem TR, Mrug S, Granger DA, Deshpande G, Knight DC. Psychosocial stress reactivity is associated with decreased whole-brain network efficiency and increased amygdala centrality. Behav Neurosci. 2018;132(6):561–72.
62. Knyazev GG, Savostyanov AN, Bocharov AV, Rimareva JM. Anxiety, depression, and oscillatory dynamics in a social interaction model. Brain Res. 2016;1644:62–9.
63. Lackner CI, Marshall WJ, Santesso DL, Dywan J, Wade T, Segalowitz SJ. Adolescent anxiety and aggression can be differentially predicted by electrocortical phase reset variables. Brain Cogn. 2014;89:90–8.
64. Bissonette GB, Powell EM, Roesch MR. Neural structures underlying set-shifting: roles of medial prefrontal cortex and anterior cingulate cortex. Behav Brain Res. 2013;250:91–101.
65. Tompson S, Chua HF, Kitayama S. Connectivity between mPFC and PCC predicts post-choice attitude change: the self-referential processing hypothesis of choice justification. Hum Brain Mapp. 2016;37(11):3810–20.
66. Reiser EM, Schulter G, Weiss EM, Fink A, Rominger C, Papousek I. Decrease of prefrontal-posterior EEG coherence: loose control during social-emotional stimulation. Brain Cogn. 2012;80:144–54.
67. Knyazev GG, Savostyanov AN, Levin EA. Anxiety and synchrony of alpha oscillations. Int J Psychophysiol. 2005;57:175–80.
68. Knyazev GG. Motivation, emotion, and their inhibitory control mirrored in brain oscillations. Neurosci Biobehav Rev. 2007;31:377–95.
69. Fries P. Neuronal gamma-band synchronization as a fundamental process in cortical computation. Annu Rev Neurosci. 2009;32:209–24.
70. De Vera L, González J, Rial RV. Reptilian waking EEG: slow waves, spindles and evoked potentials. Electroencephalogr Clin Neurophysiol. 1994;90:298–303.
71. Lopes FL, Azevedo TM, Imbiriba LA, Freire RC, Valença AM, Caldirola D, Perna G, Volchan E, Nardi AE. Freezing reaction in panic disorder patients associated with anticipatory anxiety. Depress Anxiety. 2009;26(10):917–21.
72. Porges SW. The polyvagal theory: phylogenetic contributions to social behaviour. Physiol Behav. 2003;79:503–13.
73. Knyazev GG, Slobodskaya HR. Personality trait of behavioral inhibition is associated with oscillatory systems reciprocal relationships. Int J Psychophysiol. 2003;48(3):247–61.
74. Harmony T. The functional significance of delta oscillations in cognitive processing. Front Integr Neurosci. 2013;7:83. https://doi.org/10.3389/fnint.2013.00083.
75. Kirov R, Weiss C, Siebner HR, Born J, Marshall L. Slow oscillation electrical brain stimulation during waking promotes EEG theta activity and memory encoding. PNAS. 2009;106(36):15460–5.
76. Hasselmo ME, McClelland JL. Neural models of memory. Curr Opin Neurobiol. 1999;9(2):184–8.
77. Klimesch W, Sauseng P, Hanslmayr S. EEG alpha oscillations: the inhibition-timing hypothesis. Brain Res Rev. 2007;53:63–88.
78. Pfurtscheller G, Andrew C. Event-related changes of band power and coherence: methodology and interpretation. J Clin Neurophysiol. 1999;16:512–9.
79. Palva S, Palva JM. New vistas for α-frequency band oscillations. Trends Neurosci. 2007;30:150–8.
80. Reimann MW, Nolte M, Scolamiero M, Turner K, Perin R, Chindemi G, Dłotko P, Levi R, Hess K, Markam H. Cliques of neurons bound into cavities provide a missing link between

structure and function. Front Comput Neurosci. 2017;11:48. https://doi.org/10.3389/fncom.2017.00048.

81. Shi W, Yeh C-H, Hong Y. Cross-frequency transfer entropy characterize coupling of interacting nonlinear oscillators in complex systems. IEEE Trans Biomed Eng. 2019;66(2):521–9.
82. Novelli L, Wollstadt P, Mediano P, Wibral M, Lizier JT. Large-scale directed network inference with multivariate transfer entropy and hierarchical statistical testing. Accessed via Cornell University Open Source Quantitative Biology>Neurons and Cognition 2019. arXiv:1902.06828 [q-bio.NC].
83. Jin MJ, Kim JS, Kim S, Hyun MH, Lee S-H. An integrated model of emotional problems, beta power of electroencephalography, and low frequency of heart rate variability after childhood trauma in a non-clinical sample: a path analysis study. Front Psych. 2018;8:314. https://doi.org/10.3389/fpsyt.2017.00314.
84. Tulay EE, Metin B, Tarhan N, Arikan MK. Multimodal neuroimaging: basic concepts and classifications of neuropsychiatric diseases. Clin EEG Neurosci. 2019;50(1):20–33.
85. Rabinovich MI, Muezzinoglu MK, Strigo I, Bystritsky A. Dynamical principles of emotion-cognition interaction: mathematical images of mental disorders. PLoS One. 2010;5(9):e12547. https://doi.org/10.1371/journal.pone.0012547.
86. Lee D, Kang D-H, Ha N-H, Oh C-Y, Lee U, Kang SW. Effects of an online mind-body training program on the default mode network: an EEG functional connectivity study. Sci Rep. 2018;8:16935. https://doi.org/10.1038/s41598-018-34947-x.
87. Song P, Lin H, Li S, Wang L, Liu J, Li N, Wang Y. Repetitive transcranial magnetic stimulation (rTMS) modulates time-varying electroencephalography (EEG) network in primary insomnia patients: a TMS-EEG study. Sleep Med. 2019; https://doi.org/10.1016/j.sleep.2019.01.007.
88. Xing M, GadElkarim J, Ajilore O, Wolfson O, Forbes A, Phan KL, Klumpp H, Leow A. Thought chart: tracking the thought with manifold learning during emotion regulation. Brain Inform. 2018;5:7. https://doi.org/10.1186/s40708-018-0085-y.

4 White Matter-Based Structural Brain Network of Anxiety

Kang Soo Lee and Sang Hyuk Lee

Anxiety disorders are characterized by excessive fear, intense anxiety, and related behavioral disturbances [1]. Anxiety is the emotional response in anticipation of a future threat, while fear is the psychological reaction to real or perceived imminent danger [2]. Although anxiety and fear are normal adaptive functions, anxiety disorders may develop when excessive and persistent symptoms cause clinically significant distress or impairment in functioning [3]. White matter (WM) provides the structural connectivity between gray matter regions and enables the rapid and efficient transfer of neural signals [4]. WM tracts can be divided into association pathways (which connect cerebral cortex areas within the same hemisphere), commissural pathways (connecting similar structures across hemispheres), and projection and thalamic pathways (which connect cortical and subcortical structures). Diffusion tensor imaging (DTI) enables the measurement of the restricted diffusion of water in tissue allowing the reconstruction of tracts. Each voxel has several parameters such as rate and direction of diffusion, making it sensitive to subtle pathology in the brain [5]. DTI measures have been used to characterize differences in WM microstructure for a broad spectrum of psychiatric disorders. The most commonly used DTI measure is fractional anisotropy (FA), which has been used as a quantitative indicator of WM integrity and connectivity. We aimed to offer a detailed account of the WM-based structural brain network involved in the behavioral and cognitive manifestations of fear and anxiety. We reviewed findings from DTI studies examining WM microstructural abnormalities in anxiety disorders.

K. S. Lee · S. H. Lee (✉)
Department of Psychiatry, School of Medicine, CHA Bundang Medical Center, CHA University, Seongnam, Korea

Y.-K. Kim (ed.), *Anxiety Disorders*, Advances in Experimental Medicine and Biology 1191, https://doi.org/10.1007/978-981-32-9705-0_4

DTI Studies on Anxiety Disorders

Trait Anxiety in Healthy Subjects

A study published in the *Journal of Neuroscience* found that the fractional anisotropy (FA) values of an amygdala-prefrontal cortex pathway, including the uncinate fasciculus (UF), were negatively correlated with trait anxiety scores measured by the State-Trait Anxiety Inventory (STAI) [6]. Another study showed that there was a negative association between harm avoidance and FA in the cortico-limbic pathway in a large sample of healthy subjects [7]. Higher harm avoidance was associated with lower FA in the corpus callosum (CC), the corticospinal tract, forceps major and minor, inferior fronto-occipital fasciculus (IFOF), inferior and superior longitudinal fasciculus (ILF and SLF), and UF. One study investigated the association between STAI and DTI measurements in patients with social anxiety disorder (SAD) and found a negative correlation between FA in the UF and trait anxiety [8]. Furthermore, reduced WM integrity in the UF was found in children with anxiety disorders [9]. Together, these results strengthen the idea that the neurobiological underpinnings of a state-independent characteristic of anxiety seem to be associated with the UF, which is a major pathway of communication between the amygdala and the orbitofrontal cortex (OFC).

Panic Disorder (PD)

The fear network, comprising the amygdala, hippocampus, thalamus, insula, and prefrontal areas, is the key brain circuit involved in the pathophysiology of PD. The anterior cingulate cortex (ACC) is thought to have a regulatory role in the fear network and is also involved in fear conditioning [10]. One study investigated WM integrity in cingulate regions of patients with PD and found increased FA values of the left anterior and right posterior cingulate, which were positively correlated with anxiety symptom severity [11]. Our research group aims to elucidate the WM neural correlates of anxiety in panic disorder [12–18]. Anxiety sensitivity (AS) in patients with PD was significantly correlated with the FA values of the WM regions near the insula, the splenium of the CC, the fornix, the internal capsule, the posterior thalamic radiation, and the posterior corona radiata (CR) [12]. The neural correlates of AS may be associated with the connections between the insula, thalamus, amygdala, and the entorhinal cortex, which are known to be parts of the neural circuit involved in the modulation of interoception [13]. Our study group also showed increased FA values in the cingulate gyrus in the 5-HTR1A CC genotype compared to the GG/CG genotype in PD. This suggests that the 5-HTR1A polymorphism may be associated with WM connectivity in the cingulum of PD patients [14]. Another study showed a significant association between early trauma and WM connectivity in the fornix body of major limbic structures in PD [15]. We also found that WM microstructural changes are associated with alcohol use in patients with panic disorder [16]. Aberrant WM integrity in the internal capsule and thalamic radiations was significantly associated with suicide attempts in patients with PD [17]. Studies by our group indicate altered integrity of the frontal WM in patients with PD without any comorbidity and provide evidence of altered WM integrity in the frontal lobe, suggesting its contribution to PD anxiety symptoms [18]. Taken together, apparent WM microalterations in patients with PD are present in diverse and widespread regions, although alterations vary in terms of clinical symptom severity and comorbidities.

Generalized Anxiety Disorder (GAD)

Neuroimaging studies have suggested that the amygdala and the prefrontal cortex might underlie the neural circuit of GAD. Adult and adolescent GAD patients showed lower fractional anisotropy (FA) in the uncinate fasciculus than control subjects. One study suggested that GAD patients had higher FA in the right amygdala WM and lower FA in the anterior cingulate WM compared with healthy controls [19]. The reduced FA values of the uncinate fasciculus might suggest a decreased connection between the prefrontal cortex and limbic structures [20]. The FA abnormalities in the UF are consistent with GAD studies describing abnormalities in the connectivity between limbic and frontal structures [21–23].

Social Anxiety Disorder (SAD)

It has been reported that patients with SAD had lower FA in the UF and decreased FA values in the WM of the insula, inferior frontal gyrus, middle temporal gyrus, and inferior parietal gyrus [24, 25]. The studies showed a left lateralized pattern of decreased FA in the SLF, the corticospinal tract, and the UF bilaterally. Although a limited number of DTI studies have investigated global WM volume in SAD, the current state of evidence does not support global WM volume change as a characteristic of SAD. One study showed a significant increase in the FA values of the CC in patients with SAD and interpreted it as increased connectivity in cognitive reappraisal regions. Reduced connectivity in the UF can be interpreted as underlying a diminished top-down control leading to amygdala hyperactivation in patients with SAD. In one study, patients with SAD attended a 10-week cognitive behavioral group therapy and were scanned to investigate cognitive behavioral group therapy-induced structural brain alterations of the WM. DTI analysis revealed a significant increase in FA in the bilateral uncinate fasciculus and the right inferior longitudinal fasciculus, and network-based statistics revealed a significant increase of structural connectivity in a fronto-limbic network.

Post-traumatic Stress Disorder (PTSD)

Lower FA values in PTSD subjects were reported for the ACC, prefrontal gyrus, and posterior angular gyrus and also the left cingulate bundle, in trauma-exposed subjects. The decreased FA values of the UF, cingulum, and CC in patients with PTSD are consistent with the neural circuit of PTSD. The amygdala, medial prefrontal cortex, and hippocampus are suggested as key regions in the functional neurocircuitry of PTSD. Previous studies found decreased FA in the ACC of patients with PTSD [26, 27], and decreased FA in the ACC was more specific for patients with PTSD compared with GAD patients [23]. Another study found abnormal FA values in the left anterior cingulum bundle (CB) [28, 29]. WM integrity in the left frontal gyrus has been reported to be either decreased or increased [23, 29]. One study examined WM connectivity in children with PTSD and found a decreased FA in the body of the CC [30]. Abnormalities in the structural integrity of cingulate regions, the CB, and the amygdala are interpreted as the disturbed top-down control of PTSD. Abnormal findings in the frontal lobe are also interpreted as disturbances in the recall of extinction memory or disturbed emotion processing [23, 26]. According to a meta-analysis, reduced FA is found in the left UF below the region identified in SAD and OCD.

Obsessive-Compulsive Disorder (OCD)

The thalamo-cortico-striatal circuit—which comprises the fronto-basal pathways, the OFC and ACC, the ganglia-thalamic pathways, and the striatum and medial thalamic areas—is thought to be involved in the pathophysiology of OCD. Previous ROI-based studies examined the CB, anterior thalamic radiations, the internal capsule, and the CC [31–33]. Notably, FA values in the ACC and the CB have been found to be conflicting [31–42]. Abnormalities of the structural integrity of the CC have been reported consistently, but there is disagreement regarding the direction and the specific location of such abnormalities in the CC. Taken together, DTI studies in OCD seem to implicate the ACC, CB, and the CC in the pathophysiology of OCD. One study found alterations in the forceps major and the cingulum bundle and suggested that the pathogenesis of OCD may include an abnormal myelination process in not only the fronto-striato-thalamic circuit but also the posterior and temporal regions.

DTI Findings in WM Tracts

Association Pathways

Uncinate Fasciculus (UF)

The UF reciprocally connects structures in the frontal lobe, such as the cingulate gyrus and OFC, to structures in the temporal lobe such as the amygdala [43]. DTI studies in patients with anxiety disorders have reported altered structural connectivity between temporal lobe and prefrontal cortical regions, revealing that FA in the UF is reduced in patients with anxiety disorders [44–49]. The UF is highly relevant to anxiety and emotion regulation, as it connects structures that are crucial to affective processing—such as the amygdala, entorhinal cortices, and parahippocampal gyrus—to frontal regions, including the anterior prefrontal cortex, orbital frontal cortex, ventromedial prefrontal cortex, ACC, and insula [50, 51]. Reduced UF FA is not specific to anxiety disorders, as similar WM alterations have been observed in some studies of individuals with trait anxiety [6, 52–56] as well as in patients with affective and other psychiatric disorders [57, 58]. Early-life alterations in a key WM tract involved in emotion and anxiety regulation contribute to childhood anxiety [59]. Several DTI studies have indicated that WM characteristics are heritable, which is further supported by the finding that genetic factors explain 60–80% of the variance in UF microstructure. More specifically, structural integrity of the UF is influenced by genetic variations in the brain-derived neurotrophic factor and the 5-HTTLPR genotype. Taken together, reduced integrity of the UF might be associated with SAD, GAD, trait anxiety, and heritability. Moreover, UF FA reductions have been associated with increased anxiety symptoms and deficits in fear extinction.

Cingulum, IFOF, SLF, and ILF

The cingulum is a key component of the limbic system that allows for communication between the cingulate, medial frontal lobe, and temporal lobe. It is importantly involved in emotion regulation [60, 61]. The prefrontal cortex and the ACC inhibit the activation of anxiety-related regions via their interaction with the cingulate gyrus [62]. Deficits in this circuit are associated with dysfunction of the top-down system and with activation of posterior structures and other limbic systems. One study showed that decreased FA in the left cingulum correlated with clinical symptom severity in patients with GAD [63]. Aberrant integrity of the CB contributes to emotion dysregulation and is a neural substrate of GAD. Cingulum changes in DTI studies, which appear most consistently in OCD and PTSD, are concentrated in the dorsal cingulum. Cingulum FA correlates with default-mode functional connectivity. The IFOF, which integrates the auditory and visual association cortices into the prefrontal cortex, has been implicated in the pathophysiology of OCD [64]. Reduced FA is also present in the right IFOF of patients with GAD [65]. Young healthy participants with high trait anxiety have been reported to show significantly decreased FA values in the CR, anterior thalamic radiation (ATR), IFOF, and CC, compared with subjects with low trait anxiety [66]. The altered WM may be a vulnerability marker in individuals at high risk for clinical anxiety. The inferior IFOF might connect the frontal and occipital lobes and modulate anxiety responses to environmental stimuli [67]. Decreased FA in the inferior IFOF has been reported in other anxiety disorders including panic disorder and OCD [68, 69]. Reduced FA in the inferior IFOF might disturb sensory integration and cognitive or emotional regulation of sensory perception. The ILF is an important afferent fiber of the amygdala and connects the temporal cortex with visual areas and with the fusiform gyrus. According to a meta-analysis, the ILF takes part in a common brain mechanism of anxiety disorders and plays a major role in face processing. One study found that reduced FA values in the right ILF and SLF in patients with SAD negatively correlate with the severity of anxiety [70]. These findings raise the possibility that the pattern of WM abnormalities in the right SLF and ILF may be implicated in one aspect of the pathophysiology of SAD, namely, the dysfunction of face processing, leading to deficits in social interactions. Reduced FA values in the SLF are also supported by tract-based spatial statistics studies in OCD and PTSD. The anterior CR (ACR) includes thalamic projections from the internal capsule to the cortex [37, 38]. WM abnormalities in the ACR can affect emotional processing and the executive attention network [32, 43]. In previous DTI studies on PTSD, the cortical-ACR-IC-thalamus-limbic pathway, rather than the cortical UF-limbic pathway, was suggested to play a key role [45–47]. The cortical-thalamus-limbic pathway plays a prominent role in the regulation of emotional behavior in PTSD patients [48, 49]. Alterations in the ACR-thalamus pathway are more extensive than in the UF pathway, as they result in decreased FA values in the CR and ATR in healthy individuals with high levels of anxiety.

Commissural Pathways

CC, Forceps Minor, and Forceps Major

The CC is the largest fiber bundle connecting the two hemispheres. The forceps minor passes through the CC and connects medial and lateral frontal cortices and the forceps major curves backward, linking the occipital poles of the hemispheres. The CC has many functions such as the regulation of motor, sensory, emotional, and cognitive functions, including attention and intelligence [25, 50]. Dysfunction of the CC has been implicated in the pathogenesis of OCD [31, 52]. In an OCD study, increased FA values were found in the major and minor forceps, bilateral corticospinal tract, right ATR, and bilateral SLF [44]. In contrast, a PTSD study reported decreased FA in the forceps minor [57]. WM abnormalities in populations with high anxiety involved not only the anterior CC but also extended to the forceps minor.

Projection and Thalamic Pathways

Fornix

The fornix is the main axonal output pathway from the hippocampus to the mammillary bodies and has been identified as a key region in memory and executive functions [71–73]. The fornix is also a part of the larger Papez circuit, an important network in the limbic system, and is involved in the regulation of emotions [74]. A positive correlation between trait anxiety scores and the mean FA value was obtained in the fornix and the left UF [58].

Conclusion

The white matter-based structural brain network of a trait anxiety seems to be associated with the uncinate fasciculus, a major pathway between the amygdala and orbitofrontal cortex. Apparent white matter micro-alterations in patients with panic disorder are present in diverse and widespread regions, although alterations vary in terms of clinical symptom severity and comorbidities. Social anxiety disorder is associated with structural dysconnectivity in a fronto-limbic network consistent with reduced fractional anisotropy values in uncinate fasciculus and inferior longitudinal fasciculus. The pathogenesis of obsessive-compulsive disorder may include abnormal findings in not only the fronto-striato-thalamic circuit but also the posterior and temporal regions of forceps major and cingulum bundle. Taken together, studies of white matter in anxiety elucidated overlapping patterns of fronto-cortical and fronto-limbic changes with uncinate fasciculus and cingulum alterations.

References

1. American Psychiatric Association. DSM 5. American Psychiatric Association; 2013
2. Kim MJ, Loucks RA, Palmer AL, Brown AC, Solomon KM, Marchante AN, et al. The structural and functional connectivity of the amygdala: from normal emotion to pathological anxiety. Behav Brain Res. 2011;223(2):403–10.
3. Rauch SL, Shin LM, Wright CI. Neuroimaging studies of amygdala function in anxiety disorders. Ann N Y Acad Sci. 2003;985(1):389–410.
4. Filley CM, Fields RD. White matter and cognition: making the connection. J Neurophysiol. 2016;116(5):2093–104.
5. Basser PJ, Pajevic S, Pierpaoli C, Duda J, Aldroubi A. In vivo fiber tractography using DT-MRI data. Magn Reson Med. 2000;44(4):625–32.
6. Kim MJ, Whalen PJ. The structural integrity of an amygdala-prefrontal pathway predicts trait anxiety. J Neurosci. 2009;29:11614–8.
7. Westlye LT, Bjornebekk A, Grydeland H, Fjell AM, Wallhovd KB. Linking an anxiety-related personality trait to brain white matter microstructure: diffusion tensor imaging and harm avoidance. Arch Gen Psychiatry. 2011;68:369–77.
8. Baur V, Hanggi J, Rufer M, Delsignore A, Jäncke L, Herwig U, et al. White matter alterations in social anxiety disorder. J Psychiatr Res. 2011;45:1366–72.
9. Tromp DPM, Williams LE, Fox AS, Oler JA, Gregory R, et al. 2015. White matter alterations in pre-adolescent children with anxiety disorders. Biological Psychiatry, 70th annual scientific convention and meeting, S142–286.
10. Shin LM, Liberzon I. The neurocircuitry of fear, stress, and anxiety disorders. Neuropsychopharmacology. 2010;35:169–91.
11. Han DH, Renshaw PF, Dager SR, Chung A, Hwang J, Daniels MA, et al. Altered cingulate white matter connectivity in panic disorder patients. J Psychiatr Res. 2008;42:399–407.
12. Kim MK, Kim B, Kiu Choi T, Lee SH. White matter correlates of anxiety sensitivity in panic disorder. J Affect Disord. 2017;207:148–56.
13. Paulus MP, Stein MB. An insular view of anxiety. Biol Psychiatry. 2006;60(4):383–7.
14. Yu ST, Kim MK, Kim B, Yoo E, Lee JY, Lee KS, et al. The effects of 5-HTR1A polymorphism on cingulum connectivity in patients with panic disorder. Psychiatry Investig. 2013;10(4):399–406.
15. Yu ST, Lee KS, Lee SH. Fornix microalterations might be associated with early trauma in panic disorder. J Affect Disord. 2017;220:139–46.
16. Kim B, Shin WS, Kim MK, Lee SH. White matter microstructural changes are associated with alcohol use in patients with panic disorder. J Affect Disord. 2016;199:65–72.
17. Kim B, Oh J, Kim MK, Lee S, Tae WS, Kim CM, et al. White matter alterations are associated with suicide attempt in patients with panic disorder. J Affect Disord. 2015;175:139–46.
18. Kim B, Kim JH, Kim MK, Lee KS, Kim Y, Choi TK, et al. Frontal white matter alterations in short-term medicated panic disorder patients without comorbid conditions: a diffusion tensor imaging study. PLoS One. 2014;9(4):e95279.

19. Hettema JM, Kettenmann B, Ahluwalia V, McCarthy C, Kates WR, Schmitt JE, et al. Pilot multimodal twin imaging study of generalized anxiety disorder. Depress Anxiety. 2011;29(3):202–9.
20. Ebeling U, von CD. Topography of the uncinate fascicle and adjacent temporal fiber tracts. Acta Neurochir. 1992;115:143–8.
21. Etkin A, Prater KE, Schatzberg AF, Menon V, Greicius MD. Disrupted amygdalar subregion functional connectivity and evidence of a compensatory network in generalized anxiety disorder. Arch Gen Psychiatry. 2009;66:1361–72.
22. Etkin A, Prater KE, Hoeft F, Menon V, Schatzverg AF. Failure of anterior cingulate activation and connectivity with the amygdala during implicit regulation of emotional processing in generalized anxiety disorder. Am J Psychiatry. 2010;167:545–54.
23. Zhang L, Zhang Y, Li L, Li Z, Li W, Ma N, et al. Different white matter abnormalities between the first-episode, treatment-naive patients with posttraumatic stress disorder and generalized anxiety disorder without comorbid conditions. J Affect Disord. 2011;133:294–9.
24. Liao W, Xu Q, Mantini D, Ding J, Machado-de-Sousa JP, Jallak JE, et al. Altered gray matter morphometry and resting-state functional and structural connectivity in social anxiety disorder. Brain Res. 2011;1388:167–77.
25. Phan KL, Orlichenko A, Boyd E, Angstadt M, Coccaro EF, Liberzon I, et al. Preliminary evidence of white matter abnormality in the uncinate fasciculus in generalized social anxiety disorder. Biol Psychiatry. 2009;66:691–4.
26. Schuff N, Zhang Y, Zhan W, Lenoci M, Ching C, Boreta L, et al. Patterns of altered cortical perfusion and diminished subcortical integrity in posttraumatic stress disorder: an MRI study. Neuroimage. 2011;54(Suppl 1):S62–8.20.
27. Kim MJ, Lyoo IK, Kim SJ, Sim M, Kim N, Choi N, et al. Disrupted white matter tract integrity of anterior cingulate in trauma survivors. Neuroreport. 2005;16:1049–53.
28. Kim SJ, Jeong DU, Sim ME, Bae SC, Chung A, Kim MJ, et al. Asymmetrically altered integrity of cingulum bundle in posttraumatic stress disorder. Neuropsychobiology. 2006;54:120–5.
29. Abe O, Yamasue H, Kasai K, Yamada H, Aoki S, Iwanami A, et al. Voxel-based diffusion tensor analysis reveals aberrant anterior cingulum integrity in posttraumatic stress disorder due to terrorism. Psychiatry Res. 2006;146:231–42.
30. Jackowski AP, Douglas-Palumberi H, Jackowski M, Win L, Schultz RT, Staib LW, et al. Corpus callosum in maltreated children with posttraumatic stress disorder: a diffusion tensor imaging study. Psychiatry Res. 2008;162:256–61.
31. Chiu CH, Lo YC, Tang HS, Liu IC, Chiang WY, Yeh FC, et al. White matter abnormalities of fronto-striato-thalamic circuitry in obsessive-compulsive disorder: a study using diffusion spectrum imaging tractography. Psychiatry Res. 2011;192:176–82.
32. Cannistraro PA, Makris N, Howard JD, Wedig MM, Hodge SM, Wilhelm S, et al. A diffusion tensor imaging study of white matter in obsessive-compulsive disorder. Depress Anxiety. 2007;24:440–6.
33. Saito Y, Nobuhara K, Okugawa G, Takase K, Sugimoto T, Horiuchi M, et al. Corpus callosum in patients with obsessive-compulsive disorder: diffusion-tensor imaging study. Radiology. 2008;246:536–42.
34. Li F, Huang X, Yang Y, Li B, Wu Q, Zhang T, et al. Microstructural brain abnormalities in patients with obsessive-compulsive disorder: diffusion tensor MR imaging study at 3.0 T. Radiology. 2011;260:216–23.
35. Menzies L, Williams GB, Chamberlain SR, Ooi C, Fineberg N, Suckling J, et al. White matter abnormalities in patients with obsessive-compulsive disorder and their first-degree relatives. Am J Psychiatry. 2008;165:1308–15.
36. Szeszko PR, Ardekani BA, Ashtari M, Malhotra AK, Robinson DG, Bilder RM, et al. White matter abnormalities in obsessive-compulsive disorder: a diffusion tensor imaging study. Arch Gen Psychiatry. 2005;62:782–90.
37. Nakamae T, Narumoto J, Sakai Y, Nishida S, Yamada K, Nishimura T, et al. Diffusion tensor imaging and tract-based spatial statistics in obsessive-compulsive disorder. J Psychiatr Res. 2011;45:687–90.

38. Garibotto V, Scifo P, Gorini A, Alonso CE, Brambati S, Bellodi L, et al. Disorganization of anatomical connectivity in obsessive compulsive disorder: a multi-parameter diffusion tensor imaging study in a subpopulation of patients. Neurobiol Dis. 2010;37:468–76.
39. Zarei M, Mataix-Cols D, Heyman I, Hough M¿ Doherty J, Burge L, et al. Changes in gray matter volume and white matter microstructure in adolescents with obsessive-compulsive disorder. Biol Psychiatry 2011;70:1083–1090.
40. Bora E, Harrison BJ, Fornito A, Cocchi L, Pujol J, Fontenelle LF, et al. White matter microstructure in patients with obsessive-compulsive disorder. J Psychiatry Neurosci. 2011;36:42–6.
41. Nakamae T, Narumoto J, Shibata K, Matsumoto R, Kitabayashi Y, Yoshida T, et al. Alteration of fractional anisotropy and apparent diffusion coefficient in obsessive-compulsive disorder: a diffusion tensor imaging study. Prog Neuro-Psychopharmacol Biol Psychiatry. 2008;32:1221–6.
42. Yoo SY, Jang JH, Shin YW, Kim DJ, Park HJ, Moon WJ, et al. White matter abnormalities in drug-naive patients with obsessive-compulsive disorder: a diffusion tensor study before and after citalopram treatment. Acta Psychiatr Scand. 2007;116:211–9.
43. Leng B, Han S, Bao Y, Zhang H, Wang Y, Wu Y, et al. The uncinate fasciculus as observed using diffusion spectrum imaging in the human brain. Neuroradiology. 2016;58:595–606.
44. Tromp DPM, Grupe DW, Oathes DJ, McFarlin DR, Hernandez PJ, Kral TR, et al. Reduced structural connectivity of a major frontolimbic pathway in generalized anxiety disorder. Arch Gen Psychiatry. 2012;69:925–34.
45. Baur V, Brühl AB, Herwig U, Eberle T, Rufer M, Delsignore A, et al. Evidence of frontotemporal structural hypoconnectivity in social anxiety disorder: a quantitative fiber tractography study. Hum Brain Mapp. 2013;34:437–46.
46. Hettema JM, Kettenmann B, Ahluwalia V, McCarthy C, Kates WR, Schmitt JB, et al. Pilot multimodal twin imaging study of generalized anxiety disorder. Depress Anxiety. 2012;29:202–9.
47. Baur V, Hänggi J, Rufer M, Delsignore A, Jäncke L, Herwig U, et al. White matter alterations in social anxiety disorder. J Psychiatr Res. 2011;45:1366–72.
48. Liao M, Yang F, Zhang Y, He Z, Lu L, Li L. White matter abnormalities in adolescents with generalized anxiety disorder: a diffusion tensor imaging study. BMC Psychiatry. 2014;14:41.
49. Schmahmann JD, Pandya DN, Wang R, Dai G, D'Arceuil HR, de Crespigny AJ, et al. Association fibre pathways of the brain: parallel observations from diffusion spectrum imaging and autoradiography. Brain. 2007;130:630–53.
50. Catani M, Howard RJ, Pajevic S, Jones DK. Virtual in vivo interactive dissection of white matter fasciculi in the human brain. NeuroImage. 2002;17:77–94.
51. Montag C, Reuter M, Weber B, Markett S, Schoene-Bake JC. Individual differences in trait anxiety are associated with white matter tract integrity in the left temporal lobe in healthy males but not females. Neuroscience. 2012;217:77–83.
52. Kim MJ, Brown AC, Mattek AM, Chavez SJ, Taylor JM, Palmer AL, et al. The inverse relationship between the microstructural variability of amygdala-prefrontal pathways and trait anxiety is moderated by sex. Front Syst Neurosci. 2016;10:93.
53. Kim MJ, Avinun R, Knodt AR, Radtke SR, Hariri AR. Neurogenetic plasticity and sex influence the link between corticolimbic structural connectivity and trait anxiety. Sci Rep. 2017;7:10959.
54. Westlye LT, Bjørnebekk A, Grydeland H, Fjell AM, Wallhovd KB. Linking an anxiety related personality trait to brain white matter microstructure: diffusion tensor imaging and harm avoidance. Arch Gen Psychiatry. 2011;68:369–77.
55. Baur V, Hänggi J, Jäncke L. Volumetric associations between uncinated fasciculus, amygdala, and trait anxiety. BMC Neurosci. 2012;13:4.
56. Modi S, Trivedi R, Singh K, Kumar P, Rathore RK, Tripathi RP, et al. Individual differences in trait anxiety are associated with white matter tract integrity in fornix and uncinate fasciculus: preliminary evidence from a DTI based tractography study. Behav Brain Res. 2013;238:188–92.
57. Sexton CE, Mackay CE, Ebmeier KP. A systematic review of diffusion tensor imaging studies in affective disorders. Biol Psychiatry. 2009;66:814–23.

58. Von Der Heide RJ, Skipper LM, Klobusicky E, Olson IR. Dissecting the uncinate fasciculus: disorders, controversies, and a hypothesis. Brain. 2013;136:1692–707.
59. Tromp DPM, Williams LE, Fox AS, Oler JA, Roseboom PH, Rogers GM, et al. Altered uncinate fasciculus microstructure in childhood anxiety disorders in boys but not girls. Am J Psychiatry. 2019 doi: https://doi.org/10.1176/appi.ajp.2018.18040425 [Epub ahead of print]
60. Hartley CA, Phelps EA. Changing fear: the neurocircuitry of emotion regulation. Neuropsychopharmacology. 2010;35:136–46.
61. Delgado MR, Nearing KI, Ledoux JE, Phelps EA. Neural circuitry underlying the regulation of conditioned fear and its relation to extinction. Neuron. 2008;59:829–38.
62. Mochcovitch MD, da Rocha Freire RC, Garcia RF, Nardi AE. A systematic review of fMRI studies in generalized anxiety disorder: evaluating its neural and cognitive basis. J Affect Disord. 2014;167:336–42.
63. Wang W, Qian S, Liu K, Li B, Li M, Xin K, et al. Reduced white matter integrity and its correlation with clinical symptom in first-episode, treatment-naive generalized anxiety disorder. Behav Brain Res. 2016;314:159–64.
64. Catani M. Diffusion tensor magnetic resonance imaging tractography in cognitive disorders. Curr Opin Neurol. 2006;19:599–606.
65. Liao M, Yang F, Zhang Y, He Z, Su L, Li L. White matter abnormalities in adolescents with generalized anxiety disorder: a diffusion tensor imaging study. BMC Psychiatry. 2014;14:41.
66. Lu M, Yang C, Chu T, Wu S. Cerebral white matter changes in young healthy individuals with high trait anxiety: a tract-based spatial statistics study. Front Neurol. 2018;9:704.
67. Gorman JM, Kent JM, Sullivan GM, Coplan JD. Neuroanatomical hypothesis of panic disorder, revised. Am J Psychiatry. 2000;157:493–505.
68. Lai CH, Wu YT. Fronto-occipital fasciculus, corpus callosum and superior longitudinal fasciculus tract alterations of first-episode, medication-naive and late-onset panic disorder patients. J Affect Disord. 2013;146:378–82.
69. Peng Z, Lui SS, Cheung EF, Jin Z, Miao G, Jing J, et al. Brain structural abnormalities in obsessive-compulsive disorder: converging evidence from white matter and grey matter. Asian J Psychiatr. 2012;5:290–6.
70. Tükel R, Ulasoglu Yildiz C, Ertekin E, Kurt E, Koyuncu A, Aydın K. Evidence for alterations of the right inferior and superior longitudinal fasciculi in patients with social anxiety disorder. Brain Res. 2017;1662:16–22.
71. Aggleton JP, Vann SD, Saunders RC. Projections from the hippocampal region to the mammillary bodies in macaque monkeys. Eur J Neurosci. 2005;22:2519–30.
72. Hescham S, Lim LW, Jahanshahi A, Steinbusch HW, Prickaerts J, Blokland A, et al. Deep brain stimulation of the forniceal area enhances memory functions in experimental dementia: the role of stimulation parameters. Brain Stimul. 2012;6(1):72–7.
73. Lee DY, Fletcher E, Carmichael OT, Singh B, Mungas D, Reed B, et al. Sub-regional hippocampal injury is associated with fornix degeneration in Alzheimer's disease. Front Aging Neurosci. 2012;4:1.
74. Dalgleish T. The emotional brain. Nat Rev Neurosci. 2004;5(7):583–9.

Anxiety Disorders and the Brain's Resting State Networks: From Altered Spatiotemporal Synchronization to Psychopathological Symptoms

5

Georg Northoff

Introduction

Anxiety and depression are among the most common symptoms. According to the World Health Organization (WHO), 20 million North Americans suffer from depression which also causes major economic burden with an estimated €92 billion [1]. Importantly, depression and anxiety can be associated with different disorders. Most typically, depression and anxiety occur in those psychiatric disorders that are classified as anxiety disorders in DSM V; these include mainly panic disorder (PD), generalized anxiety disorder (GAD), social anxiety disorder (SAD), agoraphobia, and specific phobias – our review focuses on GAD, PD, and SAD.

In addition to cognitive and affective symptoms, subjects suffering from PD, GAD, and SAD often also show (i) reduced heart rate variability (HRV) [2–4]; (ii) abnormal perception of their own heart beat [5–8]; (iii) somatic and especially cardiac symptoms like racing heart rate, heart palpitations, and chest pain [2, 3, 7, 8]; and (iv) increased risk for coronary artery disease and atrial fibrillation [9]. Taken together, the combination of both affective and somatic-cardiac symptoms in anxiety disorders suggests abnormal coupling of the brain's neural and the heart's cardiac activity.

The present paper focuses on how the brain's resting state and its functional connectivity process and monitor the cardiac inputs from the heart. The main and overarching hypothesis is that abnormalities in the brain's resting state activity lead to dysfunctional coupling between the heart and brain in anxiety disorders. That, as we elaborate on the basis of recent findings in both healthy and anxiety subjects, can be traced to alterations in those mechanisms that allow for spatiotemporal synchronization between neural and cardiac activity within the brain's resting state.

G. Northoff (✉)
EJLB-Michael Smith Chair for Neuroscience and Mental Health, Royal Ottawa Healthcare Group, University of Ottawa Institute of Mental Health Research, Ottawa, ON, Canada
e-mail: georg.northoff@theroyal.ca; http://www.georgnorthoff.com

Y.-K. Kim (ed.), *Anxiety Disorders*, Advances in Experimental Medicine and Biology 1191, https://doi.org/10.1007/978-981-32-9705-0_5

The first part of the paper focuses on recent findings in healthy subjects on how neural and cardiac activity are coupled and synchronized both spatially and temporally within the brain's ongoing resting state activity. We will establish various hypotheses about various mechanisms of spatiotemporal synchronization of neural and cardiac activity within the brain's spontaneous activity. That provides the basis for the second part about anxiety disorders. We here review recent findings on resting state abnormalities in various networks in anxiety disorders. That provides the basis for suggesting abnormal expression of spatiotemporal synchronization of neural and cardiac activity within the brain's spontaneous activity in these disorders. We thus establish a set of experimentally testable hypothesis about abnormal spatiotemporal brain-heart synchronization in anxiety disorders which provide direct link to the psychopathological symptoms like anxiety, unstable self, and increased bodily awareness.

The here suggested approach sets anxiety disorders within the framework of the recently introduced "Spatiotemporal Psychopathology." Spatiotemporal Psychopathology suggests primarily spatiotemporal (rather than affective or cognitive) mechanisms to provide the link between the brain's neural changes and the mind's psychopathological symptoms [10–16].

Part I: Brain-Heart Coupling – Spatiotemporal Synchronization in the Healthy Brain

Spatial Relationship Between Neural and Cardiac Variability

Initial investigations using brain imaging with functional magnetic resonance imaging (fMRI) during the so-called resting state (i.e., absence of specific tasks or stimuli) show close coupling between the heart and brain in healthy subjects [17]. observed direct relation between heart rate variability (HRV) and resting state functional connectivity in the brain, specifically in the insula, amygdala, and dorsal anterior cingulate cortex (dACC) see also [18, 19].

Others have observed close coupling of neural activity in ventromedial prefrontal cortex (VMPFC) (and adjacent perigenual anterior cingulate cortex (PACC)), somatosensory and somatomotor cortex, and subcortical regions (like periaqueductal gray (PAG), thalamus, and lentiform nucleus as part of the so-called central autonomic network (CAN)) with the high-frequency content of HRV recordings [18–20].

Other investigations of the brain using magnetoencephalography (MEG, a technique with high temporal resolution) show direct temporal coupling between the brain's oscillatory activity and heart rhythm during the brain's resting state [21–26] (see below for mechanistic details). Such coupling can specifically be observed in the insula, the somatosensory cortex, and the PACC and VMPFC – the MEG (which cannot capture subcortical regions) thus confirm the cortical regions involved in brain-heart coupling.

The very same regions and SN are also recruited during task-evoked activity related to the awareness of the heartbeat, i.e., interoceptive awareness (see [27, 28] for the recent distinction between interoceptive accuracy, sensitivity, and awareness as well as physiological interoception). Studies by others [29, 30] and our group [5, 31, 32] demonstrated strong involvement of the insula, dACC, thalamus, and somatosensory/motor cortex during heartbeat counting which requires awareness of one's own heart. Moreover, VMPFC and PACC are less deactivated (in fMRI) during interoceptive awareness of the own heartbeat [31].

Taken together, the regions mediating brain-heart coupling in the healthy brain can be subsumed under three main neural networks as they have been identified in the brain's resting state [33]. (i) The insula, amygdala, dACC, and thalamus form the salience network (SN) [34]; (ii) the somatosensory cortex and its associated regions are at the core of the somatosensory-motor network (SMN); and finally (iii) VMPFC and PACC are part of the default-mode network (DMN) [35, 36].

The involvement of these networks in brain-heart coupling raises the question for the exact mechanisms how the neural activity of these regions couples to the ongoing cardiac input. The findings suggest that neural variability in the brain's resting state activity is related to cardiac variability, e.g., heartrate variability (HRV). That is further supported by the fact that the frequency range in which fMRI resting state neuronal variability is typically measured (0.01–0.1 Hz) corresponds well to the frequency range of slow and fast HRV (0.01–0.1). Our first neuro-cardiac hypothesis is thus that there is a direct relationship between neural and cardiac variability in the brain's resting state activity in regions of SN, SMN, and DMN. The coupling between neural and cardiac variabilities raises the question for the underlying temporal mechanisms which shall be the focus in the following.

Phase-Based Temporal Synchronization of the Heart and Brain

We so far discussed the regions/networks and their resting state and task-evoked activity implicated in brain-heart coupling in anxiety disorders. This leaves open (i) the neuronal mechanisms underlying brain-heart coupling and (ii) how the latter can account for the psychopathological symptoms observed in anxiety disorders. Let us start with the first point, the neuronal mechanisms of brain-heart coupling, for which we briefly go back to the healthy brain.

As said above, the group around Tallon-Baudry conducted several MEG and intracranial EEG studies on brain-heart coupling and how that is related to perception and our sense of self (see [37] for an excellent overview). Thereby we need to understand two different mechanisms, first, the one that allows for the coupling between neuronal and cardiac activity in the brain and, second, how that transforms into mental features like perception, emotion, and self. Let us start with the first mechanism, that is, neuro-cardiac coupling.

Park et al. [21–24] conducted several studies using MEG or intracranial EEG (iEEG) while at the same time recording the heartbeat with ECG. It is well known that the heartbeat induces an event-related potential in the brain which is described as heartbeat-evoked potential (HEP) [38]. Since the HEP is time-wise related to the heartbeat, it must reflect the coupling of the brain's ongoing neuronal activity, e.g., its so-called resting state or spontaneous activity, with the heart-related rhythmic activity.

The brain's ongoing neuronal activity, that is, its spontaneous or resting state activity, can be characterized by continuous fluctuations and oscillation with varying phase onsets [39, 40]. Various investigations demonstrated that the brain can shift its phase onsets according to the onset of external stimuli as in music – this is described as entrainment [41] or alignment [42]. Phase shift can thus be seen as mechanisms by means of which the brain can actively align its own neural activity to external stimuli. This raises the question whether such active alignment also underlies the resting state's activity coupling to the continuous cardiac input. Specifically, one may now want to investigate whether the phase onsets in the brain's spontaneous activity are temporally coupled and ultimately synchronized with the onsets of the heartbeats (and ultimately the amplitude of the HEP) as they are processed by the brain.

[21–24] observed that the phase onset of the brain's ongoing neuronal fluctuations is locked to the timing of the ongoing heartbeat, e.g., the R-peak as measured in ECG. The phase onset in the brain's ongoing neuronal fluctuations was coherent across numerous heartbeats and thus temporally phase-locked to the latter – this could be measured by and was reflected in high values of the brain's intertrial coherence (ITC) (as in the frequency range between 4 and 7 Hz) relative to the heartbeat onset [24]. This strongly suggests that the timing of both the brain, e.g., phase onset as measured by ITC, and the heart, e.g., heartbeat onset including HRV, is coupled and thus synchronized with each other.

The ITC, in turn, was directly related to the amplitude of the HEP with higher ITC leading to higher amplitude in the HEP [24] therefore concluded that the amplitude of HEP can, at least in some major part, be linked to the phase shift, i.e., ITC. The ITC reflecting the coupling between the brain's phase onsets and the timing of the heartbeat (as processed by the brain) can thus be conceived a marker of brain-heart coupling. Moreover, as ITC mediates the amplitude of HEP, the latter can be conceived a second marker. That needs to be distinguished from mere power changes as measured by event-related spectral perturbation (ERSP) as these do not mediate the amplitude of HEP [24].

Taken together, the data suggest that temporal synchronization is a central mechanism in coupling neural and cardiac activity in the brain. Specifically, as such synchronization is mediated by coherence or phase, we speak of phase-based temporal synchronization between the rhythms of both the brain and the heart – this is manifested in the amplitude of HEP. Low temporal synchronization, e.g., low ITC, will reduce HEP amplitude, while high degrees of temporal synchronization with high ITC will enhance the amplitude of HEP (see Fig. 5.1a).

We can thus formulate yet another neuro-cardiac hypothesis. The temporal coupling of neural and cardiac activity in the brain's resting state activity is mediated by phase shift resulting in neuro-cardiac phase entrainment or alignment as it can be measured

a

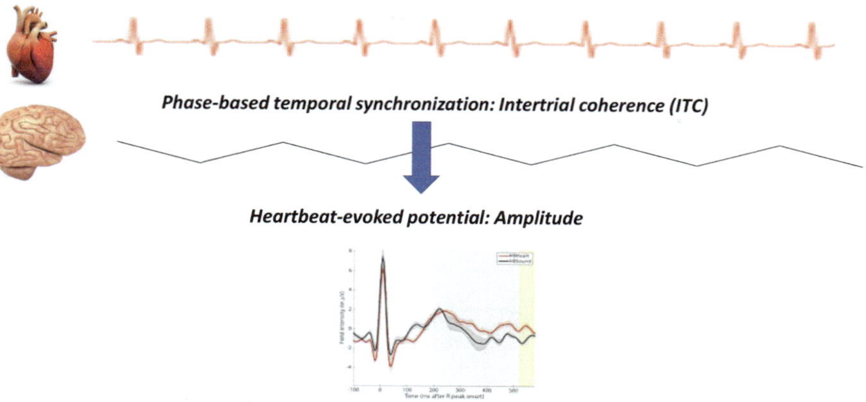

Fig. 5.1a Neuro-cardiac phase-based temporal synchronization and the heartbeat-evoked potential (HEP)

by intertrial coherence (ITC). The phase shift, in turn, mediates the amplitude of the heartbeat-evoked potential (HEP). The temporal phase alignment thus modulates the amplitude the cardiac input can induce in the brain's resting state activity.

Spatiotemporal Brain-Heart Synchronization Shapes Mental Functions (Emotion, Bodily Awareness, and the Sense of Self)

We identified phase-based temporal synchronization as core mechanism in coupling neural and cardiac activity. This allows us to proceed to our second question, namely, how temporal synchronization impacts and shapes mental features like perception, emotion, bodily awareness, and sense of self. To test that, Park included tasks for visual perception [21, 22], body awareness [24], and sense of self [23, 25, 26] and related them to neuro-cardiac coupling, namely, the HEP, as investigated in various MEG or iEEG studies (see [37] for an overview).

[21, 22] observed that visual perception, e.g., visual consciousness, is directly dependent upon the degree of neuro-cardiac temporal synchronization in specifically visual cortex and insula. The same holds for bodily awareness: different degrees in the amplitude of HEP in the insula (as region of SN) and the somatosensory cortex (as part of SMN) were related to different degrees of bodily awareness (as tested for by synchronous vs asynchronous brush stroke) [24].

Finally, [25, 26] and [23] could show that the subjects' experience of their own self as either "I" or "me" is directly related to the amplitude of the HEP and its underlying neuro-cardiac temporal synchronization in specifically the VMPFC as part of DMN. The DMN is well known to be involved in our sense of self [43, 44], the SMN has been related to bodily awareness [45], and the SN is central in mediating both sense of self [46, 47] and emotions [48]. Given the findings by the group

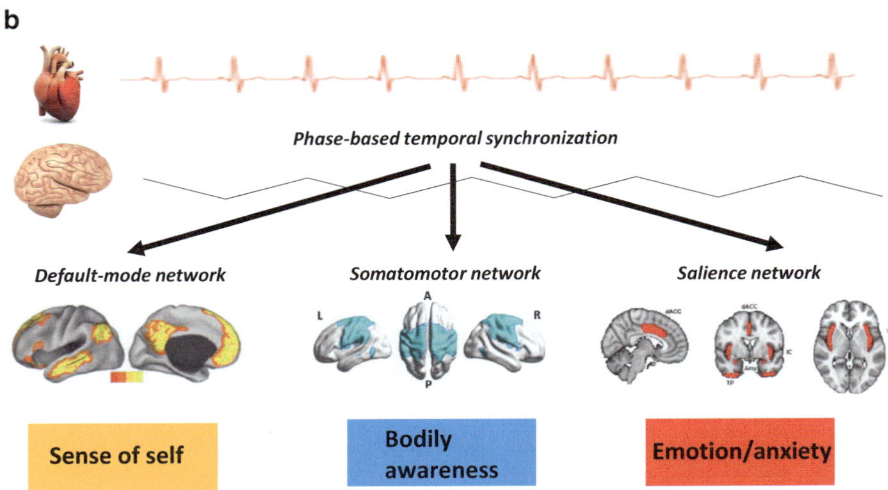

Fig. 5.1b Neuro-cardiac phase-based temporal synchronization in different networks and related psychological functions

around Tallon-Baudry, one may now want to assume that these regions'/networks' involvement in mental features like emotions, bodily awareness, and sense of self may, in part, be related to phase-based temporal synchronization between neural and cardiac activity (see [37] who makes this point) (see Fig. 5.1b).

In sum, the findings by the group around Tallon-Baudry shows that phase-based temporal synchronization in specific networks like SN, SMN, and DMN between neuronal and cardiac rhythms may not only allow for brain-heart coupling but, at the same time, also strongly shape mental features like perception, bodily awareness, and sense of self. We therefore speak of spatiotemporal synchronization as mechanism of brain-heart coupling which (i) allows connecting, e.g., synchronizing neuronal and cardiac activity, and (ii) shaping mental features like emotions, bodily awareness, and sense of self.

One would consequently expect that changes in these networks, e.g., DMN, SN, and SMN, as in anxiety disorders, are (i) based on altered phase-based temporal synchronization between neuronal and cardiac activity, which, in turn, (ii) should lead to abnormalities in the respective mental features like emotion, bodily awareness, and sense of self with the corresponding psychopathological symptoms. That is indeed the case in anxiety disorder as we will discuss in the following.

Part II: Brain-Heart Coupling – Spatiotemporal Hypo- and Hyper-synchronization and Its Psychopathology in Anxiety Disorders

Resting State Abnormalities in SN, DMN, and SMN

Several studies on resting state functional connectivity in PD have been conducted (see [3, 49–51] for reviews). [52] reported increased rsFC in the midline regions including perigenual and posterior cingulate cortex (PACC, PCC) as typical default-mode network (DMN) regions. Corresponding changes in midline regions of DMN were observed by [53] who found increased rsFC especially in the PCC and precuneus (see also [54]).

In addition to midline DMN regions, other studies observed major changes in sensory and motor regions that are included in the salience network (SN) and, in part, the sensorimotor network (SMN). A whole-brain approach observed increased rsFC between the thalamus and postcentral cortex, i.e., sensory cortex which also correlated with the degree of anxiety (as measured in the Spielberger trait-state anxiety) [7] (see also [55]).

The SN is also a major focus in PD. Various resting state fMRI studies of the brain's functional connectivity in PD demonstrate alterations in the regions of the SN with changes in the amygdala and MPFC, dACC, and insula being most prominent [7, 49, 56–58]. Moreover, [59] demonstrated abnormal change in resting state functional connectivity in the amygdala and medial prefrontal cortex during induction of perseverative cognition (i.e., excessive worry) in PD patients.

A recent review by [3] (see also [51]) compared resting state functional connectivity (rsFC) in panic and social anxiety disorders. They demonstrate that SAD and PD share abnormalities, e.g., reductions in rsFC, in DMN in specifically PACC and VMPFC that are closely connected to the insula [47]. At the same time, SAD and PD also differ from each other. In addition to DMN changes, SAD exhibits increased rsFC in SN including especially the amygdala, insula, and dorsal anterior cingulate. PD, in contrast, while sharing abnormal rsFC in DMN with SAD, exhibits additional rsFC changes, e.g., increases, in specifically the SMN with a specific focus on primary and secondary somatosensory cortex.

In sum, the resting state data in anxiety disorders like PD, SAD, and GAD show shared rsFC abnormalities, e.g., reductions, in DMN including anterior midline regions like PACC and VMPFC that are known to be closely connected to the insula [47, 48]. At the same time, rsFC changes, e.g., increases, can be observed in other networks like salience network and somatosensory network that, unlike DMN, are rather restricted to one of the anxiety disorders (see Fig. 5.2).

In sum, anxiety disorders show abnormalities in DMN, SN, and SMN. Specifically, DMN rsFC is reduced in all anxiety disorders, whereas connected networks like SN and SMN exhibit rsFC increases that are more specific to GAD, PD, or SAD. This suggests that the resting state balance of each of these networks, e.g., SMN and SN, relative to the DMN may be abnormal in the different anxiety disorders: the balance shifts away from the generally shared reduction in DMN toward increases in either SN (SAD) or SMN (PD). We will see later that such abnormal DMN/SMN or DMN/SN balances may be central in yielding both shared and differential psychopathological symptoms in GAD, SAD, and PD (see also [3]).

Most interestingly, these networks also mediate the brain's processing of the cardiac input (see above). Given the supposed relationship between neural and cardiac variability (see above), these findings in anxiety disorders raise the question whether the abnormalities in their resting state's functional connectivity go along with abnormal neuronal variability in the same regions. Following our neuro-cardiac hypothesis in healthy subjects (see above), one would expect first that neuronal

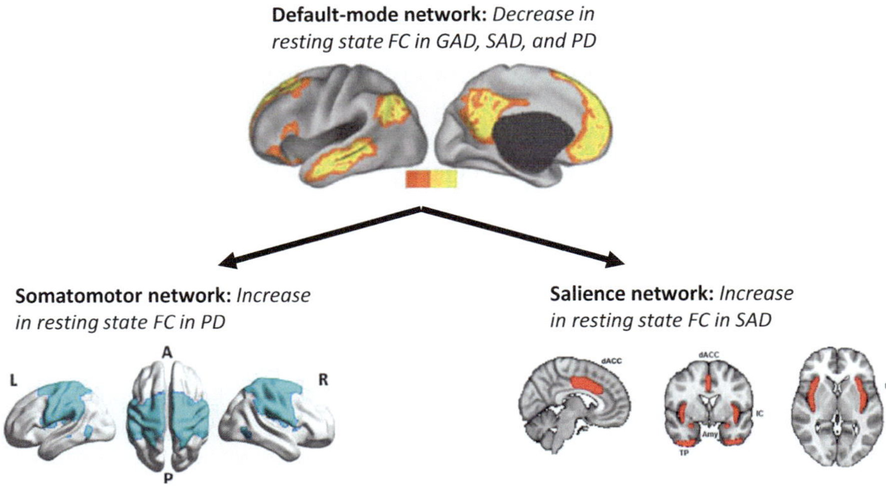

Fig. 5.2 Changes in resting state functional connectivity (FC) in different neural networks in anxiety disorders

variability in regions of DMN, SN, and SMN is abnormal and, second, that abnormal neuronal variability is related to abnormal HRV in anxiety disorders.

There is indeed some support for abnormal neuronal variability in DMN, SN, and SMN in anxiety disorders which though is tentative given the low number of studies [60–62]. Moreover, as indicated in the introduction, HRV is abnormal, e.g., reduced in anxiety disorders [2, 3]. What remains unclear is whether the potentially altered neuronal variability in DMN, SN, and SMN is related to the reduced HRV in anxiety disorders.

Hence, future investigation may want to (i) investigate neuronal variability in resting state activity of DMN, SMN, and SN; (ii) demonstrate relationship between altered rsFC and resting state neuronal variability in DMN, SMN, and SN in anxiety disorder; (iii) connect potentially altered neuronal variability to the changes in HRV; and (iv) investigate neuronal variability in different infraslow frequency ranges (slow 5, 0.01–0.027 Hz; slow 4, 0.27–0.073 Hz; slow 3, 0.073–0.198 Hz; slow 2, 0.198–0.2 Hz) as they are related to different frequency ranges of HRV (very slow, slow, fast). This may allow for detecting frequency-specific alterations in coupling between neuronal and cardiac variability in anxiety disorder (which may vary from patient to patient as well as between the different anxiety disorders).

Task-Evoked Abnormalities of Brain-Heart Coupling in SN, DMN, and SMN

Task-evoked activity during brain-heart coupling can be measured by employing the abovementioned paradigm of interoceptive awareness of the own heartbeat (see above). While this has been done in healthy subjects [29, 31, 32], there is only one recent study in anxiety disorders like GAD, SAD, and PD (see though [63] in phobia) measuring task-evoked activity during interoceptive awareness of the own heartbeat [8]. As we here only focus on task-evoked studies that specifically test for the neural correlates of interoceptive awareness of the own heart, we leave out others that investigate task-evoked activity during emotional processing in anxiety disorders (see [50] for a recent review).

Applying fMRI, [8] investigated both rsFC and task-evoked activity during interoceptive awareness in drug-naïve subjects with GAD [8]. They firstly observed that GAD subjects showed an abnormally increased sensitivity in their body perception (as measured with the Body Perception Questionnaire, BPQ) which concerned dimensions as autonomous nervous system, awareness, and stress. These findings confirm that these subjects (and that applies to PD and SAD too) suffer from an abnormally increased perception related to their own vegetative or autonomous nervous system that monitors the vegetative input to the brain from the body including the heart.

The same subjects also underwent task-evoked and resting state activity in fMRI. Task-evoked activity was measured during the abovementioned interoceptive awareness task where subjects are required to monitor and count their own heartbeat (when compared to the monitoring and counting of an external tone). Most interestingly, the GAD subjects showed increased task-evoked activity in various subregions of the insula including both the anterior insula (as closely connected to the PACC and VMPFC) and the posterior insula (as closely connected to the secondary somatosensory cortex and thus the SMN). The abnormally increased task-evoked activity in specifically left anterior insula was positively correlated with the severity of symptoms such as "psychic anxiety." In addition to task-evoked activity, the same subjects also underwent rsFC. RsFC from anterior and posterior insula to other regions such as VMPFC was reduced which was related negatively to the severity of somatic anxiety.

Taken together, this study in GAD demonstrates abnormally increased task-evoked activity in regions of SN and SMN during interoceptive awareness (of the own heartbeat) (see also Fonzo et al. 2015 for more or similar regions being involved in task-evoked activity of positive and negative emotions) while, at the same time, rsFC from these regions to those of the DMN (VMPFC) is reduced. Importantly, both increased task-evoked activity in SMN and SN regions as well as their reduced rsFC to DMN correlated with psychopathological symptoms, e.g., psychic and somatic anxiety. Confirming and extending the resting state data, these findings further point out the central role of an abnormal balance of DMN with SN and/or SMN in anxiety disorders like GAD, SAD, and PD.

Moreover, these findings support the above suggested assumption that the resting state abnormalities in these networks are related to abnormal brain-heart coupling and HRV in anxiety disorders. Future investigations may thus want to demonstrate (i) how resting state neuronal variability in SMN, DMN, and SN changes during specifically task-evoked activity related to interoceptive awareness; (ii) how HRV changes during the transition from rest to task; and (iii) how the frequency ranges of neuronal variability change during the transition from rest to task including their relationship to the HRV changes.

Spatiotemporal Hypo- and Hyper-synchronization in DMN, SMN, and SN

The findings show abnormal resting state and task-evoked activity in regions of DMN, SMN, and SN in anxiety disorders (see above). At the same time, we demonstrated that the very same regions and networks are central in mediating phase-based temporal synchronization between neuronal and cardiac activity in the healthy brain (see above). This raises the question whether, and, if so, how the changes in resting state and task-evoked activity in these networks in anxiety disorders are related to abnormal phase-based temporal synchronization between neuronal and cardiac activity.

The resting state findings in anxiety disorders are mostly based on functional connectivity, e.g., rsFC. RsFC describes the correlation between two or more regions' time series [36]. Importantly, recent investigation in both fMRI [39, 64] and iEEG [65] show that rsFC is based on phase and, more specifically, phase-based coherence or synchronization between the different regions' time series: the more, for instance, two regions' phase are coherent and thus synchronized, the higher their resulting rsFC [39] which directly probed and compared different ways of calculating rsFC.

The phase-based nature of rsFC carries far-reaching implications for interpreting the abnormalities of rsFC in anxiety disorders. All anxiety disorders share the decrease of rsFC in DMN. Given the presumably phase-based nature of rsFC, one would now assume that the DMN regions are less synchronized with each other in anxiety disorders resulting in their decreased spatiotemporal synchronization with reduced rsFC. One can thus speak of "spatiotemporal hypo-synchronization" of DMN resting state activity in anxiety disorders. The converse seems to hold in SMN and SN. The findings suggest that SN and SMN show increased rsFC and thus increased spatiotemporal synchronization in SAD and PD, respectively – we therefore speak of "spatiotemporal hyper-synchronization" of SMN and SN resting state activity in SAD and PD.

Putting all together, one may want to speak of a spectrum of different possible degrees of spatiotemporal synchronization. What we observe as "normal" degree of spatiotemporal synchronization, e.g., rsFC, in the healthy brain may reflect an intermediate or average value. In contrast, the abnormally high or low degrees of rsFC, indexing spatiotemporal hypo- or hyper-synchronization, may reflect extreme values on the spectrum of different possible degrees of spatiotemporal synchronization. Accordingly, future phase-based analysis of rsFC in DMN, SN, and SMN in anxiety disorders is needed to support our hypothesis.

Abnormal Cross-Frequency Coupling in DMN, SMN, and SN

Based on the data on healthy subjects, we now traced rsFC abnormalities in DMN, SN, and SMN anxiety disorders to an underlying abnormal phase-based coherence or synchronization. Interestingly, as pointed out above, the same networks that show rsFC in anxiety disorders are also central to processing in temporal synchronization between neuronal and cardiac activity. This raises the question whether abnormal spatiotemporal hypo- or hyper-synchronization, e.g., rsFC, in DMN, SN, and SMN is related to abnormal temporal synchronization between neuronal and cardiac activity in anxiety disorders. We thus search for connecting rsFC-related spatiotemporal (hypo- and hyper-) synchronization with neuro-cardiac temporal (de) synchronization.

We first and foremost have to say that both rsFC-based spatial synchronization and neuro-cardiac temporal synchronization operate in different frequency ranges. The rsFC findings were obtained in fMRI which operates in the infraslow frequency domain between 0.01 and 0.1 Hz, whereas the ITC and HEP findings were observed in MEG/iEEG that operate in a much faster frequency domain from 1 to 180 Hz with the ITC being observed between 4 and 7 Hz (while the slower frequencies around the heartbeat itself and its HRV (e.g., 1 Hz and lower) were filtered for methodological reasons). Due to such difference in frequency ranges, we cannot directly link and relate rsFC-related spatial (hypo- or hyper-)synchronization to neuro-cardiac temporal (de)synchronization in anxiety disorders.

However, impossible direct linkage does not preclude indirect connection. One mechanism could be cross-frequency coupling (CFC) [66]. For instance, the phase of the slower frequency around 0.01–0.1 Hz (as measured with rsFC in fMRI) could bind and thus synchronize with the amplitude of the faster frequencies (0.1–4 and 7 Hz) (as measured in MEG/iEEG) which indeed has been demonstrated in combined fMRI-EEG studies [67]. Moreover, as shown by [39], there is also cross-frequency coupling within the infraslow frequency range as measured with fMRI. Accordingly, CFC seems to be important in synchronizing neuronal activity between different frequency ranges.

Such temporal synchronization between different frequency ranges within the neuronal activity itself in terms of CFC raises the question whether analogous CFC also occurs between neuronal and cardiac activity. For instance, CFC could mediate temporal synchronization between the infraslow frequencies (0.01–0.1 Hz) of the ongoing brain's resting state activity and the faster frequencies of the heartbeat (0.1–1 Hz). This is especially important in anxiety disorders. The clearly observed resting state abnormalities in the infraslow range of 0.01–0.1 Hz may alter CFC within the brain's neuronal activity which, in turn, may alter temporal synchronization between neural and cardiac activity. The resting state FC abnormalities would then directly impact neuro-cardiac temporal synchronization. That remains to be tested though.

We therefore hypothesize that anxiety disorders can be characterized by abnormal cross-frequency coupling between infraslow and faster frequency ranges as that is important for temporal synchronization of neural and cardiac activity. Specifically, following rsFC and its spatial hypo- and hyper-synchronization, we hypothesize decreased CFC in DMN but increased CFC in SMN and SN in anxiety disorders. That, in turn, as it needs to be investigated in the future, may strongly impact neuro-cardiac temporal synchronization in the 0.1/4–7 Hz range including its measures like ITC and HEP. Specifically, one would expect that decreased CFC leads to decreases in both ITC and HEP, as we assume it to be the case in DMN, while increased CFC should lead to increases in both ITC and HEP as it may be the case in SN and SMN in anxiety disorders.

Spatiotemporal Hypo-synchronization in DMN and Instability of Self

We demonstrated that rsFC is most likely related to spatiotemporal synchronization reflecting phase-based coherence between different regions within a specific network like DMN. Decreased rsFC in DMN, as observed in anxiety disorders, may thus reflect spatiotemporal hypo-synchronization such that the various DMN regions are no longer as strongly tied and connected together as network. Due to such spatiotemporal hypo-synchronization, the DMN as one phase-based coherent network may thus become more instable, e.g., less synchronized, on neuronal grounds. This carries major neuronal and psychological implications.

Spatiotemporal hypo-synchronization means that the DMN as network becomes more unstable in its neuronal activity as its various regions, being less synchronized with each other, are "now more on their own" and independent of each other. That introduces instability into DMN neuronal activity in the infraslow range of 0.01–0.1 Hz. That very same infraslow neuronal instability may make the coupling of the infraslow to faster frequencies more difficult resulting in decreased CFC. The decrease in CFC, in turn, may also impair temporal synchronization between neuronal and cardiac activity (in faster frequencies) such that both become desynchronized from each other in regions like VMPFC and PACC as core regions of DMN. The DMN's spatiotemporal hypo-synchronization in the resting state's rsFC would then lead to reduction in neuro-cardiac temporal hypo-synchronization – the latter should be manifested in lower phase locking with reduced intertrial coherence (ITC) (in the 0.1/4–7 Hz range) in the brain's resting state activity.

Accordingly, put in a nutshell, we assume that spatiotemporal hypo-synchronization in DMN leads to neuro-cardiac temporal desynchronization in anxiety disorders. The neuronal instability of DMN may thus affect the neuro-cardiac coupling which, analogously, may then also become unstable if not resulting in neuro-cardiac temporal desynchronization. While such neuro-cardiac temporal desynchronization remains to be investigated in anxiety disorders, it has already been reported in fMRI of posttraumatic stress disorders (PTSD) [18, 19]. Since PTSD is also featured by strong anxiety, one would expect more or less analogous findings in anxiety disorders.

The instability of both DMN network neuronal activity and neuro-cardiac coupling may carry major psychological implications. As pointed out the DMN is central in mediating internal cognition like sense of self [43], episodic simulation with prospection into the future and retrospection into the past, and mind wandering [13, 68]. Moreover, the above-presented findings by the group around Tallon-Baudry clearly demonstrate that neuro-cardiac coupling in VMPFC/PACC as core regions of DMN is related to the sense of self, e.g., "I" vs "me" [25, 26] (see above for details).

Given the DMN findings, one would expect the sense of self to be less synchronized and thus instable as based on spatiotemporal hypo-synchronization within DMN as manifested in reduced rsFC. Moreover, one would expect that the self can no longer integrate different time scales as it has no access to different frequency ranges beyond itself due to reduced CFC in DMN. Finally, one would expect that the self is no longer aligned with the own body as related to decreased neuro-cardiac temporal synchronization. Together, this leads to an increased spatial and temporal instability in self as it is indeed a hallmark feature in anxiety disorders which may be manifested in abnormal internal cognition as in episodic simulation (as in abnormal foresight of the futures) [69] and mind wandering [68–70] (see Fig. 5.3a).

Spatiotemporal Hyper-synchronization in SMN/SN and Increased Emotions/Anxiety (SAD) and Bodily Awareness (PD)

In addition to rsFC decrease in DMN, the findings show increased rsFC in SMN (PD) and SN (SAD) (see above). One would consequently assume spatiotemporal hyper-synchronization in SMN or SN accompanying DMN spatiotemporal hypo-synchronization. Following the neuronal and psychological results of brain-heart coupling in the healthy brain (see above), one would assume the following hypotheses.

Spatiotemporal hyper-synchronization means that the SMN/SN as network becomes more stable in its neuronal activity as its various regions, being more synchronized with each other, are "now more with each other" and thus more dependent on each other. That increases network stability in the infraslow range of 0.01–0.1 Hz. Increased infraslow neuronal stability may make the coupling of the infraslow to faster frequencies easier resulting in increased CFC. The increase in CFC, in turn, may enhance temporal synchronization between neuronal and cardiac activity (in faster frequencies) such that both become hyper-synchronized with each other in regions like insula, amygdala, thalamus (SN), and somatosensory cortex (SMN).

The spatiotemporal hyper-synchronization of SMN/SN in the 0.01–0.1 Hz range of rsFC should lead to higher CFC and ultimately higher neuro-cardiac temporal synchronization in the 0.1/4–7 Hz range. The spatiotemporal hyper-synchronization of SMN/SN in the 0.01–0.1 Hz range of rsFC should lead to higher CFC and ultimately higher neuro-cardiac temporal synchronization in the 0.1/4–7 Hz range. The increased rsFC indexing spatiotemporal hyper-synchronization in SMN/SN should then be accompanied by increased neuro-cardiac temporal synchronization as manifested in higher phase synchronization with increased intertrial coherence (ITC) (in the 0.1/4–7 Hz range) in the brain's resting state activity.

Accordingly, put in a nutshell, we assume that spatiotemporal hyper-synchronization in SMN/SN leads to temporal hyper-synchronization between the brain and the heart in anxiety disorders. The increased neuronal stability of SMN/SN may thus affect the neuro-cardiac coupling which, analogously, may then also become increasingly stable resulting in increased neuro-cardiac coupling mediated by phase-based temporal hyper-synchronization. Such neuro-cardiac temporal

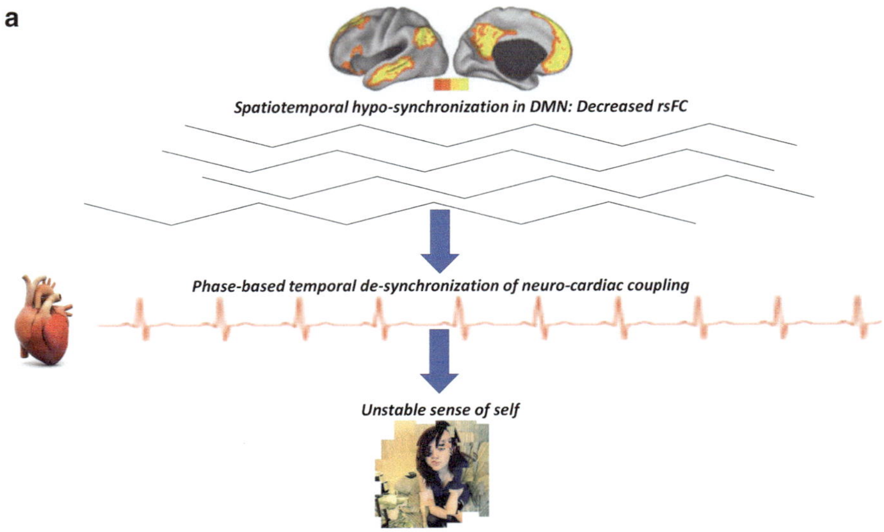

Fig. 5.3a Altered default-mode network (DMN) and its psychopathological symptoms

hyper-synchronization remains to be investigated in SMN and SN in anxiety disorders though. The increased stability of both SMN/SN network neuronal activity and neuro-cardiac coupling may carry major psychological implications.

The SN and specifically the insula, thalamus, dACC, and amygdala including their neuro-cardiac coupling are strongly associated with emotions like anxiety. Increased stability of both SN and neuro-cardiac coupling in specifically amygdala and insula may abnormally enhance emotions and especially anxiety – that is exactly what one can observe psychopathologically in specifically SAD (see also [71] and [39] as well as [57, 59, 72]) where increases in SN rsFC have been reported. Moreover, since SN and especially insula are central in processing the heartbeat (see above), one would expect abnormally increased SN function to go along with increased neuro-cardiac coupling in this region resulting in increased perception or awareness of the own heartbeat, e.g., increased interoceptive awareness (see [73, 74]), in specifically SAD (where the insula and SN are abnormal) (see Fig. 5.3b).

The SMN and its neuro-cardiac coupling are involved in proprioception and awareness of the own body (see [24, 45]). Increased stability of both SMN and its neuro-cardiac coupling may result in increased awareness of the own body's proprioceptive stimuli and ultimately in higher awareness of the own body, e.g., increased bodily awareness, and various somatic symptoms (see also [75]). This is well compatible with the psychopathological symptoms of PD (where the SMN is abnormally increased) where increased awareness of bodily sensation and the own body as a whole are core symptoms (see, as described above, 8 for support). However, the link between SMN spatiotemporal hyper-synchronization, increased neuro-cardiac coupling, and increased bodily awareness remains to be demonstrated (see Fig. 5.3c).

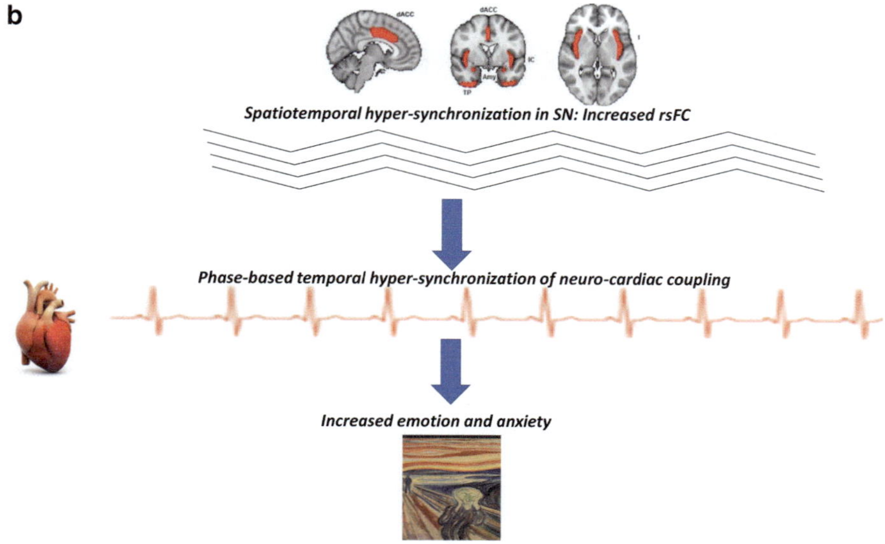

Fig. 5.3b Altered salience network (SN) and its psychopathological symptoms. Altered somatomotor network (SMN) and its psychopathological symptoms

Conclusion

We here reviewed recent rsFC and task-evoked findings, conducted mainly in fMRI, in anxiety disorders. Core findings include altered rsFC in DMN, SN, and SMN which suggests disbalances in their relationships with the tilting of DMN/SN and DMN/SMN ratios toward the non-DMN networks. These findings suggest spatiotemporal hypo-synchronization between the regions' rsFC in the infraslow frequency range in DMN (all anxiety disorders), while one may assume spatiotemporal hyper-synchronization in SMN (PD) and SN (SAD). That, as we hypothesize, may be related to increased or decreased neuro-cardiac coupling as mediated by phase-based temporal synchronization (as demonstrated in healthy subjects in MEG and iEEG).

Taken together, we assume two spatiotemporal mechanisms to play a central role in yielding psychopathological symptoms in anxiety disorders. First, the findings suggest spatiotemporal hypo- and/or hyper-synchronization in the neural activity of networks like DMN< SMN and SN. Secondly, based on clinical finding in anxiety disorders and neuronal observations in healthy subjects, we suggest these network abnormalities to lead to abnormal, e.g., increased or decreased, phase-based temporal synchronization of neuronal and cardiac activity. Together, spatiotemporal hypo- or hyper-synchronization in the different resting state networks, e.g., DMN, SMN, and SN, and associated increased/decreased neuro-cardiac temporal synchronization are shown to most likely underlie core psychopathological symptoms such as unstable self, increased emotions/anxiety, and/or increased interoceptive and/or bodily awareness in anxiety disorders.

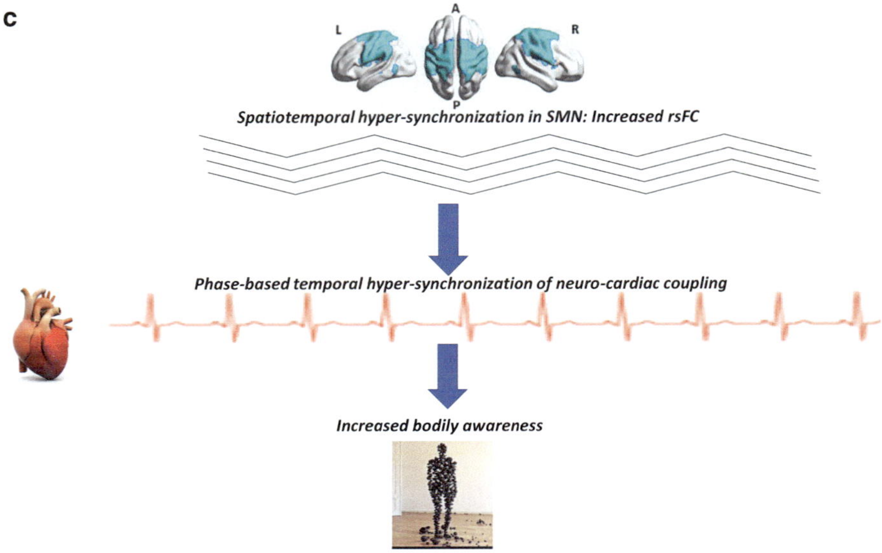

Fig. 5.3c Altered somatomotor network (SMN) and its psychopathological symptoms

Psychopathological symptoms in the different domains of self (unstable self), emotions (increased anxiety), and bodily awareness (increased interoceptive and bodily awareness) may thus be based primarily on spatiotemporal mechanisms in the brain's neuronal activity, e.g., its resting state networks' spatiotemporal synchronization, and its neuro-cardiac coupling, e.g., phase-based temporal synchronization. Such spatiotemporal (rather than primarily affective or cognitive) basis of psychopathological symptoms in anxiety disorders is well compatible with the recent suggestion of "Spatiotemporal Psychopathology" [10–13, see also 14–16].

References

1. Olesen KKW, Madsen M, Lip GYH, Egholm G, Thim T, Jensen LO, et al. Coronary artery disease and risk of adverse cardiac events and stroke. Eur J Clin Invest. 2017 Nov;47(11):819–28.
2. Kim EY, Lee MY, Kim SH, Ha K, Kim KP, Ahn YM. Diagnosis of major depressive disorder by combining multimodal information from heart rate dynamics and serum proteomics using machine-learning algorithm. Prog Neuropsychopharmacol Biol Psychiatry. 2017 Jun;76:65–71.
3. Kim Y-K, Yoon H-K. Common and distinct brain networks underlying panic and social anxiety disorders. Prog Neuropsychopharmacol Biol Psychiatry. 2018 Jan;80(Pt B):115–22.
4. Chalmers JA, Quintana DS, Abbott MJ-A, Kemp AH. Anxiety disorders are associated with reduced heart rate variability: a meta-analysis. Front Psychiatry. 2014;5:80.
5. Wiebking C, de Greck M, Duncan NW, Heinzel A, Tempelmann C, Northoff G. Are emotions associated with activity during rest or interoception? An exploratory fMRI study in healthy subjects. Neurosci Lett. 2011 Mar;491(1):87–92.

6. Avery JA, Drevets WC, Moseman SE, Bodurka J, Barcalow JC, Simmons WK. Major depressive disorder is associated with abnormal interoceptive activity and functional connectivity in the insula. Biol Psychiatry. 2014 Aug;76(3):258–66.
7. Cui H, Zhang J, Liu Y, Li Q, Li H, Zhang L, et al. Differential alterations of resting-state functional connectivity in generalized anxiety disorder and panic disorder. Hum Brain Mapp. 2016 Apr;37(4):1459–73.
8. Cui S, Len J, Zhang J, Li C, Northoff G. Alterations in interceptive awareness and the brain's resting state and task-evoked activity in generalized anxiety disorder. Biol Psychiatry Cogn Neurosci Neuroimaging. In review. 2019
9. Emdin CA, Odutayo A, Wong CX, Tran J, Hsiao AJ, Hunn BHM. Meta-analysis of anxiety as a risk factor for cardiovascular disease. Am J Cardiol. 2016 Aug;118(4):511–9.
10. Northoff G. Spatiotemporal psychopathology I: no rest for the brain's resting state activity in depression? Spatiotemporal psychopathology of depressive symptoms. J Affect Disord. 2016 Jan;190:854–66.
11. Northoff G. Spatiotemporal psychopathology II: how does a psychopathology of the brain's resting state look like? Spatiotemporal approach and the history of psychopathology. J Affect Disord. 2016 Jan;190:867–79.
12. Northoff G. "Paradox of slow frequencies" – are slow frequencies in upper cortical layers a neural predisposition of the level/state of consciousness (NPC)? Conscious Cogn. 2017 Sep;54:20–35.
13. Northoff G. The brain's spontaneous activity and its psychopathological symptoms – "spatiotemporal binding and integration". Prog Neuropsychopharmacol Biol Psychiatry. 2018 Jan;80(Pt B):81–90.
14. Northoff G, Duncan NW. How do abnormalities in the brain's spontaneous activity translate into symptoms in schizophrenia? From an overview of resting state activity findings to a proposed spatiotemporal psychopathology. Prog Neurobiol. 2016 Oct;145–146:26–45.
15. Northoff G, Stanghellini G. How to link brain and experience? Spatiotemporal psychopathology of the lived body. Front Hum Neurosci. 2016;10:76.
16. Fingelkurts AA, Fingelkurts AA. Brain space and time in mental disorders: paradigm shift in biological psychiatry. Int J Psychiatry Med. 2018 Aug;91217418791438
17. Chang C, Metzger CD, Glover GH, Duyn JH, Heinze H-J, Walter M. Association between heart rate variability and fluctuations in resting-state functional connectivity. Neuroimage. 2013 Mar;68:93–104.
18. Thome J, Densmore M, Frewen PA, McKinnon MC, Theberge J, Nicholson AA, et al. Desynchronization of autonomic response and central autonomic network connectivity in posttraumatic stress disorder. Hum Brain Mapp. 2017 Jan;38(1):27–40.
19. Rabellino D, D'Andrea W, Siegle G, Frewen PA, Minshew R, Densmore M, et al. Neural correlates of heart rate variability in PTSD during sub- and supraliminal processing of trauma-related cues. Hum Brain Mapp. 2017 Oct;38(10):4898–907.
20. Jennings JR, Sheu LK, Kuan DC-H, Manuck SB, Gianaros PJ. Resting state connectivity of the medial prefrontal cortex covaries with individual differences in high-frequency heart rate variability. Psychophysiology. 2016 Apr;53(4):444–54.
21. Park H-D, Correia S, Ducorps A, Tallon-Baudry C. Spontaneous fluctuations in neural responses to heartbeats predict visual detection. Nat Neurosci. 2014 Apr;17(4):612–8.
22. Park H-D, Tallon-Baudry C. The neural subjective frame: from bodily signals to perceptual consciousness. Philos Trans R Soc Lond B Biol Sci. 2014 May;369(1641):20130208.
23. Park H-D, Bernasconi F, Bello-Ruiz J, Pfeiffer C, Salomon R, Blanke O. Transient modulations of neural responses to heartbeats covary with bodily self-consciousness. J Neurosci. 2016 Aug;36(32):8453–60.
24. Park H-D, Bernasconi F, Salomon R, Tallon-Baudry C, Spinelli L, Seeck M, et al. Neural sources and underlying mechanisms of neural responses to heartbeats, and their role in bodily self-consciousness: an intracranial EEG study. Cereb Cortex. 2018 Jul;28(7):2351–64.

25. Babo-Rebelo M, Wolpert N, Adam C, Hasboun D, Tallon-Baudry C. Is the cardiac monitoring function related to the self in both the default network and right anterior insula? Philos Trans R Soc Lond B Biol Sci. 2016 Nov;371(1708)
26. Babo-Rebelo M, Richter CG, Tallon-Baudry C. Neural responses to heartbeats in the default network encode the self in spontaneous thoughts. J Neurosci. 2016 Jul;36(30):7829–40.
27. Garfinkel SN, Seth AK, Barrett AB, Suzuki K, Critchley HD. Knowing your own heart: distinguishing interoceptive accuracy from interoceptive awareness. Biol Psychol. 2015 Jan;104:65–74.
28. Forkmann T, Scherer A, Meessen J, Michal M, Schachinger H, Vogele C, et al. Making sense of what you sense: disentangling interoceptive awareness, sensibility and accuracy. Int J Psychophysiol. 2016 Nov;109:71–80.
29. Critchley HD, Wiens S, Rotshtein P, Ohman A, Dolan RJ. Neural systems supporting interoceptive awareness. Nat Neurosci. 2004 Feb;7(2):189–95.
30. Kuehn E, Mueller K, Lohmann G, Schuetz-Bosbach S. Interoceptive awareness changes the posterior insula functional connectivity profile. Brain Struct Funct. 2016 Apr;221(3):1555–71.
31. Wiebking C, Duncan NW, Tiret B, Hayes DJ, Marjanska M, Doyon J, et al. GABA in the insula – a predictor of the neural response to interoceptive awareness. Neuroimage. 2014 Feb;86:10–8.
32. Wiebking C, Duncan NW, Qin P, Hayes DJ, Lyttelton O, Gravel P, et al. External awareness and GABA – a multimodal imaging study combining fMRI and [18F]flumazenil-PET. Hum Brain Mapp. 2014 Jan;35(1):173–84.
33. Menon V. Large-scale brain networks and psychopathology: a unifying triple network model. Trends Cogn Sci. 2011 Oct;15(10):483–506.
34. Seeley WW, Menon V, Schatzberg AF, Keller J, Glover GH, Kenna H, et al. Dissociable intrinsic connectivity networks for salience processing and executive control. J Neurosci. 2007 Feb;27(9):2349–56.
35. Raichle ME, MacLeod AM, Snyder AZ, Powers WJ, Gusnard DA, Shulman GL. A default mode of brain function. Proc Natl Acad Sci USA. 2001 Jan;98(2):676–82.
36. Greicius MD, Krasnow B, Reiss AL, Menon V. Functional connectivity in the resting brain: a network analysis of the default mode hypothesis. Proc Natl Acad Sci USA. 2003 Jan;100(1):253–8.
37. Tallon-Baudry C, Campana F, Park H-D, Babo-Rebelo M. The neural monitoring of visceral inputs, rather than attention, accounts for first-person perspective in conscious vision. Cortex. 2018 May;102:139–49.
38. Pollatos O, Herbert BM, Mai S, Kammer T. Changes in interoceptive processes following brain stimulation. Philos Trans R Soc Lond B Biol Sci. 2016 Nov;371(1708)
39. Huang Z, Zhang J, Longtin A, Dumont G, Duncan NW, Pokorny J, et al. Is there a nonadditive interaction between spontaneous and evoked activity? Phase-dependence and its relation to the temporal structure of scale-free brain activity. Cereb Cortex. 2017 Feb;27(2):1037–59.
40. He BJ. Spontaneous and task-evoked brain activity negatively interact. J Neurosci. 2013 Mar;33(11):4672–82.
41. Lakatos P, Schroeder CE, Leitman DI, Javitt DC. Predictive suppression of cortical excitability and its deficit in schizophrenia. J Neurosci. 2013 Jul;33(28):11692–702.
42. Northoff G, Huang Z. How do the brain's time and space mediate consciousness and its different dimensions? Temporo-spatial theory of consciousness (TTC). Neurosci Biobehav Rev. 2017 Sep;80:630–45.
43. Northoff G, Heinzel A. First-Person Neuroscience: a new methodological approach for linking mental and neuronal states. Philos Ethics Humanit Med. 2006 Mar;1(1):E3.
44. Qin P, Northoff G. How is our self related to midline regions and the default-mode network? Neuroimage. 2011 Aug;57(3):1221–33.
45. Blanke O. Multisensory brain mechanisms of bodily self-consciousness. Nat Rev Neurosci. 2012 Jul;13(8):556–71.

46. Enzi B, de Greck M, Prosch U, Tempelmann C, Northoff G. Is our self nothing but reward? Neuronal overlap and distinction between reward and personal relevance and its relation to human personality. PLoS One. 2009 Dec;4(12):e8429.
47. Northoff G, Wiebking C, Feinberg T, Panksepp J. The "resting-state hypothesis" of major depressive disorder-a translational subcortical-cortical framework for a system disorder. Neurosci Biobehav Rev. 2011 Oct;35(9):1929–45.
48. Craig AD. The sentient self. Brain Struct Funct. 2010 Jun;214(5–6):563–77.
49. Peterson A, Thome J, Frewen P, Lanius RA. Resting-state neuroimaging studies: a new way of identifying differences and similarities among the anxiety disorders? Can J Psychiatry. 2014 Jun;59(6):294–300.
50. Fonzo GA, Etkin A. Affective neuroimaging in generalized anxiety disorder: an integrated review. Dialogues Clin Neurosci. 2017 Jun;19(2):169–79.
51. MacNamara A, DiGangi J, Phan KL. Aberrant spontaneous and task-dependent functional connections in the anxious brain. Biol psychiatry Cogn Neurosci neuroimaging. 2016 May;1(3):278–87.
52. Shin Y-W, Dzemidzic M, Jo HJ, Long Z, Medlock C, Dydak U, et al. Increased resting-state functional connectivity between the anterior cingulate cortex and the precuneus in panic disorder: resting-state connectivity in panic disorder. J Affect Disord. 2013 Sep;150(3):1091–5.
53. Lai C-H, Wu Y-T. The alterations in inter-hemispheric functional coordination of patients with panic disorder: the findings in the posterior sub-network of default mode network. J Affect Disord. 2014 Sep;166:279–84.
54. Pannekoek JN, van der Werff SJA, Stein DJ, van der Wee NJA. Advances in the neuroimaging of panic disorder. Hum Psychopharmacol. 2013 Nov;28(6):608–11.
55. Lai C-H, Wu Y-T. The changes in the low-frequency fluctuations of cingulate cortex and postcentral gyrus in the treatment of panic disorder: the MRI study. World J Biol Psychiatry. 2016;17(1):58–65.
56. Mochcovitch MD, da Rocha Freire RC, Garcia RF, Nardi AE. A systematic review of fMRI studies in generalized anxiety disorder: evaluating its neural and cognitive basis. J Affect Disord. 2014;167:336–42.
57. Makovac E, Meeten F, Watson DR, Herman A, Garfinkel SN, D Critchley H, et al. Alterations in amygdala-prefrontal functional connectivity account for excessive worry and autonomic dysregulation in generalized anxiety disorder. Biol Psychiatry. 2016 Nov;80(10):786–95.
58. Hilbert K, Lueken U, Beesdo-Baum K. Neural structures, functioning and connectivity in generalized anxiety disorder and interaction with neuroendocrine systems: a systematic review. J Affect Disord. 2014 Apr;158:114–26.
59. Makovac E, Watson DR, Meeten F, Garfinkel SN, Cercignani M, Critchley HD, et al. Amygdala functional connectivity as a longitudinal biomarker of symptom changes in generalized anxiety. Soc Cogn Affect Neurosci. 2016 Nov;11(11):1719–28.
60. Zhang Y, Zhu C, Chen H, Duan X, Lu F, Li M, et al. Frequency-dependent alterations in the amplitude of low-frequency fluctuations in social anxiety disorder. J Affect Disord. 2015 Mar;174:329–35.
61. Wang W, Hou J, Qian S, Liu K, Li B, Li M, et al. Aberrant regional neural fluctuations and functional connectivity in generalized anxiety disorder revealed by resting-state functional magnetic resonance imaging. Neurosci Lett. 2016 Jun;624:78–84.
62. Yuan C, Zhu H, Ren Z, Yuan M, Gao M, Zhang Y, et al. Precuneus-related regional and network functional deficits in social anxiety disorder: A resting-state functional MRI study. Compr Psychiatry. 2018 Apr;82:22–9.
63. Caseras X, Murphy K, Mataix-Cols D, Lopez-Sola M, Soriano-Mas C, Ortriz H, et al. Anatomical and functional overlap within the insula and anterior cingulate cortex during interoception and phobic symptom provocation. Hum Brain Mapp. 2013 May;34(5):1220–9.
64. Zhang J, Magioncalda P, Huang Z, Tan Z, Hu X, Hu Z, et al. Altered global signal topography and its different regional localization in motor cortex and hippocampus in mania and depression. Schizophr Bull. 2018 Oct;

65. Wen H, Liu Z. Broadband electrophysiological dynamics contribute to global resting-state fMRI signal. J Neurosci. 2016 Jun;36(22):6030–40.
66. He BJ, Zempel JM, Snyder AZ, Raichle ME. The temporal structures and functional significance of scale-free brain activity. Neuron. 2010 May;66(3):353–69.
67. Ganzetti M, Mantini D. Functional connectivity and oscillatory neuronal activity in the resting human brain. Neuroscience. 2013 Jun;240:297–309.
68. Christoff K, Irving ZC, Fox KCR, Spreng RN, Andrews-Hanna JR. Mind-wandering as spontaneous thought: a dynamic framework. Nat Rev Neurosci. 2016 Nov;17(11):718–31.
69. Miloyan B, Bulley A, Suddendorf T. Episodic foresight and anxiety: proximate and ultimate perspectives. Br J Clin Psychol. 2016 Mar;55(1):4–22.
70. Miloyan B, Pachana NA, Suddendorf T. The future is here: a review of foresight systems in anxiety and depression. Cogn Emot. 2014;28(5):795–810.
71. Paulus MP, Stein MB, Craske MG, Bookheimer S, Taylor CT, Simmons AN, et al. Latent variable analysis of positive and negative valence processing focused on symptom and behavioral units of analysis in mood and anxiety disorders. J Affect Disord. 2017 Jul;216:17–29.
72. Makovac E, Meeten F, Watson DR, Garfinkel SN, Critchley HD, Ottaviani C. Neurostructural abnormalities associated with axes of emotion dysregulation in generalized anxiety. NeuroImage Clin. 2015;10:172–81.
73. Khalsa SS, Adolphs R, Cameron OG, Critchley HD, Davenport PW, Feinstein JS, et al. Interoception and mental health: a roadmap. Biol psychiatry Cogn Neurosci Neuroimaging. 2018 Jun;3(6):501–13.
74. Quadt L, Critchley HD, Garfinkel SN. The neurobiology of interoception in health and disease. Ann NY Acad Sci. 2018 Sep;1428(1):112–28.
75. Mallorqui-Bague N, Bulbena A, Pailhez G, Garfinkel SN, Critchley HD. Mind-body interactions in anxiety and somatic symptoms. Harv Rev Psychiatry. 2016;24(1):53–60. 71.

Part II

Neurobiological Aspects

Gene-Environment Interactions and Role of Epigenetics in Anxiety Disorders

6

Eugene Lin and Shih-Jen Tsai

Introduction

One of the most prevalent and complicated mental illnesses worldwide would be anxiety disorders. As genome-wide association studies (GWASs) have not been able to identify genes related to anxiety disorders [1], it has been proposed that there is a strong association between environmental factors and the epigenome. However, only two loci were pinpointed using meta-analysis study of GWASs, namely, the rs1709393 single nucleotide polymorphism (SNP) found in the long intergenic non-protein coding RNA 2085 (*LINC02085*) gene and also the rs1067327 SNP in the calcium/calmodulin-dependent protein kinase II gamma (*CAMKMT*) gene associated with anxiety disorders [2]. In terms of clinical applications of genomic association studies, advanced methods such as epigenetics and gene-environment (G × E) interactions have been applied to investigate the involvement of genes to anxiety disorder pathogenesis. In comparison to healthy controls, anxiety disorder patients reflected relationship between environmental factors and relevant genes, as accumulating evidence suggested [3]. Although more discoveries in agreement with such hypothesis are validated, a significant amount of possible biomarkers in G × E interactions and epigenetics have been found to be related to anxiety disorders. In this

E. Lin
Department of Biostatistics, University of Washington, Seattle, WA, USA

Department of Electrical & Computer Engineering, University of Washington, Seattle, WA, USA

Graduate Institute of Biomedical Sciences, China Medical University, Taichung, Taiwan

S.-J. Tsai (✉)
Department of Psychiatry, Taipei Veterans General Hospital, Taipei, Taiwan

Division of Psychiatry, National Yang-Ming University, Taipei, Taiwan

Institute of Brain Science, National Yang-Ming University, Taipei, Taiwan

Y.-K. Kim (ed.), *Anxiety Disorders*, Advances in Experimental Medicine and Biology 1191, https://doi.org/10.1007/978-981-32-9705-0_6

chapter, we briefly examined a variety of current research investigations in relation to G × E interactions and epigenetics considering the assessment and understanding of presumed risk mechanisms in anxiety disorder pathogenesis.

Initially, we reexamined candidate genes associated with anxiety disorders in G × E interaction studies which were pinpointed as candidate biomarkers. Then, some of the probable genes evaluated in epigenetic studies and also found to be associated with anxiety disorders were assessed. Last, we suggested future perspectives of studies in G × E interactions and epigenetics, as well as their limitations. As for future research, replication studies with independent and extensive cohorts are required to determine the role of candidate biomarkers discussed in previous studies regarding G × E interactions and epigenetics for anxiety disorders.

Environmental Factors and G × E Interactions on Anxiety Disorders

Environmental Factors on Anxiety Disorders

Many environmental risk factors are associated with anxiety disorders, including early adverse childhood experiences (e.g., child maltreatment, emotional abuse, sexual abuse, physical abuse, and neglect), stressful life events (such as family and marriage conflicts, disrupted interpersonal relationships, job problems, events related to adverse physical health, loss events, financial difficulties, legal and crime matters, and perinatal conditions), as well as stress itself [4].

G × E Interactions on Anxiety Disorders

In this section we looked into several association studies which examined both multilocus interactions and single-locus effects to verify the hypothesis that selected candidate genes may play a part in anxiety disorders individually and through complex G × E interactions. Robust replications and associations with anxiety-related G × E interactions have been made for three loci, namely, *BDNF*, FKBP prolyl isomerase 5 (*FKBP5*), and serotonin transporter gene-linked polymorphic region (5-HTTLPR) [5]. This review is not intended as a comprehensive survey of all possible reports studied in the literature.

Serotonin Transporter Gene-Linked Polymorphic Region (5-HTTLPR)

Short and long alleles are commonly reported for the 5-HTTLPR variant in the *SLC6A4* (serotonin neurotransmitter transporter) gene, with the short allele having an association with lower *SLC6A4* gene expression activity [6]. In comparison to the long allele, which has 16 repeats of a sequence, the short allele in 5-HTTLPR has only 14 repeats.

Caspi et al. carried out the first study to investigate G × E interactions between stressful life events and the 5-HTTLPR variant in the *SLC6A4* gene [7]. A prospective longitudinal study was conducted by Caspi et al. involving 1037 Dunedin children by interviewing them at regular intervals for evaluations of their stressful life events occurring between the age of 21 and 26 years [7]. Caspi et al. reported an interesting G × E interaction between the 5-HTTLPR variant and stressful life events to make risk predictions for depressive symptoms among individuals carrying one or two 5-HTTLPR short alleles [7].

In addition, in order to verify the hypothesis that the 5-HTTLPR variant may affect anxiety sensitivity, which is an intermediate phenotype for anxiety, Stein et al. examined the 5-HTTLPR variant and G × E interactions through complex interactions with childhood maltreatment within a sample of ethnically diverse college undergraduates [8]. Stein et al. found that, when compared to carriers with long/long or short/long genotypes of 5-HTTLPR, the G × E interactions between childhood maltreatment and the 5-HTTLPR variant have arousing tendency of anxiety sensitivity for short-allele homozygotes [8].

On the other hand, research done by Klauke et al. indicated supporting results in G × E interactions between the 5-HTTLPR variant and childhood maltreatment showing that healthy adults with long/long genotype in the 5-HTTLPR variant expressed higher anxiety sensitivity when compared to those of short/short or short/long genotypes [9].

Brain-Derived Neurotrophic Factor (*BDNF*)

The gene known as brain-derived neurotrophic factor (*BDNF*) gene has been associated with geriatric depressive disorders, cognitive function, as well as major depressive disorder [10]. Kaufman et al. carried out an earlier G × E study and discovered a putative three-way interaction between the 5-HTTLPR variant, the rs6265 (Val66Met) SNP of *BDNF* gene, and childhood maltreatment history in influencing depression risks [11].

Verification of G × E interactions between childhood maltreatment and the BDNF rs6265 SNP in influencing anxiety in healthy young Korean adults ($n = 206$) was investigated by Min et al. [12]. In this study, combinations between the *BDNF* rs6265 SNP and childhood maltreatment in all subjects and then in gender-stratified groups were evaluated in order to assess their impacts [12]. In all subjects, G × E interactions involving childhood maltreatment and the *BDNF* rs6265 SNP were not found [12]. Interestingly, their analysis showed G × E interactions for male individuals (n = 108), but were not identified for female subjects [12].

In a group of 308 participating adolescents in mixed races ($n = 308$), Martin et al. verified the hypothesis of whether *BDNF* rs6265 may be linked to anxiety sensitivity risks through complex G × E interactions by evaluating G × E interactions between childhood treatment and the *BDNF* rs6265 SNP [13]. However, no such interactions were found, indicating that the relationship between the increased risk of anxiety sensitivity and BDNF rs6265 is not determined by childhood maltreatment [13].

Fatty Acid Amide Hydrolase (*FAAH*)

Lazary et al. carried out a study ($n = 858$) to verify the hypothesis of the possible association between the etiology of anxiety and depression and the fatty acid amide hydrolase (FAAH) gene, individually, and also by complex G × E interactions with addition of impact from childhood trauma by examining both single-locus effects and G × E interactions [14]. In this study, considerable effects of the *FAAH* rs324420 SNP with depression were found in single-locus analysis [14]. In addition, childhood trauma and the FAAH rs324420 SNP involved in G × E interactions were suggested for depression as well as anxiety [14].

RAR-Related Orphan Receptor A (*RORA*)

The RAR-related orphan receptor A (*RORA*) gene has been reported to be in association with depressive symptoms [15]. Min et al. studied whether G × E interactions exist between childhood maltreatment and the *RORA* gene in terms of their impacts on the anxiety sensitivity of healthy young Korean adults (n = 205) [16]. Gender differences were used to assess the effects of combinations between childhood maltreatment and the *RORA* gene [16]. Again, Min et al. observed a G × E interaction for male subjects but not for females [16]. However, when male and female individuals were combined, no G × E interactions with regard to childhood maltreatment and the RORA rs11071547 SNP were found [16].

Epigenetic Mechanisms and Anxiety Disorders

With the recent progressions in scientific research, results indicated that psychiatric disorders such as anxiety disorders may involve epigenetic mechanisms, including microRNAs, histone modifications, and DNA methylation [17]. Methyl group is added to the DNA molecule, especially when a cytosine is followed by a guanine (CpG dinucleotide), which is the process of DNA methylation [18]. DNA methylation is activated by the DNA methyl-transferase protein family, which now plays a major role in epigenetics and is globally linked with decreased transcriptional activity [19]. Although CpG islands, CpG-rich regions in promoter regions, have been the main focuses in DNA methylation studies, other genomic regions (e.g., intergenic regions and gene bodies) remain less understood [19].

In addition, a key player in the new arena of epigenetics is small noncoding RNAs, namely, microRNAs, in particular. Nucleotide sequence of noncoding RNAs does not produce proteins. MicroRNAs are smaller than 200 nucleotides in length [20, 21]. Histone modification involves histone proteins being covalently modified through posttranslational changes, which then form nucleosomes through their interactions with DNA [17].

This section discusses the latest developments in the realm of epigenetics in regard to anxiety disorders. As researchers pay a fair share of attention to epigenetics research, growing numbers of research studies have been conducted. This review is not intended as a comprehensive survey of all possible epigenetic reports studied in the literature.

DNA Methylation

The association between childhood maltreatment and epigenetic mechanism of DNA methylation has been proposed by both human and animal studies [22]. There are also supportive evidence showing that the risks in psychiatric disorders like anxiety disorders are affected by adverse early-life events and stressful events during adulthood [23]. In terms of animal models, it has been suggested that changes in the DNA methylation levels of genes, namely, estrogen receptor (*Esr1*), glial cell-derived neurotrophic factor (*Gdnf*), nuclear receptor subfamily 3 group C member 1 (*Nr3c1*), arginine vasopressin (*Avp*), glutamate decarboxylase 1 (*Gad1*), and *Slc6a4* (serotonin neurotransmitter transporter) genes, may be due to stressful early-life events [24–29].

A well-known potential biomarker for anxiety disorder pathophysiology is the *BDNF* gene [30]. In a study investigating animal model of childhood maltreatment, Roth et al. found that DNA methylation in the *Bdnf* gene in rats was increased by stressful early-life events [31]. Consistently, subsequent animal studies also indicated that the *Bdnf* gene is linked to changes in DNA methylation [32, 33].

In a study in older French women (aged 65 and older), Chagnon et al. tested whether four selected genes, namely, oxytocin receptor (*OXTR*; including rs53576), apolipoprotein E (*APOE*; including rs429358 and rs7412), *BDNF* (including rs6265), and *SLC6A4* (including rs25531), are involved in the association between their DNA methylation and anxiety disorders [34]. In comparison to healthy controls ($n = 19$), they found that the *BDNF* gene in patients with anxiety disorders ($n = 24$) had greater DNA methylation [34]. The same was also found for the *OXTR* gene in terms of greater DNA methylation [34]. On the other hand, no significant differences were found for the *SLC6A4* and *APOE* genes [34].

Childhood maltreatment and anxiety, substance abuse, and depressive disorders in adults ($n = 340$) were all found to be associated with decreased methylation in the *NR3C1* gene in a study done by Tyrka et al., where the *NR3C1* gene encodes glucocorticoid receptor [35]. Furthermore, comparing with healthy subjects ($n = 85$), a study done by Wang et al. indicated an association between patients with generalized anxiety disorders ($n = 64$) and increased methylation of the *NR3C1* gene [36].

Also, Ciuculete et al. investigated whether there are epigenetic differences in the promoter of the serine/threonine kinase 32B (*STK32B*) gene associated with an increased risk for generalized anxiety disorder in adolescents ($n = 221$) [37]. Their data confirmed there is 1% methylation increase in one CpG site within the *STK32B* promoter in adolescents at high risk of generalized anxiety disorder [37].

To test whether anxiety disorders are associated with epigenetic mechanisms, Emeny et al. further conducted an epigenome-wide association study using a population-based cohort ($n = 1522$) on anxiety [38]. A methylation increase of 48.5% was found in a single CpG site within the ankyrin repeat promoter and also that of the SOCS box containing 1 (*ASB*1) gene in subjects with high risks of severe anxiety, indicating that the *ASB1* gene in anxiety disorders is epigenetically regulated [38].

MicroRNAs

It has been suggested by results from many human and animal studies that miRNAs may be involved in the etiology of anxiety disorders, as well as in many central nervous system functions [39]. In the context of animal models, it has also been implicated that microRNAs may be a risk factor for anxiety disorders, such as miR-132/miR-212, miR-101a, miR-34b, and miR-135a [40–43]. It has been suggested that anxiety-induced microRNAs may regulate inflammation through suppression, as well as induce metabolic syndrome-related conditions, indicating that microRNAs may have similar regulatory networks in both metabolic syndrome-related and anxiety-related disorders [44].

Furthermore, Cohen et al. explored whether there is a causal relationship between miR-101a and anxiety disorders in the amygdala of rats. Their data indicated that an increase in the expression of miR-101a in the rat amygdala leads to also an increase in anxiety-like behavior in rats [41]. Consistently, Mannironi et al. demonstrated that the knockdown of miR-135a in the mouse amygdala may also cause an increase in anxiety-like behavior, implicating microRNA-dependent mechanism of miR-135a in the mouse amygdala for regulating anxiety-like behavior [42].

Histone Modifications

In an animal model, Moonat et al. reported that the epigenetic role of histone deacetylase isoforms may be implicated in the genetic predisposition to anxiety and alcoholism [45]. Epigenetic modulation of chronic anxiety disorders may involve histone deacetylation, as reported by Tran et al. [46].

Perspectives

It should be noted that the aforementioned studies are faced with several limitations. First, as the size of the cohort is small-scaled, conclusions made are not completely reliable [47]. Second, there is no complete replication consistency in the various biomarkers across independent studies, making results somewhat inconclusive. Third, it is also important to examine the differences in these candidate biomarkers across the various ethnic groups as diversified populations may result in different outcomes [20].

To assess G × E interactions, future studies may take advantage of leveraging novel machine learning techniques such as generalized multifactor dimensionality reduction [10, 48]. Quite a number of preferred learning methods also encompass artificial neural network algorithms, multifactor dimensionality reduction, Bayesian approaches, regression models, and generalized multifactor dimensionality reduction to weigh G × E interactions [49]. Moreover, future research can look over the contributions of genetic biomarkers by whole genome sequencing [50] or exome sequencing [51]. Due to the cost-cutting and maximized throughput of next-generation sequencing technologies, whole genome sequencing is a more complete method of genomic research and gives a wide spectrum of genetic variation in a single person [52]. Exome sequencing, which selectively sequences the nucleotides of protein-coding exons in a single person, has been utilized as an efficient and alternative method for Mendelian disorders and common diseases [51]. In summary, integrating whole genome sequencing approaches with novel machine learning tools may potentially build up an in-depth understanding of G × E interactions on anxiety disorders.

In future work, thorough evaluations may involve the use of bioinformatics pipelines, validating whether diagnostic prediction studies may replicate initial findings. In addition, in order to illustrate genetic networks at the genome level, custom machine learning and data mining pipelines should be used to investigate candidate biomarkers. Ultimately, future work requires a combination of different biomarkers, including metabolomics, clinical data, transcriptomics, genetics, imaging data, proteomics, and epigenetics in order to understand the pathogenesis of anxiety disorders and therapy [49]. Additionally, artificial intelligence approaches (such as machine learning, natural language processing, and computer vision) play a key role in wiping out the false-positive candidate genes that were found in the current association studies with meta-analysis, epistasis analysis, and pathway models [49]. Finding information missing from any data source and bridging up the gap between biological regulation models and phenotypes may be carried out by combining machine learning and deep learning models with multi-omics data [53]. While predictive tests are currently unavailable for disease states and treatment remission in anxiety disorders ahead of time, artificial intelligence and data science techniques will be explored to forecast the likelihood of drug efficacy and provide guidance on choosing medications for clinicians in future research [54].

Acknowledgments This study was supported in part by grants from the Ministry of Science and Technology of Taiwan (grant MOST 107-2634-F-075-002) and from the Taipei Veterans General Hospital (grant V105D17-002-MY2-2). We thank Emily Ting for English editing.

References

1. Bartlett AA, Singh R, Hunter RG. Anxiety and epigenetics. Adv Exp Med Biol. 2017;978:145–66.
2. Otowa T, Hek K, Lee M, Byrne EM, Mirza SS, Nivard MG, et al. Meta-analysis of genome-wide association studies of anxiety disorders. Mol Psychiatry. 2016;21(10):1391–9.
3. Schiele MA, Domschke K. Epigenetics at the crossroads between genes, environment and resilience in anxiety disorders. Genes Brain Behav. 2018;17(3):e12423.
4. Faravelli C, Lo Sauro C, Lelli L, Pietrini F, Lazzeretti L, Godini L, et al. The role of life events and HPA axis in anxiety disorders: a review. Curr Pharm Des. 2012;18(35):5663–74.
5. Sharma S, Powers A, Bradley B, Ressler KJ. Gene × environment determinants of stress- and anxiety-related disorders. Annu Rev Psychol. 2016;67:239–61.
6. Gibb BE, McGeary JE, Beevers CG, Miller IW. Serotonin transporter (5-HTTLPR) genotype, childhood abuse, and suicide attempts in adult psychiatric inpatients. Suicide Life Threat Behav. 2006;36(6):687–93.

7. Caspi A, Sugden K, Moffitt TE, Taylor A, Craig IW, Harrington H, et al. Influence of life stress on depression: moderation by a polymorphism in the 5-HTT gene. Science. 2003;301(5631):386–9.
8. Stein MB, Schork NJ, Gelernter J. Gene-by-environment (serotonin transporter and childhood maltreatment) interaction for anxiety sensitivity, an intermediate phenotype for anxiety disorders. Neuropsychopharmacology. 2008;33(2):312–9.
9. Klauke B, Deckert J, Reif A, Pauli P, Zwanzger P, Baumann C, et al. Serotonin transporter gene and childhood trauma – a G × E effect on anxiety sensitivity. Depress Anxiety. 2011;28(12):1048–57.
10. Lin E, Hong CJ, Hwang JP, Liou YJ, Yang CH, Cheng D, et al. Gene-gene interactions of the brain-derived neurotrophic-factor and neurotrophic tyrosine kinase receptor 2 genes in geriatric depression. Rejuvenation Res. 2009;12(6):387–93.
11. Kaufman J, Yang BZ, Douglas-Palumberi H, Grasso D, Lipschitz D, Houshyar S, et al. Brain-derived neurotrophic factor-5-HTTLPR gene interactions and environmental modifiers of depression in children. Biol Psychiatry. 2006;59(8):673–80.
12. Min JA, Lee HJ, Lee SH, Park YM, Kang SG, Chae JH. Gender-specific effects of brain-derived neurotrophic factor Val66Met polymorphism and childhood maltreatment on anxiety. Neuropsychobiology. 2013;67(1):6–13.
13. Martin L, Hemmings SMJ, Kidd M, Seedat S. No gene-by-environment interaction of BDNF Val66Met polymorphism and childhood maltreatment on anxiety sensitivity in a mixed race adolescent sample. Eur J Psychotraumatol. 2018;9(1):1472987.
14. Lazary J, Eszlari N, Juhasz G, Bagdy G. Genetically reduced FAAH activity may be a risk for the development of anxiety and depression in persons with repetitive childhood trauma. Eur Neuropsychopharmacol. 2016;26(6):1020–8.
15. Maglione JE, Nievergelt CM, Parimi N, Evans DS, Ancoli-Israel S, Stone KL, et al. Associations of PER3 and RORA circadian gene polymorphisms and depressive symptoms in older adults. Am J Geriatr Psychiatry. 2015;23(10):1075–87.
16. Min JA, Lee HJ, Lee SH, Park YM, Kang SG, Park YG, et al. RORA polymorphism interacts with childhood maltreatment in determining anxiety sensitivity by sex: a preliminary study in healthy young adults. Clin Psychopharmacol Neurosci. 2017;15(4):402–6.
17. Pena CJ, Bagot RC, Labonte B, Nestler EJ. Epigenetic signaling in psychiatric disorders. J Mol Biol. 2014;426(20):3389–412.
18. Klose RJ, Bird AP. Genomic DNA methylation: the mark and its mediators. Trends Biochem Sci. 2006;31(2):89–97.
19. Jones PA. Functions of DNA methylation: islands, start sites, gene bodies and beyond. Nat Rev Genet. 2012;13(7):484–92.
20. Lin E, Tsai SJ. Genome-wide microarray analysis of gene expression profiling in major depression and antidepressant therapy. Prog Neuro-Psychopharmacol Biol Psychiatry. 2016;64:334–40.
21. Nagano T, Fraser P. No-nonsense functions for long noncoding RNAs. Cell. 2011;145(2):178–81.
22. Lutz PE, Turecki G. DNA methylation and childhood maltreatment: from animal models to human studies. Neuroscience. 2014;264:142–56.
23. Gross C, Hen R. The developmental origins of anxiety. Nat Rev Neurosci. 2004;5(7):545–52.
24. Champagne FA, Weaver IC, Diorio J, Dymov S, Szyf M, Meaney MJ. Maternal care associated with methylation of the estrogen receptor-alpha1b promoter and estrogen receptor-alpha expression in the medial preoptic area of female offspring. Endocrinology. 2006;147(6):2909–15.
25. Kinnally EL, Capitanio JP, Leibel R, Deng L, LeDuc C, Haghighi F, et al. Epigenetic regulation of serotonin transporter expression and behavior in infant rhesus macaques. Genes Brain Behav. 2010;9(6):575–82.
26. Murgatroyd C, Patchev AV, Wu Y, Micale V, Bockmuhl Y, Fischer D, et al. Dynamic DNA methylation programs persistent adverse effects of early-life stress. Nat Neurosci. 2009;12(12):1559–66.

27. Uchida S, Hara K, Kobayashi A, Otsuki K, Yamagata H, Hobara T, et al. Epigenetic status of Gdnf in the ventral striatum determines susceptibility and adaptation to daily stressful events. Neuron. 2011;69(2):359–72.
28. Weaver IC, D'Alessio AC, Brown SE, Hellstrom IC, Dymov S, Sharma S, et al. The transcription factor nerve growth factor-inducible protein a mediates epigenetic programming: altering epigenetic marks by immediate-early genes. J Neurosci. 2007;27(7):1756–68.
29. Zhang TY, Hellstrom IC, Bagot RC, Wen X, Diorio J, Meaney MJ. Maternal care and DNA methylation of a glutamic acid decarboxylase 1 promoter in rat hippocampus. J Neurosci. 2010;30(39):13130–7.
30. Chen ZY, Jing D, Bath KG, Ieraci A, Khan T, Siao CJ, et al. Genetic variant BDNF (Val66Met) polymorphism alters anxiety-related behavior. Science. 2006;314(5796):140–3.
31. Roth TL, Lubin FD, Funk AJ, Sweatt JD. Lasting epigenetic influence of early-life adversity on the BDNF gene. Biol Psychiatry. 2009;65(9):760–9.
32. Blaze J, Asok A, Roth TL. Long-term effects of early-life caregiving experiences on brain-derived neurotrophic factor histone acetylation in the adult rat mPFC. Stress. 2015;18(6):607–15.
33. Doherty TS, Forster A, Roth TL. Global and gene-specific DNA methylation alterations in the adolescent amygdala and hippocampus in an animal model of caregiver maltreatment. Behav Brain Res. 2016;298(Pt A):55–61.
34. Chagnon YC, Potvin O, Hudon C, Preville M. DNA methylation and single nucleotide variants in the brain-derived neurotrophic factor (BDNF) and oxytocin receptor (OXTR) genes are associated with anxiety/depression in older women. Front Genet. 2015;6:230.
35. Tyrka AR, Parade SH, Welch ES, Ridout KK, Price LH, Marsit C, et al. Methylation of the leukocyte glucocorticoid receptor gene promoter in adults: associations with early adversity and depressive, anxiety and substance-use disorders. Transl Psychiatry. 2016;6(7):e848.
36. Wang W, Feng J, Ji C, Mu X, Ma Q, Fan Y, et al. Increased methylation of glucocorticoid receptor gene promoter 1F in peripheral blood of patients with generalized anxiety disorder. J Psychiatr Res. 2017;91:18–25.
37. Ciuculete DM, Bostrom AE, Tuunainen AK, Sohrabi F, Kular L, Jagodic M, et al. Changes in methylation within the STK32B promoter are associated with an increased risk for generalized anxiety disorder in adolescents. J Psychiatr Res. 2018;102:44–51.
38. Emeny RT, Baumert J, Zannas AS, Kunze S, Wahl S, Iurato S, et al. Anxiety associated increased CpG methylation in the promoter of Asb1: a translational approach evidenced by epidemiological and clinical studies and a murine model. Neuropsychopharmacology. 2018;43(2):342–53.
39. Malan-Muller S, Hemmings SM, Seedat S. Big effects of small RNAs: a review of microRNAs in anxiety. Mol Neurobiol. 2013;47(2):726–39.
40. Aten S, Page CE, Kalidindi A, Wheaton K, Niraula A, Godbout JP, et al. miR-132/212 is induced by stress and its dysregulation triggers anxiety-related behavior. Neuropharmacology. 2019;144:256–70.
41. Cohen JL, Jackson NL, Ballestas ME, Webb WM, Lubin FD, Clinton SM. Amygdalar expression of the microRNA miR-101a and its target Ezh2 contribute to rodent anxiety-like behaviour. Eur J Neurosci. 2017;46(7):2241–52.
42. Mannironi C, Biundo A, Rajendran S, De Vito F, Saba L, Caioli S, et al. miR-135a regulates synaptic transmission and anxiety-like behavior in amygdala. Mol Neurobiol. 2018;55(4):3301–15.
43. Zhu J, Chen Z, Tian J, Meng Z, Ju M, Wu G, et al. miR-34b attenuates trauma-induced anxiety-like behavior by targeting CRHR1. Int J Mol Med. 2017;40(1):90–100.
44. Meydan C, Shenhar-Tsarfaty S, Soreq H. MicroRNA regulators of anxiety and metabolic disorders. Trends Mol Med. 2016;22(9):798–812.
45. Moonat S, Sakharkar AJ, Zhang H, Tang L, Pandey SC. Aberrant histone deacetylase2-mediated histone modifications and synaptic plasticity in the amygdala predisposes to anxiety and alcoholism. Biol Psychiatry. 2013;73(8):763–73.

46. Tran L, Schulkin J, Ligon CO, Greenwood-Van Meerveld B. Epigenetic modulation of chronic anxiety and pain by histone deacetylation. Mol Psychiatry. 2015;20(10):1219–31.
47. Lin E, Lane HY. Genome-wide association studies in pharmacogenomics of antidepressants. Pharmacogenomics. 2015;16(5):555–66.
48. Lou XY, Chen GB, Yan L, Ma JZ, Zhu J, Elston RC, et al. A generalized combinatorial approach for detecting gene-by-gene and gene-by-environment interactions with application to nicotine dependence. Am J Hum Genet. 2007;80(6):1125–37.
49. Lin E, Lane HY. Machine learning and systems genomics approaches for multi-omics data. Biomark Res. 2017;5:2.
50. Cirulli ET, Goldstein DB. Uncovering the roles of rare variants in common disease through whole-genome sequencing. Nat Rev Genet. 2010;11(6):415–25.
51. Bamshad MJ, Ng SB, Bigham AW, Tabor HK, Emond MJ, Nickerson DA, et al. Exome sequencing as a tool for Mendelian disease gene discovery. Nat Rev Genet. 2011;12(11):745–55.
52. Hodkinson BP, Grice EA. Next-generation sequencing: a review of technologies and tools for wound microbiome research. Adv Wound Care (New Rochelle). 2015;4(1):50–8.
53. Lin E, Kuo PH, Liu YL, Yu YW, Yang AC, Tsai SJ. A deep learning approach for predicting antidepressant response in major depression using clinical and genetic biomarkers. Front Psych. 2018;9:290.
54. Lin E, Lin CH, Lai YL, Huang CH, Huang YJ, Lane HY. Combination of G72 genetic variation and G72 protein level to detect schizophrenia: machine learning approaches. Front Psych. 2018;9:566.

The Role of the Oxytocin System in Anxiety Disorders

7

Seoyoung Yoon and Yong-Ku Kim

Introduction

Oxytocin is a neuropeptide, a category that includes endogenous substances that are synthesized by, distributed within, and involve a variety of functional roles in the central nervous system. Oxytocin is synthesized in magnocellular oxytocin-containing neurons in hypothalamic nuclei, including the paraventricular (PVN), accessory, and supraoptic nuclei (SON). These nuclei project to the posterior pituitary where oxytocin is then secreted into the general circulation. Parvocellular oxytocin-containing neurons are also in the PVN, which project to the brainstem and spinal cord [1]. Oxytocin acts on peripheral targets and modulates renal water reuptake, autonomic functions, metabolism, pain, and immune responses [2, 3]. Oxytocin also acts as a key hormone in parturition and lactation [4]. As a neuropeptide, oxytocin receptors are widely distributed in the central nervous system, and oxytocin's neurobehavioral role has received attention. Immunohistochemical localization of oxytocin receptors using a monoclonal antibody determined that the distribution of oxytocin receptors includes cortical, limbic, and hypothalamic regions, especially the amygdala, cingulate cortex, and ventrolateral septum [5]. As the anterior cingulate and amygdala have been studied with respect to psychiatric disorders including depression, bipolar disorder, schizophrenia, autism, and anxiety disorders [6–12], oxytocin may affect the pathophysiology of a number of psychiatric disorders. The largest volume of behavioral studies focuses on the role of oxytocin in social affiliative behavior, between parent and newborn, and in the contexts of sexual intimacy and trust [4, 13]. Oxytocin has thereby been referred to as the "love hormone" or "cuddle hormone." As psychosocial stress heightens anxiety, against

S. Yoon
Department of Psychiatry, School of Medicine, Catholic University of Daegu, Daegu, Korea

Y.-K. Kim (✉)
Department of Psychiatry, College of Medicine, Korea University, Seoul, South Korea
e-mail: yongku@korea.edu

Y.-K. Kim (ed.), *Anxiety Disorders*, Advances in Experimental Medicine and Biology 1191, https://doi.org/10.1007/978-981-32-9705-0_7

which social support might have a protective effect, oxytocin potentially moderates anxiety via its prosocial properties [14]. Oxytocin also seems to modulate the hypothalamic-pituitary-adrenal axis response to stressful conditions [15]. Other neurobehavioral roles for oxytocin seem to be related to anxiety. Functional neuroimaging studies demonstrated the effects of intranasal oxytocin in modulating brain functions, especially connections between the amygdala and various regions including the periaqueductal gray, reticular formation, medial frontal cortex, and inferior frontal gyrus, which are related to the autonomic fear response, emotional regulation, and social cognition [16–18]. Fear learning and extinction, which are important in the pathogenesis and treatment of anxiety disorder, are also related to oxytocin [19].

As growing evidence suggested that oxytocin is related not only to the anxiety and fear response itself but also to behavioral and cognitive characteristics that moderate anxiety, such as social behavior and cognition and learning and extinction, the relationship between anxiety disorder and oxytocin has also been studied regarding genetic vulnerability, gene-environment interactions via epigenetic studies, and the impacts of oxytocin on anxiety, related behavior, and brain activity. Clinical trials of therapeutic oxytocin as monotherapy or adjunctive with psychotherapy also have been conducted. Although a previous study found a correlation between plasma oxytocin concentrations and cerebrospinal fluid [20], oxytocin has limited ability to cross the blood-brain barrier, and assessing the effects of oxytocin on the central nervous system using peripheral markers such as serum level has limitations [21]. Therefore, intranasal application has been used in clinical trials to deliver oxytocin to the central nervous system [22].

In this chapter, we review current evidence from preclinical and clinical studies to introduce neurobehavioral mechanisms by which oxytocin modulates anxiety disorder. Then, we review the interactions of oxytocin and other systems known to be related to anxiety disorder. Finally, studies restricted to anxiety disorder as defined by the DSM-5 and the potential role of oxytocin in managing anxiety disorder are discussed.

Anxiolytic Effects of Oxytocin and Possible Neurobehavioral Mechanisms

(1) Oxytocin and Anxiety

In preclinical studies, oxytocin has shown anxiolytic effects in rodent animal models that may be gender dependent. Female rats with oxytocin deficits (OT-/OT-) showed increased anxiety-related behavior compared to the wild type. Further, central infusion of oxytocin antagonists increased anxiety-related behavior in wild-type female rats. Central administration of synthetic oxytocin showed anxiolytic effects. However, oxytocin-deficient male mice did not show increased anxiety [23]. In other rodent studies, oxytocin administration showed anxiolytic effects in both female and male rodents [24–28]. Increases of endogenous oxytocin related to pregnancy, lactation, and sexual activity were also related to anxiolytic effects [29–31].

Relationships between anxiety and oxytocin were also found in humans. In children, oxytocin concentrations in both plasma and cerebrospinal fluid had negative correlations with trait anxiety scores [20]. However, this negative correlation between plasma oxytocin and trait anxiety was only found in male adults, and furthermore, in female adults higher plasma oxytocin was related to greater attachment anxiety [32]. In a genetic study, the AA genotype of oxytocin receptor (OXTR) single nucleotide polymorphism (SNP) rs53576 was related to increased harm avoidance, which reflects an anxiety-related temperament, compared to G carriers. A carriers also had smaller amygdala volumes bilaterally and reduced resting-state functional coupling between the prefrontal cortex and amygdala bilaterally in an allele load-dependent manner [33]. This suggests a role for oxytocin in anxiety along with a potential neural mechanism. Intranasal oxytocin administration in healthy volunteers reduced levels of anxiety in anxiogenic conditions including 7.5% CO_2 inhalation, a validated experimental model of generalized anxiety disorder, anxiety felt before public speaking, or the Trier social stress test [14, 34, 35].

(2) Social Cognition and Emotion Processing

Social cognition is defined as "the mental operations that underlie social interactions, including perceiving, interpreting, and generating responses to the intentions, dispositions, and behaviors of others" [36]. Social perception, emotion processing, theory of mind, and attributional style are considered four core domains of social cognition by expert consensus [37]. Although this concept arose from research on schizophrenia and has received attention in relation to autism, studies of social cognition in anxiety disorder itself also exist. A recent meta-analysis reported deficits in mentalizing and emotional cognition in PTSD, although this disorder has been excluded from the anxiety disorder category since the DSM-5, and attributional biases in social phobia and other anxiety disorders [38]. In particular, in social anxiety disorder, which is characterized by marked fear or anxiety about social situations, fear of being negatively evaluated by others can be affected by social cognitive processing such as self-focused attention, deficits in emotional knowledge, and theory of mind [39–41]. In a functional brain imaging study, patients with social anxiety disorder showed diminished medial prefrontal cortex activation compared to healthy controls during a trust game used to evaluate mentalizing [42].

Within vertebrates, studies of neuropeptides and social behavior focus primarily on the oxytocin/vasopressin family. Homologs of oxytocin and vasopressin have been identified in various species, from hydra to vertebrates, and have been shown to modulate social and reproductive behaviors [4]. In animal studies, OT knockout mice showed loss of social recognition, reduced social behaviors, and increased aggression [43–45]. Cognitive and emotional processing, including accurate appraisal of socially relevant emotional information and enhanced memory of previously encountered socially relevant emotional information, influence emotions such as fear and anxiety toward social situations as well as the relevant behavior [46]. Effects of oxytocin on the recognition of emotions through facial expression have been repeatedly studied. Intranasal oxytocin improved the recognition of facial expressions reflecting positive emotion [47]. Although this earlier study did not find improved recognition of negative facial expressions [47], other studies demonstrated similar effects toward negative facial expressions associated with fear, anger, or sadness [48–50]. A meta-analysis found that intranasal oxytocin enhanced emotional recognition of overall facial expressions, and when restricted to the type of emotion, accuracy of recognition for happy and fearful faces was significantly improved [51]. As the eye region gives the most relevant cues about emotion in facial expression, improved facial recognition might be regulated by increased reflexive gaze shifts toward the eye region, which are also increased by oxytocin [52, 53]. This increased attention to the eye region showed correlations with superior colliculi and amygdala activation [52]. However, the relevant study findings are inconsistent. In a clinical study of healthy males, intranasal oxytocin improved speed of processing in facial recognition but did not affect accuracy or eye-gaze pattern, unlike in previous studies [54]. The authors proposed that the effect of oxytocin on eye-gaze varies depending on the salience of the eye regions of the faces used in tasks.

Deficits in social cognition are important negative and cognitive symptoms of schizophrenia. Studies in schizophrenic patients found that oxytocin improves social cognition in complex and higher level processes. Detection of sarcasm, deception, and empathy was improved after a single dose of intranasal oxytocin [55]. Another clinical trial in schizophrenia found that intranasal oxytocin also improved controlled social cognition, such as comprehending indirectly expressed emotions and intentions through deliberations, although this effect was not found in healthy controls [56]. A meta-analysis of social cognition and intranasal oxytocin in schizophrenia found that features of high-level social cognition, such as mentalizing and theory of mind, were significantly improved by oxytocin, but low-level features such as social cue perception were not affected by oxytocin [57]. In another meta-analysis among patients with neurodevelopmental disorders, intranasal oxytocin significantly affected the theory of mind and led to moderate but nonsignificant effects on empathy.

As individuals with anxiety disorders often find it difficult to discriminate between threatening and non-threatening conditions and tend to overgeneralize fear to harmless situations, improving the accuracy of emotional stimuli recognition can be anxiolytic [58]. More accurate recognition of facial expressions after oxytocin treatment could therefore induce prosocial behavior. Reduced ambiguity may help individuals to feel more at ease with facial expressions and increase empathy [59, 60]. Previous studies found that oxytocin modulates early attentional bias toward facial expressions [61, 62]. Considering that maladaptive attentional bias preferentially heightens attention to threat cues and strengthens avoidance behavior, which in turn aggravates anxiety, reduced attentional bias and improved recognition can help in exploration of facial expression and promote social engagement, eventually reducing anxiety. However, not all studies demonstrate positive impacts of oxytocin on social cognition. A randomized controlled trial found that intranasal oxytocin impaired social working memory in individuals with higher levels of social anxiety [63]. Such heterogeneous results from oxytocin studies of social cognition might be explained by context and subject characteristics that warrant more study [64].

(3) Fear Learning and Fear Extinction

Fear conditioning and generalization are pathogenic mechanisms in anxiety disorder [65]. Fear conditioning happens when an aversive stimulus is paired with a neutral stimulus. After repeated pairing, the neutral stimulus alone can induce fear responses and anticipation of the aversive stimulus, resulting in avoidance behavior. Posttraumatic stress disorder (which is no longer included as an anxiety disorder according to DSM-5 criteria) and specific phobia are typical disorders related to fear conditioning. Fear responses can be elicited by a broader range of stimuli sharing similar characteristics to the original stimulus, a phenomenon known as fear generalization [66]. Generalization of fear to safe conditions due to proliferating anxiety cues in an individual's surroundings is an important psychopathology of generalized anxiety [67]. In panic disorder, fear conditioning to interoceptive stimuli also takes part in pathogenesis [68]. In a clinical study, patients with panic disorder showed stronger conditioned generalization effects than healthy counterparts [69]. After fear learning, when the neutral stimulus that elicited the fear response is repeatedly presented without the aversive stimulus, the fear response is reduced, an effect known as extinction. Delayed or reduced fear extinction has been observed in anxiety patients [70]. Fear memory and extinction memory are consolidated in long-term memory. Consolidated memory becomes temporarily unstable when recalled, and behavioral or pharmacological interventions during this unstable period can reconsolidate memory [71]. Through this mechanism, exposure-related therapy that reactivates the traumatic event and facilitates extinction learning can reduce fear and anxiety.

Several brain regions and circuits play important roles in fear learning and extinction. The amygdala is related to fear acquisition and extinction [72]. The hippocampus and prefrontal cortex modulate fear acquisition and extinction [71]. As previously described, oxytocin receptors are present in the amygdala and prefrontal cortex. Previous studies found that oxytocin plays an important role in fear memory-related processes. In a rodent study, infusion of oxytocin or a selective agonist resulted in time-dependent and brain region-dependent effects on fear acquisition and extinction. Manipulation of oxytocin after fear retrieval was related to facilitated extinction in the infralimbic region of the medial prefrontal cortex. However, in the basolateral amygdala, synthetic oxytocin impaired extinction, while the selective agonist facilitated extinction. Oxytocin manipulation before conditioning also resulted in different effects in region-specific ways, such that micro-infusion into the basolateral amygdala was related to an increased fear response, while a reduced fear response was found after micro-infusions of a selective agonist in the central amygdala [19]. Another rodent study found conflicting results that the infusion of oxytocin in both the basolateral and central amygdalae were related to fear acquisition. This difference may have resulted from variation in the level of fear among subjects [73]. A third rodent study found that the effects of oxytocin-related fear acquisition and extinction may differ according to subject age. Juvenile rats showed enhanced fear with reduced extinction after microinjection of an oxytocin agonist in the amygdala, unlike adult rats [74]. Inhibition of fear memory reconsolidation after reactivation by oxytocin administration was suggested as a possible mechanism [75]. It was also suggested that rather than the inhibition of reconsolidation, the post-retrieval process was related to extinction facilitation after oxytocin treatment [76]. In a clinical study, modulatory effects of oxytocin on fear memory and extinction were also found. Intranasal oxytocin administered after fear conditioning was related to increased prefrontal activity and dampened amygdala activity, along with an increased fear response, in early phases of extinction; however, the fear response decreased in later phases and thus oxytocin facilitated fear extinction [77]. Facilitation of fear extinction by oxytocin has also been shown in other human studies [76, 78]. Based on such findings, exposure therapy with oxytocin augmentation was studied in patients with anxiety-related disorders, including specific phobia [79], social anxiety disorder [80], and posttraumatic stress disorder [81, 82]. But for now, clinical trials of oxytocin in patients with anxiety disorder show only limited effectiveness.

Interactions with Other Neuroendocrine and Neurotransmitter Systems

(1) Hypothalamic-Pituitary-Adrenal Axis

Psychosocial stress is the major trigger or aggravating factor in various psychiatric disorders. The hypothalamic-pituitary-adrenal axis (HPA axis) directs the neuroendocrinologic response to stress, which is mediated by corticotropin-releasing factor (CRF), adrenocorticotropic hormone (ACTH), and corticosteroids [83, 84]. A meta-analysis of cortisol stress reactivity across psychiatric disorders found that females with current anxiety disorder exhibited blunted responses to stresses, whereas males with current social anxiety disorder exhibited increased response [85]. Children with anxiety disorders, including separation anxiety disorder, generalized anxiety disorder, social phobia, and specific phobia, tend to have lower basal HPA axis function, which is a psychophysiological characteristic of chronic stress [86]. The relationships between the HPA axis and oxytocin have been studied. Oxytocin is released after stress and seems to modulate HPA axis reactivity [15]. A combination of oxytocin and social support decreased salivary cortisol levels in response to stress, increased calmness, and decreased anxiety during stress, indicating that combined oxytocin treatment is more effective than social support alone [14]. Genetic variation in the oxytocin receptor (rs53576) is related to salivary cortisol levels and exhibits diurnal fluctuation [87]. Oxytocin has been suggested to reduce the stress response of the HPA axis by enhancing the negative feedback loop of the HPA axis through a potentiating effect on CRF-induced ACTH secretion [15]. Oxytocin also downregulates CRF expression in two ways, indirectly via GABA and directly via inhibition of a coactivator of CRF gene transcription, CRTC3 [15]. Discrimination between safety and threat, previously described in this chapter to be important in anxiety and fear, is also related to interactions between CRF and oxytocin. CRF output neurons of bed nuclei of the stria terminalis suppress the discrimination of safety and threat and sustain anxiety, but oxytocin receptor-mediated activation of protein kinase C delta seems to inhibit the bed nuclei of stria terminalis output [88].

(2) Serotonergic System

The serotonergic system is the major treatment target in patients with anxiety disorder. Selective serotonin reuptake inhibitors and selective serotonin norepinephrine inhibitors have been used with improved efficacy in various anxiety disorders including panic disorder, generalized anxiety disorder, and social anxiety disorder [89]. Modulation of the serotonergic system by oxytocin has been demonstrated in animal studies and is suggested to be related to anxiety and social cognition [27]. In the brains of mice, tryptophan hydroxylase immunoreactive neurons were observed to be positive for a variant of yellow fluorescent protein that replaced the oxytocin receptor gene, suggesting a close relationship between the serotonergic and oxytocin systems. In that study, oxytocin infusion was related to reduced anxiety-related behavior and facilitation of serotonin release within the median raphe nucleus. Further infusion of a 5-$HT_{2a/2c}$ receptor antagonist inhibited the anxiolytic effects of infused oxytocin, suggesting that these anxiolytic effects are mediated by the serotonergic system [27]. However, another study of oxytocin receptor expression in raphe neuron-eliminated mice showed decreased inter-male aggression, which is important for habitat protection, but not decreased anxiety-like behavior in either male or female mice [90]. The procedural difference between oxytocin infusion for anxiolytic effects versus oxytocin receptor knockout for anxiogenic effects could explain this difference. Further, the lifelong deletion of the oxytocin receptor could result in the development of a compensatory mechanism [90]. In humans, PET imaging of the brains of healthy males with $5HT_{1A}$ receptor antagonist showed that oxytocin administration modulates the serotonergic system in the dorsal raphe nucleus, amygdala, insula, and orbitofrontal cortex, which might be related to the anxiolytic action of oxytocin [91]. Effects of oxytocin on social reward also seem to be affected by the serotonergic system. In mice, oxytocin works as a social reinforcement signal within the nucleus accumbens, but the genetic deletion of the dorsal raphe nucleus, the origin of serotonergic innervation to the nucleus accumbens, abolished the social reinforcing effect [92]. Social rewards and prosocial behavior related to oxytocin have primarily been studied in autism, but as previously described this property is important in social anxiety disorders in general and in reducing maladaptive avoidance behavior.

(3) GABAergic System

The GABAergic system is important for inhibitory neurotransmission in the central nervous system via the $GABA_A$ receptor. Benzodiazepine exerts anxiolytic effects via increased $GABA_A$ receptor-mediated inhibitory postsynaptic currents (IPSCs) within the central amygdala [93]. An oxytocin agonist, TGOT, was found to enhance GABAergic transmission within the central amygdala [94]. Further study showed that although both benzodiazepine and TGOT act on IPSCs and exert anxiolytic effects, the specific mechanisms by which they act are different because the former increases the decay time of IPSCs, while the latter increases the frequency of IPSCs. The combination of the two resulted in greater decreases in spontaneous spiking frequency of the central amygdala than either alone [95]. Brain regions other than the central amygdala are related to the anxiolytic effect of oxytocin via $GABA_A$ activation. In female prairie voles, oxytocin injections within the hypothalamic paraventricular nucleus prior to exposure to stress result in reduced anxiety-like behaviors and HPA axis activity along with increased GABAergic activity in the paraventricular nucleus. This oxytocin-related anxiolytic effect was hindered by $GABA_A$ receptor antagonist [96]. The medial prefrontal cortex is also involved in oxytocin-mediated anxiolytic effects. In the rat medial prefrontal cortex, oxytocin administration within the prelimbic area, but not in the infralimbic area or anterior cingulate cortex, was related to reduced anxiety-like behavior, increased GABAergic activity, and alterations in amygdala activation [97].

Oxytocin and Anxiety Disorders

(1) Social Anxiety Disorder

Because of the efficacy of oxytocin's prosocial effects, most studies of anxiety disorder and oxytocin have focused on social anxiety disorder. These studies analyzed plasma level, genetic predispositions regarding effects of oxytocin using functional neuroimaging, and related symptoms. A study found that plasma oxytocin levels are not significantly different between generalized anxiety disorder patients and healthy controls, but within the patient group, higher anxiety symptom severity and greater dissatisfaction with social relationships are related to higher levels of plasma oxytocin. In that study, the authors assumed that increased oxytocin levels resulted from attempts to reduce anxiety and increase social behavior [98]. This finding was replicated by another recent clinical study [99], in which the same research group assessed oxytocin levels in social anxiety disorder patients and healthy controls during a trust game. They found that the levels of oxytocin were lower in the patient group than in healthy controls, which might reflect decreased ability to react prosocially to other people's indications of trust [100]. Genetic studies of social anxiety and the oxytocin system using oxytocin receptor (OXTR) or oxytocin (OXT) gene single nucleotide polymorphisms (SNP) indicate that in healthy subjects, OXTR rs53576 A allele carriers exhibit stronger negative effects of a less secure attachment style on social anxiety than GG homozygotes [101]. Adolescent girls who experienced early life adversity and are heterozygous for OXTR rs2254298 demonstrated higher social anxiety than controls, but anxiety symptoms of GG homozygous carriers who experienced early life adversity were similar to those of girls who experienced only low early adversity regardless of genotype [102]. In adolescent samples, social anxiety in OXT rs4813625 C allele carriers was affected by parenting style; in that lower perceived parental support was related to greater social anxiety, and higher perceived parental support was related to lower levels of social anxiety. However, GG homozygotes did not demonstrate any relationship between parenting style and social anxiety [103]. A polygenic study that analyzed five polygenic components (rather than single SNPs) using principal covariate regression for 223 oxytocin-related SNPs found strong associations with adolescent social anxiety symptoms, and environmental variables moderated the effect [104]. Epigenetic findings that reflect gene-environmental interactions also support the relationship between oxytocin, environmental factors such as childhood adversity, and social anxiety. A multilevel epigenetic study of OXTR gene methylation in patients with social anxiety disorder and healthy controls found that reduced methylation is related to categorical phenotypes of social anxiety disorder, increased symptom severity, increased HPA axis reactivity during the Trier social stress test, and increased amygdala reactivity during social phobia-related word processing [105].

Effects of intranasal oxytocin in social anxiety disorder patients were studied in randomized controlled trials observing brain imaging, anxiety symptoms, and related social cognition. Heightened amygdala reactivity to fearful faces was observed in social anxiety disorder patients compared to healthy controls through brain functional magnetic resonance imaging (fMRI), but this heightened reactivity in the patient group was attenuated by oxytocin administration [106]. In generalized anxiety disorder patients, medial prefrontal cortex/anterior cingulate cortex reactivity to sad faces but not happy faces was greater than in healthy controls, and this hyperreactivity was normalized by oxytocin [107]. In brain fMRI during fearful face processing task in males, functional connectivity between the amygdala and the insula and middle cingulate/dorsal anterior cingulate gyri was reduced in social anxiety disorder patients compared with healthy controls. However, intranasal oxytocin normalized the previously decreased connectivity between the amygdala and the insula and middle cingulate/dorsal anterior cingulate in the patient group [108]. This decreased functional connectivity between the amygdala and frontal connectivity which was reversed by oxytocin administration was also found in other study. Resting-state functional connectivity between the amygdala and rostral anterior cingulate cortex/medial prefrontal cortex was reduced in social anxiety disorder patients compared to healthy controls, with a relationship between higher anxiety symptoms and lower amygdala connectivity. However, this reduced resting-state functional connectivity in social anxiety disorder patients was reversed by oxytocin administration [109]. In a study of male social anxiety disorder patients, intranasal oxytocin enhanced other-oriented reward motivation in a less anxious group but not in more highly anxious patients. Although no significant effects on social anxiety itself were found in that study, this finding indicates that prosocial effects of oxytocin might help social anxiety disorder patients in a severity-dependent manner [110]. Oxytocin-augmented exposure therapy in social anxiety disorder patients did not result in significant differences in overall symptom severity compared to placebo-augmented exposure therapy, but the oxytocin-treated group showed improved mental representations of self, such as evaluations of appearance and speech performance [80].

(2) Separation Anxiety Disorder

Oxytocin plays important roles in parturition and lactation and is related to mother-infant bonding [111]. Other affiliative behaviors among other social relationships, such as relationships between spouses, are also affected by oxytocin [112]. Therefore, oxytocin is commonly referred to as the "love hormone" or "cuddle hormone." Studies of relationship attachment styles, the oxytocin system, and psychiatric disorders have therefore been performed [113–115]. Separation anxiety disorder is characterized by excessive anxiety about separation from an attachment figure. Therefore, the relationship between separation anxiety and oxytocin has been studied in both children and adults. Salivary oxytocin levels were tested in youths with DSM-5 anxiety disorders. In that study, separation anxiety disorder was related to significantly lower levels of salivary oxytocin than in other anxiety disorders, and symptom severity was negatively associated with salivary oxytocin level [116]. Children's salivary oxytocin response to interactions with their mothers was higher among children with separation anxiety disorder compared with children with DSM-5 anxiety disorders other than separation anxiety [117]. A study of pregnant women found that plasma oxytocin levels at 3 months postpartum were significantly negatively correlated with symptom severity in adult separation anxiety, depression, and state and trait anxiety [118]. A genetic study did not detect any relationships between OXT mutations and social anxiety disorder, but the authors believed this was due to the small number of subjects and lack of statistical power [119]. In the case of OXTR, the rs53576 GG genotype with T-carrier genotype of Gβ3 (G protein β3 subunit gene) rs5443 was associated with higher levels of childhood and adult separation anxiety symptoms compared to other genotype carriers [120]. Although peripheral oxytocin levels and genetic polymorphisms of the oxytocin system have repeatedly shown correlations with separation anxiety, no clinical studies have tested the effects of intranasal oxytocin in separation anxiety disorder.

(3) Other Anxiety Disorders

Although oxytocin and its relationships to general anxiety and related neurobehavioral aspects have been extensively studied, most studies are restricted to social anxiety disorder and separation anxiety disorder, while studies of other anxiety disorders are scarce. In the case of specific phobia, intranasal oxytocin administration prior to exposure therapy was performed with patients with arachnophobia as a double-blind, placebo-controlled trial. However, in this study, contrary to expectations, the oxytocin pretreatment group showed poorer outcomes and reduced credibility of the therapy and perceptions of therapeutic alliance than the placebo group [121]. In a previous section of this chapter regarding fear learning and extinction, we described heterogeneous results from animal and human studies according to severity of symptoms, age, context, and timing of oxytocin administration. It has been suggested that as oxytocin affects neural activity in the amygdala, anterior insula, anterior cingulate, and precuneus, areas that are related to conscious monitoring of the surroundings, contextual framing is crucial and should be accounted for to avoid unfavorable outcomes [122].

A genetic association study of OXTR and panic disorder was performed in Japanese patients. The AA+AG genotype had protective effects compared to the GG genotype in both rs2254298 and rs53576 [123].

Conclusion

Evidence has accumulated that oxytocin modulates anxiety and fear. In the brain, oxytocin seems to dampen phobic reactions in the amygdala and increase connectivity between the amygdala and frontal areas, which in turn seems to modulate the top-down regulation of emotional processing. Oxytocin secretion increases under stressful conditions and modulates physiologic responses such as HPA axis response. Neurotransmitter, serotonergic, and GABAergic systems are deeply related to anxiety disorder and are also related to oxytocin. Some studies have found increased ability to correctly discriminate safe and harmful phenomena due to oxytocin treatment, which in turn could reduce anxiety toward a broad range of stimuli and maladaptive avoidance that also strengthens anxiety. Further, as oxytocin also modulates social behaviors including social reward, social cognition facilitating prosocial behavior, social anxiety disorder, and separation anxiety disorder have also been studied mostly among anxiety disorder patients. Regarding the clinical use of oxytocin, intranasal oxytocin with exposure therapy has been attempted in patients with anxiety disorders, due to oxytocin's role in fear memory extinction, reconsolidation, and prosocial effects that might enhance the therapeutic synergy. But for now most clinical studies have failed to validate the superior efficacy of oxytocin augmentation in anxiety disorder and sometimes demonstrate unfavorable outcomes. These inconclusive and heterogeneous findings in studies of anxiety disorder and oxytocin are explained by complex moderators, such as gender [124, 125], age [74], childhood adversity and attachment style [113, 115], symptom severity [110], personality, and contextual variables such as situational factors, for example, dealing with familiar people or under conditions of competition or interacting with members of out-groups [64, 126].

In conclusion, oxytocin is a valuable biomarker and treatment target for anxiety disorder. However, the complexity of variables affecting the relationship between anxiety disorder and oxytocin makes it difficult to interpret the results of relevant studies and to develop treatment strategies. Further investigations of related factors and clinical studies on various anxiety disorders might help to develop strategic and precision medicine approaches for patients with anxiety disorder.

References

1. de Wied D, Diamant M, Fodor M. Central nervous system effects of the neurohypophyseal hormones and related peptides. Front Neuroendocrinol. 1993;14(4):251–302.
2. Verbalis JG, Mangione MP, Stricker EM. Oxytocin produces natriuresis in rats at physiological plasma concentrations. Endocrinology. 1991;128(3):1317–22.
3. Yang HP, Wang L, Han L, Wang SC. Nonsocial functions of hypothalamic oxytocin. ISRN Neurosci. 2013;2013:179272.
4. Donaldson ZR, Young LJ. Oxytocin, vasopressin, and the neurogenetics of sociality. Science. 2008;322(5903):900–4.
5. Boccia ML, Petrusz P, Suzuki K, Marson L, Pedersen CA. Immunohistochemical localization of oxytocin receptors in human brain. Neuroscience. 2013;253:155–64.
6. Bench CJ, Friston KJ, Brown RG, Scott LC, Frackowiak RS, Dolan RJ. The anatomy of melancholia – focal abnormalities of cerebral blood flow in major depression. Psychol Med. 1992;22(3):607–15.
7. Devinsky O, Morrell MJ, Vogt BA. Contributions of anterior cingulate cortex to behaviour. Brain. 1995;118(Pt 1):279–306.
8. Malizia AL. What do brain imaging studies tell us about anxiety disorders? J Psychopharmacol. 1999;13(4):372–8.
9. Odriozola P, Dajani DR, Burrows CA, Gabard-Durnam LJ, Goodman E, Baez AC, et al. Atypical frontoamygdala functional connectivity in youth with autism. Dev Cogn Neurosci. 2018;2018:100603.
10. Davis M. The role of the amygdala in fear and anxiety. Annu Rev Neurosci. 1992;15:353–75.
11. Rasetti R, Mattay VS, Wiedholz LM, Kolachana BS, Hariri AR, Callicott JH, et al. Evidence that altered amygdala activity in schizophrenia is related to clinical state and not genetic risk. Am J Psychiatry. 2009;166(2):216–25.
12. Pillay SS, Gruber SA, Rogowska J, Simpson N, Yurgelun-Todd DA. fMRI of fearful facial affect recognition in panic disorder: the cingulate gyrus-amygdala connection. J Affect Disord. 2006;94(1–3):173–81.
13. Porges SW. Love: an emergent property of the mammalian autonomic nervous system. Psychoneuroendocrinology. 1998;23(8):837–61.
14. Heinrichs M, Baumgartner T, Kirschbaum C, Ehlert U. Social support and oxytocin interact to suppress cortisol and subjective responses to psychosocial stress. Biol Psychiatry. 2003;54(12):1389–98.
15. Winter J, Jurek B. The interplay between oxytocin and the CRF system: regulation of the stress response. Cell Tissue Res. 2019;375(1):85–91.
16. Domes G, Heinrichs M, Glascher J, Buchel C, Braus DF, Herpertz SC. Oxytocin attenuates amygdala responses to emotional faces regardless of valence. Biol Psychiatry. 2007;62(10):1187–90.
17. Kirsch P, Esslinger C, Chen Q, Mier D, Lis S, Siddhanti S, et al. Oxytocin modulates neural circuitry for social cognition and fear in humans. J Neurosci. 2005;25(49):11489–93.
18. Sripada CS, Phan KL, Labuschagne I, Welsh R, Nathan PJ, Wood AG. Oxytocin enhances resting-state connectivity between amygdala and medial frontal cortex. Int J Neuropsychopharmacol. 2013;16(2):255–60.
19. Lahoud N, Maroun M. Oxytocinergic manipulations in corticolimbic circuit differentially affect fear acquisition and extinction. Psychoneuroendocrinology. 2013;38(10):2184–95.
20. Carson DS, Berquist SW, Trujillo TH, Garner JP, Hannah SL, Hyde SA, et al. Cerebrospinal fluid and plasma oxytocin concentrations are positively correlated and negatively predict anxiety in children. Mol Psychiatry. 2014;20:1085.
21. Ermisch A, Barth T, Ruhle HJ, Skopkova J, Hrbas P, Landgraf R. On the blood-brain barrier to peptides: accumulation of labelled vasopressin, DesGlyNH2-vasopressin and oxytocin by brain regions. Endocrinol Exp. 1985;19(1):29–37.

22. Quintana DS, Smerud KT, Andreassen OA, Djupesland PG. Evidence for intranasal oxytocin delivery to the brain: recent advances and future perspectives. Ther Deliv. 2018;9(7):515–25.
23. Mantella RC, Vollmer RR, Li X, Amico JA. Female oxytocin-deficient mice display enhanced anxiety-related behavior. Endocrinology. 2003;144(6):2291–6.
24. Ring RH, Malberg JE, Potestio L, Ping J, Boikess S, Luo B, et al. Anxiolytic-like activity of oxytocin in male mice: behavioral and autonomic evidence, therapeutic implications. Psychopharmacology. 2006;185(2):218–25.
25. McCarthy MM, McDonald CH, Brooks PJ, Goldman D. An anxiolytic action of oxytocin is enhanced by estrogen in the mouse. Physiol Behav. 1996;60(5):1209–15.
26. Uvnas-Moberg K, Ahlenius S, Hillegaart V, Alster P. High doses of oxytocin cause sedation and low doses cause an anxiolytic-like effect in male rats. Pharmacol Biochem Behav. 1994;49(1):101–6.
27. Yoshida M, Takayanagi Y, Inoue K, Kimura T, Young LJ, Onaka T, et al. Evidence that oxytocin exerts anxiolytic effects via oxytocin receptor expressed in serotonergic neurons in mice. J Neurosci. 2009;29(7):2259–71.
28. Sabihi S, Durosko NE, Dong SM, Leuner B. Oxytocin in the prelimbic medial prefrontal cortex reduces anxiety-like behavior in female and male rats. Psychoneuroendocrinology. 2014;45:31–42.
29. Jurek B, Slattery DA, Maloumby R, Hillerer K, Koszinowski S, Neumann ID, et al. Differential contribution of hypothalamic MAPK activity to anxiety-like behaviour in virgin and lactating rats. PLoS ONE. 2012;7(5):e37060.
30. Neumann ID, Torner L, Wigger A. Brain oxytocin: differential inhibition of neuroendocrine stress responses and anxiety-related behaviour in virgin, pregnant and lactating rats. Neuroscience. 1999;95(2):567–75.
31. Waldherr M, Neumann ID. Centrally released oxytocin mediates mating-induced anxiolysis in male rats. Proc Natl Acad Sci. 2007;104(42):16681.
32. Weisman O, Zagoory-Sharon O, Schneiderman I, Gordon I, Feldman R. Plasma oxytocin distributions in a large cohort of women and men and their gender-specific associations with anxiety. Psychoneuroendocrinology. 2013;38(5):694–701.
33. Wang J, Qin W, Liu B, Zhou Y, Wang D, Zhang Y, et al. Neural mechanisms of oxytocin receptor gene mediating anxiety-related temperament. Brain Struct Funct. 2014;219(5):1543–54.
34. de Oliveira DCG, Chagas MHN, Garcia LV, Crippa JAS, Zuardi AW. Oxytocin interference in the effects induced by inhalation of 7.5% CO2 in healthy volunteers. Hum Psychopharmacol Clin Exp. 2012;27(4):378–85.
35. de Oliveira DCG, Zuardi AW, Graeff FG, Queiroz RHC, Crippa JAS. Anxiolytic-like effect of oxytocin in the simulated public speaking test. J Psychopharmacol. 2011;26(4):497–504.
36. Green MF, Penn DL, Bentall R, Carpenter WT, Gaebel W, Gur RC, et al. Social cognition in schizophrenia: an NIMH workshop on definitions, assessment, and research opportunities. Schizophr Bull. 2008;34(6):1211–20.
37. Pinkham AE, Penn DL, Green MF, Buck B, Healey K, Harvey PD. The social cognition psychometric evaluation study: results of the expert survey and RAND panel. Schizophr Bull. 2013;40(4):813–23.
38. Plana I, Lavoie M-A, Battaglia M, Achim AM. A meta-analysis and scoping review of social cognition performance in social phobia, posttraumatic stress disorder and other anxiety disorders. J Anxiety Disord. 2014;28(2):169–77.
39. Gkika S, Wittkowski A, Wells A. Social cognition and metacognition in social anxiety: a systematic review. Clin Psychol Psychother. 2018;25(1):10–30.
40. O'Toole MS, Hougaard E, Mennin DS. Social anxiety and emotion knowledge: a meta-analysis. J Anxiety Disord. 2013;27(1):98–108.
41. Hezel DM, McNally RJ. Theory of mind impairments in social anxiety disorder. Behav Ther. 2014;45(4):530–40.

42. Sripada CS, Angstadt M, Banks S, Nathan PJ, Liberzon I, Phan KL. Functional neuroimaging of mentalizing during the trust game in social anxiety disorder. Neuroreport. 2009;20(11):984–9.
43. Winslow JT, Insel TR. The social deficits of the oxytocin knockout mouse. Neuropeptides. 2002;36(2–3):221–9.
44. Ferguson JN, Aldag JM, Insel TR, Young LJ. Oxytocin in the medial amygdala is essential for social recognition in the mouse. J Neurosci. 2001;21(20):8278–85.
45. Lazzari VM, Becker RO, de Azevedo MS, Morris M, Rigatto K, Almeida S, et al. Oxytocin modulates social interaction but is not essential for sexual behavior in male mice. Behav Brain Res. 2013;244:130–6.
46. Guastella AJ, MacLeod C. A critical review of the influence of oxytocin nasal spray on social cognition in humans: evidence and future directions. Horm Behav. 2012;61(3):410–8.
47. Marsh AA, Yu HH, Pine DS, Blair RJ. Oxytocin improves specific recognition of positive facial expressions. Psychopharmacology. 2010;209(3):225–32.
48. Fischer-Shofty M, Shamay-Tsoory SG, Harari H, Levkovitz Y. The effect of intranasal administration of oxytocin on fear recognition. Neuropsychologia. 2010;48(1):179–84.
49. Ellenbogen MA, Linnen AM, Grumet R, Cardoso C, Joober R. The acute effects of intranasal oxytocin on automatic and effortful attentional shifting to emotional faces. Psychophysiology. 2012;49(1):128–37.
50. Schulze L, Lischke A, Greif J, Herpertz SC, Heinrichs M, Domes G. Oxytocin increases recognition of masked emotional faces. Psychoneuroendocrinology. 2011;36(9):1378–82.
51. Shahrestani S, Kemp AH, Guastella AJ. The impact of a single administration of intranasal oxytocin on the recognition of basic emotions in humans: a meta-analysis. Neuropsychopharmacology. 2013;38(10):1929–36.
52. Gamer M, Zurowski B, Büchel C. Different amygdala subregions mediate valence-related and attentional effects of oxytocin in humans. Proc Natl Acad Sci. 2010;107(20):9400.
53. Guastella AJ, Mitchell PB, Dadds MR. Oxytocin increases gaze to the eye region of human faces. Biol Psychiatry. 2008;63(1):3–5.
54. Hubble K, Daughters K, Manstead ASR, Rees A, Thapar A, van Goozen SHM. Oxytocin reduces face processing time but leaves recognition accuracy and eye-gaze unaffected. J Int Neuropsychol Soc. 2017;23(1):23–33.
55. Davis MC, Lee J, Horan WP, Clarke AD, McGee MR, Green MF, et al. Effects of single dose intranasal oxytocin on social cognition in schizophrenia. Schizophr Res. 2013;147(2–3):393–7.
56. Woolley JD, Chuang B, Lam O, Lai W, O'Donovan A, Rankin KP, et al. Oxytocin administration enhances controlled social cognition in patients with schizophrenia. Psychoneuroendocrinology. 2014;47:116–25.
57. Burkner PC, Williams DR, Simmons TC, Woolley JD. Intranasal oxytocin may improve high-level social cognition in schizophrenia, but not social cognition or neurocognition in general: a multilevel bayesian meta-analysis. Schizophr Bull. 2017;43(6):1291–303.
58. Dunsmoor JE, Paz R. Fear generalization and anxiety: behavioral and neural mechanisms. Biol Psychiatry. 2015;78(5):336–43.
59. Di Simplicio M, Harmer CJ. Oxytocin and emotion processing. J Psychopharmacol. 2016;30(11):1156–9.
60. Geng Y, Zhao W, Zhou F, Ma X, Yao S, Hurlemann R, et al. Oxytocin enhancement of emotional empathy: generalization across cultures and effects on amygdala activity. Front Neurosci. 2018;12:512.
61. Clark-Elford R, Nathan PJ, Auyeung B, Mogg K, Bradley BP, Sule A, et al. Effects of oxytocin on attention to emotional faces in healthy volunteers and highly socially anxious males. Int J Neuropsychopharmacol. 2014;18(2):pyu012.
62. Domes G, Sibold M, Schulze L, Lischke A, Herpertz SC, Heinrichs M. Intranasal oxytocin increases covert attention to positive social cues. Psychol Med. 2013;43(8):1747–53.

63. Tabak BA, Meyer ML, Dutcher JM, Castle E, Irwin MR, Lieberman MD, et al. Oxytocin, but not vasopressin, impairs social cognitive ability among individuals with higher levels of social anxiety: a randomized controlled trial. Soc Cogn Affect Neurosci. 2016;11(8):1272–9.
64. Bartz JA, Zaki J, Bolger N, Ochsner KN. Social effects of oxytocin in humans: context and person matter. Trends Cogn Sci. 2011;15(7):301–9.
65. Lissek S, Powers AS, McClure EB, Phelps EA, Woldehawariat G, Grillon C, et al. Classical fear conditioning in the anxiety disorders: a meta-analysis. Behav Res Ther. 2005;43(11):1391–424.
66. Lissek S, Biggs AL, Rabin SJ, Cornwell BR, Alvarez RP, Pine DS, et al. Generalization of conditioned fear-potentiated startle in humans: experimental validation and clinical relevance. Behav Res Ther. 2008;46(5):678–87.
67. Lissek S, Kaczkurkin AN, Rabin S, Geraci M, Pine DS, Grillon C. Generalized anxiety disorder is associated with overgeneralization of classically conditioned fear. Biol Psychiatry. 2014;75(11):909–15.
68. Acheson DT, Forsyth JP, Prenoveau JM, Bouton ME. Interoceptive fear conditioning as a learning model of panic disorder: an experimental evaluation using 20% CO(2)-enriched air in a non-clinical sample. Behav Res Ther. 2007;45(10):2280–94.
69. Lissek S, Rabin S, Heller RE, Lukenbaugh D, Geraci M, Pine DS, et al. Overgeneralization of conditioned fear as a pathogenic marker of panic disorder. Am J Psychiatry. 2010;167(1):47–55.
70. Duits P, Cath DC, Lissek S, Hox JJ, Hamm AO, Engelhard IM, et al. Updated meta-analysis of classical fear conditioning in the anxiety disorders. Depress Anxiety. 2015;32(4):239–53.
71. Britton JC, Evans TC, Hernandez MV. Looking beyond fear and extinction learning: considering novel treatment targets for anxiety. Curr Behav Neurosci Rep. 2014;1(3):134–43.
72. LaBar KS, Gatenby JC, Gore JC, LeDoux JE, Phelps EA. Human amygdala activation during conditioned fear acquisition and extinction: a mixed-trial fMRI study. Neuron. 1998;20(5):937–45.
73. Campbell-Smith EJ, Holmes NM, Lingawi NW, Panayi MC, Westbrook RF. Oxytocin signaling in basolateral and central amygdala nuclei differentially regulates the acquisition, expression, and extinction of context-conditioned fear in rats. Learn Mem. 2015;22(5):247–57.
74. Kritman M, Lahoud N, Maroun M. Oxytocin in the amygdala and not the prefrontal cortex enhances fear and impairs extinction in the juvenile rat. Neurobiol Learn Mem. 2017;141:179–88.
75. Hou Y, Zhao L, Zhang G, Ding L. Effects of oxytocin on the fear memory reconsolidation. Neurosci Lett. 2015;594:1–5.
76. Hu J, Wang Z, Feng X, Long C, Schiller D. Post-retrieval oxytocin facilitates next day extinction of threat memory in humans. Psychopharmacology. 2019;236(1):293–301.
77. Eckstein M, Becker B, Scheele D, Scholz C, Preckel K, Schlaepfer TE, et al. Oxytocin facilitates the extinction of conditioned fear in humans. Biol Psychiatry. 2015;78(3):194–202.
78. Acheson D, Feifel D, de Wilde S, McKinney R, Lohr J, Risbrough V. The effect of intranasal oxytocin treatment on conditioned fear extinction and recall in a healthy human sample. Psychopharmacology. 2013;229(1):199–208.
79. Acheson DT, Feifel D, Kamenski M, McKinney R, Risbrough VB. Intranasal oxytocin administration prior to exposure therapy for arachnophobia impedes treatment response. Depress Anxiety. 2015;32(6):400–7.
80. Guastella AJ, Howard AL, Dadds MR, Mitchell P, Carson DS. A randomized controlled trial of intranasal oxytocin as an adjunct to exposure therapy for social anxiety disorder. Psychoneuroendocrinology. 2009;34(6):917–23.
81. Sack M, Spieler D, Wizelman L, Epple G, Stich J, Zaba M, et al. Intranasal oxytocin reduces provoked symptoms in female patients with posttraumatic stress disorder despite exerting sympathomimetic and positive chronotropic effects in a randomized controlled trial. BMC Med. 2017;15(1):40.

82. Flanagan JC, Sippel LM, Wahlquist A, Moran-Santa Maria MM, Back SE. Augmenting prolonged exposure therapy for PTSD with intranasal oxytocin: a randomized, placebo-controlled pilot trial. J Psychiatr Res. 2018;98:64–9.
83. Goldstein DS, Kopin IJ. Evolution of concepts of stress. Stress. 2007;10(2):109–20.
84. Selye H. Stress and the general adaptation syndrome. Br Med J. 1950;1(4667):1383–92.
85. Zorn JV, Schur RR, Boks MP, Kahn RS, Joels M, Vinkers CH. Cortisol stress reactivity across psychiatric disorders: a systematic review and meta-analysis. Psychoneuroendocrinology. 2017;77:25–36.
86. Dieleman GC, Huizink AC, Tulen JH, Utens EM, Creemers HE, van der Ende J, et al. Alterations in HPA-axis and autonomic nervous system functioning in childhood anxiety disorders point to a chronic stress hypothesis. Psychoneuroendocrinology. 2015;51:135–50.
87. Norman GJ, Hawkley L, Luhmann M, Ball AB, Cole SW, Berntson GG, et al. Variation in the oxytocin receptor gene influences neurocardiac reactivity to social stress and HPA function: a population based study. Horm Behav. 2012;61(1):134–9.
88. Janeček M, Dabrowska J. Oxytocin facilitates adaptive fear and attenuates anxiety responses in animal models and human studies – potential interaction with the corticotropin-releasing factor (CRF) system in the bed nucleus of the stria terminalis (BNST). Cell Tissue Res. 2019;375(1):143–72.
89. Bandelow B, Michaelis S, Wedekind D. Treatment of anxiety disorders. Dialogues Clin Neurosci. 2017;19(2):93–107.
90. Pagani JH, Williams Avram SK, Cui Z, Song J, Mezey É, Senerth JM, et al. Raphe serotonin neuron-specific oxytocin receptor knockout reduces aggression without affecting anxiety-like behavior in male mice only. Genes Brain Behav. 2015;14(2):167–76.
91. Mottolese R, Redouté J, Costes N, Le Bars D, Sirigu A. Switching brain serotonin with oxytocin. Proc Natl Acad Sci USA. 2014;111(23):8637–42.
92. Dölen G, Darvishzadeh A, Huang KW, Malenka RC. Social reward requires coordinated activity of nucleus accumbens oxytocin and serotonin. Nature. 2013;501(7466):179–84.
93. Kang-Park MH, Wilson WA, Moore SD. Differential actions of diazepam and zolpidem in basolateral and central amygdala nuclei. Neuropharmacology. 2004;46(1):1–9.
94. Huber D, Veinante P, Stoop R. Vasopressin and oxytocin excite distinct neuronal populations in the Central Amygdala. Science. 2005;308(5719):245.
95. Viviani D, Terrettaz T, Magara F, Stoop R. Oxytocin enhances the inhibitory effects of diazepam in the rat central medial amygdala. Neuropharmacology. 2010;58(1):62–8.
96. Smith AS, Tabbaa M, Lei K, Eastham P, Butler MJ, Linton L, et al. Local oxytocin tempers anxiety by activating GABAA receptors in the hypothalamic paraventricular nucleus. Psychoneuroendocrinology. 2016;63:50–8.
97. Sabihi S, Dong SM, Maurer SD, Post C, Leuner B. Oxytocin in the medial prefrontal cortex attenuates anxiety: Anatomical and receptor specificity and mechanism of action. Neuropharmacology. 2017;125:1–12.
98. Hoge EA, Pollack MH, Kaufman RE, Zak PJ, Simon NM. Oxytocin levels in social anxiety disorder. CNS Neurosci Ther. 2008;14(3):165–70.
99. Oh KS, Kim EJ, Ha JW, Woo HY, Kwon MJ, Shin DW, et al. The relationship between plasma oxytocin levels and social anxiety symptoms. Psychiatry Investig. 2018;15(11):1079–86.
100. Hoge EA, Lawson EA, Metcalf CA, Keshaviah A, Zak PJ, Pollack MH, et al. Plasma oxytocin immunoreactive products and response to trust in patients with social anxiety disorder. Depress Anxiety. 2012;29(11):924–30.
101. Notzon S, Domschke K, Holitschke K, Ziegler C, Arolt V, Pauli P, et al. Attachment style and oxytocin receptor gene variation interact in influencing social anxiety. World J Biol Psychiatry. 2016;17(1):76–83.
102. Thompson RJ, Parker KJ, Hallmayer JF, Waugh CE, Gotlib IH. Oxytocin receptor gene polymorphism (rs2254298) interacts with familial risk for psychopathology to predict symptoms of depression and anxiety in adolescent girls. Psychoneuroendocrinology. 2011;36(1):144–7.

103. Olofsdotter S, Åslund C, Furmark T, Comasco E, Nilsson KW. Differential susceptibility effects of oxytocin gene (OXT) polymorphisms and perceived parenting on social anxiety among adolescents. Dev Psychopathol. 2018;30(2):449–59.
104. Nelemans SA, van Assche E, Bijttebier P, Colpin H, van Leeuwen K, Verschueren K, et al. Parenting interacts with oxytocin polymorphisms to predict adolescent social anxiety symptom development: a novel polygenic approach. J Abnorm Child Psychol. 2018;
105. Ziegler C, Dannlowski U, Bräuer D, Stevens S, Laeger I, Wittmann H, et al. Oxytocin receptor gene methylation: converging multilevel evidence for a role in social anxiety. Neuropsychopharmacology. 2015;40(6):1528–38.
106. Labuschagne I, Phan KL, Wood A, Angstadt M, Chua P, Heinrichs M, et al. Oxytocin attenuates amygdala reactivity to fear in generalized social anxiety disorder. Neuropsychopharmacology. 2010;35(12):2403–13.
107. Labuschagne I, Phan KL, Wood A, Angstadt M, Chua P, Heinrichs M, et al. Medial frontal hyperactivity to sad faces in generalized social anxiety disorder and modulation by oxytocin. Int J Neuropsychopharmacol. 2012;15(7):883–96.
108. Gorka SM, Fitzgerald DA, Labuschagne I, Hosanagar A, Wood AG, Nathan PJ, et al. Oxytocin modulation of amygdala functional connectivity to fearful faces in generalized social anxiety disorder. Neuropsychopharmacology. 2015;40(2):278–86.
109. Dodhia S, Hosanagar A, Fitzgerald DA, Labuschagne I, Wood AG, Nathan PJ, et al. Modulation of resting-state amygdala-frontal functional connectivity by oxytocin in generalized social anxiety disorder. Neuropsychopharmacology. 2014;39(9):2061–9.
110. Fang A, Treadway MT, Hofmann SG. Working hard for oneself or others: effects of oxytocin on reward motivation in social anxiety disorder. Biol Psychol. 2017;127:157–62.
111. Nagasawa M, Okabe S, Mogi K, Kikusui T. Oxytocin and mutual communication in mother-infant bonding. Front Hum Neurosci. 2012;6:31.
112. Lieberwirth C, Wang Z. Social bonding: regulation by neuropeptides. Front Neurosci. 2014;8:171.
113. Costa B, Pini S, Gabelloni P, Abelli M, Lari L, Cardini A, et al. Oxytocin receptor polymorphisms and adult attachment style in patients with depression. Psychoneuroendocrinology. 2009;34(10):1506–14.
114. Kiss I, Levy-Gigi E, Keri S. CD 38 expression, attachment style and habituation of arousal in relation to trust-related oxytocin release. Biol Psychol. 2011;88(2-3):223–6.
115. Eapen V, Dadds M, Barnett B, Kohlhoff J, Khan F, Radom N, et al. Separation anxiety, attachment and inter-personal representations: disentangling the role of oxytocin in the perinatal period. PLoS One. 2014;9(9):e107745.
116. Lebowitz ER, Leckman JF, Feldman R, Zagoory-Sharon O, McDonald N, Silverman WK. Salivary oxytocin in clinically anxious youth: Associations with separation anxiety and family accommodation. Psychoneuroendocrinology. 2016;65:35–43.
117. Lebowitz ER, Silverman WK, Martino AM, Zagoory-Sharon O, Feldman R, Leckman JF. Oxytocin response to youth-mother interactions in clinically anxious youth is associated with separation anxiety and dyadic behavior. Depress Anxiety. 2017;34(2):127–36.
118. Eapen V, Dadds M, Barnett B, Kohlhoff J, Khan F, Radom N, et al. Separation anxiety, attachment and inter-personal representations: disentangling the role of oxytocin in the perinatal period. PLoS One. 2014;9(9):e107745-e.
119. Costa B, Pini S, Martini C, Abelli M, Gabelloni P, Ciampi O, et al. Mutation analysis of oxytocin gene in individuals with adult separation anxiety. Psychiatry Res. 2009;168(2):87–93.
120. Costa B, Pini S, Baldwin DS, Silove D, Manicavasagar V, Abelli M, et al. Oxytocin receptor and G-protein polymorphisms in patients with depression and separation anxiety. J Affect Disord. 2017;218:365–73.
121. Milrod B, Altemus M, Gross C, Busch F, Silver G, Christos P, et al. Adult separation anxiety in treatment nonresponders with anxiety disorders: delineation of the syndrome and exploration of attachment-based psychotherapy and biomarkers. Compr Psychiatry. 2016;66:139–45.

122. Hurlemann R. Oxytocin-augmented psychotherapy: beware of context. Neuropsychopharmacology. 2017;42(1):377.
123. Onodera M, Ishitobi Y, Tanaka Y, Aizawa S, Masuda K, Inoue A, et al. Genetic association of the oxytocin receptor genes with panic, major depressive disorder, and social anxiety disorder. Psychiatr Genet. 2015;25(5):212.
124. Li K, Nakajima M, Ibanez-Tallon I, Heintz N. A cortical circuit for sexually dimorphic oxytocin-dependent anxiety behaviors. Cell. 2016;167(1):60–72.e11.
125. Bredewold R, Veenema AH. Sex differences in the regulation of social and anxiety-related behaviors: insights from vasopressin and oxytocin brain systems. Curr Opin Neurobiol. 2018;49:132–40.
126. Olff M, Frijling JL, Kubzansky LD, Bradley B, Ellenbogen MA, Cardoso C, et al. The role of oxytocin in social bonding, stress regulation and mental health: an update on the moderating effects of context and interindividual differences. Psychoneuroendocrinology. 2013;38(9):1883–94.

Translational Studies in the Complex Role of Neurotransmitter Systems in Anxiety and Anxiety Disorders

8

Jocelien D. A. Olivier and Berend Olivier

Introduction

Anxiety disorders are among the most prevalent psychiatric disorders. In DSM5, anxiety disorders include separation anxiety disorder, selective mutism, specific phobia, social anxiety disorder, panic disorder, agoraphobia, and generalized anxiety disorder. The impact of anxiety disorders is high; studies in Europe [1] or the USA [2] estimate anxiety disorders at approximately 14% and 22%, respectively, of the population. Moreover, anxiety disorders are often comorbid with other psychiatric disorders, including mood and substance abuse disorders. Anxiety disorders are quite heterogeneous, and the individual disorders (e.g., panic disorder versus phobia or generalized anxiety disorder) are quite distant from each other. No laboratory blood test or brain scan or any other biomarker is yet available to distinguish between either disorder. The complex classification of the various anxiety disorders complicates or even hinders the finding and development of new psychoactive drugs, notably because one-to-one translation to preclinical animal models of such anxiety disorders is extremely difficult to realize, if not impossible. Although a classification in intermediate phenotypes [3] that are based on neurobiological mechanisms probably underlying pathological anxiety processes has been suggested, this has not led to novel, innovative treatments for anxiety yet. Another hypothesis is

J. D. A. Olivier
Department of Neurobiology, Groningen Institute for Evolutionary Life Sciences (GELIFES), University of Groningen, Groningen, The Netherlands

B. Olivier (✉)
Department of Neurobiology, Groningen Institute for Evolutionary Life Sciences (GELIFES), University of Groningen, Groningen, The Netherlands

Department of Psychopharmacology, Utrecht Institute for Pharmaceutical Sciences, Utrecht University, Utrecht, The Netherlands

Department of Psychiatry, Yale University School of Medicine, New Haven, CT, USA
e-mail: B.Olivier@uu.nl

Y.-K. Kim (ed.), *Anxiety Disorders*, Advances in Experimental Medicine and Biology 1191, https://doi.org/10.1007/978-981-32-9705-0_8

based on the assumption that dysfunctional neurotransmitter systems underlie disturbances in the regulation of anxiety processes in the CNS. However, dysfunctional neurotransmitter systems are difficult to prove, and a direct relationship between certain neurotransmitter systems (e.g., the serotonergic or GABA-ergic system) is difficult to establish. Actually, drugs influencing these systems and exerting anxiolytic properties were found serendipitously (benzodiazepines; BDZ) or developed initially for other purposes (selective serotonin reuptake inhibitors; SSRIs) many decades ago. Since 2007, no new anxiolytic agents have been approved by the US Food and Drug Administration [4], and SSRIs (and to a lesser extent serotonin-norepinephrine reuptake inhibitors; SNRIs) remain the mainstay for the treatment of anxiety disorders (often augmented or preceded by certain BDZs for a limited time period). Not all patients respond well to serotonergic therapy, either due to bothersome side effects or treatment resistance.

The last decades' intense efforts have been made both in preclinical and clinical studies, to find and test other mechanisms targeting anxiety-involved mechanisms in the CNS in order to detect new and/or improved treatments for the various anxiety disorders.

In the following parts, the existing anxiolytics and putative new promising anxiety targets will be described. First, using a translational approach, serotonergic targets, in particular the serotonergic transporter and 5-HT_{1A}- and 5-HT_2 receptors, will be outlined, followed by the involvement of the $GABA_A$ system in anxiety and anxiety disorders. Some new potential targets will be shortly introduced (CRF_1 receptor, neurokinin$_1$ (NK_1) receptors, glucocorticoid receptors, and glutamatergic receptors [5, 6].

The Serotonin Transporter (5-HTT)

There are at least 14 different serotonin receptor types (5-HT_1 ($_{A,B,D,E,F}$), 5-HT_2 ($_{A,B,C}$), 5-HT_3, 5-HT_4, 5-HT_5 ($_{A,B}$), 5-HT_6, and 5-HT_7). Except for the 5-HT_3 receptor, all 5-HT receptors are G-protein coupled. Signaling via these G-protein-coupled serotonin receptors is extremely diverse, and we are still largely in the dark how these various 5-HT receptors operate in and contribute to the extremely complex interaction in the various ongoing functions in the CNS. This diversity in combination with a complex distribution of the various receptors in the brain seems to bring the serotonergic system in an important position to modulate various functions in the brain. Disruption of various aspects of serotonergic neurotransmission contributes to changes in vulnerability or even psychopathology, including depression, anxiety disorders, schizophrenia, and others [7]. All 5-HT receptors are postsynaptically located on non-serotonergic neurons (heteroreceptors), but 5-HT_{1A} and $5\text{-HT}_{1B/1D}$ receptors are also present as autoreceptors on the soma (5-HT_{1A}) or synaptic area ($5\text{-HT}_{1B/1D}$) of serotonergic cells (Fig. 8.1). The serotonin transporter (5-HTT) is localized both in the synaptic area and on the cell bodies of 5-HT neurons (Fig. 8.2). If serotonergic neurons fire, 5-HT is released into the synaptic cleft where it exerts its action on nearby pre- and postsynaptic serotonergic receptors. The 5-HTT is a

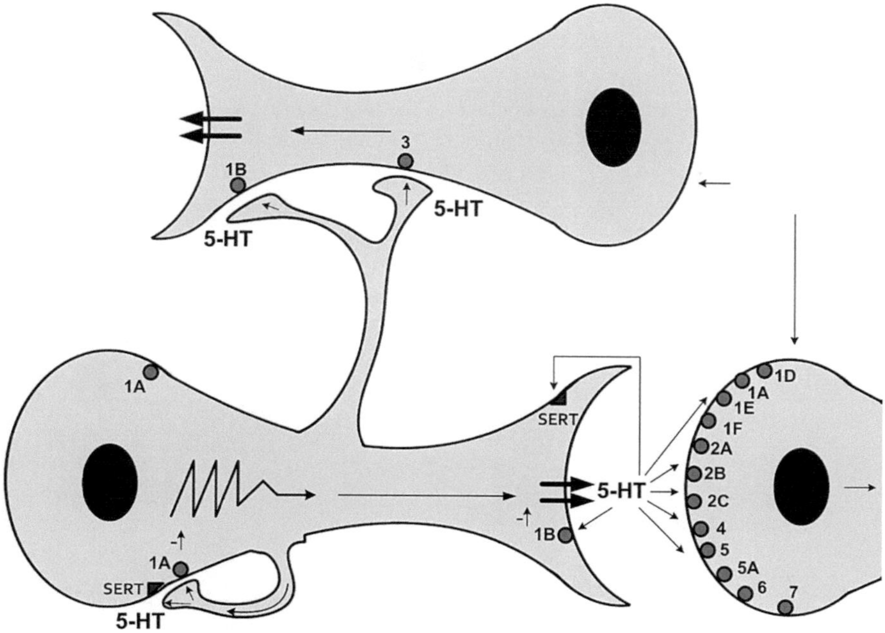

Fig. 8.1 Cartoon showing a serotonergic neuron and two postsynaptic non-serotonergic neuron (top and right). The various 5-HT receptors and the 5-HTT are schematically localized on these three neurons. 5-HT_{1A} receptors are present as somatodendritic autoreceptors and postsynaptic heteroreceptors. 5-$HT_{1B/1D}$ receptors are present as presynaptic receptors in the synaptic cleft and as postsynaptic heteroreceptors. The 5-HTT is present in the presynaptic part and at the soma of the serotonergic neuron. All other receptor(types) are postsynaptically located

key regulator of 5-HT signaling. The reuptake of serotonin by serotonergic neurons via the 5-HTT is the main mechanism for clearing extracellular 5-HT after its release; this leads to the termination of 5-HT signaling and recycles 5-HT in the neuron for subsequent release via uptake by the vesicular monoamine transporter (VMAT2). Another major route to end serotonergic activity is via uptake of synaptic 5-HT by surrounding glial cells where it is degraded to its main metabolite 5-hydroxyindole acetic acid (5-HIAA) by monoamine oxidase-A (MAO-A).

SSRIs are presently considered as the first-line pharmaceutical treatment for anxiety disorders, whereas SNRIs are considered second- or third-line treatments [8]. In a meta-analysis the differential efficacy of SSRIs and SNRIs and placebo in the treatment of anxiety disorders (together with depression, obsessive-compulsive disorder (OCD) and post-traumatic stress disorder) were estimated in children and adolescents [8]. SSRIs and SNRIs were about equally effective in treating anxiety disorders and more beneficial than placebo. Well-known, but still remarkable, is the placebo effect. This is a general observation in psychiatric treatments, both in pediatric and adult applications [9]. Notwithstanding this influential factor, it is generally accepted that psychotropic drugs that inhibit the 5-HTT in the CNS, exert anxiolytic activity in various patient groups with diverse anxiety and fear disturbances.

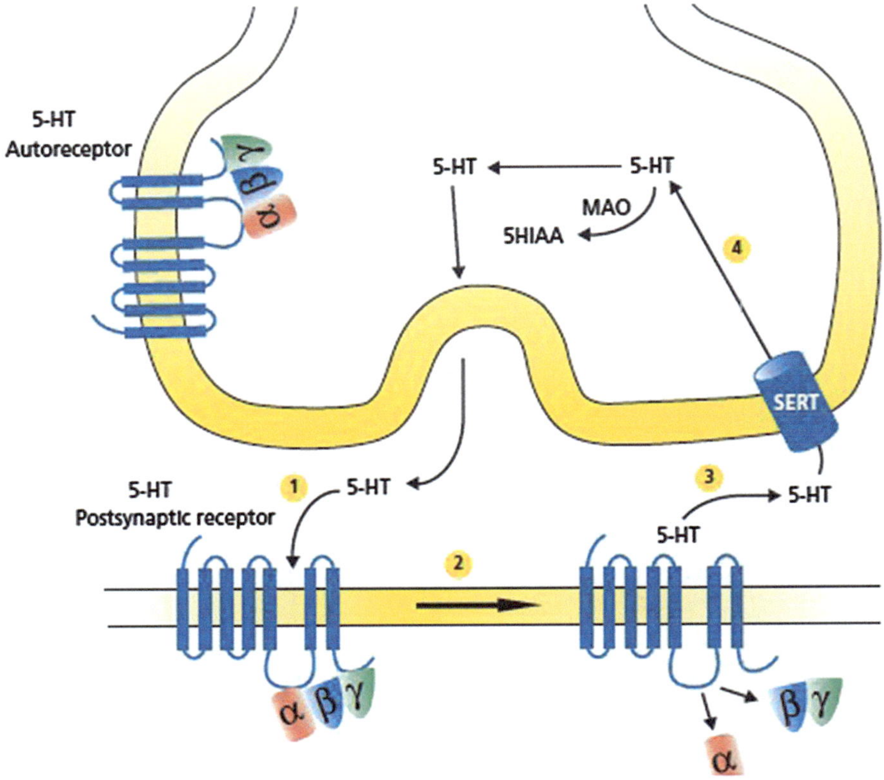

Fig. 8.2 Schematic representation of serotonin (5-HT) in the terminal and synapse. G-protein-coupled receptors are located presynaptically (5-HT autoreceptors 5-$HT_{1A/1B}$) or postsynaptically (5-HT $_{1/2/3/4/5/6/7}$ receptors). **1:** 5-HT is released from the presynaptic neuron and binds to a heterotrimeric G-protein postsynaptic receptor. Heterotrimeric G-protein complexes contain an α, β, and γ subunit, which in an inactive state are bound to GDP. **2**: 5-HT acts on postsynaptic receptors and induces a change in the conformation of the postsynaptic receptor. GDP is phosphorylated to GTP and binds to the α subunit, which subsequently becomes active. The β and γ subunits are freed. **3:** Extracellular 5-HT is taken up by the 5-HTT into the presynaptic neuron. **4:** Back in the presynaptic neuron 5-HT is broken down by MAO to 5-HIAA (occurs also extracellularly) or is being stored in vesicles for future release. MAO, monoamine oxidase; 5-HIAA, 5-hydroxyindole acetic acid; SERT, serotonin transporter. Figure adapted from Vinkers et al. [113]

For an extensive time, the 5-HT transporter has been implicated in processes involved in anxiety mechanisms. Blocking the 5-HTT leads to an acute rise in the synaptic concentration of serotonin. However, anxiolytic efficacy (similarly to depression) only emerges after longer treatment (weeks to months), indicating that long-term plasticity changes in presumed anxiety mechanisms have to take place. It is still largely unknown what those mechanisms are and how they come about. But the pathogenesis of anxiety disorders (and particularly GAD) is clearly influenced by many factors; moreover, many complex interactions between biological factors, experimental influences, and psychological mechanisms are involved [10]. In linkage studies and non-hypothesis-driven genome-wide association studies (GWAS) in large populations with various anxiety disorder problems, an association with the 5-HTT was never found [10], although some evidence emerged that a highly interconnected molecular network around GAD is part of the anxiety spectrum of disorders [11]. Candidate gene studies on putative individual nodes of this putative anxiety network have generated considerable evidence in favor of the serotonergic system and its impact on anxiety-related endophenotypes [10, 12].

Polymorphisms in the 5-HTT-gene and its associated transcriptional control area influence the activity of the serotonergic system [13]. Although focusing on anxiety, the 5-HTT gene and its variations (SLC6A4), producing a 5-HTT protein, are associated with several human behavioral and neurological traits but also with various medical disorders (e.g., myocardial infarction, pulmonary hypertension, irritable bowel syndrome, and sudden infant death syndrome) [14]. This indicates the important role of serotonergic neurotransmission in various aspects of our healthy and diseased biology. The emerging developmental role of serotonin [15] in psychiatric disorders, including anxiety-related ones, suggests that developmental disturbances in changing serotonin levels during critical periods in development (windows) may have long-lasting effects on brain function, particularly on later anxiety-related behaviors in adulthood [16, 17].

Most research involving the role of the 5-HTTs in psychiatry including anxiety disorders has been performed in adult populations. Lower 5-HTT expression and function in genotypes with the S-allele of the 5-HTTLPR (5-HTT length polymorphism) are associated with anxiety and depression [18, 19]. Subsequently, the 5-HTTPRL and associated variations in coding and noncoding regions of the 5-HTT-gene have been associated with many aspects of psychiatric disorders [14, 20]. The human 5-HTT is located at chromosome 17 (17q11.2) and consists of 14 exons spanning around 40kB; the emerging 5-HTT protein of 630 amino acids comprises 12 membrane domains (see Fig. 8.3). The expression of the human 5-HTT gene is modulated by a length variation (L and S variants) in the 5-HTTLPR and two SNPs in the same region (rs25531 and rs25332), all located upstream of the start of transcription. The single-nucleotide polymorphism rs25531 subdivides the L-allele into L_A and L_G alleles; the L_A allele is associated with increased expression of activity compared to L_G or S-alleles. A variable number of tandem repeat (VNTR) polymorphisms, called STin2, are present in intron 2, together with various SNPs influencing the activity of the 5-HTT protein. This polymorphism is tri-allelic, with 9-, 10-, or 12-repeat alleles. The latter allele is associated with enhanced 5-HTT expression [21]. It seems possible that individuals with low 5-HTT expressing genotypes, when exposed to SSRIs, have higher 5-HTT occupancy and saturation, theoretically associated with increasing central and peripheral serotonin availability. This might manifest itself as more side effects [21]. The 5-HTT is an extremely important modulator of the serotonergic neuronal system; duration and magnitude of serotonergic neurotransmission is largely steered by 5-HTT activity, thereby fine-tuning 5-HT signaling. Dysfunctions in these signaling pathways have been related to several psychiatric disorders and traits, including anxiety. Although initially, the S- and L-variants of the 5-HTTPRL were associated with differential vulnerability to neuropsychiatric disorders (or as a

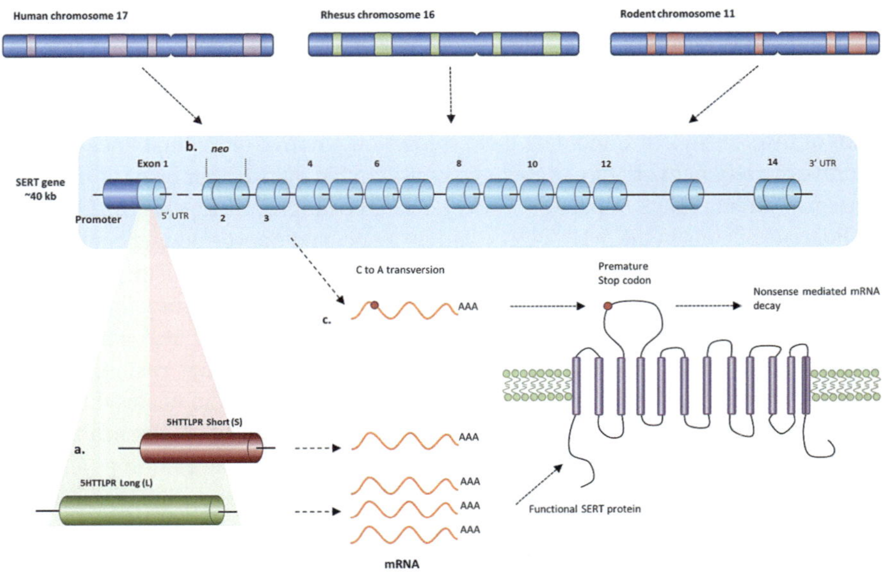

Fig. 8.3 Different alterations in the human, rhesus, and rodent 5-HTT gene resulting in changed transcription levels of 5-HTT. (**a**) Humans and rhesus macaques have short or long alleles for the 5-HTTLPR resulting in either lower or higher transcription levels, respectively. Since rodents do not carry an orthologue of this polymorphism, knockout of the 5-HTT can be achieved by (**b**) replacing exon 2 with a neo cassette (mice) or by (**c**) inducing a premature stop codon in exon 3 (rats) resulting in the absence of functional 5-HTT protein. (Figure adapted from Houwing et al. [56])

predictor of response to SSRIs), recent findings show a far more complicated pattern, where various polymorphisms inside the promoter region influence the activity of the 5-HT reuptake properties of the 5-HTT [22].

The story of modulation of serotonergic transmission via 5-HT mechanisms reveals the complexity of factors involved in the genetic-function coupling of any psychiatric disorder. The identification of specific gene variants and their involvement in anxiety is quite troublesome. It is increasingly clear that these variations may influence "intermediate biological phenotypes" in concert with other (background) genes, epigenetic factors, and environmental and developmental factors. Such complex interactions likely seem to contribute to the risk or resilience to develop certain psychiatric conditions. Using intermediate phenotypes may enable the search for associations between specific candidate genes (e.g., for 5-HTT) that mediate between a modulating allele and a complex disease phenotype [23].

Various studies on the 5-HTTPRL-promoter area are testing gene X environment and gene X gene interactions [24–27]. In general, S-alleles of 5-HTTPRL are associated with enhanced risk for a variety of psychiatric disorders, including anxiety. Caspi et al. [28] hypothesized the S-allele as a "risk" or "vulnerability" allele; the role of the L-allele is less clear although it has been suggested as risk factor for development of psychopathological traits [29]. Because L- and S-alleles are common to each human, and psychopathology does not happen in every individual, it is assumed that our genome includes "protective" alleles (e.g., CRF_1-receptor variants) rendering individuals resilient to pathology [30–32].

Genes certainly contribute to the risk to develop anxiety disorders and underlie anxious traits. No candidate genes have emerged playing a primary role in the expression for a vulnerability to anxiety or anxiety disorders [33, 34]. Meta-analysis of genotypes implicated in mental disorders with varying levels of heritability has indicated that for anxiety disorders, only three genes have been identified that contribute to anxious phenotypes (in contrast to, e.g., 50 genes for schizophrenia [33]). The former study did not implicate the 5-HTT and associated promoter area (5-HTTPRL) as a candidate gene involved in anxiety disorders. These complex findings indicate the difficulties in discovering new targets for anxiolytic drugs. This reflects that such complex neurobehavioral traits or disorders like the anxiety disorders are not caused by single genes, but probably by disturbances in extensive networks comprising of numerous parts and pathways that are created by many biological mechanisms.

Preclinical Evidence for the 5-HTT in Anxiety

Human linkage and association studies have not identified genes that contribute influentially in the modulation of anxiety or anxiety disorders; therefore, animal pathology and models might be of considerable importance in studying the involvement of genes in anxiety disorders as well. Human anxiety pathology is also present to a certain degree in animal pathology [35], and studies in animal anxiety models have considerably enhanced our knowledge of the neurobiological mechanisms involved in anxiety. Animal models can be very helpful in delineating putative genes involved in either anxiety or anxiety-associated traits [12, 36]. In order to progress in our search for fundamental mechanisms involved in anxiety processes and those in anxiety disorders, we need a translational approach; data from anxiety research, including genetic and environmental factors, should lead to fundamental research in animal models but also the reverse. The availability of cell-specific and inducible knockout and knock-ins, optogenetic technology [37], and, more recently, the Crispr/Cas 9 technology [38] enables selective operations in cellular mechanisms and circuit functions that are coupled to the gene's function [37]. These techniques are now starting to be applied in animal models of fear and anxiety.

Numerous preclinical anxiety and fear tests have been developed and used over time [39–41]. Animal tests and models of anxiety are based on face validity (is it measuring something analogous to human anxiety symptoms?), predictive validity (is it sensitive to clinically efficacious anxiolytics?), and construct validity (are the same pathophysiological mechanisms involved as in human anxiety disorders?). Till up now, none of the currently available anxiety tests or models does meet all of these criteria unequivocally. Griebel and Holmes [6] give an extensive review on preclinical models and tests used in anxiety and fear research. Many animal tests and models of anxiety are based on natural behavior patterns in rodents (often rats and mice

[42]). Such, often ethologically derived models include "approach-avoidance" tasks [40]. Animals are typically exposed to aversive situations like an open field, elevated plus maze or light/dark box. The level of anxiety is measured by the amount of avoidance of the aversive environment (open area, light). Besides the "unconditioned" procedures, several "conditioned" paradigms are also used to model anxiety disorders, including conflict procedures like the Vogel-lick test, defensive burying, four-plate test, and fear-potentiated startle. Various other paradigms are used to measure anxiety levels including physiological parameters [43], social interaction, predator stress, and stress-induced vocalizations [44]. Over the last decades, several models were introduced to investigate the role of the 5-HTT and adaptive 5-HT signaling in the in vivo action of SSRIs. In a transgenic knock-in mouse model, a single amino acid substitution (M172) in one of the 14-membrane domains led to diminished activity of SSRIs in emotional behavior, although the recognition of 5-HT and its clearance were not affected [45]. 5-HTT-overexpressing mice showed reduced anxiety levels [46, 47] and enhanced 5-$HT_{2A/2C}$ receptor functioning [48]. This 5-HTT-overexpressing model in combination with the effects associated with 5-HTT knockout (5-HTT−/−) mice might shed light on the compensatory effects observed on the life-span but also on those after chronic SSRI administration. 5-HTT−/− mice [49] and rats [50] have been created, and in both genotypes, extracellular 5-HT in the brain is enhanced [14, 51]. These enhanced extracellular 5-HT levels are associated with neurodevelopmental changes in 5-HT synthesis and metabolism [14]. The 5-HTT−/− rodent model has been regarded as a rather extreme example of the human 5-HTTLPR polymorphism, and as such, animals have difficulties to cope with stress and display anxiogenic behavior [52–55]. Heterozygous 5-HTT knockout rodents (5-HTT+/−) might be seen as a better model for the human 5-HTTPRL model, because the reduced 5-HTT expression is more or less comparable to the expression in human S-allele carriers. 5-HTT+/− animals have reduced 5-HTT expression and function, with 40–50% less 5-HTT protein levels [49, 51]. However, few neurochemical changes are present in 5-HTT+/− rodents compared to wild types (5-HTT+/+) [56]. Several serotonergic parameters, like basal extracellular levels and intracellular 5-HT levels, are not affected [51, 57], although often when challenged, 5-HTT+/− animals differ from 5-HTT+/+ animals [56]. 5-HTT+/− mice did not differ in anxiety levels from 5-HTT+/+ mice in an open field or in novelty-suppressed feeding [58]. There is evidence that the 5-HTT genotype may determine the stress response to the environment, particularly during early life stress (ELS) [56]. It has as yet to be established what mechanisms possibly underlie the 5-HTT gene variation X ELS interaction and specifically how that may lead to enhanced anxiety. Whether the 5-HTT+/− rodent is a good animal model to study the effects of anxiolytic drugs in situations where the animal is challenged toward showing anxiogenic behavior has to be awaited.

In wildtype mice and rats, often measured in different strains, SSRIs display anxiolytic activity [6]. In humans, SSRIs are also the first-line pharmaceutical agents to treat generalized anxiety disorder (GAD), post-traumatic stress disorder (PTSD), panic disorder (PD), and social anxiety disorder (SAD). Although none of the presently available animal anxiety tests and models specifically models for the specific human anxiety disorders, many tests/models have properties that predict for specific aspects present in the various anxiety disorders. Although in the early years (1960s till 1990s) of pharmacological treatment of anxiety disorders, BDZs were often first-line treatment, SSRIs have taken this position later on, mainly due to several safety issues (dependence liability, memory disturbances, tolerance, sedation). In an extensive summary of all published results in animal experiments on various pharmacological targets [6], SSRIs exerted in approx. 40% anxiolytic activity, were ineffective in 40% of the studies, and had 20% anxiogenic activity (in over 400 publications). Similar to human anxiety disorder treatment, the effects in animal tests and models are inconsistent. However, most, if not all, animal experimentation is performed using acute studies, whereas anxiolytic activity of SSRIs in humans occurs after at least 4–6 weeks of chronic treatment. Many patients do not respond to SSRIs, and disturbing side effects such as sexual dysfunction, sometimes associated with an initial transient period of enhanced anxiety, have contributed to a reduced acceptability of SSRIs in daily clinical practice.

In our own research, we mainly concurred with this preclinical anxiolytic profile of SSRIs. In a behavioral paradigm for screening putative anti-panic drugs, SSRIs like fluvoxamine and clomipramine (a TCA with predominant 5-HTT blockade) inhibited conditioned ultrasonic distress vocalizations (USV) in adult male rats [59]. SSRIs were also anxiolytic in guinea pig isolation calls [60], whereas in an approach-avoidance ambivalent situation in rats, SSRIs were inactive [61]. In a mouse model of anticipatory anxiety, stress-induced hyperthermia, antidepressants including SSRIs are inactive [62]. SSRIs are also inactive in the fear-potentiated startle [63]. In an ultrasonic vocalization model in isolated rat pups, SSRIs appeared anxiolytic [64].

The 5-HT$_{1A}$ Receptor and Anxiety

The anxiolytic activity of SSRIs that in some way is caused by the enhanced 5-HT release due to blockade of the 5-HTT must be effectuated by the effects of the released 5-HT on one or more 5-HT receptors; an important candidate is the 5-HT$_{1A}$ receptor (5-HT$_{1A}$R). Already in 1979 it was found that buspirone, a partial 5-HT$_{1A}$R agonist (and dopamine D$_2$R antagonist) exerts mild anxiolysis in human anxiety patients, notably in GAD [65]. This could be confirmed in rodent models of anxiety [7, 66]. Clinically, development of new 5-HT$_{1A}$R agonists for anxiety disorders (e.g., ipsapirone, gepirone, tandospirone, flesinoxan) failed. However, 5-HT$_{1A}$ receptors remain a clinically attractive target implied in anxiety and depression [7, 57, 66–69]. 5-HT$_{1A}$Rs are inhibitory G-protein-coupled receptors and present as autoreceptors on serotonergic neurons and as heteroreceptors (Fig. 8.1) on non-serotonergic neurons (e.g., GABA-ergic). The somatodendritic 5-HT$_{1A}$ autoreceptors regulate the serotonergic tone via feedback inhibition in concert with the 5-HTT [7]. Postsynaptic 5-HT$_{1A}$ heteroreceptors are abundantly distributed and expressed in the brain although restricted to specific and in high density in limbic areas like hippocampus and various parts of the frontal cortex. Autoreceptors are present in the dorsal and median raphé nuclei. There is evidence from genetic and imaging studies in humans that 5-HT_{1A}R density or regulation is associated with anxiety [70]. A C(−1019)G polymorphism (rs6295) in the promoter region of the 5-HT_{1A}R gene (Htr1a) was associated with mood-related variables like amygdala reactivity [71]. Increased 5-HT_{1A} autoreceptor expression was associated with enrichment of the G-allele; postsynaptically the C-allele was enhanced [72]. In panic disorder with agoraphobia patients [73], variations in the rs6295 polymorphism had functional consequences in defensive behavior, amygdala activity, and neural plasticity, suggesting that differentiation in 5-HT_{1A}R activities plays a role in anxiety and fear.

Griebel and Holmes [6] summarized the effects of 5-HT_{1A}R agonists in animal models of anxiety and fear. Of the 855 studies reported, approx. 65% of 5-HT_{1A}R agonists exerted anxiolytic activity, 30% was inactive and 5% anxiogenic. Moreover, knocking-out (KO) the 5-HT_{1A}R in three mouse strains [74–76] led to enhanced anxiety in most standard anxiety tests. This unanimous finding in different labs that inactivation of all 5-HT_{1A}R genes led to more anxious mice is rather striking. Later research found, however, that the anxious phenotype was dependent on the paradigm applied [77, 78] and one strain (but not the other two) of 5-HT_{1A}R KO mice had reduced sensitivity to anxiolytic effects of diazepam, associated with changes in particular α-subunits of the $GABA_A$- benzodiazepines (BDZ) R complex [79]. Apparently, such a dysfunction of the $GABA_A$-BDZ complex is not a prerequisite for the "anxiogenic" phenotype of 5-HT_{1A}R KO mice. The enhanced anxiety in the 5-HT_{1A}R KO mouse could not be antagonized by SSRIs [80]. Moreover, overexpression of 5-HT_{1A}Rs reduced anxiety [81]. These data strongly support a role for 5-HT_{1A}Rs in anxiety and fear processes. Further research on differential roles of pre- and postsynaptic 5-HT_{1A}Rs in anxiety, using rescue experiments on forebrain (postsynaptic) 5-HT_{1A}Rs, indicated a critical role for the latter in development of the anxiogenic phenotype [82]. Pharmacological blockade of 5-HT_{1A}Rs during early development, but not in adulthood, enhanced anxiety in normal mice [83, 84]. The complex regulation of anxiety processes during development and adulthood again indicates the complexity of the neural substrates, including genetic regulation, environmental influences, and its interactions. Simple relationships between a certain receptor (in this case 5-HT_{1A}R) and the complexity involved in the regulation of anxiety and fear but also its pathology are highly unlikely. However, rodent models (genetic and behavioral) are extremely useful in trying to untangle the different underlying parts and functions.

In our own (preclinical) research, both partial and full 5-HT_{1A}R agonists displayed anxiolytic activity. The azapirones (buspirone, ipsapirone, tandospirone, and zalospirone) are partial 5-HT_{1A}R agonists. Flesinoxan is a potent and selective full 5-HT_{1A}R agonist [66]. 5-HT_{1A}R agonists have limited anxiolytic activity in animal anxiety models based on release of suppressed behavior (both conditioned and unconditioned procedures like Geller-Seifter conflict, Vogel-lick test, four-plate, open field, elevated plus maze, light-dark box and social interaction test). In general 5-HT_{1A}R agonists are very active in animal models of stress-evoked behaviors both in conditioned and unconditioned situations like fear-potentiated startle, shock-induced USV, conditioned USV, defensive burying, stress-induced hyperthermia, rat pup, and guinea pig pup isolation calls [66]. Biased 5-HT_{1A}R agonists [85, 86] display either selectivity for 5-HT_{1A} autoreceptors (F13714) or 5-HT_{1A} heteroreceptors (F15599). However, both compounds had anxiolytic activity in several anxiety tests [87], not solving a potential role distribution between pre- and postsynaptic 5-HT_{1A}Rs in anxiety, although evidence is gathering that postsynaptic 5-$HT1_A$Rs, e.g., in the prelimbic cortex [88], may play a conclusive role.

Buspirone is worldwide marketed as anxiolytic, whereas tandospirone is only available in Japan and China. Flesinoxan was tested in phase III clinical trials as a potential anxiolytic (and antidepressant), but did not show superior activity compared to placebo and was subsequently abandoned. Although buspirone (and to a lesser extent tandospirone) are clinically effective anxiolytics, 5-HT_{1A}R agonists have not emerged as a breakthrough and successor of BDZ in the treatment of anxiety disorders.

Other 5-HT Receptors and Anxiety

Agonists and antagonists for practically all 5-HT receptors have been synthetized over the last 30 years and subsequently also tested in (mostly simple) anxiety tests. Griebel and Holmes [6] summarize results on frequently tested compounds, including 5-HT_2R antagonists, 5-HT_3R antagonists, and others. In 314 experiments 5-HT_2R antagonists were around 43% anxiolytic and 5-HT_3R antagonists (232 experiments) in 65%. Źmudzka et al. [87] essentially confirmed these data on 5-HT receptors and anxiety. Although in the 1990s the 5-HT_3R antagonists were considered the anxiolytics of the future based on animal results, they did not confirm their anxiolytic profile in human anxiety disorders [89]. Thus far, no promising new serotonin receptor ligands have been introduced to the market as new anxiolytics.

$GABA_A$ Receptors ($GABA_A$R) and Anxiety

$GABA_A$ (γ-aminobutyric acid, type A) receptors are the molecular targets of BDZ, chemical structures that were serendipitously found in the 1950s and were developed as the main anxiolytics of the ensuing decades. Besides anxiolysis, the classical BDZs also exerted sedation, epileptic seizure suppression, muscle relaxation, and hypnotic effects [90]. It took over 20 years before the molecular target of the BDZ, the $GABA_A$ receptor ($GABA_A$R) was discovered [91]. The $GABA_A$R is the target of clinically important anxiolytics, including BDZ, and is an ionotropic inhibitory receptor [90]. Although BDZs are excellent anxiolytics both in man and animals and are very safe in overdosing, it was realized that long-term use carries risks of tolerance, physical and psychological dependence, and withdrawal signs after discontinuation. Moreover, adverse side effects of sedation and cognitive slowing were found [92]. As a consequence, the SSRIs (and to a lesser extent the SNRIs) are now the first-line pharmacological treatment option for many anxiety disorders, whereas the BDZs are prescribed as second- or third-line treatment if antidepressants are ineffective and cognitive behavioral therapy has failed [4]. Notwithstanding the reluctance to prescribe BDZs, its use is still high in many countries; more than 5% of adults in the USA are prescribed BDZ each year [93].

Ionotropic $GABA_A$Rs are ubiquitous in the brain and are crucial in the temporal precise activity of neuronal circuitry and synchronized oscillatory activity of neuronal populations [94]. The diversity of this inhibitory neurotransmission regulating these networks is on one side regulated by the large variety in interneurons but also by the extremely diverse group of GABA receptors on the receiving side. GABAR consist of five subunits that surround a central pore in a pentameric composition (Fig. 8.4). When the endogenous ligand (GABA) binds to the GABAR, chloride ions flow into the neuron, leading to hyperpolarization of the cell membrane and inhibition of cell firing. $GABA_A$R are synaptically and extrasynaptically localized and molecularly very heterogeneous via variable subunit composition (Fig. 8.4). This heterogeneity determines the physiological/pharmacological properties [90, 95]. The various subunits ($\alpha_{1\text{-}6}$; $\beta_{1\text{-}3}$; $\gamma_{1\text{-}3}$; δ, ε, θ, and $\rho_{1\text{-}3}$) make up, although not randomly, the $GABA_A$R. Most $GABA_A$Rs are composed of two α, two β, and one γ subunit. GABA binds to the GABA site, which is formed by α and β subunits, and the BDZ binding site by one of the α subunits ($\alpha_{1, 2, 3, \text{or } 5}$) and a γ_2 subunit. Circa 60%

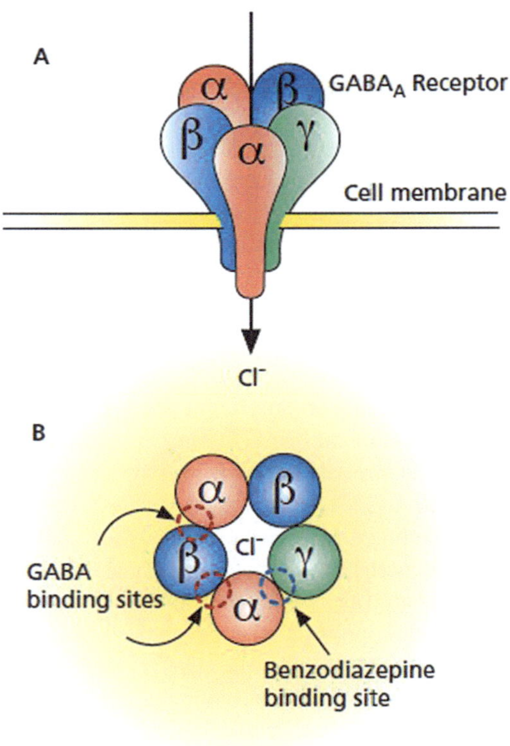

Fig. 8.4 Schematic picture of the $GABA_A$ receptor. A shows a global picture of the 2 α, the 2 β, and the γ subunits and the direction of the chloride influx. Fig. 1B shows a top view with the subunits arranged around a central pore, the chloride channel. The GABA and BDZ binding sites are indicated. (Figure adapted from Vinkers and Olivier [104])

of $GABA_A$Rs are of the $\alpha_1\beta_2\gamma_2$ subtype, 15–20% of $\alpha_2\beta_3\gamma_2$, 10–15% of $\alpha_3\beta_n\gamma_2$; the rest has variable subunit composition [96].

$GABA_A$R subtype selective compounds and mouse models with subunit point mutations have generated promising insight in the contribution of the different subunits in the different clinical effects of BDZ [90]. BDZ do not open the chloride channel in absence of the endogenous ligand GABA. Different pharmacological agents act at different $GABA_A$R sites, including GABA itself and various $GABA_A$R agonists (e.g., muscimol) or antagonists (e.g., bicuculline). Classical BDZ (e.g., chlordiazepoxide and diazepam) binds to the $GABA_A$-BDZ modulatory site. Other drug classes, including alcohol, barbiturates, and neurosteroids, also bind to the $GABA_A$R. The allosteric binding site for BDZ is always formed by two α-subunits ($\alpha_{1,\ 2,\ 3,\ \mathrm{or}\ 5}$), two β-subunits, and the γ_2 subunit. Only if the GABA receptor is activated, stimulation of the BDZ site modulates the channel opening (frequency and/or time). Ligands at the BDZ binding site are allosteric modulators and modify the efficacy and/or the affinity of GABA in a positive way (positive allosteric modification-PAM), negative way (NAM), or not (neutral), by stabilizing different three-dimensional conformations of the $GABA_A$R complex. The potency of a certain ligand is determined by its specificity for a certain receptor subtype and affinity/efficacy modulation [97]. BDZ have anxiolytic, sedative, hypnotic, muscle-relaxant, and anticonvulsant properties. If anxiolysis is the primary indication, sedative-hypnotic properties are unwanted. Classic BDZ are nonselectively activating $\alpha_{1,2,3,\ \mathrm{and}\ 5}$ subunits and bring automatically these unwanted effects. Extensive research, mainly in genetically engineered mice where certain $GABA_A$Rα subunits (GABRA) were made insensitive to diazepam binding, leaving the GABA site intact, illustrated that different subunits represent different functions. The broad therapeutic (and side) effects of BDZ are due to activation of α-subunits [90, 98]. Extensive preclinical work implicated α_2 and α_3 subunits in the modulation of anxiety, but intensive efforts (partial agonists, inverse agonists) have not yielded new and highly specific anxiolytics. Moreover, the main negative effects of all $GABA_A$R agonistic ligands (PAMs), i.e., tolerance and addiction, are still present in these new ligands [99], and none of these drugs has advanced to the market. A big outstanding question is whether activation of all α-subunits leads to tolerance/addiction or activation of only one specific subunit [100]. Also the once promising development of the so-called Z-drugs (zolpidem, zaleplon, and zopiclone) did not yield anxiolytics without tolerance/dependence problems. Interestingly, pregabalin, a structural analogue of GABA, is anxiolytic, although via a non-BDZ-like mechanism. Pregabalin is approved in 2007 for anxiety disorders in the EU, but not in the USA [101]. There is some evidence that activation of α_1-subunits is essential in the addictive potential of BDZ ligands [102, 103], but processes underlying tolerance development are complex and endpoint-dependent [104]. All these failures have probably been a major reason why pharmaceutical companies have left the development of innovative drugs for anxiety disorders that target the $GABA_A$R-BDZ complex.

Other Mechanisms and Anxiety

Although only preclinically investigated, many pharmacological mechanisms seem to be involved in the modulation of anxiety and fear, and intense efforts have tried to find new mechanisms outside the $GABA_AR$-BDZ complex and the serotonergic system; no new clinically effective drugs have emerged thus far. The situation since 2012 (described by Griebel and Holmes [6]) has not changed. Still most preclinical drug discovery is focused on neuropeptides and glutamate targets. Preclinically, CRF_1R antagonists, CCK_2R antagonists, and neurokinin$_{1,2}$R antagonists have anxiolytic-like activity, but no clinically successful drugs have emerged [6]. Several findings strongly implicated the glutamatergic system in anxiety and anxiety disorders [105]. Glutamate neurotransmission is complex due to the large number of signaling receptors and the wide and diverse distribution in the CNS. Metabotropic glutamate receptors (mGluRs) play a role in the modulation of anxiety in preclinical studies. Various ligands for mGluR subtype receptors have been tested clinically but failed to reach the market. Ketamine was recently found to have anxiolytic activity in social anxiety disorder [106] and refractory anxiety disorder patients [107] but also in animal anxiety models [108]. Ketamine, an NMDA (N-methyl-D-aspartate) R antagonist, exerts a rapid onset of antidepressant action in depressed patients and is particularly effective in anxious depression [109]. Most clinical studies utilize intravenous ketamine administration, which is a severe limitation for therapeutic use. Moreover, ketamine has serious side effects and abuse potential, which limits its widespread use as rapid onset therapeutic [110]. Further research into the efficacy of one of the enantiomers of ketamine (R,S) or one of its metabolites (e.g., 2R,6R)-hydroxynorketamine [111, 112]) may lead to a new, fast-acting anxiolytic (and antidepressant) treatment.

Conclusions

Discovery and development of innovative anxiolytics is severely hampering. Existing anxiolytics are developed decades ago and are still the therapeutics of choice. Moreover, the lack of new effective drug targets is not only disappointing but does indicate how the future treatment of the enormous population of CNS-disorder patients is in jeopardy. The reasons for this lack of progress are manyfold and are, besides, e.g., regulatory, financial, and marketing reasons, also due to our limited knowledge of the mechanisms and pathophysiology of the various anxiety disorders. We simply lack the knowledge on what is wrong in the brains of anxious people (normal and disordered). It is also not clear yet whether DSM5-based anxiety disorder classification is justifying the underlying disturbed processes in the brain. Whether the new classification of psychiatric diseases, the Research Domain Criteria (RDoC) project, as part of the 2008 NIMH strategic plan's call for new ways of classifying mental illnesses—based on dimensions of observable behavior and neurobiological measures—will be helpful in the research process has to be awaited. Translational research, with exchanging theories and data coming from clinical and preclinical research, is extremely urgent, because intrusive investigations of the human brain are impossible. In this endeavor, genetic and genomic approaches are part of the spectrum of contributing factors. The initial idea of identifying individual genes causing the disease very soon appeared an illusion. It becomes increasingly clear that anxiety disorders, probably similar to the fundamental mechanisms involved in anxiety and fear processes in the brain, are caused by the (inter)actions of hundreds of genes including environmental factors (epigenesis) and gene X gene interactions. In the present contribution, main focus has been on three druggable targets, 5-HTT, 5-HT_{1A} receptor, and $GABA_A$ receptor, whereas other approaches have been summarized. Although the therapeutics, acting via these targets were found before the mechanisms involved were unraveled, it is still uncertain whether or how these targets are involved in normal and diseased anxiety processes. In case of the serotonergic anxiolytics, the slow onset of action points to indirect effects leading to plasticity changes in certain systems in the brain that finally lead to reduced anxiety. In case of the $GABA_A$ receptors, a direct anxiolytic effect is found which indicates primary mechanisms directly influencing anxiety processes. Close translational collaboration between fundamental academic and discovery research might lead to badly needed breakthroughs in the development of drugs, not only for anxiety disorders but also other brain diseases.

References

1. Wittchen HU, Jacobi F. Size and burden of mental disorders in Europe – a critical review and appraisal of 27 studies. Eur Neuropsychopharmacol. 2005 Aug;15(4):357–76.
2. Kessler RC, Ruscio AM, Shear K, Wittchen HU. Epidemiology of anxiety disorders. Curr Top Behav Neurosci. 2010;2:21–35.
3. Ressler KJ, Mayberg HS. Targeting abnormal neural circuits in mood and anxiety disorders: from the laboratory to the clinic. Nat Neurosci. 2007 Sep;10(9):1116–24.
4. Stein MB, Craske MG. Treating anxiety In 2017: optimizing care to improve outcomes. JAMA. 2017 Jul 18;318(3):235–6.
5. Cryan JF, Sweeney FF. The age of anxiety: role of animal models of anxiolytic action in drug discovery. Br J Pharmacol. 2011 Oct;164(4):1129–61.
6. Griebel G, Holmes A. 50 years of hurdles and hope in anxiolytic drug discovery. Nat Rev Drug Discov. 2013 Sep;12(9):667–87.
7. Olivier B. Serotonin: a never-ending story. Eur J Pharmacol. 2015 Apr 15;753:2–18.
8. Locher C, Koechlin H, Sr Z, Werner C, Pine DS, Kirsch I, Et A. Efficacy and safety of selective serotonin reuptake inhibitors, serotonin-norepinephrine reuptake inhibitors, and placebo for common psychiatric disorders among children and adolescents: a systematic review and meta-analysis. JAMA Psychiatry. 2017 Oct 1;74(10):1011–20.
9. Janiaud P, Cornu C, Lajoinie A, Djemli A, Cucherat M, Kassai BI. The perceived placebo effect comparable between adults and children? A meta-regression analysis. Pediatr Res. 2017 Jan;81(1–1):11–7.
10. Gottschalk MG, Domschke K. Genetics of generalized anxiety disorder and related traits. Dialogues Clin Neurosci. 2017 Jun;19(2):159–68.
11. Okbay A, Baselmans BM, De Neve JE, Turley P, Nivard MG, Fontana MA, et al. Genetic variants associated with subjective well-being, depressive symptoms, and neuroticism identified through genome-wide analyses. Nat Genet. 2016 Jun;48(6):624–33.

12. Kas MJ, Krishnan V, Gould TD, Collier DA, Olivier B, Lesch KP, et al. Advances in multidisciplinary and cross-species approaches to examine the neurobiology of psychiatric disorders. Eur Neuropsychopharmacol. 2011 Jul;21(7):532–44.
13. Lesch KP. Molecular foundation of anxiety disorders. J Neural Transm. 2001;108(6):717–46.
14. Murphy DL, Lesch KP. Targeting the murine serotonin transporter: insights into human neurobiology. Nat Rev Neurosci. 2008 Feb;9(2):85–96.
15. Teissier A, Soiza-Reilly M, Gaspar P. Refining the role of 5-Ht In postnatal development of brain circuits. Front Cell Neurosci. 2017;11:139.
16. Marin O. Developmental timing and critical windows for the treatment of psychiatric disorders. Nat Med. 2016 Nov;22(11):1229–38.
17. Suri D, Teixeira CM, Cagliostro MK, Mahadevia D, Ansorge MS. Monoamine-sensitive developmental periods impacting adult emotional and cognitive behaviors. Neuropsychopharmacology. 2015 Jan;40(1):88–112.
18. Caspi A, Sugden K, Moffitt TE, Taylor A, Craig IW, Harrington H, et al. Influence of life stress on depression: moderation by a polymorphism in the 5-HTT gene. Science. 2003 Jul 18;301(5631):386–9.
19. Lesch KP, Bengel D, Heils A, Sabol SZ, Greenberg BD, Petri S, et al. Association of anxiety-related traits with a polymorphism in the serotonin transporter gene regulatory region. Science. 1996 Nov 29;274(5292):1527–31.
20. Haddley K, Bubb VJ, Breen G, Parades-Esquivel UM, Quinn JP. Behavioural genetics of the serotonin transporter. Curr Top Behav Neurosci. 2012 Jan 20;12:503–35.
21. Zhu J, Klein-Fedyshin M, Stevenson JM. Serotonin transporter gene polymorphisms and selective serotonin reuptake inhibitor tolerability: review of Pharmacogenetic evidence. Pharmacotherapy. 2017 Sep;37(9):1089–104.
22. Iurescia S, Seripa D, Rinaldi M. Role of the 5-HTTPRL and Snp promoter polymorphisms on serotonin transporter gene expression: a closer look at genetic architecture and in vitro functional studies of common and uncommon allelic variants. Mol Neurobiol. 2016 Oct;53(8):5510–26.
23. Murrough JW, Charney DS. The serotonin transporter and emotionality: risk, resilience, and new therapeutic opportunities. Biol Psychiatry. 2011 Mar 15;69(6):510–2.
24. Drabant EM, Ramel W, Edge MD, Hyde LW, Kuo JR, Goldin PR, et al. Neural mechanisms underlying 5-HTTPRL-related sensitivity to acute stress. Am J Psychiatry. 2012 Apr;169(4):397–405.
25. Montag C, Fiebach CJ, Kirsch P, Reuter M. Interaction of 5-HTTPRL and a variation on the oxytocin receptor gene influences negative emotionality. Biol Psychiatry. 2011 Mar 15;69(6):601–3.
26. Pluess M, Velders FP, Belsky J, Van Ijzendoorn MH, Bakermans-Kranenburg MJ, Jaddoe VW, et al. Serotonin transporter polymorphism moderates effects of prenatal maternal anxiety on infant negative emotionality. Biol Psychiatry. 2011 Mar 15;69(6):520–5.
27. Stollstorff M, Bean SE, Anderson LM, Devaney JM, Vaidya CJ. Rationality and emotionality: serotonin transporter genotype influences reasoning Bias. Soc Cogn Affect Neurosci. 2012 Mar 5;8(4):404–9.
28. Caspi A, Hariri AR, Holmes A, Uher R, Moffitt TE. Genetic sensitivity to the environment: the case of the serotonin transporter gene and its implications for studying complex diseases and traits. Am J Psychiatry. 2010 May;167(5):509–27.
29. Glenn AL. The other Allele: exploring the long allele of the serotonin transporter gene as a potential risk factor for psychopathy: a review of the parallels in findings. Neurosci Biobehav Rev. 2011 Jan;35(3):612–20.
30. Belsky J, Jonassaint C, Pluess M, Stanton M, Brummett B, Williams R. Vulnerability genes or plasticity genes? Mol Psychiatry. 2009 Aug;14(8):746–54.
31. Homberg JR, Lesch KP. Looking on the bright side of serotonin transporter gene variation. Biol Psychiatry. 2011 Mar 15;69(6):513–9.
32. Stein MB, Campbell-Sills L, Gelernter J. Genetic variation In 5-HTTPRL is associated with emotional resilience. Am J Med Genet B Neuropsychiatr Genet. 2009 Oct 5;150b(7):900–6.

33. Gatt JM, Burton KL, Williams LM, Schofield PR. Specific and common genes implicated across major mental disorders: a review of meta-analysis studies. J Psychiatr Res. 2015 Jan;60:1–13.
34. Smoller JW, Block SR, Young MM. Genetics of anxiety disorders: The complex road from Dsm To Dna. Depress Anxiety. 2009;26(11):965–75.
35. Fernando AB, Robbins TW. Animal models of neuropsychiatric disorders. Annu Rev Clin Psychol. 2011 Apr;7:39–61.
36. Flint J, Shifman S. Animal models of psychiatric disease. Curr Opin Genet Dev. 2008 Jun;18(3):235–40.
37. Tye KM, Deisseroth K. Optogenetic investigation of neural circuits underlying brain disease in animal models. Nat Rev Neurosci. 2012 Apr;13(4):251–66.
38. Walters BJ, Azam AB, Gillon CJ, Josselyn SA, Zovkic IB. Advanced in vivo use of Crispr/Cas9 and anti-sense Dna inhibition for gene manipulation in the brain. Front Genet. 2015;6:362.
39. Belzung C, Lemoine M. Criteria of validity for animal models of psychiatric disorders: focus on anxiety disorders and depression. Biol Mood Anxiety Disord. 2011 Nov 7;1(1):9.
40. Cryan JF, Holmes A. The ascent of mouse: advances in modelling human depression and anxiety. Nat Rev Drug Discov. 2005 Sep;4(9):775–90.
41. Haller J, Aliczki M, Gyimesine PK. Classical and novel approaches to the preclinical testing of anxiolytics: a critical evaluation. Neurosci Biobehav Rev. 2013 Dec;37(10 Pt 1):2318–30.
42. Rodgers R. Animal models of 'anxiety': where next? Behav Pharmacol. 1997 Nov;8(6–7):477–96.
43. Bouwknecht JA, Olivier B, Paylor RE. The stress-induced hyperthermia paradigm as a physiological animal model for anxiety: a review of pharmacological and genetic studies in the mouse. Neurosci Biobehav Rev. 2007;31(1):41–59.
44. Olivier B, Zethof T, Pattij T, Van Boogaert M, Van Oorschot R, Leahy C, et al. Stress-induced hyperthermia and anxiety: pharmacological validation. Eur J Pharmacol. 2003 Feb 28;463(1–3):117–32.
45. Thompson BJ, Jessen T, Henry LK, Field JR, Gamble KL, Gresch PJ, et al. Transgenic elimination of high-affinity antidepressant and cocaine sensitivity in the presynaptic serotonin transporter. Proc Natl Acad Sci USA. 2011 Mar 1;108(9):3785–90.
46. Jennings KA, Loder MK, Sheward WJ, Pei Q, Deacon RM, Benson MA, et al. Increased expression of the 5-HT transporter confers a low-anxiety phenotype linked to decreased 5-HT transmission. J Neurosci. 2006 Aug 30;26(35):8955–64.
47. Line SJ, Barkus C, Rawlings N, Jennings K, Mchugh S, Sharp T, et al. Reduced sensitivity to both positive and negative reinforcement in mice over-expressing the 5-hydroxytryptamine transporter. Eur J Neurosci. 2014 Dec;40(12):3735–45.
48. Dawson N, Ferrington L, Lesch KP, Kelly PA. Cerebral metabolic responses to 5-HT2A/C receptor activation in mice with genetically modified serotonin transporter (Sert) expression. Eur Neuropsychopharmacol. 2011 Jan;21(1):117–28.
49. Bengel D, Murphy DL, Andrews AM, Wichems CH, Feltner D, Heils A, et al. Altered brain serotonin homeostasis and locomotor insensitivity to 3, 4-Methylenedioxymethamphetamine ("Ecstasy") in serotonin transporter-deficient mice. Mol Pharmacol. 1998 Apr;53(4):649–55.
50. Smits BM, Mudde JB, Van De Belt J, Verheul M, Olivier J, Homberg J, et al. Generation of gene knockouts and mutant models in the laboratory rat by Enu-driven target-selected mutagenesis. Pharmacogenet Genomics. 2006 Mar;16(3):159–69.
51. Homberg JR, Olivier JD, Smits BM, Mul JD, Mudde J, Verheul M, et al. Characterization of the serotonin transporter knockout rat: a selective change in the functioning of the serotonergic system. Neuroscience. 2007 Apr 27;146(4):1662–76.
52. Adamec R, Burton P, Blundell J, Murphy DL, Holmes A. Vulnerability to mild predator stress in serotonin transporter knockout mice. Behav Brain Res. 2006 Jun 3;170(1):126–40.
53. Kalueff AV, Olivier JD, Nonkes LJ, Homberg JR. Conserved role for the serotonin transporter gene in rat and mouse neurobehavioral endophenotypes. Neurosci Biobehav Rev. 2010;34(3):373–86.

54. Olivier JD, Van Der Hart MG, Van Swelm RP, Dederen PJ, Homberg JR, Cremers T, et al. A study in male and female 5-HT transporter knockout rats: an animal model for anxiety and depression disorders. Neuroscience. 2008 Mar 27;152(3):573–84.
55. Wellman CL, Izquierdo A, Garrett JE, Martin KP, Carroll J, Millstein R, et al. Impaired stress-coping and fear extinction and abnormal corticolimbic morphology in serotonin transporter knock-out mice. J Neurosci. 2007 Jan 17;27(3):684–91.
56. Houwing DJ, Buwalda B, Van Der Zee EA, De Boer SF, Olivier JDA. The serotonin transporter and early life stress: translational perspectives. Front Cell Neurosci. 2017;11:117.
57. Holmes A, Li Q, Murphy DL, Gold E, Crawley JN. Abnormal anxiety-related behavior in serotonin transporter null mutant mice: the influence of genetic background. Genes Brain Behav. 2003 Dec;2(6):365–80.
58. Muller JM, Morelli E, Ansorge M, Gingrich JA. Serotonin transporter deficient mice are vulnerable to escape deficits following inescapable shocks. Genes Brain Behav. 2011 Mar;10(2):166–75.
59. Molewijk HE, Van Der Poel AM, Mos J, Van Der Heyden JA, Olivier B. Conditioned ultrasonic distress vocalizations in adult male rats as a behavioural paradigm for screening antipanic drugs. Psychopharmacology (Berl). 1995 Jan;117(1):32–40.
60. Molewijk HE, Hartog K, Van Der Poel AM, Mos J, Olivier B. Reduction of guinea pig pup isolation calls by anxiolytic and antidepressant drugs. Psychopharmacology (Berl). 1996 Nov;128(1):31–8.
61. Molewijk HE, Van Der Poel AM, Olivier B. The ambivalent behaviour "stretched approach posture" in the rat as a paradigm to characterize anxiolytic drugs. Psychopharmacology (Berl). 1995 Sep;121(1):81–90.
62. Zethof TJ, Van Der Heyden JA, Tolboom JT, Olivier B. Stress-induced hyperthermia as a putative anxiety model. Eur J Pharmacol. 1995 Dec 27;294(1):125–35.
63. Joordens RJ, Hijzen TH, Peeters BW, Olivier B. Fear-potentiated startle response is remarkably similar in two laboratories. Psychopharmacology (Berl). 1996 Jul;126(2):104–9.
64. Olivier B, Molewijk HE, Van Der Heyden JA, Van Oorschot R, Ronken E, Mos J, et al. Ultrasonic vocalizations in rat pups: effects of serotonergic ligands. Neurosci Biobehav Rev. 1998;23(2):215–27.
65. Goldberg HL, Finnerty RJ. The comparative efficacy of buspirone and diazepam in the treatment of anxiety. Am J Psychiatry. 1979 Sep;136(9):1184–7.
66. Olivier B, Soudijn W, van Wijngaarden I. The 5-HT1A receptor and its ligands: structure and function. Prog Drug Res. 1999;52:103–65.
67. Akimova E, Lanzenberger R, Kasper S. The serotonin-1a receptor in anxiety disorders. Biol Psychiatry. 2009 Oct 1;66(7):627–35.
68. Lanfumey L, Mongeau R, Cohen-Salmon C, Hamon M. Corticosteroid-serotonin interactions in the neurobiological mechanisms of stress-related disorders. Neurosci Biobehav Rev. 2008 Aug;32(6):1174–84.
69. Savitz J, Lucki I, Drevets WC. 5-HT(1a) receptor function in major depressive disorder. Prog Neurobiol. 2009 May;88(1):17–31.
70. Lesch KP, Gutknecht L. Focus on the 5-HT1A receptor: emerging role of a gene regulatory variant in psychopathology and pharmacogenetics. Int J Neuropsychopharmacol. 2004 Dec;7(4):381–5.
71. Fakra E, Hyde LW, Gorka A, Fisher PM, Munoz KE, Kimak M, et al. Effects of Htr1a C(-1019) G on amygdala reactivity and trait anxiety. Arch Gen Psychiatry. 2009 Jan;66(1):33–40.
72. Le François B, Czesak M, Steubl D, Albert PR. Transcriptional regulation at a Htr1a polymorphism associated with mental illness. Neuropharmacology. 2008 Nov;55(6):977–85.
73. Straube B, Reif A, Richter J, Lueken U, Weber H, Arolt V, et al. The functional -1019c/G Htr1a polymorphism and mechanisms of fear. Transl Psychiatry. 2014 Dec 16;4:E490.
74. Heisler LK, Chu HM, Brennan TJ, Danao JA, Bajwa P, Parsons LH, et al. Elevated anxiety and antidepressant-like responses in serotonin 5-HT1A receptor mutant mice. Proc Natl Acad Sci USA. 1998 Dec 8;95(25):15049–54.

75. Parks CL, Robinson PS, Sibille E, Shenk T, Toth M. Increased anxiety of mice lacking the Serotonin1a receptor. Proc Natl Acad Sci USA. 1998 Sep 1;95(18):10734–9.
76. Ramboz S, Oosting R, Amara DA, Kung HF, Blier P, Mendelsohn M, et al. Serotonin receptor 1a knockout: an animal model of anxiety-related disorder. Proc Natl Acad Sci USA. 1998 Nov;95(24):14476–81.
77. Pattij T, Hijzen TH, Groenink L, Oosting RS, Van Der Gugten J, Maes RA, et al. Stress-induced hyperthermia in the 5-HT1A receptor knockout mouse is normal. Biol Psychiatry. 2001 Apr 1;49(7):569–74.
78. Van Bogaert MJ, Oosting R, Toth M, Groenink L, Van Oorschot R, Olivier B. Effects of genetic background and null mutation of 5-HT1A receptors on basal and stress-induced body temperature: modulation by serotonergic and $Gaba_a$-Ergic drugs. Eur J Pharmacol. 2006;550(1–3):84–90.
79. Sibille E, Pavlides C, Benke D, Toth M. Genetic inactivation of the serotonin(1a) receptor in mice results in downregulation of major GABA(A) receptor alpha subunits, reduction of GABA(A) receptor binding, and benzodiazepine-resistant anxiety. J Neurosci. 2000 Apr 15;20(8):2758–65.
80. Santarelli L, Saxe M, Gross C, Surget A, Battaglia F, Dulawa S, et al. Requirement of hippocampal neurogenesis for the behavioral effects of antidepressants. Science. 2003 Aug 8;301(5634):805–9.
81. Kusserow H, Davies B, Hortnagl H, Voigt I, Stroh T, Bert B, et al. Reduced anxiety-related behaviour in transgenic mice overexpressing serotonin 1a receptors. Brain Res Mol Brain Res. 2004 Oct 22;129(1–2):104–16.
82. Gross C, Zhuang X, Stark K, Ramboz S, Oosting R, Kirby L, et al. Serotonin1a receptor acts during development to establish normal anxiety-like behaviour in the adult. Nature. 2002 Mar 28;416(6879):396–400.
83. Lo Iacono L, Gross C. Alpha-Ca2+/Calmodulin-dependent protein kinase Ii contributes to the developmental programming of anxiety in serotonin receptor 1a knock-out mice. J Neurosci. 2008 Jun 11;28(24):6250–7.
84. Vinkers CH, Oosting RS, Van Bogaert MJ, Olivier B, Groenink L. Early-life blockade of 5-HT1A receptors alters adult anxiety behaviour and benzodiazepine sensitivity. Biol Psychiatry. 2010;67(4):309–16.
85. Assié MB, Bardin L, Auclair AL, Carilla-Durand E, Depoortere R, Koek W, et al. F15599, a highly selective post-synaptic 5-HT(1a) receptor agonist: in-vivo profile in behavioural models of antidepressant and serotonergic activity. Int J Neuropsychopharmacol. 2010 Nov;13(10):1285–98.
86. Newman-Tancredi A. Biased agonism at serotonin 5-HT1A receptors: preferential postsynaptic activity for improved therapy of Cns disorders. Neuropsychiatry. 2011;1(2):149–64.
87. Żmudzka E, Salaciak K, Sapa J, Pytka K. Serotonin receptors in depression and anxiety: insights from animal studies. Life Sci. 2018 Oct 1;210:106–24.
88. Yamashita PS, Rosa DS, Lowry CA, Zangrossi H Jr. Serotonin actions within the prelimbic cortex induce anxiolysis mediated by serotonin 1a receptors. J Psychopharmacol. 2018 Dec 19:269881118817384.
89. Olivier B, van Wijngaarden I, Soudijn W. 5-HT(3) receptor antagonists and anxiety; a preclinical and clinical review. Eur Neuropsychopharmacol. 2000 Mar;10(2):77–95.
90. Rudolph U, Knoflach F. Beyond classical benzodiazepines: novel therapeutic potential of GABA(A) receptor subtypes. Nat Rev Drug Discov. 2011;10(9):685–97.
91. Möhler H, Okada T. Benzodiazepine receptor: demonstration in the central nervous system. Science. 1977 Nov 25;198(4319):849–51.
92. Lader M. benzodiazepines revisited--will we ever learn? Addiction. 2011 Dec;106(12):2086–109.
93. Olfson M, King M, Schoenbaum M. Benzodiazepine use in the United States. JAMA Psychiatry. 2015 Feb;72(2):136–42.

94. Engin E, Benham RS, Rudolph U. An emerging circuit pharmacology of GABAA receptors. Trends Pharmacol Sci. 2018 Aug;39(8):710–32.
95. Sigel E, Ernst M. The benzodiazepine binding sites of GABAA receptors. Trends Pharmacol Sci. 2018 Jul;39(7):659–71.
96. Möhler H, Fritschy JM, Rudolph U. A new benzodiazepine pharmacology. J Pharmacol Exp Ther. 2002 Jan;300(1):2–8.
97. Farb DH, Ratner MH. Targeting the modulation of neural circuitry for the treatment of anxiety disorders. Pharmacol Rev. 2014 Oct;66(4):1002–32.
98. Möhler H. GABAA receptors in central nervous system disease: anxiety, epilepsy, and insomnia. J Recept Signal Transduct Res. 2006;26(5–6):731–40.
99. Cheng T, Wallace DM, Ponteri B, Tuli M. Valium without dependence? Individual GAABAA receptor subtype contribution toward benzodiazepine addiction, tolerance, and therapeutic effects. Neuropsychiatr Dis Treat. 2018;14:1351–61.
100. Vinkers CH, Van Oorschot R, Nielsen EO, Cook JM, Hansen HH, Groenink L, et al. GABA(A) receptor alpha subunits differentially contribute to diazepam tolerance after chronic treatment. Plos One. 2012;7(8):E43054.
101. Millan MJ, Goodwin GM, Meyer-Lindenberg A, Ove OS. Learning from the past and looking to the future: emerging perspectives for improving the treatment of psychiatric disorders. Eur Neuropsychopharmacol. 2015 May;25(5):599–656.
102. Tan KR, Brown M, Labouebe G, Yvon C, Creton C, Fritschy JM, et al. Neural bases for addictive properties of benzodiazepines. Nature. 2010 Feb 11;463(7282):769–74.
103. Tan KR, Rudolph U, Luscher C. Hooked on benzodiazepines: GAABA receptor subtypes and addiction. Trends Neurosci. 2011 Apr;34(4):188–97.
104. Vinkers CH, Olivier B. Mechanisms underlying tolerance after long-term benzodiazepine use: a future for subtype-selective GABA(A) receptor modulators? Adv Pharmacol Sci. 2012;2012:416864.
105. Krystal JH, Mathew SJ, D'souza DC, Garakani A, Gunduz-Bruce H, Charney DS. Potential psychiatric applications of metabotropic glutamate receptor agonists and antagonists. Cns Drugs. 2010 Aug;24(8):669–93.
106. Taylor JH, Landeros-Weisenberger A, Coughlin C, Mulqueen J, Johnson JA, Gabriel D, et al. Ketamine for social anxiety disorder: a randomized, placebo-controlled crossover trial. Neuropsychopharmacology. 2018 Jan;43(2):325–33.
107. Glue P, Medlicott NJ, Harland S, Neehoff S, Anderson-Fahey B, Le NM, et al. Ketamine's dose-related effects on anxiety symptoms in patients with treatment refractory anxiety disorders. J Psychopharmacol. 2017 Oct;31(10):1302–5.
108. Fraga DB, Olescowicz G, Moretti M, Siteneski A, Tavares MK, Azevedo D, et al. Anxiolytic effects of ascorbic acid and ketamine in mice. J Psychiatr Res. 2018 May;100:16–23.
109. Ionescu DF, Niciu MJ, Richards EM, Zarate CA Jr. Pharmacologic treatment of dimensional anxious depression: a review. Prim Care Companion Cns Disord. 2014;16(3)
110. Sanacora G, Frye MA, Mcdonald W, Mathew SJ, Turner MS, Schatzberg AF, et al. A consensus statement on the use of ketamine in the treatment of mood disorders. JAMA Psychiatry. 2017 Apr 1;74(4):399–405.
111. Highland JN, Morris PJ, Zanos P, Lovett J, Ghosh S, Wang AQ, et al. Mouse, rat, and dog bioavailability and mouse oral antidepressant efficacy of (2r,6r)-hydroxynorketamine. J Psychopharmacol. In Press. 2018;
112. Zhang K, Hashimoto K. An update on ketamine and its two enantiomers as rapid-acting antidepressants. Expert Rev Neurother. 2018 Dec;4:1–10.
113. Vinkers CH, Olivier B, Bouwknecht JA, Groenink L, Olivier JDA. Stress-induced hyperthermia, the serotonin system and anxiety. Open Pharmacol J. 2010;4:15–29.

9 The Role of Early Life Stress in HPA Axis and Anxiety

Mario F. Juruena, Filip Eror, Anthony J. Cleare, and Allan H. Young

Introduction

It is now broadly accepted that psychological stress may change the internal homeostatic state of an individual. Anxiety disorders are among the most common of all mental disorders, and their pathogenesis is a major topic in psychiatry, both for prevention and treatment. Early stressful life (ELS) events and alterations of hypothalamic-pituitary-adrenal (HPA) axis function seem to have a significant role in the onset of anxiety.

Recent literature has demonstrated significant associations between traumatic events occurring in childhood and adolescence, called ELS, with unfavorable outcomes for the individual's health. Experiences of early life stress have been demonstrated by research to be associated with a wide number of psychological disorders including anxiety disorders. Recent studies have helped us to understand complex interaction, interplay, and bidirectional modulatory effects that these systems have, even in embryo life. They also help us to understand how ELS modulates the capacity of adaptation and how it can produce "wounds" that endure for a lifetime. Different areas of very specific subjects of research on different biological mechanisms are now beginning to be integrated, and that advance will hopefully improve our global understanding of mental disorders. It is imperative to find biological substrates and new therapeutic targets and diagnostic models in a psychiatric disease that play a pivotal role in this challenging and exciting task [1].

The effects of ELS negatively influence child development, affecting all spheres of an individual's life: behavioral, emotional, social, cognitive, and physical [2].

Parts of this chapter were previously published in Understanding Depression volume I, pp. 71–82, Springer Nature, 2018.

M. F. Juruena (✉) · F. Eror · A. J. Cleare · A. H. Young
Centre for Affective Disorders, Department of Psychological Medicine, Institute of Psychiatry, Psychology and Neuroscience-King's College London, London, UK
e-mail: mario.juruena@kcl.ac.uk

Y.-K. Kim (ed.), *Anxiety Disorders*, Advances in Experimental Medicine and Biology 1191, https://doi.org/10.1007/978-981-32-9705-0_9

Early Life Stress

The concept of early life stress (ELS) is quite broad and includes the different traumatic experiences that occur during childhood and adolescence, which may have repercussions in adulthood. Among these are a parental loss, separation from parents, childhood illness, family violence, and deprivation of food, clothing, shelter, and love.

Childhood maltreatment is a major social problem. It is a complex global phenomenon that does not respect boundaries of class, race, religion, age, or educational level and can occur both publicly and privately, resulting in serious physical injury or even death. Moreover, its psychological consequences can acutely affect a child's mental health well into adulthood [3].

ELS is an important, although non-specific, risk factor for anxiety and medical disorders, and that includes intrauterine stress as well [4, 5]. Stress in pregnancy has been shown to have a programming effect on the offspring, and one of the most consistent findings is HPA axis alterations [6]. Although GC plays a vital role in the development of the embryo, prolonged exposures may have deleterious neural effects, especially in brain areas that are rich in cortisol receptors [6].

Enduring effects were also suggested by studies of the great Dutch famine of 1944 to 1945 that showed that prenatal maternal malnutrition was associated with a variety of physical conditions in offspring when adults [7]. The offspring of fathers that were intrauterine at that time (grandchildren of the malnourished) were found to have increased risk of obesity, suggesting a transgenerational effect [8].

These studies are mentioned only to illustrate the magnitude and complexity of the effects of ELS and when that can happen, being in childhood, intrauterine, or, it seems, even before conception takes place. More studies are needed to elucidate the precise mechanisms of these effects and their relevance in human development and physical and mental health.

Studies of childhood maltreatment and their persistent effects throughout the lifespan have been more studied, and from now on, we will refer to ELS as stress that occurs in early childhood.

The childhood maltreatment may be subdivided into the following domains [9]:

(I) Physical abuse: physical aggression by someone older, with the risk of or result of injury.
(II) Emotional abuse: verbal aggression that affects the welfare or morals of the child or any conduct that humiliates, embarrasses, or threatens the child.
(III) Sexual abuse: any sexual contact or conduct between a child and someone older.
(IV) Emotional neglect: failure of caretakers to provide for basic emotional and psychological needs such as love, motivation, and support.
(V) Physical neglect: failure of caretakers to provide for basic physical needs such as feeding, a home, security, supervision, and health.

Childhood maltreatment significantly contributes to disease morbidity and mortality in adults [10], and it is essential to elucidate the mechanisms by which these early life events can elicit illnesses that become apparent decades after the presumed initial insult and why some people can adapt and others will present with an increased risk for psychiatric disorders, especially depression. It seems that a complex interaction between genes and environment are responsible for these effects.

Early life stress is associated with a diverse range of psychiatric consequences. In children and adolescents, it increases the risk of behavioral problems, including internalizing and externalizing behavior. Internalizing refers to behavioral symptoms reflected by anxiety, depression, somatic complaints, and inhibition. Externalizing refers to behavioral symptoms reflected by aggression, delinquency, and increased activity level. Sexual behavior problems most likely fall into this domain [3].

Considerable evidence from various studies suggests a preeminent role for early adverse experiences in the development of mood and anxiety disorders. Child abuse and neglect can be perceived as agents for neurodevelopmental disruption and, depending on when it occurs, can cause serious neurological "scars" in some structures, which could make some individuals vulnerable to certain types of psychopathology, including generalized anxiety disorder (GAD), panic disorder, and phobias [11].

Children and adolescents exposed to ELS experience serious consequences in their biopsychosocial constitution. The literature shows that, during early childhood and adolescence, important brain structures are being formed, so the negative consequences of traumatic events are lasting and can remain during the life of the children [12]. These children and adolescents may experience short- to long-term losses, including damage to health in general (fractures, lacerations, brain injuries) and mental health problems (GAD, panic disorder, phobias, social isolation, and, more specifically, symptoms of posttraumatic stress disorder, such as numbness, chronic anxiety, helplessness, low self-esteem, and disorders) at the entrance to adulthood [13]. Other consequences of ELS are related to cognitive developmental delay, intellectual deficit and school failure, as well as violence and crime in adolescence [2, 14–16].

In this sense, researchers point out that approximately 60% of cases of anxiety episodes are preceded by the occurrence of stressors, especially of psychosocial origin, so the influence of genetic factors in the development of anxiety disorders could be due to an increased sensitivity to stressful events [17].

Early Life Stress Abnormalities in Anxiety

Genes, ELS, adult life events, lifestyle, and stressful life experiences all add to the way the body adapts to a changing environment; and all these factors help to determine the cost to the body or the "allostatic load." Emergent data in the field of psychoneuroimmunology contributes to the understanding of the mechanisms by which traumatic events affect health. The interaction between behavior, neurobiology, and endocrine system that may cause immunosuppression is the most interesting discovery in current medicine, and its implications are important for the prevention and treatment of somatic diseases [18].

In a systematic review that we have published [11] reviewing the literature on ELS associated with psychiatric disorders in adulthood, we sought to identify whether there are independent effects between subtypes of early stress in triggering psychopathology in adults. We found the physical abuse was associated with anxiety disorders. Physical abuse contributes to the severity of anxiety disorders [19]. Sexual abuse was associated with anxiety disorders particularly with panic disorder and agoraphobia [11]. Emotional abuse was associated with anxiety disorders in two studies [20, 21] and a stronger correlation with social phobia.

While research of early life stress has focused primarily on development and outcomes relating to depression, substance abuse, and PTSD, emerging research is examining the considerable link relating to anxiety disorders. In a cohort study of individuals born between 1934 and 1944, subjects demonstrated greater late adulthood anxiety symptoms associated with emotional and physical trauma, parental divorce, and low socioeconomic status [22]. Green et al. [23] find that psychopathology risk from childhood adversities is differentiated throughout lifetime stages with risk of affective disorders being greater in ages 4–19 but superseded by anxiety disorders from ages 20 to 30+. The impact of adverse experiences in key developmental periods can have significant effect continuing throughout life with anxiety being latent until later in life.

The link with anxiety disorders and psychopathology can be differentiated through separate subtypes of early life stress as specifiers for risk, meaning particular stressors increase for certain disorders. Fergusson et al. [24] examined reporting rates of major depressive, anxiety, and conduct disorders in subjects aged 16–25 with varying degrees of physical or sexual abuse experiences. Subjects with experiences of child physical abuse reported more often major depression than anxiety disorder in all severity domains, in those who has experienced child sexual abuse reports of anxiety disorder were greater in all severity domains. Schilling et al. [25] found that sexual abuse correlates with the highest risk for psychopathology severity and dysfunction.

Early life stress has primarily been examined in the context of physical and emotion abuse/neglect as well sexual abuse, but childhood adversity such as dysfunctional family, school, and neighborhood environments and economic strain impacts on psychopathology. Significant anxiety disorder predictors included sociological dysfunctions such as low income, social support, family satisfaction, sense of mastery, and self-esteem and high stress regarding school, neighborhood, and economic perception. Cumulative burden of these risks as well as others was consistently higher for anxiety disorder compared to affective or substance use disorders for subjects aged 12–16+ [26].

While certain isolated factors and types of stress have been shown to have stronger links compared to other links with anxiety disorders, incidences of childhood adversity are complex with often there being multiple, continuous, and overlapping stressors. Cohen et al. [27] in an international sample investigation found that around half of subjects reported two or more early life stresses. They as well found sexual and emotional abuse were frequently co-occurring where when subjects report one of these abuses, they would often report the other as well. Co-occurrence of different types on abuse was reported in a sample of male juvenile offenders with many reporting emotional and physical abuse and a subsample reporting co-occurring sexual abuse [28]. Dong et al. [29] found similar reports regarding rates of two or more report early life stressors and co-occurrence of two types of child abuse but found both these to be slightly less severe than found by Cohen et al. [27].

Outcomes in psychological health, with regard to treatment efficacy and difficult adulthood features such as individual's stress, are currently evidentially limited for anxiety disorders but suggest a more dire prospect. Most studies about early life stress and its relation to adulthood difficulties currently focus on depressive disorders, but the suggestions are of a worsened course. The hammen et al. [30] study of adults with childhood adversity experiences showed that subjects had greater stress sensitivity with a focus on measuring depression. These findings are not dependent on the specific disorder and likely would be found in anxiety disorder subjects, suggesting greater risks of relapse and deterioration.

Klein et al. [31] found in a sample of chronically depressed patients, various forms of early life stress were significantly associated with lower probability of remission with pharmacotherapy. This is much in line with Nemeroff et al. [32] study of depressed patients with childhood trauma experience who had a weaker response with drug treatment options. Nemeroff et al. [32] do find that patients with childhood trauma showed greater response to psychotherapy or psychotherapy with pharmacotherapy when compared to patients without childhood trauma. Such features may be present in patients with anxiety disorder due to the major overlap between pharmacological treatment options for both anxiety and depression as well as common comorbidity of both of these disorders. Very limited studies have been found that which focus on pharmacotherapy response in subjects with anxiety disorder and early life stress experiences. One study [33] found patients with greater levels of emotional abuse predicted worse social anxiety disorder severity at baseline but a more rapid response to paroxetine compared with subjects with lower levels of emotional abuse.

Early Life Stress, HPA Axis, and Dysfunction in Anxiety

The stress response system is a complex, multilevel mechanism largely dependent on feedback regulation. The suppression of the subgenus prefrontal cortex and the activation of the amygdala lead to the stimulation of the hypothalamic-sympathetic-adrenomedullary, or autonomic sympathetic axis, and the hypothalamic-pituitary-adrenal (HPA) axis [34–36]. The autonomic sympathetic axis is responsible for the most rapid response and acts via the secretion of epinephrine by the adrenal glands; the HPA axis is activated minutes after the epinephrine surge and represents a cascade of events starting with the secretion of the corticotropin-releasing factor (CRF) from paraventricular nucleus of the hypothalamus into the portal circulation, which stimulates the synthesis and release of adrenocorticotropic hormone (ACTH) by the pituitary. ACTH further stimulates the synthesis and release of the glucocorticoid hormone cortisol by the adrenal cortex. Glucocorticoids are known to exert some functions, including gluconeogenesis, catabolic and antianabolic effects, mild inflammation, insulin resistance, and a prothrombotic state. The key role of glucocorticoids consists in maintaining homoeostasis in response to stress [37, 38].

It is believed that GCs exert their function through two types of receptors: MR receptors are high-affinity but low-specificity receptors, which means that they usually bind with basal cortisol which circulates in the blood at lower concentrations than in stress response situations following a circadian pattern, as well as with mineralocorticoids; glucocorticoid receptors (GR), on the other hand, exhibit lower affinity but higher specificity to GCs [39].

There is evidence that the expression and function of glucocorticosteroid receptors in the hippocampus, mainly MR, are regulated by the stimulation of 5-HT receptors [40]. Stressful stimuli increase 5-HT release and turnover in the hippocampus, and it seems reasonable to suggest that some of the changes in mineralocorticoid and GR expression may be mediated, in part at least, by the increase in 5-HT. These findings have led some investigators to propose that postsynaptic 5-HT1A and 5-HT2A receptors have functionally opposing effects, that a disturbed balance of these receptors may be contributing to the pathophysiology of depression, anxiety, and impulsivity, and that restoring this balance is necessary for therapeutic action [41].

Along with basal measurements, much effort has been put into assessing the functional impairments of the HPA-axis. For this aim, some challenge tests have been introduced.

More recently, an alternative test to DST was introduced– prednisolone suppression test [42]. Prednisolone's affinity to GRs has been shown to be more comparable to that of cortisol compared to dexamethasone [43, 44].

The majority of the studies show that early life stress leads to permanent changes in the HPA axis and may lead to the development of anxiety in adults. The most consistent findings in the literature show increased activity of the HPA axis associated with hypercortisolemia and reduced inhibitory feedback. These findings suggest that this dysregulation of the HPA axis is partially attributable to an imbalance between glucocorticoids (GR) and mineralocorticoid (MR) receptors; see Fig. 9.1 [38].

A study with infants experiencing social disadvantage and family adversity [45] found have higher cortisol response to awakening (CAR) than infants not exposed to social or familial adversity. Similar results were found [46] in emotional neglect and sexual abuse that strongly predicted maximal cortisol release. High levels of cortisol lead to hippocampal damage as seen in subjects with a history of emotional neglect during childhood. These patients had reduced left hippocampal white matter compared to those without a history of emotional neglect [47]. This evidence indicates that childhood trauma changes hippocampal structures during brain development leading to a higher vulnerability for stress-related psychiatric disease later in life.

Childhood trauma is associated with decreased responsiveness to pharmacological treatment [48] and a higher likelihood of relapse [49].

Evidence indicates that stress in the early phases of development can induce persistent changes in the ability of the HPA axis to respond to stress in adulthood and that this mechanism can lead to a raised susceptibility to anxiety. These

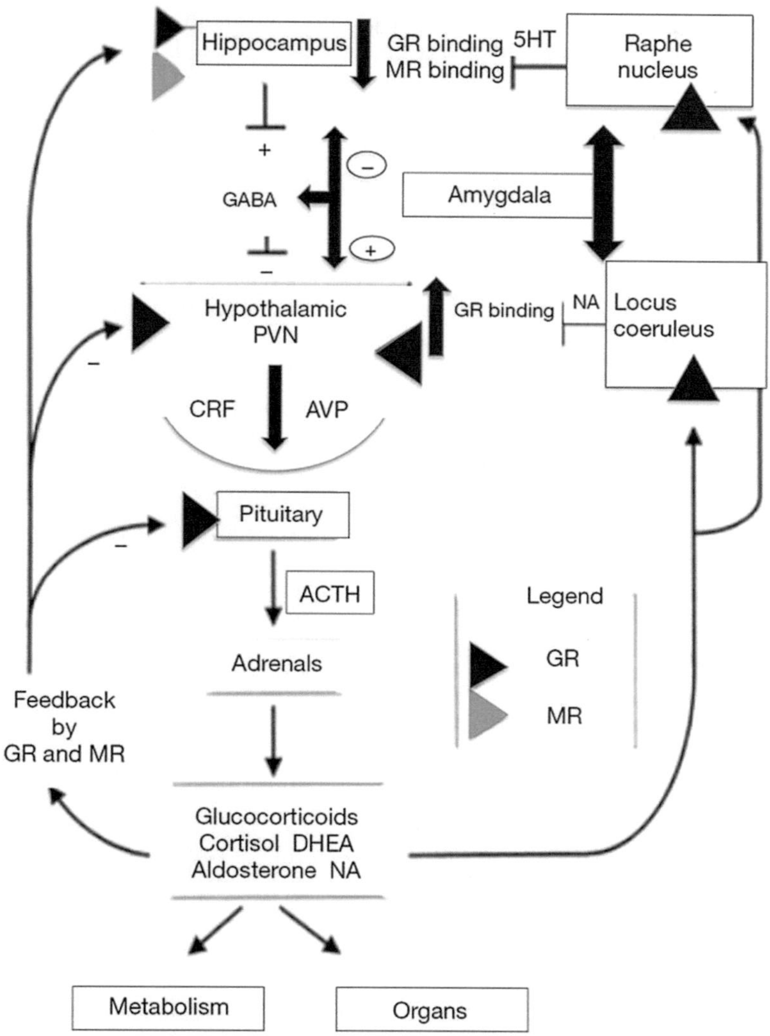

Fig. 9.1 Schematic diagram of hypothalamic-pituitary-adrenal (HPA) axis. It describes regulation and negative feedback (−) of cortisol via glucocorticoid receptors (GRs) and mineralocorticoid receptors (MRs). Including hippocampus, amygdala, raphe nucleus, locus coeruleus, and relation via serotonin (5HT) and noradrenaline (NA) with GR/MRs and adrenal hormones

abnormalities appear to be related to changes in the ability to circulate GCs to exert negative feedback on the secretion of HPA hormones through binding to GR and MR [37, 38, 43, 44].

In humans, while MRs are thought to be involved in the tonic inhibitory activity within the HPA axis, GRs appear to "switch off" cortisol production at times of stress. It seems that MRs are necessary for GC regulation of HPA axis activity during mild stressors but not during stressors that result in a stronger corticosteroid response. It is proposed that the maintenance of corticosteroid homoeostasis and the balance in MR−/ GR-mediated effects limit vulnerability to stress-related diseases in genetically predisposed individuals [37, 38, 43, 44].

Three different mechanisms of GR resistance have been considered: (I) downregulation secondary to persistent hypercortisolism, (II) a primary alteration in the genetic structure, and (III) a decrease in GR function secondary to alterations in ligand-independent pathways. It has also been proposed that the balance between MR and GR is an important factor in resilience to stress, and studies suggest that there may be an imbalance in the MR/GR ratio in depression [50]. Another possibility (that can happen concomitantly or independent) is the excessive production of corticotropin release factor (CRF) from the hypothalamus.

This structure receives fibers from some brain areas, notably the brain stem (that receives input from all sensory systems), the prefrontal cortex, and the limbic system (i.e., amygdala) [51]. The chronic overexpression of CRF in the amygdala is also associated with altered gene expression in the hippocampus and PVN, leading to increased hyperactivity [52]. These afferents play an important role in HPA responses to behavioral and emotional stimuli. The elevated CRF secretion will persistently stimulate the HPA axis, leading ultimately to increase in GC levels and to possible mechanisms of dysfunction in GR and MR already described. The prolonged exposure to GC has damaging effects on important brain structures, mainly the hippocampus, that is essential for HPA axis restraint, as well as memory consolidation. The role of GC and stress in memory was recently reviewed and linked with potential psychiatric disorders [53].

Gaining a thorough background and history during the diagnostic interview and assessment becomes important to understand the role of trauma and/or neglect in our patients. Part of this process should also reflect the fact that our interpretation and understanding of trauma may differ greatly from our patients. Children often try to protect and defend their parents; they may minimize abuse or may not view certain actions as abuse. An adult that has experienced childhood trauma may not want to revisit these events [54].

It is clear from the above data that psychotherapy should be the core component of treatment for anxious patients with a history of early childhood stress. It is also important to consider the role of different types of traumas at different developmental stages to elucidate whether there are precise developmental time periods for prevention of the adverse outcomes of childhood trauma [11, 55].

Childhood trauma in humans is associated with sensitization of the neuroendocrine stress response, glucocorticoid resistance, and increased HPA axis activity [56]. Hormones play a critical role in the development and expression of a wide range of behaviors. One aspect of the influence of hormones on behavior is their potential contribution to the pathophysiology of psychiatric disorders and the mechanism of action of psychotropic drugs. Of the endocrine axes, the HPA axis has been the most widely studied. It plays a fundamental role in response to external and internal stimuli, including psychological stressors. Abnormalities in the function of the HPA axis have been described in people who experience psychiatric disorders [57].

Studies conducted in both animals and humans suggest that stress experienced during the early phases of development can induce persistent changes in the ability of the HPA axis to respond to stress in adulthood, increasing the susceptibility to depression [58]. Evidence suggests that neurochemical and molecular changes induced by stressful situations and depression trigger changes in the HPA axis. A flaw in this system caused by factors such as excessive stress, high glucocorticoid levels, social isolation, and anxiety symptoms results in difficulty adapting to stress and can impair hippocampal serotonergic neurotransmission [37, 38, 43, 44].

It has been concluded from these studies that child maltreatment may lead to disruptions in HPA axis functioning and that factors such as the age of maltreatment, parental responsiveness, subsequent exposure to stressors, type of maltreatment, and type of psychopathology or behavioral disturbance displayed may influence the degree and pattern of HPA disturbance. However, results from studies examining the relationship between child maltreatment, psychopathology, and the HPA axis do vary. While most studies report HPA axis dysregulation, inconsistencies have been noted. Furthermore, results should be analyzed by gender and by type of stressor for maximum consistency, as the effects on the HPA axis may vary due to these factors [11, 37, 58].

Conclusion

Studies of the association between early life stress and anxiety disorders should be evaluated carefully. No consensus has been reached in the literature regarding the concept of early life stress, and the respondents in these studies likely underestimated or overestimated the frequency/intensity of events. Much descriptive work has been published on the relationship between adult psychopathology and early adversities such as parental loss in childhood, inadequate parental care, divorce, "affectionless" or dysfunctional parenting, childhood physical and sexual abuse, and other childhood traumas. Importantly, anxiety disorders, as well as depressive disorders, are associated with the history of early life stress. The results of existing studies suggest the importance of preventing early life stress and its consequences in both the short and long terms. Intervention at an early stage can reduce the likelihood of developing health problems in the long term and revictimization in adulthood. Furthermore, early interventions may reduce the burden of public spending on health care for abused individuals.

The more recent studies reviewed suggest that early life stressors are associated with an increased risk for anxiety disorders in adulthood. This review examined the emerging literature concerning the relationship between stress, HPA axis function, and GAD, panic disorder, and phobias and the role of early life stress as an important risk factor for HPA axis dysfunction and early life stress as an important risk factor for HPA axis dysregulation. These findings suggest that this dysregulation of the HPA axis is partially attributable to an imbalance between GRs and MRs [59].

Social and physical environments have an enormous impact on our physiology and behavior, and they influence the process of adaptation or allostasis. At the same time that our experiences change our brain and thoughts (i.e., changing our mind), we change our neurobiology. Although disturbances in the HPA axis are an important factor in the etiology of anxiety and treatment resistance, very little is known about the neurobiology of these disorders. Therefore, a psychometric assessment that quantifies the level of early life stress, recent stress, the evolution of anxiety symptoms and diagnosis, and neuroendocrine activity is essential. Childhood stressful events and HPA axis overactivity in adulthood are not specific to anxiety states, but several studies have linked these conditions. As demonstrated in this review, early life stress leads to permanent changes in the HPA axis and may lead to the development of anxiety disorders in adulthood. Considering the importance of early detection of violence in childhood and adolescence, to prevent the development of severe and disabling psychiatric disorders in adulthood, further research is needed to elucidate the mechanisms involved in the association between early stress and the development of psychopathology in adulthood. Stressful life experiences also play a prominent role in the development of anxiety disorders; several lines of research suggest the possibility that personality or temperament may account for some of the association between stress, anxiety, and HPA axis hyperactivity. These studies suggest that an evaluation of the HPA axis during treatment may help identify patients who are at a higher risk for relapse. These findings suggest that this dysfunction of the HPA axis is partially attributable to an imbalance between glucocorticoid and mineralocorticoid receptors. Evidence has consistently demonstrated that glucocorticoid receptor function is impaired in anxiety disorders. Moreover, normal basal cortisol levels and hyper-responsiveness of the adrenal cortex during a psychosocial stressor are observed in social phobics [60]. Finally, abnormal HPA axis activity has also been observed in generalized anxiety disordered patients. Early stressful life events may provoke alterations of the stress response and thus of the HPA axis, that can endure during adulthood, predisposing individuals to develop psychopathology.

References

1. Juruena MF, Marques AH, Mello AF, Mello MF. A paradigm for understanding and treating psychiatric illness. RBP. 2007;29(Suppl 1):s1–2.
2. Mclaughlin KA, Green JG, Gruber MJ, Sampson NA, Zaslavsky AM, Kessler RC. Childhood adversities and adult psychiatric disorders in the National Comorbidity Survey Replication associations with persistence of DSM-IV disorders. Arch Gen Psychiatry. 2010;67:113–23.
3. Friedrich W. Behavioral manifestations of child sexual abuse. Child Abuse Negl. 1998;22:523–31.
4. Cowan CS, Callaghan BL, Kan JM, et al. The lasting impact of early-life adversity on individuals and their descendants: potential mechanisms and hope for intervention. Genes Brain Behav. 2016;15:155–68.
5. Lupien SJ, McEwen BS, Gunnar MR, et al. Effects of stress throughout the lifespan on the brain, behaviour and cognition. Nat Rev Neurosci. 2009;10:434–45.
6. Egliston KA, McMahon C, Austin MP. Stress in pregnancy and infant HPA axis function: conceptual and methodological issues relating to the use of salivary cortisol as an outcome measure. Psychoneuroendocrinology. 2007;32:1–13.

7. Painter RC, Roseboom TJ, Bleker OP. Prenatal exposure to the Dutch famine and disease in later life: an overview. Reprod Toxicol. 2005;20:345–52.
8. Veenendaal MV, Painter RC, de Rooij SR, et al. Transgenerational effects of prenatal exposure to the 1944- 45 Dutch famine. BJOG. 2013;120:548–53.
9. Bernstein DP, Stein JA, Newcomb MD, et al. Development and validation of a brief screening version of the childhood trauma questionnaire. Child Abuse Negl. 2003;27:169–90.
10. Grandjean P, Heindel JJ. In utero and early-life conditions and adult health and disease. N Engl J Med. 2008;359:1523; author reply 1524
11. Carr CP, Martins CMS, Stingel AM, Lemgruber VB, Juruena MF. The role of early life stress in adult psychiatric disorders. A systematic review according to childhood trauma subtypes. J Nerv Ment Dis. 2013;201(12):1007–20.
12. Teicher MH. Scars that won t heal: the neurobiology of child abuse. Sci Am. 2002;286:68Y75.
13. Koss MP, Bailey JA, Yan NP. Depression and PTSD in survivors of male violence: research and training initiatives to facilitate recovery. Psychol Women Q. 2003;27:130–42.
14. Heim C, Nemeroff CB. The role of childhood trauma in the neurobiology of mood and anxiety disorders: preclinical and clinical studies. Biol Psychiatry. 2001;49:1023–39.
15. Neigh GN, Gillespie CF, Nemeroff CF. The neurobiological toll of child abuse and neglect. Trauma Violence Abuse. 2009;10:389–410.
16. Vitolo YLC, Fleitlich-Bilyk B, Goodman R, Bordin IAS. Parental beliefs and child-rearing attitudes and mental problems among school children. Rev Saude Publica. 2005;39:716–24.
17. Mello AF, Juruena MF, Pariante CM, Tyrka AR, Price LH, Carpenter LL, Del Porto JA. Depression and stress: this there an endophenotype? RBP. 2007;29:13–8.
18. Tosevski DL, Milovancevic MP. Stressful life events and physical health. Curr Opin Psychiatry. 2006;19:184–9.
19. Hovens JGFM, Wiersma JE, Giltay EJ, van Oppen P, Spinhoven P, Penninx BWJH, Zitman FG. Childhood life events and childhood trauma in adult patients with depressive, anxiety and comorbid disorders vs. control. Acta Psychiatr Scand. 2010;122:66–74.
20. Gibb BE, Chelminski I, Zimmerman M. Childhood emotional, physical, and sexual abuse, and diagnoses of depressive and anxiety disorders in adult psychiatric outpatients. Depress Anxiety. 2007;24:256–63.
21. Khoury L, Tang YL, Bradley B, Cubells JF, Ressler KJ. Substance use, childhood traumatic experience, and posttraumatic stress disorder in an urban civilian population. Depress Anxiety. 2010;27:1077–86.
22. Lähdepuro A, Savolainen K, Lahti-Pulkkinen M, Eriksson JG, Lahti J, Tuovinen S, et al. The impact of early life stress on anxiety symptoms in late adulthood. Sci Rep. 2019;9(1):4395.
23. Green JG, McLaughlin KA, Berglund PA, Gruber MJ, Sampson NA, Zaslavsky AM, Kessler RC. Childhood adversities and adult psychopathology in the National Comorbidity Survey Replication (NCS-R) I: associations with first onset of DSM-IV disorders. Arch Gen Psychiatry. 2010;67:113–23.
24. Fergusson DM, Boden JM, Horwood LJ. Exposure to childhood sexual and physical abuse and adjustment in early adulthood. Child Abuse Negl. 2008;32(6):607–19.
25. Schilling C, Weidner K, Schellong J, Joraschky P, Pöhlmann K. Patterns of childhood abuse and neglect as predictors of treatment outcome in inpatient psychotherapy: a typological approach. Psychopathology. 2015;48(2):91–100.
26. Roberts RE, Roberts CR, Chan W. One-year incidence of psychiatric disorders and associated risk factors among adolescents in the community. J Child Psychol Psychiatry. 2009;50(4):405–15.
27. Cohen RA, Hitsman BL, Paul RH, McCaffery J, Stroud L, Sweet L, et al. Early life stress and adult emotional experience: an international perspective. Int J Psychiatry Med. 2006;36(1):35–52.
28. Aebi M, Linhart S, Thun-Hohenstein L, Bessler C, Steinhausen H-C, Plattner B. Detained male adolescent Offender's emotional, physical and sexual maltreatment profiles and their associations to psychiatric disorders and criminal Behaviors. J Abnorm Child Psychol. 2015;43(5):999–1009.

29. Dong M, Anda RF, Felitti VJ, Dube SR, Williamson DF, Thompson TJ, et al. The interrelatedness of multiple forms of childhood abuse, neglect, and household dysfunction. Child Abuse Negl. 2004;28(7):771–84.
30. Hammen C, Henry R, Daley SE. Depression and sensitization to stressors among young women as a function of childhood adversity. J Consult Clin Psychol. 2000;68(5):782–7.
31. Klein DN, Arnow BA, Barkin JL, Dowling F, Kocsis JH, Leon AC, et al. Early adversity in chronic depression: clinical correlates and response to pharmacotherapy. Depress Anxiety. 2009;26(8):701–10.
32. Nemeroff CB, Heim CM, Thase ME, Klein DN, Rush AJ, Schatzberg AF, et al. Differential responses to psychotherapy versus pharmacotherapy in patients with chronic forms of major depression and childhood trauma. Proc Natl Acad Sci. 2003;100(24):14293–5.
33. Bruce LC, Heimberg RG, Blanco C, Schneier FR, Liebowitz MR. Childhood maltreatment and social anxiety disorder: implications for symptom severity and response to pharmacotherapy. Depress Anxiety. 2012;29(2):131–8. https://doi.org/10.1002/da.20909.
34. Diorio D, Viau V, Meaney MJ. The role of the medial prefrontal cortex (cingulate gyrus) in the regulation of hypothalamic-pituitary-adrenal responses to stress. J Neurosci. 1993;13:3839–47.
35. Phelps EA, LeDoux JE. Contributions of the amygdala to emotion processing: from animal models to human behavior. Neuron. 2005;48:175–87.
36. Gold PW. The organization of the stress system and its dysregulation in depressive illness. Mol Psychiatry. 2015;20:32–47.
37. Juruena MF. Early-life stress and HPA axis trigger recurrent adulthood depression. Epilepsy Behav. 2014;38:148–59.
38. Juruena MF, Agustini B, Cleare AJ, Young AH. A translational approach to clinical practice via stress-responsive glucocorticoid receptor signaling. Stem Cell Investig. 2017;4:13.
39. De Kloet ER, Vreugdenhil E, Oitzl MS, Joels M. Brain corticosteroid receptor balance in health and disease. Endocr Rev. 1998;19:269–301.
40. Semont A, Fache M, Ouafik L, et al. Effect of serotonin inhibition on glucocorticoid and mineralocorticoid expression in various brain structures. Neuroendocrinology. 1999;69:121–8.
41. Juruena MF, Gama CS, Berk M, et al. Improved stress response in bipolar affective disorder with adjunctive spironolactone (mineralocorticoid receptor antagonist): case series. J Psychopharmacol. 2009;23:985–7.
42. Pariante CM, Papadopoulos AS, Poon L, Checkley SA, English J, Kerwin RW, Lightman S. A novel prednisolone suppression test for the hypothalamic-pituitary—adrenal axis. Biol Psychiatry. 2002;51:922–30.
43. Juruena MF, Cleare AJ, Papadopoulos AS, Poon L, Lightman S, Pariante CM. Different responses to Dex and prednisolone in the same depressed patients. Psychopharmacology. 2006;189(2):225–35.
44. Juruena MF, Pariante CM, Papadopoulos AS, Poon L, Lightman S, Cleare AJ. Prednisolone suppression test in depression: a prospective study of the role of HPA axis dysfunction in treatment resistance. Br J Psychiatry. 2009;194:342–9.
45. Saridjan NS, Huizink AC, Koetsier JA, Jaddoe VW, Mackenbach JP, Hofman A, et al. Do social disadvantage and early family adversity affect the diurnal cortisol rhythm in infants? The generation R study. Horm Behav. 2010;57:247–54.
46. Carpenter LL, Carvalho JP, Tyrka AR, Wier LM, Mello AF, Mello MF, et al. Decreased adrenocorticotropic hormone and cortisol responses to stress in healthy adults reporting significant childhood maltreatment. Biol Psychiatry. 2007;62:1080–7.
47. Frodl T, Reinhold E, Koutsouleris N, Reiser M, Meisenzahl EM. Interaction of childhood stress with hippocampus and prefrontal cortex volume reduction in major depression. J Psychiatr Res. 2010;44:799–807.
48. Hayden EP, Klein DN. Outcome of dysthymic disorder at 5-year follow-up: the effect of familial psychopathology, early adversity, personality, comorbidity, and chronic stress. Am J Psychiatry. 2001;158:1864–70.

49. Lara ME, Klein DN, Kasch KL. Psychosocial predictors of the short-term course and outcome of major depression: a longitudinal study of a nonclinical sample with recent-onset episodes. J Abnorm Psychol. 2000;109:644–50.
50. Baes CW, Martins CM, Tofoli SM, Juruena MF. Early life stress in depressive patients: HPA axis response to GR and MR agonist. Front Psych. 2014;5:2.
51. Ulrich-Lai YM, Herman JP. Neural regulation of endocrine and autonomic stress responses. Nat Rev Neurosci. 2009;10(6):397–409.
52. Flandreau EI, Ressler KJ, Owens MJ, Nemeroff CB. Chronic overexpression of corticotropin-releasing factor from the central amygdala produces HPA axis hyperactivity and behavioral anxiety associated with gene-expression changes in the hippocampus and paraventricular nucleus of the hypothalamus. Psychoneuroendocrinology. 2012;37(1):27–38.
53. de Quervain D, Schwabe L, Roozendaal B. Stress, glucocorticoids and memory: implications for treating fearrelated disorders. Nat Rev Neurosci. 2017;18(1):7–19.
54. Martins-Monteverde CMS, Baes CVW, Reisdorfer E, Padovan T, Tofoli SMC, Juruena MF. Relationship between depression and subtypes of early life stress in adult psychiatric patients. Front Psych. 2019;10:19.
55. Heim C, Newport DJ, Mletzko T, Miller AH, Nemeroff CB. The link between childhood trauma and depression: insights from HPA axis studies in humans. Psychoneuroendocrinology. 2008;33:693–710.
56. Nemeroff CB. The corticotropin-releasing factor (CRF) hypothesis of depression: new findings and new directions. Mol Psychiatry. 1996;1(4):336–42.
57. Glover V, O'Connor TG. Effects of antenatal stress and anxiety: implications for development and psychiatry. Br J Psychiatry. 2002;180:389–91.
58. D'Elia ATD, Matsuzaka CT, Neto JBB, Mello MF, Juruena MF, Mello AF. Childhood sexual abuse and indicators of immune activity: a systematic review. Front Psych. 2018;9:354.
59. Abelson JL, Khan S, Liberzon I, Young EA. HPA axis activity in patients with panic disorder: review and synthesis of four studies. Depress Anxiety. 2007;24:66–76.
60. Faravelli C, Lo Sauro C, Godini L, Lelli L, Benni L, Pietrini F, et al. Childhood stressful events, HPA axis and anxiety disorders. World J Psychiatry. 2012;2(1):13–25.

Immune-Kynurenine Pathways and the Gut Microbiota-Brain Axis in Anxiety Disorders

10

Alper Evrensel, Barış Önen Ünsalver, and Mehmet Emin Ceylan

Introduction

Medical research has entered a new era with the turn of the millennium. It has started to be understood that not only the eukaryotic cell function but also the prokaryotic cells located in the body serve an important role in human health and diseases [1]. Understanding the effect of the microorganisms in the intestine on brain function is undoubtedly one of these discoveries. Many potential mechanisms for the role of intestinal microbiota on brain functions have been proposed. Numerous evidences have been obtained about leaky gut, vagal afferent nerve signaling, neurotransmitters, or short-chain fatty acids (SCFAs) that enter the systemic circulation after being synthesized by microorganisms, the activation of the hypothalamic-pituitary-adrenal (HPA) axis by proinflammatory cytokines, and changes in tryptophan metabolism [2].

All these mechanisms are mentioned in this article; however, it especially focuses on the kynurenine pathway in the pathogenesis of anxiety disorders. First of all, traces of intestinal microbiota after birth and even of intrauterine period are discussed. Then, the role of microbiota's neural physiology and physiopathology on the host is explained. Next, this article considers the importance of the immuno-kynurenine pathway in the etiopathogenesis of anxiety disorders. Finally, it looks at potential treatment methods that can be taken into consideration with the elucidation of these mechanisms.

A. Evrensel (✉)
Department of Psychiatry, Uskudar University, Umraniye, Istanbul, Turkey

B. Ö. Ünsalver
Vocational School of Health Services, Department of Medical Documentation and Secretariat, Uskudar University, Istanbul, Turkey

M. E. Ceylan
Departments of Psychology and Philosophy, Uskudar University, Istanbul, Turkey

Y.-K. Kim (ed.), *Anxiety Disorders*, Advances in Experimental Medicine and Biology 1191, https://doi.org/10.1007/978-981-32-9705-0_10

Formation of Gut Microbiota and Their Structural Features

The intestinal microorganisms that form a commensal relationship with the host are called "gut microbiota" [3, 4]. Until recently, the common belief has been that a fetus is sterile in the intrauterine period and that the flora bacteria colonize after birth. However, recent research shows that the fetus is not sterile and that there are nonpathogenic bacteria resident in the amniotic fluid, placenta, and meconium [5, 6]. The intestinal bacterial composition of a newborn exposed to different environmental stimuli (psychological, nutritional, and medicinal) at birth begins to evolve toward an individual-specific shape [7, 8]. The delivery mode is the most important determining factor. Babies born vaginally are exposed to the mother's vaginal and fecal microbiota, while cesarean section babies are in contact with the skin microbiota around the mother's nipples and the bacteria in the hospital [9].

Despite these individual differences and the uniqueness of the intestinal microbiota's composition, there is still a common core of microbiota in all humans, from birth to death [10]. This core microbiota consists of four phyla (*Firmicutes*, *Bacteroidetes*, *Verrucomicrobia*, and *Actinobacteria*) [11]. This composition shows distinct differences between individuals and also changes with age. *Enterobacteriaceae*, *Bifidobacterium*, and *Bacteroides* are dominant in newborns [12]. *Lactobacillus*, *Staphylococcus*, *Veillonella*, and *Lachnospiraceae* are also present in the intestinal microbiota of a vaginally born infant [13].

Although the microbial composition is quite variable in the first years of life, it starts to become more stable and adult-like beginning in a child's third year [5]. While microbial diversity is greater during middle age, it decreases with aging and *Clostridium* and *Bacteroides* rates increase [14]. This changeable content of intestinal microbiota makes it difficult to generalize and classify. The concept of an "enterotype" is objectionable due to significant differences between individuals [15, 16]. In all stages of life, dysbiosis in the microbiota's composition has a wide impact on the host's health. In this article, its effect on neuronal functions is investigated.

General Terms and Mechanisms

The communication between the brain and the intestinal microbiota defines a reticular structure called the gut-brain axis where many different mechanisms play a complex role [10]. This bidirectional axis encompasses the autonomic nervous system, the enteric nervous system, and the neuroimmune-neuroendocrine pathways [1]. The sympathetic and parasympathetic nervous systems, which function in the afferent and efferent directions between the central nervous system and the intestinal mucosa, make this interaction fast and effective [17]. A large number of studies have been carried out over the last two decades on diseases such as irritable bowel syndrome (IBS), anorexia nervosa, obesity, and inflammatory bowel diseases (Crohn's disease and ulcerative colitis), in which the relative intestinal-brain relationship is more clearly visible [18–20].

In recent years, however, there is growing evidence that the intestinal microbiota's composition also has a decisive and effective role in many neuronal functions and behaviors [7, 21, 22]. In preclinical studies, germ-free (GF) conditions, fecal microbiota transplantation (FMT), the bacterial infection of intestines, antibiotic therapy, and the effectiveness of prebiotics/probiotics have been examined [6, 7, 23]. As a result of these studies, the vagus nerve, neuroimmune and neurohormonal pathways, SCFAs and neurotransmitters produced by intestinal bacteria, and mechanisms related to the pathway of tryptophan-kynurenine metabolism have been clarified to a great degree.

Serotonergic Dysfunction and the Gut Microbiota-Brain Axis

Numerous preclinical studies have shown that intestinal microbiota modulates anxiety-like [24–28] and depression-like [29–33] behavior in experimental animals. Bercik et al. showed that anxiety-like behavior can be transferred by transplanting fecal microbiota between mice [34]. The data obtained from research on healthy people supports preclinical studies on the effects of microbiota [31, 35, 36]. Although a study in 2014 showed that the fecal microbiota's composition of patients with depression did not differ from non-depressive individuals [37], differences between two groups were found in two studies performed later [38, 39]. In parallel with these findings, one of the most important studies demonstrating the role of intestinal microbiota in the etiopathogenesis of depression is the study of Zheng et al. [39]. In this trial, fecal microbiota transplantation (FMT) from patients with depression was performed in GF mice, and a depressive behavioral pattern was observed in mice following the transplantation [39].

The most important reason for the contradictory results obtained from human studies may be the heterogeneous nature of the sample. There is a need for methodically better designed studies with a large sample to obtain more precise information on the role of intestinal microbiota in the development of anxiety and depression. One of the mechanisms that is thought to play a role in the relationship between intestinal microbiota and anxiety/depression is the immune-kynurenine pathway [40]. We will now discuss the pathogenesis of serotonin metabolism and the immune-kynurenine pathway in detail in order to understand the potential impact of intestinal microbiota on the serotonergic system.

Tryptophan-Serotonin Metabolism and the Kynurenine Pathway

Tryptophan biosynthesis is possible in bacteria and plant cells although it requires high energy [41]. This energy-intensive function seems to have been an evolutionary loss in mammals, since tryptophan is readily available from dietary proteins [42]. Tryptophan is an essential amino acid that plays a very important role in the gut-brain axis as the precursor molecule of serotonin and kynurenine [43, 44]. Only

5% of serotonin is present in the central nervous system (CNS) of an adult human body. The rest is synthesized by enterochromaffin cells located in the intestines [45–47]. Serotonin in the gastrointestinal tract plays a role in autonomic functions such as intestinal secretion, absorption, and motility under healthy conditions [48, 49]. Peripheral serotonin affects the appetite through the afferent fibers of the vagus nerve [50] and causes nausea and vomiting [51]. Serotonin in the CNS is involved in mood and cognitive functions [52].

More than 90% of a body's tryptophan, as summarized in Fig. 10.1, is converted into kynurenine by the enzymes indoleamine-2,3-dioxygenase (IDO) located in all cells and tryptophan-2,3-dioxygenase (TDO) located in the liver [52, 53]. IDO has two subtypes (IDO1 and IDO2) [54]. Inflammation plays an important role in IDO1 and TDO activity. The IDO1 pathway is stimulated by proinflammatory cytokines, especially the interferon-gamma, and the TDO pathway is stimulated by glucocorticoids [52]. Kynurenic acid and quinolinic acid, which are metabolites of kynurenine, are neuroactive. These metabolites stimulate N-methyl D-aspartate (NMDA) and alpha-7 nicotinic receptors [55]. While kynurenic acid shows neuroprotective features, quinolinic acid is excitotoxic [56]. Kynurenic acid also has anti-inflammatory features [57].

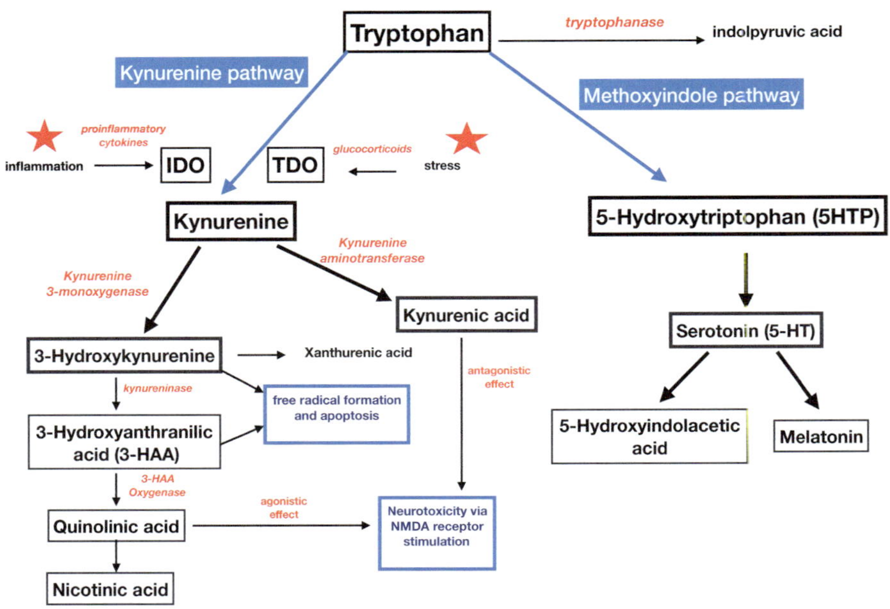

Fig. 10.1 Kynurenine pathway of the tryptophan metabolism

Anxiety, the Immune System, and the Kynurenine Pathway Interaction

Today, the relationship between intestinal microbiota and the stress response is well understood. Evidence showing that the interaction is bidirectional has been obtained. Stress can change the intestinal microbiota during the prenatal and postnatal periods [58–60]. A preliminary trial examining the effect of the change in microbiota on the stress response was conducted in 2004 by Sudo et al. [61]. In this study, it was shown that GF mice exhibited a more severe hypothalamic-pituitary-adrenal (HPA) axis response to mild stress stimuli compared to the control group and that this response was normalized following colonization with *Bifidobacterium infantis* [61]. The study of Gareau et al. supports these findings [62]. In addition, in a study conducted on healthy people, it was found that prebiotic supplementation could alter the cortisol response [63].

There is a lifelong reciprocal relationship between the immune system and the microbiota [64]. The immune system of GF mice is dysfunctional, and this dysfunction can be resolved following bacterial colonization [26]. One of the most common problems causing immune system dysfunction in GF animals is the decrease in the expression of toll-like receptors (TLRs) [65]. Recognition of bacterial components in the intestinal lumen occurs through TLRs [4]. TLRs' activation stimulates the kynurenine pathway by causing IDO1 induction [66]. Numerous studies have shown that low-level immune activation in IBS is associated with increased kynurenine pathway metabolism [67, 68].

Another factor that influences the relationship between microbiota and the kynurenine pathway may be the metabolites of microbiota bacteria (e.g., SCFAs). SCFAs stabilize the intestinal barrier permeability and, if there is a decrease in their level, may cause systemic inflammation by triggering intestinal leakage [69]. While the metabolites of the kynurenine pathway were not detected in the CNS of GF experimental animals, an increase in the levels of these metabolites was observed in mice infected with *Toxoplasma gondii* [70].

Role of Intestinal Microbiota in the Kynurenine Pathway's Metabolism

Interesting findings have been obtained in preclinical studies about kynurenine and tryptophan. For example, in GF animals, blood tryptophan levels were found to be higher and levels of serotonin (5-hydroxy tryptamine [5-HT]) and kynurenine metabolites were lower [71–73]. This may be due to the fact that microbiota-induced metabolites play a role in the 5-HT synthesis process in the lumen [74, 75]. In this respect, the 5-HT synthesis from tryptophan and systemic kynurenine metabolism are reduced when the microbiota is eliminated through antibiotic administration [32]. However, these changes can be reversed with colonization [26, 71, 76]. In another trial, tryptophan and kynurenic acid levels were found to be increased in the circulation of mice who were given *Bifidobacterium infantis* [29]. In experimental animals colonized with *Lactobacillus johnsonii*, the level of serum kynurenine decreased [77, 78].

The intestinal microbiota not only changes the levels of circulating tryptophan and kynurenine but also affects their levels in the CNS because the permeability of the blood-brain barrier is higher in GF animals [79]. While tryptophan and kynurenine can cross the blood-brain barrier through the large amino acid transporter system, kynurenic acid and quinolinic acid cannot pass through the CNS [80, 81]. It is worth mentioning here that tryptophan competes with other amino acids (tyrosine, valine, methionine, leucine, isoleucine, phenylalanine) while crossing the blood-brain barrier through the large amino acid transport system [82]. The kynurenine transported to the CNS is metabolized by microglia [83]. Considering that the microglia maturation and function are impaired under GF conditions [84], another connection between the microbiota and the kynurenine pathway metabolism can be established.

Indole compounds resulting from the metabolization of tryptophan by microbiota bacteria may cause neuroinflammation by affecting another brain cell called astrocyte [85]. Indole produced by the intestinal bacteria is the major metabolite of tryptophan obtained through a protein-rich diet [86]. Indole synthesis in bacteria is catalyzed only by tryptophanase, an enzyme found in prokaryotic cells [87]. The indole produced by bacteria in the intestinal lumen may have many effects on the host, including anti-inflammatory, oxidative, and endocrinological reflections [88, 89]. Indole is thought to increase intestinal epithelium strength and reduce permeability [90]. There is also evidence that indole regulates bacterial 5-HT synthesis in the lumen [74].

Bacteria contribute to the kynurenine pathway in various ways. For example, it is interesting that intestinal bacteria synthesize molecules in the neurotransmitter structure, such as 5-HT [91], and that these bacteria are sensitive to psychotropic drugs (tricyclic antidepressants and selective serotonin reuptake inhibitors) [92, 93]. There are also analogues of the enzymes that catalyze the quinolinic acid pathway from tryptophan in the bacterial genome of eukaryotic cells [94, 95]. Microbiota bacteria can produce quinolinic acid from aspartate [96] and kynurenic acid from kynurenine [97]. Therefore, metabolites of the tryptophan-kynurenine pathway that enter the systemic circulation after being produced in the intestinal lumen may be effective on the host's CNS functions [56, 81].

Anxiety and Tryptophan-Kynurenine Metabolism

The relationship between intestinal microbiota, tryptophan metabolism, and neuropsychiatric disorders (anxiety, depression, social avoidance, pain, and cognitive impairment) has been demonstrated by many studies [81, 98–102]. In a study conducted on healthy volunteers whose anxiety levels were induced by low-dose caffeine, the levels of plasma kynurenine were found to be high. There was a correlation between plasma kynurenine levels and anxiety scale scores [103]. A similar correlation was obtained after anxiety and depression cases were subjected to a dexamethasone suppression test [104].

Kynurenine pathway metabolism is considered to be most effective on the CNS during the prenatal and postnatal periods, especially a child's first 1000 days [105]. In animal experiments, the inhibition of kynurenine metabolism in the prenatal period causes changes in neuronal morphology and functions, which continues in adulthood as well [105–107]. Cognitive dysfunctions occur in mice with increased brain kynurenic acid levels during pregnancy and the postpartum period [108–110]. Changes during the postnatal period are of great importance as it is the time in which the composition of the microbiota is determined [52].

The microbiota-brain-immune-kynurenine pathway metabolism in human samples has mostly been studied in terms of IBS [53, 111, 112]. IBS is a special disease at the intersection of psychiatric and gastrointestinal symptoms [113]. There is a correlation between tryptophan catabolism metabolites and gastrointestinal symptom severity in women with IBS [114]. On the other hand, a correlation was found between depression and anxiety scores and the 5-HT levels of intestinal mucosa as well [112]. While acute tryptophan depletion increases plasma kynurenic acid levels in healthy individuals, it decreases plasma kynurenine levels in IBS patients [115]. In the light of these studies, it can be said that there are disruptions in tryptophan metabolism at various levels in IBS patients.

Conclusion

The serotonin hypothesis in the etiopathogenesis of anxiety disorders is current and valid. Therefore, functions and dysfunctions related to the catabolism of tryptophan, a precursor of serotonin, are of great importance. Serotonin is synthesized only from a small percentage (1–5%) of tryptophan. The rest of the tryptophan is transformed into kynurenine, kynurenic acid, 3-hydroxykynurenine, quinolinic acid, and nicotinic acid through the kynurenine pathway. Two enzymes, TDO and IDO, that catalyze the biosynthesis of kynurenine from tryptophan have been identified. TDO, located in hepatocytes, is stimulated by glucocorticoids and IDO, located outside the liver, is stimulated by proinflammatory cytokines. These tryptophan catabolites, which are activated in times of stress and inflammation and result from the kynurenine pathway, have anxiogenic and anxiolytic functions. They may also increase intestinal permeability. Bacteria-induced serotonin biosynthesis may be reduced as a result of changes in intestinal microbiota composition and dysbiosis. Also, low-grade systemic inflammation occurs due to bowel leakage. With the impact of inflammation, production of serotonin decreases, and production of kynurenine increases. This cyclic system may play a role in the etiopathogenesis of anxiety disorders. Randomized controlled studies with large samples are needed in order to shed light on the effectiveness of microbiota-based treatments (probiotics, prebiotics, and FMT) for dysbiosis.

Acknowledgments We would like to thank Dr. Barış Önen Ünsalver for preparing the image used in this review.

References

1. Evrensel A, Önen Ünsalver B, Ceylan ME. Therapeutic potential of the microbiome in the treatment of neuropsychiatric disorders. Med Sci. 2019;7(2):21.
2. Evrensel A, Önen Ünsalver B, Ceylan ME. Gut-brain axis and psychiatric disorders. Curr Psych Rev. 2018;14(3):178–86.
3. Evrensel A, Ceylan ME. Gut-microbiota-brain axis and depression. In: Kim YK, editor. Understanding depression. Singapore: Springer; 2018. p. 197–207.
4. Evrensel A, Ceylan ME. The gut-brain axis: the missing link in depression. Clin Psychopharmacol Neurosci. 2015;13(3):239–44.
5. Rodríguez JM, Murphy K, Stanton C, Ross RP, Kober OI, Juge N, et al. The composition of the gut microbiota throughout life, with an emphasis on early life. Microb Ecol Health Dis. 2015;26:26050.
6. Evrensel A, Ceylan ME. Fecal microbiota transplantation in the treatment-resistant psychiatric disorders. In: Kim YK, editor. Treatment resistance in psychiatry. Singapore: Springer; 2019. p. 369–76.
7. Cryan JF, Dinan TG. Mind-altering microorganisms: the impact of the gut microbiota on brain and behaviour. Nat Rev Neurosci. 2012;13(10):701–12.
8. Lozupone CA, Stombaugh JI, Gordon JI, Jansson JK, Knight R. Diversity, stability and resilience of the human gut microbiota. Nature. 2012;489(7415):220–30.
9. Borre YE, O'Keeffe GW, Clarke G, Stanton C, Dinan TG, Cryan JF. Microbiota and neurodevelopmental windows: implications for brain disorders. Trends Mol Med. 2014;20(9):509–18.
10. Kennedy PJ, Cryan JF, Dinan TG, Clarke G. Kynurenine pathway metabolism and the microbiota-gut-brain axis. Neuropharmacology. 2017;112(Pt B):399–412.
11. Consortium HMP. Structure, function and diversity of the healthy human microbiome. Nature. 2012;486(7402):207–14.
12. Adlerberth I, Wold AE. Establishment of the gut microbiota in Western infants. Acta Paediatr. 2009;98(2):229–38.
13. Palmer C, Bik EM, DiGiulio DB, Relman DA, Brown PO. Development of the human infant intestinal microbiota. PLoS Biol. 2007;5(7):e177.
14. Claesson MJ, Cusack S, O'Sullivan O, Greene-Diniz R, de Weerd H, Flannery E, et al. Composition, variability, and temporal stability of the intestinal microbiota of the elderly. Proc Natl Acad Sci U S A. 2011;108(Suppl 1):4586–91.
15. Falony G, Joossens M, Vieira-Silva S, Wang J, Darzi Y, Faust K, et al. Population-level analysis of gut microbiome variation. Science. 2016;352(6285):560–4.
16. Zhernakova A, Kurilshikov A, Bonder MJ, Tigchelaar EF, Schirmer M, Vatanen T, et al. Population-based metagenomics analysis reveals markers for gut microbiome composition and diversity. Science. 2016;352(6285):565–9.
17. Furness JB. The enteric nervous system and neurogastroenterology. Nat Rev Gastroenterol Hepatol. 2012;9(5):286–94.
18. Hoebel BG. Neuroscience and appetitive behavior research: 25 years. Appetite. 1997;29(2):119–33.
19. Mayer EA, Aziz Q, Coen S, Kern M, Labus JS, Lane R, et al. Brain imaging approaches to the study of functional GI disorders a Rome working team report. Neurogastroenterol Motil. 2009;21(6):579–96.
20. Schellekens H, Finger BC, Dinan TG, Cryan JF. Ghrelin signalling and obesity: at the interface of stress, mood and food reward. Pharmacol Ther. 2012;135(3):316–26.
21. Mayer E, Knight R, Mazmanian SK, Cryan JF, Tillisch K. Gut microbes and the brain: paradigm shift in neuroscience. J Neurosci. 2014;34(46):15490–6.
22. Sampson TR, Mazmanian SK. Control of brain development, function, and behavior by the microbiome. Cell Host Microbe. 2015;17(5):565–76.
23. Evrensel A, Ceylan ME. Fecal microbiota transplantation and its usage in neuropsychiatric disorders. Clin Psychopharmacol Neurosci. 2016;14(3):231–7.

24. Diaz Heijtz R, Wang S, Anuar F, Qian Y, Bjorkholm B, Samuelsson A, et al. Normal gut microbiota modulates brain development and behavior. Proc Natl Acad Sci U S A. 2011;108(7):3047–52.
25. Neufeld KM, Kang N, Bienenstock J, Foster JA. Reduced anxiety-like behavior and central neurochemical change in germ-free mice. Neurogastroenterol Motil. 2011;23(3):255–64.
26. Clarke G, Grenham S, Scully P, Fitzgerald P, Moloney RD, Shanahan F, et al. The microbiome-gut-brain axis during early life regulates the hippocampal serotonergic system in a sex-dependent manner. Mol Psychiatry. 2013;18(6):666–73.
27. Savignac HM, Kiely B, Dinan TG, Cryan JF. Bifidobacteria exert strain-specific effects on stress-related behavior and physiology in BALB/c mice. Neurogastroenterol Motil. 2014;26(11):1615–27.
28. Arentsen T, Raith H, Qian Y, Forssberg H, Heijtz RD. Host microbiota modulates development of social preference in mice. Microb Ecol Health Dis. 2015;26:29719.
29. Desbonnet L, Garrett L, Clarke G, Bienenstock J, Dinan TG. The probiotic Bifidobacteria infantis: an assessment of potential antidepressant properties in the rat. J Psychiatr Res. 2008;43(2):164–74.
30. Bravo JA, Forsythe P, Chew MV, Escaravage E, Savignac HM, Dinan TG, et al. Ingestion of Lactobacillus strain regulates emotional behavior and central GABA receptor expression in a mouse via the vagus nerve. Proc Natl Acad Sci U S A. 2011;108(38):16050–5.
31. Messaoudi M, Lalonde R, Violle N, Javelot H, Desor D, Nejdi A, et al. Assessment of psychotropic-like properties of a probiotic formulation (*Lactobacillus helveticus* R0052 and *Bifidobacterium longum* R0175) in rats and human subjects. Br J Nutr. 2011;105(5):755–64.
32. Desbonnet L, Clarke G, Traplin A, O'Sullivan O, Crispie F, Moloney RD, et al. Gut microbiota depletion from early adolescence in mice: implications for brain and behaviour. Brain Behav Immun. 2015;48:165–73.
33. Wong ML, Inserra A, Lewis MD, Mastronardi CA, Leong L, Choo J, et al. Inflammasome signaling affects anxiety- and depressive-like behavior and gut microbiome composition. Mol Psychiatry. 2016;21(6):797–805.
34. Bercik P, Denou E, Collins J, Jackson W, Lu J, Jury J, et al. The intestinal microbiota affect central levels of brain-derived neurotropic factor and behavior in mice. Gastroenterology. 2011;141(2):599–609.
35. Benton D, Williams C, Brown A. Impact of consuming a milk drink containing a probiotic on mood and cognition. Eur J Clin Nutr. 2007;61(3):355–61.
36. Steenbergen L, Sellaro R, van Hemert S, Bosch JA, Colzato LS. A randomized controlled trial to test the effect of multispecies probiotics on cognitive reactivity to sad mood. Brain Behav Immun. 2015;48:258–64.
37. Naseribafrouei A, Hestad K, Avershina E, Sekelja M, Linløkken A, Wilson R, et al. Correlation between the human fecal microbiota and depression. Neurogastroenterol Motil. 2014;26(8):1155–62.
38. Jiang H, Ling Z, Zhang Y, Mao H, Ma Z, Yin Y, et al. Altered fecal microbiota composition in patients with major depressive disorder. Brain Behav Immun. 2015;48:186–94.
39. Zheng P, Zeng B, Zhou C, Liu M, Fang Z, Xu X, et al. Gut microbiome remodeling induces depressive-like behaviors through a pathway mediated by the host's metabolism. Mol Psychiatry. 2016;21(6):786–96.
40. Kim YK, Jeon SW. Neuroinflammation and the immune-kynurenine pathway in anxiety disorders. Curr Neuropharmacol. 2018;16(5):574–82.
41. Martinez JA, Bolivar F, Escalante A. Shikimic acid production in *Escherichia coli*: from classical metabolic engineering strategies to omics applied to improve its production. Front Bioeng Biotechnol. 2015;3:145.
42. Priya VK, Sarkar S, Sinha S. Evolution of tryptophan biosynthetic pathway in microbial genomes: a comparative genetic study. Syst Synth Biol. 2014;8(1):59–72.
43. Badawy AA. Tryptophan availability for kynurenine pathway metabolism across the life span: control mechanisms and focus on aging, exercise, diet and nutritional supplements. Neuropharmacology. 2017;112(Pt B):248–63.

44. Palego L, Betti L, Rossi A, Giannaccini G. Tryptophan biochemistry: structural, nutritional, metabolic, and medical aspects in humans. J Amino Acids. 2016;2016:8952520.
45. Mayer EA, Naliboff BD, Chang L. Basic pathophysiologic mechanisms in irritable bowel syndrome. Dig Dis. 2001;19(3):212–8.
46. Camilleri M. Serotonergic modulation of visceral sensation: lower gut. Gut. 2002;51(Suppl 1):81–6.
47. Gershon MD, Tack J. The serotonin signaling system: from basic understanding to drug development for functional GI disorders. Gastroenterology. 2007;132(1):397–414.
48. Bearcroft CP, Andre EA, Farthing MJ. In vivo effects of the 5-HT3 antagonist alosetron on basal and cholera toxin-induced secretion in the human jejunum: a segmental perfusion study. Aliment Pharmacol Ther. 1997;11(6):1109–14.
49. Chial HJ, Camilleri M, Burton D, Thomforde G, Olden KW, Stephens D. Selective effects of serotonergic psychoactive agents on gastrointestinal functions in health. Am J Physiol Gastrointest Liver Physiol. 2003;284(1):G130–7.
50. Donovan MH, Tecott LH. Serotonin and the regulation of mammalian energy balance. Front Neurosci. 2013;7:36.
51. Klatt S, Bock W, Rentschler J, Beckh K, Adler G. Effects of tropisetron, a 5-HT3 receptor antagonist, on proximal gastric motor and sensory function in nonulcer dyspepsia. Digestion. 1999;60(2):147–52.
52. O'Mahony SM, Clarke G, Borre YE, Dinan TG, Cryan JF. Serotonin, tryptophan metabolism and the brain-gut-microbiome axis. Behav Brain Res. 2015;277:32–48.
53. Clarke G, McKernan DP, Gaszner G, Quigley EM, Cryan JF, Dinan TG. A distinct profile of tryptophan metabolism along the kynurenine pathway downstream of toll-like receptor activation in irritable bowel syndrome. Front Pharmacol. 2012;3:90.
54. Fatokun AA, Hunt NH, Ball HJ. Indoleamine 2,3-dioxygenase 2 (IDO2) and the kynurenine pathway: characteristics and potential roles in health and disease. Amino Acids. 2013;45(6):1319–29.
55. Forrest CM, Youd P, Kennedy A, Gould SR, Darlington LG, Stone TW. Purine, kynurenine, neopterin and lipid peroxidation levels in inflammatory bowel disease. J Biomed Sci. 2002;9(5):436–42.
56. Stone TW, Darlington LG. The kynurenine pathway as a therapeutic target in cognitive and neurodegenerative disorders. Br J Pharmacol. 2013;169(6):1211–27.
57. Kaszaki J, Erces D, Varga G, Szabo A, Vecsei L, Boros M. Kynurenines and intestinal neurotransmission: the role of N-methyl-D-aspartate receptors. J Neural Transm. 2012;119(2):211–23.
58. O'Mahony SM, Marchesi JR, Scully P, Codling C, Ceolho AM, Quigley EM, et al. Early life stress alters behavior, immunity, and microbiota in rats: implications for irritable bowel syndrome and psychiatric illnesses. Biol Psychiatry. 2009;65(3):263–7.
59. Jasarevic E, Howerton CL, Howard CD, Bale TL. Alterations in the vaginal microbiome by maternal stress are associated with metabolic reprogramming of the offspring gut and brain. Endocrinology. 2015;156(9):3265–76.
60. Reber SO, Siebler PH, Donner NC, Morton JT, Smith DG, Kopelman JM, et al. Immunization with a heat-killed preparation of the environmental bacterium Mycobacterium vaccae promotes stress resilience in mice. Proc Natl Acad Sci U S A. 2016;113(22):E3130–9.
61. Sudo N, Chida Y, Aiba Y, Sonoda J, Oyama N, Yu XN, et al. Postnatal microbial colonization programs the hypothalamic-pituitary-adrenal system for stress response in mice. J Physiol. 2004;558(Pt 1):263–75.
62. Gareau MG, Jury J, MacQueen G, Sherman PM, Perdue MH. Probiotic treatment of rat pups normalises corticosterone release and ameliorates colonic dysfunction induced by maternal separation. Gut. 2007;56(11):1522–8.
63. Schmidt K, Cowen PJ, Harmer CJ, Tzortzis G, Errington S, Burnet PW. Prebiotic intake reduces the waking cortisol response and alters emotional bias in healthy volunteers. Psychopharmacology (Berl). 2015;232(10):1793–801.

64. El Aidy S, Dinan TG, Cryan JF. Gut microbiota: the conductor in the orchestra of immune-neuroendocrine communication. Clin Ther. 2015;37(5):954–67.
65. Kawai T, Akira S. The role of pattern-recognition receptors in innate immunity: update on Toll-like receptors. Nat Immunol. 2010;11(5):373–84.
66. Campbell BM, Charych E, Lee AW, Moller T. Kynurenines in CNS disease: regulation by inflammatory cytokines. Front Neurosci. 2014;8:12.
67. Kennedy PJ, Cryan JF, Dinan TG, Clarke G. Irritable bowel syndrome: a microbiome-gut-brain axis disorder? World J Gastroenterol. 2014;20(39):14105–25.
68. Clarke G, Fitzgerald P, Cryan JF, Cassidy EM, Quigley EM, Dinan TG. Tryptophan degradation in irritable bowel syndrome: evidence of indoleamine 2,3-dioxygenase activation in a male cohort. BMC Gastroenterol. 2009;9:6.
69. Tilg H, Moschen AR. Food, immunity, and the microbiome. Gastroenterology. 2015;148(6):1107–19.
70. Notarangelo FM, Wilson EH, Horning KJ, Thomas MA, Harris TH, Fang Q, et al. Evaluation of kynurenine pathway metabolism in *Toxoplasma gondii*-infected mice: implications for schizophrenia. Schizophr Res. 2014;152(1):261–7.
71. El Aidy S, Kunze W, Bienenstock J, Kleerebezem M. The microbiota and the gut-brain axis: insights from the temporal and spatial mucosal alterations during colonisation of the germ-free mouse intestine. Benef Microbes. 2012;3(4):251–9.
72. Mardinoglu A, Shoaie S, Bergentall M, Ghaffari P, Zhang C, Larsson E. The gut microbiota modulates host amino acid and glutathione metabolism in mice. Mol Syst Biol. 2015;11(10):834.
73. Wikoff WR, Anfora AT, Liu J, Schultz PG, Lesley SA, Peters EC, et al. Metabolomics analysis reveals large effects of gut microflora on mammalian blood metabolites. Proc Natl Acad Sci U S A. 2009;106(10):3698–703.
74. Reigstad CS, Salmonson CE, Rainey JF 3rd, Szurszewski JH, Linden DR, Sonnenburg JL, et al. Gut microbes promote colonic serotonin production through an effect of short-chain fatty acids on enterochromaffin cells. FASEB J. 2015;29(4):1395–403.
75. Yano JM, Yu K, Donaldson GP, Shastri GG, Ann P, Ma L, et al. Indigenous bacteria from the gut microbiota regulate host serotonin biosynthesis. Cell. 2015;161(2):264–76.
76. El Aidy S, van Baarlen P, Derrien M, Lindenbergh-Kortleve DJ, Hooiveld G, Levenez F, et al. Temporal and spatial interplay of microbiota and intestinal mucosa drive establishment of immune homeostasis in conventionalized mice. Mucosal Immunol. 2012;5(5):567–79.
77. Freewan M, Rees MD, Plaza TS, Glaros E, Lim YJ, Wang XS, et al. Human indoleamine 2,3-dioxygenase is a catalyst of physiological heme peroxidase reactions: implications for the inhibition of dioxygenase activity by hydrogen peroxide. J Biol Chem. 2013;288(3):1548–67.
78. Valladares R, Bojilova L, Potts AH, Cameron E, Gardner C, Lorca G, et al. *Lactobacillus johnsonii* inhibits indoleamine 2,3- dioxygenase and alters tryptophan metabolite levels in BioBreeding rats. FASEB J. 2013;27(4):1711–20.
79. Braniste V, Al-Asmakh M, Kowal C, Anuar F, Abbaspour A, Toth M, et al. The gut microbiota influences blood-brain barrier permeability in mice. Sci Transl Med. 2014;6(263):263ra158.
80. Ruddick JP, Evans AK, Nutt DJ, Lightman SL, Rook GA, Lowry CA. Tryptophan metabolism in the central nervous system: medical implications. Expert Rev Mol Med. 2006;8(20):1–27.
81. Schwarcz R, Bruno JP, Muchowski PJ, Wu HQ. Kynurenines in the mammalian brain: when physiology meets pathology. Nat Rev Neurosci. 2012;13(7):465–77.
82. Silber BY, Schmitt JA. Effects of tryptophan loading on human cognition, mood, and sleep. Neurosci Biobehav Rev. 2010;34:387–407.
83. Schwarcz R, Pellicciari R. Manipulation of brain kynurenines: glial targets, neuronal effects, and clinical opportunities. J Pharmacol Exp Ther. 2002;303(1):1–10.
84. Erny D, Hrabe de Angelis AL, Jaitin D, Wieghofer P, Staszewski O, David E, et al. Host microbiota constantly control maturation and function of microglia in the CNS. Nat Neurosci. 2015;18(7):965–77.

85. Rothhammer V, Mascanfroni ID, Bunse L, Takenaka MC, Kenison JE, Mayo L, et al. Type I interferons and microbial metabolites of tryptophan modulate astrocyte activity and central nervous system inflammation via the aryl hydrocarbon receptor. Nat Med. 2016;22(6):586–97.
86. Berstad A, Raa J, Valeur J. Indole – the scent of a healthy 'inner soil'. Microb Ecol Health Dis. 2015;26:27997.
87. Scherzer R, Gdalevsky GY, Goldgur Y, Cohen-Luria R, Bittner S, Parola AH. New tryptophanase inhibitors: towards prevention of bacterial biofilm formation. J Enzyme Inhib Med Chem. 2009;24(2):350–5.
88. Lee JH, Lee J. Indole as an intercellular signal in microbial communities. FEMS Microbiol Rev. 2010;34(4):426–44.
89. Lee JH, Wood TK, Lee J. Roles of indole as an interspecies and interkingdom signaling molecule. Trends Microbiol. 2015;23(11):707–18.
90. Bansal T, Alaniz RC, Wood TK, Jayaraman A. The bacterial signal indole increases epithelial-cell tight-junction resistance and attenuates indicators of inflammation. Proc Natl Acad Sci U S A. 2010;107:228–33.
91. Clarke G, Stilling RM, Kennedy PJ, Stanton C, Cryan JF, Dinan TG. Minireview: gut microbiota: the neglected endocrine organ. Mol Endocrinol. 2014;28(8):1221–38.
92. Munoz-Bellido JL, Munoz-Criado S, Garcia-Rodriguez JA. Antimicrobial activity of psychotropic drugs: selective serotonin reuptake inhibitors. Int J Antimicrob Agents. 2000;14(3):177–80.
93. Macedo D, Filho AJ, Soares de Sousa CN, Quevedo J, Barichello T, Júnior HV. Antidepressants, antimicrobials or both? Gut microbiota dysbiosis in depression and possible implications of the antimicrobial effects of antidepressant drugs for antidepressant effectiveness. J Affect Disord. 2017;208:22–32.
94. Kurnasov O, Goral V, Colabroy K, Gerdes S, Anantha S, Osterman A, et al. NAD biosynthesis: identification of the tryptophan to quinolinate pathway in bacteria. Chem Biol. 2003;10(12):1195–204.
95. Kurnasov O, Jablonski L, Polanuyer B, Dorrestein P, Begley T, Osterman A. Aerobic tryptophan degradation pathway in bacteria: novel kynurenine formamidase. FEMS Microbiol Lett. 2003;227(2):219–27.
96. Begley TP, Kinsland C, Mehl RA, Osterman A, Dorrestein P. The biosynthesis of nicotinamide adenine dinucleotides in bacteria. Vitam Horm. 2001;61:103–19.
97. Kuc D, Zgrajka W, Parada-Turska J, Urbanik-Sypniewska T, Turski WA. Micromolar concentration of kynurenic acid in rat small intestine. Amino Acids. 2008;35(2):503–5.
98. Nestler EJ, Barrot M, DiLeone RJ, Eisch AJ, Gold SJ, Monteggia LM. Neurobiology of depression. Neuron. 2002;34(1):13–25.
99. McKernan D, Fitzgerald P, Dinan T, Cryan J. The probiotic Bifidobacterium infantis 35624 displays visceral antinociceptive effects in the rat. Neurogastroenterol Motil. 2010;22:1029–68.
100. O'Mahony S, Felice V, Nally K, Savignac H, Claesson M, Scully P, et al. Disturbance of the gut microbiota in early-life selectively affects visceral pain in adulthood without impacting cognitive or anxiety-related behaviors in male rats. Neuroscience. 2014;277:885–901.
101. Muller CP, Homberg JR. Serotonin revisited. Behav Brain Res. 2015;277:1–2.
102. Moloney RD, Johnson AC, O'Mahony SM, Dinan TG, Greenwood-Van Meerveld B, Cryan JF. Stress and the microbiota-gut-brain axis in visceral pain: relevance to irritable bowel syndrome. CNS Neurosci Ther. 2016;22(2):102–17.
103. Orlikov A, Ryzov I. Caffeine-induced anxiety and increase of kynurenine concentration in plasma of healthy subjects: a pilot study. Biol Psychiatry. 1991;29(4):391–6.
104. Orlikov AB, Prakhye IB, Ryzov IV. Kynurenine in blood plasma and DST in patients with endogenous anxiety and endogenous depression. Biol Psychiatry. 1994;36(2):97–102.
105. Clarke G, O'Mahony SM, Dinan TG, Cryan JF. Priming for health: gut microbiota acquired in early life regulates physiology, brain and behaviour. Acta Paediatr. 2014;103(8):812–9.
106. Khalil OS, Pisar M, Forrest CM, Vincenten MC, Darlington LG, Stone TW. Prenatal inhibition of the kynurenine pathway leads to structural changes in the hippocampus of adult rat offspring. Eur J Neurosci. 2014;39(10):1558–71.

107. Pisar M, Forrest CM, Khalil OS, McNair K, Vincenten MC, Qasem S, et al. Modified neocortical and cerebellar protein expression and morphology in adult rats following prenatal inhibition of the kynurenine pathway. Brain Res. 2014;1576:1–17.
108. Alexander KS, Pocivavsek A, Wu HQ, Pershing ML, Schwarcz R, Bruno JP. Early developmental elevations of brain kynurenic acid impair cognitive flexibility in adults: reversal with galantamine. Neuroscience. 2013;238:19–28.
109. Pershing ML, Bortz DM, Pocivavsek A, Fredericks PJ, Jorgensen CV, Vunck SA, et al. Elevated levels of kynurenic acid during gestation produce neurochemical, morphological, and cognitive deficits in adulthood: implications for schizophrenia. Neuropharmacology. 2015;90:33–41.
110. Pocivavsek A, Wu HQ, Elmer GI, Bruno JP, Schwarcz R. Pre- and postnatal exposure to kynurenine causes cognitive deficits in adulthood. Eur J Neurosci. 2012;35(10):1605–12.
111. Clarke G, Quigley EMM, Cryan JF, Dinan TG. Irritable bowel syndrome: towards biomarker identification. Trends Mol Med. 2009;15(10):478–89.
112. Keszthelyi D, Troost FJ, Jonkers DM, Kruimel JW, Leue C, Masclee AA. Decreased levels of kynurenic acid in the intestinal mucosa of IBS patients: relation to serotonin and psychological state. J Psychosom Res. 2013;74(6):501–4.
113. Clarke G, O'Mahony SM, Hennessy AA, Ross P, Stanton C, Cryan JF. Chain reactions: early-life stress alters the metabolic profile of plasma polyunsaturated fatty acids in adulthood. Behav Brain Res. 2009;205(1):319–21.
114. Fitzgerald P, Cassidy Eugene M, Clarke G, Scully P, Barry S, Quigley Eamonn MM, et al. Tryptophan catabolism in females with irritable bowel syndrome: relationship to interferon-gamma, severity of symptoms and psychiatric co-morbidity. Neurogastroenterol Motil. 2008;20(12):1291–7.
115. Kennedy PJ, Allen AP, O'Neill A, Quigley EM, Cryan JF, Dinan TG, et al. Acute tryptophan depletion reduces kynurenine levels: implications for treatment of impaired visuospatial memory performance in irritable bowel syndrome. Psychopharmacology (Berl). 2015;232(8):1357–71.

Experimental Anxiety Model for Anxiety Disorders: Relevance to Drug Discovery

11

Michel Bourin

Introduction

Experimental anxiety refers to induced anxiety. It is possible to induce emotions, by abusing a subject or subjecting him to stressful conversations that often induce anger. It is difficult, however, to predict the subject's reaction, which can range from anger to anxiety to fear. There are scales of evaluation to quantify the emotions. In psychopathology, many studies have identified relevant variations of this subjective dimension of sensation seeking in addictive behaviors, as well as in certain personality traits. In a more practical way, each of us knows the pleasant or unpleasant situations inducing some form of anxiety as well as anxiogenic substances. Among these, caffeine is one of the most frequently cited and rightly so. Caffeine is an alkaloid substance found in coffee beans or tea leaves. Thus, when a patient says to you "I took a cup of coffee," this unit besides the fact that the cup varies from a few milliliters to the bowl which represents a third of a liter leads to a great difficulty of appreciation. The somatic manifestations of the more or less important caffeine intake are essentially cardiovascular, that is to say a sensation of rapid pulse. Sensitivity to coffee is often associated with stress, and there is apparently a potentiation of both. Caffeine can be a genuine anxiety-provoking agent, which is causing a panic attack [1]. This acute anxiety attack is inducible by various substances including caffeine. The subject then describes what looks pretty close to a bout of angina pectoris (angina) or even myocardial infarction. He feels chest pain radiating more or less in the left arm, with difficulty breathing, sweating, and feeling of loss of sense of reality and self-control, and gastrointestinal discomfort with the feeling of imminent death.

The consumption of coffee only very rarely leads to a panic attack, especially since the anxious subjects fear the feeling of anxiety that it causes without reaching

M. Bourin (✉)
Neurobiology of anxiety and mood disorders, University of Nantes, Nantes, France
e-mail: michel.bourin@univ-nantes.fr

Y.-K. Kim (ed.), *Anxiety Disorders*, Advances in Experimental Medicine and Biology 1191, https://doi.org/10.1007/978-981-32-9705-0_11

this paroxysm. They then complain more commonly of sleep disorders. Here again we find the link between vigilance, even hypervigilance and anxiety. Coffee has long been used as a psychostimulant, but it took a very long time to demonstrate its interaction with adenosine receptors, which are currently known to be able to reduce anxiety in animals. Other substances may induce anxiety but may be considered to be only substances used for research purposes. Some of them are: sodium lactate in intravenous injection, carbonic gas and cholecystokinin, a physiological peptide substance which induced panic attack in 20 s when injected intravenously [2]. Cholecystokinin perfectly induces all the symptoms of the panic attacks, which suggests that some of us may have a central dysregulation of this peptide.

On the other hand, there are animal models of anxiety; some are more suitable for the screening of new anxiolytic drugs, while others are more based on operant conditioning and learning [3]. In this chapter we will discuss the human or animal models that seem most suitable for the development of new drugs.

Problems with the Development of Anxiolytic Drugs

Despite advances in knowledge of the anatomy and neurobiology of anxiety disorders, there are no new anxiolytics marketed for almost 20 years. It is interesting to understand the reasons, which contribute to both the lack of knowledge of psychopathology and certainly the difficulty of designing clinical trials in this area. Animal models are also problematic in that they have been designed with reference to benzodiazepines. Clearly, the developments are carried out in different directions for lack of referent molecules [4].

To date, around hundred molecules are under development exploring different modes of action:

- Drugs that act on the GABA-A receptor complex, whether they are partial agonists of acceptor sites to benzodiazepines including beta-carbolines as well as compounds that act on the complex GABA-A without our knowledge exactly at what level.
- Other drugs acting as CCK-B receptors antagonists.
- Many drugs interacting with serotonin transmission mainly agonists at the 5-HT1A receptor level, antagonists at the 5-HT2 receptor, and antagonists at the 5-HT3 receptor.
- At the end, drugs with other mechanisms.

This has been the case for many years without finding one of these derivatives on the counters of pharmacists.

Preclinical Evaluation of Anxiolytics

Before studying, in humans, a new molecule potentially anxiolytic, it is necessary to resort to animal models. These models make it possible to create and measure anxiety and thereby determine whether or not an anxiolytic effect exists. Some will criticize this way of proceeding by replying that we cannot recreate, in the animal, the anxiety that can feel the human being. Obviously, there is no model of anxiety quite satisfactory. But it is the role of a model to constitute a simple basis of reasoning in order to better understand complex phenomena [5]. Thus, animal models do not claim to faithfully reproduce the anxiety disorders observed in humans but are only tools used to better understand these disorders and to be able, perhaps, to discover the factors determining or not in anxiety and/or the anxiolytic action of the studied molecules. Existing models, even if they are not perfect, have a good correlation with the clinic and are therefore widely used for predictive or explanatory purposes. The sensitivity of these models for benzodiazepines continues to be used as a reference, thus making it possible to compare the efficacy of different anxiolytic molecules [6], whereas non-benzodiazepine molecules are now developed. Animal models of anxiety are used for two main purposes: firstly, the "screening" of new substances and, secondly, the study of neurotransmitters and their mechanism, involved in anxiety. Two objectives that are in fact not independent of each other are as follows: information on the mode of action of new molecules can lead to new approaches in the study of the biological bases of anxiety, and this knowledge can themselves lead to a more rational development of new medicines. The different models used to express and measure anxiety in animals are based on the study of spontaneous behaviors (exploration, food intake) or on those of more complex behaviors where interaction and learning play a larger role. Patterns of spontaneous behavior are based on the animal's reactions to punishment, non-reward (frustration), and novelty. Reducing or blocking these behaviors is a reflection of the animal's fear, anxiety, or frustration [7]. One can also observe the reaction of the animals, which will have been placed in a situation known to be anxiety-provoking: ultrasonic calls from neonatal rodents separated from their mother [8], rat burial behavior, and intimidating mimicry of marmoset monkey [9]. It is also possible to induce anxiety by various substances. These are the procedures of discrimination and preference of place. These are not strictly speaking models of anxiety, but they give an overall idea of the effect induced by the molecule, which effect is undoubtedly the result of the superposition of the central and peripheral effects of this substance. Recently, the number of behavioral models has increased significantly. Only two tests – the study of social interactions and the Elevated Plus Maze – were evaluated on behavioral, physiological, and pharmacological bases; the others have not been fully validated, and we will focus our attention on these two models as well as the two-compartment test, dark and illuminated.

Behavioral Models of Anxiety

Persistent and uncontrollable anxiety is the main feature of generalized anxiety. Thus, some tests based on a rat's ability to adapt to a new territory were initially proposed as models of anxiety. It is essentially neophobic anxiety such as it can be encountered in humans in new behavioral situations (airport, station, etc.), for example, the increase in locomotion in an open space, the number of entries in a Y maze, the number of steps in a staircase, or the number of holes explored in a "hole board" [10]. While anxiety is a hindrance to discovering a new territory, a number of other factors are involved in controlling exploration behavior. It is now considered that these tests are not specific enough; they can nevertheless be used as complementary tests to eliminate the behavior related to psychostimulation.

The Dark and White Box [11]

Crawley developed an anxiety test based on the number of transitions of a mouse between the two compartments (one dark, the other illuminated) of an enclosure. Mice have to choose between their natural dislike for light and their desire to explore a new environment of ambivalence that often causes anxiety in humans. The effect which increases the transitions between the two compartments without increasing spontaneous locomotion is regarded as anxiolytic. It should be noted that this effect is only observed in mice with an initial level of high spontaneous transitions. Later this test was modified with notably a change in the relative size of the two compartments (1/3 dark, 2/3 lit) but also by suppressing the neophobic anxiety by using pregnant already soiled by mice [12]. In this new model, an increase in locomotor activity in the illuminated compartment with a decrease in this same activity in the dark compartment reflects an anxiolytic effect. But even though this modified version is sensitive to the effect of benzodiazepines, it does not reflect a true decrease in anxiety. In addition, benzodiazepines do not change the transition times when the two compartments are lit identically and do not affect the number of entries in the closed arms either. In addition, they do not alter the locomotor activity of animals that have been previously trained. All these elements suggest that the effects of benzodiazepines are only observed when emotional factors limiting exploratory activity can be considered very intense. Therefore, the action of benzodiazepines may be correlated with a reduction in the emotional factors that control behavior, inversely to what is observed in humans [13].

The Elevated Plus Maze (EPM) is another model of anxiety [14]. The apparatus is constituted by a platform in cross, elevated with respect to the ground. Two opposing arms are open "dangerous arms" and may be related to the illuminated compartment of the dark and illuminated compartment box test; the other two arms are closed "safe arms," corresponding to the dark compartment. Animals can freely access each of the two types of arms. The increase in the number of entries and/or the time spent in the open arms is a parameter considered to be the translation of an anxiolytic effect [15]. The open arms of the "Elevated Plus Maze," as the clear compartment of the "black and white box," are determining factors in the induction of behavioral and physiological changes that reflect anxiety. Moreover, in these two models, the administration of benzodiazepines to the mice causes an anxiolytic effect. This effect is reflected overall in an increase in locomotion in "stressful" compartments (i.e., open arms and illuminated compartment) without any change in behavior in "safe" compartments. One could then think that the induced effect is not an anxiolytic effect but a simple increase in locomotor activity. These methods based on the study of the sensitivity of the exploration behavior to the effect of benzodiazepines raise an important problem as to the meaning of this parameter. A single test only provides limited information, and in this case the information concerning the exploration behavior is difficult to interpret in rodents.

It is now accepted that the "Elevated Plus Maze" is sensitive to a wide variety of anxiolytic or anxiogenic compounds and appears to be predictive of the activity of drugs used in generalized anxiety. The social interaction test is based on the uncertainty and anxiety induced in rats placed in an intensely illuminated unknown medium. The time during which the two rats interact with each other is measured. Two factors influence the level of interaction: the intensity of lighting and the fact that the place is known to rats. In untreated rats, a certain level of interaction is observed when the experimental area is familiar and dimly lit. This level of interaction decreases when the area is unknown and/or brightly lit. The administration of anxiolytic molecules to both rats facilitates social interaction. This test is probably the one that best reflects generalized anxiety. Indeed social interaction is the only dependent variable. The behavior varies only according to the characteristics of the experimental area. This test cannot be a model of social anxiety because there are more social interactions when the two rats do not know each other the four-plate test (FPT). The apparatus consists of a cage (18 cm × 25 cm × 16 cm) whose floor is composed of four metal plates (8 cm × 11 cm) separated from each other by a space of 8 mm. These four plates are connected to an electric shock generator. The shocks administered have an intensity of 0.6 mA and duration of 0.5 s. After a latent period of 15 s, the animal receives a plantar electric shock each time it passes from one plate to another. The number of electric shocks, or punishments, is counted for 60 s. The administration of a substance with anxiolytic activity results in an increase in the number of accepted punishments. Antidepressants used in the treatment of anxiety disorders have been shown to be effective in this test [16].

Zebrafish: A New Behavioral and Genetic Model

The zebrafish is a model organism emerging in recent years because it is accessible genetically and is characterized in the embryonic and larval stages by a high transparency. In the last 5 years, he has become a model of choice for using optics to study the organization of neural networks. Real-time observation of neural signals is an important source of information on how the brain responds to external stimuli.

To capture and analyze this activity, neurobiologists have chosen to apply the technique of imaging to the study of zebrafish brain [17]. The transparency and the small size of this fish, scarcely 4 to 5 cm, have been exploited to follow the activity of about 80,000 neurons out of a total of about 100,000 that counts their nervous system. To visualize the zebrafish brain circuits, the confocal optical microscope was used. Under this microscope, the neuronal signals are visible by fluorescence, the zebrafish neurons having been previously labeled with the GCaMP7a protein. It has been shown that the brain of zebrafish is never inactive. It has been possible to identify a region responsible for movements associated with characteristic movements of exploration [18].

In addition to anxiety models built in rodents, more recently (this is the twenty-first-century model) the zebrafish (Danio rerio) is used. With a genome of 26,000 coding genes, the zebrafish gene pool is currently the largest ever in a vertebrate. A characteristic that makes this small fish a very serious subject of study among geneticists who today claim to be able to understand certain diseases that affect the human being simply by making comparisons with the genes of the fish. Today the genome of the zebrafish has been sequenced to such a degree of precision that it is quite possible to make a comparison with that of man [19]. In an immersion test in a new aquarium, behaviors such as swimming at the bottom, latency to swim in higher levels, are important indicators of stress and anxiety, or the effects of anxiogenic or anxiolytic substances. Behavioral models with zebrafish are mostly similar to those developed in rodents [20]. It should be remembered that what is most important in terms of development, hence of "screening," is more important action size on a model, than modest action sizes on several models. Indeed, the developers are often reassured when a molecule is active on several models, thus presuming the activity on several forms of anxiety, or it proves most often that an important answer on a paradigm is the best guarantor of an activity in humans.

If these models or others not described in this present chapter have not led to successful developments, it may be due to a lack of adaptation to different forms of human anxiety.

Cholecystokinin Injection as Anxiety Model in Human?

Almost 30 years ago, it was reported that exogenous cholecystokinin (CCK-4) produced panic-like attacks in healthy volunteers and that these effects could be attenuated by pretreatment with lorazepam [21]. CCK-4 was administered to patients with a current-point diagnosis of panic disorder using a double-blind placebo-controlled methodology. Bolus injections of CCK-4 (50 mg) precipitated a panic attack, as defined by DSM-III criteria and patient self-report, within 1 minute following administration in 11 trial patients studied, whereas none of the patients panicked following placebo. CCK-4 treatment elicited an average of 12 symptoms per patient, the most common symptoms being dyspnea, palpitations/rapid heart, chest pain/

discomfort, faintness, dizziness, paresthesia, hot flushes/cold chills, nausea/abdominal distress, anxiety/fear/apprehension, and fear of losing control. It has been found that response to CCK-4 reliably differentiates panic disorder patients from healthy controls with no personal or family history of panic attacks [22]. In a double-blind placebo-controlled study, the patients with panic disorder experienced a greater number of symptoms and more intense symptoms following challenge with two doses of CCK-4 (25 and 50 mg) [23]. In addition, the incidence of panic attacks was markedly higher in patients than controls following injection of 25 mg (91% vs 17%) and 50 mg (100% vs 47%) of the peptide. Interestingly, the number and intensity of symptoms as well as the symptom profile were remarkably similar in both patients and normal subjects who panicked with the 50 mg dose of CCK-4, suggesting that the enhanced response in patients could not be readily attributed to a tendency to over endorse symptoms. These results are corroborated by other authors who used pentagastrin, a CCK agonist which incorporates the identical 4-amino-acid sequence of CCK-4 [24, 25]. Moreover, patients with panic disorder have been shown to have decreased concentrations of CCK-8 s in the cerebrospinal fluid relative to control subjects [26]. Concentrations of CCK-8 in lymphocytes were also significantly reduced in patients with panic disorder compared to healthy controls [27]. These findings are also in favor of anomalies in the CCK system due to panic disorder. It is also likely that systemic administration of CCK-4 produces prominent respiratory and cardiovascular alterations in humans through a direct action on the brainstem [28].

Antipanic drugs and no other compounds block the effect of CCK-4 [29]. It was demonstrated that the panic-induced effects of CCK-4 can be antagonized by chronic treatment with imipramine [30]. In one study, the pretreatment of patients with the selective CCK-B receptor antagonist L-365,260 dose dependently blocked CCK-4-induced panic attacks [31]. These data apparently support the role of CCK-B receptors in the mediation of panicogenic-like action of CCK-4. Nevertheless, a placebo-controlled trial of L-365,260 did not result in any clinically significant improvement in patients with panic disorder [32]. The possible reasons for lack of effect with L-365,260 are not clear, but the poor pharmacokinetic properties of the drug are the most plausible explanation.

A large body of evidence suggests that the neuropeptide cholecystokinin might be an important modulator of the neuronal networks that are involved in anxiety, in particular in panic disorders. The key regions of the fear network, such as basolateral amygdala [33], hypothalamus, periaqueductal gray, or cortical regions such as the anterior cingulate cortex, seem to be connected by CCK-ergic pathways [34, 35].

It seems that this human model that seemed promising to us three decades ago has not been exploited properly probably because of the lack of CCK-B receptors specific ligands among others [36]. The most convincing preliminary data were obtained for CCK-B [37]. The most convincing data were obtained for CCK-B receptors [38]. In addition, some cholecystokinin neurons may be able to mimic the behavioral effects of SSRIs [39].

Other Substances Inducing Anxiety

Whether it is induction of a panic attack by sodium lactate, inhalation of CO_2, or the administration of a high dose of caffeine, these models have not led to the discovery of new anxiolytics. For a long time, subjects with panic attacks were considered to be chronic hyperventilants who suddenly plunged into hypocapnia. In fact, examination of hyperventilation provocation test results, respiratory reeducation, and ambulatory PCO_2 measurement shows that there is no perfect overlap between the two phenomena. The hyperventilation syndrome is probably a separate entity. The biological models are based on experiments using lactate or carbon dioxide to reach, with Klein [40], the hypothesis of hypersensitivity to CO_2 with false suffocation. Psychological models consider various options. Lum [41] sees the consequence of a bad habit of breathing. Hyperventilation plays a role in panic attacks only if it is interpreted catastrophically [42]. Dyspnea was considered as the initiating element of an innate fear reflex. Finally, Van den Bergh has conducted promising studies on respiratory conditioning [43]. If all panic-attacked subjects are not chronic hyperventilants, it must be admitted that both phenomena coexist frequently. An integrative approach to biological and psychological models could shed light on the nature of their relationship. Is it still relevant to rely on these types of human models to develop an anxiolytic? It seems that the model of cholecystokinin is more relevant, although it essentially reproduces a panic attack with its respiratory symptoms.

Is the Theoretical Model of Deakin and Graeff (1991) Useful?

The ascending serotonergic pathways originating in the dorsal raphe nucleus (DRN) innervate the amygdala and the frontal cortex and facilitate the escape or avoidance behaviors involved in the response to a threat [44]. These behavioral strategies invoke learning and are therefore related to conditioned or anticipatory anxiety, possibly the general anxiety disorder (GAD). The 5-HT2A/5-HT2C and 5-HT3 postsynaptic receptors could be activated in this neurotransmission pathway. In return, the periventricular pathways of the DRN innervate the periventricular and periaqueductal gray matter. In these regions, 5-HT inhibits emerging reactions of combat/flight triggered by a nearby danger [45], acute pain, or asphyxia associated with panic disorder. This function could be mediated by postsynaptic 5-HT2A/5-HT2C and 5-HT1A receptors. These two paths of transmission regulate adaptive responses to acute stressors. In cases where the body has to adapt to chronic stressors, the pathways connecting the medial raphe nucleus (MRN) to the hippocampus develop resistance to chronic stimuli in disconnecting aversive events from the psychobiological processes underlying social behaviors and appetite allowing the individual to lead a normal life despite these events. Serotoninergic pathways initiated in MNR and innervating the dorsal hippocampus increase resistance and tolerance to stressful stimuli. The depression can appear when this mechanism is defeated. 5-HT1A receptors appear to be the main mediators of the hippocampus-MRN pathway whose disconnection with aversive stimuli in mitigating the behavioral consequences.

Experimental and clinical evidences supporting these hypotheses have been already reported [46–49] related to recent findings as well in animals and humans.

The Deakin/Graeff hypothesis proposes that different subpopulations of serotonergic neurons through topographically organized projections to forebrain and brainstem structures modulate the response to acute and chronic stressors and that dysfunction of these neurons increases vulnerability to affective and anxiety disorders, including panic disorder [50].

The role of 5-HT1A receptors has been well studied in the control of anxiety states [51–53]. These receptors, which are localized as inhibitory autoreceptors on the dendrites of serotoninergic cell bodies in the raphe nuclei, are also located at the postsynaptic level of serotoninergic neurons in the hippocampus, septum, amygdala, PAG, and entorhinal and frontal cortex in which they exert a negative feedback on the electrical activity of serotonergic neurons in DRN [54]. Prolonged stressors alter the brain levels of 5-HT1A sites and preferentially desensitize presynaptic 5-HT1A receptor populations rather than postsynaptic receptors [55], a process contributing to the overall potentiation of serotonergic transmission under these conditions. 5-HT1A receptor agonists induce a pronounced stimulation of the hypothalamo-corticotropic axis and act via multiple locations in the hypothalamus itself. In addition, activation of 5-HT1A receptors is involved in the induction of secretion of ACTH and corticosteroids in response to stressful stimuli [56]. The latter exert a reciprocal modulatory influence on the activity of pre- and postsynaptic 5-HT1A receptor populations and have been also involved in mood control. 5-HT1A receptor agonists, mainly via central actions, facilitate the release of noradrenaline from the neurons noradrenergic, an action that could, secondarily, stimulate the activity of the hypothalamic-pituitary-adrenal axis.

This model is widely used to develop new anxiolytics with a very relative success because it is essentially based on the serotonergic hypothesis.

The Knockout Mice

Knockout (KO) animals not only make it possible to study a concrete disease but also to answer the fundamental question: what is the gene for in the body? The idea is simple: if a gene is deficient within an organism (thus causing the corresponding protein), it means that there is some dysfunction. When, for example, the insulin gene is defective, the glucose metabolism is no longer properly controllable in the body. When the growth gene is inactivated in a fly, then there will be formation of mini-flies. In other words, by inactivating genes in a targeted way, one can conclude to the normal function of the gene. The advantage of knockout mice lies in the fact that the incidence of the gene can be observed in a living animal.

Mice genetically deprived of the 5-HT1A receptor have a generally anxiogenic phenotype and increased susceptibility to stressful stimuli in both behavioral and autonomic variables [57]. Although the precise basic mechanisms to explain the increase of this anxiety remain uncertain, it has been postulated that it reflects a loss of activity of the post-5-HT1A receptor sites rather than presynaptics, especially as extracellular levels of 5-HT are slightly modified in animals lacking this receptor. This lack of alteration of the release of 5-HT can be surprising. However, this probably reflects changes in the control of compensatory serotonergic transmission to the loss of 5-HT1A autoreceptors that exert only a limited and tonic influence on the corticolimbic release of 5-HT [58]. Studies on 5-HT1A knockout model directly support the idea that a loss of postsynaptic 5-HT1A receptors accounts for an increase in anxiety. Indeed, in the mouse, when the function of the 5-HT1A receptors is preserved in the amygdala, the hippocampus, and the frontal cortex while the 5-HT1A autoreceptors located in the raphe have been deactivated, the anxious phenotype does not occur and not express. 5-HT1A receptors exert an inhibitory influence on limbic (e.g., amygdala) and cortical release of glutamate [59].

Disinhibition of glutamatergic activity could contribute to the anxiety behavior of 5-HT1A receptor-deficient mice. In addition, a hypothesis to explain the phenotype of KO mice for 5-HT1A receptors invokes an interaction between postsynaptic 5-HT1A receptors and glutamatergic transmission at the level of pyramidal cells of the frontal cortex and possibly subcortical structures. It has been argued that the anxious phenotype of KO mice for 5-HT1A receptors may imply indirect dysregulation of amygdala and amyloid BZD/GABA-A receptor function, which are probably expressed by neurons bearing the 5-HT1A receptors and GABA-A receptors [60].

However, equivalent observations could not be obtained in all KO mouse lines for 5-HT1A receptors whose genetic backgrounds were divergent [61]. Other studies have demonstrated that there are complex functional interactions between 5-HT1A receptors and GABAergic mechanisms [62]. Indeed, whereas when they are localized, the GABA-A/GABA-B and 5-HT1A receptors exert a mutual inhibition on the activity of the neurons (including the serotoninergic cell bodies of the raphe); it has been shown that the receptors 5-HT1A can suppress the limbic release of GABA [63]. This latter action may contribute to the anxiogenic effects of postsynaptic 5-HT1A receptor agonists. The role of 5-HT1A receptors in the modulation of anxiety states is complex and reflects the contrasting roles of pre- versus postsynaptic sites and individual populations of 5-HT1A receptors in discrete supraspinal regions.

Other examples of knockout mice have been developed which may be useful in the search for new anxiolytics, such as mGluR7 [64]. Indeed, more and more evidence suggests that the glutamatergic system may be a relevant therapeutic target for anxiety disorders. Glutamate is the neurotransmitter used by the vast majority of excitatory synapses in the brain. And the metabotropic glutamate receptor subtypes (mGluR1–mGluR8) act primarily as important postsynaptic regulators of neurotransmission in the central nervous system (CNS), providing a mechanism by which rapid synaptic responses across channels glutamate-dependent cationic agents can be refined. Thus, mGluR participate in a wide variety of central nervous system functions.

In the family of metabotropic glutamate receptors, the presynaptic mGluR7 receptor exhibits the highest evolutionary conservation and is thought to act as a regulator of neurotransmitter release. The mGluR7 receptor is the most widely distributed presynaptic receptor demonstrated to be critical in the normal functioning of the central nervous system. In addition, a growing body of experimental evidence suggests that the mGluR7 receptor is not only a key player in the development of synaptic responses at glutamatergic synapses but is also a key regulator of inhibitory GABAergic transmission [65]. The development of selective pharmacological and genetic tools has allowed the dismantling of the mGluR7 receptor function in a multitude of physiological and behavioral processes. Thus, knockout mice have demonstrated a role for the mGluR7 receptor in anxiety, fear conditioning, aversion, learning, and spatial memory. In addition, these mGluR7-deficient metabotropic mice demonstrate increased susceptibility to epileptic seizures suggesting a unique role for this receptor in the regulation of neuronal excitability. Similarly, short-term alteration of synaptic plasticity in transgenic mGluR7-deficient metabotropic mice demonstrates that the absence of mGluR7 receptors causes short-term changes in synaptic plasticity in the hippocampus [66]. In addition, the discovery and recent characterization of the first allosteric antagonist acting on the Venus flytrap domain [67] of the N-terminus of the mGluR7 receptor definitively potentiate the observations made on mGluR7 knockout mice as to the function of this receptor in the anxiety [68]. Together, these data suggest that the mGluR7 receptor is an important regulator of glutamatergic function, fear, aversion, and cognition, and therefore this receptor represents an innovative therapeutic target for stress-related disorders at the cognition and anxiety.

The two examples of knockout receptors indicated above for the development of anxiolytics are not exhaustive. There are indeed many other possibilities using knockout mice for the different types of serotonin or glutamatergic receptors. However, there are limits in the use of such animals. Indeed, it is very reductive to consider that a single receptor would be at the origin of the anxiety, and especially the experiments must take place quickly to avoid reactivation of the receptor by a phenomenon of homeostasis.

Optogenetics

For the past 10 years, the development of optogenetics has revolutionized neuroscience. By making the neurons sensitive to light, this new tool makes it possible to intervene on the brain and the nervous system, such as control of memories, taste, thirst, etc. The brain is an immense network of nearly 100 billion interconnected neurons, distributed in different brain areas and classified into different subtypes according to the neurotransmitter they release in response to stimulation. Poor neuronal communication is at the root of neuropsychiatric disorders such as schizophrenia and autism. Studying how information is transmitted in neural networks is therefore essential to better understand the functioning of the brain in physiological and pathological conditions. For this, it is essential to be able to specifically control the different types of cells that make up the neural networks. This was made possible a little more than 10 years ago by the development of optogenetics [69], a veritable technological revolution in the field of neuroscience. This technique consists in genetically modifying neurons so that they become sensitive to light thanks to the expression of a protein: opsin [70].

In 2002, channelrhodopsin was identified [71]; it is a photosensitive protein derived from a green algae called *Chlamydomonas reinhardtii*. In this unicellular microorganism, channelrhodopsin is necessary for phototaxis, a process by which certain organisms move in space with respect to the light present in the environment, for example, to promote photosynthesis. In 2003, these same researchers discovered and characterized channelrhodopsin-2, which is now the most used protein for optogenetic approaches. From a mechanistic point of view, channelrhodopsin-2 is an ion channel activated by blue light. Its activation results in the opening of the channel which then allows the entry into the positive ion cell.

In 2005, it was shown that exposing neurons expressing channelrhodopsin-2 to blue light allows their activation by triggering action potentials (electrical message neurons) [72]. The power of this approach is based on spatial as well as temporal accuracy of neuron control. Spatial precision since it is possible, by genetic engineering, to target a specific neuronal population in a well-defined brain region so that the channelrhodopsin is produced only by the neurons that one wishes to activate. In addition, technological developments are underway to restrict the application of light to a much localized area of the brain. Temporal precision, too, since neuron activation is transient and only occurs when the experimenter turns on the light.

By example, optogenetic stimulation of 5-HT fibers at the terminal streak induces the release of 5-HT, an effect enhanced by SSRIs. This stimulation does not modify the acquisition of the "conditioning of fear." On the other hand, it increases the freezing response to conditional stimulus and contextual stimulus, as well as anxiety. These effects of 5-HT released in the terminal streak, or SSRIs, are blocked by a 5-HT2C receptor antagonist, in agreement with the literature. It can be concluded at this stage that the activation of the 5-HT2C receptors of the terminal streak induces an anxiogenic effect [73]. It has previously been shown that knockout mice for the 5-HT2C receptor are less anxious and that corticotropin-releasing factor neurons of the terminal streak have reduced activation after a conditioning test of fear.

Conclusion

The anxiety patterns are numerous, both animal and human; they have resulted in few new discoveries concerning anxiolytics. However, the zebrafish model may have a definite future combined with optogenetics. It also seems appropriate to use the different knockout animals for both serotoninergic and glutamatergic receptors. The fact that the pathophysiology of anxiety is complex has led to very different drug developments in this area. The first ones were the benzodiazepines discovered by serendipity. It turns out that animal models of anxiety have been developed as to their effectiveness for "screener" BZDs. BZDs decrease serotonin in the brain; they have a more disinhibiting effect than really anxiolytic. The SSRIs decrease cerebral serotonin concentrations in a less abrupt way than BZDs. That progressive decrease of serotonin contributes to their therapeutic success in anxiety disorders. Yet, it remains to be studied if more specific derivatives of serotonergic receptors subtypes could have a better benefit/risk. The glutamatergic track arises from time to time, but we do not know how to control the balance glutamate/GABA. As to neuropeptides, they are not effective in themselves but seem to be modulators of serotoninergic activity, used alone they are of a limited effectiveness. In my opinion we have tools to find and to develop new anxiolytics, but don't forget that the clinical trials in the field of anxiety disorders are very difficult regarding the high level of placebo response.

References

1. Vilarim MM, Rocha Araujo DM, Nardi AE. Caffeine challenge test and panic disorder: a systematic literature review. Expert Rev Neurother. 2011;11:1185–95.
2. Bourin M, Malinge M, Guitton B. Provocative agents in panic disorder. Therapie. 1995;50:301–6.
3. Bourin M. Animal models for screening anxiolytic-like drugs: a perspective. Dialogues Clin Neurosci. 2015;17:295–303.
4. Murrough JW, Yaqubi S, Sayed S, Charney DS. Emerging drugs for the treatment of anxiety. Expert Opin Emerg Drugs. 2015;20:393–406.
5. Bourin M. Le devenir des modèles pharmacologiques. Therapie. 1995;50:375–9.
6. Bourin M. Animal models of anxiety: are they suitable for predicting drug action in humans? Pol J Pharmacol. 1997;49:79–84.
7. Lister RG. Ethologically based animal models of anxiety disorder. In: File SE, editor. Psychopharmacology of anxiolytics and antidepressants. New York: Pergamon Press; 1991. p. 155–85.
8. Benton D, Nastiti K. The influence of psychotropic drugs on ultrasonic calling of mouse pups. Psychopharmacology (Berl). 1988;95:99–102.
9. Okano H, Hikishima K, Iriki A, Sasaki E. The common marmoset as a novel animal model system for biomedical and neuroscience research applications. Semin Fetal Neonatal Med. 2012;17(6):336–40.
10. Labots M, Van Lith HA, Ohl F, Arndt SS. The modified hole board – measuring behavior, cognition and social interaction in mice and rats. J Vis Exp. 2015;98:52529. https://doi.org/10.3791/52529.
11. Bourin M, Hascöet M. The mouse light/dark box test. Eur J Pharmacol. 2003;463:55–65.
12. Hascoët M, Bourin M. A new approach to the light/dark test procedure in mice. Pharmacol Biochem Behav. 1998;60:645–53.
13. Bourin M. The test retest model of anxiety: an appraisal of findings to explain benzodiazepine tolerance. Pharmacol Biochem Behav. 2018. pii: S0091–3057(17)30265–4.
14. Ennaceur A, Chazot PL. Preclinical animal anxiety research – flaws and prejudices. Pharmacol Res Perspect. 2016;4(2):e00223. https://doi.org/10.1002/prp2.223.
15. Bourin M, Dailly E, Hascöet M. Preclinical and clinical pharmacology of cyamemazine: anxiolytic effects and prevention of alcohol and benzodiazepine withdrawal syndrome. CNS Drug Rev. 2004;10:219–29.
16. Hascoët M, Bourin M. The mouse light-dark box test. In: Gould TD, editor. Mood and anxiety related phenotypes in mice. Characterization using behavioral tests. New York: Humana Press; 2009. p. 197–223.
17. Kettunen P. Calcium imaging in the zebrafish. Adv Exp Med Biol. 2012;740:1039–71.
18. Fetcho JR, Liu KS. Zebrafish as a model system for studying neuronal circuits and behavior. Ann N Y Acad Sci. 1998;860:333–45.

19. Hoshijima K, Jurynec MJ, Grunwald DJ. Precise editing of the zebrafish genome made simple and efficient. Dev Cell. 2016;36:654–67.
20. Stewart AM, Braubach O, Spitsbergen J, Gerlai R, Kalueff AV. Zebrafish models for translational neuroscience research: from tank to bedside. Trends Neurosci. 2014;37:264–78.
21. de Montigny C. Cholecystokinin tetrapeptide induces panic-like attacks in healthy volunteers. Preliminary findings. Arch Gen Psychiatry. 1989;46:511–7.
22. Bradwejn J, Koszycki D, Meterissian G. Cholecystokinin-tetrapeptide induces panic attacks in patients with panic disorder. Can J Psychiatry. 1990;35:83–5.
23. Bradwejn J, Koszycki D, Bourin M. Dose ranging study of the effects of cholecystokinin in healthy volunteers. J Psychiatry Neurosci. 1991;16:91–5.
24. Abelson JL, Nesse RM, Vinik A. Stimulation of corticotropin release by pentagastrin in normal subjects and patients with panic disorder. Biol Psychiatry. 1991;29:1220–3.
25. van Megen HJ, Westenberg HG, den Boer JA, Haigh JR, Traub M. Pentagastrin induced panic attacks: enhanced sensitivity in panic disorder patients. Psychopharmacology (Berl). 1994;114:449–55.
26. Lydiard RB, Ballenger JC, Laraia MT, Fossey MD, Beinfeld MC. CSF cholecystokinin concentrations in patients with panic disorder and in normal comparison subjects. Am J Psychiatry. 1992;149:691–3.
27. Brambilla F, Bellodi L, Perna G, Garberi A, Panerai A, Sacerdote P. Lymphocyte cholecystokinin concentrations in panic disorder. Am J Psychiatry. 1993;150:1111–3.
28. Bradwejn J, Koszycki D, Couëtoux du Tertre A, Paradis M, Bourin M. Effects of flumazenil on cholecystokinin-tetrapeptide-induced panic symptoms in healthy volunteers. Psychopharmacology (Berl). 1994;114:257–61.
29. Bradwejn J, Koszycki D, Paradis M, Reece P, Hinton J, Sedman A. Effect of CI-988 on cholecystokinin tetrapeptide-induced panic symptoms in healthy volunteers. Biol Psychiatry. 1995;38:742–6.
30. Bradwejn J, Koszycki D. Imipramine antagonism of the panicogenic effects of cholecystokinin tetrapeptide in panic disorder patients. Am J Psychiatry. 1994;151:261–3.
31. Bradwejn J, Koszycki D, Couëtoux du Tertre A, van Megen H, den Boer J, Westenberg H. The panicogenic effects of cholecystokinin-tetrapeptide are antagonized by L-365,260, a central cholecystokinin receptor antagonist, in patients with panic disorder. Arch Gen Psychiatry. 1994;5:486–93.
32. Kramer MS, Cutler NR, Ballenger JC, Patterson WM, Mendels J, Chenault A, Shrivastava R, Matzura-Wolfe D, Lines C, Reines S. A placebo-controlled trial of L-365,260, a CCKB antagonist, in panic disorder. Biol Psychiatry. 1995;37:462–6.
33. Del Boca C, Lutz PE, Le Merrer J, Koebel P, Kieffer BL. Cholecystokinin knock-down in the basolateral amygdala has anxiolytic and antidepressant-like effects in mice. Neuroscience. 2012;218:185–95.
34. Dieler AC, Sämann PG, Leicht G, Eser D, Kirsch V, Baghai TC, Karch S, Schüle C, Pogarell O, Czisch M, Rupprecht R, Mulert C. Independent component analysis applied to pharmacological magnetic resonance imaging (phMRI): new insights into the functional networks underlying panic attacks as induced by CCK-4. Curr Pharm Des. 2008;14:3492–507.
35. Zwanzger P, Domschke K, Bradwejn J. Neuronal network of panic disorder: the role of the neuropeptide cholecystokinin. Depress Anxiety. 2012;29:762–74.
36. Bourin M, Dailly E. Cholecystokinin and panic disorder. Acta Neuropsychiatr. 2004;16:1–9.
37. Wilson J, Markie D, Fitches A. Cholecystokinin system genes: associations with panic and other psychiatric disorders. J Affect Disord. 2012;136:902–8.
38. Domschke K, Maron E. Genetic factors in anxiety disorders. Mod Trends Pharmacopsychiatr. 2013;29:24–46.
39. Medrihan L, Sagi Y, Inde Z, Krupa O, Daniels C, Peyrache A, Greengard P. Initiation of behavioral response to antidepressants by of the dentate gyrus. Neuron. 2017;95:564–76.
40. Klein DF. False suffocation alarms, spontaneous panics, and related conditions. An integrative hypothesis. Arch Gen Psychiatry. 1993;50:306–17.
41. Lum LC. Hyperventilation and anxiety state. J R Soc Med. 1981;74:1–4.

42. Clark DM, Hemsley DR. The effects of hyperventilation; individual variability and its relation to personality. J Behav Ther Exp Psychiatry. 1982;13:41–7.
43. van den Bergh O, Kempynck PJ, van de Woestijne KP, Baeyens F, Eelen P. Respiratory learning and somatic complaints: a conditioning approach using CO_2-enriched air inhalation. Behav Res Ther. 1995;33:517–27.
44. Blanchard DC, Hori K, Rodgers RJ, Hendrie CA, Blanchard RJ. Attenuation of defensive threat and attack in wild rats (Rattus rattus) by benzodiazepines. Psychopharmacology (Berl). 1989;97:392–401.
45. Blanchard RJ, Blanchard DC. Attack and defense in rodents as ethoexperimental models for the study of emotion. Prog Neuropsychopharmacol Biol Psychiatry. 1989;13(Suppl):S3–14.
46. Deakin JF, Graeff FG. 5-HT and mechanisms of defence. J Psychopharmacol. 1991;5:305–15.
47. Graeff FG. Role of 5-HT in defensive behavior and anxiety. Rev Neurosci. 1993;4:181–211.
48. Graeff FG, Guimarães FS, De Andrade TG, Deakin JF. Role of 5-HT in stress, anxiety, and depression. Pharmacol Biochem Behav. 1996;54:129–41.
49. Graeff FG, Viana MB, Mora PO. Opposed regulation by dorsal raphe nucleus 5-HT pathways of two types of fear in the elevated T-maze. Pharmacol Biochem Behav. 1996;53:171–7.
50. Paul ED, Johnson PL, Shekhar A, Lowry CA. The Deakin/Graeff hypothesis: focus on serotonergic inhibition of panic. Neurosci Biobehav Rev. 2014;46(Pt 3):379–96.
51. Garcia-Garcia AL, Newman-Tancredi A, Leonardo ED. 5-HT (1A) [corrected] receptors in mood and anxiety: recent insights into autoreceptor versus heteroreceptor function. Psychopharmacology (Berl). 2014;231:623–36.
52. Altieri SC, Garcia-Garcia AL, Leonardo ED, Andrews AM. Rethinking 5-HT1A receptors: emerging modes of inhibitory feedback of relevance to emotion-related behavior. ACS Chem Nerosci. 2013;4:72–83.
53. Molina E, Cervilla J, Rivera M, Torres F, Bellón JA, Moreno B, King M, Nazareth I, Gutiérrez B. Polymorphic variation at the serotonin 1-A receptor gene is associated with comorbid depression and generalized anxiety. Psychiatr Genet. 2011;21:195–201.
54. Courtney NA, Ford CP. Mechanisms of 5-HT1A receptor-mediated transmission in dorsal raphe serotonin neurons. J Physiol. 2016;594:953–65.
55. Laaris N, Le Poul E, Laporte AM, Hamon M, Lanfumey L. Differential effects of stress on presynaptic and postsynaptic 5-hydroxytryptamine-1A receptors in the rat brain: an in vitro electrophysiological study. Neuroscience. 1999;91:947–58.
56. Pilar-Cuéllar F, Vidal R, Díaz Á, Garro-Martínez E, Linge R, Castro E, Haberzettl R, Fink H, Bert B, Brosda J, Romero B, Crespo-Facorro B, Pazos Á. Enhanced stress response in 5-HT1AR overexpressing mice: altered HPA function and hippocampal long-term potentiation. ACS Chem Nerosci. 2017;8:2393–401.
57. Freeman-Daniels E, Beck SG, Kirby LG. Cellular correlates of anxiety in CA1 hippocampal pyramidal cells of 5-HT1A receptor knockout mice. Psychopharmacology (Berl). 2011;213:453–63.
58. Millan MJ, Lejeune F, Gobert A. Reciprocal autoreceptor and heteroreceptor control of serotonergic, dopaminergic and noradrenergic transmission in the frontal cortex: relevance to the actions of antidepressant agents. J Psychopharmacol. 2000;14:114–38.
59. Bruening S, Oh E, Hetzenauer A, Escobar-Alvarez S, Westphalen RI, Hemmings HC Jr, Singewald N, Shippenberg T, Toth M. The anxiety-like phenotype of 5-HT receptor null mice is associated with genetic background-specific perturbations in the prefrontal cortex GABA-glutamate system. J Neurochem. 2006;99:892–9.
60. Li Q, Luo T, Jiang X, Wang J. Anxiolytic effects of 5-HT_1A receptors and anxiogenic effects of 5-HT_2C receptors in the amygdala of mice. Neuropharmacology. 2012;62:474–84.
61. Toth M. 5-HT1A receptor knockout mouse as a genetic model of anxiety. Eur J Pharmacol. 2003;463:177–84.
62. García-Oscos F, Torres-Ramírez O, Dinh L, Galindo-Charles L, Pérez Padilla EA, Pineda JC, Atzori M, Salgado H. Activation of 5-HT receptors inhibits GABAergic transmission by pre-and post-synaptic mechanisms in layer II/III of the juvenile rat auditory cortex. Synapse. 2015;69:115–27.

63. Kishimoto K, Koyama S, Akaike N. Presynaptic modulation of synaptic gamma-aminobutyric acid transmission by tandospirone in rat basolateral amygdala. Eur J Pharmacol. 2000;407:257–65.
64. Palazzo E, Marabese I, de Novellis V, Rossi F, Maione S. Metabotropic glutamate receptor 7: from synaptic function to therapeutic implications. Curr Neuropharmacol. 2016;14:504–13.
65. O'Connor RM, Finger BC, Flor PJ, Cryan JF. Metabotropic glutamate receptor 7: at the interface of cognition and emotion. Eur J Pharmacol. 2010;639:123–31.
66. Jantas D, Lech T, Gołda S, Pilc A, Lasoń W. New evidences for a role of mGluR7 in astrocyte survival: possible implications for neuroprotection. Neuropharmacology. 2018;141:223–37.
67. Spooren W, Lesage A, Lavreysen H, Gasparini F, Steckler T. Metabotropic glutamate receptors: their therapeutic potential in anxiety. Curr Top Behav Neurosci. 2010;2:391–413.
68. Lesne E, Dupré E, Lensink MF, Locht C, Antoine R, Jacob-Dubuisson F. Coiled-coil antagonism regulates activity of Venus flytrap-domain-containing sensor kinases of the BvgS family. MBio. 2018;9(1):pii: e02052–17. https://doi.org/10.1128/mBio.02052-17.
69. Duebel J, Marazova K, Sahel JA. Optogenetics. Curr Opin Ophthalmol. 2015;26:226–32.
70. Deisseroth K. Optogenetics: 10 years of microbial opsins in neuroscience. Nat Neurosci. 2015;18:1213–25.
71. Nagel G, Ollig D, Fuhrmann M, Kateriya S, Musti AM, Bamberg E, Hegemann P. Channelrhodopsin-1: a light-gated proton channel in green algae. Science. 2002;296:2395–8.
72. Boyden ES, Zhang F, Bamberg E, Nagel G, Deisseroth K. Millisecond-timescale, genetically targeted optical control of neural activity. Nat Neurosci. 2005;8:1263–8.
73. Bockaert J, Bécamel C. The anxiogenic effects of SSRI are mediated by 5-HT2c receptors of the stria terminalis. Med Sci (Paris). 2017;33:87–9.

Part III

Diagnostic and Clinical Issues of Anxiety Disorders

Anxiety Disorders in the DSM-5: Changes, Controversies, and Future Directions

12

Seon-Cheol Park and Yong-Ku Kim

Introduction

Consistent with the comment that "the Catholic Church changes its pope more than the American Psychiatric Association (APA) publishes a new version of the Diagnostic and Statistical Manual of Mental Disorder (DSM)" [1], the DSM has a tendency to resist change, and its revisions are primarily based on a pragmatic attitude rather than scientific facts. This tendency is denoted as the "DSM conservative pragmatism" [2]. The DSM-I and DSM-II were published by the APA in 1952 and 1968, respectively, based on psychoanalytic and psychodynamic theories. However, based on the Rosenhan experiment findings, the anti-psychiatry movement, and public debate on the diagnosis of homosexuality as a mental disorder, psychiatric diagnoses in the DSM-I and DSM-II were criticized for its conceptual inconsistency and arbitrariness [3–5]. To overcome the limitations of the DSM-I and DSM-II, an increase in empirical trends and biological psychiatry was proposed by the "DSM-III revolution" in 1980. Therefore, the operational diagnostic systems characterized by a descriptive approach, explicit inclusion and exclusion criteria, algorithms for the criteriology, nominalist conceptualization, etiological neutrality, and quantitative approach were used in psychiatric nosology [6–8]. Additionally, in 1994, the DSM-IV succeeded the empirical trend and biological psychiatry of the DSM-III; however, the paradigm shift from categorical approach to dimensional approach was proposed to overcome the rigorously defined diagnoses in the DSM-III and DSM-IV. The DSM-5 was published by the APA in 2013, based on the combination of categorical and dimensional approaches in psychiatric nosology [9, 10].

S.-C. Park
Department of Psychiatry, Inje University Haeundae Paik Hospital, Busan, South Korea

Y.-K. Kim (✉)
Department of Psychiatry, College of Medicine, Korea University Ansan Hospital, Ansan, Kyunggi Province, South Korea
e-mail: yongku@korea.ac.kr

Y.-K. Kim (ed.), *Anxiety Disorders*, Advances in Experimental Medicine and Biology 1191, https://doi.org/10.1007/978-981-32-9705-0_12

For the DSM-5 revision process of anxiety disorders, evidence for their reconceptualization has been evaluated based on molecular genetics and neuroimaging findings in fear circuit disorders and possibilities for their simplification based on common mechanisms and meta-structures. However, the bases for the reclassification and reordering of anxiety disorders have not been sufficiently described by molecular genetics and neuroimaging studies. Moreover, consistent biological markers and processes for anxiety disorders have not been discovered. The availability of a new dimensional meta-structure has been restricted due to the evaluation of psychiatric nosology [11–13]. Thus, the anxiety disorder revisions from the DSM-IV to DSM-5 cannot be significantly affected by the paradigm shifts in psychiatric nosology. The most dominant changes in anxiety disorders have been as follows. First, because obsessive compulsive, acute stress, and post-traumatic stress disorders were excluded, the definitions of anxiety disorders have become clearer and more unified in the DSM-5. Second, selective mutism and separation anxiety disorder were reassigned to the anxiety disorder section; however, they have been included in disorders typically diagnosed in infancy, childhood, or adolescence in the DSM-IV. Thus, anxiety disorders were reordered with a development-related axis. Third, panic disorder and agoraphobia were reclassified with radical simplification and adapted to the International Classification of Diseases, tenth revision (ICD-10) convention. Thus, panic disorder and agoraphobia were changed into separate diagnoses. Fourth, in the diagnostic criteria for agoraphobia, specific phobia, and social anxiety disorder, a diagnostic restriction rule was removed that the patient had to be 18 years of age or older. In addition, the restriction of age of onset before 18 years was removed from separation anxiety disorder [13–15]. Fifth, although several revisions have been proposed to overcome adapting the categorical concept to define generalized anxiety disorder (GAD), the DSM-5 diagnostic criteria for GAD remained unchanged from the DSM-IV criteria [16]. Sixth, mixed anxiety and depressive disorder requires symptoms of both anxiety and depression are present, whereas neither are predominant in the ICD-10. Although it has been intended to include mixed anxiety and anxiety disorder in the DSM-5, it has not been seen to fruition [17]. In this chapter, the changes, controversies, and future directions in anxiety disorders in the DSM-5 are reviewed and discussed.

Panic Disorder and Agoraphobia

The diagnostic criteria for panic disorder and agoraphobia in the DSM-5 were changed primarily by separating agoraphobia from panic disorder, distinguishing agoraphobia from specific phobia, adding a 6-month duration criterion for agoraphobia, adding panic attacks as a specifier, and changing the descriptors of panic types [17–19].

The changes from the diagnoses of panic disorder with agoraphobia, panic disorder without agoraphobia, and agoraphobia without history of panic disorder to the two independent diagnoses were adapted in the DSM-5 [17–19]. According to the suggestions of Wittchen et al. [20], the separation of agoraphobia from panic disorder has been evidenced as follows: First, agoraphobia can develop without any panic symptoms. Second, agoraphobia does not always occur secondary to panic symptoms. Third, agoraphobia without panic disorder can be characterized by its significant functional limitations, symptom persistence in clinical course, and low rates of spontaneous recovery. Fourth, agoraphobia and panic disorder evidently differ in terms of incidence, gender disparity, and treatment outcome. Fifth, agoraphobia can be regarded as an independent contributor to the severity, course, and outcome of panic disorder. Sixth, panic disorder, panic disorder with agoraphobia, and agoraphobia differ in terms of temporal progression and stability; however, the interrelation between panic disorder and agoraphobia may blur the two diagnoses.

To overcome the unclear definitions of feared situations required to diagnose agoraphobia in the DSM-IV, specific criteria were introduced into the DSM-5 diagnostic criteria. The marked fear or anxiety in at least two of five situations including using public transportation, being in open spaces, being in enclosed places, standing in line or being in a crowd, and being outside of the home alone is required to diagnose agoraphobia in the DSM-5; however, the minimum of two fearful situations is poorly evidenced, thus leading to criticism. The distinguishing feature of agoraphobia from specific phobia is the "cognitive ideation," which is associated with "dysfunctional beliefs," "catastrophic misinterpretation," and "fear consequences," as defined in the DSM-5 [17–19]. Conversely, specific phobia such as claustrophobia, acrophobia, and blood-injection-injury phobia are characterized by fears of suffocation, dizziness, and fainting, respectively [21–23]. The blurred diagnostic boundaries between agoraphobia and specific phobia are criticized but can be developed in terms of cognitive ideation (feared consequence). Thus, it is speculated that the etiologies and treatment outcomes between the two disorders can be arbitrarily differed in clinical research.

A 6-month duration criterion is newly required to diagnose agoraphobia in accordance with the diagnostic criteria for other anxiety disorders, since anxiety disorders can be distinguished from transient fear reactions based on the presence of enduring fear and anxiety [24, 25]. However, the suitability of the 6-month duration criterion for agoraphobia is poorly evidenced and needs to be researched further. Moreover, the panic attack specifier can be used in the full range of mental disorders including other anxiety disorders, mood, psychotic, and substance use disorders in the DSM-5, although panic attacks have been confined to panic disorders in the DSM-IV. More universal uses of the panic attacks specifier are evidenced by an association of panic attacks with increased symptom severity, increased suicidal ideation and behavior, and diminished treatment response [26, 27].

Additionally, in terms of panic type descriptors, simplification from unexpected (uncued), situationally bound (cued), and situationally predisposed types to unexpected and expected subtypes are offered in the DSM-5. Cultural attributions, in terms of distinguishing expected from unexpected panic attacks, are newly stressed in the DSM-5, which are evidenced in a previous finding [28]. It is expected that a foundation for diagnosis and differential diagnosis of panic attacks is provided by the clear distinctions of panic types. Moreover, because the clarification that panic attacks occur in the absence of real danger was removed in the DSM-5, panic attacks can be developed in both non-anxious and anxious states, and panic onset is indicated by the abrupt change in anxiety. However, by removal of this clarification, misclassifying the fear responses to actual threat as panic attacks may be increased in clinical practices.

Social Anxiety Disorder

The primary changes in the DSM-5 criteria for social anxiety disorder were social phobia was renamed social anxiety disorder, social situations associated with primary fear were broadened, the importance of sociocultural context in determining whether an anxious response to a social situation is out of proportion to the actual threat, diagnosis of social anxiety disorder in the context of a medical condition, and revisions in variations of social anxiety disorder [29].

Social anxiety disorder, rather than social phobia, was conceptualized as a primary name in the DSM-5, although social anxiety disorder was only an alternative name for social phobia in the DSM-IV. This renaming was caused by a mistaken impression in mental health and primary care settings that the disorder may be neither frequent nor impairing in individuals with social phobia. Moreover, the sense of pervasiveness and impairment can be conveyed more efficiently by social anxiety disorder rather than social phobia [30]. A study on the influences of the name of the disorder on perceived need for treatment in a sample of 806 community residents found that recommending a person seek treatment is more frequently labeled as social anxiety disorder rather than social phobia [31].

Additionally, the social situations are broadened from humiliating or embarrassing to being negatively evaluated by others in the changes from the DSM-IV to DSM-5. This broadening is partly supported by a proposal that not only humiliation and embarrassment but also rejection and offending others can be main sources for fear in individuals with social anxiety disorder [29]. The broadening is evidenced by an association between anger suppression and fear of negative evaluation in persons with social anxiety disorder, greater ambivalence about emotional expression, and more negative beliefs about emotional expression in persons with high social anxiety rather than those with low anxiety [32, 33].

In the DSM-5, excessive or unreasonable fear in social anxiety disorder is characterized by objective evaluation rather than subjective report. This judgment based on objective evaluation is evidenced by the well-replicated findings in individuals with social anxiety that the quality of behavioral performance is underestimated and the likelihood of negative outcomes in social situations is overestimated [34–36]. Furthermore, to evaluate the disproportion between fear of anxiety and actual risk, the sociocultural context is recommended as the judging method in the DSM-5. Cultural syndromes are defined as clusters or groups of co-occurring and relatively invariant symptoms specific to a particular culture and are represented by the influences of culture on the presentation and course of psychiatric disorders in the DSM-5. Thus, the specific influences of culture on anxiety disorders can be significant in the context of cultural syndromes [37]. Taijin kyofusho denotes a Japanese culture-specific syndrome, which is characterized by the extreme fear of embarrassing oneself. A person who fears that he or she would offend others by emitting offensive odors, blushing, staring inappropriately, and presenting an improper facial expression or physical deformity has been categorized as the offensive subtype of taijin kyofusho [38–40]. The relationship between members of social organizations is regarded as an important factor to define taijin kyofusho and other cultural syndromes, which are described in the DSM-5 [41]. Thus, it is speculated that social anxiety disorder and other emotional disorders are contributed to by a person's cultural orientation and cultural norms [42].

Distinguishing social anxiety disorder from social anxiety symptoms that are the consequence of a medical disorder has been an important boundary issue in psychiatric nosology, since social fear can be caused by medical disorders including hyperhidrosis or tremors. In the DSM-IV, the diagnosis of social anxiety disorder was given to persons whose symptoms could not be attributed to another psychiatric or medical condition, with the social fear unrelated to another condition. Conversely, in the DSM-5, the diagnosis of social anxiety disorder in the presence of a comorbid medical condition is permitted, therefore recognizing the social fear, anxiety, or avoidance is unrelated to the medical disorder or is excessive. This change partly supported the findings that individuals with Parkinson's disease, hyperhidrosis, hyperkinesis, obesity, essential tremors, or psoriasis present clinically significant and impairing social anxiety [43–48].

In the DSM-5, the "generalized" specifier is replaced by a "performance-only" specifier, which is applied to fear related to speaking or performing in public. To overcome the limitations of the generalized specifier on the quantity, rather than content, in the DSM-IV, the performance-only specifier was newly added in the DSM-5. The necessity for the performance-only specifier is supported by findings from patients who suffer exclusively from performance fears, which are considered a distinct subtype. Additionally, some patients with social anxiety disorder have reported only performance fears [25, 49]. The response to beta-adrenergic blocking agents is more frequently presented in some patients with performance-only social fears rather than patients with other types of social anxiety disorder [50].

Separation Anxiety Disorder

Separation anxiety disorder was reassigned into anxiety disorders in the DSM-5, although it had been included in disorders typically first diagnosed in infancy, childhood, or adolescence in the DSM-IV. This was accompanied by lifting the age of onset restriction and revising symptom descriptors to facilitate their application to not only children and adolescents but also adults [51]. These changes are supported by findings that early separation anxiety is a genetic risk factor for a range of anxiety disorders in adulthood [52–56]. This research on adult separation anxiety disorder has demonstrated core findings, including consistent identification of separation anxiety disorder in adults, the relatively high prevalence (20–40%) of separation anxiety disorder in clinical populations, similar comorbidity patterns to other subtypes of anxiety, relatively common onset of separation anxiety disorder in adulthood, relatively high prevalence of adult separation anxiety disorder among women, familial pattern of separation anxiety disorder, and exposure to overprotective parenting as childhood risk factor [51, 57–61]. Thus, by reducing the differences between childhood and adulthood categories by lifting the age of onset restriction and reassignment into anxiety disorders in separation anxiety disorder, more symmetrical classifications of anxiety subtypes have been created in the DSM-5.

Selective Mutism

Similar to separation anxiety disorder, selective mutism was reassigned into anxiety disorders in the DSM-5; however, the symptom descriptors are slightly changed. This reassignment of selective mutism into anxiety disorders is partly supported by the following findings: First, anxiety is identified as a prominent feature of selective mutism. Anxiety-related symptoms and behaviors are displayed by children with selective mutism. Second, many factors related to the origins of other childhood anxiety disorders contribute to the etiology of selective mutism. Third, children with selective mutism are effectively treated with behavioral and cognitive-behavioral therapies and pharmacotherapy with selective serotonin reuptake inhibitors [62].

Controversies and Future Directions

The changes from the DSM-IV to DSM-5 inevitably contributed to lowering the diagnostic threshold in anxiety disorders including agoraphobia, social anxiety disorder, and other disorders. Remarkably, in a population-based survey of 136,357 adult respondents, the World Mental Health Initiative has presented that the lifetime and 12-month prevalence of agoraphobia per the DSM-5 are 1.5% and 1.0%, respectively, whereas those of DSM-IV agoraphobia are 1.4% and 0.9%, respectively. Of those who had met the criteria for agoraphobia, 57.1% were diagnosed based on both the DSM-IV and DSM-5 criteria, whereas 24.2% met for the DSM-5 only, and 18.8% met for the DSM-IV only. In accordance with lowering the diagnostic threshold, new cases of agoraphobia have been identified by the diagnostic criteria in the DSM-5 [63]. In addition, increased prevalence rates of social anxiety disorder have been presented in the general population due to the criteria changes in the DSM-5 [64, 65].

Defining anxiety disorders is most often in terms of the categorical versus dimensional approaches. Several studies have demonstrated that GAD is more closely related to depressive disorders than other anxiety disorders including panic disorder, agoraphobia, social phobia, and specific phobia. Because of a close link between GAD and major depressive disorder at the genetic level, it is proposed that GAD cannot be appropriately defined in the dichotomy between anxiety and depression and GAD should be classified within depressive disorders. However, the proposed associated symptom criterion changes for GAD have been dropped, and the symptom descriptors for GAD were unchanged in the DSM-5. The arbitrariness in defining the diagnoses of anxiety disorders, such as GAD, remains in the DSM-5. Thus, it is proposed that transdiagnostic specifiers and dimensional assessment tools may be used as alternative methods to overcome the restrictions for the dichotomous views of the categorical and dimensional approaches in the DSM-5 diagnostic criteria for anxiety disorders [16, 66–68].

Cultural influences are regarded as a significant factor for clinical presentations of panic attacks in the DSM-5. Additionally, the cross-cultural presentations of panic attacks may be influenced by the cultural syndromes including khyal attacks

(Cambodian cultural syndrome involving dizziness, tinnitus, and neck soreness), trung gio attacks (Vietnamese cultural syndrome involving headaches), and ataque de nervios (Latin cultural syndrome involving trembling, uncontrollable screaming, crying, aggressive or suicidal behavior, and depersonalization or derealization). Moreover, the sociocultural context is proposed as a defining factor for the disproportion between the fear of anxiety and the actual risk in the DSM-5 criteria for social anxiety disorder [15]. It is speculated that the clinical significance of cultural influences on the definitions and presentations of anxiety disorders may reveal the limitations of categorical concepts.

Conclusions

The DSM-5 was revised from the DSM-IV, based on the combination of categorical and dimensional approaches. Due to the removal of obsessive-compulsive, acute stress, and post-traumatic stress disorders, clearness and unification of diagnostic concepts for anxiety disorders have increased. The symmetrical classification of anxiety subtypes in general has increased by reassigning separation and selective mutism as anxiety disorders and lifting the age of onset restrictions in agoraphobia, specific phobia, social anxiety disorder, and separation anxiety disorder. Additionally, the development-related axis is stronger in the DSM-5 criteria for anxiety disorders. Moreover, based on accumulated evidence, agoraphobia has been separated from panic disorder, and sociocultural context is regarded as a significant factor for definitions and presentations of anxiety disorders. However, the DSM-5 criteria for anxiety disorders have several debates as follows: Lowering the diagnostic thresholds is shown in the DSM-5 anxiety disorders including agoraphobia and social anxiety disorder. Partly under the influence of the categorical approach, GAD is unchanged and mixed anxiety and depressive disorder in the DSM-5. To overcome the listed limitations in the DSM-5 criteria for anxiety disorder, further studies for alternative methods including transdiagnostic specifiers and assessment tools may be necessary.

References

1. Adam D. Mental health: on the spectrum. Nature. 2013;496(7446):416–28.
2. Sadler JZ. Considering the economy of DSM alternatives. In: Paris J, Phillips J, editors. Making the DSM-5: concepts and controversies. New York: Springer; 2013. p. 21–38.
3. Shorter E. The history of DSM. In: Paris J, Phillips J, editors. Making the DSM-5: concepts and controversies. New York: Springer; 2013. p. 3–20.
4. Ghaemi SN. The perils of open-mindedness: Adolf Meyer's psychobiology. In: Ghaemi SN, editor. The rise and fall of the biopsychosocial model: Reconciling art & science in psychiatry. Baltimore: The Johns Hopkins University Press; 2010. p. 3–11.
5. Park S-C. Karl Jaspers' general psychopathology (Allgemeine Psychopathologie) and its implication for the current psychiatry. Psychiatry Investig. 2019;16:99–108.
6. Engstrom EJ, Kendler KS. Emil Kraepelin: icon and reality. Am J Psychiatry. 2015;172:1190–6.
7. Kendler KS, Engstrom EJ. Criticisms of Kraepelin's psychiatric nosology: 1896–1927. Am J Psychiatry. 2018;175:316–26.

8. Sass H, Karl Jaspers VU. Hierarchical principle and current psychiatric classification. In: Stanghellini G, Fuchs T, editors. One century of Karl Jaspers' general psychopathology. Oxford: Oxford University Press; 2013. p. 185–207.
9. Clarke DE, Narrow WE, Regier DA, Kuramoto SJ, Kupfer DJ, Kuhl EA, Greiner L, Kraemer HC. DSM-5 field trials in the United States and Canada, part I: study design, sampling strategy, implementation, and analytic approaches. Am J Psychiatry. 2013;170:43–58.
10. Kim Y-K, Park S-C. Challenges and strategies for treatment-resistant psychiatric disorders. In: Kim Y-K, editor. Research methods and interventions in psychiatry: risk factors, biology, and management. New York: Springer Nature; 2019. p. 87–96.
11. Andrews G, Charney DS, Sirovatka PJ, Stress-induced RDA. Fear circuitry disorders: refining the research agenda for DSM-V. Washington, DC: American Psychiatric Association; 2009.
12. Andrews G, Goldberg DP, Krueger RF, Carpenter WT, Hyman SE, Sachdev P, Pine DS. Exploring the feasibility of a meta-structure for DSM-V and ICD-11: could it improve utility and validity? Psychol Med. 2009;39:1993–2000.
13. Witthchen HU, Heinig I, Beesdo-Baum K. Anxiety disorders in DSM-5: an overview on changes in structure and content. Nervenarzt. 2014;85:548–52.
14. Phillips KA, Stein DJ, Rauch SL, Hollander E, Fallon BA, Barsky A, Fineberg N, Mataix-Cols D, Ferrao YA, Saxena S, Wilhelm S, Kelly MM, Clark LA, Pinto A, Bienvenu OJ, Farrow J, Leckman J. Should an obsessive-compulsive spectrum grouping of disorders be included in DSM-V? Depress Anxiety. 2010;27:528–55.
15. Mori N, Sugiyama T, Iwata Y. DSM-5 reference book for clinicians. Tokyo: Nippon Hyoron Co.; 2014. p. 86–90.
16. Starcevic V, Portman ME. The status quo as a good outcome: how the DSM-5 diagnostic criteria for generalized anxiety disorder remained unchanged from the DSM-IV criteria. Aust N Z J Psychiatry. 2013;47:995–7.
17. Bandelow B. Comparison of the DSM-5 and ICD-10: panic and other anxiety disorders. CNS Spectr. 2017;22:404–6.
18. Asmundson GJG, Taylor ST, Smits JAJ. Panic disorders and agoraphobia: an overview and commentary on DSM-5 changes. Depress Anxiety. 2014;31:480–6.
19. Park S-C, Kim Y-K. A novel bio-psychosocial-behavioral treatment model of panic disorder. Psychiatry Investig. 2019;16:4–15.
20. Wittchen HU, Gloster AT, Beesdo-Baum K, Fava GA, Craske MG. Agoraphobia: a review of the diagnostic classificatory position and criteria. Depress Anxiety. 2010;27:113–33.
21. Rachman S, Taylor S. Analyses of claustrophobia. J Anxiety Disord. 1993;7:281–91.
22. Huweler R, Kandil FI, Alpers GW, Gerlach AL. The impact of visual flow stimulation on anxiety dizziness, and body sway in individuals with and without fear of heights. Behav Res Ther. 2009;47:345–52.
23. Kleinknecht RA, Thorndike RM, Walls MM. Factorial dimensions and correlates of blood, injury, injection and related medical fears: cross validation of the medical fear survey. Behav Res Ther. 1996;34:323–31.
24. Roemer L, Molina S, Borkovec TD. An investigation of worry content among generally anxious individuals. J Nerv Ment Dis. 1997;185:314–9.
25. Stein MB, Walker JR, Forde DR. Public-speaking fears in a community sample: prevalence, impact on functioning, and diagnostic classification. Arch Gen Psychiatry. 1996;53:169–74.
26. Craske MG, Kircanski K, Epstein A, Wittchen HU, Pine DS, Lewis-Fernández R, Hinton D, DSM-V Anxiety; OC Spectrum; Posttraumatic and Dissociative Disorder Work Group. Panic disorder: a review of DSM-IV panic disorder and proposals for DSM-V. Depress Anxiety. 2010;27:93–112.
27. Goodwin RD, Pine DS. Respiratory disease and panic attacks among adults in the United States. Chest. 2002;122:645–50.
28. Lewis-Fernandez R, Hinton DE, Laria AJ, Patterson EJ, Hofmann SG, Craske MG, Stein DJ, Asnaai A, Liao B. Culture and the anxiety disorders: recommendations for DSM-V. Depress Anxiety. 2010;27:212–29.

29. Heimber RG, Hofmann SG, Liebowitz MR, Schneider FR, Smits JAJ, Stein MB, Hinton DE, Craske MG. Social anxiety disorder in DSM-5. Depress Anxiety. 2014;31:472–9.
30. Liebowitz MR, Heimberg RG, Fresco DM, Travers J, Stein MB. Social phobia or social anxiety disorder: What's in a name? Arch Gen Psychiatry. 2000;57:191–2.
31. Bruce LC, Heimberg RG, Coles ME. Social phobia versus social anxiety disorder: effect of disorder name on recommendation for treatment. Am J Psychiatry. 2012;169:538.
32. Erwin BA, Heimberg RG, Scheier FR, Liebowitz MR. Anger experience and expression in social anxiety disorder: pretreatment profile and predictors for attribution and response to cognitive behavioral treatment. Behav Ther. 2003;34:331–50.
33. Spokas M, Luterek JA, Heimberg RG. Social anxiety and emotional inhibition: the mediating role of beliefs. J Behav Ther Exp Psychiatry. 2009;40:283–91.
34. Rapee RM, Lim L. Discrepancy between self- and observer ratings of performance in social phobics. J Abnorm Psychol. 1992;101:728–31.
35. Stopa L, Clark DM. Cognitive processes in social phobia. Behav Res Ther. 1993;21:255–67.
36. Foa EB, Franklin ME, Perry KJ, Herbert JD. Cognitive biases in generalized social phobia. J Abnorm Psychol. 1996;105:433–9.
37. Hinton DE, Park L, Hsia C, Hofmann S, Pollack MH. Anxiety disorder presentations in Asian populations: a review. CNS Neurosci Ther. 2009;15:295–303.
38. Kleinknecht RA, Dinnel DL, Kleinknecht EE, Hiruma N, Harada N. Cultural factors in social anxiety: a comparison of social phobia symptoms and Taijin kyofusho. J Anxiety Disord. 1997;11:157–77.
39. Choy Y, Schneider FR, Heimberg RG, Oh KS, Liebowitz MR. Features of the offensive subtype of Taijin-Kyofu-Sho in US and Korean patients with DSM-IV social anxiety disorder. Depress Anxiety. 2008;25:230–40.
40. Takahashi T. Social phobia syndrome in Japan. Compr Psychiatry. 1989;30:45–52.
41. Argyle M. Rules for social relationships in four cultures. Aust J Psychol. 1986;38:309–18.
42. Hofmann SG, Asnaani A, Hinton DE. Cultural aspects in social anxiety and social anxiety disorder. Depress Anxiety. 2010;27:1117–27.
43. Kummer A, Cardoso F, Teixeira AL. Frequency of social phobia and psychometric properties of the Liebowitz Social Anxiety Scale in Parkinson's disease. Mov Disord. 2008;23:1739–43.
44. Schneier FR, Heimberg RG, Liebowitz MR, Chemlminski I, Young D, O'Brien E, Zimmerman M. Social anxiety and functional impairment in patients seeking surgical evaluation for hyperhidrosis. Compr Psychiatry. 2012;53:1181–6.
45. Ozel-Kizil ET, Akbostanci MC, Ozguven HD, Atbasoglu EC. Secondary social anxiety in hyperkinesis. Mov Disord. 2008;23:641–5.
46. Dalrymple KL, Galione J, Hrabosky J, Chelminski I, Young D, O'Brien E, Zimmerman M. Diagnosing social anxiety disorder in the presence of obesity: implications for a proposed change in DSM-5. Depress Anxiety. 2011;28:377–82.
47. Schneider FR, Barnes LF, Albert SM, Louis ED. Characteristics of social phobia among persons with essential tremor. J Clin Psychiatry. 2001;62:367–72.
48. Schneider G, Heuft G, Hockmann J. Determinants of social anxiety and social avoidance in psoriasis outpatients. J Eur Acad Dermatol Venereol. 2013;27:383–6.
49. Stein MB, Laine JT, Walker JR. Social phobia symptoms, subtypes, and severity: findings from a community survey. Arch Gen Psychiatry. 2000;57:1046–52.
50. Blote AW, Kint MJW, Miers AC, Westernberg MP. The relation between public speaking anxiety and social anxiety: a review. J Anxiety Disord. 2009;23:305–13.
51. Silove D, Ress S. Separation anxiety disorder across the lifespan: DSM-5 lifts age restriction on diagnosis. Asian J Psychiatry. 2014;11:98–101.
52. Flakierska-Praquin N, Lindstrom M, Gillberg C. School phobia with separation anxiety disorder: a comparative 20- to 29-year follow-up study of 35 school refusers. Compr Psychiatry. 1997;38:17–22.
53. Aschenbrand SG, Kendall PC, Webb A, Safford SM, Flannery-Schroeder E. Is childhood separation anxiety disorder a predictor of adult panic disorder and agoraphobia? A seven-year longitudinal study. J Am Acad Child Adolesc Psychiatry. 2003;42:1478–85.

54. Bruckl TM, Wittchen HU, Hofler M, Pfister H, Schneider S, Lieb R. Childhood separation anxiety and the risk of subsequent psychopathology: results from a community study. Psychother Psychosom. 2006;76:47–56.
55. Lewinsohn PM, Holm-Denoma JM, Small JW, Seeley JR, Joiner TE Jr. Separation anxiety disorder in childhood as a risk factor for future mental illness. J Am Acad Child Adolesc Psychiatry. 2008;47:548–55.
56. Kossowsky J, Pfaltz MC, Schneider S, Taeymans J, Locher C, Gaab J. The separation anxiety hypothesis of panic disorder revisited: a meta-analysis. Am J Psychiatry. 2013;170:768–81.
57. Bogels SM, Knappe S, Clark LA. Adult separation anxiety disorder in DSM-5. Clin Psychol Rev. 2013;33:663–74.
58. Pini S, Abelli M, Shear K, Cardini A, Lari L, Gesi C, Muti M, Calugi S, Galderisi S, Troisi A, Bertolino A, Cassano G. Frequency and clinical correlates of adult separation anxiety in a sample of 508 outpatients with mood and anxiety disorders. Acta Psychiatr Scand. 2010;122:40–6.
59. Manicavasagar V, Marnane C, Pini S, Abelli M, Rees S, Eapen V, Silove D. Adult separation anxiety disorder: a disorder comes of age. Curr Psychiatry Rep. 2010;12:290–7.
60. Silove DM, Marnane CL, Wagner R, Manicavasagar VL, Rees S. The prevalence and correlates of adult separation anxiety disorder in an anxiety clinic. BMC Psychiatry. 2010;10:21.
61. Silove D, Momartin S, Marnane C, Steel Z, Manicavasagar V. Adult separation anxiety disorder among war-affected Bosnian refugees: comorbidity with PTSD and associations with dimensions of trauma. J Trauma Stress. 2010;23:169–72.
62. Muris P, Ollendick TH. Children who are anxious in silence: a review on selective mutism, the new anxiety disorder in DSM-5. Clin Child Fam Psychol Rev. 2015;18:151–69.
63. Roest AM, de Vries YA, Lim CCW, Wittchen HU, Stein DJ, Adamowski T, Al-Hamzawi A, Bromet EJ, Viana MC, de Girolamo G, Demyttenaere K, Florescu S, Gureje O, Haro JM, Hu C, Karam EG, Caldas-de-Almeida JM, Kawakami N, Lépine JP, Levinson D, Medina-Mora ME, Navarro-Mateu F, O'Neill S, Piazza M, Posada-Villa JA, Slade T, Torres Y, Kessler RC, Scott KM, de Jonge P, World Mental WHO, Health Survey Collaborators. A comparison of DSM-5 and DSM-IV agoraphobia in the World Mental Health Surveys. Depress Anxiety. 2019;36(6):499–510. https://doi.org/10.1002/da.22885.
64. Crome E, Grove R, Baillie AJ, Sunderland M, Teesson M, Slade T. DSM-IV and DSM-5 social anxiety disorder in the Australian community. Aust N Z J Psychiatry. 2015;49:227–35.
65. Karlsson B, Sistrom R, Ostling S, Waern M, Borjesson-Hanson A, DSM-IV SI. DSM-5 prevalence of social anxiety disorder in a population sample of older people. Am J Geriatr Psychiatry. 2016;24:1237–45.
66. Park S-C, Kim Y-K. Depression in DSM-5: changes, controversies and future direction. In: Kim Y-K, editor. Understanding depression: Volume 2. Clinical manifestations, diagnosis and treatment. New York: Springer Nature; 2018. p. 3–14.
67. Starcevic V. Anxiety disorders no more? Australas Psychiatry. 2008;16:317–21.
68. Comer JS, Pincus DB, Hofmann SG. Generalized anxiety disorder and the proposed associated symptom criterion change for DSM-5 in a treatment-seeking sample of anxious youth. Depress Anxiety. 2012;29:994–1003.

Biological and Clinical Markers to Differentiate the Type of Anxiety Disorders

13

Fiammetta Cosci and Giovanni Mansueto

Anxiety Disorders and the Use of Markers in Psychiatry

A biomarker is an objectively measured characteristic used as an indicator of normal biologic and pathologic processes or biological responses to specific treatments [1, 2]. Some authors suggested that mental diseases might be classified according to objective markers [1–4] assuming that they may clarify the aetiology of psychiatric diseases, confirm a diagnosis, identify individuals at risk, determine the severity of mental illness, and predict the course of the illness [3–5]. The use of markers is claimed to lead to personalized psychiatric treatments informing about the type, timing, and course of interventions and monitoring the clinical response to them [3–5]. Markers, to be valid, must have at least an acceptable level of sensitivity, specificity, and predictivity [4]. They were classified as (a) susceptibility/risk markers, which indicate the potential for developing a disease in an individual who, from a clinical standpoint, does not have that disease or clinically relevant medical condition; (b) diagnostic markers, which are used to identify ill subjects or a subset of the disease; (c) prognostic markers, which allow to identify the likelihood of a clinical event, disease recurrence, or progression in ill subjects or in medical condition of clinical interest; and (d) predictive markers, which identify those who are more likely to have a favorable or unfavorable response to treatment [6–8].

We will here illustrate the markers proposed in the literature for anxiety disorders as classified by the DSM-5 [9], that is, panic disorder (PD), agoraphobia (AGO), generalized anxiety disorder (GAD), specific phobia (SP), social anxiety disorder (SAD), separation anxiety disorder (SepAD), and selective mutism (SM).

F. Cosci (✉) · G. Mansueto
Department of Health Sciences, University of Florence, Florence, Italy

Maastricht University Medical Center, Department of Psychiatry & Psychology, School for Mental Health & Neuroscience, Maastricht, The Netherlands
e-mail: fiammetta.cosci@unifi.it

Y.-K. Kim (ed.), *Anxiety Disorders*, Advances in Experimental Medicine and Biology 1191, https://doi.org/10.1007/978-981-32-9705-0_13

Biological Markers in Panic Disorder

Structural/Activity Changes in Brain Regions

Structural or activity changes in the amygdala, hippocampus, parahippocampal gyri, brainstem nuclei, insular cortices, left inferior frontal gyrus, dorsomedial prefrontal cortex, and left caudatum have been proposed as biomarkers of panic disorder given that significant differences in the volume or in functional activity of these areas were observed in PD patients and not in controls [10, 11]. Similarly, the cerebral blood flow in the left occipital cortex and the serotonin (5-TH) were suggested as biomarkers given that a higher cerebral blood flow and lower 5-TH plasma levels were found in PD patients than in controls ($P < 0.05$) [10–12]. Some evidences also reported a significant association between noradrenergic hyperactivation and symptoms in PD patients but not in patients with other psychiatric disorders (i.e., GAD, obsessive compulsive disorder, depression, schizophrenia) [10, 11, 13]. One study showed that variations in serum levels of tetranectin (TN) and creatine kinase MB (CK MB) might be associated with the severity of PD since that elevated serum of TN and CK MB was found to predict the development of depressive episodes in PD patients within 2 years [14]. There is still little evidence of neuropsychological impairment in PD patients when compared to healthy controls or to subjects with anxiety disorders other than panic [15].

Genetics

Monoamine oxidase A (MAOA) gene and genetic variants conceivably related to lower 5-HT neurotransmission might be putative biomarkers of PD [10, 16, 17]. In PD patients, the serotonin-transporter-linked polymorphic region (5-HTTLPR) L-allele, the HTR1A -1019C allele, and the catechol-O-methyltransferase (COMT) 158Val allele were found associated with a better response to antidepressant treatment and to exposure-based cognitive behavior therapy [10].

Respiratory Patterns

An aberrant respiratory regulation was proposed as a core feature of PD given that hyperventilation, higher rates of sighs, and apnea in respiratory patterns were found in PD patients and not in healthy controls or in patients with social phobia or GAD [18, 19].

Hearth Rate Variability, Cardiac Vagal Tone, Blood Cells, and Blood Stem Cells

Hearth rate variability (HRV), i.e., the extent to which the interval between beats varies with time, is a core feature of cardiovascular diseases [20] that was found positively associated with panic attacks in clinical and nonclinical samples [10, 21, 22]. PD patients have a higher risk of cardiovascular disease than healthy subjects, and the prevalence of PD among cardiac outpatients ranges from 4 to 12.5% [23].

A recent study [24] suggests that the variations on the cardiac vagal tone (CVT) may be a putative biomarker of PD since changes in CVT levels might be a useful tool to predict a favorable/unfavorable response to the exposure-focused therapy. It has been observed that PD patients with lower CVT are more likely to show residual symptoms after completing exposure-focused therapy [24].

Blood cells and peripheral blood stem cells were suggested as further biomarkers since higher platelet distribution width, red cell distribution width, and mean platelet volume were reported in PD patients and not in healthy controls [25–27].

Hypothalamic–Pituitary–Adrenal Axis

McEwen [28–30] introduced the concept of allostasis to describe the ability of the organism to detect environmental (external) and physiological (internal) changes and to activate adaptive responses by which homeostasis is achieved through change. Three highly integrated systems, i.e., the nervous, the immune, and endocrine systems, were assumed to mediate the allostatic process leading to a short-term adaptation in the face of environmental challenges [28–30]. In response to chronic stress, the prolonged and sustained activation of the allostatic systems (i.e., allostatic load-overload) might lead to detrimental physiological consequences [28–30]. When stress is persistent, the negative feedback system that serves to dampen the HPA axis activation is impaired inducing chronic cortisol release [31, 32], flattering circadian variation, and heightening daily cortisol secretion [28, 31].

Among PD patients, the role of HPA axis activity seems to be still unclear. Some evidences showed higher cortisol secretion during panic attacks compared to the same individuals at comparable times of panic-free days [33, 34]. Other studies showed unchanged or lack of HPA axis activation during spontaneous attacks [33]. Furthermore, while some studies reported higher cortisol secretion in PD patients than in healthy controls, others did not [10, 11, 32, 33, 35, 36]. These mixed findings might be due to several issues: (a) the HPA axis might be activated at the beginning of the panic attack due to an arousal in reaction to novelty cues and/or to anxiety anticipatory about other attacks and/or the avoidance of situations where having an attack might be embarrassing [32, 36, 37]; (b) after the acute phase of a panic attack, the HPA axis activation normalizes due to a successful habituation to the repeated experiences of panic [32, 36, 38]; and (c) the association between HPA axis activity and PD might be biased by comorbidities since higher cortisol levels were found also in depressed patients [39, 40].

Classification of Putative Biomarkers for Panic Disorder

Structural or activity changes in brain regions increased flow of the cerebral blood level in the left occipital cortex, monoamine oxidase gene and genetic variants (e.g., lower 5-HT neurotransmission), aberrant respiratory regulation, and variation in the HRV and in the peripheral blood stem cell distribution were observed in PD patients but not in controls [10, 11, 18, 19, 25]; thus they are susceptibility and/or diagnostic biomarkers which can be used to identify subjects with higher vulnerability for developing PD from healthy controls. On the contrary, due to the mixed results, the identification of HPA axis activation and neurocognition as susceptibility and/or diagnostic biomarkers of PD is still far.

Since longitudinal studies showed that the changes in the HPA axis activity, HRV, and serum levels of TN and CK MB predict more severe symptoms, increased vulnerability to depressive disorders and to cardiovascular diseases, and lower social and work functioning [14, 40–42], these biomarkers can be prognostic ones.

Serotonin-transporter-linked polymorphic region L-allele, the HTR1A –1019C allele, the COMT 158Val allele, and the CVT might be predictive biomarkers since they can allow to identify PD patients with a positive/negative response to treatment [10, 24].

Table 13.1 shows a possible classification of PD biomarkers. To be noted that the majority of the studies included PD subjects with agoraphobia, thus the biomarkers illustrated might be biased by the comorbidity with agoraphobia [43].

Biological Markers in Agoraphobia

To date, research aimed at identifying biomarkers of agoraphobia are lacking because a minority of studies explored patients with AG without PD [10, 11, 44]. Neurobiological and genetic studies proposed the variation in urinary lysosomal enzyme N-acetyl-b-glucosaminidase (NAG) levels as candidate biomarker of AG since NAG levels were found higher in agoraphobic patients than in PD patients (mean ± SD, 22.0 ± 21.0 vs 9.7 ± 8.0; $P < 0.05$) [45]. The HPA axis activity was

Table 13.1 Putative biomarkers for panic disorder

	Susceptibility/ risk marker	Diagnostic marker	Prognostic marker	Predictive marker
Structural or activity changes in brain regions[a]		●		
Cerebral blood level in the left occipital cortex	●	●		
Molecular serum panel			●	
Neurocognition	●[b]	●[b]		
MAOA, 5-HT	●	●		
5-HTTLPR L-allele, HTR1A -1019C allele, COMT 158Val allele				●
Aberrant respiratory regulation	●	●		
Heart rate variability–cardiac vagal tone	●	●	●	●
Blood cells and peripheral blood stem cells	●	●		
HPA axis dysregulation/ cortisol secretion	●	●[b]	●	

Note: *MAOA* monoamine oxidase, *HTR1A, HTR2A* serotonin receptor gene, *TPH2* tryptophan hydroxylase 2 genes, *CRH1* corticotropin-releasing hormone type 1 receptor, *CCK* neuropeptide cholecystokinin, *5-HTTLPR* serotonin-transporter-linked polymorphic region, *COMT* catechol-O-methyltransferase, *HPA* hypothalamic–pituitary–adrenal

[a]amygdala, hippocampus, parahippocampal gyri, brainstem nuclei, insular cortices, left inferior frontal gyrus, dorsomedial prefrontal cortex, and left caudatum

[b]is not fully confirmed

proposed as a biomarker since a higher cortisol secretion was found in agoraphobic PD patients if compared to PD patients without AG and if compared with healthy controls [46, 47].

Biological Markers in Generalized Anxiety Disorder

Structural/Activity Changes in Brain Regions

Structural changes in the amygdala, hippocampus, midbrain, dorsolateral prefrontal cortex (DPC), and basal ganglia were proposed as biomarkers of generalized anxiety disorder [10, 11, 48] since meaningful anatomical changes and significant differences in the gray or white matter volume of these regions were found in GAD patients and not in healthy controls [10, 11, 48]. However, changes in hippocampus/amygdala volumes are associated with an increased or with a decreased severity of symptoms [48]. One study showed that changes in the left hippocampal volume were positively associated with anxiety [49], while in another study, hippocampus and amygdala volumes were not found correlated with worry or intolerance of uncertainty [50].

Similarly, the functional activity changes in the prefrontal cortex, middle temporal gyrus, amygdala, and basal ganglia [10, 48, 51, 52] might be biomarkers for GAD, but results are mixed due to a heterogeneity of methods [51]. The most consistent results suggested that patients with GAD report weaker functional activity in the prefrontal cortex, anterior cingulate cortex, and middle temporal gyrus if compared to healthy controls [10, 48, 51, 52]. Moreover, lower reactivity in the anterior cingulate cortex and in the amygdala was found to predict a positive response to treatment with venlafaxine in terms of reduction of worry and anxiety severity [53, 54].

Magnetic resonance imaging and positron emission tomography studies proposed that the variation in N-acetylaspartate/creatine and dopamine transporter (DAT) levels might be biomarkers of GAD [10, 48]. Negative correlations between the low level of choline/NAA in the dorsolateral PFC and anxiety severity, higher ratio of N-acetylaspartate/creatine in the right DPC, and lower levels of DAT in the striatum were observed in GAD patients and not in healthy controls [10, 48].

The error-related negativity, i.e., a burst of electrical activity that appears as a sharp negative-going peak in the event-related potential waveform at fronto-central sites elicited when subjects make mistakes on laboratory-based reaction time tasks [55], was proposed as biomarker since a greater activation was reported in GAD patients than in healthy controls [55].

Some evidences showed that GAD patients, compared to healthy controls, score lower on cognitive domains such as attention, information processing speed, working memory, inhibition, problem-solving, and immediate and delayed memory [11, 56, 57]. It was hypothesized that GAD patients suffer from a decrement in target stimulus processing due to an attentional bias toward emotional distracters [56]. However, studies exploring neurocognitive impairments in GAD patients are still rare [11, 56]; neurocognitive performance might be influenced by state and trait anxiety [58]; and in those prone to evaluative anxiety, the results of neurocognitive test might underestimate the effective cognitive capacity or can be misinterpreted because of an impaired neurocognitive functioning [58].

Genetics

Some evidences proposed the polymorphism of the brain-derived neurotrophic factor (BDNF) gene, the catechol-O-methyltransferase gene, the MAOA 941T allele, and the 5-HTT gene-linked functional polymorphic region as possible biomarkers since they are positively associated with GAD [48]. However, these findings might be biased by the rate of psychiatric comorbidities. For instance, within the 5-HT1A gene although the G(−1019) allele carriers were found to be more frequent among GAD patients than among healthy controls (OR = 2.54; 95% CI = 1.28–4.86; p = 0.003), the association was no longer significant after adjusting for the presence of major depression (OR = 1.97; 95% CI = 0.99–3.91; P = 0.05) [59].

Heart Rate Variability

Lower heart rate variability was found in GAD patients when compared to healthy controls [60]. However, some studies did not find significant differences when HRV was evaluated during experimental task inducing worry, imagery, or hyperventilation [61–64].

Hypothalamic–Pituitary–Adrenal Axis Dysregulation

Compared to healthy controls, both increased and decreased HPA axis activities was observed in GAD patients [10, 11, 32, 48]. Variations in cortisol levels in GAD patients seem not to be associated with the medication in use [65]. However, whether the hyper–hypo activation of the HPA axis in GAD patients might be influenced by the co-occurrence of other psychiatry disorders is unclear. Indeed, higher cortisol awakening was found among those with GAD and major depression and not among those with GAD without depression [39]. On the other hand, cortisol changes seem associated with anxiety and worry [65, 66] since GAD patients with and without comorbid depression or other anxiety disorders and presenting increased levels of worry and more severe anxiety symptoms had higher cortisol secretion at waking and 30 min afterward [65]. Decreased cortisol levels were observed in GAD patients reporting low anxiety after 24 weeks of cognitive behavioral therapy [66].

Classification of Putative Biomarkers of Generalized Anxiety Disorder

Structural or activity changes in the brain regions, variation in the gray and white matter volumes, variation in N-acetylaspartate/creatine and dopamine, and error-related negativity were reported in GAD patients and not in healthy controls; thus they might be classified as susceptibility and/or diagnostic biomarkers of GAD and might be used to discriminate between GAD subjects and healthy controls. On the other hand, HPA axis activity, heart rate variability, neurocognition, and genes polymorphisms are still not identifiable as susceptibility and/or diagnostic biomarkers of GAD due to the paucity of the scientific evidence.

Anatomical changes in the right striatum, anterior cingulate cortex, prefrontal cortex, and right precentral and a higher cortisol secretion at waking were found correlated with higher worry and anxiety in GAD patients [50, 65, 67]; thus they might be prognostic biomarkers. In addition, lower reactivity in the anterior cingulate cortex and in the amygdala predicts a positive response to treatment with venlafaxine [53, 54]; thus they might be predictive biomarkers. Table 13.2 shows a proposal of biomarkers for GAD.

Biological Markers of Specific Phobia

Structural/Activity Changes in Brain Regions

Significant differences in the functional activity in the insula; ACC; amygdala; prefrontal, dorsolateral, and orbitofrontal cortex; bilateral fusiform gyrus; left cingulate cortex; and left thalamus were observed between patients with phobia related to situations and/or animals and healthy controls in relation to the exposure to phobic

Table 13.2 Putative biomarkers for generalized anxiety disorder

	Susceptibility/risk marker	Diagnostic marker	Prognostic marker	Predictive marker
Structural changes in the brain[a]	●	●	●[e]	●
Variations in gray matter volumes [b]	●	●		
Variations in white matter volumes[c]			●	
Activity of the ACC, PFC, and amygdala	●	●		●
N-Acetylaspartate/creatine	●	●		
Dopamine	●	●		
Neurocognition	●[e]	●[e]		
Genes[d]	●[e]	●[e]		
Hearth rate variability	●[e]	●[e]		
HPA axis dysregulation/cortisol secretion	●[e]	●[e]	●	
Error related negativity	●	●		

Note: *HPA* hypothalamic–pituitary–adrenal, *ACC* anterior cingulate cortex, *PFC* prefrontal cortex

[a]amygdala, hippocampus, midbrain, dorsolateral prefrontal cortex (DPC), and basal ganglia

[b]hippocampus, amygdala, dorsolateral prefrontal cortex (DPC), and basal ganglia

[c]dorsolateral prefrontal cortex (DPC)

[d]polymorphism of the brain-derived neurotrophic factor (BDNF) gene, catechol-O-methyltransferase (COMT) gene, the MAOA 941T allele, and the 5-HTT gene-linked functional polymorphic region

[e]it is not fully confirmed

stimuli [68, 69]. However, Linares et al. [68] did not show increased activity of the amygdala within SP patients.

Genetics

A study conducted on the Mainland Chinese population reported a higher frequency of A-allele of the rs10835210 in subjects with phobia if compared to healthy subjects suggesting a possible association between the BDNF gene and phobic disorders [70].

Hypothalamic–Pituitary–Adrenal Axis Dysregulation

Some studies reported hyper–hypo cortisol secretion in SP patients compared to healthy controls as well as in subjects exposed to phobic stimuli compared to subjects exposed to neutral exposures [71, 72]. Other studies did not find statistically significant differences in salivary cortisol levels between patients with SP and healthy controls or statistically significant correlations between cortisol levels and fear after the exposure to phobic stimuli in SP patients [10, 11, 73]. Other authors proposed that changes in cortisol levels in SP subjects might be associated with a better response to extinction-based psychotherapy, since the administration of cortisol enhanced extinction-based psychotherapy effects in terms of reduction of fear of heights [74].

Classification of Putative Biomarkers for Social Phobia

Given that activity changes in the brain regions and the BDNF gene was found the most prevalent in SP patients than in healthy controls [10, 11, 68, 69], these can be proposed as susceptibility and/or diagnostic biomarkers. On the contrary, the role of HPA axis activation as susceptibility and/or diagnostic biomarkers of SP is unclear, and longitudinal studies are needed to confirm it. Since the administration of cortisol was found to enhance extinction-based psychotherapy therapeutic effects [74], cortisol secretion might be classified as a predictive biomarker [11, 74, 75]. In Table 13.3 a classification of biomarkers for SP is proposed.

Table 13.3 Putative biomarkers for specific phobia

	Susceptibility/risk marker	Diagnostic marker	Prognostic marker	Predictive marker
Activity changes in the brain[a]	●	●		
BDNF gene	●	●		
HPA axis dysregulation/ cortisol secretion	●	●[b]		●

Note: *HPA* hypothalamic–pituitary–adrenal, *BDNF* brain-derived neurotrophic factor
[a]insula, ACC, amygdala, prefrontal, dorsolateral, and orbitofrontal cortex, bilateral fusiform gyrus, left cingulate cortex, and the left thalamus
[b]it is not fully confirmed

Biological Markers in Social Anxiety Disorder

Structural/Activity Changes in Brain Regions

Structural and activity changes in the amygdala, hippocampus, and insula were proposed as biomarkers of SAD given that, compared to healthy controls, altered brain functioning and lower brain volume (limited to the amygdala and hippocampus) were found in SAD patients [10, 76]. Moreover, anatomical changes in the hippocampus and amygdala as well as variations in oxytocin and serum molecular levels (i.e., AXL receptor tyrosine kinase, vascular cell adhesion molecule, collagen IV, vascular cell adhesion molecule 1, vitronectin) were found associated with more severe anxiety, onset of social fears, and depressive symptoms [76, 77].

Dopamine and serotonin were proposed as biomarkers since lower striatal density of dopamine and higher serotonin synthesis were observed in SAD patients and not in healthy controls [10].

Significant differences between SAD patients and healthy subjects were found related to the cerebral blood flow levels, with SAD subjects reporting enhanced regional blood flow in the amygdala–hippocampal region, right dorsolateral PFC, left inferior temporal cortex, and lower brain blood flow in the left temporal pole and in the cerebellum [10]. In addition, higher regional cerebral blood flow at the baseline in the anterior and lateral part of the left temporal cortex, and in the lateral part of the left mid-frontal regions, was observed in SAD patients who did not respond to selective serotonin reuptake inhibitors (SSRIs) as compared to responders [78]. However, these findings were not replicated [10].

Heart Rate Variability

Lower heart rate variability, that is, lower time domain and lower frequency domain, was observed in SAD patients when compared to healthy controls [60], and an experimental study showed lower HRV in SAD patients if compared to healthy controls when they undergo a social performance task [79]. Reduced HRV was positively correlated with anxiety symptom severity, psychological distress, and harmful alcohol use within SAD patients [80].

However, the strength of the association between HRV and SAD was smaller if compared with other anxiety disorders [60, 64, 81]; thus further studies are in need to confirm the role of HRV as biomarker of SAD.

Neurocognition

SAD patients reported decreased performance on tests of visuospatial and constructive ability than healthy controls [82, 83]. There are also evidences of poor verbal memory functions, working memory, processing speed, visuospatial memory, and word fluency [82, 83].

However, the relationship between SAD and neurocognitive functioning is debated since it might be biased by state anxiety, emotion suppression, depressive comorbidity, medication use, and cognitive style which includes cognitively vicious circles during social situations [82, 83].

Hypothalamic–Pituitary–Adrenal Axis Dysregulation

During psychosocial stress (e.g., public speech task, performed mental arithmetic, and memory tests before an audience) changes in cortisol levels were found in SAD subjects and not in healthy controls [11]. Such changes are associated with severity of the disorder since negative correlations were found between cortisol plasma levels and 5-HT1A binding in the amygdala, hippocampus, and retrosplenial cortex as well as between cortisol plasma levels and trait anxiety scores, while among healthy controls, no significant correlations were found between cortisol plasma levels and trait anxiety scores [84]. However, some studies did not confirm these results [11]. The inconsistency could be due to methodological differences across studies: (a) the non-standardization of stressors to evaluate the cortisol stress reactivity [85]; (b) the confounding effect of age which influences the cortisol response to stressor [85, 86] was not taken into account; and (c) the comorbidity with major depressive disorder, which influences the cortisol stress reactivity in SAD patients [87].

Classification of Putative Biomarkers for Social Anxiety Disorder

Structural and activity changes in the brain regions and variations in dopamine/serotonin as well as in cerebral blood flow levels and in heart rate variability were found in SAD subjects and not in controls [10, 11, 60]; thus these might be susceptibility and/or diagnostic biomarkers. On the contrary, the role of HPA axis activation and neurocognition as susceptibility and/or diagnostic biomarkers of SAD remains unclear; longitudinal studies are needed to confirm it.

Anatomical changes in the hippocampus and amygdala as well as variations in oxytocin, serum molecular levels, and heart rate variability were found associated with more severe outcomes in SAD subjects, that is, more severe anxiety, social fears co-occurrence, and comorbidity with other psychiatric disorders such as depression [76, 77, 80]. Thus, they might be suitable as prognostic biomarkers which predict unfavorable/favorable course of the social anxiety disorder.

Higher cerebral blood flow in the left temporal cortex and in the lateral part of the left mid-frontal regions and lower anterior cingulate cortex activity were suggested as prognostic biomarkers, and they might allow to identify subjects at a higher risk of unfavorable response to SSRIs [78] (Table 13.4).

Biological Markers in Separation Anxiety Disorder and in Selective Mutism

Although separation anxiety disorder and selective mutism are now held to be relevant in adulthood [88], studies exploring their biomarkers in adults are still limited in numbers and focus only on separation anxiety disorder.

Oxytocin and translocator protein (TSPO) were assumed as possible biomarkers of SepAD since the single nucleotide polymorphism rs53576 of the oxytocin receptor (OXTR) gene and lower platelet expression of TSPO density were found associated with adult separation anxiety disorder (A-SepAD) [10, 11].

Table 13.4 Putative biomarkers for social anxiety disorder

	Susceptibility/risk marker	Diagnostic marker	Prognostic marker	Predictive marker
Structural and activity changes in the brain[a]	●	●	●[b]	
Molecular serum panel			●	
Cerebral blood flow	●	●		●
Anterior cingulate cortex activity				●
Dopamine	●	●		
Serotonin	●	●		
Oxytocin			●	
Hearth rate variability	●	●	●	
HPA axis dysregulation/ cortisol secretion	●	●	●	
Neurocognition	●	●		

Note: HPA, hypothalamic–pituitary–adrenal
[a]amygdala, hippocampus, and insula
[b]amygdala, hippocampus

The Biomarker Crisis

Research on biomarkers is prone to consider single brain region/circuit or a specific neurotransmitter without ravelling one-to-one association with a specific illness [2] although psychiatry disorders are multidimensional in their description, are multifactorial in their origins, and involve nonlinear interactions in their progress [89]. Thus, it is doubtful that a single biomarker might explain the aetiology of an exclusive mental disease [4, 90]. Multimodal approach where the diagnosis and the course of illness are explained by the interaction between different biomarkers might be more appropriate [2, 4, 90]. In addition, different biomarkers might be related to a cluster of symptoms rather than to a specific diagnosis [2, 4, 90].

Some biomarkers as structural variations in the amygdala and/or hippocampus present good reliability [10, 11], and clinicians do not use them for practical and economic reasons [90]; simple and cost-effective biomarkers [4, 90] such as saliva cortisol, heart rate variability, blood cells, peripheral blood stem cells, respiratory pattern, and neurocognitive functioning [10, 11, 23, 48] are needed. Even the classifications of biomarkers proposed here have a great limitation due to the overlap of biomarkers among psychiatric disorders. Some biomarkers (e.g., structural brain morphology; lower 5-HT plasma; dopamine; serotonin; oxytocin; norepinephrine; genes; hyper–hyper secretion of cortisol; instability of the cortical arousal system; HRW, PWD, and RDW; neurocognition) do not differentiate subjects with different anxiety disorders or do not differentiate subjects with different psychiatric disorders (see Table 13.5).

As stated by Kapur et al. [91], psychiatry seems to be in a Catch-22 because the diagnostic system was not designed to facilitate biological differentiation.

Table 13.5 Summing up of putative biomarkers of anxiety disorders

	Panic disorder	Agoraphobia	Generalized anxiety disorder	Specific phobia	Social anxiety	Separation anxiety disorder	Psychiatry disease other than anxiety disorders
Structural or activity changes in brain regions	●		●	●	●		●
N-acetylaspartate/creatine			●				●
N-acetyl-b-glucosaminidase		●					
Oxytocin					●	●	●
Serotonin	●				●		●
Dopamine			●		●		●
Molecular serum panel	●				●		
Genes	●	●	●[a]	●		●	●
Aberrant respiratory regulation	●						
Hearth rate variability	●		●		●		●
Blood cells and peripheral blood stem cells	●						●
HPA axis dysregulation/ cortisol secretion	●[a]	●[a]	●[a]	●[a]	●[a]		●
Error related negativity			●				●
Neurocognition	●[a]		●[a]		●[a]		●

Note: *HPA* hypothalamic–pituitary–adrena
[a]it is not fully confirmed

Furthermore, contemporary diagnostic definitions of mental illness based on symptoms include heterogeneous populations yielding spurious results when exploring biological features of psychiatry disorders [92].

The classifications of biomarkers described here have limitations also because of the interfering effects of variables such as stress, comorbidity, physical activity, and psychotropic medications [90], which are not taken into account into the classifications. For instance, psychiatry comorbidity and exposure to stressful life events during childhood are known to contribute to HPA axis activation as well as to induce poor neurocognitive functioning [40, 46, 47, 93, 94]. Moderate and intense physical activity was found correlated with higher heart rate variability [95, 96]. Psychotropic medications, particularly after a long-term use, might modify the course, the clinical features, and the responsiveness to treatment of a psychiatric disorder; these liabilities go under the rubric of iatrogenic comorbidity [97, 98] and can be explained via the phenomenon of toxicity behavior [99].

Beyond Biological Markers in Anxiety Disorders

Within the clinimetric perspective [99], it has been proposed that clinical utility and validity of biological markers might be enhanced linking them with different stages of anxiety disorders [23]. Here, the clinimetric approach and the use of staging as clinical marker of anxiety disorders are described.

The Science of Clinimetrics

In 1982, Feinstein [99] introduced the term clinimetrics to indicate a domain regarding clinical indexes, rating scales, and other expressions which concern the assessment of symptoms, signs, and clinical phenomena which do not find room in the classical taxonomy [99]. Clinimetrics assumes a set of rules to define the structure of indexes, the choice of component variables, as well as the evaluation of consistency, validity, and responsiveness [100]. Within the clinimetric science, indexes are classified as ailment-oriented; they are referred to specific diseases, states and clinical manifestations, and general indexes; and they are referred to general health and functional states that are not disease or condition specific [100]. Fava et al. [100] proposed that the reliance to clinimetrics in the clinical practice, rather than to the psychometric approach used in the current nosography [9, 101], may help expanding the content of customary clinical information and improve outcomes in clinical research and practice.

Although the current diagnostic systems showed good reliability, their clinical utility is elusive [102, 103]: an epidemiological research involving 1764 clinicians suing ICD or DSM reported that 1123 subjects (64%) think these diagnostic systems have low utility to choose the treatment and define the prognosis [101]. The DSM [9] and the ICD [101] are influenced by the psychometric models [104, 105] according to which the severity of the disease results from the sum of the symptoms rather than from their intensity or quality [106]. This implies that all symptoms have an equivalent load in determining the diagnosis. Clinimetrics proposed that symptoms are differentiated into major and minor and have different weights in the indexes [99]. The DSM [9] and the ICD [101] do not take into account clinical issues which are instead relevant for the clinimetric science. These issues are, for instance, pattern of symptoms, severity of disease, comorbidity, timing of phenomena, staging, and responses to previous treatments and might outline prognostic and therapeutic differences among patients who otherwise seem deceptively similar because they share the same diagnosis [107, 108].

The exclusive reliance on diagnostic criteria implies that symptoms not reaching the diagnostic threshold might be not relevant from a clinical point of view [105] although they affect the quality of life of the patient and have pathophysiological and therapeutic repercussions [107].

Since the taxonomy in psychiatry derives from the traditional method of clinical medicine, which offers operating specifications for making a clinical decision regarding the subsistence of a disease, the diagnostic reasoning process ends with the identification of a disorder [109, 110]. However, the clinimetric science suggests to include also the clinical judgment which proceeds across a series of "transfer stations" where hypothetical link between symptoms and pathophysiological processes are drawn and are amenable to longitudinal validation or adjustment as long as therapeutic goals are reached [104].

Staging of Psychiatric Disorders

Psychiatry disorders are not static, sharply defined disorders with separate aetiologies and courses, but are syndromes that overlap and develop across stages [103]. Moving from a cross-sectional nosography to the longitudinal view of the development of mental diseases, Fava and Kellner [111] for the first time introduced the clinimetric concept of staging as a clinical tool to enhance the diagnostic and treatment procedure in psychiatry.

Staging was later refined [103], and it nowadays allows to define the degree of progression of a diseases at a particular point in time and where the patient is located on the continuum of the course of the disease. Staging distinguishes prodromes (i.e., early symptoms and signs that differ from the acute clinical phase) and residual symptoms (i.e., persistent symptoms and signs despite apparent remission or recovery) [111]. Prodromal and residual symptoms may associate since as the illness remits, it progressively recapitulates, although in reverse order, this is the so-called rollback phenomenon [112]. Some prodromes may be overshadowed by the acute phase of the disorder but persist as residual symptoms and progress to become prodromes of following episode of illness. Staging identifies four steps for development of a psychiatry disorder: the prodromal phase, which refers to the time interval between the onset of prodromal symptoms and the onset of the characteristic manifestations of the fully developed illness (stage 1); the acute phase (stage 2); attenuated symptoms (e.g., residual symptoms) (stage 3) that might be due to partial persistence of the disease or a worsening of a pre-existing abnormal personality trait; and the chronic phase (stage 4) [103].

Staging as Clinical Marker of Anxiety Disorders

Prevalence data estimate that about 60% of subjects with an anxiety disorder has one or more additional anxiety disorders [113]. The staging model [103] may clarify this high comorbidity referring to a longitudinal development of anxiety.

The following stages of development of anxiety were proposed. Stage 1 includes subclinical symptoms of agoraphobia (e.g., distress and avoidance of closed spaces), social phobia, generalized anxiety (e.g., excessive worries), and/or hypochondriasis (e.g., illness phobia, health anxiety, or fear of disease). These symptoms might be stable. The patient may stay on stage 1 or the disease may become more severe, and in the latter case, the patient proceeds to stage 2. This means that, according to the current nosography, the patient at stage 1 is under the diagnostic threshold of agoraphobia, social phobia, generalized anxiety, and hypochondriasis, while at stage 2 he satisfies the diagnostic criteria for agoraphobia, social anxiety, generalized anxiety disorder, and/or hypochondriasis. Afterward, panic attacks occur leading to a likely worsening of prodromal symptomatology and co-occurrence of demoralization and major depression. This is stage 3 which means, according to the current nosography, that the patient satisfies the diagnostic criteria for panic disorder which can be comorbid with major depressive disorder (MDD) or a subthreshold MDD. During stage 3, prodromal symptoms may be overshadowed and persist as residual (i.e., generalized anxiety, somatic anxiety, health anxiety, low self-esteem, agoraphobia, hypochondriasis, lower psychological and physical well-being, limited symptoms of panic attacks, anticipatory anxiety, and depression), they can also progress to become the prodromes of a coming relapse [103]. Afterward, the clinical manifestations observed at stage 3 may become chronic. This is stage 4 which, according to the current nosography, means that the patient has panic disorder with comorbid agoraphobia and/or social anxiety and/or generalized anxiety disorder and/or hypochondriasis and/or MDD. In brief, the staging model described here includes several anxiety disorder diagnoses, if we use the current nosography; alternatively we may think that the staging model described here refers to a unique syndrome which develops longitudinally and acquires different clinical features.

Under this light, the staging model might be a clinical marker for anxiety disorders: (a) it is a susceptibility/risk marker since clinicians can assess prodromal symptoms; (b) it is a diagnostic marker since it identifies the extent of progression of the anxiety syndrome across different stages; (c) it is a prognostic marker since the persistence of residual symptoms decreased the level of well-being [114] and predicts the onset of relapse, suggesting that residual symptoms might be a treatment goal [115]; (d) it is a predictive marker since clinicians should promote the regression to lower stages [103], and, for instance, in MDD, stage-specific treatments have been proposed and showed to be effective [116]; and (e) it is a specific marker for the anxiety syndrome, and its use is simple with a low economic burden for clinicians [23].

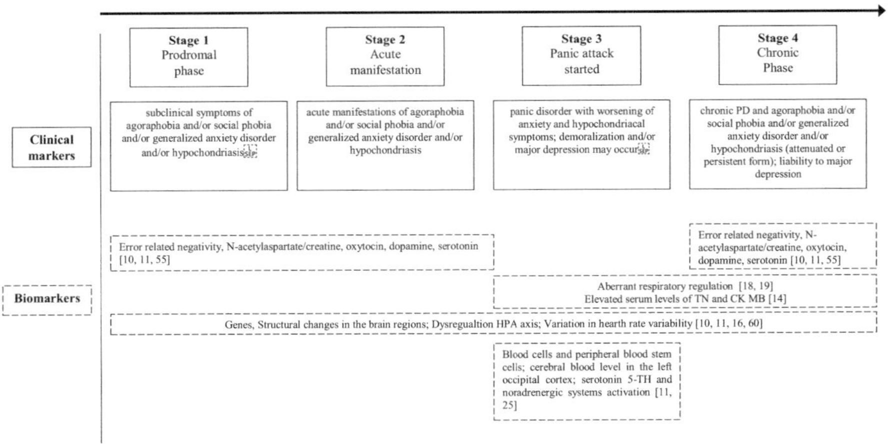

Fig. 13.1 Stage-specific biomarkers for anxiety disorders

Conclusion

We here propose a stage-specific biomarker model for anxiety disorders (Fig. 13.1) since staging seems to present good sensitivity, specificity, and predictivity [4], while biomarkers seem to have low specificity and predictive value [10, 11, 23, 90]. The stage-specific biomarker model for anxiety disorders could contribute in increasing specificity and predictivity of biomarkers.

Structural changes in the brain regions; variation in N-acetylaspartate/creatine, dopamine, serotonin, and oxytocin levels; HRV; HPA axis activity; error-related negativity; and genetic variants were found more prevalent in patients with generalized anxiety, social anxiety, and agoraphobia than in healthy controls [10, 11, 18, 19, 48, 55]. Thus, these biomarkers might be associated with stage 1 and stage 2 as well as with stage 4. Structural or activity changes in brain regions, variation in the flow of cerebral blood levels in the left occipital cortex, genetic variants, aberrant respiratory regulation, variation in the HRV, and peripheral blood stem cell distribution were found in patients with panic disorder and not in healthy controls [10, 11, 18, 19, 25, 60]. Thus, these biomarkers might be associated with stage 3 and stage 4.

However, once again, biomarkers are not stage-specific, suggesting that if biomarkers are used without the clinical marker, they have poor clinical utility.

References

1. Biomarkers Definitions Working Group. Biomarkers and surrogate endpoints. Clin Pharmacol Ther. 2001;69:89–95.
2. Venkatasubramanian G, Keshavan MS. Biomarkers in psychiatry-a critique. Ann Neurosci. 2016;23:3–5.

3. Kalia M, Silva JC. Biomarkers of psychiatric diseases: current status and future prospects. Metabolism. 2015;64:S11–5.
4. Scarr E, Millan MJ, Bahn S, Bertolino A, Turck CW, Kapur S, et al. Biomarkers for psychiatry: the journey from fantasy to fact, a report of the 2013 CINP think tank. Int J Neuropsychopharmacol. 2015;18:1–9.
5. Macaluso M, Preskorn SH. How biomarkers will change psychiatry: from clinical trials to practice. Part I. Introduction. J Psychiatr Pract. 2012;18:118–21.
6. Amur S, LaVange L, Zineh I, Buckman-Garner S, Woodcock J. Biomarker qualification: toward a multiple stakeholder framework for bio- marker development, regulatory acceptance, and utilization. Clin Pharacol Ther. 2015;98:34–46.
7. Ankeny JS, Labadie B, Luke J, Hsueh E, Messina J, Zager JS. Review of diagnostic, prognostic, and predictive biomarkers in melanoma. Clin Exp Metastasis. 2018;3(5-6):487–93.
8. FDA-NIH Biomarker Working Group. BEST (Biomarkers, Endpoints, and Other Tools) Resource. Silver Spring, MD: Food and Drug Ad- ministration (US); Bethesda, MD: National Institutes of Health (US); 2016.
9. American Psychiatric Association. Diagnostic and Statistical Manual of Mental Disorders. 5th ed. Washington, DC: American Psychiatric Association; 2013.
10. Bandelow B, Baldwin D, Abelli M, Altamura C, Dell'Osso B, Dom- schke K, et al. Biological markers for anxiety disorders, OCD and PTSD–a consensus statement. Part I: neuroimaging and genetics. World J Biol Psychiatry. 2016;17:321–65.
11. Bandelow B, Baldwin D, Abelli M, Bolea-Alamanac B, Bourin M, Chamberlain SR, et al. Biological markers for anxiety disorders, OCD and PTSD: a consensus statement. Part II: Neurochemistry, neurophysiology and neurocognition. World J Biol Psychiatry. 2017;18:162–214.
12. De Cristofaro MT, Sessarego A, Pupi A, Biondi F, Faravelli C. Brain perfusion abnormalities in drug-naive, lactate-sensitive panic patients: a SPECT study. Biol Psychiatry. 1993;33:505–12.
13. Gecici O, Acar M, Haktanir A, Emul M, Demirel R, YüCEL A, et al. Evaluation of cerebral blood flow volume using color duplex sonography in patients with untreated panic disorder. Psychiatry Clin Neurosci. 2005;59:676–82.
14. Gottschalk MG, Cooper JD, Chan MK, Bot M, Penninx BWJH, Bahn S. Serum biomarkers predictive of depressive episodes in panic disorder. J Psychiatr Res. 2016;73:53–62.
15. O'Sullivan K, Newman EF. Neuropsychological impairments in panic disorder: a systematic review. J Affect Disord. 2014;167:268–84.
16. Maron E, Lang A, Tasa G, Liivlaid L, Toru I, Must A, et al. Associations between serotonin-related gene polymorphisms and panic disorder. Int J Neuropsychopharmacol. 2005;8(2):261–6.
17. Deckert J, Catalano M, Syagailo YV, Bosi M, Okladnova O, Di Bella D, et al. Excess of high activity monoamine oxidase A gene promoter alleles in female patients with panic disorder. Hum Mol Genet. 1999;8:621–4.
18. Grassi M, Caldirola D, Vanni G, Guerriero G, Piccinni M, Valchera A, et al. Baseline respiratory parameters in panic disorder a meta-analysis. J Affect Disord. 2013;146:158–73.
19. Grassi M, Caldirola D, Di Chiaro N, Riva A, Daccò S, Pompili M, et al. Are respiratory abnormalities specific for panic disorder? A meta-analysis. Neuropsychobiology. 2014;70:52–60.
20. Hillebrand S, Gast KB, de Mutsert R, Swenne CA, Jukema JW, Middeldorp S, et al. Heart rate variability and first cardiovascular event in populations without known cardiovascular disease: meta-analysis and dose–response meta-regression. Europace. 2013;15:742–9.
21. Davies SJ, Ghahramani P, Jackson PR, Noble TW, Hardy PG, Hippisley-Cox J, et al. Association of panic disorder and panic attacks with hypertension. Am J Med. 1999;107: 310–6.
22. Davies SJ, Bjerkeset O, Nutt DJ, Lewis G. A U-shaped relationship between systolic blood pressure and panic symptoms: the HUNT study. Psychol Med. 2012;42:1969–76.

23. Cosci F, Mansueto G. Biological and clinical markers in Panic Disorder. Psychiatry Investig 2018; https://doi.org/10.30773/pi.2018.07.26
24. Wendt J, Hamm AO, Pané-Farré CA, Thayer JF, Gerlach A, Gloster AT, et al. Pretreatment cardiac vagal tone predicts dropout from and residual symptoms after exposure therapy in patients with panic disorder and agoraphobia. Psychother Psychosom. 2018;87(3):187–9.
25. Bhad R. Red blood cell and platelet indices: a potential biomarker for panic disorder. J Neurosci Rural Pract. 2017;8:164.
26. Asoglu M, Aslan M, Imre O, Kivrak Y, Akil O, Savik E, et al. Mean platelet volume and red cell distribution width levels in initial evaluation of panic disorder. Neuropsychiatr Dis Treat. 2016;12:2435–8.
27. Ransing RS, Patil B, Grigo O. Mean platelet volume and platelet distribution width level in patients with panic disorder. J Neurosci Rural Pract. 2017;8:174–8.
28. McEwen BS. Stress, adaptation, and disease: Allostasis and allostatic load. Ann N Y Acad Sci. 1998;840(1):33–44.
29. McEwen BS. Allostasis and allostatic load: implications for neuropsychopharmacology. Neuropsychopharmacology. 2000;22(2):108–24.
30. McEwen BS. Protection and damage from acute and chronic stress: allostasis and allostatic overload and relevance to the pathophysiology of psychiatric disorders. Ann N Y Acad Sci. 2004;1032(1):1–7.
31. Heim C, Shugart M, Craighead WE, Nemeroff CB. Neurobiological and psychiatric consequences of child abuse and neglect. Dev Psychobiol. 2010;52:671–90.
32. Faravelli C, Lo Sauro C, Lelli L, Pietrini F, Lazzeretti L, Godini L, et al. The role of life events and HPA axis in anxiety disorders: a review. Curr Pharm Des. 2012;18:5663–74.
33. Ising M, Hohne N, Siebertz A, Parchmann AM, Erhardt A, Keck M. Stress response regulation in panic disorder. Curr Pharm Des. 2012;18:5675–84.
34. Bandelow B, Wedekind D, Pauls J, Broocks A, Hajak G, Ruther E. Salivary cortisol in panic attacks. Am J Psychiatry. 2000;157:454–6.
35. Du X, Pang TY. Is dysregulation of the HPA-axis a core pathophysiology mediating comorbid depression in neurodegenerative diseases? Front Psychiatry. 2015;9(6):32.
36. Jakuszkowiak-Wojten K, Landowski J, Wiglusz MS, Cubała WJ. Cortisol as an indicator of hypothalamic-pituitary-adrenal axis dysregulation in patients with panic disorder: a literature review. Psychiatr Danub. 2015;27(Suppl 1):S445–51.
37. Graeff FG, Zangrossi H Jr. The hypothalamic-pituitary-adrenal axis in anxiety and panic. Psychol Neurosci. 2010;3:3–8.
38. Schommer NC, Hellhammer DH, Kirschbaum C. Dissociation between reactivity of the hypothalamus-pituitary-adrenal axis and the sympathetic-adrenal-medullary system to repeated psychosocial stress. Psychosom Med. 2003;65:450–60.
39. Vreeburg SA, Zitman FG, van Pelt J, Derijk RH, Verhagen JC, van Dyck R, et al. Salivary cortisol levels in persons with and without different anxiety disorders. Psychosom Med. 2010;72(4):340–7.
40. Coryell W, Noyes R Jr, Reich J. The prognostic significance of HPA-axis disturbance in panic disorder: a three-year follow-up. Biol Psychiatry. 1991;29:96–102.
41. Abelson JL, Curtis GC. Hypothalamic-pituitary-adrenal axis activity in panic disorder: prediction of long-term outcome by pretreatment cortisol levels. Am J Psychiatry. 1996;153:69–73.
42. Machado S, Sancassiani F, Paes F, Rocha N, Murillo-Rodriguez E, Nardi AE. Panic disorder and cardiovascular diseases: an overview. Int Rev Psychiatry. 2017;29:436–44.
43. Del Casale A, Serata D, Rapinesi CD, Kotzalidis G, Angeletti G, Tatarelli R, et al. Structural neuroimaging in patients with panic disorder: findings and limitations of recent studies. Psychiatr Danub. 2013;25(2):0–114.
44. Wittchen HU, Gloster AT, Beesdo-Baum K, Fava GA, Craske MG. Agoraphobia: a review of the diagnostic classificatory position and criteria. Depress Anxiety. 2010;27(2):113–33.

45. Garvey MJ, Noyes JR. NAG level differences in panic disorder and agoraphobia. J Anxiety Disord. 2005;19(7):818–25.
46. Westberg P, Modigh K, Lisjö P, Eriksson E. Higher postdexamethasone serum cortisol levels in agoraphobic than in nonagoraphobic panic disorder patients. Biol Psychiatry. 1991;30(3):247–56.
47. Kartalci S, Dogan M, Unal S, Ozcan AC, Ozdemir S, Atmaca M. Pituitary volume in patients with panic disorder. Prog Neuropsychopharmacol Biol Psychiatry. 2011;35(1):203–7.
48. Maron E, Nutt D. Biological markers of generalized anxiety disorder. Dialogues Clin Neurosci. 2017;19(2):147–58.
49. Abdallah CG, Coplan JD, Jackowski A, Sato JR, Mao X, Shungu DC, et al. A pilot study of hippocampal volume and N-acetylaspartate (NAA) as response biomarkers in riluzole-treated patients with GAD. Eur Neuropsychopharmacol. 2013;23(4):276–84.
50. Hilbert K, Pine DS, Muehlhan M, Lueken U, Steudete-Schmiedgen S, Beesdo-Baum K. Gray and white matter volume abnormalities in generalized anxiety disorder by categorical and dimensional characterization. Psychiatry Res. 2015;234(3):314–20.
51. Mochcovitch MD, da Rocha Freire RC, Garcia RF, Nardi AE. A systematic review of fMRI studies in generalized anxiety disorder: evaluating its neural and cognitive basis. J Affect Disord. 2014;167:336–42.
52. Qiao J, Li A, Cao C, Wang Z, Sun J, Xu G. Aberrant functional network connectivity as a biomarker of generalized anxiety disorder. Front Hum Neurosci. 2017;11:626.
53. Nitschke JB, Sarinopoulos I, Oathes DJ, Johnston T, Whalen PJ, Davidson RJ, et al. Anticipatory activation in the amygdala and anterior cingulate in generalized anxiety disorder and prediction of treatment response. Am J Psychiatry. 2009;166(3):302–10.
54. Whalen PJ, Johnstone T, Somerville LH, Nitschke JB, Polis S, Alexander AL, et al. A functional magnetic resonance imaging predictor of treatment response to venlafaxine in generalized anxiety disorder. Biol Psychiatry. 2008;63:858–63.
55. Meyer A. Developing psychiatric biomarkers: a review focusing on the error-related negativity as a biomarker for anxiety. Curr Treat Options Psychiatry. 2016;3(4):356–64.
56. Yang Y, Zhang X, Zhu Y, Dai Y, Liu T, Wang Y. Cognitive impairment in generalized anxiety disorder revealed by event-related potential N270. Neuropsychiatr Dis Treat. 2015;11:1405–11.
57. Butters MA, Bhalla RK, Andreescu C, Wetherell JL, Mantella R, Begley AE, et al. Changes in neuropsychological functioning following treatment for late-life generalised anxiety disorder. Br J Psychiatry. 2011;199(3):211–8.
58. Thames AD, Panos SE, Arentoft A, Byrd DA, Hinkin CH, Arbid N. Mild test anxiety influences neurocognitive performance among African Americans and European Americans: identifying interfering and facilitating sources. Cultur Divers Ethnic Minor Psychol. 2015;21(1):105–13.
59. Molina E, Cervilla J, Rivera M, Torres F, Bellon JA, Moreno B, et al. Polymorphic variation at the serotonin 1-A receptor gene is associated with comorbid depression and generalized anxiety. Psychiatr Genet. 2011;21:195–201.
60. Chalmers JA, Quintana DS, Abbott MJ, Kemp AH. Anxiety disorders are associated with reduced heart rate variability: a meta-analysis. Front Psychiatry. 2014;5:80.
61. Levine JC, Fleming R, Piedmont JI, Cain SM, Chen WJ. Heart rate variability and generalized anxiety disorder during laboratory-induced worry and aversive imagery. J Affect Disord. 2016;205:207–15.
62. Lyonfields JD, Borkovec TD, Thayer JF. Vagal tone in generalized anxiety disorder and the effects of aversive imagery and worrisome thinking. Behav Ther. 1995;26(3):457–66.
63. Thayer JF, Friedman BH, Borkovec TD. Autonomic characteristics of generalized anxiety disorder and worry. Biol Psychiatry. 1996;39(4):255–66.

64. Pittig A, Arch JJ, Lam CWR, Craske MG. Heart rate and heart rate variability in panic, social anxiety, obsessive–compulsive, and generalized anxiety disorders at baseline and in response to relaxation and hyperventilation. Int J Psychophysiol. 2013;87(1):19–27.
65. Mantella RC, Butters MA, Amico JA, Mazumdar S, Rollman BL. Begley et al. Salivary cortisol is associated with diagnosis and severity of late-life generalized anxiety disorder. Psychoneuroendocrinology. 2008;33(6):773–81.
66. Tafet GE, Feder DJ, Abulafia DP, Roffman SS. Regulation of hypothalamic–pituitary–adrenal activity in response to cognitive therapy in patients with generalized anxiety disorder. Cogn Affect Behav Neurosci. 2005;5:37–40.
67. Schienle A, Ebner F, Schäfer A. Localized gray matter volume abnormalities in generalized anxiety disorder. Eur Arch Psychiatry Clin Neurosci. 2011;261(4):303–7.
68. Linares IM, Trzesniak C, Chagas MH, Hallak JE, Nardi AE, Crippa JA. Neuroimaging in specific phobia disorder: a systematic review of the literature. Rev Bras Psiquiatr. 2012;34:101–11.
69. Peñate W, Fumero A, Vina C, Herrero M, Marrero RJ, Rivero F. A meta-analytic review of neuroimaging studies of specific phobia to small animals. Eur Psychiatry. 2017;31(1):23–36.
70. Xie B, Wang B, Suo P, Kou C, Wang J, Meng X, et al. Genetic association between BDNF gene polymorphisms and phobic disorders: a case-control study among mainland Han Chinese. J Affect Disord. 2011;132:239–42.
71. Alpers GW, Abelson JL, Wilhelm FH, Roth WT. Salivary cortisol response during exposure treatment in driving phobics. Psychosom Med. 2003;65:679–87.
72. Lilliecreutz C, Theodorsson E, Sydsjo G, Josefsson A. Salivary cortisol in pregnant women suffering from blood and injection phobia. Arch Womens Ment Health. 2011;14:405–11.
73. Van Duinen MA, Schruers KRJ, Griez EJL. Desynchrony of fear in phobic exposure. J Psychopharmacol. 2010;24(5):695–9.
74. de Quervain DJ, Bentz D, Michael T, Bolt OC, Wiederhold BK, Margraf J, et al. Glucocorticoids enhance extinction-based psychotherapy. Proc Natl Acad Sci USA. 2011;108(16):6621–5.
75. Soravia LM, Heinrichs M, Aerni A, Maroni C, Schelling G, Ehlert U, et al. Glucocorticoids reduce phobic fear in humans. Proc Natl Acad Sci USA. 2006;103(14):5585–90.
76. Irle E, Ruhleder M, Lange C, Seidler-Brandler U, Salzer S, Dechent P, et al. Reduced amygdalar and hippocampal size in adults with generalized social phobia. J Psychiatry Neurosci. 2010;35:126–31.
77. Hoge EA, Pollack MH, Kaufman RE, Zak PJ, Simon NM. Oxytocin levels in social anxiety disorder. CNS Neurosci Ther. 2008;14(3):165–70.
78. Van der Linden G, van Heerden B, Warwick J, Wessels C, van Kradenburg J, Zungu-Dirwayi N, et al. Functional brain imaging and pharmacotherapy in social phobia: single photon emission computed tomography before and after treatment with the selective serotonin reuptake inhibitor citalopram. Prog Neuropsychopharmacol Biol Psychiatry. 2000;24:419–38.
79. García-Rubio MJ, Espín L, Hidalgo V, Salvador A, Gómez-Amor J. Autonomic markers associated with generalized social phobia symptoms: heart rate variability and salivary alpha-amylase. Stress. 2017;20(1):61–8.
80. Alvares GA, Quintana DS, Kemp AH, Van Zwieten A, Balleine BW, Hickie IB, et al. Reduced heart rate variability in social anxiety disorder: associations with gender and symptom severity. PloS one. 2013;8(7):e70468.
81. Friedman BH. An autonomic flexibility–neurovisceral integration model of anxiety and cardiac vagal tone. Biol Psychol. 2007;74(2):185–99.
82. O'Toole MS, Pedersen AD. A systematic review of neuropsychological performance in social anxiety disorder. Nord J Psychiatry. 2011;65:147–61.
83. O'Toole MS, Pedersen AD, Hougaard E, Rosenberg NK. Neuropsychological test performance in social anxiety disorder. Nord J Psychiatry. 2015;69:444–52.

84. Lanzenberger R, Wadsak W, Spindelegger C, Mitterhauser M, Akimova E, Mien LK, et al. Cortisol plasma levels in social anxiety disorder patients correlate with serotonin-1A receptor binding in limbic brain regions. Int J Neuropsychopharmacol. 2010;13(9):1129–43.
85. Klumbies E, Braeuer D, Hoyer J, Kirschbaum C. The reaction to social stress in social phobia: discordance between physiological and subjective parameters. Plos one. 2014;9(8):e105670.
86. Strahler J, Mueller A, Rosenloecher F, Kirschbaum C, Rohleder N. Salivary alpha-amylase stress reactivity across different age groups. Psychophysiology. 2010;47:587–95.
87. Yoon KL, Joormann J. Stress reactivity in social anxiety disorder with and without comorbid depression. J Abnorm Psychol. 2012;121(1):250–5.
88. Ströhle A, Gensichen J, Domschke K. The diagnosis and treatment of anxiety disorders. Dtsch Arztebl Int. 2018;155(37):611–20.
89. Cloninger CR. Implications of comorbidity for the classification of mental disorders: the need for a psychobiology of coherence. In: Maj M, Gaebel W, Lopez-Ibor JJ, Sartorius N, editors. Psychiatric diagnosis and classification. Chichester: Wiley; 2002. p. 79–105.
90. Boksa P. A way forward for research on biomarkers for psychiatric dis- orders. J Psychiatry Neurosci JPN. 2013;38:75–7.
91. Kapur S, Phillips AG, Insel TR. Why has it taken so long for biological psychiatry to develop clinical tests and what to do about it? Mol Psychiatry. 2012;17(12):1174–9.
92. Fava GA, Guidi J, Grandi S, Hasler G. The missing link between clinical states and biomarkers in mental disorders. Psychother Psychosom. 2014;83:136–41.
93. Danese A, McEwen BS. Adverse childhood experiences, allostasis, allostatic load, and age-related disease. Physiol Behav. 2012;106:29–39.
94. Mansueto G, Schruers K, Cosci F, van Os J, GROUP Investigators. Childhood adversities and psychotic symptoms: the potential mediating or moderating role of neurocognition and social cognition. Schizophr Res. 2018; https://doi.org/10.1016/j.schres.2018.11.028.
95. May R, McBerty V, Zaky A, Gianotti M. Vigorous physical activity predicts higher heart rate variability among younger adults. J Physiol Anthropol. 2017;36(1):24.
96. Sandercock GR, Bromley PD, Brodie DA. Effects of exercise on heart rate variability: inferences from meta-analysis. Med Sci Sports Exerc. 2005;37(3):433–9.
97. Fava GA, Cosci F, Offidani E, Guidi J. Behavioral toxicity revisited: iatrogenic comorbidity in psychiatric evaluation and treatment. J Clin Psychopharmacol. 2016;36:550–3.
98. Tomba E, Guidi J, Fava GA. What psychologists need to know about psychotropic medications. Clin Psychol Psychother. 2018;25:181–7.
99. Feinstein AR. T. Duckett Jones Memorial Lecture. The Jones criteria and the challenge of clinimetrics. Circulation. 1982;66:1–5.
100. Fava GA, Rafanelli C, Tomba E. The clinical process in psychiatry: a clinimetric approach. J Clin Psychiatry. 2012;73:177–84.
101. World Health Organization. International statistical classification of diseases and related health problems 2018. (11th Revision). https://icd.who.int/browse11/l-m/en
102. First MB, Rebello TJ, Keeley JW, Bhargava R, Dai Y, Kulygina M, et al. Do mental health professionals use diagnostic classifications the way we think they do? A global survey. World Psychiatry. 2018;17:187–95.
103. Cosci F, Fava GA. Staging of mental disorders: systematic review. Psychother Psychosom. 2013;82:20–34.
104. Fava GA, Tomba E. New modalities of assessment and treatment planning in depression: the sequential approach. CNS Drugs. 2010;24:453–65.
105. Faravelli C. Assessment of psychopathology. Psychother Psychosom. 2004;73:139–41.
106. Fava GA, Ruini C, Rafanelli C. Psychometric theory is an obstacle to the progress of clinical research. Psychother Psychosom. 2004;73:145–8.
107. Tomba E, Bech P. Clinimetrics and clinical psychometrics: macro-and micro-analysis. Psychother Psychosom. 2012;81:333–43.

108. Fava GA, Tomba E, Sonino N. Clinimetrics: the science of clinical measurements. Int J Clin Pract. 2012;66(1):11–5.
109. Feinstein AR. An analysis of diagnostic reasoning: I. The domains and disorders of clinical macrobiology. Yale J Biol Med. 1973;46:212–32.
110. Feinstein AR. Clinimetrics. New Haven: Yale University Press; 1987.
111. Fava GA, Kellner R. Staging: a neglected dimension in psychiatric classification. Acta Psychiatr Scand. 1993;87:225–30.
112. Detre TP, Jarecki HG. Modern Psychiatric Treatment. Philadelphia: Lippincott; 1971.
113. Goldstein-Piekarski AN, Williams LM, Humphreys K. A trans-diagnostic review of anxiety disorder comorbidity and the impact of multiple exclusion criteria on studying clinical outcomes in anxiety disorders. Transl Psychiatry. 2016;6(6):e847.
114. Fava GA, Rafanelli C, Ottolini F, Ruini C, Cazzaro M, Grandi S. Psychological well-being and residual symptoms in remitted patients with panic disorder and agoraphobia. J Affect Disord. 2001;65:185–90.
115. Cosci F. The psychological development of panic disorder: implications for neurobiology and treatment. Braz J Psychiatry. 2012;34:9–19.
116. Guidi J, Tomba E, Cosci F, Park SK, Fava GA. The role of staging in planning psychotherapeutic interventions in depression. J Clin Psychiatry. 2017;78(4):456–63.

Comorbid Anxiety and Depression: Clinical and Conceptual Consideration and Transdiagnostic Treatment

14

Kwan Woo Choi, Yong-Ku Kim, and Hong Jin Jeon

Introduction

Although anxiety and depression have been considered as two distinct entities according to the diagnostic criteria, anxious depression (comorbid anxiety and depression) is relatively a common syndrome. It has been considered that 45–67% of patients with major depressive disorder (MDD) meet criteria for at least one comorbid anxiety disorder [1, 2]. Similarly, 30–63% of anxiety disorder patients meet criteria for concurrent MDD [3–5]. Recently there have been two kinds of attempts to define anxious depression, categorical and dimensional criteria. According to categorical criteria of anxious depression, one should meet the criteria of MDD depending on the DSM or ICD criteria and the presence of at least one comorbid anxiety disorder simultaneously [6]. Alternatively, the dimensional diagnosis of anxious depression depends on a diagnosis of MDD plus anxiety symptoms (either above threshold or subthreshold range, based on cutoff scores from psychological scales) [6, 7]. Even though there is no definition for anxious depression in the DSM-5 criteria, it uses "with anxious distress specifier" to define anxious depression in its MDD section. If one meets criteria of MDD plus at least two of five anxiety symptoms, such as "feeling keyed up or tense," "being unusually restless," "having trouble concentrating because of worry," "having fear that something awful

K. W. Choi
Department of Psychiatry, Korea University College of Medicine, Seoul, South Korea

Department of Psychiatry, Depression Center, Samsung Medical Center, Sungkyunkwan University School of Medicine, Seoul, South Korea

Y.-K. Kim
Department of Psychiatry, Korea University College of Medicine, Seoul, South Korea

H. J. Jeon (✉)
Department of Psychiatry, Depression Center, Samsung Medical Center, Sungkyunkwan University School of Medicine, Seoul, South Korea
e-mail: jeonhj@skku.edu

Y.-K. Kim (ed.), *Anxiety Disorders*, Advances in Experimental Medicine and Biology 1191, https://doi.org/10.1007/978-981-32-9705-0_14

may happen," and "feeling of that one might lose control of oneself," we can define him/her as having MDD with anxious distress [8]. As anxious depression is known to have clinically distinctive feature such as poor treatment response and lower remission rates, it is imperative to understand the current diagnostic concept, etiology, clinical course, prognosis, and current treatment strategy regarding anxious depression [9–11].

Diagnostic Consideration

Categorical Perspective

Previously, individuals with anxious depression, or comorbid anxiety and depression, were defined as having two disorders concomitantly, according to diagnostic criteria (ICD-10 or DSM diagnosis) as shown in Table 14.1 [12]. In the previous study by Young et al. [13], they approached clinical comorbid participants in the categorical way. The researcher compared the HPA axis function among four groups categorically, such as patients with MDD alone, those with anxiety disorder alone (social anxiety disorder, panic disorder, or posttraumatic stress disorder), those with comorbid anxiety and depressive disorder, and heathy controls.

Although diagnosis of anxious depression is a complicated issue, we could use hypotheses of five different psychobiological assumptions, such as "additive," "average," "single-disorder dominant," "distinctive," and "shared" [14].

A. Additive: Individuals with anxious depression show psychobiological features that are related with both of MDD and anxiety disorders in an additive manner.
B. Average: Individuals with anxious depression exhibit psychobiological patterns which are "in between" those for MDD and anxiety disorders.
C. Single-disorder dominant: Individuals with anxious depression show the characteristics which are observed by patients who are diagnosed with only one of the pure MDD or anxiety disorders alone.
D. Distinctive: Individuals with anxious depression show distinct psychobiological characteristics which are different from either of two disorders.
E. Shared: Individuals with both MDD and anxiety disorders exhibit common psychobiological characteristics.

Dimensional Perspective

Dimensional diagnosis of anxious depression is based on a MDD diagnosis plus subthreshold anxiety symptoms which might be based on cutoff scores from standardized scales as shown in Table 14.1 [6, 7]. In the previous study, subjects with anxious depression were defined as having MDD diagnosis plus concomitant high levels of anxiety which was defined as a baseline anxiety/somatization factor score ≥ 7 from the 17-item or 21-item HAM-D [12]. Recently, Seo et al. defined anxious depression,

Table 14.1 Dimensional and categorical definitions of anxious depression

Definition	Categorically defined	Dimensionally defined
Criteria	DSM (ICD) diagnosis of MDD, plus ICD or DSM diagnosis of an anxiety disorder	DSM (ICD) diagnosis of MDD, plus high levels of anxiety symptoms defined by a cutoff score on a standardized scale

using DSM-IV diagnosis of MDD, as a dysthymic disorder or depressive disorder not otherwise specified (NOS) plus a score of HAM-A of ≥20 [15].

The Research Domain Criteria (RDoC) initiative of the National Institute of Mental Health changed the current concept of psychiatric diseases [16]. The RDoC proposed dimensional and transdiagnostic approach that aims at investigating underlying neurobiological dysfunction, such as brain circuit dysfunction, for mental diseases. Recently, researchers in this field have taken dimensional approach to psychopathology in which subjects' functioning could be measured by continuous measures that operationalize core psychobiological constructs [16]. In this context, "unique," "shared," and "interactive" features of anxious depression might be needed to further understand the psychopathology for this comorbid disorder [14].

A. Unique: Unique features imply that manifestations of MDD and anxiety disorders are related to each of different psychobiological factors.
B. Shared: Shared features reflect that symptoms of MDD and anxiety disorders are correlated to similar biological factors.
C. Interactive: Interactive features mean that symptoms of MDD and anxiety disorders might interact or moderate each other, regarding psychobiological functioning.

Hypothetical Psychological Models

There are three hypothetical psychological models which explain the mechanism of anxious depression, such as the tripartite model, the approach-withdrawal model, and the valence-arousal model [17].

A. Tripartite model: It utilizes three dimensions as a framework for classifying manifestations of anxiety and depression, such as negative affectivity, positive affectivity, and physiological hyperarousal [18]. Depression might be related to the absence of positive affectivity (e.g., anhedonia), while increased psychological hyperarousal might be associated with anxiety. High level of negative affectivity could be seen in comorbid anxiety and depression state. However, this model does not identify specific neurobiological correlates that might be useful for understanding anxious depression.
B. Approach-withdrawal model: This model attempts to associate the clinical deficit of emotion and motivation observed in anxiety and depression. This model assumes two distinct systems for emotion and motivation [19]. The approach

system moderates behavioral motivation toward reward and implies left frontal regions, and it is hypothesized that this left frontal function is hypoactive in MDD. The withdrawal system moderates behavioral inhibition and is thought to be associated with right frontal lobe. In anxiety disorder, hyperactivity in right frontal lobe is hypothesized to be associated.

C. Valence-arousal model: This model expands on the approach-withdrawal model by implying hemispheric difference for arousal in anxiety and depression. Depression is associated with decreased activity in the right parietotemporal brain region for arousal, and anxiety correlates with increased activity in this region [20]. This model might explain the neurobiological mechanism of anxious depression.

Neurobiological Findings

Subjects with anxious depression are known to have different neurobiological profiles compared to those with non-anxious depression. Even with lack of research on this issue, several studies have revealed significant differences between anxious depression and non-anxious depression regarding the HPA axis function, structural and functional brain imaging findings and, inflammation markers.

Hypothalamic-Pituitary-Adrenal (HPA) Axis

Several studies showed stronger dysfunctions of the HPA axis in patients with anxious depression compared to non-anxious depression [21–24]. Meller et al. found that 14 subjects with anxious depression exhibited an attenuated adrenocorticotropic hormone (ACTH) response to exogenous corticotropin-releasing hormone (CRH) compared to 11 non-anxious depression and 27 healthy controls [21]. However, they allowed for heterogenous depression population including bipolar II disorder subjects, and they did not require to be medication free. Rao et al. showed more impaired suppression (50%) of cortisol following the dexamethasone suppression test (DST) in female subjects with anxious depression compared to female subjects with anxiety disorders (37%) and female subjects with MDD (18%) [24]. Also, Cameron et al. found that 18 subjects with anxious depression were the only group found to have increased ACTH and cortisol levels after Trier social stress test (TSST), compared to 15 subjects with anxiety disorder, 15 subjects with MDD, and 48 healthy controls [22]. Even though the researchers defined anxious depression as having diagnosis of MDD plus an anxiety disorder, there were limitations in that the definition included not only concurrent comorbidity of anxiety disorder but also diagnosis of anxiety disorder in the past year [22]. These results imply the HPA axis dysfunction in subjects with anxious depression. Recent review also implicated that chronic stress and HPA hyperactivity might initiate a cascade change involving the serotonergic system, which may be related to the pathophysiology of anxiety and depression [23]. However, more elaborate studies will be needed to overcome these previous limitations.

Structural Brain Imaging

Several studies used structural brain magnetic resonance imaging (MRI) to examine the difference between anxious depression and non-anxious depression. One structural MRI compared subjects with MDD, anxiety disorders, categorically defined anxious depression (MDD plus one of anxiety disorders including panic disorder, social anxiety disorder (SAD), and generalized anxiety disorder (GAD)), and healthy controls [25]. The researcher found that all patient groups had decreased gray matter (GM) volume of the rostral anterior cingulate cortex (ACC) (extending into the dorsal ACC) than healthy controls. This result was independent of severity of illness and suggested a shared neurobiological mechanism of dysfunctions in emotional processing and regulation between anxiety and depressive disorders. However, the researcher did not find any significant differences between subjects with anxious depression and those with non-anxious depression (25). In another structural brain imaging study, 96 MDD subjects were compared to 49 subjects with dimensionally defined anxious depression and 183 healthy controls [26]. The researcher defined anxious depression as having MDD with at least one of the following symptoms occurring concurrently: (1) general rating of anxiety, (2) general rating of phobia, (3) free-floating anxiety, and (4) anxious foreboding with autonomic symptoms. The researchers found increased GM volumes in the superior temporal gyrus, extending to the posterior middle temporal gyrus and inferior temporal gyrus in the right hemisphere in patients with anxious depression compared to those with non-anxious depression. However, no differences were found in these regions compared to healthy controls. These results implicate a plausible diagnosis-dependent change in GM thickness in the global pattern of sulcal or gyral structures [26]. Future research will be needed to find whether these changes can be explained by the valence-arousal or the approach-withdrawal mechanisms, which imply a hyperactive right hemisphere in subjects with anxious depression.

Functional Brain Imaging

There have been several functional brain MRI studies to investigate the difference between subjects with anxious depression and those with non-anxious depression. Using categorical criteria of MDD plus comorbid GAD to define anxious depression, Etkin et al. compared four groups of subjects with currently experiencing MDD, anxiety disorder, anxious depression, and healthy subjects during an emotional conflict identification task [27]. In the emotional conflict identification task, participants had to identify whether happy or fearful faces were labeled correctly while examined under fMRI. During incongruent stimuli, all patient groups showed deficits in both activation and connectivity of the ventral ACC and amygdala, which were known to be involved in the regulation of emotional conflict. These results suggest a shared pathophysiology between anxiety and depression. However, unlike the anxious depression group, the non-anxious depression group showed compensating findings for these deficits by also activating regions bilateral anterior lateral prefrontal cortices, improving their ability to regulate their emotional conflict.

Another fMRI study examined the differences between subjects with MDD, SAD, anxious depression (MDD plus SAD), and healthy controls [28]. Subjects completed a social evaluative threat task. In this study, subjects with anxious depression showed similar pattern to the other two patients group, except for an intermediate level of activation of the middle cingulate cortex and precentral gyrus (less than the MDD group and more than the SAD group) and posterior cingulate (more than the MDD group and less than the SAD group). Additionally, patients with anxious depression and healthy controls revealed similar activation patterns in several regions: greater activation of the insula and middle temporal gyrus and less activation of the cerebellum and cuneus [28]. There were also several fMRI studies to investigate the difference between anxious depression and non-anxious depression, following the cognitive stimulating tasks. Andreescu et al. examined subjects which were older than 65 years old. The researchers found that subjects with anxious depression showed a greater and more prolonged activation of the dorsal ACC, prefrontal cortex supplementary motor area, and the posterior cingulate during an executive control task [29]. In their other study, Andreescu et al. found that subjects with anxious depression had a dissociative pattern in default mode network, specifically an increased connectivity in the posterior regions (occipital and parietal association areas) and a decreased connectivity in the anterior regions (rostral ACC, medial frontal and orbitofrontal cortex) [30]. According to one recently published study, the researcher compared functional connectivity between subjects with MDD only, with SAD only, and with MDD plus SAD [31]. Hamilton et al. attempted to examine anomalies in neural response to positive and negative self- and other-related stimuli in these subjects. In subjects with MDD, the researchers found a cortically diffuse reduction in the height of neural response to self-related positive stimuli, which implies cortical hyporesponse to positive stimuli in MDD. In subjects with SAD, increased insula response (a part of salience network) was found, and decreased default mode network (DMN) response to negative self-related cues was shown. These findings suggest neural underlying mechanisms of vigilance-avoidance in SAD or anxiety disorders. Especially, MDD plus SAD subjects uniquely exhibited decreased response in the dorsal ACC to positive self-related cues [31]. Meta-analyses of anxiety and of depression have shown that abnormal dorsal ACC response to affective stimuli is a consistent aspect of these disorders [32, 33]. Given the reduced dorsal ACC response to positive self-related stimuli in the MDD plus SAD group, this region might have acted as a hub area for the neural effects of negative self-image that are associated with MDD and anxiety regarding social interactions [31]. However, further studies will be needed to reveal more consistent and sophisticated structural and functional neuroimaging biomarkers to differentiate the anxious depression subtype from non-anxious depression group.

Neuropsychiatric and Sensory Testing

Several studies found significant association between neuropsychiatric and sensory testing and anxious depression. Fixed-response design fluency tasks required participants to draw as many as novel designs as possible during 5 min. These tasks were intended to test the neuronal circuit underlying withdrawal in the right hemisphere [34]. Nelson et al. revealed that 30 anxious depression subjects (categorically defined anxious depression; MDD plus lifetime anxiety disorder including SAD, panic disorder, specific phobia, posttraumatic stress disorder (PTSD), or obsessive compulsive disorder (OCD)) showed significantly lower performance levels on fixed response design fluency tasks than 34 MDD patients and 33 healthy controls [34]. These findings implicate hemispheric-specific deficit of the right frontal lobe in subjects with anxious depression accords with the approach-withdrawal hypothesis, which anxiety is associated with right frontal lobe dysfunction. The researchers in this study replicated their results using several different design fluency tasks in subjects with MDD plus current panic disorders. Subjects with anxious depression showed better performance on the verbal tasks relative to the design fluency, which might confirm hemispheric asymmetries [34].

Electroencephalography

Electroencephalography (EEG) examines hemispheric asymmetries directly. Bruder et al. compared patients with anxious depression, patients with MDD, and healthy controls on resting EEG. In this study, patients with anxious depression showed less alpha activity (greater activation) over the right anterior hemisphere than the left hemisphere [35]. More recently conducted study investigated task-related EEG differences among depressed patients with high anxiety, low anxiety, and healthy controls during word finding (verbal) and dot localization (spatial) neurocognitive tasks [36]. Although the researcher did not find any group differences between these three groups, the high-anxiety group showed more pronounced activation pattern in the right central and parietal regions during spatial task compared to the left. Otherwise, the low-anxiety group showed more pronounced left frontal and central activation during verbal task [36]. These electrophysiologic data are consistent with the approach-withdrawal theory and the valence-arousal mechanism, which are thought to be underlying hypothetical mechanisms in anxious depression [17].

Inflammation

Recently, inflammation has acquired interest as an important neurobiological mechanism in anxious depression. Several meta-analyses revealed that inflammatory markers, such as interleukin (IL) -1, IL-6, C-reactive protein (CRP), tumor necrosis factor alpha (TNF-α), etc., were more often present in subjects with MDD compared to healthy controls [37–39]. A number of studies also suggested neuroinflammation as a pathophysiologic marker of anxiety disorders [40–43]. To our knowledge, few studies have investigated the association between inflammation and anxious depression. Shim et al. found higher monocyte counts in subjects with MDD and moderate-severe to severe anxious distress compared to in those with mild to moderate group [44]. Baek et al. revealed reduced venous blood basophil counts and elevated fragmented neutrophils in subjects with MDD and high level of anxiety [45]. These findings implicated alterations in white blood cell subset counts and suggest alterations in the immune system indirectly, especially in anxious depression. Other than these two studies, Gaspersz et al. investigated not only basal inflammation markers but also markers of innate cytokine production capacity such as inflammatory markers stimulated by lipopolysaccharide (LPS) [46]. Such markers may give us more insight into the functioning of the immune system [47]. The researchers compared these inflammatory markers between 585 MDD subjects with anxious distress and 494 MDD subjects without anxious distress. According to this study, anxious distress was associated with higher LPS-stimulated inflammation levels (including interferon-gamma, IL-6, monocyte chemotactic protein-1, macrophage inflammatory protein-1 alpha, matrix metalloproteinase-2, TNF-alpha, etc.) [46]. These results support that anxious depression has distinct neurobiological inflammatory processes form MDD alone. However, since these previous studies had all cross-sectional design, large-scale multicenter well-designed longitudinal research will be needed to find more definite findings regarding neuroinflammation in anxious depression.

Clinical Characteristics

Comorbid anxiety and depressive disorder, also known as anxious depression, have demographical and clinical characteristics which are distinct from non-anxious depression. Demographically, according to the Sequenced Treatment Alternatives to Relieve Depression (STAR∗D) project, patients with anxious depression were significantly more likely to be in primary care setting and more likely to be associated with female gender, non-single, unemployed, less educated, and severe depression [48]. Fava et al. also found that anxious depression patients were more likely to meet diagnostic threshold for GAD and somatoform disorder [48]. Hovens et al. and de Graaf et al. also found that patients with anxious depression were less educated and less often employed and more often had a positive parental psychiatric history and child trauma [49, 50]. Comorbidity of depressive and anxiety disorders has consistently shown to have earlier age of onset compared to non-anxious MDD both in cross-sectional surveys and prospective studies [3, 51, 52]. Previous epidemiologic studies also showed that MDD with comorbid anxiety disorders tended to be more severe and persistent compared to non-anxious depression [3, 53–55]. Previous reports also showed that patients with anxious depression had more frequent episodes of major depression and a higher risk of suicidal ideation and previous suicide attempts than those with non-anxious depression [1, 48]. According to the WHO world mental health surveys, which were the nationally or representative epidemiological interview from 24 countries, Kessler et al. found that higher proportions of respondents with 12-month anxious depression reported severe role impairment and suicidal ideation [56]. Similarly, several studies found significantly increased risk of suicidal ideation in patients with anxious depression [15, 57, 58]. Patients with anxious depression are more likely to seek mental health-care services compared to non-anxious depression or anxiety disorder patients [55, 56]. However, patients with anxious depression showed significantly lower response to treatment compared to those with non-anxious depression [59, 60]. For these reasons, different therapeutic approaches are needed to treat patients with anxious depression more efficiently.

Treatment

Anxious depression is known to be associated with greater symptom severity and poor treatment outcomes in several studies [1, 61, 62]. One naturalistic study showed that subjects with anxious depression revealed worse responses to a wide range of antidepressants such as selective serotonin reuptake inhibitors (SSRI) and serotonin norepinephrine reuptake inhibitors (SNRI), longer treatment response time, and smaller decrease in symptoms [63]. Another naturalistic study found that subjects with anxious depression took double the time to recovery from depressive episode than those with non-anxious depression [64].

SSRI

Several studies revealed that SSRIs were effective to treat anxious depression. One previous study compared the efficacy and tolerability of three different SSRIs including fluoxetine, sertraline, and paroxetine for the treatment of anxious depression [65]. Response rates and remission rates were more than 50% for all three treatment groups. Another study pooled results of five studies investigating the efficacy of sertraline or fluoxetine in anxious depression patients [66]. The researchers found that both antidepressants improved several depressive symptoms similarly; 70% of patients achieved response, and 47% achieved remission. A previous study reanalyzed the data from five different studies investigating the efficacy of escitalopram, citalopram, and sertraline in 1690 MDD patients, of which 756 with anxious depression [67]. In this study, SSRI treatment was superior to placebo for treating those with severe depression on the basis of the response rates, but not the remission rates, whereas patients with non-anxious depression showed significant improvement on both of the response rates and remission rates. When patients with severe depression were analyzed separately, anxiety status was considered to serve as a treatment moderator, so the SSRI treatment could be more effective than placebo only for severe non-anxious depression [67].

SNRI

Duloxetine has been examined by many researchers as a potent treatment medication for anxious depression. An 8-week duloxetine study found that anxious depression patients showed greater improvements at all points (HAM-D scores), and the improvements increased proportionally over time compared to non-anxious depression patients [68]. Another open-label duloxetine trial also found that subjects with anxious depression improved significantly especially in the later course of treatment measured by HAM-D compared to those with non-anxious depression. In their duloxetine study, subjects with anxious depression also showed significantly shorter median time to respond to the duloxetine therapy, although remission and response rates at the end point were similar [69]. According to another post hoc study from 11 duloxetine trials, response rates to duloxetine were higher than placebo in both anxious and non-depression groups. Anxious depression group showed significantly higher response rates than non-anxious depression group, whereas anxious depression group were less likely to experience remission than non-anxious group [70]. In one study comparing the efficacy of venlafaxine extended release (XR) to reboxetine, no significant differences were observed in response and remission rates from depression between treatment groups in patients with anxious depression [71].

Benzodiazepine Augmentation

In clinical practice, benzodiazepines are usually used as augmenting medications [72]. In the previous double-blind study, clonazepam augmentation of fluoxetine treatment was found to be superior to fluoxetine monotherapy for treating depressive symptoms in week 3 [73]. A review article also supported that benzodiazepine augmentation of SSRI might result in rapid control of baseline anxiety, as well as SSRI-induced anxiety, and might improve adherence to antidepressant treatment [72]. However, it should be noted that the benefits of treatment must be outweighed against the risk of side effects such as cognitive slowing in elderly patients, transient sedation, abuse potentials, and withdrawal symptoms.

SSRI Versus Bupropion

Papakostas et al. conducted a meta-analysis of the previous 10 studies comparing the efficacy of bupropion, dopamine, and norepinephrine reuptake inhibitor (DNRI), with SSRIs (escitalopram, fluoxetine, sertraline, and paroxetine) for the treatment of anxious depression [74]. Response rates were greater with SSRI treatment than with bupropion on the basis of total scores from HAM-D and HAM-A. However, remission rates were not different [74].

SSRI Versus SNRI

A post hoc study of pooled data from five different studies examined the difference of efficacy of venlafaxine to fluoxetine or placebo in 1454 outpatients with MDD [75]. Remission rates in the severe anxious depression group were significantly greater for venlafaxine compared to placebo from week 3 (17% vs. 6%, $p = 0.003$) to week 6 (34% vs. 15%, $p < 0.001$). There was no significant improvement in the fluoxetine group, and remission rates for venlafaxine were higher than those in the fluoxetine group at week 3 and week 6 [75]. However, a previous post hoc subgroup analysis of 8-week study comparing venlafaxine XR to sertraline found that there were no significant differences in their treatment outcomes of anxious depression between two groups [76]. Another post hoc study investigated the efficacy of escitalopram in treating anxious depression from 13 trials [77]. In this study, researchers found that escitalopram was more effective than placebo and same as effective as the SNRI or the other SSRIs regarding treating anxious depression. There were no significant differences in response rates, adherence, tolerability, or side effect profiles between anxious depression group and non-anxious depression group. However, anxious depression group had lower remission rates than non-anxious depression group [77].

Atypical Antipsychotics

Atypical antipsychotics, specifically quetiapine and aripiprazole, have been approved by the Food and Drug Administration as augmentation agents in conjunction with serotonergic antidepressants for the treatment of MDD. Augmentation treatment with atypical antipsychotics showed approximately double the odds of achieving remission compared to augmentation with placebo in patients with treatment-resistant depression (TRD) [78–80]. Several studies examined data of atypical antipsychotics aripiprazole and quetiapine for the treatment of anxious depression, and atypical antipsychotics are usually used as augmentation agents to improve symptoms among TRD patients with anxious depression [81, 82]. Ionescu et al. recently showed that ziprasidone augmentation of escitalopram was equally effective in treating depression in patients with anxious depression and patents with non-anxious depression [83].

Recent Medication Studies

Ketamine, an N-methyl-D-aspartate (NMDA) receptor antagonist and glutaminergic modulator, has gained considerable attention in the past decade for the treatment of TRD [84]. Several controlled clinical trials supported the rapid and robust therapeutic effect of ketamine for the treatment of TRD. Two post hoc studies published and suggested that ketamine was efficacious in both anxious and non-anxious depression. One study showed relatively higher ketamine efficacy in the anxious depression group [85], whereas another study revealed that there was similar response to ketamine in both anxious and non-anxious depression groups [86]. Further investigation will be needed to clarify these controversial results.

Dysregulation of the endogenous opioid system is one of the candidate mechanisms to contribute to the pathophysiology of MDD [87]. Recently, Richards et al. examined AZD2327, a selective delta-opioid receptor agonist, for the treatment of anxious depression [88]. In the double-blind, placebo-controlled pilot study in humans (n = 22), patients were supposed to take 4 weeks of either AZD2327 3 mg bid or placebo. Although there were no significant differences between the two arms based on the score of HAM-D or HAM-A, one of the major metabolites of AZD 2327 had significantly higher levels in patients who had an anxiolytic response compared to non-responders [88]. This study might provide preliminary evidence for the potential anxiolytic effects of opioid receptor-associated medications.

Brexpiprazole, an atypical antipsychotic, was recently approved in the USA, Saudi Arabia, Honduras, and Mexico as an adjunctive treatment to antidepressants for the treatment of MDD in adults, in October 2018. Recently published one post hoc analysis of three brexpiprazole studies revealed that 2–3 mg/day of brexpiprazole as an augmentation treatment to antidepressant was efficacious in patients with anxious depression [89].

Conclusion

MDD is a heterogenous and complicated mental disorder. RDoC and recent transdiagnostic approaches have attempted to investigate distinct neurobiological mechanisms of different types of MDD. Anxious depression or comorbid anxiety and depressive disorder is currently diagnosed with MDD with anxious distress specifier according to the DSM-5. We reviewed possible neurobiological correlates including structural and functional brain imaging and also examined demographical and clinical course in the patients with anxious depression. Although treatment response and treatment were worse in the patients with anxious depression than in those with non-anxious depression, recent research has sought to find better treatment strategy to improve anxious depression.

References

1. Fava M, Alpert JE, Carmin CN, Wisniewski SR, Trivedi MH, Biggs MM, et al. Clinical correlates and symptom patterns of anxious depression among patients with major depressive disorder in STAR*D. Psychol Med. 2004;34(7):1299–308.
2. Kessler RC, Berglund P, Demler O, Jin R, Koretz D, Merikangas KR, et al. The epidemiology of major depressive disorder: results from the National Comorbidity Survey Replication (NCS-R). JAMA. 2003;289(23):3095–105.
3. Lamers F, van Oppen P, Comijs HC, Smit JH, Spinhoven P, van Balkom AJ, et al. Comorbidity patterns of anxiety and depressive disorders in a large cohort study: the Netherlands study of depression and anxiety (NESDA). J Clin Psychiatry. 2011;72(3):341–8.
4. Fava M, Rankin MA, Wright EC, Alpert JE, Nierenberg AA, Pava J, et al. Anxiety disorders in major depression. Compr Psychiatry. 2000;41(2):97–102.
5. Brown TA, Campbell LA, Lehman CL, Grisham JR, Mancill RB. Current and lifetime comorbidity of the DSM-IV anxiety and mood disorders in a large clinical sample. J Abnorm Psychol. 2001;110(4):585–99.
6. Rao S, Zisook S. Anxious depression: clinical features and treatment. Curr Psychiatry Rep. 2009;11(6):429–36.
7. Rush AJ. The varied clinical presentations of major depressive disorder. J Clin Psychiatry. 2007;68(Suppl 8):4–10.

8. American Psychiatric Association. Diagnostic and statistical manual of mental disorders. 5th ed. Washington, DC: American Psychiatric Association; 2013.
9. Wu Z, Chen J, Yuan C, Hong W, Peng D, Zhang C, et al. Difference in remission in a Chinese population with anxious versus nonanxious treatment-resistant depression: a report of OPERATION study. J Affect Disord. 2013;150(3):834–9.
10. Papakostas GI, McGrath P, Stewart J, Charles D, Chen Y, Mischoulon D, et al. Psychic and somatic anxiety symptoms as predictors of response to fluoxetine in major depressive disorder. Psychiatry Res. 2008;161(1):116–20.
11. Souery D, Oswald P, Massat I, Bailer U, Bollen J, Demyttenaere K, et al. Clinical factors associated with treatment resistance in major depressive disorder: results from a European multicenter study. J Clin Psychiatry. 2007;68(7):1062–70.
12. Ionescu DF, Niciu MJ, Henter ID, Zarate CA. Defining anxious depression: a review of the literature. CNS Spectr. 2013;18(5):252–60.
13. Young EA, Abelson JL, Cameron OG. Effect of comorbid anxiety disorders on the hypothalamic-pituitary-adrenal axis response to a social stressor in major depression. Biol Psychiatry. 2004;56(2):113–20.
14. Kircanski K, LeMoult J, Ordaz S, Gotlib IH. Investigating the nature of co-occurring depression and anxiety: comparing diagnostic and dimensional research approaches. J Affect Disord. 2017;216:123–35.
15. Seo HJ, Jung YE, Kim TS, Kim JB, Lee MS, Kim JM, et al. Distinctive clinical characteristics and suicidal tendencies of patients with anxious depression. J Nerv Ment Dis. 2011;199(1):42–8.
16. Insel T, Cuthbert B, Garvey M, Heinssen R, Pine DS, Quinn K, et al. Research domain criteria (RDoC): toward a new classification framework for research on mental disorders. Am J Psychiatry. 2010;167(7):748–51.
17. Ionescu DF, Niciu MJ, Mathews DC, Richards EM, Zarate CA Jr. Neurobiology of anxious depression: a review. Depress Anxiety. 2013;30(4):374–85.
18. Clark LA, Watson D. Tripartite model of anxiety and depression: psychometric evidence and taxonomic implications. J Abnorm Psychol. 1991;100(3):316–36.
19. Davidson RJ. Anterior cerebral asymmetry and the nature of emotion. Brain Cogn. 1992;20(1):125–51.
20. Heller W, Etienne MA, Miller GA. Patterns of perceptual asymmetry in depression and anxiety: implications for neuropsychological models of emotion and psychopathology. J Abnorm Psychol. 1995;104(2):327–33.
21. Meller WH, Kathol RG, Samuelson SD, Gehris TL, Carroll BT, Pitts AF, et al. CRH challenge test in anxious depression. Biol Psychiatry. 1995;37(6):376–82.
22. Cameron OG. Anxious-depressive comorbidity: effects on HPA axis and CNS noradrenergic functions. Essent Psychopharmacol. 2006;7(1):24–34.
23. Leonard BE, Myint A. The psychoneuroimmunology of depression. Hum Psychopharmacol. 2009;24(3):165–75.
24. Rao ML, Vartzopoulos D, Fels K. Thyroid function in anxious and depressed patients. Pharmacopsychiatry. 1989;22(2):66–70.
25. van Tol MJ, van der Wee NJ, van den Heuvel OA, Nielen MM, Demenescu LR, Aleman A, et al. Regional brain volume in depression and anxiety disorders. Arch Gen Psychiatry. 2010;67(10):1002–11.
26. Inkster B, Rao AW, Ridler K, Nichols TE, Saemann PG, Auer DP, et al. Structural brain changes in patients with recurrent major depressive disorder presenting with anxiety symptoms. J Neuroimaging. 2011;21(4):375–82.
27. Etkin A, Schatzberg AF. Common abnormalities and disorder-specific compensation during implicit regulation of emotional processing in generalized anxiety and major depressive disorders. Am J Psychiatry. 2011;168(9):968–78.
28. Waugh CE, Hamilton JP, Chen MC, Joormann J, Gotlib IH. Neural temporal dynamics of stress in comorbid major depressive disorder and social anxiety disorder. Biol Mood Anxiety Disord. 2012;2:11.

29. Andreescu C, Butters M, Lenze EJ, Venkatraman VK, Nable M, Reynolds CF 3rd, et al. fMRI activation in late-life anxious depression: a potential biomarker. Int J Geriatr Psychiatry. 2009;24(8):820–8.
30. Andreescu C, Wu M, Butters MA, Figurski J, Reynolds CF 3rd, Aizenstein HJ. The default mode network in late-life anxious depression. Am J Geriatr Psychiatry. 2011;19(11):980–3.
31. Hamilton JP, Chen MC, Waugh CE, Joormann J, Gotlib IH. Distinctive and common neural underpinnings of major depression, social anxiety, and their comorbidity. Soc Cogn Affect Neurosci. 2015;10(4):552–60.
32. Etkin A, Wager TD. Functional neuroimaging of anxiety: a meta-analysis of emotional processing in PTSD, social anxiety disorder, and specific phobia. Am J Psychiatry. 2007;164(10):1476–88.
33. Hamilton JP, Etkin A, Furman DJ, Lemus MG, Johnson RF, Gotlib IH. Functional neuroimaging of major depressive disorder: a meta-analysis and new integration of base line activation and neural response data. Am J Psychiatry. 2012;169(7):693–703.
34. Nelson BD, Sarapas C, Robison-Andrew EJ, Altman SE, Campbell ML, Shankman SA. Frontal brain asymmetry in depression with comorbid anxiety: a neuropsychological investigation. J Abnorm Psychol. 2012;121(3):579–91.
35. Bruder GE, Fong R, Tenke CE, Leite P, Towey JP, Stewart JE, et al. Regional brain asymmetries in major depression with or without an anxiety disorder: a quantitative electroencephalographic study. Biol Psychiatry. 1997;41(9):939–48.
36. Manna CB, Tenke CE, Gates NA, Kayser J, Borod JC, Stewart JW, et al. EEG hemispheric asymmetries during cognitive tasks in depressed patients with high versus low trait anxiety. Clin EEG Neurosci. 2010;41(4):196–202.
37. Howren MB, Lamkin DM, Suls J. Associations of depression with C-reactive protein, IL-1, and IL-6: a meta-analysis. Psychosom Med. 2009;71(2):171–86.
38. Dowlati Y, Herrmann N, Swardfager W, Liu H, Sham L, Reim EK, et al. A meta-analysis of cytokines in major depression. Biol Psychiatry. 2010;67(5):446–57.
39. Liu Y, Ho RC, Mak A. Interleukin (IL)-6, tumour necrosis factor alpha (TNF-alpha) and soluble interleukin-2 receptors (sIL-2R) are elevated in patients with major depressive disorder: a meta-analysis and meta-regression. J Affect Disord. 2012;139(3):230–9.
40. Vogelzangs N, de Jonge P, Smit JH, Bahn S, Penninx BW. Cytokine production capacity in depression and anxiety. Transl Psychiatry. 2016;6(5):e825.
41. Vogelzangs N, Beekman AT, de Jonge P, Penninx BW. Anxiety disorders and inflammation in a large adult cohort. Transl Psychiatry. 2013;3:e249.
42. Furtado M, Katzman MA. Neuroinflammatory pathways in anxiety, posttraumatic stress, and obsessive compulsive disorders. Psychiatry Res. 2015;229(1–2):37–48.
43. Liukkonen T, Rasanen P, Jokelainen J, Leinonen M, Jarvelin MR, Meyer-Rochow VB, et al. The association between anxiety and C-reactive protein (CRP) levels: results from the Northern Finland 1966 birth cohort study. Eur Psychiatry. 2011;26(6):363–9.
44. Shim IH, Woo YS, Bahk WM. Associations between immune activation and the current severity of the "with anxious distress" specifier in patients with depressive disorders. Gen Hosp Psychiatry. 2016;42:27–31.
45. Baek JH, Kim HJ, Fava M, Mischoulon D, Papakostas GI, Nierenberg A, et al. Reduced venous blood basophil count and anxious depression in patients with major depressive disorder. Psychiatry Investig. 2016;13(3):321–6.
46. Gaspersz R, Lamers F, Wittenberg G, Beekman ATF, van Hemert AM, Schoevers RA, et al. The role of anxious distress in immune dysregulation in patients with major depressive disorder. Transl Psychiatry. 2017;7(12):1268.
47. van der Linden MW, Huizinga TW, Stoeken DJ, Sturk A, Westendorp RG. Determination of tumour necrosis factor-alpha and interleukin-10 production in a whole blood stimulation system: assessment of laboratory error and individual variation. J Immunol Methods. 1998;218(1–2):63–71.

48. Fava M, Rush AJ, Alpert JE, Carmin CN, Balasubramani GK, Wisniewski SR, et al. What clinical and symptom features and comorbid disorders characterize outpatients with anxious major depressive disorder: a replication and extension. Can J Psychiatry. 2006;51(13):823–35.
49. de Graaf R, Bijl RV, Smit F, Vollebergh WA, Spijker J. Risk factors for 12-month comorbidity of mood, anxiety, and substance use disorders: findings from the Netherlands mental health survey and incidence study. Am J Psychiatry. 2002;159(4):620–9.
50. Hovens JG, Wiersma JE, Giltay EJ, van Oppen P, Spinhoven P, Penninx BW, et al. Childhood life events and childhood trauma in adult patients with depressive, anxiety and comorbid disorders vs. controls. Acta Psychiatr Scand. 2010;122(1):66–74.
51. Kessler RC, Ormel J, Petukhova M, McLaughlin KA, Green JG, Russo LJ, et al. Development of lifetime comorbidity in the World Health Organization world mental health surveys. Arch Gen Psychiatry. 2011;68(1):90–100.
52. Klein DN, Glenn CR, Kosty DB, Seeley JR, Rohde P, Lewinsohn PM. Predictors of first lifetime onset of major depressive disorder in young adulthood. J Abnorm Psychol. 2013;122(1):1–6.
53. Fichter MM, Quadflieg N, Fischer UC, Kohlboeck G. Twenty-five-year course and outcome in anxiety and depression in the upper Bavarian longitudinal community study. Acta Psychiatr Scand. 2010;122(1):75–85.
54. Roy-Byrne PP, Stang P, Wittchen HU, Ustun B, Walters EE, Kessler RC. Lifetime panic-depression comorbidity in the National Comorbidity Survey. Association with symptoms, impairment, course and help-seeking. Br J Psychiatry. 2000;176:229–35.
55. McLaughlin TP, Khandker RK, Kruzikas DT, Tummala R. Overlap of anxiety and depression in a managed care population: prevalence and association with resource utilization. J Clin Psychiatry. 2006;67(8):1187–93.
56. Kessler RC, Sampson NA, Berglund P, Gruber MJ, Al-Hamzawi A, Andrade L, et al. Anxious and non-anxious major depressive disorder in the World Health Organization world mental health surveys. Epidemiol Psychiatr Sci. 2015;24(3):210–26.
57. Baek JH, Heo JY, Fava M, Mischoulon D, Nierenberg A, Hong JP, et al. Anxiety symptoms are linked to new-onset suicidal ideation after six months of follow-up in outpatients with major depressive disorder. J Affect Disord. 2015;187:183–7.
58. McIntyre RS, Woldeyohannes HO, Soczynska JK, Vinberg M, Cha DS, Lee Y, et al. The prevalence and clinical characteristics associated with diagnostic and statistical manual Version-5-defined anxious distress specifier in adults with major depressive disorder: results from the international mood disorders collaborative project. Ther Adv Chronic Dis. 2016;7(3):153–9.
59. Jakubovski E, Bloch MH. Prognostic subgroups for citalopram response in the STAR∗D trial. J Clin Psychiatry. 2014;75(7):738–47.
60. Saveanu R, Etkin A, Duchemin AM, Goldstein-Piekarski A, Gyurak A, Debattista C, et al. The international study to predict optimized treatment in depression (iSPOT-D): outcomes from the acute phase of antidepressant treatment. J Psychiatr Res. 2015;61:1–12.
61. Fava M, Rush AJ, Alpert JE, Balasubramani GK, Wisniewski SR, Carmin CN, et al. Difference in treatment outcome in outpatients with anxious versus nonanxious depression: a STAR∗D report. Am J Psychiatry. 2008;165(3):342–51.
62. Domschke K, Deckert J, Arolt V, Baune BT. Anxious versus non-anxious depression: difference in treatment outcome. J Psychopharmacol. 2010;24(4):621–2.
63. Altamura AC, Montresor C, Salvadori D, Mundo E. Does comorbid subthreshold anxiety affect clinical presentation and treatment response in depression? A preliminary 12-month naturalistic study. Int J Neuropsychopharmacol. 2004;7(4):481–7.
64. Clayton PJ, Grove WM, Coryell W, Keller M, Hirschfeld R, Fawcett J. Follow-up and family study of anxious depression. Am J Psychiatry. 1991;148(11):1512–7.
65. Fava M, Rosenbaum JF, Hoog SL, Tepner RG, Kopp JB, Nilsson ME. Fluoxetine versus sertraline and paroxetine in major depression: tolerability and efficacy in anxious depression. J Affect Disord. 2000;59(2):119–26.

66. Feiger AD, Flament MF, Boyer P, Gillespie JA. Sertraline versus fluoxetine in the treatment of major depression: a combined analysis of five double-blind comparator studies. Int Clin Psychopharmacol. 2003;18(4):203–10.
67. Papakostas GI, Fan H, Tedeschini E. Severe and anxious depression: combining definitions of clinical sub-types to identify patients differentially responsive to selective serotonin reuptake inhibitors. Eur Neuropsychopharmacol. 2012;22(5):347–55.
68. Nasso ED, Chiesa A, Serretti A, De Ronchi D, Mencacci C. Clinical and demographic predictors of improvement during duloxetine treatment in patients with major depression: an open-label study. Clin Drug Investig. 2011;31(6):385–405.
69. Fava M, Martinez JM, Greist J, Marangell LB, Brown E, Chen L, et al. The efficacy and tolerability of duloxetine in the treatment of anxious versus non-anxious depression: a post-hoc analysis of an open-label outpatient study. Ann Clin Psychiatry. 2007;19(3):187–95.
70. Nelson JC. Effects of baseline depression severity on remission rates with duloxetine and placebo in anxious and nonanxious patients with major depression. J Clin Psychopharmacol. 2011;31(5):682–4.
71. Akkaya C, Sivrioglu EY, Akgoz S, Eker SS, Kirli S. Comparison of efficacy and tolerability of reboxetine and venlafaxine XR in major depression and major depression with anxiety features: an open label study. Hum Psychopharmacol. 2006;21(5):337–45.
72. Dunlop BW, Davis PG. Combination treatment with benzodiazepines and SSRIs for comorbid anxiety and depression: a review. Prim Care Companion J Clin Psychiatry. 2008;10(3):222–8.
73. Smith WT, Londborg PD, Glaudin V, Painter JR. Short-term augmentation of fluoxetine with clonazepam in the treatment of depression: a double-blind study. Am J Psychiatry. 1998;155(10):1339–45.
74. Papakostas GI, Stahl SM, Krishen A, Seifert CA, Tucker VL, Goodale EP, et al. Efficacy of bupropion and the selective serotonin reuptake inhibitors in the treatment of major depressive disorder with high levels of anxiety (anxious depression): a pooled analysis of 10 studies. J Clin Psychiatry. 2008;69(8):1287–92.
75. Davidson JR, Meoni P, Haudiquet V, Cantillon M, Hackett D. Achieving remission with venlafaxine and fluoxetine in major depression: its relationship to anxiety symptoms. Depress Anxiety. 2002;16(1):4–13.
76. Sir A, D'Souza RF, Uguz S, George T, Vahip S, Hopwood M, et al. Randomized trial of sertraline versus venlafaxine XR in major depression: efficacy and discontinuation symptoms. J Clin Psychiatry. 2005;66(10):1312–20.
77. Papakostas GI, Larsen K. Testing anxious depression as a predictor and moderator of symptom improvement in major depressive disorder during treatment with escitalopram. Eur Arch Psychiatry Clin Neurosci. 2011;261(3):147–56.
78. Papakostas GI, Shelton RC, Smith J, Fava M. Augmentation of antidepressants with atypical antipsychotic medications for treatment-resistant major depressive disorder: a meta-analysis. J Clin Psychiatry. 2007;68(6):826–31.
79. Nelson JC, Papakostas GI. Atypical antipsychotic augmentation in major depressive disorder: a meta-analysis of placebo-controlled randomized trials. Am J Psychiatry. 2009;166(9):980–91.
80. Spielmans GI, Berman MI, Linardatos E, Rosenlicht NZ, Perry A, Tsai AC. Adjunctive atypical antipsychotic treatment for major depressive disorder: a meta-analysis of depression, quality of life, and safety outcomes. PLoS Med. 2013;10(3):e1001403.
81. Thase ME, Demyttenaere K, Earley WR, Gustafsson U, Udd M, Eriksson H. Extended release quetiapine fumarate in major depressive disorder: analysis in patients with anxious depression. Depress Anxiety. 2012;29(7):574–86.
82. Trivedi MH, Thase ME, Fava M, Nelson CJ, Yang H, Qi Y, et al. Adjunctive aripiprazole in major depressive disorder: analysis of efficacy and safety in patients with anxious and atypical features. J Clin Psychiatry. 2008;69(12):1928–36.
83. Ionescu DF, Shelton RC, Baer L, Meade KH, Swee MB, Fava M, et al. Ziprasidone augmentation for anxious depression. Int Clin Psychopharmacol. 2016;31(6):341–6.

84. Sanacora G, Frye MA, McDonald W, Mathew SJ, Turner MS, Schatzberg AF, et al. A consensus statement on the use of ketamine in the treatment of mood disorders. JAMA Psychiat. 2017;74(4):399–405.
85. Ionescu DF, Luckenbaugh DA, Niciu MJ, Richards EM, Slonena EE, Vande Voort JL, et al. Effect of baseline anxious depression on initial and sustained antidepressant response to ketamine. J Clin Psychiatry. 2014;75(9):e932–8.
86. Salloum NC, Fava M, Freeman MP, Flynn M, Hoeppner B, Hock RS, et al. Efficacy of intravenous ketamine treatment in anxious versus nonanxious unipolar treatment-resistant depression. Depress Anxiety. 2019;36(3):235–43.
87. Lutz PE, Kieffer BL. Opioid receptors: distinct roles in mood disorders. Trends Neurosci. 2013;36(3):195–206.
88. Richards EM, Mathews DC, Luckenbaugh DA, Ionescu DF, Machado-Vieira R, Niciu MJ, et al. A randomized, placebo-controlled pilot trial of the delta opioid receptor agonist AZD2327 in anxious depression. Psychopharmacology (Berl). 2016;233(6):1119–30.
89. Thase ME, Weiller E, Zhang P, Weiss C, McIntyre RS. Adjunctive brexpiprazole in patients with major depressive disorder and anxiety symptoms: post hoc analyses of three placebo-controlled studies. Neuropsychiatr Dis Treat. 2019;15:37–45.

Anxiety Disorders and Medical Comorbidity: Treatment Implications

15

Alicia E. Meuret, Natalie Tunnell, and Andres Roque

This chapter identifies medical illnesses with high co-occurrence with anxiety disorders and offers an analysis of implications for treatment of both types of conditions. We concentrate on medical conditions with high associations to anxiety and panic by aspects of symptomatology, specifically neurological disorders (fibromyalgia, epilepsy, cerebral palsy), non-cardiac chest pain, diabetes, gastrointestinal illness (irritable bowel syndrome, gastroesophageal reflux disease), and cardiovascular and respiratory illnesses (asthma).

Comorbid Neurological, Pain, and Anxiety

Fibromyalgia Symptoms of anxiety depression are cardinal for individuals with fibromyalgia (FM). FM is a chronic neurological disorder characterized by pervasive musculoskeletal pain and tenderness, with secondary symptoms including fatigue, memory impairment, and gastrointestinal disorders, among others [1]. It occurs in up to 4% of the general population, with women being the majority of individuals diagnosed [2]. A large percentage of individuals with FM meet diagnostic criteria for at least one anxiety or depressive disorder, making the treatment of FM a complex task [2]. Specifically, 27–60% of patients with FM report a current anxiety disorder, with 62% reporting lifetime diagnoses of any anxiety disorder [3]. Research indicates that anxiety in FM patients is associated with higher levels of perceived pain and more severe symptom [4, 5]. Because FM symptoms and associated psychiatric symptoms can substantially reduce daily functioning, the illness is of significant economic burden. Among chronic pain conditions, FM is associated with the most unemployment, financial disability, and days off [2, 6]. Because there is no cure, physicians and patients focus on managing and coping with existing pain.

A. E. Meuret (✉) · N. Tunnell · A. Roque
Department of Psychology, Southern Methodist University, Dallas, TX, USA
e-mail: ameuret@smu.edu

Y.-K. Kim (ed.), *Anxiety Disorders*, Advances in Experimental Medicine and Biology 1191, https://doi.org/10.1007/978-981-32-9705-0_15

The first line of treatment for FM patients with comorbid anxiety is antidepressants or psychotherapy [7]. It is widely accepted that cognitive, behavioral, and affective factors play a pivotal role in the level of pain, psychological distress, and impairment associated with chronic pain conditions [5]. Cognitive factors such as catastrophizing, pain-related fear or anxiety, attentional bias toward threatening or general negative cues, and helplessness all contribute to how a patient experiences and copes with their chronic pain [5, 8, 9].

Cognitive behavioral therapies (CBT) are frontline psychological treatments for chronic pain, including fibromyalgia [10], with findings of controlled trials providing supporting treatment benefits for comorbid pain and anxiety [11, 12]. For instance, in a controlled trial by Garcia and colleagues [12], the authors examined the differential efficacy of CBT, a muscle relaxant (cyclobenzaprine), combined treatment, or a waitlist control condition. The CBT and combined group conditions resulted in significantly greater long-term improvements in anxiety compared to the drug-only or control condition [13]. Other variations of CBT, including CBT-Insomnia, CBT-Pain, and their hybrid (CBT-IP), showed mixed success for reducing anxiety [13, 14].

Interventions aimed at facilitating change in attentional biases also show promise for reducing pain, anxiety sensitivity, and pain-related fear and anxiety. In a small, double-blind, controlled trial, 17 women with FM received an attentional modification group (AMP) or an active, placebo control group [8]. Participants in the AMP group completed two weekly 15-min sessions over 4 weeks targeting attentional biases toward catastrophizing and pain-related anxiety. AMP resulted in significantly greater reductions for pain, anxiety sensitivity, and pain-related fear supporting attention modification as a viable psychological treatment option [8]. Mindfulness-based stress reduction and Acceptance and Commitment Therapy (ACT) provide good grounds for treating comorbid chronic pain, fibromyalgia, and anxiety [15, 16]. For example, comparing 12 weeks of ACT to a waitlist control, Wicksell and colleagues showed significant reductions in anxiety in FM patient [17]. Additionally, a variety of intervention strategies including guided imagery/hypnosis, attachment-based compassion therapy, and meditation have shown success in controlled trials for reducing pain and psychological distress [18–20]. Lastly, studies have highlighted the efficacy of biofeedback broadly within neurological disorders, some focusing directly on its effect within FM populations [21, 22]. For example, EEG biofeedback, compared to escitalopram, resulted in greater anxiety and depression reductions, with gains being maintained throughout follow-up [22]. The initial evidence for neurofeedback intervention for comorbid fibromyalgia syndrome and affective symptoms is promising and inspired further research [23].

Epilepsy Epilepsy is a rare but highly debilitating neurological disorder. Epilepsy is characterized by unprovoked seizures with a wide range of symptoms. Epidemiological research has found anxiety disorders to be more than twice as likely in patients with epilepsy compared to the general population, with some studies reporting anxiety disorders in up to 30% of epileptic patients [24, 25]. Comorbid anxiety and epilepsy reduce health-related quality of life [26, 27].

Increasing seizure control with medical procedures, such as epilepsy surgery or medication, is the first line of treatment [25]. Once seizure control is improved, second-line treatments targeting anxiety can be employed. Antidepressants, such as buspirone and selective serotonin reuptake inhibitors (SSRIs), are commonly used to treat comorbid epilepsy and anxiety disorders [25], though special attention and monitoring are required to seize the elevated risk of inducing or exacerbating seizures. A 2017 Cochrane review [28] found that psychological treatments including CBT or ACT were beneficial for individuals with epilepsy, though only one of the five studies included anxiety as outcome measures. Overall, those studies reported significant reductions in anxiety levels post-intervention [28]. Apart from reducing anxiety, CBT and ACT have shown promise in reducing depression and improving quality of life, for people with epilepsy, particularly when combined with antidepressants [29–32]. For instance, Macrodimitris and colleagues [33] tested a ten-session group CBT pilot trial for epilepsy and comorbid depression or anxiety and observed significant improvements in anxiety, depression, and negative automatic thoughts. The generic CBT protocol targeting anxiety and depression was well accepted and provided evidence reducing affective symptoms. Following up with a controlled trial, Gandy and colleagues [34], however, failed to demonstrate significant improvements in anxiety when comparing CBT to a waitlist control for patients with epilepsy. An uncontrolled prospective ACT intervention involving 60 patients with epilepsy found the treatment had medium to large positive effects on several psychological outcomes, including anxiety, at 6-months follow-up [31]. Comparing mindfulness-based therapy (MT) and social support in drug-resistant patients with epilepsy also resulted in significantly greater reduction anxiety and depression symptoms and frequency of seizures in the MT group [35]. Lastly, preliminary results for biofeedback for epilepsy show promise, providing a low-cost and easy-to-implement therapy approach for seizure control [36–38].

Cerebral Palsy Cerebral palsy (CP) refers to a group of developmental disorders defined by impairment in motor function, muscle tone, balance, and coordination. CP is associated with varying degrees of functioning across domains (i.e., speech, cognitive functioning, eating). Research indicates that adults with CP are at a 40% increased risk of being diagnosed with an anxiety disorder, with the risk even higher for those without intellectual disability [39]. Mothers of children with CP also reported elevated levels of parenting stress [40], and up to 30% report experiencing affective symptoms at clinical levels [41]. Both are negatively associated with parenting satisfaction and child quality of life [42]. In a study involving 31 caregivers of children or adolescents with CP parents' ratings of their characteristics and children's behavior problems, parenting depression, stress, and state anxiety were significantly related to children's quality of life [43]. As a result, interventions targeting parent-child relations and parenting strategies are becoming more common.

Roux and colleagues [44] investigated the effects of a nine-session group family intervention (Stepping Stones Triple P; SSTP) on child behavior problems, parenting styles, parent-child conflicts, and parental satisfaction. Fifty-two families with a child diagnosed with autism spectrum disorder, CP, Down syndrome, or intellectual disability were randomized into a waitlist group or immediate treatment conditions, designed to facilitate positive parenting strategies, children's competence, positive relationships, and behavior management. The authors found significant improvements in child behaviors, relationship quality, and parental satisfaction [44]. Another controlled study involving 67 parents of children with CP compared the differential effects of SSTP, SSTP+ACT (Acceptance and Commitment Therapy) and a waitlist control on improvements in quality of life, functional performance, parenting style, and behavioral and emotional problems [45, 46]. Compared to the waitlist, the SSTP + ACT group demonstrated improved quality of life and functional performance as well as decreased behavioral problems and parental psychological symptoms, over-reactivity, and verbosity. SSTP alone was associated with decreased behavioral problems and emotional symptoms [45, 46].

However, there is relatively limited literature focused on psychotherapy, specifically targeting anxiety in people with CP. Three case studies, one involving systematic desensitization, one biofeedback relaxation training, and the other involving CBT, discuss that the suitability of these interventions for CP is far from providing adequate evidence for their efficacy [47–49]. Additionally, two controlled studies involving interventions for individuals with CP indicate that mindfulness yoga interventions and horse therapy could be beneficial for various physical, behavioral, and cognitive outcomes, though neither study discusses anxiety as an outcome [50, 51].

Non-cardiac Chest Pain

Non-cardiac chest pain (NCCP) is experienced as a squeezing, pressure-like, or burning substernal sensation which radiates to the spine, neck, left arm, or jaw [52]. NCCP is difficult to distinguish from malign cardiac chest pain. Consequently, it is a highly prevalent symptom reported within emergency departments and primary care clinics, with an incidence rate of 20–25% [52, 53]. Medical literature cites multiple causes of NCCP, both biological (e.g., gastrointestinal reflux disease, microvascular ischemia) or psychological (e.g., abnormal pain perception, passive pain coping strategies) [54]. Though NCCP is a benign condition with an excellent prognosis [55–57], it is often chronic and is associated with high disability and cost [58]. Research has shown higher prevalence rates of psychiatric disorders within NCCP, with up to half of the patients eventually meeting diagnostic criteria of an anxiety disorder, including generalized anxiety disorder, panic disorder, and social phobia [59, 60]. Inaccurate interoceptive sensitivity, or hypervigilance of somatic sensations, is hypothesized to play an important role in the development of anxiety, particularly within these populations. White and colleagues [61] found that patients with NCCP and comorbid anxiety or depression symptoms were prone to hypervigilance of their cardiorespiratory symptoms. This inaccurate and oversensitive perception, particularly with cardiac anxiety (the fear of sensations associated with the heart), has been linked to excessive healthcare-seeking behaviors [62]. NCCP is associated with exceedingly high healthcare costs (up to 315 million dollars annually), likely due to the various invasive procedures, incorrect diagnoses, and failed treatments associated with assessing and treating chest pain [63, 64]. Thus, interventions targeting symptom differentiation and relieve are of high priority.

Treatment for NCCP varies depending on the etiology of the pain (i.e., biological or psychological). The majority of patients who seek medical attention for chest pain are eventually diagnosed with medically based disorders such as gastroesophageal reflux disease or esophageal motility disorders and thus treated with medications, minor procedures, or surgery. However, those whose chest pain has no biological or medical explanation are thought to benefit from psychotherapy.

A 2015 Cochrane review [65] concluded that CBT-type interventions are beneficial for individuals with NCCP. By restructuring anxiogenic thoughts, providing education about symptom origination, and exposing patients to feared bodily sensation, CBT aims to reduce distress produced when experiencing non-cardiac chest pain [66, 67]. Specifically, in the studies that included anxiety as an outcome variable (four CBT interventions, two relaxation interventions, one cardiac rehabilitation, and one coping skills training), psychological conditions indicated significant improvements over control conditions [65]. Spinhoven and colleagues [68] compared the efficacy of 12 weeks of CBT, paroxetine (a selective serotonin reuptake inhibitor), or placebo in reducing pain, anxiety, and heart-related anxiety for 69 individuals with NCCP. Analyses indicated that CBT was superior to paroxetine and placebo in reducing pain, while both CBT and paroxetine were superior in reducing heart-related anxiety compared to placebo [68]. Jonsbu and colleagues [66] did not observe significant improvements in the frequency of bodily symptoms in a brief three-session CBT-controlled trial for NCCP. However, there was a reduction in avoidance and catastrophic interpretation of those symptoms, which may be the more significant finding for psychologically based NCCP [66]. Finally a recent, large controlled study demonstrate short-term success (at 3 months) but failed to show long-term (1-year) improvements in psychological health, including anxiety, in a brief BT intervention for NCCP [69].

An alternative treatment approach for NCCP is relaxation training. For example, Lahmann and colleagues [70] assigned sufferers to either functional relaxation (FR) or an enhanced medical care control group. The intervention was 6 weeks in duration during which the FR treatment group received ten group sessions focused on creating awareness of various bodily sensations by physical movements and breathing [65, 70]. FR group resulted in significantly greater anxiety and somatization symptom reductions compared to the control group [70]. Furthermore, a study recognizing the unique needs of NCCP, particularly within an emergency department, pilot tested a brief self-help psychoeducational intervention for NCCP patients with elevated anxiety [71]. The psychoeducational materials included (1) information regarding potential causes of NCCP, (2) coping techniques for stress and pain management, and (3) directions on how to implement coping techniques. The trial provided a strong preliminary signal for need, feasibility, and acceptability [71]. Other forms of psychosocial interventions for NCCP include hypnosis, coping skills training, and guided breathing [65, 72, 73].

Lastly, pharmacological treatments designed to reduce symptoms of anxiety and depression, as well as the perception of pain, are tested, but results are mixed [68, 73]. Comparing sertraline to placebo resulted in significant reductions in daily pain, but not depressive symptoms [74]. In a study by Keefe et al. [73], sertraline only did not result in reductions in anxiety. However, the combination of coping skill training (relaxation, imagery, distraction, activity-rest cycling, pleasant activity scheduling, and cognitive restructuring) was successful.

Comorbid Diabetes and Anxiety

Diabetes has become a rising health concern, with a prevalence of 8.3% in the population [75]. Diabetes complications include cardiovascular disease, circulatory difficulties, kidney failure, and amputations [76]. Clinical anxiety has been diagnosed in up to 40% of diabetes patients and is linked to worse glycemic control [77–80]. Numbers may be even higher due to healthcare professionals overlooking anxiety symptoms and attributing distress to patients having maladjustment to their diabetes diagnosis [81]. Like in other chronic somatic disorders, comorbid anxiety disorders can exacerbate disease symptoms by interfering with treatment adherence, such as having a fear of injection and needles may interfere with glycemic control [82, 83].

Symptom overlap between diabetes and anxiety, especially panic, is common and is associated with greater distress, disease worsening, and a greater burden on the healthcare system [84]. For example, feelings of tingling and numbness in the extremities are a common symptom of both panic disorder [85] and diabetes [76]. Hypoglycemia or low blood sugar can lead to shakiness, sweating, and rapid heartbeat [81], symptoms that mimic the ones of a panic attack [86]. Phobic fears include fear of needles and injection [83, 87]. Patients with comorbid diabetes and blood-injection-injury phobia [85] are more likely to have macrovascular complications and insulin non-compliance [88]. Lastly, a specific diabetes-related fear, fear of hypoglycemia, can cause patients to maintain levels above the recommended dose [84, 89].

Interventions targeting the complex interaction of diabetes-related health and affective disorders are of great interest and need. One major goal of psychotherapeutic treatment of diabetes is to improve glycemic control by helping patients manage their medications and be more adherent to better healthier lifestyles [90]. A meta-analysis on psychosocial interventions for comorbid mental health disorders and diabetes has found that a combination of lifestyle and psychosocial interventions (problem-solving, cognitive-behavior therapy, and social support) are efficacious in reducing physical or mental health symptoms, but possibly not both [91]. Psychosocial interventions that included education and skills training had better outcomes on glycosylated hemoglobin (HbA1c) [91]. A recent meta-analysis on psychosocial interventions (ranging from CBT to illness management interventions) concluded that psychosocial interventions specifically appear to reduce anxiety symptoms short-term, but not long-term, and failed to improve self-efficacy of diabetes management [92].

Results from CBT trials show promise in reducing HbA(1c), anxiety, and prevention of hypoglycemic states compared to a treatment-as-usual control group [93]. Several other psychotherapeutic intervention modalities are recommended as diabetes treatment augmenters. For instance, techniques to enhance mindfulness and nonjudgmental approach (components of ACT) resulted in normalizing glycated hemoglobin and improving coping strategies [94], but more controlled studies are needed [95, 96]. Combining CBT and motivational enhancement therapy has also shown promise in glycemic control management [97], so has motivational interviewing [98] and problem-solving therapy [99]. The latter, a pilot trial, led to reductions in psychological distress and normalization in glycemic markers.

Comorbid Gastrointestinal and Esophageal Disorders and Anxiety

Gastrointestinal (GI) and esophageal symptoms like diarrhea, abdominal distress, and nausea are commonly reported by anxiety sufferers especially those diagnosed with generalized anxiety disorder and panic disorder [100–104]. Historically, medical and psychological models independently examined GI symptoms. Current research acknowledges more strongly a brain-gut connection or axis, where psychosocial stressors and anxiety may negatively impact the gastrointestinal biome and exacerbate existing biological vulnerabilities [105–108]. Stress and the endocrine response of corticotropin-releasing factor (CRF) has been proposed to affect IBS through multiple pathways including CRF role in colon secretion, visceral hypersensitivity, hypervigilance, and effects of CRF on low-grade activation of the immune system cells in the GI tract caused by inflammation [109, 110].

An example of this interaction is irritable bowel syndrome (IBS), a condition highly comorbid with anxiety disorders [107]. Primary symptoms of IBS include high levels of recurrent abdominal pain, present at least 1 day per week over at least 3 months. IBS is associated with changes in bowel habits, including diarrhea, constipation, or both [111, 112]. Epidemiological studies indicated that 10–25% of the population suffers from IBS, with twice as many women than men [113]. Likewise, more than half of those with IBS have been reported to have a comorbid psychiatric disorder, with the most common being depression (25%), panic disorder (21–46%), and generalized disorder (25%) [107, 114–117]. Lee et al. [118] found that IBS was 4.7 times more common in GAD than in non-patients.

Targeted intervention is of high relevance and need due to the documented heightened comorbidity, high costs in the utilization of healthcare services, and negative impact on patients' overall work productivity [119–121]. Four primary psychotherapeutic interventions for IBS have been studied: cognitive behavioral therapy (CBT), relaxation techniques, hypnosis, and psychodynamic therapy [122, 123]. A recent meta-analysis by Laird et al. [124] of 41 controlled trials of psychological interventions for IBS showed that CBT, hypnosis, psychodynamic, and relaxation were beneficial in improving mental health symptoms, including anxiety and depression. Except for relaxation techniques, interventions also increased improvements in daily functioning. CBT was the most studied modality (21 studies compared to less than 5 for their other modalities) showing the most evidence for its efficacy in improving daily functioning and reducing avoidance behavior likely due to teaching exposure-based techniques. In support, systematic exposures in an Internet-delivered CBT dismantling study were associated with greater GI symptoms and anxiety decrease than CBT without [125]. Results on long-term effects (12-month follow-up) are mixed with some showing enduring effects [123, 126], while others such as gut-directed hypnotherapy do not [127].

Treatment options tailored to targeting anxiety specific to IBS symptoms, coined gastrointestinal (GI)-specific anxiety [128–130], have also been studied. GI-specific anxiety is linked to increased hypervigilance, poor coping responses, and increased pain sensitivity [128] and can lead to symptom exacerbation [131]. Targeting hypervigilance and hypersensitivity to visceral sensations through exposure-based interventions is, therefore, a promising approach [107, 124]. In a meta-analysis on CBT-controlled trials for IBS, Altayar et al. [132] found reductions in IBS symptom severity, IBS quality of life, and abdominal pain, as did a recent meta-analysis on the effects of antidepressants, relaxation therapy, multicomponent psychological therapy, hypnotherapy, and dynamic psychotherapy [133]. The psychological interventions of CBT, hypnotherapy, multicomponent psychological therapy, which included a mix of relaxation, thermal biofeedback, and cognitive therapy techniques, were more efficacious in reducing symptoms than control therapy conditions [133].

Challenges to treatment engagement in IBS sufferers can be due to lack of motivation or denials that their symptoms are connected with their psychosocial functioning [108, 134]. Therefore, clinicians and medical providers should establish good patient rapport and provide clear education on the interaction of mental health and IBS to increase treatment participation and adherence [135].

As an alternative to psychotherapy, psychotropic medications are investigated for functional gastrointestinal disorders (FGID). Tricyclic antidepressants and SRRI show promise in lowering pain and anxiety symptoms in patients with FGID [136, 137]. Despite greater efficacy over placebo, caution should be used before administering antidepressants as shown in a recent meta-analysis due to intolerability issues [138]. Corticotropin-releasing hormone antagonist also shows promise in reducing inflammation [139] and anxiety and depression [110], yet no large-scale trials are published yet.

Combined medication and drug treatments can also be considered. In a recent controlled trial for patients with nonerosive reflux disease, Li et al. [140] found the combination of GI medication (omeprazole and domperidone) and CBT was more efficacious in reducing self-report gastroesophageal reflux and quality of life than medication or CBT alone. Anxious or depressed symptoms were equally lower in CBT only or the combined treatment compared to medication only.

Lastly, research on treating psychological distress in gastroesophageal reflux disease (GERD), a condition where stomach content begins to leak back into the esophagus, causing discomfort or pain, has garnered interest. GERD, a diagnosis confirmed by endoscopic esophageal mucosal damage or erosion [141, 142], poses a heavy burden on the healthcare system with seven million visits a year attributed to GERD and reflux esophagitis [143]. First-line medical treatments, including proton pump inhibitors, pharmacology, and behavioral changes like weight loss and diet [144], have high numbers of non-responders (40%) [145–148]. Psychogastroenterology highlights the integration of psychological interventions and GI interventions [149], as psychosocial stressors appear to worsen symptoms of GERD [148]. Emerging psychosocial interventions include relaxation [150], diaphragmatic breathing [151], and speech therapy [152]. These treatments show promise for reducing reflux, GI and GERD symptoms, and stress, but objective 12 measures of acid exposure are often lacking. CBT for excessive supragastric belching, a condition related to GERD, found relief in both objective symptoms of GI as well as overall increased quality of life [153]. The sessions consisted of psychoeducation about the disease, identification of physical sensations, restructuring maladaptive thinking patterns about their belching, diaphragmatic breathing, and mouth opening/tongue position to better control belching.

Comorbid Cardiorespiratory Disorders and Anxiety

Decades of research support the heightened prevalence of anxiety disorders, particularly panic disorder, within individuals with respiratory and cardiovascular diseases [154–156]. With more than 80% of panic sufferers complaining of such physical symptoms as chest pain, tachycardia, or dyspnea [157, 158], PD is associated with substantial health decrements [159] and ranked among the top three mental and physical illnesses associated with perceived health decrements [160–162].

Cardiovascular Disorders Interactions between psychological stress and activation of the cardiovascular system have long been a focus of research [163–165]. Epidemiological studies report consistent links between anxiety disorders and cardiovascular disease, including cerebrovascular events, such as strokes [166], coronary heart disease (ischemic heart disease, myocardial infarction, angina pectoris), chronic heart failure [167], and hypertension [112, 155, 168–172]. Heart racing or palpitations are central symptoms in both cardiovascular illness conditions and anxiety. Notably, they are also the most prominent and widely reported physical symptoms in panic disorder (PD) [173]. Dysfunctional activity of the autonomic nervous system and hypothalamic-pituitary-adrenal axis, which affects the cardiovascular system, is implemented in both anxiety and depression [156, 174, 175]. Consequently, studies highlight the bidirectional nature of cardiovascular illness augmenting the vulnerability for a subsequent anxiety disorder and vice versa. In support of the earlier, hypotensive compared to normotensive individuals have elevated rates of PD, and PD onset was typically followed diagnosis [176]. Likewise, discrete cardiac events such as myocardial infarctions (MI) are associated to increased risk of posttraumatic stress disorder (PTSD), worsened adverse illness management outcomes, and subsequent major cardiac events and mortality [177]. In a recent 2019 study, 54% of cardiac disease patients met criteria for severe depression [178]. However, there is also an elevated risk for anxiety disorders, with 19% meeting criteria severe-to-very severe anxiety [178] and the manifestations of cardiovascular illness, including mortality from cardiac events [179, 180]. Reversely, several longitudinal studies show that panic onset before the age of 21 years relates to a 1.3-fold risk for subsequent hypertension [181] and panic sufferers had a 1.5-fold risk of atrial fibrillations in one study [182] and a 2-fold risk of subsequent cardiovascular illness in another [183]. Finally, postmenopausal women with PD were at more than the fourfold risk for developing coronary heart disease [184].

Cardiac rehabilitation represents the first-line treatment for coronary heart disease (CHD). It is a multicomponent treatment incorporating health education, physical exercise, and diet but also includes psychosocial (e.g., CBT, hypnotherapy, stress management such as relaxation, coping and problem-solving skill training) and psychopharmacological interventions. CHD and anxiety disorders share contributing factors that strongly link to problematic health behaviors, such as physical inactivity, smoking, and obesity.

Development of catastrophic thoughts and fear of dying following illness onset and subsequent avoidance behavior further fuel sedentary lifestyle, and foster symptoms at benign levels of physical activity. Furthermore, somatizing medical patients likely present with a lower threshold for experiencing subtle bodily sensations as aversive and fear-provoking (heart-focused anxiety), particularly if the sensation mimics those of serious medical conditions [156]. Stress resulting from heart-focused anxiety may, in turn, increase the occurrence of angina attacks and the probability of cardiac death as discussed above. Notably, while CHD patients predominantly report physical symptoms, they rarely fear them or have a sense of catastrophe. By contrast, chest pain experiences in NCCP are dominated by catastrophizing thoughts. Palpitation, the most common reason for cardiologist referrals, is indeed only weakly related to arrhythmia and is largely asymptomatic as are heart rate symptom perceptions.

Furthermore, preventative cardiac treatments such as the use of implantable cardioverter defibrillators (ICDs) to treat life-threatening ventricular fibrillation can, in and of themselves, increase the risk for clinical anxiety. The painful and uncontrollable shocks are linked to adverse psychological outcomes, with up to 20% of ICD users developing severe anxiety and panic [185, 186]. Strikingly, rates of PD and agoraphobia are 60% for those with more than two electrical discharges per year compared to 10% for one discharge [187].

Based on the latest Cochrane study [188] on the effects of p*sychological* interventions for coronary *heart* disease, no effect was found for reduced total deaths (any cause), risk of cardiac surgery, or having another heart attack. However, they showed significant reductions in stress, depression, and anxiety [188]. Particularly promising was the small but significant effect for cardiac mortality. Similarly, a recent meta-analysis on patients with myocardial infarction [189] attested to the benefits of exercise-based cardiac rehabilitation on alleviating anxiety and depression symptoms. Findings from CBT-based trials demonstrate anxiety reductions up to 60% in patients with ICDs [190] as well as for CHD more generally, including acute coronary syndrome, atrial fibrillation, and postmyocardial infarction [191]. A controlled trial testing anxiety-focused CBT for comorbid generalized anxiety disorder and chronic heart failure sufferers resulted in significant reductions in anxiety and unplanned hospital admissions [192]. Finally, promising new interventions are under investigation, such as the UNWIND study which is examining the benefits of exercise and escitalopram in anxious patients with coronary heart disease [193].

Asthma Strong and consistent associations have been found between asthma and anxiety disorders [112, 194], in particular, panic disorder, panic attacks, generalized anxiety disorder, and phobias [195–199], with prevalence rates for anxiety disorders of up to 45% in asthmatic samples [200]. Self-reported respiratory disease is associated with a 70% greater likelihood of panic attacks [195]. Sixty-three percent of asthma patients presenting to the emergency room for acute exacerbations suffered from an anxiety disorder [201], a phenomenon which is likely due to the frightening nature of asthma symptoms such as extreme dyspnea, chest tightness, or feelings of suffocation. Among the anxiety disorders, PD has often presented a specifically strong association with asthma [202], with prevalence rates up to 24% of adults and 4.7% of children/adolescents [112, 203]. Childhood asthma symptoms are associated with increased levels of shyness/anxiety [204] and predict later-life development of panic disorder and agoraphobia [198, 205]. The illness may lead to subsequent anxiety/panic through the emotional burden of chronic illness and excessive monitoring of illness-related symptoms that are interpreted as impending signals of physical catastrophe (e.g., shortness of breath interpreted as respiratory arrest) [206]. Early adulthood PD increases odds by more than six times for asthma later in life, possibly through additional problematic health behaviors, such as smoking, lack of physical activity, or dysfunctional dietary habits leading to obesity [207].

A substantial body of evidence suggests that comorbid anxiety complicates the management of asthma and is a risk factor for greater asthma morbidity, independent of objective measures of pulmonary function [200, 208]. Strong negative emotions and stress contribute to bronchoconstriction [209–212] and airway inflammation [213, 214], thereby exacerbating asthma. Asthma exacerbations lead to symptoms greatly feared by comorbid panic sufferers, thus exacerbating panic attacks. Anxiety and asthma comorbidity are associated with mutual complications in diagnosis and management. Similar symptoms may cause errors in diagnosis and treatment, leading to additional costs for the healthcare system [215, 216]. Asthma patients with higher anxiety levels are more likely to use healthcare providers, hospitalization, and emergency visits [217–219]. Comorbidity has been associated with reduced quality of life [220, 221] and elevated medication use [222].

Although medication is undisputedly the first line of treatment for asthma patients, some (e.g., oral corticosteroids, β-agonists [205]) provoke the very symptoms anxious patients fear, thus exacerbating anxiety. Likewise, psychotropic medication for anxiety can cause respiratory side effects that can complicate asthma symptoms. Thus, psychosocial interventions that are equally effective as psychotropic medication in managing anxiety may be preferable.

On the other hand, caution is warranted since psychosocial interventions for anxiety carry other risks for the asthma patient. For example, CBT for anxiety often includes interoceptive exposure exercises, such as voluntary hyperventilation, that would lead to bronchoconstriction [223, 224]. Similarly, slow abdominal breathing training without control of PCO_2 levels, a common technique included in CBT, can lead to hyperpnea or hyperventilation [225]. The efficacy of relaxation for asthma patients also is debated [226, 227] since it may encourage bronchoconstriction through enhancing parasympathetic activity.

Despite the repeated calls for standardized and evaluated interventions for comorbid asthma and anxiety [112, 196, 228], interventions remain remarkably absent, even though anxiety symptoms are viewed as modifiable risk factors. There have been attempts to improve self-management behaviors [229, 230], such as self-monitoring (symptom and peak flow diaries), allergen and trigger avoidance, and correct use of medication [231–235]. Interventions, including CBT, yoga, scriptography, and biofeedback techniques [226, 236], have been devised that target aspect of asthma pathophysiology, with assumptions derived from biobehavioral and psychophysiological models of the disease. While efficacies ranges from limited [237] to promising [238, 239]; most interventions have failed to address anxiety direct.

Notwithstanding, a few interventions are targeting comorbid asthma and anxiety. In a small trial, the investigators (Ross et al. [240]) found that an 8-week group treatment that combined CBT for panic disorder with asthma education led to reductions in panic and anxiety that endured for 6 months, in addition to short-term improvements in morning PEF and asthma-related quality of life. Lehrer et al. [224] tested the benefits of a multimodal intervention for comorbid asthma and panic disorder, combining panic control therapy [241] with asthma education, smoking reduction, and assertiveness training. Findings of the uncontrolled pilot study reported promising reductions in panic and asthma symptoms, improvement in asthma quality of life, and decreased albuterol use. The intervention resulted in clinically significant reductions in panic and asthma symptoms and albuterol use as well as improved asthma stability and quality of life. Similarly, Yorke et al. [242] tested the feasibility and acceptability of group CBT in severe asthma, which provided a moderate signal for utility due to the high attrition. Given the particularly high comorbidity of asthma and PD in Latinos [200], Feldman and colleagues [243] developed a culturally adapted behavior psychophysiological therapy. The 8-week treatment was comprised of CBT for panic disorder, asthma education, differentiation between panic and asthma symptoms, and heart rate variability biofeedback. A control group received music and relaxation therapy (MRT). While both interventions showed improvements in anxiety and asthma outcome measures, only CBT led to improvement in adherence to inhaled corticosteroids. Finally, respiratory training using biofeedback of capnometry to reduce hyperventilation may likely benefit both panic symptomatology [244, 245] and asthma outcomes [246], but evidence has yet to be established.

Conclusion

This chapter examined the link between comorbid anxiety and medical illness through the physical character of its symptoms. They overlap with the symptomatology of a range of chronic somatic illnesses and the presence of maladaptive cognitions and behaviors. Because of the highly complex nature of bodily sensations, attempts to assign "typical" versus "nontypical" symptoms to medical or psychiatric diagnoses are overly simplistic. Strictly speaking, the diagnostic nomenclature (DSM-5) prohibits a diagnosis of PD if the origin of the symptoms is the direct cause of a medical condition. However, given that anxiety and panic/fear symptoms mimic those of several a critical medical condition, differential diagnosis is far from straightforward. The most common factor identified in research on nonspecific, medically unexplained symptoms is the persistence of bodily perceptions paired with catastrophic beliefs about the nature of these symptoms. While several promising interventions for comorbid anxiety and medical conditions have emerged, research and dissemination of evidence-based, tailored interventions with an impact on both psychiatric and chronic disease health are still in its infancy.

References

1. Martínez-Poles J, Nedkova-Hristova V, Escribano-Paredes JB, García-Madrona S, Natera-Villalba E, Estévez-Fraga C, et al. Incobotulinumtoxin A for sialorrhea in neurological disorders: a real-life experience. Toxins (Basel). 2018 June;10(6):217.
2. Montesó-Curto P, García-Martinez M, Romaguera S, Mateu ML, Cubí-Guillén MT, Sarrió-Colas L, et al. Problems and solutions for patients with fibromyalgia: building new helping relationships. J Adv Nurs. 2018 Feb;74(2):339–49.
3. Arnold L, Goldenberg D, Stanford S, Lalonde J, Sandhu H, Keck P, et al. Gabapentin in the treatment of fibromyalgia: a randomized, double-blind, placebo-controlled, multicenter trial. Arthritis Rheum. 2007 Apr;56(4):1336–44.
4. Córdoba-Torrecilla S, Aparicio VA, Soriano-Maldonado A, Estévez-López F, Segura-Jiménez V, Álvarez-Gallardo I, et al. Physical fitness is associated with anxiety levels in women with fibromyalgia: the al-Ándalus project. Qual Life Res. 2016 Apr;25(4):1053–8.
5. Lami MJ, Martínez MP, Miró E, Sánchez AI, Guzmán MA. Catastrophizing, acceptance, and coping as mediators between pain and emotional distress and disability in fibromyalgia. J Clin Psychol Med Settings. 2018 Mar;25(1):80–92.
6. Luciano J, Guallar J, Aguado J, López-del-Hoyo Y, Olivan B, Magallón R, et al. Effectiveness of group acceptance and commitment therapy for fibromyalgia: a 6-month randomized controlled trial (EFFIGACT study). Pain. 2014 Apr;155(4):693–702.
7. Falcão D, Sales L, Leite J, Feldman D, Valim V, Natour J. Cognitive behavioral therapy for the treatment of fibromyalgia syndrome: a randomized controlled trial. J Musculoskelet Pain. 2008 Jan;16(3):133–40.
8. Carleton R, Richter A, Asmundson G. Attention modification in persons with fibromyalgia: a double blind, randomized clinical trial. Cogn Behav Ther. 2011 Dec;40(4):279–90.
9. Duschek S, Werner N, Limbert N, Winkelmann A, Montoya P. Attentional bias toward negative information in patients with fibromyalgia syndrome. Pain Med. 2014 Apr;15(4):603–12.
10. Bernardy K, Klose P, Welsch P, Häuser W. Efficacy, acceptability and safety of cognitive behavioural therapies in fibromyalgia syndrome – a systematic review and meta-analysis of randomized controlled trials. Eur J Pain. 2017 Feb;22(2):242–60.

11. Vallejo M, Ortega J, Rivera J, Comeche M, Vallejo-Slocker L. Internet versus face-to-face group cognitive-behavioral therapy for fibromyalgia: a randomized control trial. J Psychiatr Res. 2015 Sep;68:106–13.
12. García J, Simón MA, Durán M, Canceller J, Aneiros FJ. Differential efficacy of a cognitive-behavioral intervention versus pharmacological treatment in the management of fibromyalgic syndrome. Psychol Health Med. 2006 Nov;11(4):498–506.
13. Martínez MP, Miró E, Sánchez AI, Díaz-Piedra C, Cáliz R, Vlaeyen JW, et al. Cognitive-behavioral therapy for insomnia and sleep hygiene in fibromyalgia: a randomized controlled trial. J Behav Med. 2014 Aug;37(4):683–97.
14. Lami M, Martínez M, Miró E, Sánchez A, Prados G, Cáliz R, et al. Efficacy of combined cognitive-behavioral therapy for insomnia and pain in patients with fibromyalgia: a randomized controlled trial. Cognit Ther Res. 2018 Feb;42(1):63–79.
15. Grossman P, Niemann L, Schmidt S, Walach H. Mindfulness-based stress reduction and health benefits: a meta-analysis. J Psychosom Res. 2004 Jul;57(1):35–43.
16. Veehof M, Oskam M-J, Schreurs K, Bohlmeijer E. Acceptance-based interventions for the treatment of chronic pain: a systematic review and meta-analysis. Pain. 2011 Mar;152(3):533–42.
17. Wicksell RK, Kemani M, Jensen K, Kosek E, Kadetoff D, Sorjonen K, et al. Acceptance and commitment therapy for fibromyalgia: a randomized controlled trial. Eur J Pain. 2013 Apr;17(4):599–611.
18. Zech N, Hansen E, Bernardy K, Häuser W. Efficacy, acceptability and safety of guided imagery/hypnosis in fibromyalgia – a systematic review and meta-analysis of randomized controlled trials. Eur J Pain. 2017 Feb;21(2):217–27.
19. Montero-Marín J, Navarro-Gil M, Puebla-Guedea M, Luciano JV, Van Gordon W, Shonin E, et al. Efficacy of "attachment-based compassion therapy" in the treatment of fibromyalgia: a randomized controlled trial. Front Psychiatry. 2018 Jan;8(307)
20. Vago DR, Nakamura Y. Selective attentional bias towards pain-related threat in fibromyalgia: preliminary evidence for effects of mindfulness meditation training. Cognit Ther Res. 2011 Dec;35(6):581–94.
21. Mur E, Drexler A, Gruber J, Hartig F, Gunther V. Electromyography biofeedback therapy in fibromyalgia. Wien Med Wochenschr. 1999;149(19-20):561–3.
22. Kayıran S, Dursun E, Dursun N, Ermutlu N, Karamürsel S. Neurofeedback intervention in fibromyalgia syndrome; a randomized, controlled, rater blind clinical trial. Appl Psychophysiol Biofeedback. 2010 Dec;35(4):293–302.
23. Santoro M, Cronan T. A systematic review of neurofeedback as a treatment for fibromyalgia syndrome symptoms. J Musculoskelet Pain. 2014 Sep;22(3):286–300.
24. Jones JE, Hermann BP, Barry JJ, Gilliam F, Kanner AM, Meador KJ. Clinical assessment of axis I psychiatric morbidity in chronic epilepsy: a multicenter investigation. J Neuropsychiatry Clin Neurosci. 2005 Spr;17(2):172–9.
25. Kimiskidis VK, Valeta T. Epilepsy and anxiety: epidemiology, classification, aetiology, and treatment. Epileptic Disord. 2012 Sep;14(3):248–56.
26. Johnson EK, Jones JE, Seidenberg M, Hermann BP. The relative impact of anxiety, depression, and clinical seizure features on health-related quality of life in epilepsy. Epilepsia. 2004 May;45(5):544–50.
27. Kanner AM, Barry JJ, Gilliam F, Hermann B, Meador KJ. Anxiety disorders, subsyndromic depressive episodes, and major depressive episodes: do they differ on their impact on the quality of life of patients with epilepsy? Epilepsia. 2010 Jul;51(7):1152–8.
28. Michaelis R, Tang V, Wagner J, Modi A, LaFrance W Jr, Goldstein L, et al. Psychological treatments for people with epilepsy. Cochrane Database Syst Rev. 2017;10
29. Lundgren T, Dahl J, Hayes SC. Evaluation of mediators of change in the treatment of epilepsy with acceptance and commitment therapy. J Behav Med. 2008 Jun;31(3):225–35.

30. Gundy JM, Woidneck MR, Pratt KM, Christian AW, Twohig MP. Acceptance and commitment therapy: state of evidence in the field of health psychology. Sci Rev Ment Health Pract. 2011;8(2):23–35.
31. Dewhurst E, Novakova B, Reuber M. A prospective service evaluation of acceptance and commitment therapy for patients with refractory epilepsy. Epilepsy Behav. 2015 May;46:234–41.
32. Mula M. Treatment of anxiety disorders in epilepsy: an evidence-based approach. Epilepsia. 2013 Mar;54(Suppl 1):13–8.
33. Macrodimitris S, Wershler J, Hatfield M, Hamilton K, Backs-Dermott B, Mothersill K, et al. Group cognitive-behavioral therapy for patients with epilepsy and comorbid depression and anxiety. Epilepsy Behav. 2011 Jan;20(1):83–8.
34. Gandy M, Sharpe L, Nicholson Perry K, Thayer Z, Miller L, Boserio J, et al. Cognitive behaviour therapy to improve mood in people with epilepsy: a randomised controlled trial. Cogn Behav Ther. 2014 Apr;43(2):153–66.
35. Tang V, Poon WS, Kwan P. Mindfulness-based therapy for drug-resistant epilepsy: an assessor-blinded randomized trial. Neurology. 2015 Sep;85(13):1100–7.
36. Fried R, Fox MC, Carlton RM. Effect of diaphragmatic respiration with end-tidal CO2 biofeedback on respiration, EEG, and seizure frequency in idiopathic epilepsy. Ann NY Acad Sci. 1990 Sep;602(1):67–96.
37. Nagai Y, Trimble MR. Long-term effects of electrodermal biofeedback training on seizure control in patients with drug-resistant epilepsy: two case reports. Epilepsy Res. 2014 Jan;108(1):149–52.
38. Uhlmann C, Fröscher W. Biofeedback treatment in patients with refractory epilepsy: changes in depression and control orientation. Seizure. 2001 Jan;10(1):34–8.
39. Smith KJ, Peterson MD, O'Connell NE, Victor C, Liverani S, Anokye N, et al. Risk of depression and anxiety in adults with cerebral palsy. JAMA Neurol. 2019;76(3):294–300.
40. Britner PA, Morog MC, Pianta RC, Marvin RS. Stress and coping: a comparison of self-report measures of functioning in families of young children with cerebral palsy or no medical diagnosis. J Child Fam Stud. 2003 Sep;12(3):335–48.
41. Manuel J, Naughton MJ, Balkrishnan R, Smith BP, Koman LA. Stress and adaptation in mothers of children with cerebral palsy. J Pediatr Psychol. 2003 Apr;28(3):197–201.
42. Wanamaker CE, Glenwick DS. Stress, coping, and perceptions of child behavior in parents of preschoolers with cerebral palsy. Rehabil Psychol. 1998 Win;43(4):297–312.
43. Wiley R, Renk K. Psychological correlates of quality of life in children with cerebral palsy. J Dev Phys Disabil. 2007 Oct;19(5):427–47.
44. Roux G, Sofronoff K, Sanders M. A randomized controlled trial of group Stepping Stones Triple P: a mixed-disability trial. Fam Process. 2013 Sep;52(3):411–24.
45. Whittingham K, Sanders M, McKinlay L, Boyd RN. Interventions to reduce behavioral problems in children with cerebral palsy: an RCT. Pediatrics. 2014 May;133(5):e1249–57.
46. Whittingham K, Sanders MR, McKinlay L, Boyd RN. Parenting intervention combined with acceptance and commitment therapy: a trial with families of children with cerebral palsy. J Pediatr Psychol. 2016 Jun;41(5):531–42.
47. Peterman JS, Hoff AL, Gosch E, Kendall PC. Cognitive-behavioral therapy for anxious youth with a physical disability: a case study. Clin Case Stud. 2015 Jun;14(3):210–26.
48. Silber LD, Howard J. Systematic desensitization with a cerebral palsied college student. Psychotherapy. 1972 Spr;9(1):17.
49. Engel JM, Jensen MP, Schwartz L. Outcome of biofeedback-assisted relaxation for pain in adults with cerebral palsy: preliminary findings. Appl Psychophysiol Biofeedback. 2004 Jun;29(2):135–40.
50. Davis E, Davies B, Wolfe R, Raadsveld R, Heine B, Thomason P, et al. A randomized controlled trial of the impact of therapeutic horse riding on the quality of life, health, and function of children with cerebral palsy. Dev Med Child Neurol. 2009 Feb;51(2):111–9.

51. Mak C, Whittingham K, Cunnington R, Boyd RN. Effect of mindfulness yoga programme MiYoga on attention, behaviour, and physical outcomes in cerebral palsy: a randomized controlled trial. Dev Med Child Neurol. 2018 Sep;60(9):922–32.
52. Fass R, Achem S. Noncardiac chest pain: epidemiology, natural course and pathogenesis. J Neurogastroenterol Motil. 2011 Apr;17(2):110–23.
53. Pope C, Ziebland S, Mays N. Qualitative research in health care: analysing qualitative data. BMJ. 2000 Jan;320(7227):114–6.
54. Thurston RC, Keefe FJ, Bradley L, Rama Krishnan KR, Caldwell DS. Chest pain in the absence of coronary artery disease: a biopsychosocial perspective. Pain. 2001 Aug;93(2):95–100.
55. Eslick G, Talley N. Non-cardiac chest pain: predictors of health care seeking, the types of health care professional consulted, work absenteeism and interruption of daily activities. Aliment Pharmacol Ther. 2004 Oct;20(8):909–15.
56. Eslick G, Talley N. Gastroesophageal reflux disease (GERD): risk factors, and impact on quality of life – a population-based study. J Clin Gastroenterol. 2009 Feb;43(2):111–7.
57. Eifert G, Forsyth J, Zvolensky M, Lejuez C. Moving from the laboratory to the real world and back again: increasing the relevance of laboratory examinations of anxiety sensitivity. Behav Ther. 1999 Spr;30(2):273–83.
58. Potts SG, Bass CM. Psychological morbidity in patients with chest pain and normal or near-normal coronary arteries: a long-term follow-up study. Psychol Med. 1995 Mar;25(2):339–47.
59. Barsky AJ. Palpitations, cardiac awareness and panic disorder. Am J Med. 1992 Jan;92(1):S31–4.
60. White KS, Raffa SD, Jakle KR, Stoddard JA, Barlow DH, Brown TA, et al. Morbidity of DSM-IV Axis I disorders in patients with noncardiac chest pain: psychiatric morbidity linked with increased pain and health care utilization. J Consult Clin Psychol. 2008 Jun;76(3):422–30.
61. White K, Craft J, Gervino E. Anxiety and hypervigilance to cardiopulmonary sensations in non-cardiac chest pain patients with and without psychiatric disorders. Behav Res Ther. 2010 May;48(5):394–401.
62. Mourad G, Strömberg A, Johansson P, Jaarsma T. Depressive symptoms, cardiac anxiety, and fear of body sensations in patients with non-cardiac chest pain, and their relation to healthcare-seeking behavior: a cross-sectional study. Patient. 2016 Feb;9(1):69–77.
63. Richter JE, Bradley LA, Castell DO. Esophageal chest pain: current controversies in pathogenesis, diagnosis, and therapy. Ann Intern Med. 1989 Jan;110(1):66–78.
64. Barsky AJ, Cleary PD, Sarnie MK, Ruskin JN. Panic disorder, palpitations, and the awareness of cardiac activity. J Nerv Ment Dis. 1994 Feb;182(2):63–71.
65. Kisely S, Campbell LA, Yelland MJ, Paydar A. Psychological interventions for symptomatic management of non-specific chest pain in patients with normal coronary anatomy. Cochrane Database Syst Rev. 2015 Jun;6:CD004101.
66. Jonsbu E, Dammen T, Morken G, Moum T, Martinsen EW. Short-term cognitive behavioral therapy for non-cardiac chest pain and benign palpitations: a randomized controlled trial. J Psychosom Res. 2011 Feb;70(2):117–23.
67. Beck JS. Cognitive behavior therapy: basics and beyond. 2nd ed. New York: Guilford Press; 2011.
68. Spinhoven P, Van der Does A, Van Dijk E, Van Rood Y. Heart-focused anxiety as a mediating variable in the treatment of noncardiac chest pain by cognitive-behavioral therapy and paroxetine. J Psychosom Res. 2010 Sep;69(3):227–35.
69. Mulder R, Zarifeh J, Boden J, Lacey C, Tyrer P, Tyrer H. An RCT of brief cognitive therapy versus treatment as usual in patients with non-cardiac chest pain. Int J Cardiol. 2019 Feb 5; https://doi.org/10.1016/j.ijcard.2019.01.067.
70. Lahmann C, Loew T, Tritt K, Nickel M. Efficacy of functional relaxation and patient education in the treatment of somatoform heart disorders: a randomized, controlled clinical investigation. Psychosomatics. 2008 Sep–Oct;49(5):378–85.

71. Webster R, Thompson AR, Norman P, Goodacre S. The acceptability and feasibility of an anxiety reduction intervention for emergency department patients with non-cardiac chest pain. Psychol Health Med. 2017 Jan;22(1):1–11.
72. Schey R, Villarreal A, Fass R. Noncardiac chest pain: current treatment. Gastroenterol Hepatol (N Y). 2007 Apr;3(4):255–62.
73. Keefe F, Shelby R, Somers T, Varia I, Blazing M, Waters S, et al. Effects of coping skills training and sertraline in patients with non-cardiac chest pain: a randomized controlled study. Pain. 2011 Apr;152(4):730–41.
74. Varia I, Logue E, O'Connor C, Newby K, Wagner H, Davenport C, et al. Randomized trial of sertraline in patients with unexplained chest pain of noncardiac origin. Am Heart J. 2000 Sep;140(3):367–72.
75. Kharroubi AT, Darwish HM. Diabetes mellitus: the epidemic of the century. World J Diabetes. 2015 Jun;6(6):850–67.
76. American Diabetes Association. Standards of medical care in diabetes-2018. Diabetes Care. 2018;41
77. Anderson RJ, Grigsby AB, Freedland KE, Groot MD, Mcgill JB, Clouse RE, et al. Anxiety and poor glycemic control: a meta-analytic review of the literature. Int J Psychiatry Med. 2002;32(3):235–47.
78. Buchberger B, Huppertz H, Krabbe L, Lux B, Mattivi JT, Siafarikas A. Symptoms of depression and anxiety in youth with type 1 diabetes: a systematic review and meta-analysis. Psychoneuroendocrinology. 2016 Aug;70:70–84.
79. Grigsby AB, Anderson RJ, Freedland KE, Clouse RE, Lustman PJ. Prevalence of anxiety in adults with diabetes: a systematic review. J Psychosom Res. 2002 Dec;53(6):1053–60.
80. Smith KJ, Béland M, Clyde M, Gariépy G, Pagé V, Badawi G, et al. Association of diabetes with anxiety: a systematic review and meta-analysis. J Psychosom Res. 2013 Feb;74(2):89–99.
81. Rubin RR, Peyrot M. Psychological issues and treatments for people with diabetes. J Clin Psychol. 2001 Apr;57(4):457–78.
82. Cramer JA. A systematic review of adherence with medications for diabetes. Diabetes Care. 2004 May;27(5):1218–24.
83. Wani AL, Ara A, Bhat SA. Blood injury and injection phobia: the neglected one. Behav Neurol. 2014;2014:471340.
84. Ducat L, Philipson LH, Anderson BJ. The mental health comorbidities of diabetes. JAMA. 2014 Aug;312(7):691–2.
85. American Psychiatric Association. Diagnostic and statistical manual of mental disorders. 5th ed. Arlington: American Psychiatric Publishing; 2013.
86. Meuret AE, Rosenfield D, Wilhelm FH, Zhou E, Conrad A, Ritz T, et al. Do unexpected panic attacks occur spontaneously? Biol Psychiatry. 2011 Nov;70(10):985–91.
87. Ayala ES, Meuret AE, Ritz T. Treatments for blood-injury-injection phobia: a critical review of current evidence. J Psychiatr Res. 2009 Oct;43(15):1235–42.
88. Bienvenu OJ, Eaton WW. The epidemiology of blood-injection-injury phobia. Psychol Med. 1998 Sep;28(5):1129–36.
89. Alam U, Asghar O, Azmi S, Malik RA. General aspects of diabetes mellitus. Handb Clin Neurol. 2014;126:211–22.
90. Asche C, LaFleur J, Conner C. A review of diabetes treatment adherence and the association with clinical and economic outcomes. Clin Ther. 2011 Jan;33(1):74–109.
91. Harkness E, Macdonald W, Valderas J, Coventry P, Gask L, Bower P. Identifying psychosocial interventions that improve both physical and mental health in patients with diabetes: a systematic review and meta-analysis. Diabetes Care. 2010 Apr;33(4):926–30.
92. Pascoe MC, Thompson DR, Castle DJ, Jenkins ZM, Ski CF. Psychosocial interventions and wellbeing in individuals with diabetes mellitus: a systematic review and meta-analysis. Front Psychol. 2017 Dec;8:2063.

93. Amsberg S, Anderbro T, Wredling R, Lisspers J, Lins P-E, Adamson U, et al. A cognitive behavior therapy-based intervention among poorly controlled adult type 1 diabetes patients – a randomized controlled trial. Patient Educ Couns. 2009 Oct;77(1):72–80.
94. Gregg JA, Callaghan GM, Hayes SC, Glenn-Lawson JL. Improving diabetes self-management through acceptance, mindfulness, and values: a randomized controlled trial. J Consult Clin Psychol. 2007 Apr;75(2):336–43.
95. O'Donohue W, Snipes C, Soto C. A case study of overselling psychotherapy: an ACT intervention for diabetes management. J Contemp Psychother. 2016 Mar;46(1):15–25.
96. Rosen GM, Lilienfeld SO. On the failure of psychology to advance self-help: acceptance and commitment therapy (ACT) as a case example. J Contemp Psychother. 2016;46(2):71–7.
97. Ismail K, Winkley K, Rabe-Hesketh S. Systematic review and meta-analysis of randomised controlled trials of psychological interventions to improve glycaemic control in patients with type 2 diabetes. Lancet. 2004 May;363(9421):1589–97.
98. Welch G, Shayne R. Interactive behavioral technologies and diabetes self-management support: recent research findings from clinical trials. Curr Diab Rep. 2006 Apr;6(2):130–6.
99. Villamil-Salcedo V, Vargas-Terrez BE, Caraveo-Anduaga J, González-Olvera J, Díaz-Anzaldúa A, Cortés-Sotres J, et al. Glucose and cholesterol stabilization in patients with type 2 diabetes mellitus with depressive and anxiety symptoms by problem-solving therapy in primary care centers in Mexico City. Prim Health Care Res Dev. 2018 Jan;19(1):33–41.
100. Fadgyas-Stanculete M, Buga AM, Popa-Wagner A, Dumitrascu DL. The relationship between irritable bowel syndrome and psychiatric disorders: from molecular changes to clinical manifestations. J Mol Psychiatry. 2014 Jun;2(1):4.
101. Lydiard RB. Irritable bowel syndrome, anxiety, and depression: what are the links? J Clin Psychiatry. 2001;62(Suppl 8):38–45.
102. Lydiard RB, Laraia MT, Howell EF, Ballenger JC. Can panic disorder present as irritable bowel syndrome? J Clin Psychiatry. 1986 Sep;47(9):470–3.
103. Lydiard B, Greenwald S, Weissman M, Johnson J, Drossman D, Ballenger J. Panic disorder and gastrointestinal symptoms: findings from the NIMH Epidemiologic Catchment Area project. Am J Psychiatry. 1994 Jan;151(1):64–70.
104. Tollefson GD, Luxenberg M, Valentine R, Dunsmore G, Tollefson SL. An open label trial of alprazolam in comorbid irritable bowel syndrome and generalized anxiety disorder. J Clin Psychiatry. 1991 Dec;52(12):502–8.
105. Miller I. The gut–brain axis: historical reflections. Microb Ecol Health Dis. 2018;29(1):1542921.
106. Halpert A, Drossman D. Biopsychosocial issues in irritable bowel syndrome. J Clin Gastroenterol. 2005 Sep;39(8):665–9.
107. Oudenhove LV, Levy RL, Crowell MD, Drossman DA, Halpert AD, Keefer L, et al. Biopsychosocial aspects of functional gastrointestinal disorders: how central and environmental processes contribute to the development and expression of functional gastrointestinal disorders. Gastroenterology. 2016 May;150(6):1355–67.
108. Tanaka Y, Kanazawa M, Fukudo S, Drossman DA. Biopsychosocial model of irritable bowel syndrome. J Neurogastroenterol Motil. 2011 Apr;17(2):131–9.
109. Chatoo M, Li Y, Ma Z, Coote J, Du J, Chen X. Involvement of corticotropin-releasing factor and receptors in immune cells in irritable bowel syndrome. Front Endocrinol (Lausanne). 2018 Feb;9:21.
110. Taché Y, Kiank C, Stengel A. A role for corticotropin-releasing factor in functional gastrointestinal disorders. Curr Gastroenterol Rep. 2009 Aug;11(4):270–7.
111. Lacy B, Patel N. Rome criteria and a diagnostic approach to irritable bowel syndrome. J Clin Med. 2017 Nov;6(11):99.
112. Roy-Byrne PP, Davidson KW, Kessler RC, Asmundson GJ, Goodwin RD, Kubzansky L, et al. Anxiety disorders and comorbid medical illness. Gen Hosp Psychiatry. 2008 May-Jun;30(3):208–25.

113. Thakur ER, Gurtman MB, Keefer L, Brenner DM, Lackner JM. Gender differences in irritable bowel syndrome: the interpersonal connection. Neurogastroenterol Motil. 2015 Oct;27(10):1478–86.
114. Garakani A, Win T, Virk S, Gupta S, Kaplan D, Masand PS. Comorbidity of irritable bowel syndrome in psychiatric patients: a review. Am J Ther. 2003 Jan-Feb;10(1):61–7.
115. Gros DF, Antony MM, Mccabe RE, Swinson RP. Frequency and severity of the symptoms of irritable bowel syndrome across the anxiety disorders and depression. J Anxiety Disord. 2009 Mar;23(2):290–6.
116. Kenwright M, Mcdonald J, Talbot J, Janjua K. Do symptoms of irritable bowel syndrome improve when patients receive cognitive behavioural therapy for co-morbid anxiety disorders in a primary care psychological therapy (IAPT) service? Behav Cogn Psychother. 2017 Sep;45(5):433–47.
117. Whitehead WE, Palsson O, Jones KR. Systematic review of the comorbidity of irritable bowel syndrome with other disorders: what are the causes and implications? Gastroenterology. 2002 Apr;122(4):1140–56.
118. Lee S, Wu J, Ma YL, Tsang A, Guo W-J, Sung J. Irritable bowel syndrome is strongly associated with generalized anxiety disorder: a community study. Aliment Pharmacol Ther. 2009 Sep;30(6):643–51.
119. Inadomi JM, Fennerty MB, Bjorkman D. Systematic review: the economic impact of irritable bowel syndrome. Aliment Pharmacol Ther. 2003 Oct;18(7):671–82.
120. Longstreth G, Wilson A, Knight K, Wong J, Chiou CF, Barghout V, et al. Irritable bowel syndrome, health care use, and costs: a U.S. managed care perspective. Am J Gastroenterol. 2003 Mar;98(3):600–7.
121. Paré P, Gray J, Lam S, Balshaw R, Khorasheh S, Barbeau M, et al. Health-related quality of life, work productivity, and health care resource utilization of subjects with irritable bowel syndrome: baseline results from LOGIC (Longitudinal Outcomes Study of Gastrointestinal Symptoms in Canada), a naturalistic study. Clin Ther. 2006 Oct;28(10):1726–35.
122. Lackner JM, Mesmer C, Morley S, Dowzer C, Hamilton S. Psychological treatments for irritable bowel syndrome: a systematic review and meta-analysis. J Consult Clin Psychol. 2004 Dec;72(6):1100–13.
123. Laird KT, Tanner-Smith EE, Russell AC, Hollon SD, Walker LS. Short-term and long-term efficacy of psychological therapies for irritable bowel syndrome: a systematic review and meta-analysis. Clin Gastroenterol Hepatol. 2016 Jul;14(7):937–47.
124. Laird KT, Tanner-Smith EE, Russell AC, Hollon SD, Walker LS. Comparative efficacy of psychological therapies for improving mental health and daily functioning in irritable bowel syndrome: a systematic review and meta-analysis. Clin Psychol Rev. 2017 Feb;51:142–52.
125. Ljótsson B, Andréewitch S, Hedman E, Rück C, Andersson G, Lindefors N. Exposure and mindfulness based therapy for irritable bowel syndrome – an open pilot study. J Behav Ther Exp Psychiatry. 2010 Sep;41(3):185–90.
126. Kinsinger SW. Cognitive-behavioral therapy for patients with irritable bowel syndrome: current insights. Psychol Res Behav Manag. 2017 Jul;10:231–7.
127. Lee HH, Choi YY, Choi MG. The efficacy of hypnotherapy in the treatment of irritable bowel syndrome: a systematic review and meta-analysis. J Neurogastroenterol Motil. 2014 Apr;20(2):152–62.
128. Hazlett-Stevens H, Craske MG, Mayer EA, Chang L, Naliboff BD. Prevalence of irritable bowel syndrome among university students: the roles of worry, neuroticism, anxiety sensitivity and visceral anxiety. J Psychosom Res. 2003 Dec;55(6):501–5.
129. Jerndal P, Ringström G, Agerforz P, Karpefors M, Akkermans LM, Bayati A, et al. Gastrointestinal-specific anxiety: an important factor for severity of GI symptoms and quality of life in IBS. Neurogastroenterol Motil. 2010 Jun;22(6):646–e179.

130. Labus JS, Mayer EA, Chang L, Bolus R, Naliboff BD. The central role of gastrointestinal-specific anxiety in irritable bowel syndrome: further validation of the visceral sensitivity index. Psychosom Med. 2007 Jan;69(1):89–98.
131. Craske MG, Wolitzky-Taylor KB, Labus J, Wu S, Frese M, Mayer EA, et al. A cognitive-behavioral treatment for irritable bowel syndrome using interoceptive exposure to visceral sensations. Behav Res Ther. 2011 Jun;49(6–7):413–21.
132. Altayar O, Sharma V, Prokop LJ, Sood A, Murad MH. Psychological therapies in patients with irritable bowel syndrome: a systematic review and meta-analysis of randomized controlled trials. Gastroenterol Res Pract. 2015;2015:549308.
133. Ford AC, Lacy BE, Harris LA, Quigley EMM, Moayyedi P. Effect of antidepressants and psychological therapies in irritable bowel syndrome: an updated systematic review and meta-analysis. Am J Gastroenterol. 2019 Jan;114(1):21–39.
134. Martens U, Enck P, Matheis A, Herzog W, Klosterhalfen S, Ruhl A, et al. Motivation for psychotherapy in patients with functional gastrointestinal disorders. Psychosomatics. 2010 May-Jun;51(3):225–9.
135. Palsson OS, Whitehead WE. Psychological treatments in functional gastrointestinal disorders: a primer for the gastroenterologist. Clin Gastroenterol Hepatol. 2013 Mar;11(3):208–16.
136. Sobin WH, Heinrich TW, Drossman DA. Central neuromodulators for treating functional GI disorders: a primer. Am J Gastroenterol. 2017 May;112(5):693–702.
137. Thiwan SI, Drossman DA. Treatment of functional GI disorders with psychotropic medicines: a review of evidence with a practical approach. Gastroenterol Hepatol (N Y). 2006 Sep;2(9):678–88.
138. Xiong N, Duan Y, Wei J, Mewes R, Leonhart R. Antidepressants vs. placebo for the treatment of functional gastrointestinal disorders in adults: a systematic review and meta-analysis. Front Psychiatry. 2018 Dec;9:659.
139. Taché Y, Perdue MH. Role of peripheral CRF signalling pathways in stress-related alterations of gut motility and mucosal function. Neurogastroenterol Motil. 2004 Apr;16(Suppl 1):137–42.
140. Li X, Ding F, Luo P, Yang J, Liu Z, Liu J, et al. Study on the therapeutic effects of drug and cognitive-behavioral therapy on non-erosive reflux disease patients with emotional disorders. Front Psychiatry. 2018 May;9:115.
141. Modlin I, Hunt RH, Malfertheiner P, Moayyedi P, Quigley EM, Tytgat GN, et al. Diagnosis and management of non-erosive reflux disease--the Vevey NERD Consensus Group. Digestion. 2009;80(2):74–88.
142. Vakil N, van Zanten SV, Kahrilas P, Dent J, Jones R. The Montreal definition and classification of gastroesophageal reflux disease: a global evidence-based consensus. Am J Gastroenterol. 2006 Aug;101(8):1900–20.
143. Peery AF, Crockett SD, Barritt AS, Dellon ES, Eluri S, Gangarosa LM, et al. Burden of gastrointestinal, liver, and pancreatic diseases in the United States. Gastroenterology. 2015 Dec;149(7):1731–41.
144. Katz PO, Gerson LB, Vela MF. Guidelines for the diagnosis and management of gastroesophageal reflux disease. Am J Gastroenterol. 2013 Mar;108(3):308–28.
145. Martinez SD, Malagon IB, Garewal HS, Cui H, Fass R. Non-erosive reflux disease (NERD) – acid reflux and symptom patterns. Aliment Pharmacol Ther. 2003 Feb;17(4):537–45.
146. Mdi P, Fitzgerald R. Research advances in esophageal diseases: bench to bedside. F1000Prime Rep. 2013 Oct;5:44.
147. Riehl ME, Kinsinger S, Kahrilas PJ, Pandolfino JE, Keefer L. Role of a health psychologist in the management of functional esophageal complaints. Dis Esophagus. 2015;28(5):428–36.
148. Riehl ME, Chen JW. The proton pump inhibitor nonresponder: a behavioral approach to improvement and wellness. Curr Gastroenterol Rep. 2018 Jun;20(7):34.
149. Keefer L, Palsson OS, Pandolfino JE. Best practice update: incorporating psychogastroenterology into management of digestive disorders. Gastroenterology. 2018 Apr;154(5):1249–57.

150. Mcdonald-Haile J, Bradley LA, Bailey MA, Schan CA, Richter JE. Relaxation training reduces symptom reports and acid exposure in patients with gastroesophageal reflux disease. Gastroenterology. 1994 Jul;107(1):61–9.
151. Katzka DA. Simple office-based behavioral approach to patients with chronic belching. Dis Esophagus. 2013 Aug;26(6):570–3.
152. Hemmink GJ, Ten Cate L, Bredenoord AJ, Timmer R, Weusten BL, Smout AJ. Speech therapy in patients with excessive supragastric belching – a pilot study. Neurogastroenterol Motil. 2010 Jan;22(1):24–8.
153. Glasinovic E, Wynter E, Arguero J, Ooi J, Nakagawa K, Yazaki E, et al. Treatment of supragastric belching with cognitive behavioral therapy improves quality of life and reduces acid gastroesophageal reflux. Am J Gastroenterol. 2018 Apr;113(4):539–47.
154. Meuret AE, Ritz T. Hyperventilation in panic disorder and asthma: empirical evidence and clinical strategies. Int J Psychophysiol. 2010 Oct;78(1):68–79.
155. Davies SJ, Allgulander C. Anxiety and cardiovascular disease. Mod Trends Pharmacopsychiatry. 2013;29:85–97.
156. Meuret A, Kroll J, Ritz T. Panic disorder comorbidity with medical conditions and treatment implications. Annu Rev Clin Psychol. 2017 May;13(1):209–40.
157. Deacon B, Lickel J, Abramowitz JS. Medical utilization across the anxiety disorders. J Anxiety Disord. 2008;22(2):344–50.
158. Katon W. Panic disorder in the medical setting. Rockville: Department of Health and Human Services, Public Health Service, Alcohol, Drug Abuse and Mental Health Administration, National Institute of Mental Health; 1989.
159. Gadermann AM, Alonso J, Vilagut G, Zaslavsky AM, Kessler RC. Comorbidity and disease burden in the National Comorbidity Survey Replication (NCS-R). Depress Anxiety. 2012 Sep;29(9):797–806.
160. Alonso J, Vilagut G, Chatterji S, Heeringa S, Schoenbaum M, Bedirhan Üstün T, et al. Including information about co-morbidity in estimates of disease burden: results from the World Health Organization World Mental Health Surveys. Psychol Med. 2011 Apr;41(4):873–86.
161. Kinley DJ, Cox BJ, Clara I, Goodwin RD, Sareen J. Panic attacks and their relation to psychological and physical functioning in Canadians: results from a nationally representative sample. Can J Psychiatry. 2009 Feb;54(2):113–22.
162. Klerman GL, Weissman MM, Ouellette R, Johnson J, Greenwald S. Panic attacks in the community. Social morbidity and health care utilization. JAMA. 1991 Feb;265(6):742–6.
163. Obrist PA. Presidential address, 1975. The cardiovascular-behavioral interaction – as it appears today. Psychophysiology. 1976 Mar;13(2):95–107.
164. Dimsdale JE. Psychological stress and cardiovascular disease. J Am Coll Cardiol. 2008 Apr;51(13):1237–46.
165. Esler M. Depressive illness, the sympathetic nervous system and cardiac risk. J Hypertens. 2009 Dec;27(12):2349–50.
166. Tully PJ, Baune BT. Comorbid anxiety disorders alter the association between cardiovascular diseases and depression: the German National Health Interview and Examination Survey. Soc Psychiatry Psychiatr Epidemiol. 2014 May;49(5):683–91.
167. Müller-Tasch T, Frankenstein L, Holzapfel N, Schellberg D, Löwe B, Nelles M, et al. Panic disorder in patients with chronic heart failure. J Psychosom Res. 2008 Mar;64(3):299–303.
168. Emdin CA, Odutayo A, Wong CX, Tran J, Hsiao AJ, Hunn BH. Meta-analysis of anxiety as a risk factor for cardiovascular disease. Am J Cardiol. 2016 Aug;118(4):511–9.
169. Katerndahl DA. The association between panic disorder and coronary artery disease among primary care patients presenting with chest pain: an updated literature review. Prim Care Companion J Clin Psychiatry. 2008;10(4):276–85.
170. Player MS, Peterson LE. Anxiety disorders, hypertension, and cardiovascular risk: a review. Int J Psychiatry Med. 2011;41(4):365–77.

171. Gomez-Caminero A, Blumentals WA, Russo LJ, Brown RR, Castilla-Puentes R. Does panic disorder increase the risk of coronary heart disease? A cohort study of a national managed care database. Psychosom Med. 2005 Sep–Oct;67(5):688–91.
172. Goodwin RD, Davidson KW, Keyes K. Mental disorders and cardiovascular disease among adults in the United States. J Psychiatr Res. 2009 Jan;43(3):239–46.
173. Meuret AE, White KS, Ritz T, Roth WT, Hofmann SG, Brown TA. Panic attack symptom dimensions and their relationship to illness characteristics in panic disorder. J Psychiatr Res. 2006 Sep;40(6):520–7.
174. Celano CM, Daunis DJ, Lokko HN, Campbell KA, Huffman JC. Anxiety disorders and cardiovascular disease. Curr Psychiatry Rep. 2016 Nov;18(11):101.
175. Cohen BE, Edmondson D, Kronish IM. State of the art review: depression, stress, anxiety, and cardiovascular disease. Am J Hypertens. 2015 Nov;28(11):1295–302.
176. Davies SJ, Ghahramani P, Jackson PR, Noble TW, Hardy PG, Hippisley-Cox J, et al. Association of panic disorder and panic attacks with hypertension. Am J Med. 1999 Oct;107(4):310–6.
177. Edmondson D, Richardson S, Falzon L, Davidson KW, Mills MA, Neria Y. Posttraumatic stress disorder prevalence and risk of recurrence in acute coronary syndrome patients: a meta-analytic review. PLoS One. 2012;7(6):e38915.
178. Allabadi H, Alkaiyat A, Alkhayyat A, Hammoudi A, Odeh H, Shtayeh J, et al. Depression and anxiety symptoms in cardiac patients: a cross-sectional hospital-based study in a Palestinian population. BMC Public Health. 2019 Feb;19(1):232.
179. Roest AM, Martens EJ, de Jonge P, Denollet J. Anxiety and risk of incident coronary heart disease: a meta-analysis. J Am Coll Cardiol. 2010 Jun;56(1):38–46.
180. Walters K, Rait G, Petersen I, Williams R, Nazareth I. Panic disorder and risk of new onset coronary heart disease, acute myocardial infarction, and cardiac mortality: cohort study using the general practice research database. Eur Heart J. 2008 Dec;29(24):2981–8.
181. Stein DJ, Aguilar-Gaxiola S, Alonso J, Bruffaerts R, de Jonge P, Liu Z, et al. Associations between mental disorders and subsequent onset of hypertension. Gen Hosp Psychiatry. 2014 Mar-Apr;36(2):142–9.
182. Cheng YF, Leu HB, Su CC, Huang CC, Chiang CH, Huang PH, et al. Association between panic disorder and risk of atrial fibrillation: a nationwide study. Psychosom Med. 2013 Jan;75(1):30–5.
183. Seldenrijk A, Vogelzangs N, Batelaan NM, Wieman I, van Schaik DJ, Penninx BJ. Depression, anxiety and 6-year risk of cardiovascular disease. J Psychosom Res. 2015 Feb;78(2):123–9.
184. Smoller JW, Pollack MH, Wassertheil-Smoller S, Jackson RD, Oberman A, Wong ND, et al. Panic attacks and risk of incident cardiovascular events among postmenopausal women in the Women's Health Initiative Observational Study. Arch Gen Psychiatry. 2007 Oct;64(10):1153–60.
185. de Ornelas Maia AC, Soares-Filho G, Pereira V, Nardi AE, Silva AC. Psychiatric disorders and quality of life in patients with implantable cardioverter defibrillators: a systematic review. Prim Care Companion CNS Disord. 2013;15(2)
186. Magyar-Russell G, Thombs BD, Cai JX, Baveja T, Kuhl EA, Singh PP, et al. The prevalence of anxiety and depression in adults with implantable cardioverter defibrillators: a systematic review. J Psychosom Res. 2011 Oct;71(4):223–31.
187. Godemann F, Butter C, Lampe F, Linden M, Schlegl M, Schultheiss HP, et al. Panic disorders and agoraphobia: side effects of treatment with an implantable cardioverter/defibrillator. Clin Cardiol. 2004 Jun;27:321–6.
188. Richards SH, Anderson L, Jenkinson C, Whalley B, Rees K, Davies P, Bennett P, Liu Z, West R, Thompson DR, Taylor RS. Psychological interventions for coronary heart disease. Cochrane Database Syst Rev. 2017 Apr;28(4):CD002902.
189. Zheng X, Zheng Y, Ma J, Zhang M, Zhang Y, Liu X, et al. Effect of exercise-based cardiac rehabilitation on anxiety and depression in patients with myocardial infarction: a systematic review and meta-analysis. Heart Lung. 2019 Jan;48(1):1–7.

190. Maia AC, Braga AA, Soares-Filho G, Pereira V, Nardi AE, Silva AC. Efficacy of cognitive behavioral therapy in reducing psychiatric symptoms in patients with implantable cardioverter defibrillator: an integrative review. Braz J Med Biol Res. 2014 Apr;47(4):265–72.
191. Reavell J, Hopkinson M, Clarkesmith D, Lane DA. Effectiveness of cognitive behavioral therapy for depression and anxiety in patients with cardiovascular disease: a systematic review and meta-analysis. Psychosom Med. 2018 Oct;80(8):742–53.
192. Tully PJ, Selkow T, Bengel J, Rafanelli C. A dynamic view of comorbid depression and generalized anxiety disorder symptom change in chronic heart failure: the discrete effects of cognitive behavioral therapy, exercise, and psychotropic medication. Disabil Rehabil. 2015;37(7):585–92.
193. Blumenthal JA, Feger BJ, Smith PJ, Watkins LL, Jiang W, Davidson J, et al. Treatment of anxiety in patients with coronary heart disease: rationale and design of the UNderstanding the benefits of exercise and escitalopram in anxious patients WIth coroNary heart Disease (UNWIND) randomized clinical trial. Am Heart J. 2016 Jun;176:53–62.
194. Oh H, Stickley A, Singh F, Koyanagi A. Self-reported asthma diagnosis and mental health: findings from the Collaborative Psychiatric Epidemiology Surveys. Psychiatry Res. 2019 Jan;271:721–5.
195. Goodwin RD, Pine DS. Respiratory disease and panic attacks among adults in the United States. Chest. 2002 Aug;122(2):645–50.
196. Goodwin RD, Olfson M, Shea S, Latingua RA, Carrasquilo O, Gameroff MJ, et al. Asthma and mental disorders in primary care. Gen Hosp Psychiatry. 2003 Nov–Dec;25(6):479–83.
197. Goodwin RD, Jacobi F, Thefeld W. Mental disorders and asthma in the community. Arch Gen Psychiatry. 2003 Nov;60(11):1125–30.
198. Hasler G, Gergen PJ, Kleinbaum DG, Ajdacic V, Gamma A, Eich D, et al. Asthma and panic in young adults: a 20-year prospective community study. Am J Respir Crit Care Med. 2005 Jun;171(11):1224–30.
199. Niles AN, Dour HJ, Stanton AL, Roy-Byrne PP, Stein MB, Sullivan G, et al. Anxiety and depressive symptoms and medical illness among adults with anxiety disorders. J Psychosom Res. 2015 Feb;78(2):109–15.
200. Feldman JM, Siddique MI, Morales E, Kaminski B, Lu SE, Lehrer PM. Psychiatric disorder and asthma outcomes among high-risk inner-city patients. Psychosom Med. 2005 Nov-Dec;67(6):989–96.
201. Bouchard A, Ouellet K, Bacon SL, Lavoie, KL. Psychiatric morbidity in COPD and asthma patients presenting to the emergency room for acute exacerbations. Paper presented at: 17th Annual Meeting of the International Society for Advancement of Respiratory Psychophysiology; 2010 September 25-27; New York City, NY.
202. Goodwin RD, Wamboldt MZ, Pine DS. Lung disease and internalizing disorders. Is childhood abuse a shared etiologic factor? J Psychosom Res. 2003 Sep;55(3):215–9.
203. Katon WJ, Richardson L, Lozano P, McCauley E. The relationship of asthma and anxiety disorders. Psychosom Med. 2004 May-Jun;66(3):349–55.
204. Halterman JS, Conn KM, Forbes-Jones E, Fagnano M, Hightower AD, Szilagyi PG. Behavior problems among inner-city children with asthma: findings from a community-based sample. Pediatrics. 2006 Feb;117(2):e192–9.
205. Craske MG, Poulton R, Tsao JC, Plotkin D. Paths to panic disorder/agoraphobia: An exploratory analysis from age 3 to 21 in an unselected birth cohort. J Am Acad Child Adolesc Psychiatry. 2001 May;40(5):556–63.
206. Meuret AE, Ehrenreich JT, Pincus DB, Ritz T. Prevalence and correlates of asthma in children with internalizing psychopathology. Depress Anxiety. 2006;23(8):502–8.
207. Battaglia M, Ogliari A. Anxiety and panic: from human studies to animal research and back. Neurosci Biobehav Rev. 2005 Feb;29(1):169–79.
208. Dirks JF, Schraa JC, Brown EL, Kinsman RA. Psycho-maintenance in asthma: hospitalization rates and financial impact. Br J Med Psychol. 1980 Dec;53(4):349–54.
209. Ritz T, Steptoe A, DeWilde S, Costa M. Emotions and stress increase respiratory resistance in asthma. Psychosom Med. 2000 May–Jun;62(3):401–12.

210. Ritz T, Kullowatz A, Goldman MD, Smith HJ, Kanniess F, Dahme B, et al. Airway response to emotional stimuli in asthma: the role of the cholinergic pathway. J Appl Physiol (1985). 2010 Jun;108(6):1542–9.
211. Ritz T, Wilhelm FH, Meuret AE, Gerlach AL, Roth WT. Airway response to emotion- and disease-specific films in asthma, blood phobia, and health. Psychophysiology. 2011 Jan;48(1):121–35.
212. Sandberg S, Paton JY, Ahola S, McCann DC, McGuinness D, Hillary CR, et al. The role of acute and chronic stress in asthma attacks in children. Lancet. 2000 Sep;356(9234):982–7.
213. Kullowaltz A, Rosefield D, Dahme B, Magnussen H, Kanniess F, Ritz T. Stress effects on lung function in asthma are mediated by changes in airway inflammation. Psychosom Med. 2008 May;70(4):468–75.
214. Ritz T, Ayala ES, Trueba AF, Vance CD, Auchus RJ. Acute stress-induced increases in exhaled nitric oxide in asthma and their association with endogenous cortisol. Am J Respir Crit Care Med. 2011 Jan;183(1):26–30.
215. Potokar JP, Nutt DJ. Chest pain: panic attack or heart attack? Int J Clin Pract. 2000 Mar;54(2):110–4.
216. Greenberg PE, Sisitsky T, Kessler RC, Finkelstein SN, Berndt ER, Davidson JR, et al. The economic burden of anxiety disorders in the 1990s. J Clin Psychiatry. 1999 Jul;60(7):427–35.
217. ten Brinke A, Ouwerkerk ME, Zwinderman AH, Spinhoven P, Bel EH. Psychopathology in patients with severe asthma is associated with increased health care utilization. Am J Respir Crit Care Med. 2001 Apr;163(5):1093–6.
218. Greaves CJ, Eiser C, Seamark D, Halpin DM. Attack context: an important mediator of the relationship between psychological status and asthma outcomes. Thorax. 2002 Mar;57(3):217–21.
219. Pilipenko N, Karekla M, Georgiou A, Feldman J. Impact of psychiatric illness upon asthma patients' health care utilization and illness control. Are all psychiatric comorbidities created equal? Psychol Health Med. 2016 Oct;21(7):787–99.
220. Johansson P, Dahlström U, Broström A. Factors and interventions influencing health-related quality of life in patients with heart failure: a review of the literature. Eur J Cardiovasc Nurs. 2006 Mar;5(1):5–15.
221. Sundh J, Wireklint P, Hasselgren M, Montgomery S, Ställberg B, Lisspers K, et al. Health-related quality of life in asthma patients - a comparison of two cohorts from 2005 and 2015. Respir Med. 2017 Nov;132:154–60.
222. Jaunay E, Consoli A, Greenfield B, Guilé JM, Mazet P, Cohen D. Treatment refusal in adolescents with severe chronic illness and borderline personality disorder. J Can Acad Child Adolesc Psychiatry. 2006 Aug;15(3):135–42.
223. Lehrer PM, Isenberg S, Hochron SM. Asthma and emotion: a review. J Asthma. 1993;30(1):5–21.
224. Lehrer PM, Karavidas MK, Lu SE, Feldman J, Kranitz L, Abraham S, et al. Psychological treatment of comorbid asthma and panic disorder: a pilot study. J Anxiety Disord. 2008 May;22(4):671–83.
225. Lehrer P, Carr RE, Smetankine A, Vaschillo E, Peper E, Porges S, et al. Respiratory sinus arrhythmia vs neck/trapezius EMG and incentive inspirometry biofeedback for asthma: a pilot study. Appl Psychophysiol Biofeedback. 1997 Jun;22(2):95–109.
226. Lehrer P, Feldman J, Giardino N, Song HS, Schmaling K. Psychological aspects of asthma. J Consult Clin Psychol. 2002 Jun;70(3):691–711.
227. Ritz T. Relaxation therapy in adult asthma. Is there new evidence for its effectiveness? Behav Modif. 2001 Sep;25(4):640–66.
228. Carr RE. Panic disorder and asthma: causes, effects and research implications. J Psychosom Res. 1998 Jan;44(1):43–52.
229. Global Initiative for Asthma (GINA). Global strategy for asthma management and prevention: NHLIB/WHO workshop report. Bethesda: National Institutes of Health, National Heart, Lung and Blood Institute; 2005.

230. National Heart, Lung, and Blood Institute/National Asthma Education and Prevention Program. Expert panel report: guidelines for the diagnosis and management of asthma. Full report 2007. NIH Publication No. 07-4051. Bethesda: National Institutes of Health; 2007.
231. Bobb C, Ritz T. Do asthma patients in general practice profit from a structured allergy evaluation and skin testing? Respir Med. 2003 Nov;97(11):1180–7.
232. Bobb C, Ritz T, Rowlands G, Griffiths C. Effects of allergen and trigger factor avoidance advice in primary care on asthma control: a randomized-controlled trial. Clin Exp Allergy. 2010 Jan;40(1):143–52.
233. Gibson PG, Powell H, Coughlan J, Wilson AJ, Abramson M, Haywood P, et al. Self-management education and regular practitioner review for adults with asthma. Cochrane Database Syst Rev. 2003;1:CD001117.
234. Kotses H, Bernstein IL, Bernstein DI, Reynolds RV, Korbee L, Wigal JK, et al. A self-management program for adult asthma. Part I: development and evaluation. J Allergy Clin Immunol. 1995 Feb;95(2):529–40.
235. Kotses H, Stout C, McConnaughy K, Winder JA, Creer TL. Evaluation of individualized asthma self-management programs. J Asthma. 1996;33(2):113–8.
236. Smyth JM, Stone AA, Hurewitz A, Kaell A. Effects of writing about stressful experiences on symptom reduction in patients with asthma or rheumatoid arthritis: a randomized trial. JAMA. 1999 Apr;281(14):1304–9.
237. Ritz T, Dahme B, Roth WT. Behavioral interventions in asthma: biofeedback techniques. J Psychosom Res. 2004 Jun;56(6):711–20.
238. Kew KM, Nashed M, Dulay V, Yorke J. Cognitive behavioural therapy (CBT) for adults and adolescents with asthma. Cochrane Database Syst Rev. 2016 Sep;9:CD011818.
239. Lehrer PM, Vaschillo E, Vaschillo B, Lu SE, Scardella A, Siddique M, et al. Biofeedback treatment for asthma. Chest. 2004 Aug;126(2):352–61.
240. Ross CJ, Davis TM, MacDonald GF. Cognitive-behavioral treatment combined with asthma education for adults with asthma and coexisting panic disorder. Clin Nurs Res. 2005 May;14(2):131–57.
241. Barlow DH, Craske MG. Mastery of your anxiety and panic (MAP-3): client workbook for anxiety and panic. Oxford: Oxford University Press; 2000.
242. Yorke J, Adair P, Doyle AM, Dubrow-Marshall L, Fleming S, Holmes L, et al. A randomised controlled feasibility trial of Group Cognitive Behavioural Therapy for people with severe asthma. J Asthma. 2017 Jun;54(5):543–54.
243. Feldman JM, Matte L, Interian A, Lehrer PM, Lu SE, Scheckner B, et al. Psychological treatment of comorbid asthma and panic disorder in Latino adults: results from a randomized controlled trial. Behav Res Ther. 2016 Dec;87:142–54.
244. Meuret AE, Wilhelm FH, Ritz T, Roth WT. Feedback of end-tidal pCO2 as a therapeutic approach for panic disorder. J Psychiatr Res. 2008 Jun;42(7):560–8.
245. Meuret AE, Rosenfield D, Seidel A, Bhaskara L, Hofmann SG. Respiratory and cognitive mediators of treatment change in panic disorder: evidence for intervention specificity. J Consult Clin Psychol. 2010 Oct;78(5):691–704.
246. Ritz T, Rosenfield D, Steele AM, Millard MW, Meuret AE. Controlling asthma by training of Capnometry-Assisted Hypoventilation (CATCH) vs slow breathing: a randomized controlled trial. Chest. 2014 Nov;146(5):1237–47.

Part IV

Therapeutic Issues

Biofeedback and Neurofeedback for Anxiety Disorders: A Quantitative and Qualitative Systematic Review

16

David F. Tolin, Carolyn D. Davies, Danielle M. Moskow, and Stefan G. Hofmann

Introduction

It has long been understood that anxiety disorders are associated with physiological arousal [1, 2]. Signs of sympathetic nervous system arousal in anxiety disorders include increased muscle tension, as evidenced by electromyography (EMG) [3]; increased electrodermal activity (EDA) [4]; and increased heart rate (HR) [5]. Individuals with anxiety disorders also show decreased vagal tone, reflecting poorer parasympathetic control over sympathetic arousal; findings in this domain include decreased heart rate variability (HRV) [6] and respiratory sinus arrhythmia (RSA) [7]. Respiration is affected (specifically hyperventilation) in anxiety disorders, as evidenced by decreased end-tidal CO_2 ($ETCO_2$) [8]. Finally, anxiety-related arousal can be detected centrally using electroencephalography (EEG), with some evidence that attenuated alpha activity is associated with anxiety [9]. Given the close association between physiological arousal and anxiety disorders, over the past several decades, there has been significant interest in the use of biofeedback, based on the notion that autonomic nervous system responses can be instrumentally conditioned [10]. In this section, we will briefly review the biofeedback strategies that have been utilized for individuals with anxiety disorders.

An EMG uses electrodes to translate electrical signals from motor neurons that cause muscles to contract. *EMG biofeedback* training is conceptualized as a method

D. F. Tolin (✉)
The Institute of Living, Hartford, CT, USA

Yale University School of Medicine, New Haven, CT, USA
e-mail: david.tolin@hhchealth.org

C. D. Davies
The Institute of Living, Hartford, CT, USA

D. M. Moskow · S. G. Hofmann
Boston University, Boston, MA, USA

Y.-K. Kim (ed.), *Anxiety Disorders*, Advances in Experimental Medicine and Biology 1191, https://doi.org/10.1007/978-981-32-9705-0_16

for reducing muscle tension commonly associated with anxiety. The frontalis (forehead) muscle is the most frequently targeted muscle group in EMG biofeedback [11–13], though the trapezius (upper back) and sternomastoid (neck) muscles may also be used [13]. Electrodes are placed on the target areas, and patients receive auditory or visual feedback over the course of an approximately 20-min session, such as the sound of a tone [11, 13] or an image of bars on a computer screen turning green when EMG levels decrease [12]. EDA measures skin conductance, or what has been historically known as galvanic skin response (GSR). Sweating is controlled by the sympathetic nervous system, and thus the goal of *EDA biofeedback* is to train patients to reduce EDA levels, thereby reducing associated autonomic arousal and anxiety. During biofeedback sessions, electrodes are typically attached to the fingers or palm, and patients are taught to lower EDA levels, typically through the use of visual feedback from an EDA recording device [14, 15]. *Thermal biofeedback* is based on the notion that raising skin temperature (e.g., of the hand) is associated with vasodilation, which results from decreased sympathetic nervous system activity [16]. Relaxing imagery or relaxing self-statements may be used to facilitate warming.

HR biofeedback involves encouraging patients to decrease HR toward a target; such procedures have been demonstrated not only to decrease HR but also to engage prefrontal cortical regions [17]. In anxiety, such procedures have been conducted at rest [18], while viewing negative pictures [19], or during in vivo exposure [20, 21]. HR feedback may be auditory [21] or visual (e.g., in one study, HR feedback was given via the shrinking of a pictorial stimulus, or slowing of a moving stimulus, on the computer screen) [18]. *HRV biofeedback* consists of feeding back HR data during slow breathing exercises with the aim of increasing RSA, the cardiac pattern that occurs when heart rate increases during inhalation and decreases during exhalation [22, 23]. Patients alter their breathing rate according to a pacing stimulus, such as a light display that moves up and down on a computer screen. Patients are provided with feedback of cardiac variability and are instructed to increase the amplitude of HR fluctuations that occur in conjunction with respiration [24]. The feedback can include a beat-to-beat cardiotachometer, superimposed on a measure of respiratory activity, a moving frequency analysis of HR, or a light bar display, showing the amplitude of RSA with each breath. There are several methods for conducting HRV biofeedback, with RSA HR wave as one measure. RSA wave training involves having individuals slow down their breath to a rate in which the amplitude of RSA is maximized. When the "resonant frequency" or proper breathing rate is determined for that individual, HR and respiration will covary in a way that the person will breathe in until their HR peaks and exhale until it starts to rise once again [25].

Biofeedback of $ETCO_2$ is predicated on the concept of chronic hyperventilation, which rapidly decreases the partial pressure of carbon dioxide in arterial blood, measured indirectly via exhaled $ETCO_2$. A capnometry-guided respiratory intervention [26, 27] includes a CO_2 sensor that samples the patient's exhaled breath and measures $ETCO_2$ and respiration rate (RR). The CO_2 sensor then transmits the $ETCO_2$ and RR values to a computer or tablet, which instructs the patient visually and audibly to slow his/her breathing, and increase $ETCO_2$ levels, progressively across sessions and during home-based practice.

Although there are many variants of *EEG neurofeedback*, the most frequently studied of these in the anxiety disorders have focused on increasing alpha waves. Alpha is the dominant EEG rhythm in healthy adults at rest and is associated with a calm, relaxed state [28]. Among patients with panic disorder, alpha is attenuated [9], though in GAD patients, alpha is increased [29]. Increasing alpha magnitude can produce a calming effect in high-anxious individuals [30]. One style of EEG neurofeedback is designed to train patients to increase their alpha productivity (often in contrast to beta or theta wave activity) with the aim of achieving relaxation. Alpha training, as described by Watson et al. [28], involved having the patient watch a display that showed lights that would brighten as alpha activity increased. In subsequent trials, the patient tried to match the brightness of the feedback light to a reference light, set by the operator. After several training sessions, patients are asked to practice at home daily, attempting to replicate the enhanced alpha state without biofeedback. Subsequent variations of this protocol included using pleasant sounds to reflect alpha activity in the desired zone [31] and a computer game in which participants "walk" through a scary house and light their way using greater alpha power [32].

Neurofeedback using functional magnetic resonance imaging (fMRI) signal is comparatively new. For emotional disorders, the model of training is to teach participants to downregulate limbic structures by engaging frontal regions [33]. In one trial for anxiety disorders (spider phobia), participants received training in cognitive reappraisal while looking at spider photographs (e.g., thinking about the aesthetics of the spider, focusing on the spider's powerlessness, humanizing the spider, or imagining approaching the spider in a safe environment). In the fMRI scanner, they were encouraged to use their reappraisal strategies while increasing prefrontal activity (associated with emotion regulation) and decreasing insula activity (associated with fear) [34].

Much of the earlier research on biofeedback was in the area of behavioral medicine and general stress reduction. However, biofeedback interventions have also been applied to a broad range of psychiatric disorders, including anxiety disorders, obsessive-compulsive disorder, posttraumatic stress disorder, depression, substance use disorders, dissociative disorders, eating disorders, and schizophrenia [35]. In this chapter, we will review the literature specific to biofeedback for anxiety disorders according to the 3rd, 4th, or 5th edition of the *Diagnostic and Statistical Manual of Mental Disorders* (DSM) [36–38], using both quantitative (meta-analytic) and qualitative review strategies.

Method

Data Sources

Journal articles were identified using searches of the PsycINFO and Medline electronic databases from 1987 to February 2019. The following search terms were used: *biofeedback training* or *biofeedback* or *neurofeedback* or *neurotherapy* and *anxiety disorders*, *panic disorder*, *phobia*, *social anxiety*, *generalized anxiety disorder*, *obsessive-compulsive disorder*, *posttraumatic stress disorder*, or *agoraphobia*. Studies were also identified through the reference lists of originally obtained articles, review papers, and meta-analyses.

Study Selection and Data Extraction

We included studies that met the following inclusion criteria: used a randomized controlled trial design; included an identified biofeedback intervention in at least one arm of the study; and tested a DSM-defined anxiety disorder sample (DSM-III or later). We excluded studies that examined child samples or reported only subanalyses of another study, or if it was not possible to isolate the effects of the biofeedback intervention (e.g., biofeedback + CBT compared to a wait list condition).

The second and third authors independently reviewed all abstracts from the initial search and coded whether or not they met initial inclusion criteria. Any disagreements were resolved by mutual discussion until 100% agreement was reached. This process resulted in 47 abstracts that met initial inclusion criteria (see Fig. 16.1). Full-text articles were obtained for these abstracts and divided between the second and third authors, who independently coded their assigned articles. From the initial 151 abstracts, 21 full-text articles were retained for analyses (see Table 16.1), representing a total of 779 patients.

Coding Variables

From the selected articles, we coded the diagnosis of sample, as well as the type of biofeedback: EMG, EDA, thermal, HR, HRV, $ETCO_2$, EEG, or fMRI. We also coded the nature of the control group: wait list, active treatment control, or reverse biofeedback (i.e., training patients to modify their physiological responses in a theoretically countertherapeutic direction).

Following the Effective Public Health Practice Project (EPHPP) rating system [39], we coded six variables for each study related to methodological quality as strong, moderate, or weak. Variables were as follows:

Selection bias. A "strong" rating was given when the sample was deemed very likely to be representative of the target population, with an 80% or higher participation rate. A "moderate" rating was given when the sample was judged to be

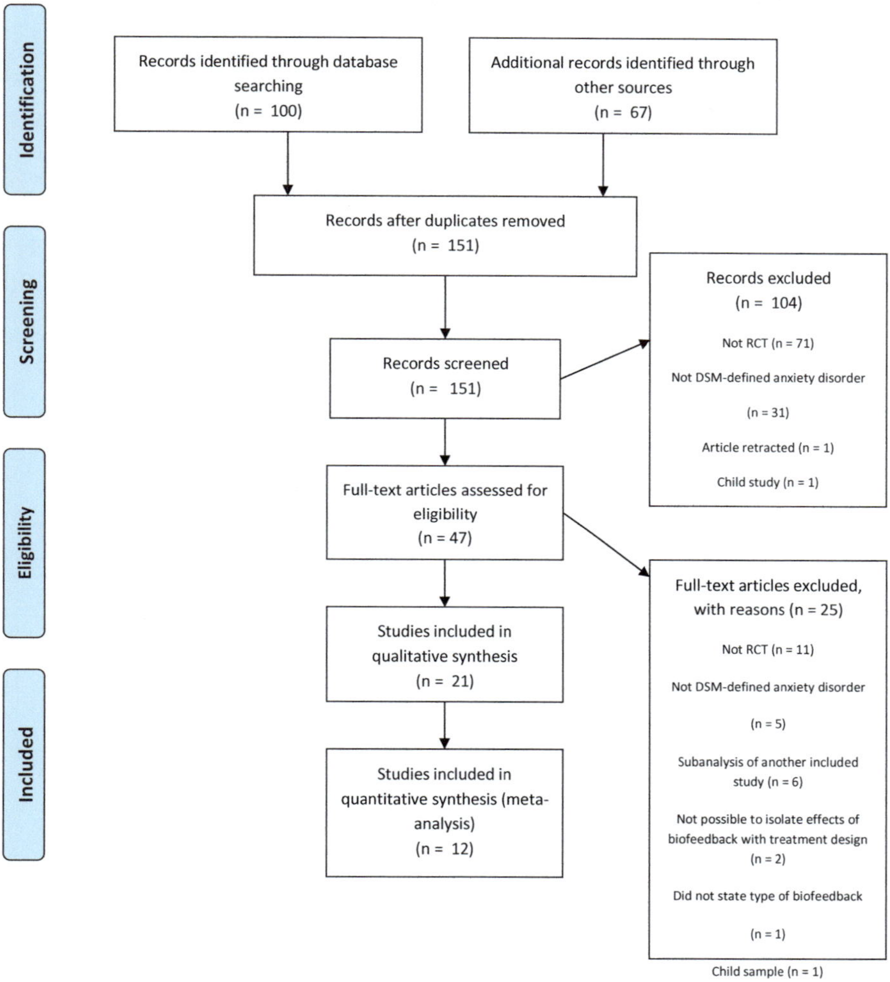

Fig. 16.1 PRISMA diagram

somewhat likely to be representative of the target population, with a 60–79% participation rate. A "weak" rating was given for all other responses, or when selection procedures were not described.

Design. A "strong" rating was given for randomized controlled trials. A "moderate" rating was given for cohort analytic, case-control, cohort, and an interrupted time series designs. A "weak" rating was given for all other designs, or when the design was not described.

Confounders. A "strong" rating was given when the researchers controlled for at least 80% of confounders. A "moderate" rating was given when the researchers controlled for 60–79% of confounders. A "weak" rating was given when confounders were not controlled for, or not described.

Table 16.1 Articles in the systematic review

Study	Target problem	Assessments	N	Comparison	Posttreatment	Follow-up	Selection bias rating	Design rating	Confounders rating	Blinding rating	Data collection methods rating	Withdrawals and dropouts rating	Global rating
Biofeedback type: EEG													
Agnihotri et al. (2007)	GAD	Semi-structured interview for DSM-IV-TR	15	EMG biofeedback	Increase in GSR and reduction in STAI	2-week follow-up: results maintained	Strong	Strong	Strong	Weak	Moderate	Weak	3
			15	EEG biofeedback	Increase in GSR and reduction in STAI								
		STAI	15	No treatment	No change								
Rice et al. (1993)	GAD	STAI-T	9	EMG biofeedback	Improvement on STAI and PSC; no changes in EMG, SCL, temperature	4-week follow-up: no change from post except for additional reduction in PSC in alpha suppression and alpha-increase BF groups	Moderate	Strong	Moderate	Moderate	Weak	Strong	2
		Welsh-A	9	Alpha enhancement biofeedback	Improvement on STAI and PSC; improvement on Welsh-A in EMG-BF; reduced HR response to stressor; no changes in EMG, SCL, temperature								
		PSC	9	Alpha suppression biofeedback (active control)	Improvement on STAI and PSC; increased HR response to stressor; no changes in EMG, SCL, temperature								
			9	Pseudo-meditation (placebo control)	Improvement on STAI and PSC; no changes in EMG, SCL, temperature								
			9	WL	No change								
Dadashi et al. (2015)	GAD	GAD-7	28 total	EEG biofeedback	GAD-7 decreased, GAFs decreased	None	Weak	Strong	Strong	Weak	Moderate	Weak	3
		GAFS		WL	No change								

Agnihotri et al. (2008)	GAD	DSM-IV-TR criteria	15	EEG biofeedback	Sig. decreased CAT, sig. increase in alpha-EEG, sig. reduction in BP	2 weeks later repeated parameters, found changes in EEG group in alpha-EEG activity	Weak	Strong	Weak	Weak	Strong	Weak	3
		CAT	15	EMG biofeedback	Sig. decreased CAT, sig. increase in alpha-EEG, most effective in reducing frontalis-EMG activity, sig. reduction in BP	Found changes in frontalis-EMG activity							
			15	No treatment	No sig. changes								
Pangotra et al. (2018)	Other anxiety disorders	HAM-A, ICD-10-DCR	10	L-Theanine	HAM-A sig. reduced	None	Weak	Strong	Strong	Weak	Strong	Weak	3
			10	PMR	HAM-A sig. reduced								
			10	EEG biofeedback	HAM-A sig. reduced								
van der Kolk et al. (2016)	PTSD	Traumatic Events Screening Inventory, CAPS, DTS, IASC	28	EEG neurofeedback	CAPS sig. decreased (more than control), DTS sig. decreased, IASC sig. decreased		Moderate	Strong	Strong	Moderate	Weak	Strong	2
			24	Wait list	Higher proportion met criteria for PTSD than participants receiving NF; CAPS sig. decreased	At 1-month follow-up, a higher proportion of WL met criteria for PTSD than neurofeedback							
Deng et al. (2014)	OCD	Y-BOCS, RBANS	40	Meds, CBT, EEG	Y-BOCS sig. decreased (more than control), sig. difference in effectiveness compared to control, sig. RBANS group main effect and between group differences	None	Strong	Strong	Strong	Strong	Strong	Strong	1
			39	Meds, CBT	Y-BOCS sig. decreased	None							

(continued)

Table 16.1 (continued)

Study	Target problem	Assessments	N	Comparison	Posttreatment	Follow-up	Selection bias rating	Design rating	Confounders rating	Blinding rating	Data collection methods rating	Withdrawals and dropouts rating	Global rating
Peniston and Kulkosky (1991)	PTSD	MMPI	15	EEG biofeedback	Decrease in MMPI scores in biofeedback compared to control group, including significant reduction on MMPI PTSD scale in biofeedback compared to control	Self-reported return of PTSD symptoms in 3/15 participants	Weak	Strong	Weak	Moderate	Weak	Strong	3
			14	TAU	No significant change in MMPI scores	Self-reported return of PTSD symptoms in all 14 participants							
Vanathy et al. (1998)	GAD	STAI, GQL, HAM-A	6	Alpha NF	Sig. decrease in self-report state and trait anxiety and observer manifest anxiety, non sig. improved quality of life	None	Weak	Strong	Strong	Weak	Weak	Weak	3
			6	Theta NF	Sig. decrease in self-report state anxiety and observer manifest anxiety, sig. improved quality of life								
			6	Wait list	Sig. increase in state anxiety								
Koprivova et al. (2013)	OCD	Y-BOCS	8	EEG NF	Y-BOCS improved; no change on BAI or BDI	No follow-up	Weak	Strong	Strong	Moderate	Moderate	Strong	2
		BAI											
		BDI	10	Sham feedback	Y-BOCS improved; BAI and BDI improved								
Biofeedback type: HR/HRV													
Pallavicini et al. (2009)	GAD	Semi-structured interview for DSM-IV TR	4	Biofeedback (HR) virtual reality and mobile phone	Improvements on GAD-7, BAI, and STAI-S	No follow-up	Strong	Strong	Strong	Weak	Moderate	Strong	2
		GAD-7	4	Virtual reality and mobile phone	Improvements on PSWQ								
		PSWQ											
		BAI	4	WL	Improvements on GAD-7								
		STAI											
		HAM-A			No differences between groups								
		VAS-A											

Lande et al. (2010)	PTSD	PCL	22	HR and respiration biofeedback + TAU	Improvement on PCL and Zung across groups, no differences between groups	No follow-up	Moderate	Strong	Weak	Weak	Moderate	Strong	3
		Zung	17	TAU									
Tan et al. (2010)	PTSD	CAPS	10	HRV biofeedback + TAU	Improved on CAPS and PCL-S; change scores not significantly different than TAU	No follow-up	Weak	Strong	Strong	Weak	Strong	Strong	3
		PCL-S	10	TAU	No change on CAPS or PCL-S								
Biofeedback type: EDA													
Biswas et al. (1995)	GAD	HAM-A	16 total	CBT	Within-group improvements on all outcome measures	Results largely maintained through 4-month follow-up	Weak	Strong	Moderate	Weak	Weak	Weak	3
		STAI			Greater improvement on SCS compared to other conditions								
		DAS											
		SCS		GSR biofeedback	Within-group improvements on 4/6 outcome measures								
		LOC		Pharmacotherapy	Within-group improvements on 3/6 outcome measures								
Biofeedback type: EMG													
Agnihotri et al. (2007)	GAD	Semi-structured interview for DSM-IV TR	15	EMG biofeedback	Increase in GSR and reduction in STAI	2-week follow-up: results maintained	Strong	Strong	Strong	Weak	Moderate	Weak	3
			15	EEG biofeedback	Increase in GSR and reduction in STAI								
		STAI	15	No treatment	No change								
Rice et al. (1993)	GAD	STAI-T	9	Frontal EMG-BF	Improvement on STAI and PSC; no changes in EMG, SCL, temperature	4-week follow-up: no change from post except for additional reduction in PSC in alpha suppression and alpha-increase BF groups	Moderate	Strong	Moderate	Moderate	Weak	Strong	2
		Welsh-A	9	Alpha enhancement BF	Improvement on STAI and PSC; improvement on Welsh-A in EMG-BF; reduced HR response to stressor; no changes in EMG, SCL, temperature								

(continued)

Table 16.1 (continued)

Study	Target problem	Assessments	N	Comparison	Posttreatment	Follow-up	Selection bias rating	Design rating	Confounders rating	Blinding rating	Data collection methods rating	Withdrawals and dropouts rating	Global rating
		PSC	9	Alpha suppression BF (active control)	Improvement on STAI and PSC; increased HR response to stressor; no changes in EMG, SCL, temperature								
			9	Pseudo-meditation (placebo control)	Improvement on STAI and PSC; no changes in EMG, SCL, temperature								
			9	WL	No change								
Agnihotri et al. (2008)	GAD	DSM-IV-TR criteria, CAT questionnaire	15	EEG biofeedback	Sig. decreased CAT, sig. increase in alpha-EEG, sig. reduction in bp	2 weeks later repeated parameters, found changes in EEG group in alpha-EEG activity	Weak	Strong	Weak	Weak	Strong	Weak	3
			15	EMG biofeedback	Sig. decreased CAT, sig. increase in alpha-EEG, most effective in reducing frontalis-EMG activity, sig. reduction in bp	Found changes in frontalis-EMG activity							
			15	No treatment	No sig. changes								

Carlson et al. (1998)	PTSD	CAPS-1	13	EMG biofeedback/ relaxation	Improved on 1 PTSD scale and STAI but not significantly more than WL; improvement on BDI and EMG measures	3-month follow-up: improvement on 1 PTSD scale, STAI, BDI, and EMG maintained	Moderate	Strong	Moderate	Weak	Moderate	Strong	2
		Mississippi Scale											
		IES	10	EMDR	Improved on 2 PTSD scales compared to WL; improvement on STAI but not significantly more than WL; reduction on BDI compared to WL; reduction on EMG measures	Improvements on 2 PTSD scales maintained; improved on CAPS-1 compared to biofeedback; improvements on STAI, BDI, and EMG maintained							
		PSS											
		BDI	12	Wait list control	Improved on 1 PTSD scale; reduction on EMG measures	EMG reduction maintained							
		STAI											
Biofeedback type: Respiratory													
Meuret et al. (2008)	PD	PDSS, CGI, ASI, SDS, MI-AAL, BDI	20	Raise-CO_2 breathing training	Psychopathology and RR decreased and pCO2 increased	40% reduction in initial PDSS scores achieved by 68% of participants at posttreatment, 79% at 2-month follow-up, and 93% at 12-month follow-up	Strong	Strong	Strong	Weak	Moderate	Strong	2
			17	Wait list	No sig. changes								

(continued)

Table 16.1 (continued)

Study	Target problem	Assessments	N	Comparison	Posttreatment	Follow-up	Selection bias rating	Design rating	Confounders rating	Blinding rating	Data collection methods rating	Withdrawals and dropouts rating	Global rating
Kim et al. (2012)	PD	SCID-IV-TR	19	Lower-CO_2 breathing training	Improved on PDSS compared to WL; end-tidal PCO_2 and RR reduced during quiet sitting and breathing exercise; large effect sizes for improvements in Agor Cogn Q, ASI, BAI, and BSQ	6 months: improvements on PDSS and RR maintained; PCO_2 reduction maintained during quiet sitting and PCO_2 reduced further during breathing exercise; effects for other measures not reported	Moderate	Strong	Strong	Weak	Moderate	Moderate	2
		PDSS											
		ASI											
		BDI	28	Raise-CO_2 breathing training	Improved on PDSS scores compared to WL; end-tidal PCO_2 increased and RR reduced during quiet sitting and breathing exercise; large effect sizes for improvements in Agor Cogn Q, ASI, BDI, BSQ, and MI	6 months: improvements on PDSS and RR maintained; PCO_2 increase maintained; effects for other measures maintained							
		MI											
		Agor Cogn Q											
		ACQ	27	Wait list	No change in PDSS; no change in end-tidal PCO_2; increase in RR	No follow-up							
		BAI											
		BSQ											

Meuret et al. (2010)	PD	ADIS-IV or SCID-IV	21	CART	Improvements on PDSS, ACQ, and BSQ/ASI for CART and CT, no differences between groups; PCO_2 increased and RR decreased in CART but not CT	No follow-up reported	Strong	Strong	Strong	Weak	Strong	Strong	2
		BSQ/ASI composite	20	CT									
		ACQ											
		PDSS											
Polak et al. (2015)	PTSD	CAPS	4	TF-CBT + respiratory biofeedback	IES-R scores improved in both conditions but faster in biofeedback condition	No follow-up	Moderate	Strong	Moderate	Weak	Strong	Strong	2
		MINI											
		IES-R	4	TF-CBT									
Biofeedback type: fMRI													
Zilverstand et al. (2015)	Spider phobia	SPQ, SBQ, FSQ, DSM-IV TR, MINI, ERQ-R, QCM	9	Neurofeedback + cognitive restructuring and exposure	Lower average anxiety levels than control, no physiology differences, sig. lower insula activation levels during regulate trials, sig. better downregulation during regulate trials	Sig. spider phobia decreases up to 3-month follow-up long-term reduction in spider fear correlated with ability to downregulate insula activation	Weak	Strong	Strong	Moderate	Weak	Weak	3
			9	Cognitive restructuring and exposure	No physiology differences	Sig. spider phobia decreases up to 3-month follow-up							

(continued)

Table 16.1 (continued)

Study	Target problem	Assessments	N	Comparison	Posttreatment	Follow-up	Selection bias rating	Design rating	Confounders rating	Blinding rating	Data collection methods rating	Withdrawals and dropouts rating	Global rating
Biofeedback type: Thermal													
Watson et al. (1997)	PTSD	PTSD-I	90 total	Relaxation + deep breathing + thermal biofeedback	No improvements in total scale scores; reductions in temperature and EMG across all groups	No follow-up	Weak	Strong	Moderate	Weak	Moderate	Weak	3
		Mississippi Scale		Relaxation only									
				Relaxation + deep breathing									

Note: Global rating 1 = strong, 2 = moderate, 3 = weak. *GAD* generalized anxiety disorder, *PTSD* posttraumatic stress disorder, *STAI-T/S* State-Trait Anxiety Inventory-Trait or State, *EMG* electromyography, *EEG* electroencephalography, *GSR* galvanic skin response, *Welsh-A* Welsh Anxiety Scale, *PSC* Psychosomatic Symptom Checklist, *SCL* skin conductance level, *BF* biofeedback, *NF* neurofeedback, *GAD-7* generalized anxiety disorder-7, *GAFS* global assessment of functioning scale, *CAT* Comprehensive Anxiety Test, *BP* blood pressure, *HAM-A* Hamilton Anxiety Rating Scale, *CAPS* Clinician-Administered PTSD Scale, *DTS* Davidson Trauma Scale, *IASC* Inventory of Altered Self-Capacities, *Y-BOCS* Yale-Brown Obsessive-Compulsive Scale, *RBANS* Repeatable Battery for the Assessment of Neuropsychological Status, *MMPI* Minnesota Multiphasic Personality Inventory, *TAU* treatment as usual, *GQL* Global Quality of Life Questionnaire, *BAI* Beck Anxiety Inventory, *BDI* Beck Depression Inventory, *PSWQ* Penn State Worry Questionnaire, *WL* wait list, *VAS-A* Visual Analogue Scale for Anxiety, *HR* heart rate, *HRV* heart rate variability, *PCL(S)* PTSD Checklist (Specific), *Zung* Zung Depression Scale, *DAS* Dysfunctional Attitude Scale, *SCS* Self-Control Schedule, *LOC* locus of control, *IES* Impact of Events Scale, *PSS* PTSD Symptom Scale, *EMDR* eye movement desensitization and reprocessing, *PDSS* Panic Disorder Severity Scale, *ASI* Anxiety Sensitivity Inventory, *SDS* Sheehan Disability Scale, *MI-AAL* Mobility Inventory (alone), PCO_2 partial pressure of end-tidal carbon dioxide, *Agor Cogn Q* Agoraphobic Cognitions Questionnaire, *ASI* Anxiety Sensitivity Index, *ACQ* Anxiety Control Questionnaire, *BSQ* Body Sensations Questionnaire, *MINI* Mini-International Neuropsychiatric Interview, *SPQ* Spider Phobia Questionnaire, *SBQ* Spider Belief Questionnaire, *FSQ* Fear of Spider Questionnaire, *ERQ-R* Emotion Regulation Questionnaire Reappraisal Score, *QCM* Questionnaire of Current Motivation, *PTSD-I* PTSD Inventory

Blinding. A "strong" rating was given when the outcome assessor and study participants were blinded to intervention status and/or the research question. A "moderate" rating was given when there was blinding of either the outcome assessor or study participants, but not both. A "weak" rating was given when the outcome assessor and study participants were aware of the intervention condition.

Data collection methods. A "strong" rating was given when the study measures were judged to be reliable and valid. A "moderate" rating was given when the measures were considered valid, but the reliability was not described. A "weak" rating was given when there was no evidence of validity or reliability of the study measures.

Withdrawals and dropouts. A "strong" rating was given when over 80% of participants had follow-up assessments. A "moderate" rating was given when 60–79% of participants had follow-up assessments. A "weak" rating was given when less than < 60% of participants had follow-up assessments, or when withdrawals and dropouts were not described.

From these ratings, an overall study quality rating was determined, with a score of 1 being the highest quality and reflecting no "weak" ratings; a score of 2 reflecting one "weak" rating; and a score of 3 reflecting two or more "weak" ratings.

Coding Reliability

Eleven articles (52.4%) were randomly selected from the pool of included studies for inter-rater reliability analyses of study quality ratings. The second and third authors independently coded each of these articles, and then an intraclass correlation coefficient was calculated for the global rating of study quality. Intraclass correlation was .80, indicating good inter-rater reliability.

Statistical Analyses

We used random effects models with Comprehensive Meta-Analysis software. Study was the unit of analysis, resulting in pooled event rates. Effect sizes (Hedges's *g*) were weighted by inverse variance (therefore, larger samples were weighted more strongly than were smaller samples). Publication bias was examined using the trim and fill procedure [40], which estimates missing studies to the left side of the mean event rate (indicating lower effect size). Heterogeneity was examined using the Q statistic and its p value. Heterogeneity was examined further using the I^2 statistic, which reflects the percentage of variation due to true heterogeneity rather than chance and is interpreted as follows: 25% = little heterogeneity, 50% = moderate heterogeneity, and 75% = high heterogeneity [41]. Meta-regression was used to examine the relationship between global quality rating and effect size.

Results

Overall Results

When compared to WL control, biofeedback (10 comparisons, N = 175) was associated with a large effect size (g = 1.10, 95% CI = −0.55–1.64). Heterogeneity was high (Q = 40.47, p < .001; I^2 = 77.76). Visual inspection of Fig. 16.2 suggests the presence of an outlying study (g = 5.25). When that study was removed from analysis, the effect remained large (g = 0.86, 95% CI = 0.57–1.16), and heterogeneity was low (Q = 10.87, p = 0.21, I^2 = 26.40). The trim and fill procedure suggested no missing studies. Meta-regression did not reveal a relationship between study quality rating and effect size (z = 0.17, p = 0.55).

When compared to active control conditions, biofeedback (9 comparisons, N = 194) was associated with a small and negative effect size (g = −0.25, 95% CI = −0.79–0.28). Heterogeneity was moderate (Q = 28.83, p < 0.001; I^2 = 72.25); see Fig. 16.3. The trim and fill procedure suggested no missing studies. A significant relationship between global quality rating and effect size was found (z = −2.26,

Study name	Statistics for each study						
	Hedges's g	Standard error	Variance	Lower limit	Upper limit	Z-Value	p-Value
Dadashi et al. (2014)	5.254	0.792	0.628	3.701	6.806	6.632	0.000
Rice et al. (1993)	0.863	0.472	0.223	-0.061	1.788	1.830	0.067
Pallavicini et al. (2009)	0.125	0.629	0.396	-1.108	1.358	0.198	0.843
Carlson et al. (1998)	0.037	0.390	0.152	-0.727	0.801	0.095	0.925
Meuret et al. (2008)	1.217	0.364	0.133	0.503	1.932	3.341	0.001
Kim et al. (2012)	1.224	0.302	0.091	0.631	1.816	4.046	0.000
van der Kolk et al. (2016)	1.313	0.312	0.097	0.702	1.923	4.211	0.000
Agnihoutri et al. (2007)	0.712	0.373	0.139	-0.020	1.443	1.907	0.057
Agnihoutri et al. (2007)	0.629	0.371	0.138	-0.099	1.357	1.694	0.090
Rice et al. (1993)	0.945	0.476	0.227	0.011	1.879	1.984	0.047
	1.096	0.277	0.077	0.553	1.640	3.957	0.000

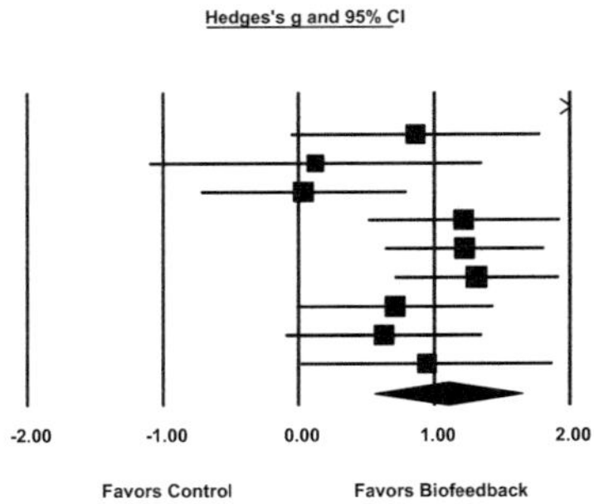

Fig. 16.2 Effect sizes for studies of biofeedback compared to wait list

Study name	Statistics for each study						
	Hedges's g	Standard error	Variance	Lower limit	Upper limit	Z-Value	p-Value
Pallavicini et al. (2009)	0.336	0.645	0.416	-0.928	1.601	0.521	0.602
Carlson et al. (1998)	-0.741	0.426	0.181	-1.576	0.093	-1.741	0.082
Deng et al. (2014)	0.742	0.241	0.058	0.269	1.215	3.076	0.002
Biswas et al. (1995)	-0.567	0.588	0.346	-1.720	0.585	-0.965	0.335
Biswas et al. (1995)	-1.360	0.685	0.470	-2.703	-0.016	-1.984	0.047
Pangotra et al. (2018)	-0.395	0.433	0.187	-1.244	0.453	-0.913	0.361
Pangotra et al. (2018)	-1.284	0.474	0.225	-2.213	-0.355	-2.709	0.007
Tan et al. (2011)	0.704	0.445	0.198	-0.169	1.577	1.581	0.114
Meuret et al. (2010)	-0.233	0.305	0.093	-0.831	0.365	-0.764	0.445
	-0.251	0.273	0.074	-0.786	0.283	-0.921	0.357

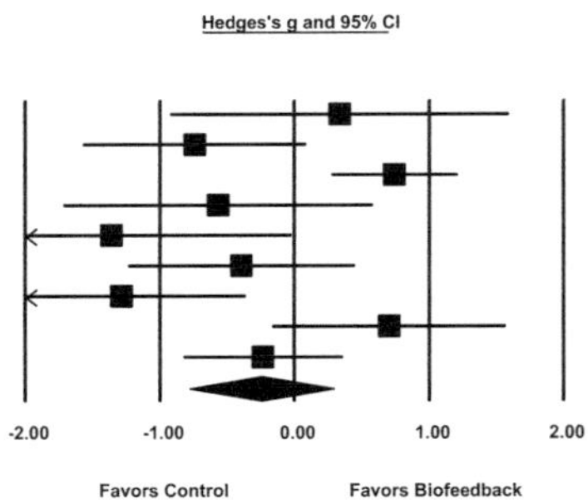

Fig. 16.3 Effect sizes for studies of biofeedback compared to active control

Study name	Statistics for each study						
	Hedges's g	Standard error	Variance	Lower limit	Upper limit	Z-Value	p-Value
Kim et al. (2012)	-0.614	0.344	0.118	-1.287	0.059	-1.787	0.074
Rice et al. (1993)	0.194	0.450	0.203	-0.688	1.077	0.432	0.666
	-0.262	0.401	0.161	-1.048	0.523	-0.654	0.513

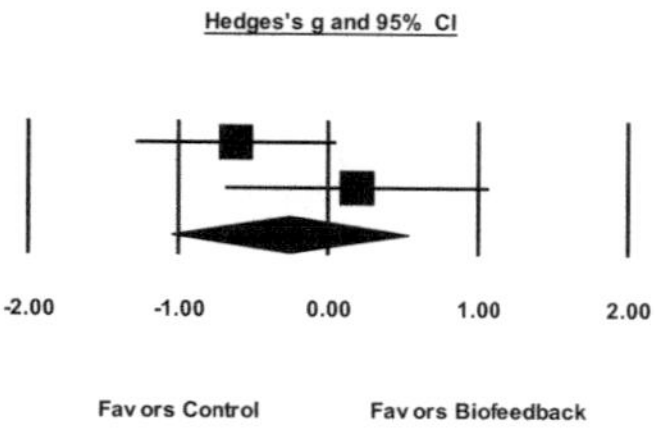

Fig. 16.4 Effect sizes for studies of biofeedback compared to reverse biofeedback control

$p = 0.02$), showing that higher-quality studies (lower number) had superior outcomes.

When compared to reverse control conditions, biofeedback (2 comparisons, N = 73) was associated with a small and negative effect size ($g = -0.26$, 95% CI = −1.05–0.52). Heterogeneity was moderate ($Q = 2.07$, $p = .0.154$; $I^2 = 50.91$); see Fig. 16.4. The trim and fill procedure and meta-regression could not be performed due to the small number of studies.

Electromyography (EMG) Biofeedback

Qualitative Review Most of the extant RCTs of EMG biofeedback have been with GAD patients. Rice et al. [11] assigned GAD patients to one of five conditions: EMG biofeedback, EEG neurofeedback to increase alpha rhythms, neurofeedback to decrease alpha rhythms, a pseudo-meditation control condition, or wait list. Whereas patients in the wait list condition did not improve, those in the EMG biofeedback condition showed improvement in trait anxiety and psychosomatic symptoms. Gains were maintained at 4-week follow-up. Agnihotri et al. [12] assigned patients to EMG biofeedback, EEG biofeedback, or no treatment. Compared to no treatment, patients receiving EMG biofeedback showed reductions in state and trait anxiety, though some increase in trait anxiety was noted at 2-week follow-up. In a follow-up study, Agnihotri et al. [42] assigned GAD patients to EMG biofeedback, EEG biofeedback, or no treatment. Patients receiving EMG biofeedback, in comparison with no treatment, showed significant decreases in anxiety symptoms, with some return of anxiety at 2-week follow-up. In one study of PTSD patients, Carlson et al. [13] randomized patients to EMG biofeedback, wait list, or an active treatment (eye movement desensitization). EMG biofeedback did not outperform wait list or active treatment in terms of PTSD symptoms or trait anxiety.

Effect size Analysis When compared to various control conditions, EMG biofeedback (4 comparisons, N = 83) was associated with a small effect size ($g = 0.23$, 95% CI = −0.48–0.95). Heterogeneity was moderate ($Q = 9.37$, $p = .0.025$; $I^2 = 67.97$). The trim and fill procedure suggested no missing studies. Meta-regression showed no relationship between study quality and effect size ($z = 0.69$, $p = 0.12$). Of the 4 comparisons, 3 were compared to wait list; these were associated with a medium effect size ($g = 0.53$, 95% CI = −0.17–0.63). One was compared to an active control, showing a medium negative effect ($g = -0.74$, 95% CI = -1.58–0.09).

Electrodermal Activity (EDA) Biofeedback

Qualitative Review Only one RCT of EDA biofeedback for anxiety disorders was identified. Biswas et al. [15] randomly assigned patients with GAD to a biofeedback condition designed to lower skin conductance or to one of the two active control treatments: cognitive behavioral therapy or pharmacotherapy (apparently unspecified benzodiazepines). Though improvement in anxiety was seen in all three groups, biofeedback did not outperform either of the two active treatments.

Effect Size Analysis EDA biofeedback (1 comparison, N = 16) was associated with a large and negative effect size ($g = -0.90$, 95% CI = 1.78–0.03). The trim and fill and meta-regression procedures could not be calculated due to the small number of studies.

Thermal Biofeedback

Qualitative Review Only one RCT of thermal biofeedback for anxiety disorders was identified. Watson et al. [43] assigned patients with PTSD to thermal biofeedback vs. two relaxation conditions. No significant differences were obtained across the 3 groups for 21 different outcome measures.

Effect Size Analysis Effect size for the single study of thermal biofeedback could not be calculated due to the lack of useable data.

Heart Rate (HR) Biofeedback

Qualitative Review Pallavicini et al. [18] assigned patients with GAD to HR biofeedback along with a virtual reality program designed to provide relaxation and exposure exercises, the virtual reality condition without biofeedback, or wait list. The biofeedback group showed a significant reduction on several measures of anxiety, though the two treated groups did not differ from one another. Lande et al. [44] assigned patients with PTSD to a HR and respiration biofeedback condition vs. treatment as usual (TAU). Participants in both conditions showed a reduction in PTSD and depressive symptoms, but did not differ from one another.

Effect Size Analysis When compared to various control conditions, HR biofeedback (2 comparisons, N = 12) was associated with a small effect size ($g = 0.23$, 95% CI = – 0.66–1.11). Heterogeneity was low ($Q = 0.06$, $p = .0.815$; $I^2 = 0.00$). The trim and fill and meta-regression procedure could not be performed due to the small number of studies. One comparison of HR biofeedback to wait list showed a negligible effect ($g = 0.12$, 95% CI = −1.11–1.36), whereas another comparing biofeedback to active treatment showed a small effect ($g = 0.34$, 95% CI = −0.93–1.60).

Heart Rate Variability (HRV) Biofeedback

Qualitative Review Only one RCT of HRV biofeedback for anxiety disorders was identified. Tan et al. [45] assigned patients with PTSD to HRV biofeedback vs. TAU. PTSD symptoms decreased in the biofeedback group, but not the TAU group; however, change scores did not differ significantly between the two groups.

Effect Size Analysis Only one study of HRV biofeedback was identified (N = 20); that study was associated with a medium effect size (g = 0.70, 95% CI = −0.17–1.58). The trim and fill and meta-regression procedures could not be performed due to the small number of studies.

End-Tidal CO_2 ($ETCO_2$) Biofeedback

Qualitative Review Meuret et al. [26] compared a capnometry-assisted breathing retraining procedure against wait list for patients with panic disorder. In this procedure, patients were trained to raise their levels of end-tidal CO_2 ($ETCO_2$). Compared to the wait list group, patients receiving biofeedback showed significant improvement in panic severity, agoraphobic avoidance, anxiety sensitivity, and functional impairment. Gains were largely maintained at 2- and 12-month follow-up. In a follow-up study, Meuret et al. [46] compared their respiratory biofeedback procedure to cognitive therapy for patients with panic disorder. The two groups showed equivalent reductions in panic severity, anxiety, and anxiety sensitivity. No follow-up was reported. Kim et al. [47] assigned panic disorder patients to two respiratory biofeedback procedures: in one, patients were instructed to raise their $ETCO_2$ levels, as was the case in the Meuret et al. [26, 46] studies. In the other condition, patients were instructed to lower their $ETCO_2$ levels. Both methods, compared to wait list, yielded significant decreases in panic disorder severity, with gains maintained at 6-month follow-up. This study raises questions about the extent to which raising $ETCO_2$ levels is an active mechanism of the intervention. Finally, in a study of patients with PTSD, Polak et al. [48] assigned patients to trauma-focused CBT with or without a respiratory biofeedback that instructed patients to correct breathing rate but not $ETCO_2$ levels. PTSD symptoms improved in each condition, though response was more rapid in patients receiving biofeedback. No follow-up was reported.

Effect Size Analysis When compared to various control conditions, $ETCO_2$ biofeedback (4 comparisons, N = 160) was associated with a small effect size ($g = 0.39$, 95% CI = −0.08–0.71). Heterogeneity was high ($Q = 25.46$, $p < .0.001$; $I^2 = 88.22$). The trim and fill procedure suggested no missing studies. Meta-regression could not be calculated due to lack of variance in quality ratings. Of the 4 comparisons, 2 compared $ETCO_2$ biofeedback to wait list. These were associated with a large effect size ($g = 1.22$, 95% CI = 0.76–1.68). One compared biofeedback to an active treatment; that study was associated with a small negative effect ($g = -0.23$, 95% CI = −0.83–0.36). The final study compared $ETCO_2$ biofeedback to a reverse biofeedback condition; that study was associated with a medium negative effect size ($g = -0.61$, 95% CI = −1.29–0.06).

Electroencephalography (EEG) Neurofeedback

Qualitative Review Several RCTs have examined the use of EEG neurofeedback in patients with GAD. Agnihotri et al. [12] assigned patients to EEG biofeedback, EMG biofeedback, or no treatment. In the neurofeedback group, patients were instructed to increase alpha rhythms. Compared to no treatment, patients receiving neurofeedback showed reductions in state and trait anxiety, though some increase in trait anxiety was noted at 2-week follow-up. In a follow-up study, Agnihotri et al. [42] assigned GAD patients to EEG biofeedback, EMG biofeedback, or no treatment. Patients receiving neurofeedback, in comparison with no treatment, showed significant decreases in anxiety symptoms, with some return of anxiety at 2-week follow-up. Rice et al. [11] assigned GAD patients to one of five conditions: EEG neurofeedback to increase alpha rhythms, neurofeedback to decrease alpha rhythms, EMG biofeedback, a pseudo-meditation control condition, or wait list. Compared to wait list, participants in both neurofeedback conditions showed decreased trait anxiety, with gains maintained at follow-up. This study raises questions about the extent to which alpha enhancement is the critical mechanism of treatment. Vanathy et al. [49] assigned GAD patients to neurofeedback designed to increase alpha rhythms, neurofeedback designed to increase theta rhythms, or wait list. Compared to wait list, both neurofeedback groups showed decreased state anxiety as well as improved quality of life, with no difference between the two neurofeedback groups. Dadashi et al. [50] employed a combined alpha- and theta-enhancing neurofeedback. Compared to wait list, neurofeedback patients showed decreased GAD symptoms and global impairment indices.

In patients with OCD, Deng et al. [51] examined the addition of EEG neurofeedback (increasing alpha and theta rhythms) to medications and CBT. OCD symptoms decreased in both groups, with a greater decrease in the group that received EEG. Koprivova et al. [52] assigned OCD patients to an individualized EEG neurofeedback (based on obtained abnormal frequencies) vs. sham biofeedback. OCD symptoms decreased in both conditions, and the neurofeedback group showed a greater reduction in compulsions than did the sham biofeedback group.

EEG neurofeedback has also been investigated in patients with PTSD. Peniston and Kulkosky [53] assigned veterans with PTSD to treatment as usual (TAU) or to EEG biofeedback (designed to increase alpha rhythm). Neurofeedback patients, compared to TAU patients, showed decreased scores on several subscales of the Minnesota Multiphasic Personality Inventory (MMPI); gains were largely maintained at 30-month follow-up. van der Kolk et al. [54] assigned PTSD patients to wait list or to EEG neurofeedback designed to increase alpha activity. Though PTSD severity decreased in both groups, a greater decrease was seen in the neurofeedback group. At 1-month follow-up, a greater proportion of wait list patients continued to meet criteria for PTSD, compared to neurofeedback patients.

Effect Size Analysis When compared to various control conditions, EEG neurofeedback (8 comparisons, N = 130) was associated with a moderate effect size ($g = 0.79$, 95% CI = −0.03–1.62). Heterogeneity was high (Q = 62.15, $p < .0.001$; $I^2 = 88.74$). The trim and fill procedure suggested no missing studies. Meta-regression showed a significant relationship between study quality and effect size ($z = -0.57$, $p = 0.045$), with higher-quality studies (lower number) showing superior effects. Of the 8 comparisons, 4 compared EEG neurofeedback to wait list and were associated with a large effect size ($g = 1.84$, 95% CI = 0.49–3.18). Three compared neurofeedback to active treatment and were associated with a small negative effect ($g = -0.26$, 95% CI = −1.50–0.97). One compared neurofeedback to a reverse neurofeedback condition; this study was associated with a negligible to small effect ($g = 0.19$, 95% CI = -0.69–1.08).

Functional Magnetic Resonance Imaging (fMRI) Neurofeedback

Qualitative Review Only one RCT of fMRI neurofeedback for anxiety disorders was identified. Zilverstand et al. [34] assigned patients with specific phobia to a neurofeedback condition, in which they were instructed to increase prefrontal activity and reduce insular activation, or to a control condition in which patients were instructed to "adapt their strategy based on intuition throughout the session" (p. 3). During exposure to phobic stimuli, neurofeedback patients reported lower fear levels than did control patients. At 3-month follow-up, patients in both conditions showed significant reductions in spider fear on a standardized measure; it was noted that fear reductions were correlated with insular downregulation in the neurofeedback group but not in the control group.

Effect Size Analysis Effect size for the single study of fMRI neurofeedback could not be calculated due to the lack of useable data.

Conclusions

In this chapter, we conducted a qualitative and quantitative systematic review of randomized controlled trials of biofeedback for anxiety disorders. Perhaps the most surprising aspect of the present review is the relatively small number of randomized controlled trials for anxiety disorders. Despite the long history of biofeedback [55], our review (see Fig. 16.1) found that of 151 articles reviewed, 54% of these were not RCTs, and 24% were not for DSM-defined anxiety disorders. Thus, there is a clear need for more well-controlled studies of biofeedback interventions for anxiety disorders. Based on the EPHPP criteria [39], the extant RCTs largely suffer from significant methodological limitations. As shown in Table 16.1, of 24 articles included in the qualitative review, 54% received an overall rating of "weak." Considering the long history of biofeedback studies, this is a surprising finding. We recommend the following for future research:

(1) *The use of representative samples.* It is important for studies to insure a high participation rate among participants who are likely to be representative of the target population.
(2) *Controlling for potential confounders.* Research should be mindful of imbalanced prognostic factors by, for example, stratifying or matching patients based on these factors.
(3) *Adequate blinding of research participants and assessors.* Ideally, sham treatments could be used, allowing for both the patient and the assessor to remain unaware of the intervention condition and/or the specific research question.
(4) *The use of reliable and valid measures of anxiety.* The reliability of measures should be verified within the treated sample, and only measures with established validity should be used.
(5) *Follow-up with withdrawals and dropouts.* Efforts should be made to insure that follow-up data are collected from all patients, including those who withdrew or dropped out.

Across studies, the meta-analytic result indicates that biofeedback (broadly defined) is superior to wait list. In particular, EMG biofeedback shows a medium advantage over wait list, and $ETCO_2$ biofeedback and EEG neurofeedback show a large advantage over wait list. Thus, these results affirm the efficacy of at least some forms of biofeedback for patients with DSM anxiety disorders.

The limitations of our review are determined by the quantity and quality of the studies we identified. Although we clearly observed that studies demonstrate the efficacy of biofeedback, there are significant limitations with wait list control. First, although a wait list condition can control for the mere effects of time, it cannot control for placebo or other nonspecific effects of treatment [56]. Second, there is evidence suggesting that wait list can actually serve as a "nocebo," creating negative expectations in patients and hampering naturally occurring recovery mechanisms [57].

It is noteworthy that our meta-analysis did not demonstrate a clear advantage of biofeedback over active treatment, or over reverse biofeedback. Moreover, the number of studies that examined reverse biofeedback was too small to provide conclusive evidence for or against it. Though one study of HRV biofeedback did exhibit a medium effect over treatment as usual, such a design still may not adequately control for nonspecific effects of treatment. The use of active treatment conditions (e.g., randomizing patients to biofeedback versus other forms of therapy) would provide an effective control for expectancy, attention, and other nonspecific effects of treatment. Even more powerful are reverse conditions, in which patients are trained to change their physiological responses in a countertherapeutic direction. Such designs not only provide an excellent control for nonspecific effects but also help to address whether physiological changes in the desired direction are indeed an active mechanism of treatment. To date, this has not yet been demonstrated conclusively in controlled trials for anxiety disorders.

References

1. Bond AJ, James DC, Lader MH. Physiological and psychological measures in anxious patients. Psychol Med. 1974;4(4):364–73.
2. Hoehn-Saric R, McLeod DR. The peripheral sympathetic nervous system. Its role in normal and pathologic anxiety. Psychiatr Clin North Am. 1988;11(2):375–86.
3. Hazlett RL, McLeod DR, Hoehn-Saric R. Muscle tension in generalized anxiety disorder: elevated muscle tonus or agitated movement? Psychophysiology. 1994;31(2):189–95.
4. Globisch J, Hamm AO, Esteves F, Ohman A. Fear appears fast: temporal course of startle reflex potentiation in animal fearful subjects. Psychophysiology. 1999;36(1):66–75.
5. Cuthbert BN, Lang PJ, Strauss C, Drobes D, Patrick CJ, Bradley MM. The psychophysiology of anxiety disorder: fear memory imagery. Psychophysiology. 2003;40(3):407–22.
6. Thayer JF, Friedman BH, Borkovec TD. Autonomic characteristics of generalized anxiety disorder and worry. Biol Psychiatry. 1996;39:255–66.
7. Ahs F, Sollers JJ 3rd, Furmark T, Fredrikson M, Thayer JF. High-frequency heart rate variability and cortico-striatal activity in men and women with social phobia. NeuroImage. 2009;47(3):815–20.
8. Wilhelm FH, Gevirtz R, Roth WT. Respiratory dysregulation in anxiety, functional cardiac, and pain disorders. Assessment, phenomenology, and treatment. Behav Modif. 2001;25(4):513–45.
9. Wise V, McFarlane AC, Clark CR, Battersby M. An integrative assessment of brain and body function 'at rest' in panic disorder: a combined quantitative EEG/autonomic function study. Int J Psychophysiol. 2011;79(2):155–65.
10. Taub E. What psychology as a science owes Neal miller: the example of his biofeedback approach. Biofeedback. 2010;38(3):108–17.
11. Rice KM, Blanchard EB, Purcell M. Biofeedback treatments of generalized anxiety disorder: preliminary results. Biofeedback Self Regul. 1993;18(2):93–105.
12. Agnihotri H, Paul M, Sandhu JS. Biofeedback approach in the treatment of generalized anxiety disorder. Iran J Psychiatry. 2007;2:90–5.
13. Carlson JG, Chemtob CM, Rusnak K, Hedlund NL, Muraoka MY. Eye movement desensitization and reprocessing (EDMR) treatment for combat-related posttraumatic stress disorder. J Trauma Stress. 1998;11(1):3–24.
14. Khanna A, Paul M, Sandhu JS. Efficacy of two relaxation techniques in reducing pulse rate among highly stressed females. Calicut Med J. 2007;5(5):e3.
15. Biswas A, Biswas D, Chattopadhyay PK. Cognitive behaviour therapy in generalised anxiety disorder. Indian J Clin Psychol. 1995;22(2):1–10.
16. Penzien DB, Holroyd KA. Psychosocial interventions in the management of recurrent headache disorders. 2: description of treatment techniques. Behav Med. 1994;20(2):64–73.

17. Jones CL, Minati L, Nagai Y, Medford N, Harrison NA, Gray M, et al. Neuroanatomical substrates for the volitional regulation of heart rate. Front Psychol. 2015;6:300.
18. Pallavicini F, Algeri D, Repetto C, Gorini A, Riva G. Biofeedback, virtual reality and mobile phones in the treatment of generalized anxiety disorder (GAD): a phase-2 controlled clinical trial. J Cyberther Rehabil. 2009;2(4):315–27.
19. Peira N, Fredrikson M, Pourtois G. Controlling the emotional heart: heart rate biofeedback improves cardiac control during emotional reactions. Int J Psychophysiol. 2014;91(3):225–31.
20. Nunes JS, Marks IM. Feedback of true heart rate during exposure in vivo. Arch Gen Psychiatry. 1975;32(7):933–6.
21. Telch MJ, Valentiner DP, Ilai D, Petruzzi D, Hehmsoth M. The facilitative effects of heart-rate feedback in the emotional processing of claustrophobic fear. Behav Res Ther. 2000;38(4):373–87.
22. Lehrer PM, Gevirtz R. Heart rate variability biofeedback: how and why does it work? Front Psychol. 2014;5:756.
23. Lehrer P, Vaschillo B, Zucker T, Graves J, Katsamanis M, Aviles M, et al. Protocol for heart rate variability biofeedback training. Biofeedback. 2013;41(3):98–109.
24. Lehrer PM, Vaschillo E, Vaschillo B. Resonant frequency biofeedback training to increase cardiac variability: rationale and manual for training. Appl Psychophysiol Biofeedback. 2000;25(3):177–91.
25. Reiner R. Integrating a portable biofeedback device into clinical practice for patients with anxiety disorders: results of a pilot study. Appl Psychophysiol Biofeedback. 2008;33(1):55–61.
26. Meuret AE, Wilhelm FH, Ritz T, Roth WT. Feedback of end-tidal pCO2 as a therapeutic approach for panic disorder. J Psychiatr Res. 2008;42(7):560–8.
27. Meuret AE, Wilhelm FH, Roth WT. Respiratory biofeedback-assisted therapy in panic disorder. Behav Modif. 2001;25(4):584–605.
28. Watson BW, Woolley-Hart A, Timmons BH. Biofeedback instruments for the management of anxiety and for relaxation training. J Biomed Eng. 1979;1(1):58–62.
29. Saletu-Zyhlarz G, Saletu B, Anderer P, Brandstatter N, Frey R, Gruber G, et al. Nonorganic insomnia in generalized anxiety disorder. 1. Controlled studies on sleep, awakening and daytime vigilance utilizing polysomnography and EEG mapping. Neuropsychobiology. 1997;36(3):117–29.
30. Hardt JV, Kamiya J. Anxiety change through electroencephalographic alpha feedback seen only in high anxiety subjects. Science. 1978;201(4350):79–81.
31. Wang S, Zhao Y, Chen S, Lin G, Sun P, Wang T. EEG biofeedback improves attentional bias in high trait anxiety individuals. BMC Neurosci. 2013;14:115.
32. Schoneveld EA, Malmberg M, Lichtwarck-Ashoff A, Verheijen GP, ENgels RCME, Granic I. A neurofeedback video game (*Mindlight*) to prevent anxiety in children: a randomized controlled trial. Comput Hum Behav. 2016;63:321–33.
33. Koush Y, Meskaldji DE, Pichon S, Rey G, Rieger SW, Linden DE, et al. Learning control over emotion networks through connectivity-based neurofeedback. Cereb Cortex. 2017;27(2):1193–202.
34. Zilverstand A, Sorger B, Sarkheil P, Goebel R. fMRI neurofeedback facilitates anxiety regulation in females with spider phobia. Front Behav Neurosci. 2015;9:148.
35. Schoenberg PL, David AS. Biofeedback for psychiatric disorders: a systematic review. Appl Psychophysiol Biofeedback. 2014;39(2):109–35.
36. American Psychiatric Association. Diagnostic and statistical manual of mental disorders. 5th ed. Washington, DC: Author; 2013.
37. American Psychiatric Association. Diagnostic and statistical manual of mental disorders. 3rd ed. Washington, DC: Author; 1980.
38. American Psychiatric Association. Diagnostic and statistical manual of mental disorders. 4th ed. Washington, DC: Author; 1994.
39. Thomas BH, Ciliska D, Dobbins M, Micucci S. A process for systematically reviewing the literature: providing the research evidence for public health nursing interventions. Worldviews Evid-Based Nurs. 2004;1(3):176–84.

40. Duval S, Tweedie R. Trim and fill: a simple funnel-plot–based method of testing and adjusting for publication bias in meta-analysis. Biometrics. 2000;56(2):455–63.
41. Higgins JP, Thompson SG, Deeks JJ, Altman DG. Measuring inconsistency in meta-analyses. BMJ. 2003;327(7414):557–60.
42. Agnihotri H, Paul M, Sandhu JS. The comparative efficacy of two biofeedback techniques in the treatment of generalized anxiety disorder. Pakistan J Soc Clin Psychol. 2008;6:35.
43. Watson CG, Tuorila JR, Vickers KS, Gearhart LP, Mendez CM. The efficacies of three relaxation regimens in the treatment of PTSD in Vietnam war veterans. J Clin Psychol. 1997;53(8):917–23.
44. Lande RG, Williams LB, Francis JL, Gragnani C, Morin ML. Efficacy of biofeedback for post-traumatic stress disorder. Comp Thr Med. 2010;18(6):256–9.
45. Tan G, Dao TK, Farmer L, Sutherland RJ, Gevirtz R. Heart rate variability (HRV) and posttraumatic stress disorder (PTSD): a pilot study. Appl Psychophysiol Biofeedback. 2011;36(1):27–35.
46. Meuret AE, Rosenfield D, Seidel A, Bhaskara L, Hofmann SG. Respiratory and cognitive mediators of treatment change in panic disorder: evidence for intervention specificity. J Consult Clin Psychol. 2010;78(5):691–704.
47. Kim S, Wollburg E, Roth WT. Opposing breathing therapies for panic disorder: a randomized controlled trial of lowering vs raising end-tidal P(CO(2)). J Clin Psychiatry. 2012;73(7):931–9.
48. Polak AR, Witteveen AB, Denys D, Olff M. Breathing biofeedback as an adjunct to exposure in cognitive behavioral therapy hastens the reduction of PTSD symptoms: a pilot study. Appl Psychophysiol Biofeedback. 2015;40(1):25–31.
49. Vanathy S, Sharma PSVN, Kumar KB. The efficacy of alpha and theta neurofeedback training in treatment of generalized anxiety disorder. Indian J Clin Psychol. 1998;25(2):136–43.
50. Dadashi M, Birashk B, Taremian F, Asgarnejad AA, Momtazi S. Effects of increase in amplitude of occipital alpha & theta brain waves on global functioning level of patients with GAD. Basic Clin Neurosci. 2015;6(1):14–20.
51. Deng X, Wang G, Zhou L, Zhang X, Yang M, Han G, et al. Randomized controlled trial of adjunctive EEG-biofeedback treatment of obsessive-compulsive disorder. Shanghai Arch Psychiatry. 2014;26(5):272–9.
52. Koprivova J, Congedo M, Raszka M, Prasko J, Brunovsky M, Horacek J. Prediction of treatment response and the effect of independent component neurofeedback in obsessive-compulsive disorder: a randomized, sham-controlled, double-blind study. Neuropsychobiology. 2013;67(4):210–23.
53. Peniston EG, Kulkosky PJ. Alpha-theta brainwave neuro-feedback for Vietnam veterans with combat-related post-traumatic stress disorder. Med Psychotherapy. 1991;4:47–60.
54. van der Kolk BA, Hodgdon H, Gapen M, Musicaro R, Suvak MK, Hamlin E, et al. A randomized controlled study of neurofeedback for chronic PTSD. PLoS One. 2016;11(12):e0166752.
55. Schwartz MS, Collura TF, Kamiya J, Schwartz NM. The history and definitions of biofeedback and applied psychophysiology. In: Schwartz MS, Andrasik F, editors. Biofeedback: a practitioner's guide. 4th ed. New York: Guilford Press; 2016. p. 3–23.
56. Campbell DT, Stanley JC. Experimental and quasi-experimental designs for research. Chicago: Rand-McNally; 1963.
57. Furukawa TA, Noma H, Caldwell DM, Honyashiki M, Shinohara K, Imai H, et al. Waiting list may be a nocebo condition in psychotherapy trials: a contribution from network meta-analysis. Acta Psychiatr Scand. 2014;130(3):181–92.

Cognitive Behavioral Therapy, Mindfulness-Based Cognitive Therapy and Acceptance Commitment Therapy for Anxiety Disorders: Integrating Traditional with Digital Treatment Approaches

17

Jennifer Apolinário-Hagen, Marie Drüge, and Lara Fritsche

Introduction

Anxiety disorders are a tremendous burden for individuals and societies given the relatively high lifetime prevalence of up to 33.7% in population-based surveys [1]. Several pharmacological and psychological interventions have been established as effective in the treatment of anxiety disorders [2]. However, pharmacological treatments may have side effects, which can impede the treatment adherence and the maintenance of therapeutic effects on the long run [3]. Hence, it appears crucial to inform patients with anxiety disorders about evidence-based psychological interventions that can be provided either as adjunctive to pharmacotherapy or as standalone treatment [3].

Among psychotherapeutic approaches, the most robust evidence base exists for cognitive behavioral therapy (CBT) that is like pharmacotherapy a first-line treatment for different anxiety disorders [4]. CBT is a directive, problem-orientated, structured, time-limited therapeutic approach that emphasizes the importance of a collaborative therapeutic relationship [5]. It is assumed that maladaptive behaviors and cognitions are learned and thus can be unlearned. For instance, in behavioral

J. Apolinário-Hagen (✉)
Institute of Occupational, Social and Environmental Medicine, Faculty of Medicine, Heinrich Heine University Düsseldorf, Düsseldorf, Germany
e-mail: jennifer.apolinario.hagen@hhu.de

M. Drüge
Department of Clinical Psychology/Psychotherapy Research, Institute of Psychology, University of Zurich, Zurich, Switzerland

L. Fritsche
Faculty of Psychology, Department of Health Psychology, FernUniversität in Hagen (University of Hagen), Hagen, Germany

Y.-K. Kim (ed.), *Anxiety Disorders*, Advances in Experimental Medicine and Biology 1191, https://doi.org/10.1007/978-981-32-9705-0_17

therapy anxiety disorders are assumed to be developed and maintained based on associative learning in terms of classical and operant conditioning [6, 7]. The core strategy of cognitive approaches in mental disorders is to identify learned maladaptive cognitions, to test their validity and to replace them with more realistic or functional cognitions. This represents a starting point for emotional symptom relief and more adaptive behaviors [8]. In other words, CBT assumes that it is not the context or circumstances that make a person suffer emotionally, but one's perceptions, beliefs, and assumptions about the situation. In the treatment of anxiety disorders, cognitive and behavioral techniques such as behavioral experiments, especially exposure to feared stimuli, are usually combined [5]. The principles of CBT are well-established in clinical practice for a range of disorder-specific treatment protocols, whereas there are newer and less frequently studied mindfulness-based approaches that try to add a broader contextual perspective to CBT [9, 10].

Hayes [11] suggested to classify developments in CBT using the metaphor of waves: The "first wave" of CBT in the 1950s was dominated by neo-behavioral principles using classical and operant conditioning techniques of behavior therapy in order to directly change problematic behavior. In the late 1960s, the "second wave" of CBT aimed to focus more on the "black box" of the human mind in order to challenge and change the content of maladaptive cognitions related to complex behavior patterns (e.g., by using cognitive restructuring [11, 12]). While the first behavioral therapies assumed that changing behavior results in changes in emotions and cognitions, cognitive therapy sees the change of cognitions as the initial step in order to change emotions and behavior [7].

"Third wave" therapies are proposed to address the meta-cognitive level by changing the function and context of maladaptive psychological processes in terms of one's association with thoughts and feelings (e.g., by fostering the momentary awareness and acceptance of these events) [9, 13]. Treatments of the "new generation" of CBT have been characterized to put more emphasis on contextual and less on pathological aspects than behavioral therapy and CBT [9, 12], for instance, by not debating dysfunctional thoughts and by focusing more on indirect and experiential avoidance techniques [14]. That means that, while the "first" and the "second" waves of CBT have focused on the elimination of specific problematic or unhealthy behaviors, thoughts, and emotions, "third wave" CBT approaches have a broader scope on the cultivation of new skills such as mindfulness and acceptance [11]. The "third wave" of CBT includes various approaches such as mindfulness-based cognitive therapy (MBCT) [15], acceptance and commitment therapy (ACT) [16], dialectical behavior therapy (DBT) [17], and meta-cognitive therapy (MCT) [18]. However, the wording "third wave" has been subject to controversy [12]. Hofmann et al. [9] argued that this terminology is misleading, because only slight differences exist in the theory base and procedures between "new wave" contextual and traditional CBT that both rather belong to a CBT family of treatments.

In the past two decades, mindfulness-based interventions (MBIs) have received broad attention among health professionals, and there is empirical support for their efficacy in clinical [10, 19–23] and nonclinical settings [24–26]. The most common MBIs are mindfulness-based stress reduction (MBSR), MBCT, and ACT [13]. Based on mindfulness meditation practice adapted from Buddhist tradition,

Kabat-Zinn developed MBSR, an 8-week group program, as the first Western MBI in the late 1970s to adjunctively treat patients with chronic pain [13]. Over the past decades, MBSR has been also applied to broader spectrum of health conditions and populations [27, 28]. As an amalgamation of MBSR and CBT principles, MBCT was developed as a disorder-specific mindfulness intervention for the relapse prevention of major depressive disorder (MDD) [29], but has been also adapted to treat anxiety [30]. Another differentiation of MBIs is ACT that is grounded on the assumption that the suppression of negative perceptions result in an increased frequency and severity of aversive experiences; ACT emphasizes the importance of accepting these perceptions as a precondition to behavior change [16]. ACT puts emphasis on experience-orientated techniques to train self-acceptance and a non-judgmental attitude toward in order to increase psychological flexibility [24]. Both MBCT and ACT are grounded on the concept of mindfulness and aim to target experiential avoidance strategies. However, in contrast to ACT, which is not considered a direct extension of CBT by Hayes [11], MBCT has a greater scope on formal mindfulness exercises and integrates elements of cognitive therapy [20]. In view of the compatible background and techniques but different stages of clinical testing, a selective overview on the current status of (1) CBT as well as potential applications of (2) MBCT and (3) ACT in the treatment of anxiety disorders will be provided. Furthermore, besides the traditional face-to-face individual and group format, innovative delivery modes of the "second" and "third wave" have been developed in recent years. Given the scarcity of treatment places in routine care, the dissemination of evidence-based Internet-delivered CBT (iCBT) (c.f., [31]) and mindfulness- and acceptance interventions (c.f. [32]) has been proposed to improve the access to professional help for people with anxiety disorders. Therefore, digital treatment approaches will be presented.

Specifically, this chapter describes the application of CBT-compatible interventions for the common DMS V anxiety disorders specific phobia, panic disorder, agoraphobia, social anxiety disorder, and generalized anxiety disorder in adults in order to derive implications for clinical practice and research.

"Second Wave" of CBT in the Treatment of Anxiety Disorders in Adults

Cognitive Behavioral Therapy

CBT is one of the most applied and best researched psychological treatments for anxiety disorders [33]. This "second wave" psychological treatment supposes that maladaptive cognitions play a key role in the development and maintenance of mental disorders [4]. CBT originates partly from the theories by Skinner and Wolpe that were the groundwork for behavioral therapy in the 1950s, which assumes that behavior changes lead to changes in emotions and cognitions [7]. It is assumed that fear is maintained through negative reinforcement such as avoidance behavior. Behavioral therapy was the starting point for the theories of cognitive therapy pioneered by Beck and Ellis. These theories postulate that changing cognitions result in changes in emotions and behavior [7]. The general purpose of CBT is to guide patients to learn modifying the content of problematic cognitions and, thereby, break the cycle of maladaptive thoughts, emotions, and behavior [8]. CBT is often used as an umbrella term for behavioral and cognitive therapy approaches, which vary in terms of a different emphasis either on behavioral or cognitive techniques [7]. CBT interventions share common principles, such as action orientation (CBT Principle 5, Table 17.1).

Traditional CBT for Anxiety Disorders

As a theoretical groundwork for CBT, Beck's cognitive model of depression [35] proposes that patients are vulnerable to negatively biased information processing based on which they develop negative core beliefs or cognitive schemas, dysfunctional assumptions, and negative automatic thoughts that contribute to dysfunctional behavior and emotional distress. The negative cognitive triad model proposes that cognitions, emotions, and behavior are interrelated [36]. The cognitive perspective has been also extended to treat anxiety and phobias [37]. Cognitive therapy applies techniques that help patients to recognize and correct dysfunctional cognitions such as negative automatic thoughts and distorted beliefs in order to reduce fear and worry [38]. Techniques include psychoeducation, cognitive restructuring, and homework assignments [7]. Guided discovery (e.g., Socratic questioning) and the collaborative therapeutic relationship are further components of CBT [5]. Behavioral experiments (i.e., exposure) are especially relevant for the treatment of anxiety [5]. Exposure is a commonly used behavioral technique in CBT for anxiety disorders that can be applied imaginal, in vivo, or interoceptive. Other exposure-based treatments involve, for instance, systematic desensitization and virtual reality (VR) exposure [39]. It is assumed that the confrontation with the pathologically feared stimulus alongside with the integration of corrective information in the fear memory will result in the decrease of fear [7]. Other behavioral techniques involve activity scheduling, progressive muscle relaxation, and breathing exercises [5].

CBT for anxiety disorders can be delivered in individual, group, or guided self-help format. Usually, CBT programs are disorder-specific for treating a single diagnosis within the anxiety disorder spectrum. In view of the high prevalence of comorbidity and overlapping symptoms in anxiety disorders with each other and other disorders [1], transdiagnostic CBT protocols have gained an increasing interest among clinicians and researchers [7]. However, many people with anxiety disorders remain untreated or seek help delayed [40]. Therefore, innovative strategies to improve help-seeking and the uptake of psychological services are investigated. One of these service types is iCBT.

Internet-Delivered CBT for Anxiety Disorders

As the most commonly applied framework, iCBT has been suggested to have additional benefits to routine care such as scalability, cost-effectiveness, and time/

Table 17.1 9 Core principles of cognitive behavioral therapy

Main principles	Key elements
Principle 1 CBT is oriented on empirical psychology	Operationalization and empirical testing of theoretical concepts and therapeutic methods with objective, reliable, and valid measures
Principle 2	Addressing the present problem
CBT is problem-oriented	Tailoring the treatment to the individual and the mental disorder (disorder-specific)
	Improving general problem-solving skills
Principle 3	Focusing on maintaining conditions due to their relevance for future conditions and durable problem solutions
CBT addresses predisposing, triggering, and maintaining problematic conditions	Addressing the present effects of predisposing conditions and triggers, although they often cannot be changed
Principle 4	Collaborative problem identification and definition of therapeutic goals by the therapist and patient in order to prevent conflicting goals or unrealistic expectations
CBT is goal-oriented	The problem is the starting point for the therapy
	The solution of the problem is the envisioned goal and the sufficient reason for completing the therapy
Principal 5 CBT is action-oriented	Active participation of the patient as a sufficient condition: motivation of the patient to actively attempt new behavior and experiences as well as problem-solving strategies
Principle 6	Generalization of the achieved changes to the patient's everyday life (beyond of the therapeutic setting)
CBT is not limited to a specific setting	Encouraging patients to test and practice new strategies between the treatment sessions
	Successful coping of the condition without therapeutic support as a treatment gain
Principle 7 CBT is transparent	Explaining the disorder and therapy increases problem-solving skills and acceptance and thus promotes adherence as well as relapse prevention
Principle 8 CBT should be help for self-help	Improving general problem-solving and learning new self-help skills in order to analyze and cope with future problems autonomously without therapist guidance
Principle 9 CBT strives for constant advancement	Ensuring a constant empirical evaluation and progress of theoretical concepts and practical methods

Note: The principles were adapted and modified from Margraf ([34], pp. 6–7)

location flexibility [41–43]. Given the high prevalence of anxiety disorders, the limited treatment places in routine care and delayed help-seeking, iCBT increase the uptake of self-help services or represent a low-treshold, early intervention for anxious patients (see CBT Principle 8, Table 17.1). General barrier to access iCBT involve, for instance, the "digital divide," new challenges for therapeutic routines, and data security concerns [44]. The latter appears especially relevant given the emergence of commercial mobile health (mHealth) apps [42].

Support formats in iCBT vary between unguided self-help and guided interventions with either real-time (synchronous) or delayed (asynchronous) therapist contact [42].

While therapist-guided iCBT programs usually include structured self-help contents and scheduled therapist feedback, for example, via text messages or telephone contact, modern self-guided iCBT programs for anxiety disorders often provide some degree of support such as automated messages, reminders, or feedback on demand (c.f. [42, 45, 46]). Another promising approach is blended care, which combines the strengths of face-to-face and Internet-based treatments. In a systematic review of 44 studies, Erbe et al. [47] identified 8 studies targeting blended interventions for anxiety disorders. Different models of blended care were identified: integrated blended care interventions with either face-to-face or Internet focus and sequential blended intervention starting with either face-to-face or Internet-based sessions [47].

Efficacy of CBT for Anxiety Disorders

Efficacy of Traditional CBT for Anxiety Disorders

Reviews of meta-analyses on the efficacy of CBT in various mental health problems [4, 48, 49] demonstrated a strong evidence base for anxiety disorders. For example, a review of 16 meta-analyses by Butler et al. [48] indicated large effect sizes of CBT in the treatment of panic disorder (PD) with or without agoraphobia, social anxiety disorder (SAD), and generalized anxiety disorder (GAD). The more recent review by Hofmann et al. [4] identified 269 meta-analyses and reviewed of representative sample of 106 meta-analyses of CBT for different common mental health problems (published since 2000); regarding DSM-V anxiety disorders, they found at least moderate effect sizes for SAD and PD. Another review of meta-analyses [49] on CBT for anxiety disorders showed that CBT also outperformed both waitlist controls and placebo controls, the latter indicating specific effects of CBT. This is in accordance with a recent meta-analysis of randomized controlled trials (RCTs) by Carpenter et al. [33] that confirmed the efficacy of CBT for different anxiety disorders compared to placebo control conditions ($g = 0.56$). Meta-analytic evidence indicates that the combination of cognitive and behavioral therapy approaches is generally effective in treating different anxiety disorders, but in some conditions, pure cognitive or behavioral therapy can work as well [38]. For instance, a meta-analytic review by Norton and Price [50] of 108 trials indicated that different treatment components of CBT (i.e., cognitive therapy and exposure therapy alone or in combination, or both combined with applied relaxation training) showed a comparable efficacy across different anxiety disorders. However, a meta-analysis by Bandelow et al. [2] found higher effect sizes for combined CBT and pharmacotherapy ($d = 2.12$) as for individual CBT/exposure ($d = 1.30$) and group CBT ($d = 1.22$). Generally, pharmacotherapy achieved higher effect sizes and gains in shorter time than psychological treatments, but given the potential side effects of medication and the potentially more sustainable effects of CBT, the authors suggest to make the decisions on choosing psychotherapy, pharmacotherapy, or a combination of both based on patients' preferences [2]. However, it has been also argued that there are no substantial and sustainable treatment benefits of combining pharmacotherapy with CBT and that there may be interference effects given the prerequisite of arousal to achieve successful fear exposure [51].

Nonetheless, CBT appears at least superior in the treatment of anxiety disorders compared to other psychotherapeutic approaches [52]. Furthermore, a meta-analysis of 11 trials by Reinholt and Krogh [53] suggested the efficacy of transdiagnostic CBT for anxiety disorders with moderate effect sizes, but the authors report a large heterogeneity and risk of bias on the study level.

Efficacy of iCBT for Anxiety Disorders

Over the past two decades, an exponentially growing number of RCTs suggest the efficacy of iCBT programs in treating anxiety disorders [54–57]. Moreover, iCBT can achieve comparable short-term effects in anxiety disorders like individual and group CBT [54, 58–60]. Therapist guidance appears to be an important active ingredient of iCBT programs for anxiety disorders in terms of improved symptom severity and adherence in comparison to unguided interventions [61]. In a meta-analysis of 38 RCTs (3214 participants) on therapist-guided iCBT for DSM-IV anxiety disorders, Olthuis et al. [31] found preliminary evidence for clinically relevant improvements in anxiety at posttreatment, favoring therapist-guided iCBT over passive and active control condition groups. Interestingly, there is currently no evidence from RCTs indicating the impact of therapist qualification on the efficacy of guided self-help iCBT programs [45, 59, 61]. Furthermore, in studies both therapist-guided and self-guided transdiagnostic iCBT programs achieve significant improvements, with no difference to more studied disorder-specific treatments [46, 59, 62, 63]. Regarding the acceptability of iCBT, a meta-analysis by Andrews et al. [54] found high rates of participant satisfaction, while adherence rates in iCBT for anxiety and depression were moderate with only 10 out of 50 trials reporting an adherence rate lower than 50% (range 6–100%). There is also evidence that iCBT can be effectively implemented into routine care settings [64] and provided as blended interventions [47].

Concerning exposure-based approaches transferred to VR, a review paper by Maples-Keller et al. [65] found preliminary evidence for its efficacy, for instance, for specific phobia, PD with or without agoraphobia, SAD, and GAD, but more research is needed to derive definitive conclusions about the comparability with in vivo exposure. In a systematic review, Sucala et al. [66] found a high discrepancy between the number of commercial apps for anxiety and the evidence for their efficacy and effectiveness: only 3.8% of the 52 reviewed apps have been rigorously tested. In line with this, a review by van Ameringen et al. [67] confirmed that the few existing clinically tested apps for anxiety are usually not publicly available.

Specific Efficacy of CBT for Different Anxiety Disorders

CBT for Specific Phobia

Rationale of CBT for Specific Phobia

The DSM-V [68] classifies four subtypes of specific phobias, including animals (e.g., spiders), natural environments (e.g., fear of heights), blood-injection-injury, and situations (e.g., fear of flying), while the fifth "other" category includes unclassified phobias. Phobias are maintained through avoiding the feared stimulus. Specific phobias are common mental health problems that can be easily and best treated with exposure-based therapies [69]. Current treatments for specific phobias include exposure approaches such as in vivo exposure, systematic desensitization, imaginal exposure, and VR or computer-assisted exposure, whereas cognitive therapy is considered as an alternative approach to exposure [39].

Efficacy of Traditional CBT for Specific Phobia

The treatment of specific phobia focuses on behavioral strategies. For instance, in vivo exposure is recommended for adults with high fear of needles, while non-in vivo types may be an alternative for people who are not willing to undergo in vivo exposure [70, 71]. A meta-analysis of 16 studies on treatments for acrophobia (fear of heights) by Arroll et al. [72] indicated that a wide range of therapies such as desensitization, in vivo exposure, and VR are effective for acrophobia only in the short term. A comprehensive review of short-term and long-term efficacy studies (published between 1960 and 2005) by Choy et al. [73] indicated that the efficacy of different treatments varied between the phobia subtypes: in vivo exposure was found to be a robust treatment in most phobias, but is associated with high attrition and poorer treatment acceptance than systematic desensitization; cognitive therapy was identified to be most effective in claustrophobia; preliminary evidence suggests VR as a promising approach for flying and height phobia, whereas applied tension was found to work best for blood-injury phobia.

However, another meta-analysis by Wolitzky-Taylor et al. [39] could not confirm the assumption that certain subtypes of specific phobia respond more favorably to specific cognitive or behavioral treatments. Wolitzky-Taylor et al. [39] analyzed 33 RCTs and found that most (88%) used an exposure-based treatment. At posttreatment, exposure treatments (18 studies) were more effective (with large effect sizes) than waiting list (behavioral measures, $d = 1.00$; questionnaires, $d = 1.16$). The effect sizes were moderate when exposure treatments were compared to a placebo treatment in five studies (behavioral measures, $d = 0.42$; questionnaires, $d = 0.61$). The effects were increased at the follow-up, favoring exposure ($d = 0.68$). Compared to non-exposure treatments, exposure treatments led to significantly higher improvements at post- and follow-up-assessments with moderate effect sizes. Another interesting finding of this meta-analysis [39] was that only the treatment length (i.e., more improvements after multiple sessions versus one-session exposure) was identified as a significant predictor of outcome, but the authors mentioned the small number of studies for comparisons as a limitation.

Efficacy of iCBT for Specific Phobia

Little iCBT research has focused on specific phobia. Animal phobia is usually treated with exposure-based treatments, but there are few pilot studies on Internet-based self-help interventions. For example, a RCT by Andersson et al. [74] compared a 5-week guided Internet-based self-help program for spider phobia with one-session live-exposure treatment in 30 patients. While no differences in anxiety measures were found between the groups at post- and follow-up assessments, significant improvements in the Behavioral Approach Test (BAT) were less in the Internet group (46.2%) than in the live-exposure group (85.7%) at posttreatment. In another study with 30 snake phobic patients, Andersson et al. [75] found that 61.5% of the group receiving guided Internet-based treatment and 84.6% of the one-session exposure group achieved a clinically significant improvement on the BAT at posttreatment – again, the within-group effect sizes were large for both groups, but higher in the live-exposure group (d = 2.31) than in the Internet group (d = 1.63).

Flying phobia is a common situational phobia. Over a decade ago, a review of 40 studies by da Costa et al. [76] indicated the efficacy of VR exposure to flying with or without additional CBT components or psychoeducation, but they suggest to combine VR exposure with cognitive therapy elements. In a recent study, 46 patients with flying phobia received "NO-FEAR Airlines" [77] either with (n = 23) or without (n = 23) therapist guidance in a RCT by Campos et al. [78] that showed that completers in both groups (n = 33, attrition rate of 28.26%) reported positive expectations, satisfaction, opinions, and usability ratings, but participants in general preferred the therapist-guided condition.

CBT for Panic Disorder and Agoraphobia

Rationale of CBT for Panic Disorder and Agoraphobia

PD is characterized by recurrent panic attacks that occur frequently, involving persistent worries about future panic attacks and their personal costs, and maladaptive behavior [3, 79]. According to Clark's cognitive model of panic [80], panic attacks are the result of catastrophic misinterpretation of specific bodily sensations. These patterns contribute to preserving the vicious cycle of recurrent panic attacks [3]. There is a high comorbidity of PD with agoraphobia [2]. Both are associated with anticipated panic attacks and subsequent safety or avoidance behavior. Agoraphobia-specific situations are difficult or embarrassing and characterized by absent possibility of immediate escape and being alone or outside of the safety zone [81]. Although PD and agoraphobia are unconnected in the DSM-V [82], treatment protocols often combine both, whereas the treatment of agoraphobia puts more emphasis on exposure-based techniques [81].

Casey et al. [83] proposed an integrated cognitive model according to which catastrophic misinterpretations and panic-related self-efficacy beliefs independently contribute to cognitions involved in the maintenance of recurrent panic attacks. CBT can help breaking the persistent cycle of anticipatory anxiety, panic, and agoraphobic avoidance behavior by educating patients about PD and by helping them to cultivate more adaptive cognitive and behavioral strategies to address the fear of panic attacks [84]. In order to eliminate catastrophic misinterpretations in PD and agoraphobia, education, cognitive techniques, interoceptive exposure (bodily sensations in PD), in vivo exposure (agoraphobic situations), and training of coping skills have been proven to be effective techniques [38].

Efficacy of Traditional CBT for Panic Disorder With or Without Agoraphobia

For PD, CBT and pharmacotherapy are considered as first-line treatments [85, 86]. In RCTs, CBT achieves rates of 85% panic-free patients at posttreatment [87]. However, about one-third of treated patients continue to suffer from treatment-resistant panic attacks [88]. In a meta-analysis by Pompoli et al. [89], CBT was the most studied psychological treatment for PD with or without agoraphobia and often associated with a greater treatment response when compared to other psychological treatments, albeit with small effect sizes. Another meta-analysis by Carpenter et al. [33] also showed small effect sizes of CBT for PD when compared to placebo controls ($g = 0.39$), which is in accordance with an earlier meta-analysis ([90], $g = 0.35$). A meta-analysis by Ougrin [91] of different CBT components showed no difference in the efficacy between cognitive therapy and exposure in the treatment of PD with or without agoraphobia (seven studies). This is in line with a component network meta-analysis by Pompoli et al. [92] that indicated that effective CBT programs for PD should include both face-to-face and interoceptive exposure components, while muscle relaxation and VR exposure showed no additional benefit. In addition, there is preliminary evidence that brief CBT for PD is an effective treatment option [93]. Regarding mediators and moderators of outcome, a systematic review of 52 papers by Porter et al. [94] identified agoraphobic avoidance as an important negative predictor of improvement in CBT for PD with or without agoraphobia. However, more research is needed to clarify the benefits of CBT in combination with pharmacotherapy on the long run for patients with treatment-resistant PD [88]. In addition, another systematic review of 16 studies by Imai et al. [95] could not identify clear benefits of psychological or pharmaceutical therapies over the other, which, however, may be due to methodological shortcomings on the study level.

Efficacy of iCBT for Panic Disorder With or Without Agoraphobia

A recent review [96] of eight RCTs on Internet-delivered psychological treatments (7 RCTs on iCBT, 1 RCT on iACT) for PD indicated the effectiveness of both disorder-specific and transdiagnostic iCBT approaches and different support formats (guided vs. unguided) in reducing the symptoms of PD. Attrition (9.8–42.1%) and adherence rates (7.8–75%) were moderate to high, while participant satisfaction of intervention completers was overall high (five studies) [96]. These findings are in line with other reviews and meta-analyses showing the efficacy of different formats of iCBT for anxiety disorders, including PD [54, 97–101]. In a meta-analysis by Andrews et al. [54], iCBT yielded a large combined effect size ($g = 1.31$) in the treatment of PD (12 studies) in comparison to control conditions. There are also pilot studies on PD showing the effectiveness of iCBT in routine care settings (c.f. [102, 103]), the feasibility of blended CBT (c.f. [104]), and comparable outcomes to face-to-face CBT (c.f. [105]). Regarding the helpfulness of mobile apps, a RCT by Christoforou et al. [106] with 170 people suffering from agoraphobia indicated the efficacy of both a disorder-specific app ("Agoraphobia Free") and a transdiagnostic app ("Stress Free") on the symptom severity at posttreatment. To date, there are only few scientifically tested apps available. The potential of publicly available apps for PD and agoraphobia developed outside of clinical studies is largely unclear.

For instance, a study on the quality of 52 Android apps for PD by van Singer et al. [107] revealed a poor content quality and capability to foster self-help of PD smartphone apps.

CBT for Social Anxiety Disorder

Rationale of CBT for Social Anxiety Disorder

SAD is characterized by excessive and persistent fear related to social situations in terms of performance as well as social interactions and has often a chronic course [81]. Cognitive behavioral models assume that people with SAD see themselves as they think others would see them, but in a highly distorted negative way. Furthermore, they become hypervigilant to social threats and tend to not pay attention to what is actually happening in the social situation [108]. Hofmann [109] suggested a comprehensive psychological maintenance model of SAD: Social worry is related to unrealistic social standards and goals. In challenging social situations, people with SAD tend to focus their attention toward their anxiety and negative self-perceptions. They overestimate the negative consequences and believe to have poor control over their emotional response as well as insufficient social skills for successful coping. As a maladaptive coping strategy, people with SAD manifest avoidance and safety behaviors to avoid negative social outcomes, which contribute to rumination and anticipatory social worry [109]. Taken together, Hofmann [110] identified negative cognitive appraisal (estimated social costs), perceived self-efficacy (perceived social skills), and perceived emotional control as potential mediators in the CBT treatment of SAD. Therapeutic strategies for SAD commonly include multiple techniques such as exposure, cognitive restructuring, social skills training, and applied tension [81].

Efficacy of Traditional CBT for Social Anxiety Disorder

Currently, CBT represents the most effective psychological treatment for SAD, with more enduring effects than pharmacotherapy [111]. For instance, a meta-analysis by Gould et al. [112] demonstrated that exposure-based interventions achieved the largest effect size, whether alone (effect size = 0.89) or combined with cognitive restructuring (effect size = 0.80). A network meta-analysis by Mayo-Wilson et al. [113] showed that different formats of psychological interventions were effective when compared to waitlist controls, including individual CBT, group CBT, exposure, social skills training, and self-help interventions with or without support. Generally, CBT interventions for SAD achieve larger effects when they are compared to a passive control condition and not to active controls [114]. A meta-analysis by Carpenter et al. [33] revealed a small-to-moderate effect size of CBT for SAD versus placebo controls ($g = 0.41$). A previous meta-analysis [90] found a moderate effect size ($g = 0.62$) of CBT for anxiety disorders when compared to placebo-controlled trials. Concerning the relative efficacy of different components of CBT for SAD, the aforementioned meta-analysis by Ougrin [91] indicated a statistically significant difference (three studies), favoring cognitive therapy over exposure in social phobia outcomes in both the short and long term.

Efficacy of iCBT for Social Anxiety Disorder

A meta-analysis by Andrews et al. [54] showed a large effect size (g = 0.92) for iCBT for SAD (11 studies) when compared to a control condition. Kampmann et al. [115] performed a meta-analysis of 37 RCTs (2,991 participants) on different types of technology-assisted psychological interventions for SAD with 21 trials of iCBT. They showed that iCBT was effective in treating SAD symptoms when compared to both passive controls (g = 0.84) and, albeit with less advantage, when compared to active controls (g = 0.38) [115]. This finding is in line with a previous review on iCBT for SAD by Boettcher et al. [116]. Transdiagnostic iCBT for SAD and comorbid disorders has also been shown to be effective (c.f. [62]). Furthermore, two meta-analyses (with six and seven studies) by Cesham et al. [117] indicated the usefulness of VR exposure therapy in reducing the symptoms of SAD. Apps may represent another useful tool to promote self-help activities. In a systematic a systematic search across the main app stores, Alyami et al. [118] identified 38 eligible apps (60% focused solely on social anxiety). However, they critically critically noted lacking information on the content source or organizational affiliations in the apps as well as on the evidence base for these apps. Stolz et al. [119] showed the efficacy of both a guided 12-week computer-delivered (d = 0.74) and app-delivered self-help iCBT program (d = 0.89) compared to a waiting-list control group among 150 people with SAD, with no difference between the treatment arms (d = 0.07).

CBT for Generalized Anxiety Disorder

Rationale of CBT for Generalized Anxiety Disorder

Excessive worry and internal avoidance strategies are core features for the maintenance of GAD. Theoretical models that can be applied to CBT-based treatments of GAD involve, for instance, the cognitive model of GAD [18], the cognitive model of pathological worry [120], the avoidance model of worry in GAD [121], and the contrast avoidance model [122, 123]. Worry is negatively reinforced by avoiding experience with the aversive stimulus as well as dysfunctional positive beliefs (e.g., that worry is useful for problem-solving when these expected negative events do not occur in the future) [124]. The intolerance of uncertainty model (IUM [125]) is another theory for the maintenance of GAD. The IUM proposes four mechanisms of intolerance of uncertainty (IU) in terms of a cognitive vulnerability and potentially a key mediator for reducing worry, positive worry beliefs, cognitive avoidance, and negative orientation. Several cognitive behavioral techniques can be applied to promote habituation and extinction learning in GAD, such as self-monitoring, self-control desensitization, and gradual stimulus control [124].

Efficacy of Traditional CBT for Generalized Anxiety Disorder

A meta-analysis of 41 studies (2132 patients) on psychological treatments for GAD by Cuijpers et al. [126] confirmed the efficacy of CBT on GAD and comorbid symptoms of depression. Most of these studies compared the effects of CBT against a waiting-list control condition. With 28 of the included studies, Cuijpers et al. [126]

found a large pooled effect size (g = 0.84) in the 38 comparisons of psychotherapy versus a control group for GAD (with low-to-moderate heterogeneity). The effect sizes for individual therapy (g = 0.80, 23 studies) and group format (g = 0.70, 8 studies) were comparable: CBT (g = 0.90, 28 studies) and behavioral therapy only (g = 0.57, 3 studies) were effective in reducing GAD symptoms [126]. Furthermore, a meta-analysis of 14 RCTs of CBT for GAD in older adults by Hall et al. [127] indicated a superiority of CBT over no treatment and treatment as usual (TAU) at posttreatment and 6-month follow-up. However, both meta-analyses [126, 127] identified a research gap concerning the advantage of CBT compared to active controls or other treatments and the long-term assessment of CBT in GAD, which is important because of its frequently chronic course. When compared to placebo controls, a meta-analysis by Carpenter et al. [33] revealed large effect sizes for CBT for GAD (g = 1.01), which was higher than in a previous meta-analysis ([90], g = 0.51).

Efficacy of iCBT for Generalized Anxiety Disorder

Given that GAD is considered as the least successfully treated anxiety disorder in routine care, novel strategies such as iCBT may represent a promising way. Preliminary results from therapist-guided iCBT have yielded large effect sizes up to 2 years follow-up, when analyzing the data of completers [128]. A meta-analysis by Andrews et al. [54] showed a moderate effect size (g = 0.70) for iCBT for GAD (nine studies) when compared to a control condition. In a meta-analysis by Cuijpers et al. [126] on psychological treatments for GAD, Internet-based self-help therapies (seven studies) achieved a large effect size (g = 1.05). An example for the use of iCBT in GAD in clinical practice is a study by Mewton et al. [129] in which they found moderate-to-large effect sizes on different outcomes among program completers. However, only half (55.1%) of the included 588 patients completed the entire six-lesson program. Another example is a RCT by Jones et al. [130] with older people with GAD in which they compared the effects of a guided iCBT program (n = 24) with a waiting-list control condition (n = 22) and found large effect sizes favoring iCBT (d = 0.85–1.17) at the posttreatment.

"New Wave" of CBT: Mindfulness and Acceptance-Based Approaches for the Treatment of Anxiety Disorders in Adults

Mindfulness-Based Interventions

In recent years, clinicians have expressed a growing interest toward integrating mindfulness-based techniques into their therapeutic practice. Therefore, it appears important to raise the understanding of the nature of mindful meditation and the applications as well as limitations of MBIs [131].

Kabat-Zinn [13] defines mindfulness as "...the awareness that emerges through paying attention on purpose, in the present moment, and nonjudgmentally to the unfolding of experience moment by moment" [13], p. 145). Bishop et al. [132] operationalize "... mindfulness as a process or regulating attention in order to bring a quality of nonelaborative awareness to current experience and a quality of relating to one's experience within an orientation of curiosity, experiential openness, and acceptance" ([132], p. 234). Bishop et al. [132] assume that two processes are involved in the training of mindfulness: the regulation of mindfulness at the present moment and an orientation to the experience within daily activities. Teasdale et al. [133] understand mindfulness as a generic skill that can be practiced in everyday experience across different situations and propose mindfulness training as a way to make people fully aware of their thoughts and feelings, regardless of the evaluation of these experiences (e.g., as unpleasant). "Mindfulness training appears to be associated with a reduction in the tendency to 'float away' into ruminative, elaborative thoughts throughout streams. (…)" ([133], p. 34). Consequently, rumination and worry (past and future orientation vs. awareness of the "here and now") are commonly proposed mechanisms of action in the treatment of depression and anxiety in MBIs [134]. MBIs aim at fostering the skill of decentering (similar to distancing in CBT), acceptance, increased exposure to cognitions and feelings, greater self-awareness, self-motivation, and reduced autonomic arousal [135].

Shapiro et al. [136] postulated that manifold mechanisms enable behavior changes in MBIs, such as self-regulation, values clarification, cognitive behavioral flexibility, and exposure. There are different MBI programs available for health problems, such as chronic pain, smoking cessation, substance abuse, and recurrent MDD [137]. Like MBSR and MBCT, ACT has Buddhists roots (Fig. 17.1).

MBIs have usually 8-week duration with 2 h of face-to-face group training plus home exercises and homework. All MBIs start from the premise that it is mandatory to develop mindfulness systematically through regular practice in daily life [13]. Preliminary evidence also indicates that mindfulness skills can be effectively trained using self-help interventions [27]. MBIs are usually embedded in a broader therapeutic framework (multifaceted interventions), but a recent meta-analysis of 18 studies has shown that even stand-alone mindfulness exercises can work for the reduction of anxiety, with small-to-medium effect sizes [139]. Mindful exposure

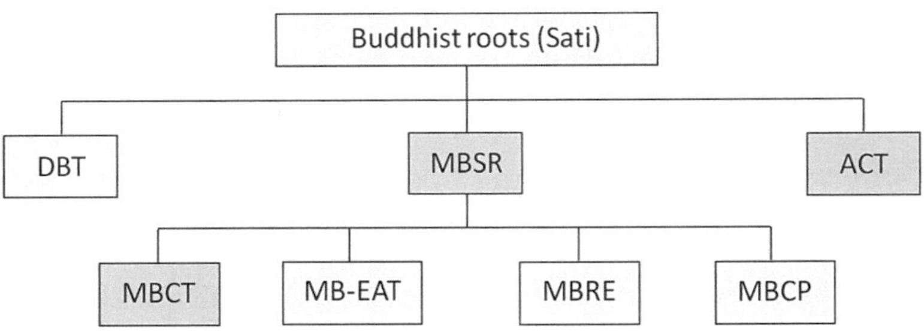

Fig. 17.1 Mindfulness-based intervention types: Subsumption of MBSR, MBCT, and ACT. Abbreviations: DBT (dialectic behavioral therapy), MBSR (mindfulness-based stress reduction), ACT (acceptance commitment therapy), MBCT (mindfulness-based cognitive therapy), MB-EAT (MBSR for anxiety and eating disorders), MBRE (mindfulness-based relationship enhancement), and MBCP (mindfulness-based childbirth and parenting). (Modified and translated from Michalak et al. [138], p. 12)

combining mindfulness practice with exposure procedures might be an approach to promote extinction learning in anxiety disorders [140], but interestingly there are only few pilot studies on this method (e.g., [141]).

Contraindications for MBIs have been rarely reported [134]. Common adverse events of MBIs involve agitation and discomfort during formal meditation exercises in mindfulness trainings [142]. Furthermore, anecdotal evidence suggests that increased distress and rumination are common side effects of mindfulness trainings [134]. However, although there are only few empirical studies reporting severe adverse effects and contraindications of MBSR and MBCT, it appears plausible that people suffering from some types of severe mental illnesses can have harmful experiences while practicing meditation and should receive other treatments [143]. Farias and Wikholm [144] raised concerns about the under-researched "dark side of meditation" that involves individuals who may suffer from a number of moderate-to-severe adverse events, including exercitations in emotional problems, anxiety, depression, psychosis, and confrontation with contents related to childhood trauma during mindfulness practice. Therefore, for vulnerable patients it is vital that MM is part of a therapeutic context in order to address these events, which is not possible in a group setting without a therapist or in home practice [144].

Mindfulness-Based Cognitive Therapy

MBCT adds cognitive therapy elements for disorder-specific treatments to MBSR and is thus a direct extension of CBT [9]. Traditional face-to-face MBSR and MBCT share the 8-week group format as well as most mindful meditation components, such as sitting meditation, body scan, and hatha yoga. Additionally, MBCT includes disorder-specific psychoeducation [145]. Based on attentional control training [133], MBCT was developed by Segal et al. [15] as a specific MBI for the relapse prevention of recurrent MDD [29]. The aim of MBCT is to learn controlling the shift between unfocused attention (autopilot modus) and focused attention (mindfulness) in daily activities [133]. The mechanisms of change in MBCT are explained based on the cognitive vulnerability model, stating that repeated automated negative thoughts become associated with the depressed state that triggers and increases the risk of relapse with each new depressive episode [29, 133, 146].

In contrast to CBT, MBCT aims to foster the awareness regarding the personal relationships to own thoughts and feelings (meta-cognitive level) and not regarding the change of the content of thoughts [147]. The traditional 8-week course structure of MBCT, with about 2-hour training per week, involves several formal and informal MM practices: while early sessions have a greater scope on guided meditations in order to bring attention to breathing and bodily sensations, later sessions focus more on the development and cultivation of independent practice and mindful awareness to emotional and cognitive events. Participants are encouraged to spend at least 45 min per day with practicing mindful activities [147]. It is important to regularly practice mindfulness outside the group setting. For instance, Parsons et al. [148] showed a significant relationship between the home practice of MBCT and mental health outcomes. The proposed mechanisms of change in MBIs like MBCT involve, for instance, mindfulness [149, 150], compassion [150], meta-cognitive awareness [150], reactivity/emotion regulation [151], rumination [134, 149–152], and worry [134, 151, 152].

MBCT for Anxiety Disorders

According to Sipe and Eisendrath [147], there is a strong theoretical rationale for applying MBCT to anxiety disorders that was initially developed for MDD relapse prevention [147]: worry as future orientation in anxiety disorders that results in the avoidance of unpleasant outcomes can be addressed with moment-to-moment awareness. MBCT protocols have been adapted to treat comorbidity in anxiety disorders (transdiagnostic, c.f. [153]) and for the disorder-specific treatment of GAD (c.f. [154]), SAD (c.f., [155]), and PD (c.f. [156]). Elements of cognitive therapy and psychoeducation are disorder-specific for anxiety disorders, but, as outlined above, MBCT differs from CBT in distinctive ways: for instance, MBCT promotes being with unpleasant states and encourages viewing thoughts as thoughts not as facts (or dysfunctional thoughts), noticing and allowing thoughts without changing or avoiding them (not challenging them), and developing a nonjudgmental moment-to-moment awareness (not on reinforcing more adaptive responses) [147].

Efficacy of MBCT and Related Approaches for Anxiety Disorders

For different health-related conditions, Khoury et al. [28] identified moderate effect sizes ($g = 0.53$) of MBIs when compared to waitlist control conditions in a meta-analysis of 209 studies ($n = 12{,}145$ participants). Concerning anxiety disorders, preliminary evidence suggests that MBIs can produce short-term treatment effects comparable to traditional CBT [9, 30]. MBIs demonstrated to be efficacious in reducing the symptoms of anxiety in clinical and healthy populations [25, 28, 30].

A meta-analysis of 39 studies (1,140 participants) by Hofmann et al. [30] found that MBIs were moderately effective at reducing symptoms of anxiety ($g = 0.63$) in clinical samples with medical and psychiatric conditions. These effect sizes are significantly larger than the effect sizes of psychological placebo conditions in RCTs targeting anxiety disorders ($g = 0.45$). MBIs demonstrated to be most effective in reducing anxiety among patients with anxiety disorders ($g = 0.97$), across a wide range of severity levels and patient populations. The moderate effect sizes for MBIs on anxiety symptoms are in line with another meta-analysis by Goyal et al. [157]. However, another meta-analysis of 12 RCTs on MBIs for people with a current episode of an anxiety or mood disorder by Strauss et al. [158] showed no significant benefit of MBIs in terms of anxiety severity ($g = -0.55$) compared to passive and active control groups, while at least MBCT was found being more effective than MBSR for anxiety severity. In a meta-analysis of 19 studies on mindfulness- and acceptance-based interventions (MABIs), Vøllestad et al. [159] found a large effect size ($g = 0.83$) for anxiety symptoms in controlled trials, with robust reductions of both anxiety and comorbid depressive symptoms. In a meta-analysis of transdiagnostic treatments, Newby et al. [160] showed that CBT and MABIs were both effective in reducing anxiety, but CBT had a higher effect size ($g = 0.88$, 40 studies) than MABIs ($g = 0.61$, 7 studies).

To date, there is support from a small number of mainly pilot studies for the efficacy of MBCT in anxiety disorders [161]. Currently, the best evidence for MBCT exists for the relapse prevention of recurrent MDD [162–164]. In addition, MBCT for anxiety disorders is often applied in addition to pharmacotherapy [165, 166]. Principally, MBSR and MBCT can help to significantly reduce the symptoms of anxiety in different medical conditions, such as pain disorders, diabetes, chronic fatigue, heart disease, and cancer [30]. For instance, a meta-analysis by Zhang et al. [167] on the efficacy of MBIs for cancer patients confirmed positive effects regarding the reduction of anxiety symptoms. There is also meta-analytic evidence that MBSR and MBCT can reduce anxiety in people with vascular disease [168]. Additionally, adults with intellectual disabilities with anxiety may also benefit from adapted briefer MBCT programs, which can be cognitively less demanding than pure CBT [169]. Furthermore, there are pilot studies suggesting positive effects of MBCT in reducing anxiety symptoms in elderly (e.g., [170, 171]), insomnic patients [172], pregnant women [173, 174], and women with postpartum depression and anxiety undergoing pharmacotherapy [175] as well as in health/illness anxiety (hypochondriasis) [176–179]. According to a systematic review by Xie et al. [145], MBCT can be recommended for patients with anxiety in the context of rehabilitation or for the long-term improvement of subjective quality of life.

Given that several review articles report heterogeneity regarding the outcome assessment, types of MBIs [142], populations, and evaluation conditions [9], it appears important to take a deeper look at the effects of MBCT for anxiety disorders in order to derive disorder-specific recommendations for clinical practice and research.

Specific Efficacy of MBCT and Related Approaches for Different Anxiety Disorders

MCBT for Specific Phobia

In contrast to exposure therapy [3, 81], there is a scarcity of studies on the application of MBCT for specific phobias. Techniques from MBIs such as focused attention can help to become aware of phobic thoughts and feelings at the present moment and foster emotion regulation strategies and exposure. Specific phobias and their negative impact are highly heterogeneous [81]. For instance, fear of flying can become impairing when it begins to affect people's lives (e.g., when the new job requires someone to travel by airplane) [81]. Clark and Rock [180] suggested that trait mindfulness may be a protective factor against flying anxiety, given that a nonjudgmental and accepting attitude (i.e., decentering stance) can serve as a coping strategy during flying. In line with this, a study by Arch and Craske [181] with 90 anxious or non-anxious people indicated a moderating role of trait mindfulness in terms of diminishing effects in response to laboratory stressors.

Few interventional studies include at least elements of mindful meditation, such as mindful breathing and focused attention training in specific phobias. For example, Hooper et al. [182] allocated 60 spider fearful students to 3 groups receiving either focused attention (mindful breathing exercise), thought suppression, or unfocused attention prior to exposing them to the BAT. They showed that participants exposed to focused attention (mindfulness) induction were more likely to approach a spider post-induction than the other groups showing more avoidance behavior and anxiety [182].

MCBT for Panic Disorder and Agoraphobia

People with PD become hypervigilant to bodily sensations and contextual details related to their panic attacks and often try to reduce their worries by seeking reassurance that causes of their physical symptoms are of medical nature [1, 3]. Intolerance of uncertainty (IU) has been suggested to be a key contributor of fear and avoidance related to panic attacks [183]. Uncertainty regarding bodily sensations and the occurrence of panic attacks is associated with anticipatory anxiety and the development of avoidance behavior, resulting in a self-perpetuating cycle in which bodily sensations induce catastrophic misinterpretations [184]. MBIs can be seen as a novel approach to improve the adherence to treatment and long-term outcomes in PD [185]. MBCT can help patients with PD to become more aware of their thoughts (from a decentering stance), to effectively deal with catastrophizing misinterpretations using mindfulness techniques (e.g., attention to physical sensations using breathing exercises), and to better tolerate uncertainties or accept misbeliefs without judgment or elaboration [184].

There are few, mainly pilot studies on MBCT for PD. For instance, a study by Sado et al. [153] showed the feasibility of an adapted MBCT program for anxiety disorders in Japanese patients with PD or SAD. Another study by MK Kim et al. [184] demonstrated positive effects of a 8-week MBCT on IU in patients with PD taking medication. In their study with 69 patients with PD, they found significant decreases of IU and PD severity, which were correlated [184]. In another study, B Kim et al. [186] investigated predictors of outcomes of 8-week MBCT among 65 PD patients. They found that a comorbid personality disorder was significantly related to MBCT non-completion, whereas improvements of anxiety sensitivity were associated with treatment response and PD remission after completing the course. As an adjuvant to pharmacotherapy, YW Kim et al. [187] allocated 46 patients with PD or GAD for 8 weeks either to MBCT or psychoeducation. Patients receiving MBCT showed significantly more improvements in anxiety measures than patients in the active control condition [187]. However, the interoceptive component of formal mindful exercises may provoke panic attacks. Thus, MBCT should be professionally guided by a trained clinician. In addition, MBCT may be contraindicated in some patients with PD, which should be carefully and individually decided by the psychotherapist.

MBCT for Social Anxiety Disorder

SAD is often associated with residual symptoms after initially successful treatment using traditional CBT techniques focusing fear extinction and habituation, which makes the exploration of extended CBT methods like MBCT worthwhile [155]. People with SAD are vulnerable to bias for negative social events. MBIs can target cognitive reappraisal, rumination, worry, cognitive and emotional reactivity by training attention control, self-awareness, and self-regulation [188]. Hence, increasing mindfulness skills might be associated with positive changes in SAD symptoms [189].

MBCT may improve the awareness of daily positive events and, thereby, may contribute to an increased positive affect and a curiosity as well as openness toward emotional experiences related to social situations. Furthermore, MBCT may help replacing ruminative thinking patterns with mindful attention by disconnecting cognitions from their emotional content [155]. There are some pilot studies that suggest positive effects of group MBSR [181, 188, 190, 191], MBCT [155, 192], MBCT plus Task Concentration Training [193], mindful exposure [194], as well as mindfulness and acceptance-based group therapy (MAGT) [189, 195] on SAD symptoms. In a recent pilot RCT on a 12-week MBI for SAD, Koszycki et al. [194] combined compassion meditation and mindful exposure and indicated the feasibility of this approach. However, a systematic review by Norton et al. [196] of 9 MABIs for SAD suggests that the evidence base is not sufficient to recommend MBSR or MBCT as an alternative to the first-line treatment with CBT [197].

MBCT for Generalized Anxiety Disorder

GAD that is characterized by pathological worries and is associated with hyperarousal and high rates of comorbidity, with depression, has often a chronic course and is the least successfully treated anxiety disorder [81, 154]. MBCT may provide an alternative approach to CBT. Evans [154] proposed an 8-week protocol specific to GAD based on the original MBCT manual for recurrent MDD [198]. As outlined above, MBCT aims to break the vicious cycle of negative thinking by promoting a decentered association between cognitions and feelings, including ruminative and catastrophizing thoughts [134]. Increased mindfulness may help reducing symptoms of GAD by fostering the meta-cognitive capacity of decentering (e.g., observe and accept thoughts as thoughts), which might have a positive impact on reducing pathological worry and rumination [199].

There are mainly uncontrolled pilot studies showing beneficial effects of MBCT in treating GAD (c.f. [200, 201]). Another example is “CALM Pregnancy,” a MBCT program that was tested in 24 pregnant women: of the 17 women with GAD, only 1 still met the diagnosis criteria for GAD at posttreatment [202]. There are only few RCTs on the effectiveness of MBIs in GAD. In an 8-week RCT with 182 people with GAD, Wong et al. [203] showed that both MBCT and group psychoeducation were significantly more effective than care as usual (CAU) in reducing anxiety symptoms. As mentioned earlier, another RCT by YW Kim et al. [187] also showed the efficacy of MBCT in the adjunctive treatment of GAD and PD in patients receiving pharmacotherapy. However, due to the common combination of MBCT with other treatments, it is difficult to determine specific effects of MBCT.

Internet-Based Mindfulness-Based Interventions for Anxiety

As outlined before, the therapeutic benefits of MBIs depend on regular exercises in daily life, which could be supported by using the Internet and modern technology [204]. Internet-based MBIs (iMBIs) have been suggested as a well-accepted self-help alternative to the traditional MBI group format (e.g., [205]). However, most iMBIs have been tested in nonclinical samples like students (e.g., [206]) and have a scope on prevention or health promotion [207], rather than on medical conditions and anxiety [208]. Another difference to traditional 8-week MBIs is the development of iMBIs with a briefer duration (e.g., [209]). According to a review by Spijkerman et al. [32], iMBIs have a duration between 2 and 12 weeks and provide at least a minimal support, including individualized feedback and coaching by therapists. Because most of these studies investigate transdiagnostic therapies, the self-reported outcomes are often assessed with general anxiety measures in different health conditions [210].

Preliminary evidence on the efficacy of iMBIs (incl. ACT) for anxiety was shown by the aforementioned meta-analysis by Spijkermans et al. [32] of 15 RCTs (n = 11 targeting anxiety) in mainly nonclinical studies; they found significant reductions of symptoms of anxiety, albeit with a small effect size (g = 0.22). Furthermore, adherence rates varied largely between 39.5% and 92% in 6- to 8-week programs [32]. A recent review and meta-analysis of a wide range of 21 Internet-based third wave or contextual CBT treatments (e.g., MBCT, ACT, and DBT) by O'Connor et al. [211] indicated the efficacy and acceptability (attrition rate of 23%) of these interventions across different conditions in reducing anxiety compared to active controls. Another review by Fish et al. [212] identified ten studies on the efficacy of iMBIs, with four studies targeting anxiety, the two studies that tested for significant effects (see [213, 214]) revealed positive findings for anxiety, but generally this review reported methodological issues due to the early nature of iMBIs. One of these studies of higher quality used a transdiagnostic approach to address the high proportion of comorbidity in anxiety disorders [3]: The RCT by Boettcher et al. [213] on an 8-week MBI for anxiety disorders with 91 patients (with SAD, GAD, PD, or an unclassified anxiety disorder) revealed large effect sizes for different outcomes (d = 0.82–1.58) in the MBI group and when compared to an online discussion forum (d = 0.45–0.76), indicating the feasibility of this stand-alone Internet-based MBI. Smartphone apps may provide a flexible delivery mode for MBIs, but the quality of publicly available, self-proclaimed mindfulness apps and their helpfulness in improving mindfulness skills remains overall unclear [215, 216]. To date, there are only few studies on the efficacy of evidence-based mindfulness apps in reducing symptoms of anxiety (e.g., [217]).

Acceptance and Commitment Therapy

ACT is another treatment approach of the "third wave" of CBT [11] and was developed in 1986 by Steven C. Hayes and co-founders. Besides its connection with the philosophy of Buddhism, ACT is also rooted in the Relational Frame Theory (RTF) [218] that integrates cognition and language into a behavioral analytic framework. On the one hand, ACT aims to foster acceptance of unwanted thoughts and feelings (e.g., by discouraging experiential avoidance and cognitive fusion); on the other hand, ACT tries to stimulate action tendencies (moving toward value-based directions) that may contribute to an improvement in the circumstances of life [219]. ACT is trying to fulfill these two general goals by the following six main principles (Box 17.1).

Box 17.1 The Six Main Principles of Acceptance and Commitment Therapy

1. *Defusion* (lowering the impact of thoughts by creating a distance between the person and the cognition)
2. *Acceptance* (trying to lower the threat of unpleasant feelings by accepting them instead of trying to control them)
3. *Contact with the present moment* (training unpurposeful attention by using mindfulness meditation)
4. *Self-as-context* (realizing that a person can observe thoughts, emotions, and behavior)
5. *Values* (clarifying what is important to a person and how the person can move towards directions of a value-driven life)
6. *Committed action* (committing to some action that is in line with values although it may cause some distress in the short term)

All six ACT principles lead to specific techniques such as mindfulness exercises that aim at increasing psychological flexibility, defined as "the ability to contact the present moment more fully as a human being, and to change or persist in behavior when doing so serves valued ends" ([16], p. 7). ACT treatments may be delivered in a face-to-face group setting or individually. In addition, there are mHealth interventions such as the "ACT Daily" app [220] and Internet-delivered ACT (iACT) interventions for different mental health outcomes available and undergoing clinical testing [221].

Apart from the differences in the philosophical foundation of CBT and ACT, ACT techniques are still compatible with the techniques of "second wave" CBT interventions by adding *mindfulness* and *acceptance* of the inner experience [222] and may lead to an improvement of the overall treatment. However, there are differences in the role of cognitions and emotions between CBT and ACT [219]. In order to focus on the *role of cognitions,* it is important to know that cognitive processes, such as automatic thoughts and schemas about the world, the self and the future, are the centerpiece of the CBT. But where CBT is distinguishing between behaviors or actions, emotions or subjective experience, and cognitions or thought processes, in ACT the cognition takes the role as a private behavior [219]. In ACT, the content of specific thoughts is not meant to be changed. Instead, ACT encourages patients to accept the thought, but aims at changing the function of the thought. Also, the *role of emotion* and emotion regulation is different between CBT and ACT. Both approaches target different stages in the emotion-generative process [219]. CBT encourages cognitive reappraisal, for instance, by exposing the patient to the emotional trigger. In terms of Gross' process model of emotions [223], this purpose of CBT can be classified as adaptive antecedent-focused emotion regulation strategy. In contrast, ACT targets acceptance as the counterpart to experiential avoidance, for instance, in the form of suppression or control of negative states, which can be classified as counteracting of maladaptive response-focusing emotion regulation strategies [219]. ACT assumes that the suppression of unpleasant states results in a worsening of emotional outcomes. Antecedent-focused strategies occur before the emotional response takes place, whereas response-focused strategies try to work with the expression or experience of emotions [219].

ACT for Anxiety Disorders

In general, the ACT approach leads to a different view on anxiety: anxiety is a collection of thoughts, feelings, urges, and body sensations as well as a normal part of life. It is not the anxiety that has the huge impact on one's life; it is the avoidance strategies to control those feelings, thoughts, urges, and body sensations that are feeding the anxiety. The struggle to avoid anxiety by unsuccessfully attempting to control it may make it a problem. ACT techniques focus on decreasing the behavior-regulatory function of anxiety and related cognitions [224]. Eifert and Forsyth [225] have developed a practitioner's guide called "ACT for anxiety disorders" that focuses on anxiety disorders such as PD, GAD, and specific phobias. This guide is used by practitioners as well as by researchers to create and evaluate disorder-specific ACT treatments. One of the latter is a 12-session intervention evaluated by Craske et al. [226] with patients suffering from social phobia: session 1 includes psychoeducation, experiential exercises, and a discussion of acceptance and valued action, while the sessions 2 and 3 explore whether previous efforts to control anxiety have worked and how some of them had led to the reduction of valued life activities as well as encouraged acceptance; sessions 4 and 5 aim to establish mindfulness, acceptance, and cognitive defusion; sessions 6–11 have the purpose to continue to deepen the practice and to understand acceptance, mindfulness, and cognitive defusion. Also, these later sessions intend to lead to the clarification of values with the goal of increasing valued life activities. By using behavioral exposures (in vivo, interoceptive, and imaginal), participants are encouraged to practice mindfully observing and accepting anxiety and to further pursuing valued activities while experiencing anxiety. Session 12 is conducted as a roundup in order to summarize how to move forward (c.f. [225]).

Efficacy of ACT for Anxiety Disorders

A meta-analysis of 39 RCTs on the efficacy of ACT for mental disorders and somatic health problems showed that ACT outperformed the waitlist ($g = 0.82$), the psychological placebo ($g = 0.51$), and the treatment as usual ($g = 0.64$) [227]. A meta-analysis of 19 studies on ACT for anxiety disorders indicated the efficacy of ACT at posttreatment, with a small-to-moderate effect size ($g = 0.45$) and low-to-moderate attrition rate (14.2% [228]). In a meta-analysis of 10 RCTs on iACT for different mental health outcomes by Brown et al. [221], none of the trials had an anxiety disorder as primary condition and, although the effect size for anxiety measures was significant, it did not reach the level for a small effect size ($g = 0.18$).

Specific Efficacy of ACT for Different Anxiety Disorders

ACT for Specific Phobia

Specific phobias include phobias on a huge variety of specific objects or situations (e.g., flying, heights, seeing blood) but have in common that the situation or object almost always provokes fear or anxiety that is out of proportion and leads to avoiding or enduring behavior [81]. An ACT-based treatment for specific phobias may follow some of the principles of ACT. For instance, the principles 5 "Values" and 6 "Committed Action" could be applied to work on avoidance behavior and to increase the quality of a value-driven life. The patient is asked to commit oneself to action that is in line with the clarified values, such as being able to visit the grandmother on the 12th floor, even if it causes some distress. If someone is approaching the fear and working on the avoidance behavior, it is helpful to focus on the values and directions someone tries to achieve (e.g., being a reliable grandchild).

Only few studies have applied ACT to treat specific phobia. An example is a study on math anxiety. Zettle [229] treated 24 college students suffering from math anxiety individually for 6 weeks with either ACT (n = 12) or systematic desensitization as the "benchmark" CBT component for specific phobia (n = 12). Only systematic desensitization was associated with statistically and clinically significant reductions in trait anxiety at posttreatment. Interestingly, the baseline levels of experiential avoidance were more strongly related to therapeutic change among participants receiving ACT than in the CBT condition.

ACT for Panic Disorder and Agoraphobia

PD is characterized by panic attacks in terms of abrupt surges of intense fear that might be associated with agoraphobia. Agoraphobia is related to fearing and avoiding situations or places that are expected to cause panic attacks, including anticipated embarrassment, helplessness, and no possibility to escape [3]. The second ACT principle "Acceptance" may be applied for PD with or without agoraphobia by training to accept discomfort in panic-related or agoraphobic situations rather than fighting against it. Accepting to have a panic attack might actually attenuate the negative impact of the panic attack on someone's life (e.g., avoidance strategies that may lead to an impaired social functioning). It is helpful to use the metaphor of quicksand to create an understanding: the more someone keeps calm in quicksand, the easier she or he will get out of it; if someone tries to fight against quicksand, she or he panics, moves quickly, and the deeper he or she sinks into it.

There are pilot studies on the efficacy of ACT in PD with or without agoraphobia. Gloster et al. [230] conducted a RCT on an eight-session ACT with 43 patients with primary treatment-resistant PD and/or agoraphobia. The patients were allocated to three groups: immediate 4-week ACT treatment (n = 33) or a 4-week waiting list (n = 10), of which n = 8 received a delayed treatment. Patients that were treated immediately with ACT reported significantly more improvements on the Panic and Agoraphobia Scale (d = 0.72) and the Clinical Global Impression Scale (d = 0.89) than patients in the waiting-list group. Treatment gains were maintained at the 6-month follow-up assessment. Another example is a pilot study by Meuret et al. [231] that has demonstrated the feasibility of a combined brief ACT and exposure treatment for PD.

ACT for Social Anxiety Disorder

SAD's core symptom is the fear in one or more social situations in which the person feels exposed to the possible scrutiny by others [81]. Interventions for SAD may use different principles of ACT, such as "Defusion": SAD may cause thoughts like "Everyone thinks I am boring" or "Everyone is staring at me and thinks that I am a loser"; ACT adopts a different perspective by adding "I am having the thought that …" to the beginning of these sentences. The third ACT principle "Mindfulness" stresses the importance of living in the "here and now" by practicing to engage in the present moment instead of becoming lost in thoughts about other peoples' opinions.

Dalrymple and Herbert [232] used a 12-week ACT treatment in a pilot study with 19 individuals diagnosed with SAD. They found significant improvements in social anxiety symptoms and quality of life with large effect sizes. Craske et al. [226] conducted a three-arm RCT with participants with social phobia (n = 87) comparing the effects of either 12 weeks of CBT, ACT, or a waiting-list control condition. Their findings showed that both treatment groups scored significantly better in the main outcomes than the control group, with no differences observed between CBT and ACT. Remarkably, the self-reported fear of negative evaluation moderated the treatment outcomes, with trends for both extremes in terms of superior outcomes for participants receiving CBT and inferior outcomes for participants receiving ACT [226]. Another RCT was conducted by Herbert et al. [233] comparing the efficacy of an ACT treatment (n = 49) with a CBT treatment (n = 53) among patients with SAD. The authors found greater improvements in self-reported social anxiety symptoms and overall functioning in the CBT group on the one hand, but on the other hand, they have found indications for greater improvements in observer-rated social behavior in the ACT group.

Concerning Internet-delivered approaches, Ivanova et al. [234] created and evaluated an iACT-treatment with additional app for SAD and PD. 152 participants were randomized either to a therapist-guided or therapist-unguided treatment or to a waiting-list control group. Both treatment groups showed significantly reduced general anxiety (d = 0.39) and social anxiety (d = 0.70), but not panic symptoms (d = 0.05) compared to the control group, with no significant differences between the guided and the unguided intervention. In another RCT on an iACT intervention, Gershkovich et al. [235] randomized 42 participants either to an eight-module self-help intervention with minimal therapist support in terms of 10–15 min per week using videoconferencing and daily text messages (n = 20) or to the same intervention, but without therapist support (n = 22). Their results showed significant reductions in SAD symptoms, with no significant differences between the both support formats. The only difference was a lower attrition rate in the guided group (20%) than in the unguided group (50%).

ACT for Generalized Anxiety Disorder

Excessive anxiety and worries that are difficult to control and have a huge impact on the patient's life, including physical complaints, are diagnostic criteria of GAD [81].

Theoretically, all six ACT principles can be applied for the treatment of GAD. For example, the fourth ACT principle, "Self-as-context," may be helpful for patients diagnosed with GAD in order to create a distance between the worries and the self and to realize that there is more to observe than worries and anxiety. In ACT, the patient is in charge of the thoughts and their content, making them not dangerous or life-threatening anymore.

Hasheminasab et al. [236] reported findings from a ten-session ACT treatment with three consecutive referrals for GAD patients at a private practice. In those sessions, different methods were used to introduce the assumption that struggle and control may actually interfere with the patient's quality of life and values as well as the aim to stay with the anxiety and live in the present moment. In addition, the treatment aimed to foster the fully perception of anxiety-related responses without judgment or elaboration (i.e., perceiving thoughts as thoughts, physical sensations as physical sensations, images as images, feelings as feelings, etc.). Participants showed clinically significant changes from pre- to posttreatment in the severity of anxiety symptoms. In another study, Avdagic et al. [237] randomly allocated 51 patients (diagnosed with GAD) to either a 6-week intervention based on ACT or on CBT. The authors found significant improvements on all anxiety-related measures for both conditions at post- and follow-up-treatment. Moreover, they found a significant interaction indicating a steeper reduction from pre- to posttreatment in worrying symptoms ($d = 0.79$) between both groups.

Furthermore, iACT has been also applied to treat GAD. For example, a RCT Dahlin et al. [238] randomly assigned 103 participants diagnosed with GAD either to a therapist-guided iACT or to a waiting-list control group. Their findings indicated the efficacy of the intervention with moderate-to-high effect sizes ($d = 0.70$–0.98) with maintained effects at the 6-month follow-up.

Conclusions

There is strong evidence for the efficacy and effectiveness of traditional CBT in treating the symptoms of PD with or without agoraphobia, SAD, and GAD (c.f. [239]), while the best evidence for specific phobia exists for exposure-based therapies [69]. There is also evidence that different delivery modes of CBT, such as individual or group formats as well as iCBT with minimal therapist guidance, work similarly well across different anxiety disorders (e.g., [240]). Given the small number of studies with diverse populations, Hofmann et al. [4] found no meta-analytic evidence for the efficacy of CBT for specific subgroups like minorities, except for children and elderly. Another drawback is the scarcity of long-term assessments of CBT outcomes. Overall, the current evidence base suggests that treatment decisions on CBT and/or pharmacotherapy can be based on patients' preferences [2].

Although the rationale for using MBCT to treat anxiety is credible [147] and the availability of adapted MBCT protocols for anxiety disorders like GAD [154], the clinical testing is at an early stage compared to traditional CBT [14]. Preliminary evidence suggests that patients with PD and GAD may benefit from MBCT in reducing pathological worry and intolerance of uncertainty [184, 187]. However, for anxiety disorders there is a need for more high-quality RCTs to draw conclusions on the effectiveness of traditional MBIs/MBCT [9, 21, 30, 142, 196] and iMBIs [32, 207, 211, 212]. Finally, few publicly available apps using the term mindfulness fulfill criteria for MBIs, which makes it difficult for patients and clinicians to identify useful apps [215, 216]. Like MBCT, ACT is at an early stage of clinical testing regarding the efficacy for anxiety disorders (c.f. [9, 219, 227]). In addition, contra-indications and potential adverse events of MBIs need to be evaluated and should be individually evaluated by clinicians [144].

References

1. Bandelow B, Michaelis S. Epidemiology of anxiety disorders in the 21st century. Dialogues Clin Neurosci. 2015;17:327–35.
2. Bandelow B, Reitt M, Röver C, Michaelis S, Görlich Y, Wedekind D. Efficacy of treatments for anxiety disorders: a meta-analysis. Int Clin Psychopharmacol. 2015;30:183–92. https://doi.org/10.1097/YIC.0000000000000078.
3. Bystritsky A, Khalsa SS, Cameron ME, Schiffman J. Current diagnosis and treatment of anxiety disorders. P T. 2013;38:30–57.
4. Hofmann SG, Asnaani A, Vonk IJJ, Sawyer AT, Fang A. The efficacy of cognitive behavioral therapy: a review of meta-analyses. Cogn Ther Res. 2012;36:427–40. https://doi.org/10.1007/s10608-012-9476-1.
5. Fenn K, Byrne M. The key principles of cognitive behavioural therapy. InnovAiT. 2013;6:579–85. https://doi.org/10.1177/1755738012471029.
6. Dymond S, Roche B. A contemporary behavior analysis of anxiety and avoidance. Behav Anal. 2009;32:7–27.
7. Kaczkurkin AN, Foa EB. Cognitive-behavioral therapy for anxiety disorders: an update on the empirical evidence. Dialogues Clin Neurosci. 2015;17:337–46.
8. Beck AT. Cognitive therapy and the emotional disorders. New York: Penguin; 1976.
9. Hofmann SG, Sawyer AT, Fang A. The empirical status of the "New Wave" of CBT. Psychiatr Clin North Am. 2010;33:701–10. https://doi.org/10.1016/j.psc.2010.04.006.
10. Hofmann SG, Gómez AF. Mindfulness-based interventions for anxiety and depression. Psychiatr Clin North Am. 2017;40:739–49. https://doi.org/10.1016/j.psc.2017.08.008.
11. Hayes SC. Acceptance and commitment therapy, relational frame theory, and the third wave of behavioral and cognitive therapies. Behav Ther. 2004;35:639–65. https://doi.org/10.1016/S0005-7894(04)80013-3.
12. Hayes SC, Hofmann SG. The third wave of cognitive behavioral therapy and the rise of process-based care. World Psych. 2017;16:245–6. https://doi.org/10.1002/wps.20442.
13. Kabat-Zinn J. Mindfulness-based interventions in context: past, present, and future. Clin Psychol Sci Pract. 2003;10:144–56. https://doi.org/10.1093/clipsy.bpg016.
14. Feliu-Soler A, Cebolla A, McCracken LM, D'Amico F, Knapp M, López-Montoyo A, et al. Economic impact of third-wave cognitive behavioral therapies: a systematic review and quality assessment of economic evaluations in randomized controlled trials. Behav Ther. 2018;49:124–47. https://doi.org/10.1016/j.beth.2017.07.001.
15. Segal ZV, Williams JMG, Teasdale JD, Kabat-Zinn J. Mindfulness-based cognitive therapy for depression: a new approach to preventing relapse. 2nd ed. New York: The Guilford Press; 2018.
16. Hayes SC, Luoma JB, Bond FW, Masuda A, Lillis J. Acceptance and commitment therapy: model, processes and outcomes. Behav Res Ther. 2006;44:1–25. https://doi.org/10.1016/j.brat.2005.06.006.
17. Linehan M. DBT skills training manual. New York: The Guilford Press; 2015.

18. Wells A. Meta-cognition and worry: a cognitive model of generalized anxiety disorder. Behav Cogn Psychother. 1995;23:301. https://doi.org/10.1017/S1352465800015897.
19. Apolinário-Hagen J, Salewski C. Internet-based mindfulness-based cognitive therapy for the adjunctive treatment of major depressive disorder. In: Kim Y-K, editor. Understanding depression: Volume 2. Clinical manifestations, diagnosis and treatment: Springer; 2018. p. 299–309. https://doi.org/10.1007/978-981-10-6577-4_22.
20. Edenfield TM, Saeed SA. An update on mindfulness meditation as a self-help treatment for anxiety and depression. Psychol Res Behav Manag. 2012;5:131–41. https://doi.org/10.2147/PRBM.S34937.
21. Goldberg SB, Tucker RP, Greene PA, Davidson RJ, Wampold BE, Kearney DJ, Simpson TL. Mindfulness-based interventions for psychiatric disorders: a systematic review and meta-analysis. Clin Psychol Rev. 2018;59:52–60. https://doi.org/10.1016/j.cpr.2017.10.011.
22. Shapero BG, Greenberg J, Pedrelli P, de Jong M, Desbordes G. Mindfulness-based interventions in psychiatry. Focus Am Psychiatr Publ. 2018;16:32–9. https://doi.org/10.1176/appi.focus.20170039.
23. Demarzo MMP, Montero-Marin J, Cuijpers P, Zabaleta-del-Olmo E, Mahtani KR, Vellinga A, et al. The efficacy of mindfulness-based interventions in primary care: a meta-analytic review. Ann Fam Med. 2015;13:573–82. https://doi.org/10.1370/afm.1863.
24. Keng S-L, Smoski MJ, Robins CJ. Effects of mindfulness on psychological health: a review of empirical studies. Clin Psychol Rev. 2011;31:1041–56. https://doi.org/10.1016/j.cpr.2011.04.006.
25. Khoury B, Sharma M, Rush SE, Fournier C. Mindfulness-based stress reduction for healthy individuals: A meta-analysis. J Psychosom Res. 2015;78:519–28. https://doi.org/10.1016/j.jpsychores.2015.03.009.
26. Janssen M, Heerkens Y, Kuijer W, van der Heijden B, Engels J. Effects of mindfulness-based stress reduction on employees' mental health: a systematic review. PLoS One. 2018;13:e0191332. https://doi.org/10.1371/journal.pone.0191332.
27. Cavanagh K, Strauss C, Forder L, Jones F. Can mindfulness and acceptance be learnt by self-help?: a systematic review and meta-analysis of mindfulness and acceptance-based self-help interventions. Clin Psychol Rev. 2014;34:118–29. https://doi.org/10.1016/j.cpr.2014.01.001.
28. Khoury B, Lecomte T, Fortin G, Masse M, Therien P, Bouchard V, et al. Mindfulness-based therapy: a comprehensive meta-analysis. Clin Psychol Rev. 2013;33:763–71. https://doi.org/10.1016/j.cpr.2013.05.005.
29. Teasdale JD, Segal ZV, Williams JMG, Ridgeway VA, Soulsby JM, Lau MA. Prevention of relapse/recurrence in major depression by mindfulness-based cognitive therapy. J Consult Clin Psychol. 2000;68:615–23. https://doi.org/10.1037/0022-006X.68.4.615.
30. Hofmann SG, Sawyer AT, Witt AA, Oh D. The effect of mindfulness-based therapy on anxiety and depression: a meta-analytic review. J Consult Clin Psychol. 2010;78:169–83. https://doi.org/10.1037/a0018555.
31. Olthuis JV, Watt MC, Bailey K, Hayden JA, Stewart SH. Therapist-supported Internet cognitive behavioural therapy for anxiety disorders in adults. Cochrane Database Syst Rev. 2016;3:CD011565. https://doi.org/10.1002/14651858.CD011565.pub2.
32. Spijkerman MPJ, Pots WTM, Bohlmeijer ET. Effectiveness of online mindfulness-based interventions in improving mental health: a review and meta-analysis of randomised controlled trials. Clin Psychol Rev. 2016;45:102–14. https://doi.org/10.1016/j.cpr.2016.03.009.
33. Carpenter JK, Andrews LA, Witcraft SM, Powers MB, Smits JAJ, Hofmann SG. Cognitive behavioral therapy for anxiety and related disorders: a meta-analysis of randomized placebo-controlled trials. Depress Anxiety. 2018;35:502–14. https://doi.org/10.1002/da.22728.
34. Margraf J. Hintergründe und Entwicklung. In: Lehrbuch der Verhaltenstherapie. p. 3–45. doi:https://doi.org/10.1007/978-3-540-79541-4.
35. Beck AT. Thinking and depression. Arch Gen Psychiatry. 1963;9:324. https://doi.org/10.1001/archpsyc.1963.01720160014002.
36. Beck AT. Cognitive therapy and the emotional disorders. New York: Penguin; 1979, 1976.

37. Beck AT, Emery G, Greenberg RL. Anxiety disorders and phobias: a cognitive perspective/ Aaron T. Beck and Gary Emery with Ruth L. Greenberg; [with a new preface by Aaron T. Beck]. 15th ed. Cambridge, MA: Basic Books; 2005.
38. Deacon BJ, Abramowitz JS. Cognitive and behavioral treatments for anxiety disorders: a review of meta-analytic findings. J Clin Psychol. 2004;60:429–41. https://doi.org/10.1002/jclp.10255.
39. Wolitzky-Taylor KB, Horowitz JD, Powers MB, Telch MJ. Psychological approaches in the treatment of specific phobias: a meta-analysis. Clin Psychol Rev. 2008;28:1021–37. https://doi.org/10.1016/j.cpr.2008.02.007.
40. Kasper S. Anxiety disorders: under-diagnosed and insufficiently treated. Int J Psychiatry Clin Pract. 2006;10(Suppl 1):3–9. https://doi.org/10.1080/13651500600552297.
41. Musiat P, Tarrier N. Collateral outcomes in e-mental health: a systematic review of the evidence for added benefits of computerized cognitive behavior therapy interventions for mental health. Psychol Med. 2014;44:3137–50. https://doi.org/10.1017/S0033291714000245.
42. Andersson G, Titov N. Advantages and limitations of internet-based interventions for common mental disorders. World Psych. 2014;13:4–11. https://doi.org/10.1002/wps.20083.
43. Apolinrio-Hagen J, Harrer M, Kählke F, Fritsche L, Salewski C, Ebert DD. Public attitudes toward guided internet-based therapies: web-based survey study. JMIR Ment Health. 2018;5:e10735. https://doi.org/10.2196/10735.
44. Apolinario-Hagen JA, Tasseit S. Access to psychotherapy in the era of web 2.0 – New Media, old inequalities?: Zugang zur Psychotherapie in der Ära des Web 2.0 – Neue Medien, Alte Ungleichheiten? Int J Heal Prof. 2015;2:179. https://doi.org/10.1515/ijhp-2015-0010.
45. Baumeister H, Reichler L, Munzinger M, Lin J. The impact of guidance on Internet-based mental health interventions — a systematic review. Internet Interv. 2014;1:205–15. https://doi.org/10.1016/j.invent.2014.08.003.
46. Fogliati VJ, Dear BF, Staples LG, Terides MD, Sheehan J, Johnston L, et al. Disorder-specific versus transdiagnostic and clinician-guided versus self-guided internet-delivered treatment for panic disorder and comorbid disorders: a randomized controlled trial. J Anxiety Disord. 2016;39:88–102. https://doi.org/10.1016/j.janxdis.2016.03.005.
47. Erbe D, Eichert H-C, Riper H, Ebert DD. Blending face-to-face and internet-based interventions for the treatment of mental disorders in adults: systematic review. J Med Internet Res. 2017;19:e306. https://doi.org/10.2196/jmir.6588.
48. Butler AC, Chapman JE, Forman EM, Beck AT. The empirical status of cognitive-behavioral therapy: a review of meta-analyses. Clin Psychol Rev. 2006;26:17–31. https://doi.org/10.1016/j.cpr.2005.07.003.
49. Olatunji BO, Cisler JM, Deacon BJ. Efficacy of cognitive behavioral therapy for anxiety disorders: a review of meta-analytic findings. Psychiatr Clin North Am. 2010;33:557–77. https://doi.org/10.1016/j.psc.2010.04.002.
50. Norton PJ, Price EC. A meta-analytic review of adult cognitive-behavioral treatment outcome across the anxiety disorders. J Nerv Ment Dis. 2007;195:521–31. https://doi.org/10.1097/01.nmd.0000253843.70149.9a.
51. Otto MW, McHugh RK, Kantak KM. Combined pharmacotherapy and cognitive-behavioral therapy for anxiety disorders: medication effects, glucocorticoids, and attenuated treatment outcomes. Clin Psychol NY. 2010;17:91–103. https://doi.org/10.1111/j.1468-2850.2010.01198.x.
52. Tolin DF. Is cognitive-behavioral therapy more effective than other therapies? A meta-analytic review. Clin Psychol Rev. 2010;30:710–20. https://doi.org/10.1016/j.cpr.2010.05.003.
53. Reinholt N, Krogh J. Efficacy of transdiagnostic cognitive behaviour therapy for anxiety disorders: a systematic review and meta-analysis of published outcome studies. Cogn Behav Ther. 2014;43:171–84. https://doi.org/10.1080/16506073.2014.897367.
54. Andrews G, Basu A, Cuijpers P, Craske MG, McEvoy P, English CL, Newby JM. Computer therapy for the anxiety and depression disorders is effective, acceptable and practical health care: an updated meta-analysis. J Anxiety Disord. 2018; https://doi.org/10.1016/j.janxdis.2018.01.001.

55. Andrews G, Cuijpers P, Craske MG, McEvoy P, Titov N. Computer therapy for the anxiety and depressive disorders is effective, acceptable and practical health care: a meta-analysis. PLoS One. 2010;5:e13196. https://doi.org/10.1371/journal.pone.0013196.
56. Peñate W, Fumero A. A meta-review of Internet computer-based psychological treatments for anxiety disorders. J Telemed Telecare. 2016;22:3–11. https://doi.org/10.1177/1357633X15586491.
57. Mayo-Wilson E, Montgomery P. Media-delivered cognitive behavioural therapy and behavioural therapy (self-help) for anxiety disorders in adults. Cochrane Datab Syst Rev. 2013:CD005330. https://doi.org/10.1002/14651858.CD005330.pub4.
58. Andersson G, Cuijpers P, Carlbring P, Riper H, Hedman E. Guided internet-based vs. face-to-face cognitive behavior therapy for psychiatric and somatic disorders: a systematic review and meta-analysis. World Psych. 2014;13:288–95. https://doi.org/10.1002/wps.20151.
59. Mewton L, Smith J, Rossouw P, Andrews G. Current perspectives on Internet-delivered cognitive behavioral therapy for adults with anxiety and related disorders. Psychol Res Behav Manag. 2014;7:37–46. https://doi.org/10.2147/PRBM.S40879.
60. Carlbring P, Andersson G, Cuijpers P, Riper H, Hedman-Lagerlöf E. Internet-based vs face-to-face cognitive behavior therapy for psychiatric and somatic disorders: An updated systematic review and meta-analysis. Cogn Behav Ther. 2018;47:1–18. https://doi.org/10.1080/16506073.2017.1401115.
61. Domhardt M, Geßlein H, von Rezori RE, Baumeister H. Internet- and mobile-based interventions for anxiety disorders: a meta-analytic review of intervention components. Depress Anxiety. 2018; https://doi.org/10.1002/da.22860.
62. Dear BF, Staples LG, Terides MD, Fogliati VJ, Sheehan J, Johnston L, et al. Transdiagnostic versus disorder-specific and clinician-guided versus self-guided internet-delivered treatment for Social Anxiety Disorder and comorbid disorders: a randomized controlled trial. J Anxiety Disord. 2016;42:30–44. https://doi.org/10.1016/j.janxdis.2016.05.004.
63. Dear BF, Staples LG, Terides MD, Karin E, Zou J, Johnston L, et al. Transdiagnostic versus disorder-specific and clinician-guided versus self-guided internet-delivered treatment for generalized anxiety disorder and comorbid disorders: a randomized controlled trial. J Anxiety Disord. 2015;36:63–77. https://doi.org/10.1016/j.janxdis.2015.09.003.
64. Titov N, Dear B, Nielssen O, Staples L, Hadjistavropoulos H, Nugent M, et al. ICBT in routine care: a descriptive analysis of successful clinics in five countries. Internet Interv. 2018;13:108–15. https://doi.org/10.1016/j.invent.2018.07.006.
65. Maples-Keller JL, Bunnell BE, Kim S-J, Rothbaum BO. The use of virtual reality technology in the treatment of anxiety and other psychiatric disorders. Harv Rev Psychiatry. 2017;25:103–13. https://doi.org/10.1097/HRP.0000000000000138.
66. Sucala M, Cuijpers P, Muench F, Cardoş R, Soflau R, Dobrean A, et al. Anxiety: there is an app for that. A systematic review of anxiety apps. Depress Anxiety. 2017;34:518–25. https://doi.org/10.1002/da.22654.
67. van Ameringen M, Turna J, Khalesi Z, Pullia K, Patterson B. There is an app for that! The current state of mobile applications (apps) for DSM-5 obsessive-compulsive disorder, posttraumatic stress disorder, anxiety and mood disorders. Depress Anxiety. 2017;34:526–39. https://doi.org/10.1002/da.22657.
68. Association AP. Diagnostic and statistical manual of mental disorders: American Psychiatric Association; 2013.
69. Grös DF, Antony MM. The assessment and treatment of specific phobias: a review. Curr Psychiatry Rep. 2006;8:298–303.
70. McMurtry CM, Noel M, Taddio A, Antony MM, Asmundson GJG, Riddell RP, et al. Interventions for individuals with high levels of needle fear: systematic review of randomized controlled trials and quasi-randomized controlled trials. Clin J Pain. 2015;31:S109–23. https://doi.org/10.1097/AJP.0000000000000273.
71. McMurtry CM, Taddio A, Noel M, Antony MM, Chambers CT, Asmundson GJG, et al. Exposure-based Interventions for the management of individuals with high levels of needle

fear across the lifespan: a clinical practice guideline and call for further research. Cogn Behav Ther. 2016;45:217–35. https://doi.org/10.1080/16506073.2016.1157204.
72. Arroll B, Wallace HB, Mount V, Humm SP, Kingsford DW. A systematic review and meta-analysis of treatments for acrophobia. Med J Aust. 2017;206:263–7.
73. Choy Y, Fyer AJ, Lipsitz JD. Treatment of specific phobia in adults. Clin Psychol Rev. 2007;27:266–86. https://doi.org/10.1016/j.cpr.2006.10.002.
74. Andersson G, Waara J, Jonsson U, Malmaeus F, Carlbring P, Ost L-G. Internet-based self-help versus one-session exposure in the treatment of spider phobia: a randomized controlled trial. Cogn Behav Ther. 2009;38:114–20. https://doi.org/10.1080/16506070902931326.
75. Andersson G, Waara J, Jonsson U, Malmaeus F, Carlbring P, Ost L-G. Internet-based exposure treatment versus one-session exposure treatment of snake phobia: a randomized controlled trial. Cogn Behav Ther. 2013;42:284–91. https://doi.org/10.1080/16506073.2013.844202.
76. da Costa RT, Sardinha A, Nardi AE. Virtual reality exposure in the treatment of fear of flying. Aviat Space Environ Med. 2008;79:899–903.
77. Quero S, Campos D, Riera Del Amo A, Bretón-López J, Tortella-Feliu M, Baños RM, Botella C. NO-FEAR airlines: a computer-aided self-help treatment for flying phobia. Stud Health Technol Inform. 2015;219:197–201.
78. Campos D, Mira A, Bretón-López J, Castilla D, Botella C, Baños RM, Quero S. The acceptability of an Internet-based exposure treatment for flying phobia with and without therapist guidance: patients' expectations, satisfaction, treatment preferences, and usability. Neuropsychiatr Dis Treat. 2018;14:879–92. https://doi.org/10.2147/NDT.S153041.
79. Bandelow B. Comparison of the DSM-5 and ICD-10: panic and other anxiety disorders. CNS Spectr. 2017;22:404–6. https://doi.org/10.1017/S1092852917000116.
80. Clark DM. A cognitive approach to panic. Behav Res Ther. 1986;24:461–70. https://doi.org/10.1016/0005-7967(86)90011-2.
81. Starcevic V, Castle DJ. Anxiety disorders. In: Fink G, editor. Stress: concepts cognition emotion and behavior: Elsevier; 2016. p. 203–11. https://doi.org/10.1016/B978-0-12-800951-2.00024-8.
82. Kupfer DJ. Anxiety and DSM-5. Dialogues Clin Neurosci. 2015;17:245–6.
83. Casey LM, Oei TPS, Newcombe PA. An integrated cognitive model of panic disorder: the role of positive and negative cognitions. Clin Psychol Rev. 2004;24:529–55. https://doi.org/10.1016/j.cpr.2004.01.005.
84. Rayburn NR, Otto MW. Cognitive-behavioral therapy for panic disorder: a review of treatment elements, strategies, and outcomes. CNS Spectr. 2003;8:356–62.
85. Otto MW, Deveney C. Cognitive-behavioral therapy and the treatment of panic disorder: efficacy and strategies. J Clin Psychiatry. 2005;66(Suppl 4):28–32.
86. McHugh RK, Smits JAJ, Otto MW. Empirically supported treatments for panic disorder. Psychiatr Clin North Am. 2009;32:593–610. https://doi.org/10.1016/j.psc.2009.05.005.
87. Goldberg C. Cognitive-behavioral therapy for panic: effectiveness and limitations. Psychiatry Q. 1998;69:23–44.
88. Freire RC, Zugliani MM, Garcia RF, Nardi AE. Treatment-resistant panic disorder: a systematic review. Expert Opin Pharmacother. 2016;17:159–68. https://doi.org/10.1517/14656566.2016.1109628.
89. Pompoli A, Furukawa TA, Imai H, Tajika A, Efthimiou O, Salanti G. Psychological therapies for panic disorder with or without agoraphobia in adults: a network meta-analysis. Cochrane Database Syst Rev. 2016;4:CD011004. https://doi.org/10.1002/14651858.CD011004.pub2.
90. Hofmann SG, Smits JAJ. Cognitive-behavioral therapy for adult anxiety disorders: a meta-analysis of randomized placebo-controlled trials. J Clin Psychiatry. 2008;69:621–32.
91. Ougrin D. Efficacy of exposure versus cognitive therapy in anxiety disorders: systematic review and meta-analysis. BMC Psychiatry. 2011;11:200. https://doi.org/10.1186/1471-244X-11-200.

92. Pompoli A, Furukawa TA, Efthimiou O, Imai H, Tajika A, Salanti G. Dismantling cognitive-behaviour therapy for panic disorder: a systematic review and component network meta-analysis. Psychol Med. 2018;48:1945–53. https://doi.org/10.1017/S0033291717003919.
93. Otto MW, Tolin DF, Nations KR, Utschig AC, Rothbaum BO, Hofmann SG, Smits JAJ. Five sessions and counting: considering ultra-brief treatment for panic disorder. Depress Anxiety. 2012;29:465–70. https://doi.org/10.1002/da.21910.
94. Porter E, Chambless DL. A systematic review of predictors and moderators of improvement in cognitive-behavioral therapy for panic disorder and agoraphobia. Clin Psychol Rev. 2015;42:179–92. https://doi.org/10.1016/j.cpr.2015.09.004.
95. Imai H, Tajika A, Chen P, Pompoli A, Furukawa TA. Psychological therapies versus pharmacological interventions for panic disorder with or without agoraphobia in adults. England; 2016 Oct 12.
96. Apolinário-Hagen J. Internet-delivered psychological treatment options for panic disorder: a review on their efficacy and acceptability. Psych Invest. 2018; https://doi.org/10.30773/pi.2018.06.26.
97. Andersen P, Toner P, Bland M, McMillan D. Effectiveness of transdiagnostic cognitive behaviour therapy for anxiety and depression in adults: a systematic review and meta-analysis. Behav Cogn Psychother. 2016;44:673–90. https://doi.org/10.1017/S1352465816000229.
98. Arnberg FK, Linton SJ, Hultcrantz M, Heintz E, Jonsson U. Internet-delivered psychological treatments for mood and anxiety disorders: a systematic review of their efficacy, safety, and cost-effectiveness. PLoS One. 2014;9:e98118. https://doi.org/10.1371/journal.pone.0098118.
99. Newby JM, Twomey C, Yuan Li SS, Andrews G. Transdiagnostic computerised cognitive behavioural therapy for depression and anxiety: a systematic review and meta-analysis. J Affect Disord. 2016;199:30–41. https://doi.org/10.1016/j.jad.2016.03.018.
100. Newby JM, Mewton L, Andrews G. Transdiagnostic versus disorder-specific internet-delivered cognitive behaviour therapy for anxiety and depression in primary care. J Anxiety Disord. 2017;46:25–34. https://doi.org/10.1016/j.janxdis.2016.06.002.
101. Păsărelu CR, Andersson G, Bergman Nordgren L, Dobrean A. Internet-delivered transdiagnostic and tailored cognitive behavioral therapy for anxiety and depression: a systematic review and meta-analysis of randomized controlled trials. Cogn Behav Ther. 2017;46:1–28. https://doi.org/10.1080/16506073.2016.1231219.
102. Bergström J, Andersson G, Ljótsson B, Rück C, Andréewitch S, Karlsson A, et al. Internet-versus group-administered cognitive behaviour therapy for panic disorder in a psychiatric setting: a randomised trial. BMC Psychiatry. 2010;10:54. https://doi.org/10.1186/1471-244X-10-54.
103. Bergström J, Andersson G, Karlsson A, Andréewitch S, Rück C, Carlbring P, Lindefors N. An open study of the effectiveness of Internet treatment for panic disorder delivered in a psychiatric setting. Nord J Psychiatry. 2009;63:44–50. https://doi.org/10.1080/08039480802191132.
104. Bruinsma A, Kampman M, Exterkate CC, Hendriks GJ. Een verkennende studie naar blended cognitieve gedragstherapie voor paniek-stoornis: Resultaten en patiëntervaringen. Tijdschr Psychiatr. 2016;58:361–70.
105. Kiropoulos LA, Klein B, Austin DW, Gilson K, Pier C, Mitchell J, Ciechomski L. Is internet-based CBT for panic disorder and agoraphobia as effective as face-to-face CBT? J Anxiety Disord. 2008;22:1273–84. https://doi.org/10.1016/j.janxdis.2008.01.008.
106. Christoforou M, Sáez Fonseca JA, Tsakanikos E. Two novel cognitive behavioral therapy-based mobile apps for agoraphobia: randomized controlled trial. J Med Internet Res. 2017;19:e398. https://doi.org/10.2196/jmir.7747.
107. van Singer M, Chatton A, Khazaal Y. Quality of smartphone apps related to panic disorder. Front Psych. 2015;6:96. https://doi.org/10.3389/fpsyt.2015.00096.
108. Roth DA, Heimberg RG. Cognitive-behavioral models of social anxiety disorder. Psychiatr Clin North Am. 2001;24:753–71.
109. Hofmann SG. Cognitive factors that maintain social anxiety disorder: a comprehensive model and its treatment implications. Cogn Behav Ther. 2007;36:193–209. https://doi.org/10.1080/16506070701421313.

110. Hofmann SG. Treatment of social phobia: potential mediators and moderators. Clin Psychol NY. 2000;7:3–16.
111. Canton J, Scott KM, Glue P. Optimal treatment of social phobia: systematic review and meta-analysis. Neuropsychiatr Dis Treat. 2012;8:203–15. https://doi.org/10.2147/NDT.S23317.
112. Gould RA, Buckminster S, Pollack MH, Otto MW, Massachusetts LY. Cognitive-behavioral and pharmacological treatment for social phobia: a meta-analysis. Clin Psychol Sci Pract. 1997;4:291–306. https://doi.org/10.1111/j.1468-2850.1997.tb00123.x.
113. Mayo-Wilson E, Dias S, Mavranezouli I, Kew K, Clark DM, Ades AE, Pilling S. Psychological and pharmacological interventions for social anxiety disorder in adults: a systematic review and network meta-analysis. Lancet Psychiatry. 2014;1:368–76. https://doi.org/10.1016/S2215-0366(14)70329-3.
114. Acarturk C, Cuijpers P, van Straten A, de Graaf R. Psychological treatment of social anxiety disorder: a meta-analysis. Psychol Med. 2009;39:241–54. https://doi.org/10.1017/S0033291708003590.
115. Kampmann IL, Emmelkamp PMG, Morina N. Meta-analysis of technology-assisted interventions for social anxiety disorder. J Anxiety Disord. 2016;42:71–84. https://doi.org/10.1016/j.janxdis.2016.06.007.
116. Boettcher J, Carlbring P, Renneberg B, Berger T. Internet-based interventions for social anxiety disorder - an overview. Verhaltenstherapie. 2013;23:160–8. https://doi.org/10.1159/000354747.
117. Chesham RK, Malouff JM, Schutte NS. Meta-analysis of the efficacy of virtual reality exposure therapy for social anxiety. Behav Chang. 2018;35:152–66. https://doi.org/10.1017/bec.2018.15.
118. Alyami M, Giri B, Alyami H, Sundram F. Social anxiety apps: a systematic review and assessment of app descriptors across mobile store platforms. Evid Based Ment Health. 2017;20:65–70. https://doi.org/10.1136/eb-2017-102664.
119. Stolz T, Schulz A, Krieger T, Vincent A, Urech A, Moser C, et al. A mobile app for social anxiety disorder: A three-arm randomized controlled trial comparing mobile and PC-based guided self-help interventions. J Consult Clin Psychol. 2018;86:493–504. https://doi.org/10.1037/ccp0000301.
120. Hirsch CR, Mathews A. A cognitive model of pathological worry. Behav Res Ther. 2012;50:636–46. https://doi.org/10.1016/j.brat.2012.06.007.
121. Borkovec TD, Alcaine OM, Behar E. Avoidance theory of worry and generalized anxiety disorder. In: Generalized anxiety disorder: Advances in research and practice. New York: Guilford Press; 2004. p. 77–108.
122. Newman MG, Llera SJ. A novel theory of experiential avoidance in generalized anxiety disorder: a review and synthesis of research supporting a contrast avoidance model of worry. Clin Psychol Rev. 2011;31:371–82. https://doi.org/10.1016/j.cpr.2011.01.008.
123. Newman MG, Llera SJ, Erickson TM, Przeworski A, Castonguay LG. Worry and generalized anxiety disorder: a review and theoretical synthesis of evidence on nature, etiology, mechanisms, and treatment. Annu Rev Clin Psychol. 2013;9:275–97. https://doi.org/10.1146/annurev-clinpsy-050212-185544.
124. Behar E, DiMarco ID, Hekler EB, Mohlman J, Staples AM. Current theoretical models of generalized anxiety disorder (GAD): conceptual review and treatment implications. J Anxiety Disord. 2009;23:1011–23. https://doi.org/10.1016/j.janxdis.2009.07.006.
125. Buhr K, Dugas MJ. The intolerance of uncertainty scale: psychometric properties of the English version. Behav Res Ther. 2002;40:931–45.
126. Cuijpers P, Sijbrandij M, Koole S, Huibers M, Berking M, Andersson G. Psychological treatment of generalized anxiety disorder: a meta-analysis. Clin Psychol Rev. 2014;34:130–40. https://doi.org/10.1016/j.cpr.2014.01.002.
127. Hall J, Kellett S, Berrios R, Bains MK, Scott S. Efficacy of cognitive behavioral therapy for generalized anxiety disorder in older adults: systematic review, meta-analysis, and meta-regression. Am J Geriatr Psychiatry. 2016;24:1063–73. https://doi.org/10.1016/j.jagp.2016.06.006.

128. Titov N, Andersson G, Paxling B. ICBT in psychiatry: generalised anxiety disorder. In: Lindefors N, Andersson G, editors. Guided internet-based treatments in psychiatry. Cham: Springer; 2016. p. 79–100. https://doi.org/10.1007/978-3-319-06083-5_5.
129. Mewton L, Wong N, Andrews G. The effectiveness of internet cognitive behavioural therapy for generalized anxiety disorder in clinical practice. Depress Anxiety. 2012;29:843–9. https://doi.org/10.1002/da.21995.
130. Jones SL, Hadjistavropoulos HD, Soucy JN. A randomized controlled trial of guided internet-delivered cognitive behaviour therapy for older adults with generalized anxiety. J Anxiety Disord. 2016;37:1–9. https://doi.org/10.1016/j.janxdis.2015.10.006.
131. Allen NB, Chambers R, Knight W. Mindfulness-based psychotherapies: a review of conceptual foundations, empirical evidence and practical considerations. Aust N Z J Psychiatry. 2006;40:285–94. https://doi.org/10.1080/j.1440-1614.2006.01794.x.
132. Bishop SR, Lau M, Shapiro S, Carlson L, Anderson ND, Carmody J, et al. Mindfulness: a proposed operational definition. Clin Psychol Sci Pract. 2004;11:230–41. https://doi.org/10.1093/clipsy.bph077.
133. Teasdale JD, Segal Z, Williams JM. How does cognitive therapy prevent depressive relapse and why should attentional control (mindfulness) training help? Behav Res Ther. 1995;33:25–39.
134. Gaynor K. A critical review of mindfulness-based psychological treatments for worry and rumination. OA Behav Med. 2014;2
135. Shonin E, van Gordon W, Griffiths MD. Mindfulness-based interventions: towards mindful clinical integration. Front Psychol. 2013; https://doi.org/10.3389/fpsyg.2013.00194.
136. Shapiro SL, Carlson LE, Astin JA, Freedman B. Mechanisms of mindfulness. J Clin Psychol. 2006;62:373–86. https://doi.org/10.1002/jclp.20237.
137. Khusid MA, Vythilingam M. The emerging role of mindfulness meditation as effective self-management strategy, Part 1: Clinical implications for depression, post-traumatic stress disorder, and anxiety. Mil Med. 2016;181:961–8. https://doi.org/10.7205/MILMED-D-14-00677.
138. Michalak J, Heidenreich T, Williams JMG. Achtsamkeit: Fortschritte der Psychotherapie 48. 1st ed. Hogrefe Verlag: Göttingen; 2012.
139. Blanck P, Perleth S, Heidenreich T, Kröger P, Ditzen B, Bents H, Mander J. Effects of mindfulness exercises as stand-alone intervention on symptoms of anxiety and depression: Systematic review and meta-analysis. Behav Res Ther. 2018;102:25–35. https://doi.org/10.1016/j.brat.2017.12.002.
140. Treanor M. The potential impact of mindfulness on exposure and extinction learning in anxiety disorders. Clin Psychol Rev. 2011;31:617–25. https://doi.org/10.1016/j.cpr.2011.02.003.
141. Brake CA, Sauer-Zavala S, Boswell JF, Gallagher MW, Farchione TJ, Barlow DH. Mindfulness-based exposure strategies as a transdiagnostic mechanism of change: an exploratory alternating treatment design. Behav Ther. 2016;47:225–38. https://doi.org/10.1016/j.beth.2015.10.008.
142. Creswell JD. Mindfulness interventions. Annu Rev Psychol. 2017;68:491–516. https://doi.org/10.1146/annurev-psych-042716-051139.
143. Dobkin PL, Irving JA, Amar S. For whom may participation in a mindfulness-based stress reduction program be contraindicated? Mindfulness. 2012;3:44–50. https://doi.org/10.1007/s12671-011-0079-9.
144. Farias M, Wikholm C. Has the science of mindfulness lost its mind? BJPsych Bull. 2016;40:329–32. https://doi.org/10.1192/pb.bp.116.053686.
145. Xie J-F, Zhou J-D, Gong L-N, Iennaco JD, Ding S-Q. Mindfulness-based cognitive therapy in the intervention of psychiatric disorders: a review. Int J Nurs Sci. 2014;1:232–9. https://doi.org/10.1016/j.ijnss.2014.05.015.
146. MacKenzie MB, Kocovski NL. Mindfulness-based cognitive therapy for depression: trends and developments. Psychol Res Behav Manag. 2016;9:125–32. https://doi.org/10.2147/PRBM.S63949.
147. Sipe WEB, Eisendrath SJ. Mindfulness-based cognitive therapy: theory and practice. Can J Psychiatr. 2012;57:63–9. https://doi.org/10.1177/070674371205700202.

148. Parsons CE, Crane C, Parsons LJ, Fjorback LO, Kuyken W. Home practice in mindfulness-based cognitive therapy and mindfulness-based stress reduction: a systematic review and meta-analysis of participants' mindfulness practice and its association with outcomes. Behav Res Ther. 2017;95:29–41. https://doi.org/10.1016/j.brat.2017.05.004.
149. Alsubaie M, Abbott R, Dunn B, Dickens C, Keil TF, Henley W, Kuyken W. Mechanisms of action in mindfulness-based cognitive therapy (MBCT) and mindfulness-based stress reduction (MBSR) in people with physical and/or psychological conditions: a systematic review. Clin Psychol Rev. 2017;55:74–91. https://doi.org/10.1016/j.cpr.2017.04.008.
150. van der Velden AM, Kuyken W, Wattar U, Crane C, Pallesen KJ, Dahlgaard J, et al. A systematic review of mechanisms of change in mindfulness-based cognitive therapy in the treatment of recurrent major depressive disorder. Clin Psychol Rev. 2015;37:26–39. https://doi.org/10.1016/j.cpr.2015.02.001.
151. Gu J, Strauss C, Bond R, Cavanagh K. How do mindfulness-based cognitive therapy and mindfulness-based stress reduction improve mental health and wellbeing? A systematic review and meta-analysis of mediation studies. Clin Psychol Rev. 2015;37:1–12. https://doi.org/10.1016/j.cpr.2015.01.006.
152. Desrosiers A, Vine V, Klemanski DH, Nolen-Hoeksema S. Mindfulness and emotion regulation in depression and anxiety: common and distinct mechanisms of action. Depress Anxiety. 2013;30:654–61. https://doi.org/10.1002/da.22124.
153. Sado M, Park S, Ninomiya A, Sato Y, Fujisawa D, Shirahase J, Mimura M. Feasibility study of mindfulness-based cognitive therapy for anxiety disorders in a Japanese setting. BMC Res Notes. 2018;11:653. https://doi.org/10.1186/s13104-018-3744-4.
154. Evans S. Mindfulness-based cognitive therapy for generalized anxiety disorder. In: Eisendrath SJ, editor. Mindfulness-based cognitive therapy: Innovative applications. Cham: Springer; 2016. p. 145–54. https://doi.org/10.1007/978-3-319-29866-5_13.
155. Strege MV, Swain D, Bochicchio L, Valdespino A, Richey JA. A pilot study of the effects of mindfulness-based cognitive therapy on positive affect and social anxiety symptoms. Front Psychol. 2018;9:866. https://doi.org/10.3389/fpsyg.2018.00866.
156. Kim B, Lee S-H, Kim YW, Choi TK, Yook K, Suh SY, et al. Effectiveness of a mindfulness-based cognitive therapy program as an adjunct to pharmacotherapy in patients with panic disorder. J Anxiety Disord. 2010;24:590–5. https://doi.org/10.1016/j.janxdis.2010.03.019.
157. Goyal M, Singh S, Sibinga EMS, Gould NF, Rowland-Seymour A, Sharma R, et al. Meditation programs for psychological stress and well-being: a systematic review and meta-analysis. JAMA Intern Med. 2014;174:357–68. https://doi.org/10.1001/jamainternmed.2013.13018.
158. Strauss C, Cavanagh K, Oliver A, Pettman D. Mindfulness-based interventions for people diagnosed with a current episode of an anxiety or depressive disorder: a meta-analysis of randomised controlled trials. PLoS One. 2014;9:e96110. https://doi.org/10.1371/journal.pone.0096110.
159. Vøllestad J, Nielsen MB, Nielsen GH. Mindfulness- and acceptance-based interventions for anxiety disorders: a systematic review and meta-analysis. Br J Clin Psychol. 2012;51:239–60. https://doi.org/10.1111/j.2044-8260.2011.02024.x.
160. Newby JM, Mewton L, Williams AD, Andrews G. Effectiveness of transdiagnostic Internet cognitive behavioural treatment for mixed anxiety and depression in primary care. J Affect Disord. 2014;165:45–52. https://doi.org/10.1016/j.jad.2014.04.037.
161. Chiesa A, Serretti A. Mindfulness based cognitive therapy for psychiatric disorders: a systematic review and meta-analysis. Psychiatry Res. 2011;187:441–53. https://doi.org/10.1016/j.psychres.2010.08.011.
162. Fjorback LO, Arendt M, Ornbøl E, Fink P, Walach H. Mindfulness-based stress reduction and mindfulness-based cognitive therapy: a systematic review of randomized controlled trials. Acta Psychiatr Scand. 2011;124:102–19. https://doi.org/10.1111/j.1600-0447.2011.01704.x.
163. Rodrigues MF, Nardi AE, Levitan M. Mindfulness in mood and anxiety disorders: a review of the literature. Trends Psychiatry Psychother. 2017;39:207–15. https://doi.org/10.1590/2237-6089-2016-0051.

164. Galante J, Iribarren SJ, Pearce PF. Effects of mindfulness-based cognitive therapy on mental disorders: a systematic review and meta-analysis of randomised controlled trials. J Res Nurs. 2013;18:133–55. https://doi.org/10.1177/1744987112466087.
165. Marchand WR. Mindfulness-based stress reduction, mindfulness-based cognitive therapy, and Zen meditation for depression, anxiety, pain, and psychological distress. J Psychiatr Pract. 2012;18:233–52. https://doi.org/10.1097/01.pra.0000416014.53215.86.
166. Marchand WR. Mindfulness meditation practices as adjunctive treatments for psychiatric disorders. Psychiatr Clin North Am. 2013;36:141–52. https://doi.org/10.1016/j.psc.2013.01.002.
167. Zhang M-F, Wen Y-S, Liu W-Y, Peng L-F, Wu X-D, Liu Q-W. Effectiveness of mindfulness-based therapy for reducing anxiety and depression in patients with cancer: a meta-analysis. Medicine (Baltimore). 2015;94:e0897-0. https://doi.org/10.1097/MD.0000000000000897.
168. Abbott RA, Whear R, Rodgers LR, Bethel A, Thompson Coon J, Kuyken W, et al. Effectiveness of mindfulness-based stress reduction and mindfulness based cognitive therapy in vascular disease: A systematic review and meta-analysis of randomised controlled trials. J Psychosom Res. 2014;76:341–51. https://doi.org/10.1016/j.jpsychores.2014.02.012.
169. Idusohan-Moizer H, Sawicka A, Dendle J, Albany M. Mindfulness-based cognitive therapy for adults with intellectual disabilities: an evaluation of the effectiveness of mindfulness in reducing symptoms of depression and anxiety. J Intellect Disabil Res. 2015;59:93–104. https://doi.org/10.1111/jir.12082.
170. Helmes E, Ward BG. Mindfulness-based cognitive therapy for anxiety symptoms in older adults in residential care. Aging Ment Health. 2017;21:272–8. https://doi.org/10.1080/13607863.2015.1111862.
171. Kishita N, Takei Y, Stewart I. A meta-analysis of third wave mindfulness-based cognitive behavioral therapies for older people. Int J Geriatr Psychiatry. 2017;32:1352–61. https://doi.org/10.1002/gps.4621.
172. Yook K, Lee S-H, Ryu M, Kim K-H, Choi TK, Suh SY, et al. Usefulness of mindfulness-based cognitive therapy for treating insomnia in patients with anxiety disorders: a pilot study. J Nerv Ment Dis. 2008;196:501–3. https://doi.org/10.1097/NMD.0b013e31817762ac.
173. Shi Z, MacBeth A. The effectiveness of mindfulness-based interventions on maternal perinatal mental health outcomes: a systematic review. Mindfulness. 2017;8:823–47. https://doi.org/10.1007/s12671-016-0673-y.
174. Lever Taylor B, Cavanagh K, Strauss C. The effectiveness of mindfulness-based interventions in the perinatal period: a systematic review and meta-analysis. PLoS One. 2016;11:e0155720. https://doi.org/10.1371/journal.pone.0155720.
175. Shulman B, Dueck R, Ryan D, Breau G, Sadowski I, Misri S. Feasibility of a mindfulness-based cognitive therapy group intervention as an adjunctive treatment for postpartum depression and anxiety. J Affect Disord. 2018;235:61–7. https://doi.org/10.1016/j.jad.2017.12.065.
176. Williams MJ, McManus F, Muse K, Williams JMG. Mindfulness-based cognitive therapy for severe health anxiety (hypochondriasis): an interpretative phenomenological analysis of patients' experiences. Br J Clin Psychol. 2011;50:379–97. https://doi.org/10.1111/j.2044-8260.2010.02000.x.
177. Surawy C, McManus F, Muse K, Williams JMG. Mindfulness-Based Cognitive Therapy (MBCT) for health anxiety (Hypochondriasis): rationale, implementation and case illustration. Mindfulness. 2015;6:382–92. https://doi.org/10.1007/s12671-013-0271-1.
178. McManus F, Surawy C, Muse K, Vazquez-Montes M, Williams JMG. A randomized clinical trial of mindfulness-based cognitive therapy versus unrestricted services for health anxiety (hypochondriasis). J Consult Clin Psychol. 2012;80:817–28. https://doi.org/10.1037/a0028782.
179. Lovas DA, Barsky AJ. Mindfulness-based cognitive therapy for hypochondriasis, or severe health anxiety: a pilot study. J Anxiety Disord. 2010;24:931–5. https://doi.org/10.1016/j.janxdis.2010.06.019.
180. Clark GI, Rock AJ. Processes contributing to the maintenance of flying phobia: a narrative review. Front Psychol. 2016; https://doi.org/10.3389/fpsyg.2016.00754.

181. Arch JJ, Craske MG. Laboratory stressors in clinically anxious and non-anxious individuals: the moderating role of mindfulness. Behav Res Ther. 2010;48:495–505. https://doi.org/10.1016/j.brat.2010.02.005.
182. Hooper N, Davies N, Davies L, McHugh L. Comparing thought suppression and mindfulness as coping techniques for spider fear. Conscious Cogn. 2011;20:1824–30. https://doi.org/10.1016/j.concog.2011.05.013.
183. Carleton RN, Fetzner MG, Hackl JL, McEvoy P. Intolerance of uncertainty as a contributor to fear and avoidance symptoms of panic attacks. Cogn Behav Ther. 2013;42:328–41. https://doi.org/10.1080/16506073.2013.792100.
184. Kim MK, Lee KS, Kim B, Choi TK, Lee S-H. Impact of mindfulness-based cognitive therapy on intolerance of uncertainty in patients with panic disorder. Psychiatry Investig. 2016;13:196–202. https://doi.org/10.4306/pi.2016.13.2.196.
185. Levitt JT, Karekla M. Integrating acceptance and mindfulness with cognitive behavioral treatment for panic disorder. In: Orsillo SM, Roemer L, editors. Acceptance and mindfulness-based approaches to anxiety: conceptualization and treatment. Boston, MA: Springer; 2005. p. 165–88. https://doi.org/10.1007/0-387-25989-9_7.
186. Kim B, Cho SJ, Lee KS, Lee J-Y, Choe AY, Lee JE, et al. Factors associated with treatment outcomes in mindfulness-based cognitive therapy for panic disorder. Yonsei Med J. 2013;54:1454–62. https://doi.org/10.3349/ymj.2013.54.6.1454.
187. Kim YW, Lee S-H, Choi TK, Suh SY, Kim B, Kim CM, et al. Effectiveness of mindfulness-based cognitive therapy as an adjuvant to pharmacotherapy in patients with panic disorder or generalized anxiety disorder. Depress Anxiety. 2009;26:601–6. https://doi.org/10.1002/da.20552.
188. Goldin PR, Morrison A, Jazaieri H, Brozovich F, Heimberg R, Gross JJ. Group CBT versus MBSR for social anxiety disorder: a randomized controlled trial. J Consult Clin Psychol. 2016;84:427–37. https://doi.org/10.1037/ccp0000092.
189. Kocovski NL, Fleming JE, Hawley LL, Ho M-HR, Antony MM. Mindfulness and acceptance-based group therapy and traditional cognitive behavioral group therapy for social anxiety disorder: Mechanisms of change. Behav Res Ther. 2015;70:11–22. https://doi.org/10.1016/j.brat.2015.04.005.
190. Koszycki D, Benger M, Shlik J, Bradwejn J. Randomized trial of a meditation-based stress reduction program and cognitive behavior therapy in generalized social anxiety disorder. Behav Res Ther. 2007;45:2518–26. https://doi.org/10.1016/j.brat.2007.04.011.
191. Hjeltnes A, Molde H, Schanche E, Vøllestad J, Lillebostad Svendsen J, Moltu C, Binder P-E. An open trial of mindfulness-based stress reduction for young adults with social anxiety disorder. Scand J Psychol. 2017;58:80–90. https://doi.org/10.1111/sjop.12342.
192. Piet J, Hougaard E, Hecksher MS, Rosenberg NK. A randomized pilot study of mindfulness-based cognitive therapy and group cognitive-behavioral therapy for young adults with social phobia. Scand J Psychol. 2010;51:403–10. https://doi.org/10.1111/j.1467-9450.2009.00801.x.
193. Bögels SM, Sijbers GFVM, Voncken M. Mindfulness and task concentration training for social phobia: a pilot study. J Cogn Psychother. 2006;20:33–44. https://doi.org/10.1891/jcop.20.1.33.
194. Koszycki D, Thake J, Mavounza C, Daoust J-P, Taljaard M, Bradwejn J. Preliminary investigation of a mindfulness-based intervention for social anxiety disorder that integrates compassion meditation and mindful exposure. J Altern Complement Med. 2016;22:363–74. https://doi.org/10.1089/acm.2015.0108.
195. Kocovski NL, Fleming JE, Hawley LL, Huta V, Antony MM. Mindfulness and acceptance-based group therapy versus traditional cognitive behavioral group therapy for social anxiety disorder: a randomized controlled trial. Behav Res Ther. 2013;51:889–98. https://doi.org/10.1016/j.brat.2013.10.007.
196. Norton AR, Abbott MJ, Norberg MM, Hunt C. A systematic review of mindfulness and acceptance-based treatments for social anxiety disorder. J Clin Psychol. 2015;71:283–301. https://doi.org/10.1002/jclp.22144.

197. The National Institute for Health and Care Excellence (NICE). Surveillance report 2017 - Social anxiety disorder: recognition, assessment and treatment (2013) NICE guideline CG159. 2017. https://www.nice.org.uk/guidance/cg159/resources/surveillance-report-2017-social-anxiety-disorder-recognition-assessment-and-treatment-2013-nice-guideline-cg159-pdf-5812324559557. Accessed 28 Dec 2018.
198. Segal ZV, Williams JMG, Teasdale JD. Mindfulness-based cognitive therapy for depression: A new approach to preventing relapse. New York: Guilford Press; 2002.
199. Hoge EA, Bui E, Goetter E, Robinaugh DJ, Ojserkis RA, Fresco DM, Simon NM. Change in decentering mediates improvement in anxiety in mindfulness-based stress reduction for generalized anxiety disorder. Cogn Ther Res. 2014;39:228–35. https://doi.org/10.1007/s10608-014-9646-4.
200. Craigie MA, Rees CS, Marsh A, Nathan P. Mindfulness-based cognitive therapy for generalized anxiety disorder: a preliminary evaluation. Behav Cogn Psychother. 2008;36:553. https://doi.org/10.1017/S135246580800458X.
201. Evans S, Ferrando S, Findler M, Stowell C, Smart C, Haglin D. Mindfulness-based cognitive therapy for generalized anxiety disorder. J Anxiety Disord. 2008;22:716–21. https://doi.org/10.1016/j.janxdis.2007.07.005.
202. Goodman JH, Guarino A, Chenausky K, Klein L, Prager J, Petersen R, et al. CALM pregnancy: results of a pilot study of mindfulness-based cognitive therapy for perinatal anxiety. Arch Womens Ment Health. 2014;17:373–87. https://doi.org/10.1007/s00737-013-0402-7.
203. Wong SYS, Yip BHK, Mak WWS, Mercer S, Cheung EYL, CYM L, et al. Mindfulness-based cognitive therapy v group psychoeducation for people with generalised anxiety disorder: randomised controlled trial. Br J Psychiatry. 2016;209:68–75. https://doi.org/10.1192/bjp.bp.115.166124.
204. Kvillemo P, Brandberg Y, Bränström R. Feasibility and outcomes of an internet-based mindfulness training program: a pilot randomized controlled trial. JMIR Ment Health. 2016;3:e33. https://doi.org/10.2196/mental.5457.
205. Wahbeh H, Svalina MN, Oken BS. Group, one-on-one, or internet? Preferences for mindfulness meditation delivery format and their predictors. Open Med J. 2014;1:66–74. https://doi.org/10.2174/1874220301401010066.
206. Mak WWS, Chan ATY, Cheung EYL, Lin CLY, Ngai KCS. Enhancing web-based mindfulness training for mental health promotion with the health action process approach: randomized controlled trial. J Med Internet Res. 2015;17:e8. https://doi.org/10.2196/jmir.3746.
207. Jayewardene WP, Lohrmann DK, Erbe RG, Torabi MR. Effects of preventive online mindfulness interventions on stress and mindfulness: a meta-analysis of randomized controlled trials. Prev Med Rep. 2017;5:150–9. https://doi.org/10.1016/j.pmedr.2016.11.013.
208. Mikolasek M, Berg J, Witt CM, Barth J. Effectiveness of mindfulness- and relaxation-based eHealth interventions for patients with medical conditions: a systematic review and synthesis. Int J Behav Med. 2018;25:1–16. https://doi.org/10.1007/s12529-017-9679-7.
209. Cavanagh K, Strauss C, Cicconi F, Griffiths N, Wyper A, Jones F. A randomised controlled trial of a brief online mindfulness-based intervention. Behav Res Ther. 2013;51:573–8. https://doi.org/10.1016/j.brat.2013.06.003.
210. Apolinário-Hagen J, Salewski C. Internet-based cognitive behavioral therapy and online mindfulness training for panic disorder. In: Kim Y-K, editor. Panic disorder: assessment, management and research insights. New York: NOVA SCIENCE; 2018. p. 221–46.
211. O'Connor M, Munnelly A, Whelan R, McHugh L. The efficacy and acceptability of third-wave behavioral and cognitive eHealth treatments: a systematic review and meta-analysis of randomized controlled trials. Behav Ther. 2018;49:459–75. https://doi.org/10.1016/j.beth.2017.07.007.
212. Fish J, Brimson J, Lynch S. Mindfulness interventions delivered by technology without facilitator involvement: what research exists and what are the clinical outcomes? Mindfulness. 2016;7:1011–23. https://doi.org/10.1007/s12671-016-0548-2.

213. Boettcher J, Aström V, Påhlsson D, Schenström O, Andersson G, Carlbring P. Internet-based mindfulness treatment for anxiety disorders: a randomized controlled trial. Behav Ther. 2014;45:241–53. https://doi.org/10.1016/j.beth.2013.11.003.
214. Krusche A, Cyhlarova E, Williams JMG. Mindfulness online: an evaluation of the feasibility of a web-based mindfulness course for stress, anxiety and depression. BMJ Open. 2013;3:e003498. https://doi.org/10.1136/bmjopen-2013-003498.
215. Plaza I, Demarzo MMP, Herrera-Mercadal P, García-Campayo J. Mindfulness-based mobile applications: literature review and analysis of current features. JMIR Mhealth Uhealth. 2013;1:e24. https://doi.org/10.2196/mhealth.2733.
216. Mani M, Kavanagh DJ, Hides L, Stoyanov SR. Review and evaluation of mindfulness-based iPhone apps. JMIR Mhealth Uhealth. 2015;3:e82. https://doi.org/10.2196/mhealth.4328.
217. Mak WW, Tong AC, Yip SY, Lui WW, Chio FH, Chan AT, Wong CC. Efficacy and moderation of mobile app-based programs for mindfulness-based training, self-compassion training, and cognitive behavioral psychoeducation on mental health: randomized controlled noninferiority trial. JMIR Ment Health. 2018;5:e60. https://doi.org/10.2196/mental.8597.
218. Hayes SC, editor. Relational frame theory: a post-Skinnerian account of human language and cognition. New York: Kluwer; 2001.
219. Hofmann SG, Asmundson GJG. Acceptance and mindfulness-based therapy: new wave or old hat? Clin Psychol Rev. 2008;28:1–16. https://doi.org/10.1016/j.cpr.2007.09.003.
220. Levin ME, Haeger J, Pierce B, Cruz RA. Evaluating an adjunctive mobile app to enhance psychological flexibility in acceptance and commitment therapy. Behav Modif. 2017;41:846–67. https://doi.org/10.1177/0145445517719661.
221. Brown M, Glendenning A, Hoon AE, John A. Effectiveness of web-delivered acceptance and commitment therapy in relation to mental health and well-being: a systematic review and meta-analysis. J Med Internet Res. 2016;18:e221. https://doi.org/10.2196/jmir.6200.
222. Eifert GH. Akzeptanz- und Commitment-Therapie (ACT). Göttingen: Hogrefe; 2011.
223. Gross JJ. Antecedent- and response-focused emotion regulation: divergent consequences for experience, expression, and physiology. J Pers Soc Psychol. 1998;74:224–37.
224. Eifert GH, Forsyth JP, Arch J, Espejo E, Keller M, Langer D. Acceptance and commitment therapy for anxiety disorders: three case studies exemplifying a unified treatment protocol. Cogn Behav Pract. 2009;16:368–85. https://doi.org/10.1016/j.cbpra.2009.06.001.
225. Eifert GH, Forsyth JP. Acceptance & commitment therapy for anxiety disorders: a practitioner's treatment guide to using mindfulness, acceptance, and values-based behavior change strategies. Oakland, CA: New Harbinger Publications; 2005.
226. Craske MG, Niles AN, Burklund LJ, Wolitzky-Taylor KB, Vilardaga JCP, Arch JJ, et al. Randomized controlled trial of cognitive behavioral therapy and acceptance and commitment therapy for social phobia: outcomes and moderators. J Consult Clin Psychol. 2014;82:1034–48. https://doi.org/10.1037/a0037212.
227. A-Tjak JGL, Davis ML, Morina N, Powers MB, Smits JAJ, Emmelkamp PMG. A meta-analysis of the efficacy of acceptance and commitment therapy for clinically relevant mental and physical health problems. Psychother Psychosom. 2015;84:30–6. https://doi.org/10.1159/000365764.
228. Smits JAJ, Hofmann SG. A meta-analytic review of the effects of psychotherapy control conditions for anxiety disorders. Psychol Med. 2009;39:229–39. https://doi.org/10.1017/S0033291708003498.
229. Zettle RD. Acceptance and Commitment Therapy (ACT) vs. systematic desensitization in treatment of mathematics anxiety. Psychol Rec. 2003;53:197–215. https://doi.org/10.1007/BF03395440.
230. Gloster AT, Sonntag R, Hoyer J, Meyer AH, Heinze S, Ströhle A, et al. Treating treatment-resistant patients with panic disorder and agoraphobia using psychotherapy: a randomized controlled switching trial. Psychother Psychosom. 2015;84:100–9. https://doi.org/10.1159/000370162.

231. Meuret AE, Twohig MP, Rosenfield D, Hayes SC, Craske MG. Brief acceptance and commitment therapy and exposure for panic disorder: a pilot study. Cogn Behav Pract. 2012;19:606–18. https://doi.org/10.1016/j.cbpra.2012.05.004.
232. Dalrymple KL, Herbert JD. Acceptance and commitment therapy for generalized social anxiety disorder: a pilot study. Behav Modif. 2007;31:543–68. https://doi.org/10.1177/0145445507302037.
233. Herbert JD, Forman EM, Kaye JL, Gershkovich M, Goetter E, Yuen EK, et al. Randomized controlled trial of acceptance and commitment therapy versus traditional cognitive behavior therapy for social anxiety disorder: Symptomatic and behavioral outcomes. J Contextual Behav Sci. 2018;9:88–96. https://doi.org/10.1016/j.jcbs.2018.07.008.
234. Ivanova E, Lindner P, Ly KH, Dahlin M, Vernmark K, Andersson G, Carlbring P. Guided and unguided Acceptance and Commitment Therapy for social anxiety disorder and/or panic disorder provided via the Internet and a smartphone application: A randomized controlled trial. J Anxiety Disord. 2016;44:27–35. https://doi.org/10.1016/j.janxdis.2016.09.012.
235. Gershkovich M, Herbert JD, Forman EM, Schumacher LM, Fischer LE. Internet-delivered acceptance-based cognitive-behavioral intervention for social anxiety disorder With and without therapist support: a randomized trial. Behav Modif. 2017;41:583–608. https://doi.org/10.1177/0145445517694457.
236. Hasheminasab M, Babapour Kheiroddin J, Mahmood Aliloo M, Fakhari A. Acceptance and Commitment Therapy (ACT) for generalized anxiety disorder. Iran J Public Health. 2015;44:718–9.
237. Avdagic E, Morrissey SA, Boschen MJ. A randomised controlled trial of acceptance and commitment therapy and cognitive-behaviour therapy for generalised anxiety disorder. Behav Chang. 2014;31:110–30. https://doi.org/10.1017/bec.2014.5.
238. Dahlin M, Andersson G, Magnusson K, Johansson T, Sjögren J, Håkansson A, et al. Internet-delivered acceptance-based behaviour therapy for generalized anxiety disorder: a randomized controlled trial. Behav Res Ther. 2016;77:86–95. https://doi.org/10.1016/j.brat.2015.12.007.
239. Chambless DL, Gillis MM. Cognitive therapy of anxiety disorders. J Consult Clin Psychol. 1993;61:248–60.
240. Cuijpers P, Cristea IA, Karyotaki E, Reijnders M, Huibers MJH. How effective are cognitive behavior therapies for major depression and anxiety disorders? A meta-analytic update of the evidence. World Psychiatry. 2016;15:245–58. https://doi.org/10.1002/wps.20346.

18 Neurostimulation in Anxiety Disorders, Post-traumatic Stress Disorder, and Obsessive-Compulsive Disorder

Rafael Christophe Freire, Casimiro Cabrera-Abreu, and Roumen Milev

Abbreviation

ALIC	Anterior limb of the internal capsule
BOLD	Blood-oxygen-level-dependent
BST	Bed nucleus of the stria terminalis
CANMAT	Canadian Network for Mood and Anxiety Treatments
CAPS	Clinician-administered post-traumatic stress disorder scale
CES	Cranial electrotherapy stimulation
DBS	Deep brain stimulation
DLPFC	Dorsolateral prefrontal cortex
ECT	Electroconvulsive therapy
GAD	Generalized anxiety disorder
HAM-A	Hamilton Anxiety Rating Scale
IPG	Implantable pulse generator
iTBS	Intermittent theta burst stimulation

Parts of the chapter were previously published in *Understanding Depression Volume II*, pp. 277–288, Springer Nature, 2018.

R. C. Freire (✉)
Institute of Psychiatry, Federal University of Rio de Janeiro, Rio de Janeiro, Brazil

Department of Psychiatry, Queen's University, Kingston, ON, Canada

C. Cabrera-Abreu
Department of Psychiatry, Queen's University and Providence Care Hospital, Kingston, ON, Canada
e-mail: cabrerac@providencecare.ca

R. Milev
Department of Psychiatry, Queen's University, Kingston, ON, Canada

Centre for Neuroscience Studies, Queen's University, Kingston, ON, Canada
e-mail: roumen.milev@queensu.ca

Y.-K. Kim (ed.), *Anxiety Disorders*, Advances in Experimental Medicine and Biology 1191, https://doi.org/10.1007/978-981-32-9705-0_18

ITP	Inferior thalamic peduncle
MDD	Major depressive disorder
MFB	Medial forebrain bundle
MPFC	Medial prefrontal cortex
MRI	Magnetic resonance imaging
NAc	Nucleus accumbens
NS	Neurostimulator.
NTS	Nucleus tractus solitarius
OCD	Obsessive-compulsive disorder
OFC	Orbitofrontal cortex
PD	Panic disorder
PDSS	Panic Disorder Severity Scale
PTSD	Post-traumatic stress disorder
REAC	Radio electric asymmetric conveyor
RMT	Resting motor threshold
rTMS	Repetitive transcranial magnetic stimulation
SAD	Social anxiety disorder
SCC	Subcallosal cingulate
SMA	Supplementary motor area
STN	Subthalamic nucleus
tDCS	Transcranial direct current stimulation
TNS	External Trigeminal Nerve Stimulation
VC/VS	Ventral internal capsule/ventral striatum
VNS	Vagus nerve stimulation
Y-BOCS	Yale-Brown Obsessive Compulsive Scale

Introduction

In the treatment of panic disorder (PD), generalized anxiety disorder (GAD), social anxiety disorder (SAD), post-traumatic stress disorder (PTSD), and obsessive-compulsive disorder (OCD), several medications demonstrated effectiveness, including selective serotonin reuptake inhibitors, serotonin-norepinephrine reuptake inhibitors, tricyclic antidepressants, and benzodiazepines. However, many patients do not achieve remission with pharmacological treatments. Psychotherapy is also effective for most anxiety disorder patients, but issues with availability, cost, and commitment limit its use. Even more, a significant portion of patients has little or no response to this form of treatment. In addition, the relapse rate is high among those who do achieve remission. From a clinical perspective, there is still a strong unmet need for effective, fast-acting, and tolerable treatments for anxiety disorders, PTSD, and OCD. Neurostimulation techniques have been studied as promising alternatives or augmentation treatments to pharmacological and psychological therapies [1].

The administration of electrical charges to the brain as an attempt to treat neurological and mental disorders has been studied since the classical antiquity, but the major turning point was the invention of electroconvulsive therapy (ECT) in the 1930s [2]. Despite being still considered an invaluable treatment option for depressive disorders [3], due to its side effects, high cost, and anesthesia-related risks, ECT has been scarcely studied in the treatment of anxiety disorders. Repetitive transcranial magnetic stimulation (rTMS) is the most studied neurostimulation technique in the treatment of anxiety disorders, PTSD, and OCD, but direct electric (nonmagnetic) neuromodulation techniques have also been studied in the treatment of these disorders. Electrodes may be placed directly into the brain (deep brain stimulation, DBS), on the scalp (ECT; transcranial direct current stimulation, tDCS), or on peripheral nerves (vagus nerve stimulation, VNS; External Trigeminal Nerve Stimulation, TNS) [3].

Cortical activation patterns can be selectively modified by means of rTMS via electromagnetic induction [4]. Transcranial magnetic stimulation uses powerful but extremely brief magnetic fields, which induce a current in the brain. It has been applied for evaluation of the motor system, functional research of cerebral regions, and pathophysiological mechanisms of mental disorders. Administration of magnetic stimuli in a rhythmic and repetitive form became an intervention tool [5]. In rTMS, trains of magnetic pulses temporarily summate to cause greater changes in neural activity than a single pulse, which can modulate cortical excitability. While high-frequency rTMS is considered to increase the cortical excitability in certain regions, low-frequency rTMS is postulated to [5] inhibit the cortical excitability of stimulated area. This way, rTMS demonstrated to modulate neurotransmitter release and – depending on its stimulation frequency – normalize prefrontal hypoactivity [4].

DBS is a neurostimulation technique that involves stereotactic neurosurgical implantation of approximately 1.3 mm diameter platinum-iridium electrodes bilaterally into a brain target. Multiple stimulating contacts at the electrode tip allow precise targeting of subregions within a target brain structure. Wires connect the stimulating electrodes to a pulse generator implanted subcutaneously below the clavicle, and a handheld wireless programmer allows ongoing noninvasive adjustment of stimulus parameters. DBS is the most invasive neuromodulation strategy, and thus patients enrolled in studies are severely ill and treatment refractory [6].

The therapeutic effect of tDCS is under investigation but has an advantage of increasing or decreasing the activity of disorder-related cortical areas. It is possible to produce excitatory or inhibitory effects depending on whether the stimulating electrode is an anode or a cathode. tDCS has a favorable safety profile, is portable, has low cost, and is easy to use, thus showing clinical and practical advantages over the other brain stimulation techniques [7].

VNS targets forebrain limbic neurocircuitry by stimulation of ascending fibers in the vagus nerve. Electric stimulation of afferent nerve fibers projecting to nucleus tractus solitarius (NTS) in the brain stem leads to stimulation of the medial temporal and prefrontal cortex which is thought to mediate psychotropic effects. In conventional VNS the electrodes are surgically implanted, wrapped around the cervical vagus nerve, while in transcutaneous VNS, the electrodes are applied at the auricular branch of the vagus nerve at the concha of the outer ear. In both techniques the electrodes are connected to a pulse generator similar to the one used in DBS [6]. VNS demonstrated efficacy in treatment-resistant depression [3]. TNS, like VNS, targets forebrain circuitry by stimulating afferent cranial nerve fibers and targeting transynaptic alteration of limbic circuitry. TNS employs cutaneous stimulation of afferent cranial nerve fibers with the intent to affect forebrain neurocircuitry via projections of brain stem nuclei. Electrodes are taped to the forehead, and stimulation occurs at night during sleep, from an attached stimulating device [6].

ECT is the oldest somatic treatment among those currently used in psychiatric practice, and it is also the most stigmatizing. ECT was a popular treatment for mental disorders between the 1940s and 1960s. After the 1960s, the use of ECT met resistance, and it was no longer a treatment option for many psychiatrists and psychiatry services. It was seen as a psychiatric asylum practice, and there was an erroneous association with punishment and torture. The prejudice against this technique was probably due to ECT applications without the patient's consent and its indiscriminate use. Rudimentary ECT devices, lack of anesthesia, and lack of muscle relaxants were associated with higher risks and side effects, reinforcing the negative public perception of ECT. In the 1990s, a new interest on ECT emerged with a great increase of clinical trials and publications. The efficacy of ECT in the treatment of mental disorders has been confirmed by several studies in the last two decades. In addition, modern devices and advanced anesthesia methods made ECT an extremely safe method that does not produce discomfort for the patients. Despite all that, ECT is frequently considered as the last therapeutic resource, reserved for very severe and refractory cases, demonstrating that the stigma about this method still exists [3].

Repetitive Transcranial Magnetic Stimulation

In rTMS, repetitive magnetic pulses are delivered over cortical areas through a coil positioned on the scalp. The magnetic field produces electrical currents in the brain, stimulating or inhibiting brain structures. There are different types of rTMS protocols; depending on the employed devices and parameters, rTMS may be deep or superficial. In conventional rTMS treatment, the electromagnetic waves have a reach of 3 cm from que coil surface to the cortex, while in deep rTMS the reach is approximately 5 cm. Several types of coils are used in rTMS, and they vary in size, format, focality, and reach. Intermittent theta burst stimulation (iTBS) is a new form of rTMS, which consists of three bursts of pulses at 50 Hz every 200 ms. This form of rTMS is as effective as rTMS with 10 or 20 Hz, but the duration of the session decreases about 5 times, which allows services to treat a larger number of patients [8].

All kinds of rTMS can be inhibitory or excitatory. These treatments can be an add-on treatment or monotherapy for several mental disorders. Commonly, psychotropic medications are maintained while performing neurostimulation treatment. The patient sits in an upright sitting position and is conscious during the whole procedure, and there is no need for anesthesia.

Usually the sites of stimulation are the left and right dorsolateral prefrontal cortex (DLPFC) and dorsomedial prefrontal cortex. Regarding DLPFC, excitatory stimuli are applied on the left hemisphere, while inhibitory stimuli are applied on the right hemisphere. Some researchers had attempted to combine both excitatory stimulation on the left DLPFC and inhibitory stimulation on the right DLPFC, but it did not show additional benefits, compared to unilateral stimulation [9] (Fig. 18.1).

Stimulus intensity ranges from 1.5 to 3.0 tesla, but intensity is based on resting motor threshold (RMT), which is the minimum intensity to contract a patient's thumb. Patients have their own RMT, and the stimulus intensity is calculated as a percentage of this RMT. The stimulus intensity applied to patients is predetermined; generally 110% to 120% of RMT in rTMS and 70–80% for iTBS. Less intense stimuli are probably not effective [9].

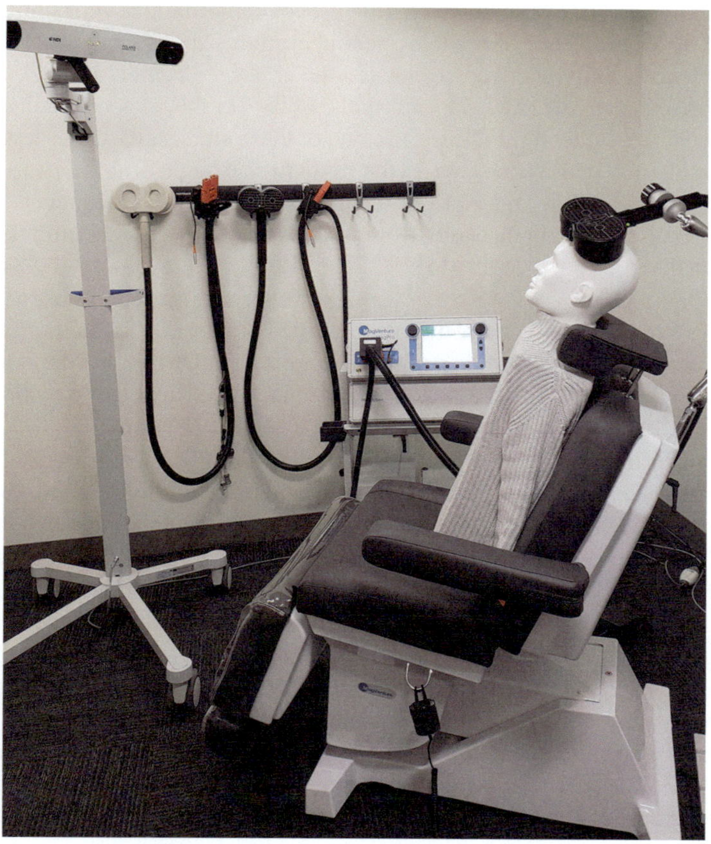

Fig. 18.1 Repetitive transcranial magnetic stimulation (rTMS) device MagPro X100 MagOption, from Magventure. Picture taken at the Laboratory for Neuropsychiatry & Neuromodulation, Division of Neuropsychiatry, Department of Psychiatry, Massachusetts General Hospital, Courtesy of Dr. Patricia Cirillo and Prof. Joan Camprodon

In a recent meta-analysis, Cirillo et al. [10] identified four studies that used rTMS to treat GAD, of which two were randomized, double-blind, and sham-controlled and two are uncontrolled open-trials [11–14]. Three studies applied low-frequency (1 Hz) rTMS over the right DLPFC [12, 14, 15], and one study evaluated bilateral rTMS treatment in patients with comorbid GAD and major depressive disorder (MDD) employing 1 Hz over the right DLPFC followed by 10 Hz over the left DLPFC [13]. rTMS demonstrated effectiveness in all four studies, and the overall effect size was −2.06 (95%CI: −2.64, −1.48), favoring active rTMS treatment [10]. In a sample of patients with GAD and insomnia, low-frequency rTMS in the right posterior parietal cortex produced significant improvement of both anxiety and insomnia [16]. Bystritsky et al. [14, 17] reported maintenance of the improvement after a 6-month follow-up without deterioration of questionnaires scores when compared to the end of rTMS treatment.

Nine studies on rTMS in PTSD were found and included in the meta-analysis from Cirillo et al. [10]. In some studies there were comorbidities with substance use disorders or MDD. There were three open-label studies and six randomized, sham-controlled trials. Only one study evaluated the effect of deep rTMS [18], while all others applied regular rTMS. In four studies [19–22] the protocol included low-frequency rTMS in the right DLPFC; in one of these studies, they also tested high-frequency (10 Hz) in the same area. In the study from Boggio et al. [23], the protocol was stimulation of either the right or left DLPFC with 20 Hz frequency. The frequency was also 20 Hz in the study from Isserles et al. [18], but the stimulated area was the medial prefrontal cortex (MPFC). The overall effect size was −0.88 (95%IC: −1.42, −0.34), favoring rTMS over sham. There were significant improvements in depressive symptoms too [10]. High-frequency rTMS seemed to be more effective than low-frequency rTMS, while stimulation of the right DLPFC was more effective than stimulation of the left DLPFC. Some studies demonstrated symptom worsening in the follow-up, which included PTSD and MDD symptoms [10].

There is little evidence of effectiveness of rTMS in PD; there are only two augmentation studies. In one randomized, controlled double-blind study [24], low-frequency stimulation on the right DLPFC produced significant improvement in PD patients, with a response rate of 50% and 25% remission rate in the active group. The second study [25] was a randomized, controlled, double-blind trial with iTBS over the left DLPFC. There were no significant differences between active and sham iTBS regarding PD symptom improvement in this study.

rTMS has been tested in different brain areas in the treatment of OCD, including the supplementary motor area (SMA), orbitofrontal cortex (OFC), and right, left, and bilateral DLPFC [26]. All trials with rTMS over the SMA used 1 Hz frequency and 100% of RMT. There were case series, open-label studies, and randomized controlled trials. Some of these studies indicated efficacy of this technique, but overall, the findings were inconsistent. There were also studies with low-frequency stimulation and high-frequency stimulation (10 or 20 Hz) over the DLPFC. The results for DLPFC trials were mainly positive but not very consistent either. Stimulation of the right side of this area had a larger effect size compared to the stimulation of the left area. There were three trials of rTMS over the OFC, one retrospective study and two randomized controlled trials, all with low-frequency stimulation. Two of these trials had positive results, while in the third trial the active treatment was not superior to sham [26].

Transcranial Direct Current Stimulation

tDCS is an electrical neurostimulation method with constant low-amplitude current focalized in specific cortical areas through electrodes placed on the scalp. This neurostimulation method increases cortical excitability in cortical areas under the anodal electrode, while it decreases cortical excitability where the cathodal electrode is placed. This effect is produced by the neuronal depolarization and hyperpolarization, respectively. There are basically two options of electrode placing in the treatment of MDD: [1] anodal electrode over the left DLPFC and cathodal electrode grounded in a noncortical area or [2] anodal electrode over the left DLPFC and the cathodal over the right DLPFC. The stimulus intensity ranges from 1 to 2 mA, and the sessions last for 30 min or more. In trials with MDD, daily sessions for at least 2 weeks are needed to obtain an antidepressant effect. Six-week treatments were more effective than shorter treatments, and tDCS combined with antidepressants was more effective than tDCS alone [8]. Usually, in the treatment of anxiety disorders, the objective is to decrease cortical excitability; for this reason the electrode placement and protocols are different from those used to treat depression (Fig. 18.2).

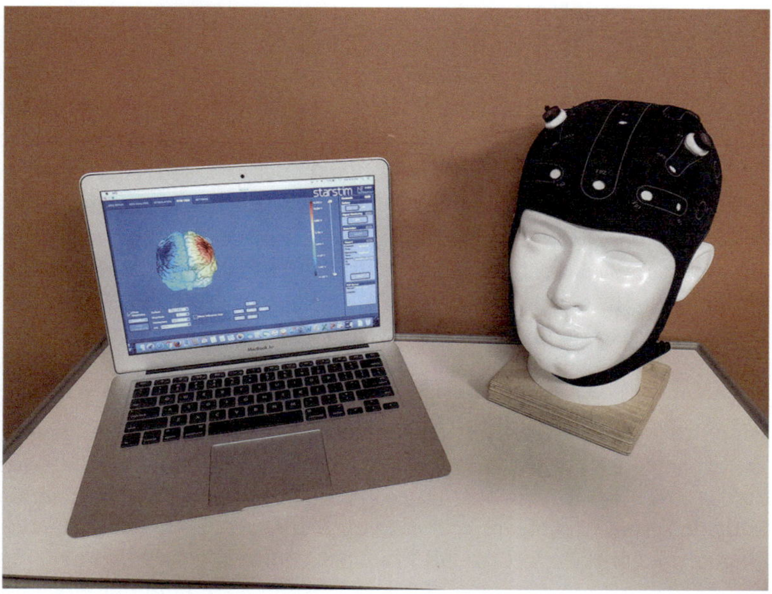

Fig. 18.2 Transcranial Direct Current Stimulation (tDCS) device Starstim 8 5G, from Neuroelectrics. Picture taken at the Laboratory for Neuropsychiatry & Neuromodulation, Division of Neuropsychiatry, Department of Psychiatry, Massachusetts General Hospital. Courtesy of Dr. Patricia Cirillo and Prof. Joan Camprodon

Cathodal stimulation was applied to the right DLPFC of one GAD patient in a case study [27], with excellent effect on anxiety decrease (90%). There are no other studies with tDCS in GAD.

There were only two studies with tDCS in PTSD, both focused on the cognitive features of this disorder. In one of them [28], the combination of working memory training and tDCS produced significant improvement of working memory performance. A fear-conditioning paradigm was used in the second study [29]. In this trial tDCS facilitated the consolidation of extinction memory. Currently there is no evidence pointing to effectiveness of tDCS in the treatment of this disorder because PTSD symptom improvement with tDCS was not assessed in any of these studies.

In a case study [30], one PD patient received cathodal tDCS over the right DLPFC and had significant improvement of anxiety symptoms. There were no other studies with tDCS in PD.

There is no evidence of effectiveness for tDCS in the treatment of SAD, but one study demonstrated that anodal stimulation on the left DLPFC decreased attentional bias in most of SAD patients [31].

Most of the studies with tDCS and OCD were case, case series, and open-label studies; there was only one randomized controlled trial with 12 patients. Besides the low quality of the studies, the results were unimpressive; in most of the studies, there was small positive effect. Stimulation of the OFC or SMA had more significant effects than stimulation of DLPFC [26].

Deep Brain Stimulation

DBS consists of neurosurgical implantation of electrodes under magnetic resonance imaging (MRI) guidance in selected brain areas connected by a wire to a neurostimulator (NS) or an implantable pulse generator (IPG). The NS/IPG, usually placed into the right chest below the clavicle, sends electrical pulses to brain electrodes in order to modulate an adjacent neural network. DBS parameters can be monitored and programmed remotely with a handheld device in a similar way to pacemakers and VNS. There are also a patient controller to turn it on and off and check battery status and self-adjust parameters provided by the DBS programmer [8].

DBS is mainly used to improve motor symptoms of Parkinson's disease, but it is also been studied in treatment-resistant depression, dystonia, essential tremor, epilepsy, and Alzheimer's disease. Studies using DBS to treat treatment-resistant depression are increasing; but this technique still needs to find more appropriate brain areas related to this disorder for the implantation of electrodes to have more consistent results. The psychopharmacological and psychotherapeutic treatments are usually performed in tandem with each other [8].

The main anatomical target for DBS in most studies is the subcallosal cingulate (SCC) white matter, but ventral internal capsule/ventral striatum (VC/VS), nucleus accumbens (NAc), medial forebrain bundle (MFB), inferior thalamic peduncle (ITP), the lateral habenular complex, and the rostral cingulate gyrus are also targeted in DBS [8, 32].

Currently there is no consensus on what the optimal stimulation parameters for MDD are. However, trials in animals concluded that some parameters are more effective for ventromedial prefrontal cortex/SCC such as high-frequency (130 Hz) and current intensity (100–300 mA). These studies also identified that prelimbic stimulation was more effective than infralimbic stimulation and that left unilateral and bilateral stimulation had similar results [8].

In the treatment of OCD, DBS target areas consisted of VC/VS, NAc, the subthalamic nucleus (STN), the anterior limb of the internal capsule (ALIC), and the ITP. A recent meta-analysis [33] showed positive findings for DBS in OCD, but there were no significant differences between the five DBS targets regarding treatment response. From 16 studies analyzed by Alonso et al. [33], the average percentage of responders to DBS was 60.0% (95% CI =49.0–69.0%), with response being defined as an OCD symptom decrease of at least 35% measured on the Yale-Brown Obsessive Compulsive Scale (Y-BOCS). The mean reduction of Y-BOCS score after DBS was available for 66 patients and was estimated to be 45.1% (95% CI =29.4–60.8%) [33]. Besides reducing OCD symptoms, DBS targeted at NAc, VC/VS, or ALIC improved mood and decreased anxiety [34].

In a case report of a patient with MDD, PD, and substance use disorder, DBS targeting the NAc produced significant decrease of alcohol consumption, but did not improve PD or MDD symptoms [58]. In fact, three studies [35–37] demonstrated that DBS into the NAc might induce panic attacks even in patients who never had them before.

DBS in the basolateral nucleus of the amygdala was effective for a combat veteran with treatment-resistant PTSD in a single case report [38]. After 8 months of treatment, there was a decrease of 37.8% of the clinician-administered post-traumatic stress disorder scale (CAPS) score, and the quality of life increased considerably.

Vagus Nerve Stimulation

In VNS, an IPG is implanted subcutaneously in the left chest, and the electrodes from the IPG deliver low-frequency, intermittent pulses to the left vagus nerve [39]. VNS demonstrated to be safe in pregnant women. VNS can be used concomitantly with psychotropic drugs and ECT [40].

The electrical stimulation through the vagus nerve provides stimulation to the NTS that modulates several subcortical and cortical regions of the brain through the neural networks [8]. The paraventricular nucleus of the hypothalamus, central nucleus of the amygdala, parabrachial nucleus, locus coeruleus, and dorsal raphe nucleus receive direct input from the NTS [41]. Afferents from the NTS and parabrachial nucleus carry input to the bed nucleus of the stria terminalis (BST). Output from the ventrolateral region of BST heavily targets the substantia innominata; the parvocellular paraventricular hypothalamic nucleus; the dorsomedial, perifornical, and lateral hypothalamus; the central nucleus of the amygdala; the periaqueductal gray matter; the pre-locus coeruleus; the lateral parabrachial area; and the NTS [42].

These interconnected brain areas are excellent targets for neurostimulation because they are heavily implicated in the pathophysiology of anxiety disorders.

Treatment parameters are similar to other devices which include the intensity of the electrical stimulus (mA), pulse width (microseconds), frequency (Hz), duration of the stimulus, and interval between them (sec or min) [40]. There is still no consensus on what the optimal stimulation parameters for MDD would be. Most studies used the output current beginning at 0.25 mA, 500-μs pulse width, and 20- or 30-Hz frequency for 30 sec of stimulation and a non-stimulation interval of 5 min. In one long-term study, the electric current was increased to 1 mA [39]. There seem to be better response rates and decreased suicide attempts in VNS with higher electrical charges, compared to low charges [8].

It was demonstrated by functional MRI that transdermal VNS, besides activating the NTS, also activates the locus coeruleus in the brain stem. In limbic networks, specifically the amygdala, hippocampus, posterior cingulate gyrus, parahippocampal gyrus, and middle and superior temporal gyrus, a reduction of blood-oxygen-level-dependent (BOLD) signal was also observed. Changes in NTS activity have been demonstrated to increase high-frequency heart rate variability and reduce sympathetic activity. As PTSD symptom severity is often correlated with elevated distress autonomic arousal, transdermal VNS may function to modulate this response [6].

In an open-label study, George et al. [43] assessed the efficacy of VNS in a sample of treatment-resistant patients, including OCD, PD, and PTSD patients. In the acute phase (12 weeks), three out of nine patients had a 50% or greater improvement in the Hamilton Anxiety Rating Scale (HAM-A) scores (response). The PD patient achieved response before completing 24 weeks of treatment. One PTSD patient achieved response only after 2 years of treatment; the other PTSD patient did not achieve response after 4 years of treatment. Regarding the six OCD patients, three had improvements of 25% or greater in Y-BOCS scores (response) in the acute phase, and four achieved response after 24 weeks of treatment, but the response rate decreased gradually during the 4-year treatment. At the end of the study, there were only two OCD patients; both of them were responders.

Trigeminal Nerve Stimulation

In TNS, small electrical currents stimulate cutaneous branches of the trigeminal nerve to stimulate the brain. The trigeminal is the largest cranial nerve and carries afferent impulses from the face to locations throughout the brain stem, limbic system, and cortex. As the vagus nerve, the trigeminal projects to the NTS, which is interconnected with the thalamus, hypothalamus, amygdala, locus coeruleus, raphe nuclei, reticular activating system, limbic forebrain, anterior cingulate, and insula. These regions are implicated in the pathophysiology of anxiety disorders. This new and noninvasive intervention has shown therapeutic effects in epilepsy, MDD, and attention deficit/hyperactivity disorder [44].

There were two small open-label studies with TNS in PTSD, and both had positive results. In the study from Trevisol et al. [45], five PTSD patients who were under antidepressants received add-on TNS in a 10-day protocol. Four in five subjects had 50% or more decrease in depression and PTSD scale scores; this improvement persisted at the 1-month follow-up. In an 8-week trial with 12 subjects with PTSD and MDD [44], TNS was also used in addition to pharmacological treatment. At the end of the study, there were significant decreases in depression and PTSD symptoms.

The same researchers published two case reports with treatment-resistant anxiety disorder patients, one with GAD [46] and the other one with SAD [47]. In both studies TNS was administered for 10 consecutive days (except weekends); the electric stimulation was performed at 120 Hz, with a pulse wave duration of 250 ms, for 30 min per day. Both patients had significant improvements.

Trevisol et al. [48] applied the same methods tested in PTSD, GAD, and SAD in a 10-day trial with a sample of seven PD patients. At the end of the treatment, there was a significant reduction of PD symptoms from 18.1 (SD = 1.8) to 6.4 (SD = 5.8) on the Panic Disorder Severity Scale (PDSS). Six in seven patients had a 50% or greater improvement in PDSS scores and were considered responders. There were also significant improvements in HAM-A and Hamilton Depression Rating Scale scores. There were no significant changes in cognitive function.

Electroconvulsive Therapy

Recent studies indicate that methods used in ECT applications, such as the position of the electrodes, current intensity, wavelength, frequency, session duration, time between applications, and number of sessions, may influence both positive and negative outcomes of this therapy. In order to produce effective seizures, unilateral ECT requires higher electrical current doses, bifrontal ECT requires lower doses, and the lowest doses are those applied in bitemporal ECT. The last is associated with more cognitive side effects, especially memory deficits, compared to unilateral and bifrontal positions. However, it is still under debate if bifrontal and right unilateral ECT are as effective as bitemporal ECT in the treatment of mood disorders [3]. The Canadian Network for Mood and Anxiety Treatments (CANMAT) classified right unilateral ECT and bifrontal ECT as a first-line treatment for MDD and bitemporal ECT as a second-line treatment, due to its cognitive side effects [8].

There are no randomized controlled trials with ECT in PTSD. In a retrospective chart review with 26 military veterans with MDD and PTSD, Watts et al. [49] found significant improvements of depression symptoms after right unilateral ECT or bitemporal ECT. However, only 35% of the patients had at least a small improvement (20%) of PTSD symptoms. In another retrospective study with patients with PTSD and MDD [50], bilateral ECT produced significant improvements of MDD and PTSD symptoms. In a recent open-label study [51], 20 treatment-resistant PTSD patients received six bilateral ECT treatments during a 3-week trial. The improvement of anxiety and depression symptoms was significant, with a mean decrease CAPS scores of 34% and a mean decrease of MDD scores of 51%. The current level of evidence for ECT in PTSD is still low and insufficient to support routine use of this treatment [52].

Fontenelle et al. [53] reviewed the evidence regarding efficacy of ECT in the treatment of OCD. The authors found 50 relevant articles, with a total number of 279 patients with OCD and related constructs, but there were no randomized controlled trials. In one non-randomized controlled trial [54], 17 OCD patients received ECT, while 49 patients received pharmacological treatment. Sixty percent of the patients who received ECT showed marked improvements, and a similar result was found in the pharmacotherapy group. The authors did not explain the allocation methods and if there were differences between the two groups in baseline regarding severity of depression or OCD symptoms. In addition, pill placebo or sham ECT were not mentioned in this article, adding up to the other methodological limitations. In the study from Dubois et al. [55], the allocation to treatments was decided based on clinical criteria; the treatment arms were defined a posteriori. The 43 patients received either oral antidepressants, intravenous followed by oral antidepressants, oral antidepressant plus ECT, or oral plus intravenous antidepressant plus ECT. The remission rates ranged from 60% to 85% depending on the arm, and the overall remission rate was 69.7%. Fontenelle et al. [53] concluded that ECT has no role in routine treatment of OCD because there are too many limitations in the studies done so far. According to the authors, OCD patients resistant to pharmacological treatment are probably resistant to ECT too; thus ECT offers no advantage over pharmacological treatments.

Other Neurostimulation Techniques

Cranial electrotherapy stimulation (CES) [56] is a noninvasive neurostimulation technique in which a pulsed, low-amplitude electrical current is applied to the head through electrodes placed on the earlobes. Recently, the effect of CES was examined in pain, headaches, fibromyalgia, smoking cessation, and opiates withdrawal. Several small low-quality studies reported anxiety decrease with CES in diverse patient samples, but the study from Bystritsky et al. [56] was the first study with anxiety disorder patients. In this open-label study with 12 GAD patients (9 completers), a significant improvement of HAM-A scores was observed in the end of the 6-week trial. Six patients were considered responders to CES.

The radio electric asymmetric conveyor (REAC) is a device used for pulsed low electric current stimulation applied to the ear pavilion [57], similar to CES. The protocol for SAD [57] consisted of seven radiofrequency bursts of 500 ms each at a frequency of 10.5 GHz and a specific absorption rate of 7 μW/kg. The same group of Italian researchers conducted clinical trials with several diseases/disorders including stress-related disorders, anxiety, depression, bipolar disorder, "antiaging" treatment, aging-related chondral damage or injuries, surgical wounds, and behavioral and psychiatric symptoms in Alzheimer's disease. None of these studies were a randomized controlled trial, and most had several methodological shortcomings; nevertheless, REAC was always considered effective by the authors.

Conclusion

The most studied neurostimulation method for anxiety disorders, PTSD, and OCD was rTMS. This neurostimulation technique had the highest level of evidence for GAD. There were also randomized sham-controlled trials indicating that rTMS may be effective in the treatment of PTSD and OCD, but there were conflicting findings regarding these two disorders. There is indication that rTMS may be effective in the treatment of panic disorder, but the level of evidence is low. DBS was most studied for treatment of OCD, but the randomized sham-controlled trials had mixed findings. Preliminary findings indicate that DBS is probably not effective for panic disorder treatment but could be affective for PTSD. There is weak evidence indicating that ECT, tDCS, VNS, and TNS could be effective in the treatment of anxiety disorders, PTSD, and OCD (Table 18.1).

The growth of research on neurostimulation techniques for treatment of anxiety disorders, OCD, and PTSD is exciting; however, the current literature does not support the use of these techniques in clinical practice. One important issue is the lack of standardization of devices and protocols; in addition, most studies are small, not sham-controlled, and riddled with methodological flaws. Large high-quality studies are warranted to clarify the role of neurostimulation in anxiety disorders, OCD, and PTSD.

Table 18.1 Neurostimulation in anxiety disorders, post-traumatic stress disorder, and obsessive-compulsive disorder: quality of evidence

	rTMS	DBS	tDCS	ECT	VNS	TNS
Panic disorder	++[a]	0[a]	+	0	+	+
Generalized anxiety disorder	+++	0	+	0	0	+
Social anxiety disorder	0	0	0	0	0	+
Obsessive-compulsive disorder	+++[a]	+++[a]	++[a]	++	+	0
Post-traumatic stress disorder	+++[a]	+	0	++	+	+

DBS Deep brain stimulation, *ECT* electroconvulsive therapy, *rTMS* repetitive transcranial magnetic stimulation, *tDCS* transcranial direct current stimulation, *TNS* trigeminal nerve stimulation, *VNS* vagus nerve stimulation

0 No evidence

+ At least one case report with positive findings

++ At least one open-label study **or** randomized sham-controlled trial with more than 10 patients (receiving active treatment) with positive findings

+++ At least two randomized sham-controlled trial with more than 10 patients (receiving active treatment) with positive findings

[a]At least one study without positive findings

References

1. Zugliani MM, Cabo MC, Nardi AE, Perna G, Freire RC. Pharmacological and neuromodulatory treatments for panic disorder: clinical trials from 2010 to 2018. Psychiatry Investig. 2019;16(1):50–8.
2. Freire RC, Nardi AE. The effect of neurostimulation in depression. In: Kim YK, editor. Understanding depression: contemporary issues, vol. 1. Singapore: Springer Singapore; 2018. p. 177–87.
3. Freire RC, Cirillo PC, Nardi AE. Clinical application of neurostimulation in depression. In: Kim YK, editor. Understanding depression: contemporary issues, vol. 2. Singapore: Springer Singapore; 2018. p. 271–82.
4. Deppermann S, Vennewald N, Diemer J, Sickinger S, Haeussinger F, Notzon S, et al. Does rTMS alter neurocognitive functioning in patients with panic disorder/agoraphobia? An fNIRS-based investigation of prefrontal activation during a cognitive task and its modulation via sham-controlled rTMS. Biomed Res Int [Internet]. 2014; 2014:542526 p. Available from: http://cochranelibrary-wiley.com/o/cochrane/clcentral/articles/909/CN-01047909/frame.html.
5. Li H, Wang J, Li C, Xiao Z. Repetitive transcranial magnetic stimulation (rTMS) for panic disorder in adults. Cochrane Database Syst Rev. 2014;9:CD009083.
6. Koek RJ, Roach J, Athanasiou N, Van 't Wout-Frank M, Philip NS. Neuromodulatory treatments for post-traumatic stress disorder (PTSD). Prog Neuro-Psychopharmacol Biol Psychiatry. 2019;92:148–60.
7. D'Urso G, Mantovani A, Patti S, Toscano E, de Bartolomeis A. Transcranial direct current stimulation in obsessive-compulsive disorder, posttraumatic stress disorder, and anxiety disorders. J ECT. 2018;34(3):172–81.
8. Milev RV, Giacobbe P, Kennedy SH, Blumberger DM, Daskalakis ZJ, Downar J, et al. Canadian network for mood and anxiety treatments (CANMAT) 2016 clinical guidelines for the Management of Adults with major depressive disorder: section 4. Neurostimul Treat Can J Psychiat. 2016;61(9):561–75.
9. Janicak PG, Dokucu ME. Transcranial magnetic stimulation for the treatment of major depression. Neuropsychiatr Dis Treat. 2015;11:1549–60.
10. Cirillo PC, Gold AK, Nardi AE, Ornelas AC, Nierenberg AA, Campodron J, et al. Transcranial magnetic stimulation in anxiety and trauma-related disorders: a systematic review and meta-analysis. Brain Behav. 2019; https://doi.org/10.1002/brb3.1284.
11. Diefenbach GJ, Bragdon LB, Zertuche L, Hyatt CJ, Hallion LS, Tolin DF, et al. Repetitive transcranial magnetic stimulation for generalised anxiety disorder: a pilot randomised, double-blind, sham-controlled trial. Br J Psychiatry. 2016;209(3):222–8.
12. Dilkov D, Hawken ER, Kaludiev E, Milev R. Repetitive transcranial magnetic stimulation of the right dorsal lateral prefrontal cortex in the treatment of generalized anxiety disorder: a randomized, double-blind sham controlled clinical trial. Prog Neuro-Psychopharmacol Biol Psychiatry. 2017;78:61–5.
13. White D, Tavakoli S. Repetitive transcranial magnetic stimulation for treatment of major depressive disorder with comorbid generalized anxiety disorder. Ann Clin Psychiatry. 2015;27(3):192–6.
14. Bystritsky A, Kaplan JT, Feusner JD, Kerwin LE, Wadekar M, Burock M, et al. A preliminary study of fMRI-guided rTMS in the treatment of generalized anxiety disorder. J Clin Psychiatry. 2008;69(7):1092–8.
15. Diefenbach GJ, Assaf M, Goethe JW, Gueorguieva R, Tolin DF. Improvements in emotion regulation following repetitive transcranial magnetic stimulation for generalized anxiety disorder. J Anxiety Disord. 2016;43:1–7.
16. Huang Z, Li Y, Bianchi MT, Zhan S, Jiang F, Li N, et al. Repetitive transcranial magnetic stimulation of the right parietal cortex for comorbid generalized anxiety disorder and insomnia: a randomized, double-blind, sham-controlled pilot study. Brain Stimul. 2018;11(5):1103–9.
17. Bystritsky A, Kerwin LE, Feusner JD. A preliminary study of fMRI-guided rTMS in the treatment of generalized anxiety disorder: 6-month follow-up. J Clin Psychiatry. 2009;70(3):431–2.

18. Isserles M, Shalev AY, Roth Y, Peri T, Kutz I, Zlotnick E, et al. Effectiveness of deep transcranial magnetic stimulation combined with a brief exposure procedure in post-traumatic stress disorder – a pilot study. Brain Stimul. 2013;6(3):377–83.
19. Nam DH, Pae CU, Chae JH. Low-frequency, repetitive Transcranial magnetic stimulation for the treatment of patients with posttraumatic stress disorder: a double-blind, sham-controlled study. Clin Psychopharmacol Neurosci. 2013;11(2):96–102.
20. Watts BV, Landon B, Groft A, Young-Xu Y. A sham controlled study of repetitive transcranial magnetic stimulation for posttraumatic stress disorder. Brain Stimul. 2012;5(1):38–43.
21. Cohen H, Kaplan Z, Kotler M, Kouperman I, Moisa R, Grisaru N. Repetitive transcranial magnetic stimulation of the right dorsolateral prefrontal cortex in posttraumatic stress disorder: a double-blind, placebo-controlled study. Am J Psychiatry. 2004;161(3):515–24.
22. Osuch EA, Benson BE, Luckenbaugh DA, Geraci M, Post RM, McCann U. Repetitive TMS combined with exposure therapy for PTSD: a preliminary study. J Anxiety Disord. 2009;23(1):54–9.
23. Boggio PS, Rocha M, Oliveira MO, Fecteau S, Cohen RB, Campanha C, et al. Noninvasive brain stimulation with high-frequency and low-intensity repetitive transcranial magnetic stimulation treatment for posttraumatic stress disorder. J Clin Psychiatry. 2010;71(8):992–9.
24. Mantovani A, Aly M, Dagan Y, Allart A, Lisanby S. Randomized sham controlled trial of repetitive transcranial magnetic stimulation to the dorsolateral prefrontal cortex for the treatment of panic disorder with comorbid major depression. J Affect Disord [Internet]. 2013; 144(1–2):153–9. Available from: http://cochranelibrary-wiley.com/o/cochrane/clcentral/articles/340/CN-00968340/frame.html.
25. Deppermann S, Vennewald N, Diemer J, Sickinger S, Haeussinger F, Dresler T, et al. Neurobiological and clinical effects of fNIRS-controlled rTMS in patients with panic disorder/agoraphobia during cognitive-behavioural therapy. Neuroimage: clinical [Internet] 2017; 16:668–77. Available from: http://cochranelibrary-wiley.com/o/cochrane/clcentral/articles/406/CN-01423406/frame.html.
26. Shivakumar V, Dinakaran D, Narayanaswamy JC, Venkatasubramanian G. Noninvasive brain stimulation in obsessive-compulsive disorder. Indian J Psychiatry. 2019;61(Suppl 1):S66–76.
27. Shiozawa P, Leiva AP, Castro CD, da Silva ME, Cordeiro Q, Fregni F, et al. Transcranial direct current stimulation for generalized anxiety disorder: a case study. Biol Psychiatry. 2014;75(11):e17–8.
28. Saunders N, Downham R, Turman B, Kropotov J, Clark R, Yumash R, et al. Working memory training with tDCS improves behavioral and neurophysiological symptoms in pilot group with post-traumatic stress disorder (PTSD) and with poor working memory. Neurocase. 2015;21(3):271–8.
29. Van't Wout M, Longo SM, Reddy MK, Philip NS, Bowker MT, Greenberg BD. Transcranial direct current stimulation may modulate extinction memory in posttraumatic stress disorder. Brain Behav. 2017;7(5):e00681.
30. Shiozawa P, Enokibara da Silva M, Cordeiro Q. Transcranial direct current stimulation (tDCS) for panic disorder: a case study. J Depress Anxiety. 2014;3(3):158.
31. Heeren A, Billieux J, Philippot P, De Raedt R, Baeken C, de Timary P, et al. Impact of transcranial direct current stimulation on attentional bias for threat: a proof-of-concept study among individuals with social anxiety disorder. Soc Cogn Affect Neurosci. 2017;12(2):251–60.
32. Delaloye S, Holtzheimer PE. Deep brain stimulation in the treatment of depression. Dialogues Clin Neurosci. 2014;16(1):83–91.
33. Alonso P, Cuadras D, Gabriels L, Denys D, Goodman W, Greenberg BD, et al. Deep brain stimulation for obsessive-compulsive disorder: a meta-analysis of treatment outcome and predictors of response. PLoS One. 2015;10(7):e0133591.
34. Graat I, Figee M, Denys D. The application of deep brain stimulation in the treatment of psychiatric disorders. Int Rev Psychiatry. 2017;29(2):178–90.
35. Sousa MB, Reis T, Reis A, Belmonte-de-Abreu P. New-onset panic attacks after deep brain stimulation of the nucleus accumbens in a patient with refractory obsessive-compulsive and bipolar disorders: a case report. Revista brasileira de psiquiatria (Sao Paulo, Brazil: 1999). 2015;37(2):182–3.

36. Shapira NA, Okun MS, Wint D, Foote KD, Byars JA, Bowers D, et al. Panic and fear induced by deep brain stimulation. J Neurol Neurosurg Psychiatry. 2006;77(3):410–2.
37. Okun MS, Mann G, Foote KD, Shapira NA, Bowers D, Springer U, et al. Deep brain stimulation in the internal capsule and nucleus accumbens region: responses observed during active and sham programming. J Neurol Neurosurg Psychiatry. 2007;78(3):310–4.
38. Langevin JP, Koek RJ, Schwartz HN, Chen JWY, Sultzer DL, Mandelkern MA, et al. Deep brain stimulation of the Basolateral amygdala for treatment-refractory posttraumatic stress disorder. Biol Psychiatry. 2016;79(10):e82–e4.
39. Daban C, Martinez-Aran A, Cruz N, Vieta E. Safety and efficacy of Vagus nerve stimulation in treatment-resistant depression. A systematic review. J Affect Disord. 2008;110(1–2):1–15.
40. Howland RH. New developments with vagus nerve stimulation therapy. J Psychosoc Nurs Ment Health Serv. 2014;52(3):11–4.
41. Carlson NR. Physiology of behavior. 10th ed. Boston: Allyn & Bacon; 2010.
42. Shin JW, Geerling JC, Loewy AD. Inputs to the ventrolateral bed nucleus of the stria terminalis. J Comp Neurol. 2008;511(5):628–57.
43. George MS, Ward HE Jr, Ninan PT, Pollack M, Nahas Z, Anderson B, et al. A pilot study of vagus nerve stimulation (VNS) for treatment-resistant anxiety disorders. Brain Stimul. 2008;1(2):112–21.
44. Cook IA, Abrams M, Leuchter AF. Trigeminal nerve stimulation for comorbid posttraumatic stress disorder and major depressive disorder. Neuromodulation. 2016;19(3):299–305.
45. Trevizol AP, Shiozawa P, Albuquerque Sato I, da Silva ME, de Barros Calfat EL, Alberto RL, et al. Trigeminal nerve stimulation (TNS) for Post-traumatic stress disorder: a case study. Brain Stimul. 2015;8(3):676–8.
46. Trevizol AP, Shiozawa P, Sato IA, Calfat EL, Alberto RL, Cook IA, et al. Trigeminal nerve stimulation (TNS) for generalized anxiety disorder: a case study. Brain Stimul. 2015;8(3):659–60.
47. Trevizol AP, Taiar I, Malta RC, Sato IA, Bonadia B, Cordeiro Q, et al. Trigeminal nerve stimulation (TNS) for social anxiety disorder: a case study. Epilepsy Behav. 2016;56:170–1.
48. Trevizol AP, Sato IA, Cook IA, Lowenthal R, Barros MD, Cordeiro Q, et al. Trigeminal nerve stimulation (TNS) for panic disorder: an open label proof-of-concept trial. Brain Stimul. 2016;9(1):161–2.
49. Watts BV. Electroconvulsive therapy for comorbid major depressive disorder and posttraumatic stress disorder. J ECT. 2007;23(2):93–5.
50. Ahmadi N, Moss L, Simon E, Nemeroff CB, Atre-Vaidya N. Efficacy and long-term clinical outcome of comorbid posttraumatic stress disorder and major depressive disorder after electroconvulsive therapy. Depress Anxiety. 2016;33(7):640–7.
51. Margoob MA, Ali Z, Andrade C. Efficacy of ECT in chronic, severe, antidepressant- and CBT-refractory PTSD: an open, prospective study. Brain Stimul. 2010;3(1):28–35.
52. Rosenquist PB, Youssef NA, Surya S, McCall WV. When all else fails: the use of electroconvulsive therapy for conditions other than major depressive episode. Psychiatr Clin North Am. 2018;41(3):355–71.
53. Fontenelle LF, Coutinho ES, Lins-Martins NM, Fitzgerald PB, Fujiwara H, Yucel M. Electroconvulsive therapy for obsessive-compulsive disorder: a systematic review. J Clin Psychiatry. 2015;76(7):949–57.
54. Garrido A. Electroconvulsive therapy in severe obsessive-compulsive disorder. Eur Psychiatry. 1998;13(Suppl 4):236s–7s.
55. Dubois JC. Obsessions and mood: apropos of 43 cases of obsessive neurosis treated with antidepressive chemotherapy and electroshock. Ann Med Psychol (Paris). 1984;142(1):141–51.
56. Bystritsky A, Kerwin L, Feusner J. A pilot study of cranial electrotherapy stimulation for generalized anxiety disorder. J Clin Psychiatry. 2008;69(3):412–7.
57. Fontani V, Mannu P, Castagna A, Rinaldi S. Social anxiety disorder: radio electric asymmetric conveyor brain stimulation versus sertraline. Patient Prefer Adherence. 2011;5:581–6.
58. Kuhn J, Lenartz D, Huff W, Lee S, Koulousakis A, Klosterkoetter J, Sturm V. Remission of alcohol dependency following deep brain stimulation of the nucleus accumbens: valuable therapeutic implications? J Neurol Neurosurg Psychiatry. 2007;78(10):1152–3.

Current and Novel Psychopharmacological Drugs for Anxiety Disorders

19

Borwin Bandelow

Introduction

Anxiety disorders, including panic disorder/agoraphobia (PDA), generalized anxiety disorder (GAD), social anxiety disorder (SAD), and others, are the most prevalent psychiatric disorders and are associated with a very high burden of illness [1]. They are more common in females [2].

The current concept of the etiology of anxiety disorders assumes an interaction of psychosocial factors, e.g., childhood adversity, trauma, or stress, and a heritable vulnerability, which manifests in dysfunctions of brain neurotransmitter systems. The available data for potential biomarkers for anxiety disorders in the fields of neuroimaging, genetics, neurochemistry, neurophysiology, and neurocognition have been summarized in two articles [3, 4]. However, according to these reviews, no specific biomarkers for anxiety disorders have been identified yet, despite high-quality neurobiological research in the field.

Patients with different anxiety disorders show varying degrees of health-care utilization. For example, in the United States, 54% of patients with PDA, but only 27% of individuals with specific phobias, utilized health-care providers in 1 year [5]. While patients with PDA often fear that they suffer from a severe somatic disorder, such as a coronary heart disease or a brain tumor, and need immediate medical help in the case of a panic attack, people with specific phobias often have the feeling that they can cope with the disorder and even think it is "normal" to have fear of cats or insects.

There is evidence for substantial undertreatment of anxiety disorders. In a large study conducted in European countries [6], only 21% of respondents with an anxiety disorder sought professional help. Of those participants who contacted

B. Bandelow (✉)
Department of Psychiatry and Psychotherapy, University Medical Center, Göttingen, Germany
e-mail: bbandel@gwdg.de

Y.-K. Kim (ed.), *Anxiety Disorders*, Advances in Experimental Medicine and Biology 1191, https://doi.org/10.1007/978-981-32-9705-0_19

health-care services, 23% were given no treatment at all, 31% received only medication treatment, 27% were treated with both medications and psychotherapy, and 20% received only psychological treatment.

Current Pharmacological Treatment

In clinical settings, the majority of patients seeking professional help for anxiety disorders suffer from PDA, followed by SAD and GAD [1]. For these disorders, a substantial number of randomized controlled trials with medications are available. Although specific phobia is the most frequent anxiety disorder, almost no studies exist for the psychopharmacological treatment of this disorder. This may be due to the fact that most people with specific phobia can cope with their fears and do not feel they have to undergo treatment for the condition. In severe cases, specific phobia can effectively be treated with exposure therapy. As two other anxiety disorders, separation anxiety disorder in adults and selective mutism, have only recently been introduced in the DSM-5, almost no medication studies exist for these disorders.

Not all patients with mild anxiety disorders have to be treated. However, treatment is indicated when a patient shows marked impairments in quality of life or suffers from complications resulting from the disorder (e.g., secondary depression, suicidality, or substance abuse). Patients with anxiety disorders can mostly be treated in outpatient settings. Only in the case of suicidal ideation, unresponsiveness to standard treatments, or relevant comorbidity, e.g., with major depression or personality disorders, hospitalization may be indicated.

The recommendations in this article are based on treatment guidelines for anxiety disorders [7–9]. For these guidelines, a systematic literature search for randomized clinical trials had been performed. Studies were analyzed by using internationally acknowledged quality assessment tools. The recommendations were reviewed by expert panels.

Patients treated with medications should receive information about their diagnostic considerations, the physiological backgrounds of anxiety symptoms, the possible etiology of pathological fears, and the putative mechanisms of action of the available treatment approaches. The treatment plan should include psychotherapy, pharmacotherapy, and other interventions which should be selected after consideration of individual conditions, e.g., the patient's history with previous treatment attempts, illness severity, comorbidities such as personality disorders, suicidality, local availability of treatment methods, wait time for psychotherapy appointments, costs, and last but not least the patient's preference.

A large number of studies have shown the efficacy of psychopharmacological drugs for PDA, SAD, and GAD. For specific phobias, however, there are only very few studies on drug treatment, e.g., a small study suggesting the efficacy of paroxetine [10].

The available drugs and dose recommendations are listed in Table 19.1. Some of the medications mentioned here are not licensed for anxiety indications in some countries. The reader is referred to the local summary of product characteristic. The

Table 19.1 Pharmacological treatment recommendations for anxiety disorders in adults

Medications		Efficacy shown in RCTs for				
	Drug	PDA	GAD	SAD	Daily dose	Adverse effects
SSRIs	Citalopram[1]	x		x	20–40 mg	Jitteriness, nausea, restlessness, headache, fatigue, increased or decreased appetite, weight gain, weight loss, tremor, sweating, QT_C prolongation, sexual dysfunction, diarrhea, constipation, and other side effects
	Escitalopram[2]	x	x	x	10–20 mg	
	Fluoxetine	x				
	Fluvoxamine	x		x		
	Paroxetine	x	x	x	20–50 mg	
	Sertraline	x	x	x	50–150 mg	
SNRIs	Duloxetine		x		60–120 mg	Jitteriness, nausea, restlessness, headache, fatigue, increased or decreased appetite, weight gain, weight loss, tremor, sweating, sexual dysfunction, diarrhea, constipation, urination problems, and other side effects
	Venlafaxine	x	x	x	75–225 mg	
Tricyclic antidepressant	Clomipramine	x			75–250 mg	Anticholinergic effects, somnolence, dizziness, cardiovascular side effects, weight gain, nausea, headache, sexual dysfunction, and other side effects
Calcium modulator	Pregabalin		x	x	150–600 mg	Dizziness, somnolence, dry mouth, edema, blurred vision, weight gain, constipation, euphoric mood, balance disorder, increased appetite, difficulty with concentration/attention, withdrawal symptoms after abrupt discontinuation, and other side effects
Azapirone	Buspirone		x		15–60 mg	Dizziness, nausea, headache, nervousness, light-headedness, excitement, insomnia, and other side effects
RIMA	Moclobemide			x	300–600 mg	Restlessness, insomnia, dry mouth, headache, dizziness, gastrointestinal symptoms, nausea, and other side effects

Not all drugs are licensed for these indications in all countries

PDA panic disorder/agoraphobia, *GAD* generalized anxiety disorder, *SAD* social phobia, *RIMA* reversible monoamine oxidase A inhibitor, *RCT* randomized controlled study

[a]Do not exceed recommended dose (possible QT_C interval prolongation). Maximal dose with diminished hepatic function 30 mg/day, for older patients 20 mg/day

[b]Do not exceed recommended dose (possible QT_C interval prolongation). Maximal dose for persons over age 65 10 mg/day

side effects of the drugs are listed in Table 19.1. For a detailed list of available randomized controlled studies, the reader is referred to guidelines which included a systematic evaluation of the available studies [7–9].

Selective Serotonin Reuptake Inhibitors (SSRIs)/Selective Serotonin-Norepinephrine Reuptake Inhibitors (SNRIs)

Due to their positive benefit/risk balance, SSRIs and SNRIs antidepressants are recommended as standard drugs for anxiety disorders. The onset of the anxiolytic effect of these antidepressants has a latency of 2 to 4 weeks (or even up to 6 weeks). During the first weeks of treatment, adverse effects may be more pronounced. Initial jitteriness or agitation may occur in this period, which may reduce the patients' treatment compliance. Starting with a lower dose of the drug may reduce these adverse effects. A review of studies in depressed patients suggested that the tolerability of SNRIs may be less when compared to the SSRIs [11]. However, according to clinical experience, tolerability may vary among patients, and it is also possible that an individual patient may experience less adverse effects when switched from an SSRI to an SNRI.

Some SSRIs and SNRIs may inhibit enzymes in the cytochrome P450 system and may interact with other medications for mental or medical illnesses [12].

After stopping treatment with an SSRI, withdrawal reactions may occur. However, these are much less frequent and severe than the withdrawal reactions observed after terminating treatment with benzodiazepines. These adverse reactions may be more frequent with paroxetine than with sertraline or fluoxetine [13].

Pregabalin

Pregabalin is a calcium modulator, acting at the alpha-2-delta subunit of voltage-gated calcium channel. The drug has more sedating effects than SSRIs/SNRIs. Sleep disorders which are common in patients with anxiety disorders may improve earlier with pregabalin than with the SSRIs or SNRIs. Onset of efficacy is earlier with pregabalin than with antidepressants. Pregabalin is not subject to hepatic metabolism and hence does not interact with inhibitors or inducers of cytochrome P450 enzymes. However, there have been concerns about the abuse of pregabalin in individuals suffering from substance abuse and withdrawal syndromes after abrupt discontinuation [14].

Tricyclic Antidepressants (TCAs)

Classic tricyclic antidepressant drugs, including imipramine and clomipramine, are effective medications for anxiety disorders. However, as the frequency of side effects is higher for TCAs than for SSRIs or SNRIs, these antidepressants should be tried first before tricyclics are used. The dosage should be on the same level as in the treatment of major depression. Slow up-titration is recommended. TCAs should be used with caution in suicidal patients, due to their potential fatal toxicity after overdose [15].

Benzodiazepines

The anxiolytic effects of benzodiazepines begin soon after intake. In contrast to SSRIs/SNRIs, benzodiazepines do not lead to initially increased nervousness and insomnia. In the United States, 55–94% of patients with anxiety disorders are treated with benzodiazepines [16]. Likewise, European studies have shown a high rate of long-term benzodiazepine use [17]. However, treatment with benzodiazepines may be associated with CNS depression, resulting in fatigue, dizziness, increased reaction time, impaired driving skills, and other side effects. Cognitive functions may be impaired, mainly in elderly patients. After prolonged use of benzodiazepines (e.g., for 4 to 8 months), addiction may occur in some patients [18–24], in particular in individuals predisposed for alcohol or substance abuse [25]. Tolerance (resulting in the constant wish of the patient to increase the dose) seems to be rare [26], as the majority of patients with prolonged use of benzodiazepines show "low-dose dependency." Most anxiety guidelines do not recommend benzodiazepines as first-line treatments [27]. The recommendations to prefer newer antidepressants are not based on direct comparison studies but rather on the known risks of benzodiazepines [28].

Benzodiazepines can be used for a limited time period in carefully selected patients, for example, in those with suicidality, severe cardiac disease, or contraindications for antidepressants, or in whom standard drugs were not effective or not tolerated. However, patients with a history of benzodiazepine, alcohol, or substance abuse should be excluded from treatment. Benzodiazepines may also be used as add-on SSRIs/SNRIs during the first weeks before the onset of efficacy of the antidepressants or to treat insomnia [29]. Additional cognitive behavioral therapy (CBT) may facilitate benzodiazepine withdrawal [30, 31].

Only in few cases, acute panic attacks may require immediate drug treatment. Then, a benzodiazepine, e.g., lorazepam 1.0–2.5 mg melting tablets, may be given as needed (up to a dose of not higher than 7.5 mg per day). In most cases, talking with the affected patient and assuring that the attack is not due to a dangerous medical condition are sufficient.

In contrast to SSRIs and SNRIs, benzodiazepines do not treat depression, which is a common comorbid condition in anxiety disorders.

Other Drugs

Buspirone. According to some randomized controlled studies, the 5-HT_{1A} agonist buspirone, an azapirone, is effective in the treatment of GAD. However, not all studies showed superiority to placebo and/or equivalence to standard drugs.

Moclobemide is a selective and reversible inhibitor of monoamine oxidase A. It was only licensed for the treatment of SAD. Because not all studies showed evidence for superiority to placebo, the drug is not recommended as first-line drug.

Opipramol is an anxiolytic with the chemical structure similar to the tricyclic antidepressants. Only one double-blind study showing the efficacy of opipramol in GAD has been published [32].

Phenelzine is an irreversible inhibitor of monoamine oxidase (MAOI). Decades ago, it was one of the mostly used drugs in the treatment of panic disorder or social phobia. However, due to a high number of side effects and serious interactions with food components or other medications, its use has decreased in the past decades.

Some other drugs have demonstrated efficacy in randomized controlled trials but are not approved for the treatment of these disorders in most countries. Medicolegal issues have to be considered whenever medications that have not been licensed for anxiety indications are prescribed off-label.

- *Agomelatine* is an antidepressant which acts as an agonist at melatonin MT_1 and MT_2 receptors and as an antagonist at serotonin 5-HT_{2C} receptors. It was shown to be effective in four studies in GAD [33–36]. However, the drug is only licensed for the treatment of depression and not for anxiety disorders.
- *Mirtazapine*, a noradrenergic and specific serotonergic drug, is licensed for depression but not for anxiety disorders in most countries. The drug has more sedating properties than SSRIs/SNRIs. According to a meta-analysis of ten Chinese trials, the drug is effective in GAD [37].
- *Quetiapine*, an atypical antipsychotic, was demonstrated to be effective in a number of trials for GAD. It is usually prescribed in the treatment of schizophrenia and other psychoses in dosages between 150 and 800 mg/d or more. For the treatment of anxiety disorder, lower doses (50–300 mg/day) are required. However, probably due to adverse effects such as the metabolic syndrome, the drug was not licensed for anxiety disorders in most countries. In general, typical side effects such as sedation or increased weight were less frequent in patients receiving lower doses [38, 39]. Quetiapine can only be used off-label in treatment-refractory patients. The onset of efficacy is earlier than with antidepressants.
- *Vilazodone* acts as a selective serotonin reuptake inhibitor and 5-HT_{1A} receptor partial agonist. It is only licensed for the treatment of major depression in the United States. It has shown efficacy in GAD [40, 41] and SAD [42] but was not licensed for anxiety disorders. The drug is available in the United States.

General Treatment Principles

Patients must be informed about relative efficacy of treatment alternatives, possible adverse effects, interactions, safety warnings, and contraindications, following the up-to-date summary of product characteristic. When patients are educated about the possibility that some early side effects might later decrease in intensity, compliance may be improved. Patients with anxiety disorders are often hesitant to take psychotropic drugs because they are afraid of adverse effects. In particular, patients with PDA may easily discontinue antidepressants because of nervousness or even an increase in panic attacks in the first days or weeks after starting treatment.

Recommended doses are shown in Table 19.2. For around three quarters of the patients, doses in lower part of the therapeutic range are sufficient (e.g., escitalopram 10 mg or sertraline 50 mg per day). However, when treatment response is insufficient, a dose increase to the upper limit of the standard dose range should be tried. Dose-finding studies are often made with samples containing twice as much women than men and are based on persons with an average weight. That means that the dose for male patients, especially those with overweight, should be adjusted accordingly. In patients with severe hepatic impairment, a dosage adjustment may be required. Pregabalin, a drug with primarily renal clearance, may be an alternative.

Relative Efficacy of Medications for Anxiety Disorders

In a meta-analysis of all controlled studies for anxiety disorders [2], the pre-post effect sizes of available antianxiety agents were determined. We simply looked at the absolute difference of anxiety scale scores before and after treatment, without regard to the relative efficacy compared to placebo. This approach makes it possible to include hundreds of studies in comparisons of differential efficacy of all available drugs and not only the few direct head-to-head comparisons. From the patients' point of view, the improvement of anxiety symptoms as measured by the change from baseline to endpoint is more relevant than the difference to a control group.

The available medications for anxiety disorders showed considerably large differences in pre-post effect sizes. For example, the improvement achieved with the most efficacious drug (quetiapine, Cohen's $d = 3.39$) was almost three times higher than what was accomplished with the drug with the weakest efficacy (buspirone, $d = 1.35$). Quetiapine, however, is not licensed for the treatment of any anxiety disorder in most countries. Among the drugs, showing high effect sizes that are licensed for anxiety disorders and recommended by guidelines, were the SSRIs escitalopram ($d = 2.75$) and paroxetine ($d = 2.42$) and the SNRIs venlafaxine ($d = 2.32$) and pregabalin (2.30). Also, some benzodiazepines, e.g., diazepam ($d = 2.46$) and lorazepam (2.44), showed high effect sizes. These drugs, however, are not recommended for routine treatment.

Similar results were found in a network analysis of drug treatment for GAD [37, 43].

Duration of Treatment

For all drugs recommended in this article, relapse prevention studies for at least one anxiety disorder have been conducted in patients who were responders to open treatment with a certain drug and were then randomized to ongoing blind treatment with the same drug to or placebo for long-term treatment for another 6–18 months.

All of these trials showed an advantage for staying on active medication when compared with switching to placebo. Based on the findings from these relapse prevention studies, clinical experiences, and expert opinions, drug treatment should be continued for 6–12 months after remission, or even more, when relapses occur after tapering the drug. To avoid withdrawal effects, the drug should be slowly tapered off over 2 weeks at treatment termination.

Drug-Drug Interactions

When treating anxiety disorders with drugs, possible drug interactions have to be monitored. SSRIs like fluoxetine, fluvoxamine, and paroxetine are particularly liable to be potentially involved in pharmacokinetic interactions such as enzyme inhibition in the cytochrome P450 system [12]. Pregabalin can be combined with SSRIs or SNRIs without causing interactions in the cytochrome system. This may be helpful for patients complaining about sleep problems which do not result in the first weeks of SSRI/SNRI treatment.

Increased CNS depression may occur when drugs with sedating properties are combined, e.g., tricyclic antidepressants, benzodiazepines, or pregabalin, resulting in unwanted sedation, drowsiness, or increased reaction time. Additive effects at neurotransmitter level can occur when medications are combined that have antagonistic effects at the same receptors, e.g., two drugs with anticholinergic effects.

Unresponsiveness to Standard Treatments

When a patient shows insufficient response to treatment, it should be made certain that the diagnostic considerations were correct, adherence to the treatment plan was sufficient, the dose prescribed had covered the full range, and there had been a trial period of adequate duration. When patients report previous treatment failures, it often turns out that a drug was only prescribed in the lowest dose or was stopped within the first 2 weeks due to side effects, which occurred in the initial phase, before the patient could experience improvement. Concurrent medications may interfere with efficacy, e.g., enzyme inhibitors. Adverse psychosocial conditions may also affect response. Comorbid personality disorders or alcohol/substance abuse may complicate anxiety disorders.

If a patient shows no response to treatment in adequate dose after 4 weeks, medication should be changed. In one analysis, it was found that the chance of responding beyond the fourth week was 20% or less, if no effect had occurred by the second week of treatment [44]. Elderly patients may take longer to respond. If partial response is seen after 4 weeks, there is still a chance that the patient will respond after another 4–6 weeks of therapy with increased dosages. There have been only few controlled trials of treatment-refractory patients with anxiety disorders, and "switching studies" are lacking. However, many treatment-unresponsive patients are reported to improve after a switch to a different class of antidepressants

Table 19.2 Stepwise plan for drug treatment if the initial standard drug treatment was not effective or poorly tolerated*

Switch from one standard drug to another	Switch from one SSRI to another Switch from an SSRI to an SNRI, or vice versa Switch to a TCA Switch to pregabalin (only in GAD)	
Switch to a second-line drug	Switch to a benzodiazepine (only when clinically justified) Switch to phenelzine, opipramol, buspirone, moclobemide	
Switch to a drug that is approved for other anxiety disorders	PDA	Pregabalin, duloxetine, phenelzine, moclobemide, buspirone
	SAD	Pregabalin, duloxetine, phenelzine, buspirone
Switch to a drug that is not approved for the anxiety disorder in question but has been found effective in RCTs	PDA	Mirtazapine, quetiapine
	GAD	Quetiapine, agomelatine, vilazodone

Abbreviations see Table 19.1

*Medicolegal issues should be considered whenever drugs that have not been approved for the treatment of a certain anxiety disorder are given off label

(e.g., change from one SSRI to another SSRI, or to an SNRI). Studies on a potential dose-response relationship of antidepressants are often inconclusive, perhaps due to the lack of statistical power for showing a difference between lower and higher doses. According to clinical experience, however, a trial with a higher dose in patients with insufficient response is warranted.

Table 19.2 contains alternatives for patients who have not responded to or have not tolerated standard medications. For all patients who are unresponsive to psychopharmacological drugs, additional CBT is generally recommended, although only few randomized studies have been conducted that support this strategy [45, 46].

A combination of antidepressants and benzodiazepines is sometimes used in treatment-refractory cases. Moreover, SSRIs/SNRIs can be combined with pregabalin due to the low interaction potential of this combination. When all standard treatments have failed, the off-label use of drugs may be considered, which are licensed for another anxiety disorder or are not licensed but have shown efficacy in clinical studies. These include quetiapine, agomelatine, and vilazodone.

Treating Older Patients

Most anxiety disorders are less common in patients over 65 years, with the exception of GAD. The mean age of patients with PDA, GAD, and SAD in clinical studies are 37, 41, and 35, respectively [47]. Only few studies for the treatment of GAD have been performed with older patients. Randomized trials have shown the efficacy of duloxetine, pregabalin, quetiapine, and venlafaxine in patients over 65 years [9].

In the elderly, increased sensitivity to adverse effects and interactions has been observed, including anticholinergic effects, orthostatic hypotension and risk of cardiovascular events, risk of falling, and paradoxical agitation with benzodiazepines.

Treating Children and Adolescents

While specific phobias, SAD, and separation anxiety disorder are common in children and youths, PDA and GAD are relatively rare before the age of 18. There are some randomized placebo-controlled studies of pharmacotherapy of anxiety disorders in children and youths, showing efficacy of sertraline, fluoxetine, and duloxetine in young patients with GAD, of venlafaxine and paroxetine in SAD, and of sertraline, fluvoxamine, and fluoxetine in mixed samples, including patients with separation anxiety disorder, SAD, and GAD [48].

The combination of cognitive therapy and sertraline was found to be more efficacious than both treatment modalities alone in children with separation anxiety disorder, GAD, or SAD [49].

In a very small study, in which children with separation anxiety disorder who did not respond sufficiently to behavioral therapy were randomized to imipramine and placebo, the drug failed to show a difference to placebo [50]; however, the sample size was perhaps too small to detect a difference.

School refusal is not a DSM-5 diagnostic category; however, the condition is mostly a consequence of separation anxiety disorder. In an early study in children with school refusal, imipramine was superior to placebo [51]. In another study, alprazolam and imipramine were not superior to placebo in children with school refusal, but the simple size was too small to detect a difference [52]. The combination of imipramine and CBT was significantly more efficacious than placebo plus CBT with regard to school attendance and depressive symptoms in adolescents with school refusal comorbid with anxiety and depression [53]. A study comparing clomipramine and placebo in the treatment of youths with school refusal did not find a difference between the two conditions [54].

There had been concerns about increased risk for suicidal ideation (not completed suicides) in children and adolescents treated for major depression with SSRIs (escitalopram, citalopram, paroxetine, and sertraline), mirtazapine, and venlafaxine [55]. According to a meta-analysis, the risk/benefit ratio seemed to be most favorable with the SSRI fluoxetine [56]. Although suicidal ideation is less common in anxiety disorders than in depression, the risks of pharmacological treatment have to be weighed carefully against the risks of nontreatment. It was reported that antidepressant prescriptions for children and adolescents decreased substantially after the European Medicines Agency and the US Food and Drugs Administration issued warnings about a possible suicide risk with antidepressant use in pediatric patients in 2003/2004 [57] and that the reduced use of antidepressants was associated with increases in rates of completed suicides in children and adolescents (although a causal relationship is not proven) [58].

Separation Anxiety Disorder in Adults

Since 2013, separation anxiety disorder can also be diagnosed in adults, according to DSM-5. There is only one controlled study with drug treatment for this condition. A small pilot trial comparing vilazodone with placebo in adults with separation anxiety disorder did not show a difference in response rates but greater improvement in separation anxiety symptoms [59].

Selective Mutism

There is also a paucity of treatment studies for children with selective mutism. Small studies have shown that psychotherapeutic approaches were at least better than wait list controls [60, 61]. Only two very small placebo-controlled drug studies were found according to a review [62], showing the efficacy of the SSRIs fluoxetine and sertraline.

Pregnancy and Breastfeeding

When pregnant women suffer from an anxiety disorder, the risk of an untreated anxiety disorder must be weighed against the risk of damage to the unborn child as a result of drug treatment. A large study suggested no substantial increase in the risk of cardiac malformations attributable to use of antidepressants during the first 3 months of pregnancy [63]. Antidepressants have been associated with increased risk of spontaneous abortions, stillbirths, early deliveries, respiratory distress, and endocrine and metabolic dysfunctions [64]. However, the current evidence suggests that the use of many antidepressants, especially the SSRIs, is favorable compared to exposing the mother to the risks of untreated depression or anxiety disorders [65].

Likewise, a careful assessment of the risk/benefit balance has to be done when a mother is breastfeeding. In such cases, CBT should be considered as an alternative to medication treatment.

Combining Psychotherapy and Medication

Patients require supportive talks and attention to the emotional problems that are associated with the anxiety disorder. Psychoeducation includes information about the physiology of the bodily symptoms of anxiety reactions and the rationale of available treatment possibilities. Many patients may require formal psychological treatment interventions, which are mostly done on an outpatient basis. In the case of phobic avoidance, exposure techniques should be included in the treatment schedule, in which patients are exposed to their feared situations.

Both psychotherapy and pharmacotherapy have been shown to be more effective than control groups. However, while drugs are mostly compared with placebo controls, the evidence for psychotherapy in anxiety disorders is mainly based on comparisons with a "wait list," a control method that was used in 70–75% of the studies in adults and children [2, 66, 67]. Because pill placebos have much higher effect sizes than wait list controls, the effect size differences between active and control conditions cannot be compared between psychotherapy and medication studies. Therefore, our research group conducted a large meta-analysis of all available controlled short-term studies for anxiety disorders and compared the pre-post effect size differences (before and after treatment) between medications and psychotherapies [2]. In this meta-analysis, which was based on studies with around 35,000 patients, antianxiety drugs were associated with a significantly higher average pre-post effect size (Cohen's $d = 2.02$) than psychotherapies ($d = 1.22$; $p < 0.0001$). It was also found that patients included in psychotherapy studies were less severely ill than those recruited for medication trials. Patients should be informed about the relative efficacy of the treatment options they are offered.

The meta-analysis also showed that combinations of psychotherapy and pharmacotherapy had a relatively high effect size ($d = 2.12$). However, only few combination studies were available for this comparison, and some of these have not been conducted with the most powerful drugs.

It is a common opinion that patients treated with medications show immediate relapse after stopping the treatment, whereas gains of psychotherapies are maintained for years after treatment termination. This would offer psychological therapies considerable advantage over pharmacological treatment. In a meta-analysis of follow-up studies with psychotherapy and medication, it was found that patients who have terminated psychotherapy maintain their gains for up to 2 years; however, the same enduring effects have been found within patients who had stopped medication [68].

Novel Treatment Approaches

In the 1980s and 1990s, many new pharmacological drugs have been developed for the treatment of anxiety disorders. However, since then, there has been substantial decline in the number of clinical studies with developmental drugs for anxiety disorders. The first reason is that we now already have a large number of drugs that are effective and well tolerated for the treatment of these disorders. The second reason is that more and more pharmaceutical companies have withdrawn from the development of new drugs because an increasing number of national and international regulations for new drug application have made the process increasingly expensive and time-consuming. Moreover, any novel drug developed for anxiety disorders has to compete with medications which are already on the market and are very inexpensive because the patents of all licensed antianxiety drugs have expired and, therefore, generic drugs are widely available.

In the past decades, research on the neurobiological backgrounds of the anxiety disorders has substantially increased our knowledge of the etiology of these disorders. While the boom of anxiety drugs in the 1980s and 1990s has stimulated a huge number of studies on the neurochemistry of anxiety disorders [4], there has been a substantial shift to genetic studies and neuroimaging [3]. However, we still do not have evidence that a specific dysfunction of a neurotransmitter system is the main cause for pathological anxiety. Still, probably the most robust evidence exists for an involvement of the serotonin system, which is based on the observation that a large number of drugs that are effective treatments for anxiety disorders have a common denominator, i.e., that they modulate serotonergic neurotransmission. The main mechanism of action of these antidepressants is serotonin reuptake inhibition, but other serotonergic drugs have agonist or antagonist properties at various serotonin receptors.

Serotonergic Drugs

In the past years, few new serotonergic drugs came to the market. Unfortunately, pharmaceutical companies are reluctant to develop new serotonergic drugs, although this might be the most promising research line, given the efficacy of so many serotonergic antidepressants in anxiety disorders. If a manufacturer would spend enormous development costs for such a new drug, it would most likely be treated as a "me-too" drug and would compete with low-price SSRIs.

GABAergic Drugs

Other medications that are effective treatments for anxiety act at the γ-aminobutyric acid (GABA) binding site by reinforcing the effects of GABA. Recent research has focused on developing drugs that target the GABA receptor but do not have the unwanted side effects of benzodiazepines, including sedation and addiction. Flavonoids, which are widely distributed in plants and fungi, have selective affinity for $GABA_A$ receptors and demonstrated anxiolytic-like activity in laboratory animals, but they are devoid of the adverse side effects of benzodiazepines. Methoxyflavanone, a positive allosteric modulator of $GABA_A$, exerted an anxiolytic-like effect in rodents [69]. ADX71441, a novel, potent, and selective $GABA_B$ positive allosteric modulator, had anxiolytic properties in mice [70]. Also, the GABA transporter (GAT) may be a target for antianxiety agents. A novel, potent, and selective inhibitor of the GAT_1, the guvacine derivative DDPM-2571, had anxiolytic effects in rodents [71].

Additionally, drugs that target the translocator protein (18 kDa), formerly called the peripheral or mitochondrial benzodiazepine receptor, may have anxiolytic properties. The translocator protein ligand XBD173 reinforced GABA-mediated neurotransmission and blunted induced panic attacks in rodents without causing sedation and tolerance [72]. Another novel ligand of translocator protein (18 kDa), ZBD-2, had anxiolytic-like effects in mice [73].

However, there is a long way from rodent studies to a new class of drugs that could replace the benzodiazepines. To date, there have been no successful trials in humans showing the efficacy of such medications. The $GABA_A$ positive allosteric modulator PF-06372865 was not superior to placebo in a small study with GAD patients [74].

Glutamatergic Drugs

Glutamate is the major excitatory neurotransmitter in the brain. It was thought that mechanisms that suppress glutamate hyperexcitability might lead to novel anxiolytics. However, the glutamatergic drugs that have been studied so far often have problematic side effects [75], and therefore their development was terminated. However, recent preclinical studies have opened new perspectives for drugs targeting metabotropic glutamate (mGlu) receptors, showing the efficacy of mGlu5 receptor negative allosteric modulators, group II mGlu receptor orthosteric agonists, and positive allosteric modulators as anxiolytic agents [76].

Other

Other research lines in the search for biomarkers of anxiety focused on the role of neurotransmitters such as serotonin, norepinephrine, dopamine, neuropeptides (e.g., cholecystokinin), neurokinins, atrial natriuretic peptide (ANP), or oxytocin. Moreover, research on the hypothalamic-pituitary-adrenal (HPA) axis or neurotrophic factors such as nerve growth factor (NGF) and brain-derived neurotrophic factor (BDNF) may result in the development of putative antianxiety medications [4]. A novel μ-opioid agonist, PILAB 8, was shown to have anxiolytic properties in mice [77]. Also, drugs that target the endocannabinoid system have been investigated in animals [78]. For example, the drug O-1602 which acts at the newly detected cannabinoid receptor GPR55 mediates anxiolytic-like effects in mice [79]. Curcumin is a substance contained in the natural spice turmeric (curcuma). It is the "natural" COX-2 inhibitor, because it can downregulate COX-2 expression. It has shown a potential to reduce depressive and anxious symptoms in patients with depression [80].

Conclusions

Anxiety disorders are the most prevalent mental disorders. A large database of randomized controlled trials permits the formulation of evidence-based recommendations for the treatment of PDA, GAD, and SAD. In most cases, drug treatment and CBT may substantially improve quality of life in GAD patients. Unfortunately, the development of novel medications for anxiety disorders has stagnated in the recent years.

References

1. Bandelow B, Michaelis S. Epidemiology of anxiety disorders in the 21st century. Dialogues Clin Neurosci. 2015;17(3):327–35. Epub 2015/10/22
2. Bandelow B, Reitt M, Rover C, Michaelis S, Gorlich Y, Wedekind D. Efficacy of treatments for anxiety disorders: a meta-analysis. Int Clin Psychopharmacol. 2015;30(4):183–92. Epub 2015/05/02
3. Bandelow B, Baldwin D, Abelli M, Altamura C, Dell'Osso B, Domschke K, et al. Biological markers for anxiety disorders, OCD and PTSD – a consensus statement. Part I: Neuroimaging and genetics. World J Biol Psychiatry. 2016;17(5):321–65. Epub 2016/07/13
4. Bandelow B, Baldwin D, Abelli M, Bolea-Alamanac B, Bourin M, Chamberlain SR, et al. Biological markers for anxiety disorders, OCD and PTSD: a consensus statement. Part II: Neurochemistry, neurophysiology and neurocognition. World J Biol Psychiatry. 2017;18(3):162–214. Epub 2016/07/16
5. Regier DA, Narrow WE, Rae DS, Manderscheid RW, Locke BZ, Goodwin FK. The de facto US mental and addictive disorders service system. Epidemiologic catchment area prospective 1-year prevalence rates of disorders and services. Arch Gen Psychiatry. 1993;50(2):85–94.
6. Alonso J, Lepine JP, Committee ESMS. Overview of key data from the European Study of the Epidemiology of Mental Disorders (ESEMeD). J Clin Psychiatry. 2007;68(Suppl 2):3–9.
7. Bandelow B, Sher L, Bunevicius R, Hollander E, Kasper S, Zohar J, et al. Guidelines for the pharmacological treatment of anxiety disorders, obsessive-compulsive disorder and post-traumatic stress disorder in primary care. Int J Psychiatry Clin Pract. 2012;16(2):77–84. Epub 2012/05/01
8. Baldwin DS, Anderson IM, Nutt DJ, Allgulander C, Bandelow B, den Boer JA, et al. Evidence-based pharmacological treatment of anxiety disorders, post-traumatic stress disorder and obsessive-compulsive disorder: a revision of the 2005 guidelines from the British Association for Psychopharmacology. J Psychopharmacol. 2014;28(5):403–39. Epub 2014/04/10
9. Bandelow B, Lichte T, Rudolf S, Wiltink J, Beutel ME. The German guidelines for the treatment of anxiety disorders. Eur Arch Psychiatry Clin Neurosci. 2015;265(5):363–73. Epub 2014/11/19
10. Benjamin J, Ben-Zion IZ, Karbofsky E, Dannon P. Double-blind placebo-controlled pilot study of paroxetine for specific phobia. Psychopharmacology. 2000;149(2):194–6. Epub 2000/05/11
11. Cipriani A, Furukawa TA, Salanti G, Geddes JR, Higgins JP, Churchill R, et al. Comparative efficacy and acceptability of 12 new-generation antidepressants: a multiple-treatments meta-analysis. Lancet. 2009;373(9665):746–58.
12. Muscatello MR, Spina E, Bandelow B, Baldwin DS. Clinically relevant drug interactions in anxiety disorders. Hum Psychopharmacol. 2012;27(3):239–53. Epub 2012/02/09
13. Stahl MMS, Lindquist M, Pettersson M, Edwards IR, Sanderson JH, Taylor NFA, et al. Withdrawal reactions with selective serotonin re-uptake inhibitors as reported to the WHO system. Eur J Clin Pharmacol. 1997;53(3–4):163–9.
14. Baldwin DS, Ajel K, Masdrakis VG, Nowak M, Rafiq R. Pregabalin for the treatment of generalized anxiety disorder: an update. Neuropsychiatr Dis Treat. 2013;9:883–92.
15. Thanacoody HK, Thomas SH. Tricyclic antidepressant poisoning: cardiovascular toxicity. Toxicol Rev. 2005;24(3):205–14.
16. Stahl SM. Don't ask, don't tell, but benzodiazepines are still the leading treatments for anxiety disorder. J Clin Psychiatry. 2002;63(9):756–7.
17. Starcevic V. The reappraisal of benzodiazepines in the treatment of anxiety and related disorders. Expert Rev Neurother. 2014;14(11):1275–86.
18. Schweizer E, Rickels K, Case WG, Greenblatt DJ. Long-term therapeutic use of benzodiazepines. II. Effects of gradual taper. Arch Gen Psychiatry. 1990;47(10):908–15.
19. Rickels K, Schweizer E, Case WG, Greenblatt DJ. Long-term therapeutic use of benzodiazepines. I. Effects of abrupt discontinuation [published erratum appears in Arch Gen Psychiatry 1991; 48(1):51]. Arch Gen Psychiatry. 1990;47(10):899–907.

20. Smith DE, Landry MJ. Benzodiazepine dependency discontinuation: focus on the chemical dependency detoxification setting and benzodiazepine-polydrug abuse. J Psychiatr Res. 1990;24(Suppl 2):145–56. Epub 1990/01/01
21. Bradwejn J. Benzodiazepines for the treatment of panic disorder and generalized anxiety disorder: clinical issues and future directions. Can J Psychiatry. 1993;38(Suppl 4):S109–13. Epub 1993/11/01
22. Shader RI, Greenblatt DJ. Use of benzodiazepines in anxiety disorders. New Engl J Med. 1993;13:1398–405.
23. Livingston MG. Benzodiazepine dependence. Br J Hosp Med. 1994;51(6):281–6. Epub 1994/03/05
24. Nelson J, Chouinard G. Guidelines for the clinical use of benzodiazepines: pharmacokinetics, dependency, rebound and withdrawal. Can J Clin Pharmacol. 1999;6(2):69–83. Epub 1999/10/16
25. Schweizer E, Rickels K, De Martinis N, Case G, Garcia-Espana F. The effect of personality on withdrawal severity and taper outcome in benzodiazepine dependent patients. Psychol Med. 1998;28(3):713–20. Epub 1998/06/17
26. Rickels K. Benzodiazepines in the treatment of anxiety. Am J Psychother. 1982;36(3):358–70. Epub 1982/07/01
27. Bandelow B, Lichte T, Rudolf S, Wiltink J, Beutel ME. The diagnosis of and treatment recommendations for anxiety disorders. Deutsch Arztebl Int. 2014;111(27–28):473–80. Epub 2014/08/21
28. Berney P, Halperin D, Tango R, Daeniker-Dayer I, Schulz P. A major change of prescribing pattern in absence of adequate evidence: benzodiazepines versus newer antidepressants in anxiety disorders. Psychopharmacol Bull. 2008;41(3):39–47.
29. Goddard AW, Brouette T, Almai A, Jetty P, Woods SW, Charney D. Early coadministration of clonazepam with sertraline for panic disorder. Arch Gen Psychiatry. 2001;58(7):681–6. Epub 2001/07/31
30. Otto MW, Pollack MH, Sachs GS, Reiter SR, Meltzer-Brody S, Rosenbaum JF. Discontinuation of benzodiazepine treatment: efficacy of cognitive-behavioral therapy for patients with panic disorder. Am J Psychiatry. 1993;150(10):1485–90. Epub 1993/10/01
31. Spiegel DA. Psychological strategies for discontinuing benzodiazepine treatment. J Clin Psychopharmacol. 1999;19(6 Suppl 2):17S–22S. Epub 1999/12/10
32. Möller HJ, Volz HP, Reimann IW, Stoll KD. Opipramol for the treatment of generalized anxiety disorder: a placebo-controlled trial including an alprazolam-treated group. J Clin Psychopharmacol. 2001;21(1):59–65. Epub 2001/02/24
33. Stein DJ, Ahokas AA, de Bodinat C. Efficacy of agomelatine in generalized anxiety disorder: a randomized, double-blind, placebo-controlled study. J Clin Psychopharmacol. 2008;28(5):561–6. Epub 2008/09/17
34. Stein DJ, Ahokas A, Albarran C, Olivier V, Allgulander C. Agomelatine prevents relapse in generalized anxiety disorder: a 6-month randomized, double-blind, placebo-controlled discontinuation study. J Clin Psychiatry. 2012;73(7):1002–8. Epub 2012/08/21
35. Stein DJ, Ahokas A, Marquez MS, Hoschl C, Oh KS, Jarema M, et al. Agomelatine in generalized anxiety disorder: an active comparator and placebo-controlled study. J Clin Psychiatry. 2014;75(4):362–8.
36. Stein DJ, Ahokas A, Jarema M, Avedisova AS, Vavrusova L, Chaban O, et al. Efficacy and safety of agomelatine (10 or 25 mg/day) in non-depressed out-patients with generalized anxiety disorder: a 12-week, double-blind, placebo-controlled study. Eur Neuropsychopharmacol. 2017;27(5):526–37.
37. Slee A, Nazareth I, Bondaronek P, Liu Y, Cheng Z, Freemantle N. Pharmacological treatments for generalised anxiety disorder: a systematic review and network meta-analysis. Lancet 2019. Epub 2019/02/05.
38. Bandelow B, Chouinard G, Bobes J, Ahokas A, Eggens I, Liu S, et al. Extended-release quetiapine fumarate (quetiapine XR): a once-daily monotherapy effective in generalized anxiety disorder. Data from a randomized, double-blind, placebo- and active-controlled study. Int J Neuropsychopharmacol. 2010;13(3):305–20. Epub 2009/08/21

39. Khan A, Joyce M, Atkinson S, Eggens I, Baldytcheva I, Eriksson H. A randomized, double-blind study of once-daily extended release quetiapine fumarate (quetiapine XR) monotherapy in patients with generalized anxiety disorder. J Clin Psychopharmacol. 2011;31(4):418–28. Epub 2011/06/23
40. Gommoll C, Durgam S, Mathews M, Forero G, Nunez R, Tang X, et al. A double-blind, randomized, placebo-controlled, fixed-dose phase III study of vilazodone in patients with generalized anxiety disorder. Depress Anxiety. 2015;32(6):451–9.
41. Gommoll C, Forero G, Mathews M, Nunez R, Tang X, Durgam S, et al. Vilazodone in patients with generalized anxiety disorder: a double-blind, randomized, placebo-controlled, flexible-dose study. Int Clin Psychopharmacol. 2015;30(6):297–306.
42. Careri JM, Draine AE, Hanover R, Liebowitz MR. A 12-week double-blind, placebo-controlled, flexible-dose trial of vilazodone in generalized social anxiety disorder. Prim Care Companion CNS Disord. 2015;17(6)
43. Bandelow B. Wedekind D. Lancet: Network analyses to rank pharmacological treatments for generalised anxiety disorder; 2019. Epub 2019/02/05
44. Baldwin DS, Stein DJ, Dolberg OT, Bandelow B. How long should a trial of escitalopram treatment be in patients with major depressive disorder, generalised anxiety disorder or social anxiety disorder? An exploration of the randomised controlled trial database. Hum Psychopharmacol. 2009;24(4):269–75. Epub 2009/04/01
45. Yoshinaga N, Matsuki S, Niitsu T, Sato Y, Tanaka M, Ibuki H, et al. Cognitive behavioral therapy for patients with social anxiety disorder who remain symptomatic following antidepressant treatment: a randomized, assessor-blinded, controlled trial. Psychother Psychosom. 2016;85(4):208–17.
46. Campbell-Sills L, Roy-Byrne PP, Craske MG, Bystritsky A, Sullivan G, Stein MB. Improving outcomes for patients with medication-resistant anxiety: effects of collaborative care with cognitive behavioral therapy. Depress Anxiety. 2016;
47. Bandelow B, Schuller K. Mean age and gender distribution of patients with major mental disorders participating in clinical trials. Eur Arch Psychiatry Clin Neurosci. 2019. Epub 2019/01/03
48. Hussain FS, Dobson ET, Strawn JR. Pharmacologic treatment of pediatric anxiety disorders. Curr Treat Options Psychiatry. 2016;3(2):151–60.
49. Walkup JT, Albano AM, Piacentini J, Birmaher B, Compton SN, Sherrill JT, et al. Cognitive behavioral therapy, sertraline, or a combination in childhood anxiety. N Engl J Med. 2008;359(26):2753–66.
50. Klein RG, Koplewicz HS, Kanner A. Imipramine treatment of children with separation anxiety disorder. J Am Acad Child Adolesc Psychiatry. 1992;31(1):21–8. Epub 1992/01/01
51. Gittelman-Klein R, Klein DF. School phobia: controlled imipramine treatment. Calif Med. 1971;115(3):42. Epub 1971/09/01
52. Bernstein GA, Garfinkel BD, Borchardt CM. Comparative studies of pharmacotherapy for school refusal. J Am Acad Child Adolesc Psychiatry. 1990;29(5):773–81. Epub 1990/09/01
53. Bernstein GA, Borchardt CM, Perwien AR, Crosby RD, Kushner MG, Thuras PD, et al. Imipramine plus cognitive-behavioral therapy in the treatment of school refusal. J Am Acad Child Adolesc Psychiatry. 2000;39(3):276–83. Epub 2000/03/14
54. Berney T, Kolvin I, Bhate SR, Garside RF, Jeans J, Kay B, et al. School phobia: a therapeutic trial with clomipramine and short-term outcome. Br J Psychiatry. 1981;138:110–8. Epub 1981/02/01
55. Selective serotonin reuptake inhibitors (SSRIs) and serotonin and noradrenaline reuptake inhibitors (SNRIs): use and safety [database on the Internet]. Medicines & Healthcare products Regulatory Agency (MHRA). 2014 [cited 18/05/2017]. Available from: https://www.gov.uk/government/publications/ssris-and-snris-use-and-safety/selective-serotonin-reuptake-inhibitors-ssris-and-serotonin-and-noradrenaline-reuptake-inhibitors-snris-use-and-safety
56. Hetrick SE, McKenzie JE, Cox GR, Simmons MB, Merry SN. Newer generation antidepressants for depressive disorders in children and adolescents. Cochrane Database Syst Rev. 2012;11:CD004851.

57. Libby AM, Brent DA, Morrato EH, Orton HD, Allen R, Valuck RJ. Decline in treatment of pediatric depression after FDA advisory on risk of suicidality with SSRIs. Am J Psychiatry. 2007;164(6):884–91.
58. Gibbons RD, Brown CH, Hur K, Marcus SM, Bhaumik DK, Erkens JA, et al. Early evidence on the effects of regulators' suicidality warnings on SSRI prescriptions and suicide in children and adolescents. Am J Psychiatry. 2007;164(9):1356–63. Epub 2007/08/31
59. Schneier FR, Moskow DM, Choo TH, Galfalvy H, Campeas R, Sanchez-Lacay A. A randomized controlled pilot trial of vilazodone for adult separation anxiety disorder. Depress Anxiety. 2017;34(12):1085–95. Epub 2017/10/27
60. Bergman RL, Gonzalez A, Piacentini J, Keller ML. Integrated behavior therapy for selective mutism: a randomized controlled pilot study. Behav Res Ther. 2013;51(10):680–9.
61. Oerbeck B, Stein MB, Wentzel-Larsen T, Langsrud O, Kristensen H. A randomized controlled trial of a home and school-based intervention for selective mutism – defocused communication and behavioural techniques. Child Adolesc Mental Health. 2014;19(3):192–8.
62. Manassis K, Oerbeck B, Overgaard KR. The use of medication in selective mutism: a systematic review. Eur Child Adolesc Psychiatry. 2016;25(6):571–8.
63. Huybrechts KF, Palmsten K, Avorn J, Cohen LS, Holmes LB, Franklin JM, et al. Antidepressant use in pregnancy and the risk of cardiac defects. N Engl J Med. 2014;370(25):2397–407.
64. Oyebode F, Rastogi A, Berrisford G, Coccia F. Psychotropics in pregnancy: safety and other considerations. Pharmacol Ther. 2012;135(1):71–7. Epub 2012/04/10
65. Muzik M, Hamilton SE. Use of antidepressants during pregnancy?: what to consider when weighing treatment with antidepressants against untreated depression. Matern Child Health J. 2016;20(11):2268–79.
66. Patterson B, Boyle MH, Kivlenieks M, Van Ameringen M. The use of waitlists as control conditions in anxiety disorders research. J Psychiatr Res. 2016;83:112–20.
67. James AC, James G, Cowdrey FA, Soler A, Choke A. Cognitive behavioural therapy for anxiety disorders in children and adolescents. Cochrane Database Syst Rev. 2015;2:CD004690.
68. Bandelow B, Sagebiel A, Belz M, Gorlich Y, Michaelis S, Wedekind D. Enduring effects of psychological treatments for anxiety disorders: meta-analysis of follow-up studies. Br J Psychiatry. 2018;212(6):333–8. Epub 2018/05/01
69. Akbar S, Subhan F, Karim N, Aman U, Ullah S, Shahid M, et al. Characterization of 6-methoxyflavanone as a novel anxiolytic agent: a behavioral and pharmacokinetic approach. Eur J Pharmacol. 2017;801:19–27. Epub 2017/03/05
70. Kalinichev M, Girard F, Haddouk H, Rouillier M, Riguet E, Royer-Urios I, et al. The drug candidate, ADX71441, is a novel, potent and selective positive allosteric modulator of the GABAB receptor with a potential for treatment of anxiety, pain and spasticity. Neuropharmacology. 2017;114:34–47. Epub 2016/11/28
71. Salat K, Podkowa A, Malikowska N, Kern F, Pabel J, Wojcieszak E, et al. Novel, highly potent and in vivo active inhibitor of GABA transporter subtype 1 with anticonvulsant, anxiolytic, antidepressant and antinociceptive properties. Neuropharmacology. 2017;113(Pt A):331–42. Epub 2016/11/05
72. Rupprecht R, Rammes G, Eser D, Baghai TC, Schule C, Nothdurfter C, et al. Translocator protein (18 kD) as target for anxiolytics without benzodiazepine-like side effects. Science. 2009;325(5939):490–3.
73. Wang DS, Han J, Li S, Sun T, Guo YY, Kang WB, et al. Antidepressant-like and anxiolytic-like effects of ZBD-2, a novel ligand for the translocator protein (18 kDa). Neuromolecular Med. 2017;19(1):57–68. Epub 2016/08/22
74. Simen A, Whitlock M, Qiu R, Miceli J, Zumpano L, Du Metz M, et al. An 8-week, randomized, phase 2, double-blind, sequential parallel-group comparison study of two dose levels of the GABAA positive allosteric modulator PF-06372865 compared with placebo as an adjunctive treatment in outpatients with inadequate response to standard of care for generalized anxiety disorder. J Clin Psychopharmacol. 2019;39(1):20–7. Epub 2018/12/12

75. Swanson CJ, Bures M, Johnson MP, Linden A-M, Monn JA, Schoepp DD. Metabotropic glutamate receptors as novel targets for anxiety and stress disorders. Nat Rev Drug Discov. 2005;4:131.
76. Ferraguti F. Metabotropic glutamate receptors as targets for novel anxiolytics. Curr Opin Pharmacol. 2018;38:37–42. Epub 2018/03/02
77. Montes GC, da Silva BNM, Rezende B, Sudo RT, Ferreira VF, de Carvalho da Silva F, et al. The hypnotic, anxiolytic, and antinociceptive profile of a novel micro-opioid agonist. Molecules. 2017;22(5):800. Epub 2017/05/17
78. Patel S, Hill MN, Cheer JF, Wotjak CT, Holmes A. The endocannabinoid system as a target for novel anxiolytic drugs. Neurosci Biobehav Rev. 2017;76(Pt A):56–66. Epub 2017/04/25
79. Shi QX, Yang LK, Shi WL, Wang L, Zhou SM, Guan SY, et al. The novel cannabinoid receptor GPR55 mediates anxiolytic-like effects in the medial orbital cortex of mice with acute stress. Mol Brain. 2017;10(1):38. Epub 2017/08/13
80. Baldwin DS, Hou R, Gordon R, Huneke NT, Garner M. Pharmacotherapy in generalized anxiety disorder: novel experimental medicine models and emerging drug targets. CNS Drugs. 2017;31(4):307–17. Epub 2017/03/18

Role of Benzodiazepines in Anxiety Disorders

20

Richard Balon and Vladan Starcevic

Introduction

Benzodiazepines (BZs) have been used for treatment of anxiety and anxiety disorders for over half of a century. The first BZ, chlordiazepoxide, was synthesized by Leo Sternbach in the mid-1950s, and the first reports of its clinical use were published in 1960 (e.g., 1). Sternbach subsequently fairly quickly discovered several other BZs, such as diazepam, clonazepam, flurazepam, nitrazepam, and flunitrazepam. In addition to antipsychotics, antidepressants, and stimulants, psychiatry's armamentarium was thus enriched by a new group of medications that were effective and easy to use, acted quickly, had no unpleasant side effects, and made anxious people feel better.

BZs started to be widely used by physicians not just because of their ability to quickly alleviate anxiety but also for their anticonvulsants, hypnotic, muscle-relaxing, and sedative properties. BZs became one of the most prescribed classes of psychotropic medications [2]. For instance, in 2008, approximately 5.2% of US adults aged 18–80 years used BZs [2], and there were 85 million BZ prescriptions issued in 2007 for outpatients with anxiety and mood disorders [3]. Psychiatrists have been historically apprehensive about BZ abuse potential, withdrawal effects, and possible side effects associated with long-term use. Thus, BZs have been more frequently prescribed by primary care physicians. With the arrival of selective serotonin reuptake inhibitors (SSRIs) during the 1990s and their subsequent approval for use in anxiety disorders, the use of BZs by psychiatrists has become even more

R. Balon (✉)
Department of Psychiatry and Behavioral Neurosciences and Anesthesiology, Wayne State University, Detroit, MI, USA
e-mail: rbalon@wayne.edu

V. Starcevic
Discipline of Psychiatry, Nepean Clinical School, Sydney Medical School, Faculty of Medicine and Health, University of Sydney, Sydney, NSW, Australia

Y.-K. Kim (ed.), *Anxiety Disorders*, Advances in Experimental Medicine and Biology 1191, https://doi.org/10.1007/978-981-32-9705-0_20

limited. SSRIs and other antidepressants (ADs) became a preferable pharmacological choice for treatment of anxiety disorders among psychiatrists, with their use being promoted in various practice guidelines.

Interestingly, as Rickels [3] pointed out, "no evidence for the superiority of the newer ADs over BZs, both in terms of efficacy or safety, exists for either short-term or long-term treatment. BZ toxicity, adverse events, and withdrawal symptoms, not better efficacy, are usually cited in support of the use of ADs over BZs in anxiety disorders. Yet ADs are not better tolerated than BZs and they also cause withdrawal symptoms." In a systematic review and meta-analysis, Offidani and colleagues [4] demonstrated that treatment with BZs resulted in comparable or greater improvements and fewer adverse events in patients suffering from generalized anxiety disorder or panic disorder and that BZs were more effective in reducing panic attacks than tricyclic antidepressants.

BZs are clearly useful, efficacious, and effective medications for the treatment of anxiety disorders. As within the framework of this book, we are rethinking anxiety disorders, and the time has come to rethink the role of BZs in the treatment of these disorders. For the purpose of this book, we are discussing the use of BZs in anxiety disorders in terms of the *Diagnostic and Statistical Manual of Mental Disorders*, Fifth Edition (DSM-5) [5]. Thus, we will focus mostly on the use of BZs in panic disorder, generalized anxiety disorder, agoraphobia, social anxiety disorder, specific phobia, and anxiety disorder due to another medical condition. However, we would like to acknowledge that BZs are effective in many other disorders, such as trauma- and stressor-related disorders, obsessive-compulsive and related disorders, sleep disorders, and anxious depression and as adjunctive treatment for anxiety within the frame of mood, psychotic, and other disorders.

Benzodiazepines: Basic Pharmacology and Classification

Mechanism of Action

All BZs have a similar structure: their molecules include a 1,4 benzodiazepine ring, but they differ by the rest of the molecule (2-keto; 3-hydroxy, 7-nitro, triazolo, and imidazo benzodiazepines).

The mechanism of action of BZs involves the potentiation of gamma-aminobutyric acid (GABA), the main inhibitory neurotransmitter in the central nervous system (CNS). BZs bind to the BZ receptors which are a part of the GABA-A-benzodiazepine receptor complex connected to the chloride channel. The GABA-A receptor is basically a chloride channel regulated by GABA binding. GABA "opens" the chloride channels by binding to these receptors and thus inhibits neuronal excitability by the influx of chlorine ions into the cell, which helps to stabilize the membrane potential close to the resting level. GABA actually works on the GABA-A receptor alone. While GABA binds to the receptor, the opening and closing of the chloride channel occurs more frequently, which results in inhibition. However, when BZs bind to this receptor complex, the GABA-A receptor is allosterically modulated, and the action of GABA is potentiated with the greater influx of chloride ions into the cell (BZs alone, without GABA, cannot influence chloride flow). It is important to note that the GABA-A receptor has several subunits (α_1, α_2, and α_3 and several more). Receptors with different subunits seem to regulate different activities – e.g., α_1 is more implicated in sedative-hypnotic activity, while α_2 is more involved in anxiolytic activity. The distribution of these receptors in the CNS also varies (e.g., α_1 can be found mostly in the cortex and cerebellum, while α_2 is more prominent in the cortex, limbic system, and spinal cord and α_3 in the periphery). Most BZs bind to all three subunits, though BZs binding mainly to the α_1 subunit have also been developed (halazepam, quazepam).

As far as pharmacokinetics of BZs is concerned, BZs differ in their speed of absorption from the gastrointestinal tract, which also determines the speed of their action (diazepam is absorbed more quickly than some others). BZs are highly lipophilic, which helps crossing the blood-brain barrier. They differ in their lipophilicity, though, and drugs that are more lipophilic (e.g., diazepam) have quicker onset of action.

BZs also differ in their half-life, existence of metabolites (e.g., oxazepam has none, while diazepam has several, e.g., desmethyldiazepam), duration of action, and metabolism. BZs are metabolized in the liver through two principal pathways [6], microsomal oxidation or glucuronide conjugation, followed by excretion through the kidneys. The glucuronide conjugation is considered less susceptible to and impaired by various disease processes and medication than oxidation, and thus BZs metabolized through glucuronide conjugation are considered safer in some subpopulations (e.g., elderly). Some BZs are prodrugs, i.e., they are not active, but their metabolites are.

Classification

There are various ways to classify BZs: according to their structure, pharmacokinetics (half-life), pathway of metabolism, and intensity of hypnotic-sedative effect. We list the best known examples in each classification, as the list of all BZs is beyond the scope of this chapter. There are well over 60 BZs with hundreds of brand names available around the world. Most BZs are indicated for treatment of anxiety, but some (estazolam, flurazepam, quazepam, temazepam, and triazolam) are also indicated (in the USA) for insomnia (among other indications).

A. *Classification based on structure.*
   2-Keto benzodiazepines: chlordiazepoxide, clorazepate, diazepam, flurazepam, halazepam, prazepam.
   3-Hydroxybenzodiazepines: lorazepam, lormetazepam, oxazepam, temazepam.
   7-Nitro benzodiazepines: clonazepam, flunitrazepam, nimetazepam, nitrazepam.
   Triazolo benzodiazepines: adinazolam, alprazolam, estazolam, triazolam.
   Imidazo benzodiazepines: climazolam, loprazolam, midazolam

B. *Classification based on pharmacokinetics.*
   1. Benzodiazepines with short half-life: midazolam, oxazepam, triazolam.
   2. Benzodiazepines with intermediate half-life: alprazolam, bromazepam, estazolam, lorazepam, lormetazepam, temazepam.
   3. Benzodiazepines with long half-life: chlordiazepoxide, clobazam, clonazepam, clorazepate, diazepam, flurazepam, halazepam, medazepam, prazepam, quazepam
C. *Classification based on metabolism.*
   1. By oxidation: alprazolam, bromazepam, chlordiazepoxide, clonazepam, clorazepate, diazepam, estazolam, flurazepam, midazolam, prazepam, quazepam, triazolam, and others.
   2. 2. By glucuronide conjugation: lorazepam, lormetazepam, oxazepam, and temazepam only
D. *Classification based on relative intensity of sedative-hypnotic effect.*

Some benzodiazepines, such as alprazolam, clonazepam, clorazepate, diazepam, and oxazepam, have a fairly low sedative-hypnotic effect, while this effect is much more pronounced with other benzodiazepines, such as flurazepam, midazolam, nitrazepam, and triazolam.

Evidence of Efficacy of Benzodiazepines in Anxiety Disorders

BZs arrived during the 1960s when the regulatory rules, clinical trials, and diagnostic system were different from those of today. The three anxiety disorder diagnoses used before 1980 (arrival of DSM-III [7]) were anxiety neurosis, phobic neurosis, and obsessive-compulsive neurosis. The delineation of panic disorder, agoraphobia with and without panic attacks, and generalized anxiety disorder came with DSM-III [7] during the time when most BZs (with the main exception of alprazolam) were already a part of the psychiatric armamentarium and frequently out of patent. For obvious reasons, the pharmaceutical industry had not much interest in investing in the evaluation of many BZs in "new" indications delineated in DSM-III. Further changes in the classification of anxiety disorders were introduced in DSM-III-R [8]. Thus, BZ registration trials for use in diagnostic categories developed since the DSM-III have not been conducted, except for alprazolam for panic disorder. Nevertheless, there exist data from clinical trials supporting the use of BZs in anxiety disorders.

Panic Disorder with and Without Agoraphobia

The first trial examining BZs in this indication compared diazepam and propranolol in 21 patients with panic disorder and agoraphobia [9]. Panic attacks and phobic symptoms responded to diazepam, but not to propranolol.

The strongest evidence for efficacy of BZs in anxiety disorders comes from evaluation of alprazolam in panic disorder. Alprazolam arrived during the late 1980s and was examined in a number of clinical trials. Ballenger and colleagues [10, 11] studied alprazolam in a placebo-controlled, 8-week, flexible dose trial of 526 patients (481 completed 3 weeks of treatment) with agoraphobia with panic attacks and panic disorder. Alprazolam was significantly more effective than placebo in improving spontaneous and situational panic attacks, phobic fears, avoidance behavior, anxiety, and secondary disability already at week 1. At week 4, 83% of patients on alprazolam vs. 43% of patients on placebo were moderately improved or better; and 50% of alprazolam recipients vs. 28% of placebo recipients were free of panic attacks. Alprazolam was well tolerated, and 84% of alprazolam patients vs. 50% of placebo patients completed the study. In an interesting report [12] analyzing the data from this study [10, 11], the authors examined Donald Klein's theory [13] that the initial event in panic disorder is an unexpected panic attack followed by anticipatory anxiety and agoraphobia. Their finding [12] that panic attacks remitted before phobias seems to indirectly support Klein's theory [13].

As the efficacy of alprazolam in panic disorder and agoraphobia with panic attacks was established, a question was raised whether alprazolam is unique among antianxiety agents as to its efficacy. Thus, alprazolam was also compared to several other medications. In a double-blind, placebo-controlled study of 55 patients with panic disorder and agoraphobia with panic attacks, alprazolam was superior to propranolol [14]. Two trials [15, 16] found lorazepam as effective as alprazolam in this indication. Some interesting observations regarding the doses were made: in one study [15], the mean dose of lorazepam was 7 mg vs. 3 mg alprazolam daily; in the other study [16], the dose of lorazepam required for antipanic efficacy was twice that of alprazolam, a ration that was considered consistent with the relative potency of these drugs for generalized anxiety. In another study [17], clonazepam was found to be similarly effective to alprazolam in the treatment of panic disorder.

As it became clear that BZs are probably all effective in the treatment of panic disorder and agoraphobia with panic attacks, the question remained whether BZs are similarly effective as antidepressants in this indication. In a large multicenter, double-blind, placebo-controlled study of 1168 panic disorder patients comparing alprazolam and imipramine, improvement with alprazolam occurred by week 1 and 2 while with imipramine by week 4. By the end of week 8, the effects of both medications were similar and superior to placebo [18]. In a 6-month maintenance treatment study [19] comparing alprazolam with imipramine or placebo, all patients who completed the maintenance phase were panic free, both medications produced significant panic relief, but imipramine was associated with poorer patient acceptance.

Interestingly, as pointed out by Offidani and colleagues [4], there have not been many comparisons of efficacy of BZs and SSRIs. In a series of reports [20–22], Nardi and colleagues compared clonazepam and paroxetine in patients with panic disorder with and without agoraphobia. The first report [20] from a randomized, open-label, naturalistic 8-week study of 63 patients on clonazepam (target dose 2 mg) and 57 patients on paroxetine (target dose 40 mg) demonstrated the efficacy of both medications, with clonazepam performing better on several measures of panic and anxiety and being better tolerated. The second report [21], a 3-year follow-up of the same cohort of patients (47 on clonazepam and 37 on paroxetine), showed that both medications remained effective in reducing the frequency of panic attacks and anxiety, with clonazepam showing a small but significantly better improvement on the Clinical Global Impressions improvement rating. After 3 years, treatment was discontinued in patients who had achieved remission. The last report [22] focused on relapse rate in these patients with follow-up at years 1, 2, 3, 4, 5, and 6 after treatment discontinuation. Cumulative relapse rate in 85 patients who completed follow-up was 50% at year 1 and 89.4% at year 6. One-year relapse rates were lower in patients previously treated with clonazepam than in those previously treated with paroxetine, and low 6-year relapse rates were associated with high anxiety rating scores before treatment and previous treatment with clonazepam.

BZs are clearly effective and efficacious in treatment of panic disorder with or without agoraphobia, are well tolerated, and it seems that they are suitable for a long treatment in this indication, too.

Generalized Anxiety Disorder

BZs have been used in treatment of symptomatology akin to symptomatology of generalized anxiety disorder (GAD) (e.g., 1) since their inception. Similar to panic disorder, double-blind, placebo-controlled, mostly short-term trials examining their efficacy started to appear during the 1980s. In 1982, Chouinard and colleagues [23] reported on the efficacy of alprazolam in GAD and panic disorder in a small double-blind study. Alprazolam up to 3 mg/day was effective in both disorders. Interestingly, Elie and Lamontagne [24] found that both alprazolam (average dose 2 mg/day) and diazepam (average dose 15.8 mg/day) were effective in GAD, with diazepam being more effective than alprazolam in the reduction of several symptoms of anxiety and depression.

Several other BZs were tested in GAD. Two placebo-controlled studies [25, 26] demonstrated the efficacy of lorazepam in GAD, one in comparison with bromazepam (both equally effective) [25] and another one [26] finding both oral and sublingual forms of lorazepam effective in GAD. Bromazepam was also found equally helpful in GAD as chlorprothixene [27]. Etizolam displayed anxiolytic activity equivalent to those of alprazolam and bromazepam and possessed a more antidepressant effect than alprazolam or bromazepam [28]. The long-acting chlordesmethyldiazepam was a more effective therapy for GAD than lorazepam in a trial by Berlin and colleagues [29]. Finally, adinazolam-SR was superior to placebo in the treatment of GAD in another trial [30].

To further explore their usefulness and place among antianxiety medications, BZs were compared to other drugs with anxiolytic properties. As mentioned, bromazepam and chlorprothixene were equally effective in a multicenter study of 245 GAD patients in a general practice [27]. Two studies [31, 32] compared diazepam and one [33] alprazolam to modest doses of buspirone. Diazepam [32] and alprazolam [33] produced a more rapid improvement, but BZs were equally effective at the endpoints. One of these studies [31] suggested that buspirone may be particularly indicated for anxious patients with depression. In an interesting study by Rickels and colleagues [34], abercanil (anxioselective β-carboline) and diazepam provided more symptom relief than placebo at the end of week 1; however, only diazepam differed from placebo at week 6.

Similar to panic disorder, not many trials compared BZs in GAD to antidepressants. Rickels and colleagues [35] compared diazepam to imipramine and trazodone in a placebo-controlled trial of 230 GAD patients (depression and panic disorder were excluded). At the end of week 8, moderate to marked improvement was reported by 73% of patients treated with imipramine, 69% of patients treated with trazodone, 66% patients treated with diazepam, and only 47% of patients on placebo. Imipramine (max 143 mg/day) had somewhat better anxiolytic efficacy than diazepam (22 mg/day). Only two studies [36, 37] provided any data on comparisons of BZs and newer antidepressants. There were no differences in response rate between venlafaxine XR, diazepam, and placebo in one of these studies [36], and both lorazepam and paroxetine were significantly better than placebo, with lorazepam separating from placebo earlier in the other one [37].

A recent meta-analytic review [38] of BZs, SSRIs, and serotonin-norepinephrine reuptake inhibitors (SNRIs) of 54 articles reporting the results of 56 studies on the use of these medications in GAD concluded that "the most common forms of pharmacotherapy for adult GAD are moderately effective, with BZs being the most effective drugs." Reinhold and Rickels [39] also wrote that "Evaluation of the literature suggests consistent, reliable efficacy of BZs in improving the central features of GAD – both the psychiatric and somatic. BZs elicit an earlier response than the ADs and provided that a response occurs by the eight week, it tends to be sustained throughout the length of treatment."

Social Anxiety Disorder (Social Phobia)

Two observations [40, 41] noted a positive response to alprazolam in a small number of patients with social phobia. Subsequently, two reports [42, 43] described a positive response to clonazepam in small groups of patients suffering from social phobia.

Clonazepam was found effective in relieving anxiety, phobic avoidance, and social phobic symptoms in 23 patients with social phobia in a study comparing clonazepam to no treatment [44]. Finally, in a double-blind, placebo-controlled trial of clonazepam in 75 outpatients with social phobia, clonazepam was found significantly more effective than placebo. Response rates were 78.3% for clonazepam and 20% for placebo [45].

Clonazepam was also compared to other treatments for social phobia. Patients treated either with clonazepam or cognitive-behavioral group therapy improved significantly, and the differences between treatment conditions were absent, except for improvement with clonazepam on several measures at week 12 in a study by Otto and colleagues [46]. A small study by Seedat and Stein [47] with a complicated design found a trend favoring the combination of clonazepam and paroxetine over paroxetine/placebo group. Global outcome measures also favored the combination of clonazepam with paroxetine over paroxetine alone.

Clonazepam thus seems to be a suitable, effective, and well-tolerated treatment for social phobia.

Other Anxiety Disorders

Alprazolam in combination with house calls was described as helpful in a small placebo-controlled study of 12 patients with agoraphobia [48]. Intranasal midazolam was effective in claustrophobia induced by MR imaging in a randomized, placebo-controlled study of 54 patients scheduled for MR imaging [49]. BZs are frequently used in fear of flying, but no study have been done in this indication.

Dependence and Withdrawal Symptoms

BZ dependence is arguably the most controversial aspect of the use of these pharmacological agents. To a large extent, this is due to the negative connotations of the concept of dependence. In addition, this concept has often been confused with the notions of abuse and addiction.

BZ dependence is a physical or pharmacological dependence that denotes a physiological adaptation to the presence of BZs that is required to maintain their use [50]. As such, dependence develops in all patients who use BZs long-term, even after only a few months. It does not reflect pathology and is similar to dependence that develops during administration of other drugs. The usual manifestations of dependence, including BZ dependence, are tolerance and/or withdrawal symptoms. Thus, tolerance and withdrawal symptoms are regarded as an "evidence of normal adaptation" [51] to a long-term substance use.

Tolerance refers to a need for greater amounts of the substance to achieve desired effect or a markedly decreased effect if the substance continues to be administered in the same dose. Tolerance does occur when BZs are abused, and there have been reports of tolerance to the therapeutic (anxiolytic) effects of BZs. However, a convincing evidence now exists that tolerance to the anxiolytic effects of BZs usually does not develop during long treatment of anxiety disorders, especially panic disorder [21, 52–56]. Consequently, escalation of doses of BZs and loss of their therapeutic benefit are rarely seen when patients with anxiety disorders use BZs long-term for therapeutic reasons – as anxiolytic agents – and in the absence of other substance abuse issues.

The main implication of BZ dependence is the likelihood of the withdrawal symptoms if BZs are ceased abruptly after long-term use. These withdrawal symptoms are common, although not inevitable. One early study reported that withdrawal problems occurred in only 40% of those who took BZs regularly [57]. BZ withdrawal often resembles a recurrence of an anxiety disorder and consists of various nonspecific symptoms such as restlessness, irritability, insomnia, feelings of weakness or fatigue, numbness or tingling sensations, nausea, stomach cramps, flu-like symptoms, muscle cramps, involuntary movements, and unsteady gait. The relatively specific BZ withdrawal symptoms include hypersensitivity to various stimuli, perceptual disturbances, "metallic taste," distorted body image, depersonalization and derealization, confusion, ringing in the ears, and a sense that things are moving as if being on a boat. Although unpleasant and distressing, BZ withdrawal symptoms are rarely serious or life-threatening (e.g., seizures). They usually last from several days to 4 weeks but can last longer. Withdrawal symptoms may abate without specific treatment and usually produce no long-lasting consequences.

Withdrawal symptoms should be distinguished from rebound symptoms upon BZ cessation. The latter represent an exacerbation of the primary condition, usually anxiety disorder, for which BZs were originally prescribed. Rebound symptoms may be more severe than those experienced before the medication was commenced. Given the manifestations of BZ withdrawal, it may be difficult to make a distinction between the withdrawal and rebound symptoms in clinical practice. However, their implications are very different, as rebound symptoms call for BZs to be reinstated and perhaps for another treatment modality to be added.

An effort should be made to prevent and alleviate BZ withdrawal symptoms. The key consideration in this regard is engaging patients in the treatment planning and decision-making process. This increases the chances of the right timing for BZ cessation, so that it is suggested only when patients are ready for it, i.e., able to cope with anxiety or distress without relying on medication and feeling relatively comfortable when facing the symptoms [58]. Coercion of any kind, including a threat to stop prescribing BZs if the patient is unwilling to discontinue them or pressuring patients to complete taper within a rigidly set time limit, is likely to lead to more difficulties and should be avoided.

Once a decision has been made to discontinue BZs after long-term treatment, this should be done gradually and in an individualized manner, discussing and negotiating with patients the rate of taper that they feel comfortable with. This rate can be changed during the taper, depending on patients' response to a decreasing dose of the medication and level of discomfort. As a result, the duration of taper varies substantially – between several weeks and more than 12 months. BZs with a shorter half-life (e.g., alprazolam) are generally more likely to be associated with more intense withdrawal symptoms than BZs with a longer half-life (e.g., clonazepam). Techniques of CBT can also facilitate the taper.

Considering the nature and the course of BZ withdrawal symptoms, it is somewhat paradoxical that there is such a widespread fear of them. Several reasons can account for it. The first has to do with sheer ignorance and misinformation. Secondly, there are terminological issues and inherently negative connotations of the term "withdrawal symptoms" and the accompanying sinister expectations. Moreover, media have tended to portray BZ withdrawal symptoms in a negative, sensationalist manner. Still another reason is a deliberate exaggeration of the severity and consequences of BZ withdrawal. This has occurred because of the conflict of interest, when promoting the alleged or real advantages of alternative pharmacotherapy (e.g., SSRIs) or psychological treatment (e.g., CBT). The fear of BZ withdrawal symptoms motivates some patients to continue taking BZs even when they do not seem to benefit from treatment. Such patients are often hypervigilant about any bodily changes and likely to misinterpret minor symptoms during minimal dose reductions as signs of withdrawal, which reinforces the erroneous notion that they are "addicted" to BZs and that they will never be able to cease them. These considerations underscore a need for proper education of both BZ users and BZ prescribers.

The characteristics of BZ dependence that develops in the context of long-term treatment, especially its occurrence even with relatively low doses of the medication and the combination of little or no tolerance with the likelihood of withdrawal symptoms upon abrupt discontinuation, have led to various terms in an effort to describe what may be relatively specific for BZ dependence and how it differs from dependence that is encountered in the context of drug abuse or addiction. These terms include "therapeutic dose dependence" [59, 60], "therapeutic dependence" [61, 62], "nonaddictive dependence" [63], "low-dose iatrogenic dependence" [64], and "low-dose dependence" [65].

"Psychological dependence" is a term that has been used somewhat loosely with reference to BZ dependence. This phenomenon has no basis in the pharmacological properties of BZs and is not a part of the physiological adaptation to the presence of BZs. It is not a form of BZ abuse and does not suggest addiction. Due to its propensity to be misused or misinterpreted, it should best be avoided. One of its manifestations is "talisman dependence," which denotes a need to constantly carry tablets of BZs and have them close at hand in case the person needs the medication. This is a safety behavior that should best be addressed during psychological therapy of the underlying condition (usually an anxiety disorder). "Last dose dependence" is another form of "psychological dependence," which refers to an inability to complete the hitherto successful BZ taper. This suggests an overreliance on medication and may mean that the patient is not quite ready for medication cessation.

An approach to BZ dependence should be rational, without appealing to its possible emotional connotations. Consequently, BZ dependence should neither be overestimated nor trivialized and needs to be approached like any other pharmacotherapy-related issue. A statement made by Dell'Osso et al. [66] reflects this position succinctly: "Dependence is neither a valid reason to continue prescribing nor a sufficient reason, on its own, to refuse to prescribe BZs."

Abuse and Addiction

Substance abuse is no longer an official diagnosis in the DSM system. Although there are various definitions of substance abuse, they all have in common two elements: [1] a pattern of excessive, indiscriminate or inappropriate substance use and [2] various negative consequences of such substance use. These consequences pertain to physical and mental health problems, difficulties in terms of social or interpersonal functioning, and/or legal issues. Features that are often associated with substance abuse include craving (intense or abnormal desire or longing for the substance), unnecessarily prolonged use, dose escalation, and tolerance.

BZs are often considered to have a high abuse potential, but this is true only in the context of other substance abuse. The mood-modulating and/or euphoria-like effects of BZs in such a context are the main reason for craving and BZ abuse. Consequently, these agents can be used as "downers" or mild euphoriants, usually in combination with other drugs or while abstaining from them.

Patients who are prescribed BZs for their anxiety disorder rarely abuse them in the absence of other substance abuse issues. This is related to the findings that neither craving for BZs nor BZ-induced euphoria have been consistently reported in anxiety disorder patients taking BZs long-term. In this context, it is important to emphasize that interpreting as euphoria a BZ-induced relief from anxiety, tension, distress, and/or misery among patients with anxiety disorders and the subsequent "good" and "calming" feeling is erroneous.

In 1990, the American Psychiatric Association Task Force Report on benzodiazepine dependency concluded that "BZs … are not widely abused drugs. When abuse does occur, it is almost always among persons who are also actively abusing alcohol, opiates, or other sedative hypnotics. In these people, diazepam and alprazolam – the most commonly used benzodiazepines – are the most abused benzodiazepines." [67]. This report also noted that cocaine abusers use BZs to ease the "crash" of the rapid decline in euphoria. This assessment continues to be relevant and is a concise summary of what is known about BZ abuse.

Like the term "abuse," the term "addiction" does not relate to an official DSM diagnosis and has also suffered from too many definitions and a tendency to be equated with dependence. In recent times, the emphasis in the conceptualization of addiction has shifted from substances to behaviors. Thus, one influential approach to the definition of addiction [68, 69] focuses on "behavioral engagement" and posits that the core features of addiction include an urge or a craving that immediately precedes behavioral engagement, poor self-control over behavioral engagement, continued behavioral engagement despite its adverse consequences, and "compulsive" behavioral engagement, which refers to a continued use of the substance to avoid withdrawal symptoms. This approach is very much in line with other definitions that consider impairment in behavioral control (over substance use) and related inability to consistently abstain the key components of addiction [70].

In view of the above definitions, it is clear that addiction and dependence must not be considered interchangeable terms. Moreover, dependence can exist without addiction [71], and this is perhaps nowhere more evident than with BZs [72]. Therefore, a response to the frequently posed question of whether patients dependent on BDZs are inevitably addicted must not be an affirmative one. As already noted and in contrast to substance addiction, BZ dependence is not characterized by craving and an all-encompassing preoccupation with BZs, and there is no compulsive or uncontrolled, drug-seeking behavior. Tolerance and adverse health and/or social consequences are usually not associated with BZ dependence, whereas they are a part of substance addiction. Withdrawal upon abrupt cessation is the only feature that BZ dependence and substance addiction have in common, although withdrawal symptoms are not inevitable as part of BZ dependence. All these considerations mandate avoidance of the term "addiction" in the context of therapeutic BZ use and dependence.

Adverse Effects

While BZs are generally well tolerated and their adverse effects are relatively rarely a reason for discontinuation, they do produce adverse effects that may limit their utility. The most common adverse effect of BZs is sedation – a general mental and motoric slowing down. Sedation is dose-dependent and is often experienced and described as a difficulty remaining focused or feeling tired, drowsy, or sleepy. While sedative effects of BZs are desirable when these agents are used as hypnotics, they can be problematic or even impairing during the day, when patients wish to remain awake and alert. Sedation is not directly related to anxiolytic properties of BZs, but there is a common perception that sedative drugs of any kind also have antianxiety effects.

Sedation usually occurs when the medication is commenced or after its dose has been increased, and patients taking BZs need to be warned about this. It is advisable to avoid driving, operating machinery, or performing other complex tasks at that time, at least until patients adapt themselves to the medication or its higher dose. This adaptation usually occurs after several days because tolerance to sedative effects of BZs tends to develop rapidly. Therefore, the dose of the medication usually does not need to be reduced if sedation occurs, and avoiding activities in which sedation might be troublesome or dangerous is all that needs to be done in that situation. Additional doses of BZs prior to driving or performing complex tasks should be avoided. Patients who are on a constant dose of a BZ for longer periods of time usually do not experience sedation, but they should remain cautious because their driving ability can still be affected. In case of severe sedation or a need to perform activities that might be affected adversely by BZ-induced sedation, the dose of a BZ medication can be decreased, or the medication can be discontinued.

An impaired psychomotor performance is a related and common adverse effect of BZs. Besides its impact on activities that require complex psychomotor coordination, this adverse effect has been implicated in the falls and fractures among the elderly. Due to the propensity of BDZs to cause sedation and psychomotor impairment, the usual recommendation is to avoid their long-term administration to the elderly and frail patients. This recommendation pertains particularly to BZs with a long half-life (e.g., diazepam) because of their slower metabolism and tendency to accumulate in the body. Consequently, BZs with a shorter half-life (e.g., lorazepam) are preferred for use in this population. Some guidelines consider use of any BZs in the elderly risky, regardless of their duration of action and half-life. If BZs are used in the elderly, this should be done with the lowest possible dose, generally avoiding increases in dosing. Moreover, a need for administration of BZs to the elderly should be frequently reassessed.

Many elderly patients are resistant to a suggestion to cease BZs or at least decrease their use substantially. This occurs for various reasons, including the effectiveness of BZs for treating insomnia, anxiety, or distress, long-term use, and fear of a "life without BZs" and fear of the withdrawal symptoms. Also, some elderly patients find the calming and soothing effects of BZs more important than any BZ-induced cognitive problems [73]. In these situations, it is crucial to carefully weigh the risks and benefits of BZ use and have appropriate discussions with patients.

The assessment of the risk of BZ use in old age in terms of falls and fractures may be confounded by various other factors, including a concomitant or sequential use of other medications (e.g., antidepressants or antipsychotics) that have been associated even more strongly with this adverse effect [74]. Therefore, attribution of the risk solely or mainly to BZs may be misleading [75], and all the relevant risk factors need to be taken into consideration.

Another problem with use of BZs in the elderly is their association with cognitive impairment. The cognitive effects of BZs in the elderly include problems with concentration, decreased speed of processing and verbal learning, and alterations in visuospatial ability. Reports of the effects of BZs on memory in the elderly have been conflicting, with studies finding memory difficulties of varying magnitude that were both reversible and irreversible, with some of them persisting after BZ cessation [76]. The clinical significance of these cognitive effects has been controversial and seems to vary from one person to another; some reports suggest that daily functioning is not significantly affected by the cognitive effects of chronic BDZ use [77]. Anterograde amnesia (difficulty recalling events that occurred during the period of several hours after taking a BZ medication) is common with use of BZs and is seen in patients of all ages. Although BZs have been linked with dementia, a direct causal relationship between BZ use and development of dementia has not been demonstrated.

The cognitive and motor impairment associated with BZ use becomes more prominent in the context of alcohol consumption. Therefore, alcohol should generally be avoided during treatment with BZs and especially after taking the BZ medication.

BZ use has been associated with irritability, disinhibition, "out of character" or inappropriate behavior, anger, and aggression, but the frequency of these adverse effects varies considerably. One systematic review has confirmed this association to be "moderate" [78], but the circumstances under which aggressive behavior follows BZ use and the underlying mechanisms remain unclear. It is often assumed that these behavioral and emotional adverse effects of BZs are more likely in individuals with severe and emotionally unstable (i.e., borderline) personality disorders, impulse control disorders, emotional or intellectual immaturity, neurodevelopmental disorders, brain damage, and substance abuse. Evidence for such associations is inconclusive, but BZs may need to be avoided in the presence of these conditions.

Positioning Benzodiazepines in the Treatment of Anxiety Disorders

Treatment Guidelines, Benzodiazepines, and Antidepressants

Virtually all clinical practice and treatment guidelines consider BZs as second- or third-line pharmacotherapy in the treatment of anxiety disorders (i.e., panic disorder, generalized anxiety disorder, and social anxiety disorder), with SSRIs and less often SNRIs being considered the pharmacological treatment of choice. Long-term treatment with BZs is generally not recommended or may be reserved for severely ill and functionally impaired patients who failed to respond to several other treatments. Only a brief use of BZs in selected circumstances is deemed appropriate by most guidelines. It has been argued that these recommendations are not based on good evidence, that they are misleading because of exaggerating the dangers of BZs and downplaying the risks with SSRIs and SNRIs, and that they are too restrictive, depriving patients of a valuable treatment option [79–83]. Therefore, a reappraisal of BDZs and their adequate positioning in the treatment of anxiety disorders are essential.

The first issue concerns the efficacy of BZs for anxiety disorders. Treatment guidelines often assume or suggest that BZs are either less effective than SSRIs and SNRIs or equally effective, at best. However, very few studies directly comparing BZs with SSRIs or SNRIs have been conducted. In one such study, clonazepam was compared with paroxetine in the treatment of panic disorder; a greater clinical improvement with clonazepam and its faster onset of action were reported during short-term treatment [21, 56], whereas treatment with clonazepam predicted a lower relapse rate after long-term treatment [22]. One systematic review and meta-analysis comparing BDZs with antidepressants (though mainly tricyclic antidepressants) for anxiety disorders found no support for the primacy given to antidepressants on the grounds of efficacy [4]. A meta-analysis of the efficacy of medications for generalized anxiety disorder reported significantly greater effect sizes (Hedges' *g*) for BZs (0.497) than for SNRIs (0.357) and SSRIs (0.325) [38]. Although more studies that directly compare BZs with SSRIs and SNRIs are needed, current evidence does not support the notion that SSRIs and SNRIs are more efficacious than BZs, especially for panic disorder and generalized anxiety disorder.

With regard to adverse effects, BZs appear to be better tolerated than SSRIs and SNRIs in the treatment of anxiety disorders [4, 84–86]. Thus, clonazepam was better tolerated than paroxetine over the course of both short-term and long-term treatment of panic disorder [21, 56]. Numerous reports suggest that early adverse effects of SSRIs and SNRIs during the treatment of anxiety disorders, especially increased anxiety and agitation, insomnia, headache, dizziness, and gastrointestinal symptoms, lead to a premature discontinuation of these agents. Adverse effects of SSRIs and SNRIs that may be more prominent in the long run, especially sexual dysfunction, also contribute to their poor tolerability.

There is abundant evidence that the cessation of SSRIs and SNRIs is associated with the withdrawal symptoms [87, 88] and that, in this respect, SSRIs and SNRIs do not differ significantly from BDZs [89]. For reasons that have much more to do with marketing and commercial interests than science, withdrawal symptoms that occur with SSRIs and SNRIs have been labelled as "discontinuation symptoms." This terminological ploy, along with conflicts of interest and a generally negative attitude toward BZs, has contributed to a biased portrayal in the treatment guidelines of the importance and magnitude of the withdrawal symptoms caused by the cessation of antidepressants [90]. As a result, treatment guidelines tend to minimize these withdrawal symptoms and make them look less severe and less clinically important than the BZ withdrawal symptoms.

Considering the comparative efficacy and tolerability data, there is no reason to deny BZs the status of the first-line pharmacotherapy for anxiety disorders. In addition, BZs have a significant advantage over antidepressants in terms of their quick onset of action. This is particularly important in an acute clinical setting, crisis situation, and whenever there is a need to quickly alleviate distress, restlessness, agitation, autonomic hyperarousal, muscle tension, and other symptoms of anxiety or panic. The fast onset of action of BZs remains one of the key reasons for their ongoing popularity among both patients and prescribers. This feature of BZs also allows them to be used on an "as needed" (PRN) basis. Such use of BZs is sometimes frowned upon by clinicians and researchers, especially when it is interpreted as a safety behavior, but many patients prefer to use BZs in this manner rather than take them continuously and for longer periods of time.

Choice of Pharmacotherapy for Anxiety Disorders

Medications with calming effects, including BZs, will always have a role in the pharmacological treatment of pathological anxiety because of the perennial human need to alleviate anxiety-associated distress and suffering [91]. Pharmacotherapy versus psychotherapy for anxiety disorders is a forced and false dichotomy, and both clinical practice and research show that a careful, well-planned combination of both modalities can contribute to favorable outcomes. BZs have often been considered unsuitable for combination with psychological interventions, especially CBT, presumably due to their interference with CBT. However, this assumption has not been tested adequately, and there is emerging evidence that BDZs can be combined with CBT safely and effectively [92].

A decision to use BZs or antidepressants as the initial treatment for anxiety disorders depends on several factors. If the key factor is the speed of onset of antianxiety effects, BZs have a clear advantage. Prominent physical symptoms of anxiety and tension may respond more reliably and more consistently to BZs than to antidepressants. Evidence is mixed about the preferential response of cognitive symptoms of anxiety, with a potential advantage of SSRIs in this realm. If the patient has a history of alcohol or other substance abuse, antidepressants are usually preferred over BZs, but this suggestion has been controversial in light of the reports that in such clinical situations, use of BZs may not be as risky [93, 94]. A history of severe adverse effects of the previously administered SSRIs or SNRIs, including sexual dysfunction, strongly suggests that these medications should be avoided and that BZs may be preferred. The presence of depressive illness or a history of depressive episodes makes SSRIs or SNRIs a logical pharmacotherapy choice, although BZs can be co-administered for their anxiolytic properties or even used as a monotherapy for anxious depression [95]. Anxiety disorders have been associated with a higher risk of suicide, and prescribing medications with a low lethality potential in an overdose constitutes good clinical practice. If used alone, both BZs and SSRIs are relatively safe in an overdose. Unfortunately, overdosing on more than one drug is common, with the outcome depending on the particular combination of pharmacological agents and their quantities.

Combining BZs with SSRIs or SNRIs is common in clinical practice. However, this approach to the pharmacotherapy of anxiety disorders should not be haphazard, and its purpose should be clear. For example, one goal of this combination is to minimize adverse effects of SSRIs and SNRIs, especially at the beginning of treatment. A better tolerability of sertraline and clonazepam than of sertraline alone in the treatment of social anxiety disorder [86] supports such use. Another reason for combining BZs with SSRIs or SNRIs is to avoid waiting for too long for antidepressants to start "working." In other words, a combination of BZs with SSRIs or SNRIs tends to achieve a faster response compared to the response to an antidepressant alone, as demonstrated in panic disorder using various combinations of medications [96, 97]. Finally, there is some evidence that combining SSRIs with BDZs may produce better outcomes than treatment with an SSRI alone [47].

How to Select and Use Benzodiazepines in Anxiety Disorders

The treating physician should carefully evaluate the patient and rule out anxiety disorder due to other illnesses or causes (e.g., excessive intake of caffeine) first. The next management step should be the consideration of an initial trial of non-pharmacological treatment which may include short-term counseling, CBT and other psychotherapies, stress management, exercise, or meditation [98]. The decision to use medication may follow the failure of non-pharmacological treatments or may be the first choice in cases of severe symptomatology or when the availability of effective non-pharmacologic treatments is limited.

As Shader and Greenblatt wrote (99, p 1399–1400), "The ultimate decision to prescribe a benzodiazepine derivative or any other medication should be based on the assessment of the patient's degree of emotional distress and level of functional disability, the potential hazards of nontreatment in relation to the probable success of pharmacologic treatment, and the hazards of the medication." BZs should always be considered as a possible first medication choice considering their efficacy and safety. BZs can offer quick symptomatic relief and, e.g., in GAD, may reduce somatic symptoms and hyperarousal fairly quickly [99, 100]. Substance abuse could, of course, be a limiting factor in selecting BZs.

The decision as to which specific BZ to select may be based on various clinical factors and on BZ characteristics, e.g., their pharmacokinetic properties (short- vs. long half-life; metabolic pathway) and the degree of sedation.

Following the advice of Shader and Greenblatt [98] again, "Approaches to initiating benzodiazepine therapy are based largely on clinical experience. Therapy is initiated with a low dose that is based on patient's age, sex, body size, and medication history, and the dose is increased every few days until therapeutic benefit is achieved or side effects supervene. When side effects are encountered, further increases in the dose should be delayed or the dose should be reduced. Many patients who have drowsiness or other sedative effects soon after the initiation of therapy report that these symptoms diminish with continued therapy" (p 1400). "The duration of benzodiazepine treatment should be tailored to the character of the underlying illness. Patients with intermittent symptoms or symptoms that are triggered by identifiable anxiety-provoking situations are candidates for intermittent therapy. Those with persistent unremitting symptoms may require more continuous treatment, but the appropriate duration of therapy has not been clearly established" [98]. Solid evidence regarding the length of treatment from long-term trials does not exist. It is our clinical experience that treatment could continue indefinitely, though decrease and/or discontinuation of BZs should be attempted from time to time, depending on clinical status.

Conclusion

BZs are effective, efficacious, and safe medications indicated for the treatment of anxiety and anxiety disorders. They may be considered the treatment of choice in a number of patients and preferred to ADs in a number of clinical situations. BZs are quite versatile agents that could be used intermittently, for short-term treatment and for long-term, or even indefinite treatment. It is clearly time to rethink their role in the treatment of anxiety disorders in the upcoming era of personalized medicine and more specifically targeted treatment.

References

1. Bowes HA. The role of Librium in an out-patient psychiatric setting. Dis Nerv Syst. 1960;21(Suppl 3):20–2.
2. Olfson M, King M, Schoenbaum M. Benzodiazepine use in the United States. JAMA Psychiat. 2015;72:136–42.
3. Rickels K. Should benzodiazepines be replaced by antidepressants in the treatment of anxiety disorders? Facts of fiction? Psychother Psychosom. 2013;82:351–2.
4. Offidani E, Guidi J, Tomba E, Fava GA. Efficacy and tolerability of benzodiazepines versus antidepressants in anxiety disorders: a systematic review and meta-analysis. Psychother Psychosom. 2013;82:355–62.
5. American Psychiatric Association. Diagnostic and statistical manual of mental disorders. 5th ed. Arlington, VA: American Psychiatric Association; 2013.

6. Greenblatt DJ, Shader RI, Abernethy DR. Drug therapy: Current status of benzodiazepines (first of two parts). N Engl J Med. 1983;309:354–8.
7. American Psychiatric Association. Diagnostic and statistical manual of mental disorders. 3rd ed. Washington, D.C.: American Psychiatric Association; 1980.
8. American Psychiatric Association. Diagnostic and statistical manual of mental disorders. 3rd ed. Revised. Washington, D.C., American Psychiatric Association, 1987.
9. Noyes R Jr, Anderson DJ, Clancy J, Crowe RR, Slymen DJ, Ghoneim MM, et al. Diazepam and propranolol in panic disorder and agoraphobia. Arch Gen Psychiatry. 1984;41:287–92.
10. Ballenger JC, Burrows GD, DuPont RL Jr, Lesser IM, Noyes R Jr, Pecknold JC, et al. Alprazolam in panic disorder and agoraphobia: results from a multicenter trial. I. Efficacy in short-term treatment. Arch Gen Psychiatry. 1988;45:413–22.
11. Noyes R Jr, DuPont RL Jr, Pecknold JC, Rifkin A, Rubin RT, Swinson RP, et al. Alprazolam in panic disorder and agoraphobia: results from a multicenter trial. II. Patient acceptance, side effects, and safety. Arch Gen Psychiatry. 1988;45:423–8.
12. Rifkin A, Pecknold JC, Swinson RP, Ballenger JC, Burrows GD, Noyes R, et al. Sequence of improvement in agoraphobia with panic attacks. J Psychiatr Res. 1990;24:1–8.
13. Klein DF. Anxiety reconceptualized. In: Klein DF, Raskin J, editors. Anxiety: new research and changing concepts. New York: Raven Press; 1981. p. 235–63.
14. Munjack DJ, Crocker B, Cabe D, Brown R, Usigli R, Zulueta A, et al. Alprazolam, propranolol, and placebo in the treatment of panic disorder and agoraphobia with panic attacks. J Clin Psychopharmacol. 1989;9:22–7.
15. Schweizer E, Pohl R, Balon R, Fox I, Rickels K, Yeragani VK. Lorazepam vs. alprazolam in the treatment of panic disorder. Pharmacopsychiatry. 1990;23:90–3.
16. Charney DS, Woods SW. Benzodiazepine treatment of panic disorder: a comparison of alprazolam and lorazepam. J Clin Psychiatry. 1989;50:418–23.
17. Tesar GE, Rosenbaum JF, Pollack MH, Otto MW, Sachs GS, Herman JB, et al. Double-blind, placebo-controlled comparison of clonazepam and alprazolam for panic disorder. J Clin Psychiatry. 1991;52:69–76.
18. Drug treatment of panic disorder. Comparative efficacy of alprazolam, imipramine and placebo. Cross-National Collaborative Panic Study. Second Phase Investigators. (No authors listed) Br J Psychiatry. 1992;160:191–202.
19. Schweizer E, Rickels K, Weiss S, Zavodnick S. Maintenance drug treatment of panic disorder. I. Results of a prospective, placebo-controlled comparison of alprazolam and imipramine. Arch Gen Psychiatry. 1993;50:51–60.
20. Nardi AE, Valenca AM, Freire RC, Mochcovitch MD, Amrein R, Sardinha A, et al. Psychopharmacotherapy of panic disorder: 8 week randomized trial with clonazepam and paroxetine. Braz J Med Biol Res. 2011;44:366–73.
21. Nardi AE, Freire RC, Mochcovitch MD, Amrein R, Levitan MN, King AL, et al. A randomized naturalistic, parallel-group study for the long-term treatment of panic disorder with clonazepam or paroxetine. J Clin Psychopharmacol. 2012;32:120–6.
22. Freire RC, Amrein R, Mochcovitch MD, Dias GP, Machado S, Versiani M, et al. A 6-year posttreatment follow-up of panic disorder patients: treatment with clonazepam predicts lower recurrence than treatment with paroxetine. J Clin Psychopharmacol. 2017;37:429–34.
23. Chouinard G, Annable L, Fontaine R, Solyom L. Alprazolam in the treatment of generalized anxiety and panic disorders: a double-blind placebo-controlled studt. Pharmacopsychiatry. 1982;77:229–33.
24. Elie R, Lamontagne Y. Alprazolam and diazepam in the treatment of generalized anxiety. J Clin Psychopharmacol. 1984;4:125–9.
25. Fontaine M, Mercier P, Beaudry P, Annable L, Chouinard G. Bromazepam and lorazepam in generalized anxiety: a placebo-controlled study with measurements of drug plasma concentrations. Acta Psychiatr Scand. 1986;74:451–8.
26. Spenard J, Caille G, de Montigny C, Vezina M, Oulette J, Lariviere L, et al. Placebo-controlled comparative study of the anxiolytic activity and pharmacokinetics of oral and sublingual lorazepam in generalized anxiety. Biopharm Drug Dispos. 1988;9:457–64.

27. Kragh-Sorensen P, Holm P, Fynboe C, Schaumburg E, Andersen B, Bech P, et al. Bromazepam in generalized anxiety. Randomized, multi-practice comparisons with both chlorprothixene and placebo. Psychopharmacology (Berl). 1990;100:383–6.
28. Bertolino A, Mastucci E, Porro V, Corfiati L, Palermo M, Ecari U, et al. Etizolam in the treatment of generalized anxiety disorder: a controlled clinical trial. J Int Med Res. 1989;17:455–60.
29. Berlin I, Colombo G, Furlanut M, Benetello P. Double-blind placebo cross-over study of long-acting (chlordesmethyldiazepam) versus short-acting (lorazepam) benzodiazepines in generalized anxiety disorders. Int J Clin Pharmacol Res. 1989;9:203–8.
30. Wilcox CS, Ryan PJ, Morrissey JL, Cohn JB, DeFrancisco DF, Linden RD, Heiser JF. A fixed-dose study of adinazolam-SR tablets in generalized anxiety disorder. Prog Neuropsychopharmacol Biol Psychiatry. 1994;18:979–93.
31. Feighner JP, Merideth CH, Hendrickson GA. A double-blind comparison of buspirone and diazepam in outpatients with generalized anxiety disorder. J Clin Psychiatry. 1982;43:103–8.
32. Jacobson AF, Dominguez RE, Goldstein BJ, Steinbook RM. Comparison of buspirone and diazepam in generalized anxiety disorder. Pharmacotherapy. 1985;5:290–6.
33. Enkelmann R. Alprazolam versus buspirone in the treatment of outpatients with generalized anxiety disorder. Psychopharmacology (Berl). 1991;105:428–32.
34. Rickels K, DeMartinis N, Aufdembrinke B. A double-blind, placebo-controlled trial of abercanil and diazepam in the treatment of patients with generalized anxiety disorder. J Clin Psychopharmacol. 2000;20:12–8.
35. Rickels K, Downing R, Schweizer E, Hassman H. Antidepressants for the treatment of generalized anxiety disorder. A placebo-controlled comparison of imipramine, trazodone, and diazepam. Arch Gen Psychiatry. 1993;50:884–95.
36. Hackett D, Haudiquet V, Salinas E. A method for controlling for a high placebo response rate in comparison of venlafaxine XR and diazepam in the short-term treatment of patients with generalized anxiety disorder. Eur Psychiatry. 2003;18:182–7.
37. Feltner DE, Harness J, Brock J, Sambunaris A, Cappelleri JC, Morlock R. Clinical evaluation of daily assessment of symptoms-anxiety (DAS-A): a new instrument to assess the onset of symptomatic improvement in generalized anxiety disorder. CNS Neurosci Ther. 2009;15:12–8.
38. Gomez AF, Barthel AL, Hofmann SG. Comparing the efficacy of benzodiazepines and serotonergic anti-depressants for adults with generalized anxiety disorder: a meta-analytic review. Expert Opin Pharmacother. 2018;19:883–94.
39. Reinhold JA, Rickels K. Pharmacological treatment for generalized anxiety disorder in adults: an update. Expert Opin Pharmacother. 2015;16:1669–81.
40. Lydiard RB, Laraia MT, Howell EF, Ballenger JC. Alprazolam in the treatment of social phobia. J Clin Psychiatry. 1988;49:17–9.
41. Reich J, Yates W. A pilot study of treatment of social phobia with alprazolam. Am J Psychiatry. 1988;145:590–4.
42. Reiter SR, Pollack MH, Rosenbaum JF, Cohen LS. Clonazepam for the treatment of social phobia. J Clin Psychiatry. 1990;51:470–2.
43. Ontiveros A, Fontaine R. Social phobia and clonazepam. Can J Psychiatry. 1990;35:439–41.
44. Munjack DJ, Baltazar PJ, Bohn PG, Cabe DD, Appleton AA. Clonazepam in the treatment of social phobia: a pilot study. J Clin Psychiatry. 1990;51(Suppl):35–40.
45. Davidson JR, Potts N, Richichi E, Krishnan R, Ford SM, Smith R, et al. Treatment of social phobia with clonazepam and placebo. J Clin Psychopharmacol. 1993;13:423–8.
46. Otto MW, Pollack MH, Gould RA, Worthington JJ 3rd, McArdle ET, Rosenbaum JF. A comparison of the efficacy of clonazepam and cognitive-behavioral group therapy for the treatment of social phobia. J Anxiety Disord. 2000;14:345–58.
47. Seedat S, Stein MB. Double-blind, placebo-controlled assessment of combined clonazepam with paroxetine compared with paroxetine monotherapy for generalized social anxiety disorder. J Clin Psychiatry. 2004;65:244–8.
48. Hooper JF. House calls and alprazolam for agoraphobia. Am Fam Physician. 1991;44:1132–4.

49. Hollenhorst J, Munte S, Friedrich L, Heine J, Leuwer M, Becker H, et al. Using intranasal midazolam spray to prevent claustrophobia. AJR Am J Roentgenol. 2001;176:865–8.
50. Haefely W. Biological basis of drug-induced tolerance, rebound, and dependence: contribution of recent research on benzodiazepines. Pharmacopsychiatry. 1986;19:353–61.
51. O'Brien CP. Response to commentaries. Addiction. 2011;106:895–7.
52. Nagy LM, Krystal JH, Woods SW, Charney DS. Clinical and medication outcome after short-term alprazolam and behavioral group treatment in panic disorder: 2.5 year naturalistic follow-up study. Arch Gen Psychiatry. 1989;46:993–9.
53. Pollack MH, Otto MW, Tesar GE, Cohen LS, Meltzer-Brody S, Rosenbaum JF. Long-term outcome after acute treatment with alprazolam or clonazepam for panic disorder. J Clin Psychopharmacol. 1993;13:257–63.
54. Worthington JJ, Pollack MH, Otto MW, McLean RYS, Moroz G, Rosenbaum JF. Long-term experience with clonazepam in patients with a primary diagnosis of panic disorder. Psychopharmacol Bull. 1998;34:199–205.
55. Soumerai SB, Simoni-Wastila L, Singer C, Mah C, Gao X, Salzman C, et al. Lack of relationship between long-term use of benzodiazepines and escalation to high dosages. Psychiatr Serv. 2003;54:1006–11.
56. Nardi AE, Valença AM, Freire RC, Amrein R, Sardinha A, Levitan MN, et al. Randomized, open naturalistic, acute treatment of panic disorder with clonazepam or paroxetine. J Clin Psychopharmacol. 2011;31:259–61.
57. Tyrer P, Rutherford D, Huggett T. Benzodiazepine withdrawal symptoms and propranolol. Lancet. 1981;317:520–2.
58. Starcevic V. Benzodiazepines for anxiety disorders: maximising the benefits and minimising the risks. Adv Psychiatr Treat. 2012;18:250–8.
59. Balter MB, Ban TA, Uhlenhuth EH. International study of expert judgment on therapeutic use of benzodiazepines and other psychotherapeutic medications: I. Current concerns. Hum Psychopharmacol Clin Exp. 1993;8:253–61.
60. Uhlenhuth EH, Balter MB, Ban TA, Yang KMS. International study of expert judgment on therapeutic use of benzodiazepines and other psychotherapeutic medications: IV. Therapeutic dose dependence and abuse liability of benzodiazepines in the long-term treatment of anxiety disorders. J Clin Psychopharmacol. 1999;19(Suppl 2):23S–9S.
61. Labelle A, Lapierre YD. Anxiety disorders. Part 2: pharmacotherapy with benzodiazepines. Can Fam Physician. 1993;39:2205–08, 2011–13.
62. Starcevic V. Issues in the pharmacological treatment of anxiety disorders. Australas Psychiatry. 2005;13:371–4.
63. Bühler K-E. Euphoria, ecstasy, inebriation, abuse, dependence, and addiction: a conceptual analysis. Med Health Care Philos. 2005;8:79–87.
64. Lader M. Benzodiazepines revisited – will we ever learn? Addiction. 2011;106:2086–109.
65. Tyrer P. Why benzodiazepines are not going away. Commentary on… benzodiazepines for anxiety disorders. Adv Psychiatr Treat. 2012;18:259–62.
66. Dell'Osso B, Albert U, Atti AR, Carmassi C, Carra G, Cosci F, et al. Bridging the gap between education and appropriate use of benzodiazepines in psychiatric clinical practice. Neuropsychiatr Dis Treat. 2015;11:1885–909.
67. The American Psychiatric Association Task Forcew on Benzodiazepine Dependency. Benzodiazepine dependence, toxicity, and abuse. The Task Force Report of the American Psychiatric Association. Washington, D.C.: American Psychiatric Association; 1990.
68. Shaffer HJ. Strange bedfellows: a critical view of pathological gambling and addiction. Addiction. 1999;94:1445–8.
69. Potenza MN. Should addictive disorders include non-substance-related conditions? Addiction. 2006;101(Suppl 1):142–51.
70. American Society of Addiction Medicine. Public Policy Statement: Definition of Addiction. 2011. Available at: https://www.asam.org/advocacy/find-a-policy-statement/view-policy-statement/public-policy-statements/2011/12/15/the-definition-of-addiction

71. O'Brien CP, Volkow N, Li T-K. What's in a word? Addiction versus dependence in DSM-V. Am J Psychiatry. 2006;163:764–5.
72. Starcevic V. The importance of distinguishing between dependence and addiction in the context of long-term benzodiazepine use. Aust NZ J Psychiatry. 2016;50:1111–2.
73. Salzman C, Fisher J, Nobel K, Glassman R, Wolfson A, Kelley M. Cognitive improvement following benzodiazepine discontinuation in elderly nursing home residents. Int J Geriatr Psychiatry. 1992;7:89–93.
74. Woolcott JC, Richardson KJ, Wiens MO, et al. Meta-analysis of the impact of 9 medication classes on falls in elderly persons. Arch Intern Med. 2009;169:1952–60.
75. Wagner AK, Ross-Degnan D, Gurwitz JH, Zhang F, Gilden DB, Cosler L, et al. Effect of New York state regulatory action on benzodiazepine prescribing and hip fracture rates. Ann Intern Med. 2007;146:96–103.
76. Bierman EJM, Comijs HC, Gundy CM, Sonnenberg C, Jonker C, Beekman AT. The effect of chronic benzodiazepine use on cognitive functioning in older persons: good, bad or indifferent? Int J Geriatr Psychiatry. 2007;22:1194–200.
77. Stewart SA. The effects of benzodiazepines on cognition. J Clin Psychiatry. 2005;66(Suppl 2):9–13.
78. Albrecht B, Staiger PK, Hall K, Miller P, Best D, Lubman DI. Benzodiazepine use and aggressive behaviour: a systematic review. Aust NZ J Psychiatry. 2014;48:1096–114.
79. Balon R. Benzodiazepines revisited. Psychother Psychosom. 2013;82:353–4.
80. Starcevic V. The reappraisal of benzodiazepines in the treatment of anxiety and related disorders. Exp Rev Neurotherap. 2014;14:1275–86.
81. Balon R, Fava GA, Rickels K. Need for a realistic appraisal of benzodiazepines. World Psychiatry. 2015;14:243–4.
82. Balon R, Chouinard G, Cosci F, Dubovsky SL, Fava GA, Freire RC, et al. International task force on benzodiazepines. Psychother Psychosom. 2018;87:193–4.
83. Starcevic V. Will the RANZCP clinical practice guidelines for the treatment of anxiety disorders assist in making adequate treatment decisions for panic disorder? Aust NZ J Psychiatry. 2019;53:362–3.
84. Cowley DS, Ha EH, Roy-Byrne PP. Determinants of pharmacologic treatment failure in panic disorder. J Clin Psychiatry. 1997;58:555–61.
85. Bruce SE, Vasile RG, Goisman RM, Salzman C, Spencer M, Machan JT, et al. Are benzodiazepines still the medication of choice for patients with panic disorder with or without agoraphobia? Am J Psychiatry. 2003;160:1432–8.
86. Pollack MH, Van Ameringen M, Simon NM, Worthington JW, Hoge EA, Keshaviah A, et al. A double-blind randomized controlled trial of augmentation and switch strategies for refractory social anxiety disorder. Am J Psychiatry. 2014;171:44–53.
87. Fava GA, Gatti A, Belaise C, Guidi J, Offidani E. Withdrawal symptoms after selective serotonin reuptake inhibitor discontinuation: a systematic review. Psychother Psychosom. 2015;84:72–81.
88. Fava GA, Benasi G, Lucente M, Offidani E, Cosci F, Guidi J. Withdrawal symptoms after serotonin-noradrenaline reuptake inhibitor discontinuation: systematic review. Psychother Psychosom. 2018;87:195–203.
89. Nielsen M, Hansen EH, Gøtzsche PC. What is the difference between dependence and withdrawal reactions? A comparison of benzodiazepines and selective serotonin re-uptake inhibitors. Addiction. 2012;107:900–8.
90. Starcevic V, Brakoulias V, Viswasam K, Berle D. Inconsistent portrayal of medication dependence, withdrawal and discontinuation symptoms in treatment guidelines for anxiety disorders. Psychother Psychosom. 2015;84:379–80.
91. Starcevic V. Is the need for medications with calming effects ever going to disappear? Aust NZ J Psychiatry. 2013;47:971.
92. Rosen CS, Greenbaum MA, Schnurr PP, Holmes TH, Brennan PL, Friedman MJ. Do benzodiazepines reduce the effectiveness of exposure therapy for posttraumatic stress disorder? J Clin Psychiatry. 2013;74:1241–8.

93. Posternak MA, Mueller TI. Assessing the risks and benefits of benzodiazepines for anxiety disorders in patients with a history of substance abuse or dependence. Am J Addict. 2001;10:48–68.
94. Mueller TI, Pagano ME, Rodriguez BF, Bruce SE, Stout RL, Keller MB. Long-term use of benzodiazepines in participants with comorbid anxiety and alcohol use disorders. Alcohol Clin Exp Res. 2005;29:1411–8.
95. Benasi G, Guidi J, Offidani E, Balon R, Rickels K, Fava GA. Benzodiazepines as a monotherapy in depressive disorders: a systematic review. Psychother Psychosom. 2018;87:65–74.
96. Goddard AW, Brouette T, Almai A, Jetty P, Woods SW, Charney D. Early coadministration of clonazepam with sertraline for panic disorder. Arch Gen Psychiatry. 2001;58:681–6.
97. Pollack MH, Simon NM, Worthington JJ, Doyle AL, Peters P, Toshkov F. Combined paroxetine and clonazepam treatment strategies compared to paroxetine monotherapy for panic disorder. J Psychopharmacol. 2003;17:276–82.
98. Shader RI, Greenblatt DJ. Use of benzodiazepines in anxiety disorders. N Engl J Med. 1993;328:1398–405.
99. Hoehn-Saric R, McLeod DR, Zimerli WD. Differential effects of alprazolam and imipramine in generalized anxiety disorder: somatic versus psychic symptoms. J Clin Psychiatry. 1988;49:293–301.
100. Rickels K, Rynn M. Pharmacotherapy of generalized anxiety disorder. J Clin Psychiatry. 2002;63(Suppl 14):9–16.

Virtual Reality for Anxiety Disorders: Rethinking a Field in Expansion

21

Javier Fernández-Álvarez, Daniele Di Lernia, and Giuseppe Riva

Introduction

In the last two decades, virtual reality (VR) emerged as an attractive alternative for the treatment of anxiety and related disorders. Ample evidence supports its efficacy and effectiveness, which has shown to be similar to face-to-face interventions [1–4]. The results were confirmed for all anxiety disorders, particularly for specific phobias [5], post-traumatic stress disorder [6], and social anxiety disorder [7]. Besides, these therapeutic gains through VR have revealed to be applicable to real-life situations [8]. VR has also yielded similar attrition rates [9] and deterioration rates [10] when compared to other active interventions and superior to waiting lists groups.

The abovementioned results entirely derive from virtual reality exposure treatments (VRETs). That is, VR for anxiety disorders has principally emerged as a convenient alternative to imaginal and in vivo *exposure*, a technique that despite its undoubted effectiveness is not always implemented given the aversion and invasiveness experienced by patients [11] and even by therapists [12]. Hence, the attractiveness of VR so far has resided in the possibility it brings to enhance patients' acceptability of exposure [11], therapists' controllability of the stimuli, or the representativeness by means of providing ecological situations to cope with those fearful stimuli [13].

The predominance of studies of VR as an exposure tool derived in many researchers using VR therapy and VRET indistinguishably. However, we consider that the potential of VR in clinical psychology and psychiatry goes beyond the

J. Fernández-Álvarez · D. Di Lernia
Universita Cattolica del Sacro Cuore, Milan, Italy
e-mail: j.fernandezkirszman@unicatt.it; daniele.dilernia@unicatt.it

G. Riva (✉)
Universita Cattolica del Sacro Cuore, Milan, Italy

ATN-P Lab, Istituto Auxologico Italiano, Milan, Italy
e-mail: giuseppe.riva@unicatt.it

Y.-K. Kim (ed.), *Anxiety Disorders*, Advances in Experimental Medicine and Biology 1191, https://doi.org/10.1007/978-981-32-9705-0_21

augmentation of exposure. Hence, in this chapter we want to start describing the latest advancements in VRET, marked by a great proliferation of clinical trials. A second section will present the contributions that VR has for the understanding, assessment, and intervention of anxiety disorders in a general context in which psychopathology is experiencing a profound rethinking.

Latest Advancements in VR as an Exposure Tool: VRET

Doubtlessly, VRET has experienced a significant growth as a research area in the last years. Not only has the outcome-focused research been greatly expanded, with many clinical trials conducted for different conditions, but also has started to emerge more process-focused research seeking to unravel mechanisms of change in VRET. In addition, recent trials have begun to include larger sample sizes, thus solving one of the main weaknesses of clinical research in virtual reality [14]. In the next section, we will summarize the most outstanding progresses in VRET achieved in the last years, both in outcome-focused research and in process-focused research.

Outcome-Focused Research Progresses

The most recent VRET meta-analysis for anxiety disorders identified 30 studies that fulfilled the inclusion criteria [15]. Confirming previous meta-analyses, this study yielded a large effect size when comparing VRET to waiting list control conditions and no significant differences when compared to in vivo *exposure*.

Besides, more specific reviews and meta-analyses started to be published, which is a pristine proof of the maturity of the field. Among the examples to mention, a meta-analysis on fear of flight revealed a larger effect size of VRET versus control conditions both at posttreatment and follow-up and also, surprisingly, a larger effect of VRET versus exposure-based interventions [16]. Another meta-analysis for social anxiety disorder showed that VRET is more effective than waiting list control groups and equally effective than imaginal and in vivo exposure, in line with the great majority of studies in the field [7]. Finally, a considerable number of trials have been conducted on post-traumatic stress disorder (PTSD), but no specific meta-analysis is conducted yet. However, in the meta-analysis previously described by Carl and colleagues [1], a pooled effect size was obtained for five studies of VRET versus placebo or waiting list and of VRET against in vivo exposure, yielding a medium effect in the first comparison and no differences in the second comparison. Besides, two comprehensive systematic reviews on the topic were elaborated [6, 17].

From this large body of evidence, some conclusion can be drawn. First and foremost, the situation is changing dramatically regarding the small sample sizes that had characterized VR research in the first 20 years [14], greatly due to the significant decrease of VR equipment [18]. In line with the first point and the natural evolution of every research field, the research quality has also greatly improved. This means that studies started to present lower risk of bias.

With regard to the researched topics, new specific phobias have been studied as therapeutic targets, being fear of falling [19], driving anxiety [20], music performance anxiety [21], and dental phobia [22, 23] as some of the salient examples.

The fact that a number of trials were conducted with off-the-shelf consumer technology and without the presence of a therapist deserves to be particularly emphasized. For example, Lindner and colleagues [24] showed not only that a one-session VRET for public speaking anxiety can be delivered with very basic portable hardware like Samsung Gear VR but also that a self-led intervention revealed to be equally efficacious than the therapist-led intervention with sustained effects at 6 months. This study yielded a large effect size, which constitutes a promising aspect for the future dissemination of VRET. In this vein, another recent study showed that a self-applied app-based VR significantly reduces acrophobia symptoms [25]. Congruent results were found for fear of heights with another application [26]. All together, these examples reflect the increasing incorporation of self-guided apps, which, in adjunction to the much lower prices of the VR devices, lead to believe that VR will be greatly disseminated in the next years.

Besides, a range of open-source VR environments is starting to appear, although not many have been tested in rigorous research studies. A good example is precisely the aforementioned study by Lindner and colleagues, who used VirtualSpeech for fear of speaking, or Kim and colleagues [27] who used the environments provided by Samsung to deliver exposure therapy in social anxiety patients. A recent example has been developed in Lithuania, in which Šalkevičius and colleagues [28] presented the elaboration of a cloud-based VRET for public speaking anxiety, which potentially could be accessed through the Internet and using also low-cost VR devices.

Another important issue for the further incorporation of VR in clinical practices revolves around the possibility of personalizing the stimuli to the needs of the patients. In this line, a pioneering example was the "EMMA's World" [29], which enables to individualize the therapeutic content (music, pictures, and even the virtual environments) for mood induction and trauma re-elaboration. A recent example in this line was developed by Loucks and colleagues, who carried out a feasibility study with tailored content for military sexual trauma using a very advanced VR environment called BRAVEMIND [30].

Last but not least, it must be mentioned that the first empirical results of augmented reality (AR) are starting to emerge. Not only has a RCT showed the equal efficacy with regard to in vivo exposure [31], but also a study comparing AR, VR, and in vivo exposure has indicated that the three modalities are equally efficacious [32]. Augmented reality has experienced a massive penetration in commercial apps (e.g., Pokémon Go), and it can be a very powerful way of bridging virtual and real worlds. Precisely, its combination with mobiles has been also proven to be efficacious in an experimental single case [33].

Going Beyond Pre-post Positive Outcomes

A decade ago, Meyerbröker and Emmelkamp [34] concluded a systematic review on VRET, stating that: "In summary, it would be a valuable addition when research into VRET would not only focus on outcome but also on the underlying processes. Understanding the processes would help implementing VRET into clinical practice given that treatment outcome would be better predictable." Accordingly, the scientific community working on VR has started to carry out a range of studies that have helped to shed light on the way therapeutic processes are deployed.

The dearth of studies regarding the mechanisms of change of VR therapy has started to be reverted in the last 10 years. The three main moments in which a therapeutic process can be divided are before a treatment starts, throughout the treatment, and once it has finished. In the following section, we aim to summarize the most relevant results that have been published in this regard.

Process-based research in virtual reality		
Before	During	After
Predictors of outcome	Therapeutic alliance	Acceptability and experiences
Expectations	Presence	Negative effects
Preferences/attitudes	Dynamic of change	Follow-up results
Feasibility	Physiological concomitants	Attrition

Before the Treatment

Predictors of Outcome

In total, four studies had the principal aim of exploring how expectations predicted the outcome of VRET. Price, Anderson, Henrich, and Rothbaum [35] identified that higher outcome expectancy is associated with a better improvement throughout the treatment for flying phobia using hierarchical linear modeling. Likewise, Price and Anderson [36] used the same methodological strategy to establish that outcome expectancy constituted also a predictor in public speaking fear within social anxiety disorder. A third study explored how outcome expectancy impacted in PTSD, comparing clinician, self-report, and biological measures. While results indicated a significant relationship for clinician measurement and self-reports, in the case of biological, measured by trauma-potentiated startle and cortisol reactivity, no significance was found [37]. Finally, Norrholm and colleagues [38] conducted a study on physiological measures (startle EMG, SCL, HR) as predictors of outcome. They found that when EMG, SCL, and HR were higher, a better outcome was identified; and in the case of neuroendocrine markers (cortisol) when were higher, worse outcome was detected. Besides, physiological responses predicted better outcome at 6-month follow-up assessment in one of the groups and cortisol in others immediately after treatment.

Attitudes, Preferences, and Perceptions of Patients and Therapists

The dissemination of VR greatly hinges on the attitudes of therapists, patients, and the general community in general toward VR as a clinical tool. In the case of users or patients, there has been a range of studies throughout the years, which overall suggest that VR is a more attractive exposure tool than in vivo *exposure*. As an illustrative example, 76% of the participants in a study expressed to prefer VRET than in vivo *exposure* [39]. These results were even higher among college students, from which 81–89% preferred VRET over in vivo exposure [40]. In this regard, VR would be preferable for the accomplishment of client's preferences and values, which constitutes one of the three aspects in the appropriate delivery of evidence-based practices according to the American Psychological Association and has shown to play an important role in the treatment satisfaction, completion, and clinical outcome [41].

However, the most defining aspect in order to incorporate a principle or technique in the clinical practice revolves around the therapists' preferences and perceptions. In that sense, it is undoubted that VR and technologies in general may be perceived as threatening by clinicians who were mainly trained, regardless the theoretical school, to implement the long-lasting Freud's statement of *talking cure*. Accordingly, some initial studies suggested that therapists found VR potentially difficult to be applied in the clinical practice due to their lack of specific training and particularly because of the technical and financial obstacles that incorporating VR may entail [42–44].

The latest results in this regard, however, suggest that the situation may be changing. This could be explained due to the advent of consumer VR technology and its respective massive penetration in many spheres of the society. In this line, Lindner and colleagues [45] revealed that attitudes toward VRET are overall positive and that familiarity with VR was higher in comparison with previous studies, although no direct comparison can be established for being completely different populations.

During the Treatment

Therapeutic Alliance

Although in the last years more attention has been paid to the deployment of therapeutic alliance (TA) in VR or AR treatments, only three empirical studies had the primary aim of studying it [46–48]. Besides, the work carried out by Ngai, Tully, and Anderson [48], the unique study that explored the course of working alliance in a VR treatment, revealing that unlike usual quadratic growth, there was a positive linear growth. Besides, three RCTs [49–51] and one pilot study [52] included a measure of TA within their secondary analyses. The current evidence suggests that there is a positive association between alliance levels and outcome and TA as a predictor of outcome. Despite the scarcity of studies on TA, the current available evidence yields similar results to face-to-face approaches, although some studies have posed the question on the possible difficulties that working with technology devices could entail for the therapeutic alliance [53]. Besides, none of the studies have posed the question yet on alliance ruptures, which is a topic of particular relevance in the literature of TA in traditional approaches. Overall, much more research is needed in order to better understand how TA behaves in VR treatments.

Attrition

Although for many years it has been thought that VR provoked more dropouts than non-VR approaches, only recently the first synthetizing study on attrition rates in VRET has been published. Benbow and Anderson [9] conducted a meta-analysis of 46 studies, showing that among the 1057 participants that were included in the study, only 16% dropped out. These results are similar to in vivo *exposure*. Besides, the inclusion of homework assignments is the most significant predictor of non-attrition, result that has direct practical implications for the implementation of VR treatments.

Engagement

One of the advantages of including VR revolves around the possibility of improving the engagement of the intervention, which in turn would permit to enhance the adherence of participants to the interventions. In that sense, the development of gamified VR applications and interventions has started to be developed [33, 54].

However, studies focused on exploring the impact of engagement on VRET outcomes show mixed results. Price and colleagues [55] posed the question on the concept of presence (spatial presence, involvement, and realness) to explore if it was a predictor of outcome. Despite the fact that the experience of fear was significantly associated with all the facets of presence, only involvement was associated with treatment response. Likewise, Reger and colleagues [56] determined that there were no differences of engagement between a prolonged exposure condition and a VRET condition for the treatment of PTSD. Overall, more research is needed in this realm to draw definite conclusions regarding the role of engagement in VRET.

Sense of Presence

The sense of presence constitutes a long-standing topic in the VRET research domain. Although its definition is a matter of controversy, the most used definition is the user's sense of "being there" in the virtual environment [57], that is, the subjective experience of the user, which can be conceptualized in three facets: spatial reference, involvement, and realness [58, 59].

In that sense, it has been suggested to play an important role in the fear and anxiety elicitation along with immersion, which represents the objective component of the sense of presence [60]. However, there is an unclear relationship between presence and emotions. While some studies found a positive association, others found a negative association and other even no association [61].

The first meta-analytical study exploring the relationship between anxiety and presence in VRET reveals the existing positive association between anxiety and presence, particularly in clinical populations and in certain technological devices (with more degrees of freedom) which constituted moderating factors [62]. However it is not clear the causal effect, that is, if more fear or more anxiety leads to more presence, or the other way around. The existing works tend to be in favor of bidirectional relationship [63, 64]. In this sense, although there is no doubt that immersion [65] and user characteristics [63] are relevant aspects that need to be further explored in order to draw definitive conclusions.

In the same vein, the relationship between the sense of presence and therapeutic outcomes in VRET remains unclear given that research yielding mixed results. While some studies did not find any relationship [66–68], others did find a significant association between presence and VRET outcome [55].

Taken together, the wealth of focusing on the relationship between VRET and presence is mixed, and thus no definite conclusions can be drawn yet. However, there is a rule of thumb supported by the large mounting evidence suggesting that presence constitutes a necessary but not sufficient condition in order to achieve positive outcomes in VRET.

After the Treatment

Mechanisms of Change

In line with the advent of the process-based CBT movement that Stefan Hofmann and Steven Hayes have recently promoted [69], more studies on the mechanisms of change have appeared in the last years. As an example, Norr and colleagues [70] performed moderator analyses to explore the characteristics of the patients who improved the most for prolonged exposure versus VR exposure. Their results indicate that being young, not to take antidepressants and having greater PTSD hyperarousal symptoms, predicted a greater reduction in PTSD symptoms undergoing VR exposure. In another study, mediational analyses revealed that the reduction of PTSD symptoms leads to the change of depressive symptoms and not the other way around [71]. Likewise, Maples-Keller and colleagues [72] investigated the role of reexperiencing symptoms in VRET for PTSD, showing its importance in line with emotional processing theory.

Another research that has been explored revolves around the role of cognitive mechanisms. Specifically, two studies have revealed that the phobic stimuli and self-efficacy are predictors of change in VRET for spider phobias [73, 74].

Finally, a very interesting group of studies were brought forth by Shiban and colleagues [75–78] shedding light on the contextual factors that determine the efficacy of VRET. In particular, some of these studies [77, 79, 80] revealed that presenting multiple contexts increments the efficacy of the VRET for spider phobia and claustrophobia.

Negative Effects

On the one hand, there are studies revolving around the potential adverse effects of VR as a technology (i.e., cybersickness). However, in the last years, this has proven to be a residual aspect that may affect only a small proportion of the users. In any case, it is an important ethical aspect to be always considered.

Instead, a new avenue that has been recently explored is the study of deterioration rates in VRET interventions. An individual patient data meta-analysis has shown that VRET was not more deleterious than other active treatments but yielded significantly lower deterioration rates compared to the waiting list group [10].

In this sense, it is of utmost importance that new trials have started to report the potential negative effects of the interventions, as some of the latest clinical trials have already done [25, 26, 31, 50, 81, 82]. In the latest available randomized control trial in the field, the Negative Effects Questionnaire was even included, allowing to identify potential experiences of unwanted and adverse effects [82].

Exposure Enhancers

Although exposure has shown to be effective, a large number of patients do not recover or suffer from relapses. Hence, there is a large body of research exploring potential enhancers, particularly combining psychosocial and pharmacological interventions. The rationale behind this approach is that the medication may catalyze the neurological circuitry of fear and thus augment the effect of the psychological intervention.

As an illustrative example, d-cycloserine has been implemented for being a partial agonist of NMDA receptors, which modulate the amygdala response and accordingly the emotional activity, including fear. An individual patient data analysis has revealed that d-cycloserine exerts a small augmenting effect on exposure-based therapy in anxiety and related disorders [83].

Given the already described advantages of VR for the delivery of exposure, several researchers exploring exposure enhancement started to incorporate VR in their studies. Precisely with d-cycloserine, a double-blind RCT was conducted by Barbara Rothbaum and her team, in which they explored differences between d-cycloserine and alprazolam [84]. Although no outcome differences were found between conditions, d-cycloserine condition yielded lower cortisol reactivity and startle response. Besides, de Quervain and colleagues [85] explored the potential augmentation of glucocorticoids, suggesting that the administering of cortisol orally 1 hour before the exposure resulted in a significant improvement in comparison with the placebo condition. Likewise, Meyerbröker and associates [86] showed that yohimbine hydrochloride, a noradrenaline agonist, may facilitate the fear extinction response in VR for aviophobia after four sessions.

Finally, it is of particular interest a study in this domain that was recently done by Maples-Keller and colleagues [72]. The authors tried to implement dexamethasone as a treatment enhancer, which has proven to suppress cortisol release via negative feedback in animal and human models and showed to suppress HPA axis reactivity in PTSD patients. However, the authors not only did not find any augmentation in comparison with placebo but also identified an increase in the drop-out rates. This study suggests that much more has to be studied in this interesting domain.

Acceptability and Opinions

A number of studies have explored the treatment satisfaction, and in general VR is well accepted by participants undergoing this kind of treatment [11, 40, 87–89]. Fortunately, qualitative research is starting to be conducted in the field of VRET. This approach is essential in order to explore acceptability and opinions and will complement the results obtained with survey and other quantitative methodologies, which may not grasp the complete picture of the experience of people undergoing VR treatments. An illustrative example has been recently done with PTSD, indicating that the group undergoing a VR treatment showed a range of self-perceived benefits of the intervention. Likewise, therapists using a VR intervention showed to incorporate the treatment in a positive way, although they were part of a clinical center in which information and communication technologies are developed and applied to the clinical practice, and thus it may not be totally representative of the acceptability of the treatment of VR [90].

Follow-Ups

One of the key aspects of any health intervention revolves around the stability of the results in the mid and long term. The great majority of the interventions in clinical psychology are measured at a posttreatment point, despite the key importance of determining the effects beyond that moment. VRET is not the exception to the rule. While a range of studies has explored follow-up results at 3, 6, or maximum 12 months, only scant examples have focused on the stability of the results in the long term. An exception was the study by Wiederhold and Wiederhold [91] who explored the maintenance of the results 3 years later. Despite the small sample, it was a first study in this direction suggesting that VR could also maintain the therapeutic gains. More recently, a very interesting work has been carried out by Anderson and colleagues [92], in which the 4- to 6-year follow-up results of an intervention for social anxiety disorder were sustained. The same research group, in a different study but based on the same sample of patients, revealed that there was a long-term improvement in probability and cognitive biases, showing that the stability of the results are beyond symptomatology [93].

Software and Hardware Progresses

In the almost 25 years of existence of VR for anxiety disorders, the most concerning problem has definitely been the lack of dissemination in real settings. Although the evidence was undisputable showing the efficacy of the VRET, there were a number of reasons that prevent VR from scaling. In this regard, the high costs of VR hardware and software have been probably the most challenging obstacle to tackle.

In the last 3 to 5 years, the situation has substantially changed. The quality of the images has improved [94], and the cost of the devices has significantly decreased [18]. Recent findings have proven that low-cost devices of VR are equally capable of inducing immersive states and thus sense of presence, essential feature to address exposure through VR [95]. Despite the appearance of new devices, the great majority of the clinical trials have still been conducted with the classical devices [96], but this is starting to change in light of the latest studies in the field [25, 81, 82].

Major Challenges in VRET Exposure Research

Despite the undoubted progress in VRET, there are still considerable challenges to face with. First, it must be said that the great majority of the studies are for adults, while children and adolescents are very understudied populations. Besides, it is very relevant to increment the simple sizes of the clinical trials, although, as described, this is an ongoing phenomenon. The latest trials are showing promising progresses in this regard, partially explained by the decline of costs of VR devices and the possibility of delivering self-applied treatments. Another key aspect is to explore the stability of results in follow-up studies. Undoubtedly, process research has grown in the last years, but many further aspects remain unknown yet, principally regarding aspects that face-to-face therapy has already identified as key aspects, such as therapist effects, patient characteristics, or process-outcome studies like shapes of change throughout VRET interventions. However, the most urgent revolves around the implementation of VRET in clinical practices. Therefore, it is of utmost importance both to conduct research in real clinical contexts in order to study how the usual barriers of translating research to practice can be hurdled.

360 Degrees Cameras: A Low-Cost Solution to Widely Disseminate VR Interventions

A way VR is starting to be widely disseminated is through the incorporation of 360 degrees cameras, which can be defined as immersive or spherical videos giving the user the possibility of viewing in every angle. These cameras are much closer to the way humans are used to perceiving reality. Currently, there are softwares that enable to create 360 videos easily (e.g., InstaVR or Google toolkit creator). In the same vein that building websites or editing pictures and films have become accessible tasks for laypeople, these novel softwares currently permit to create videos without any programming skill.

For clinical purposes, 360 cameras have many advantages. First and foremost, the reduced prices in comparison with other type of VR devices are significantly lower. Besides, as they do not need programming skills, clinical researchers and even practitioners have the chance to develop and implement tools both for assessment and intervention purposes. Of course, unlike classical VR or AR, there is no interaction with the environment but in terms of providing with multiple contexts to evaluate skills and elicit specific feelings and emotions can be of enormous help. Besides, there is no risk of inducing the uncanny valley phenomena, as all environments do not feature computer-generated graphical presentations.

A range of classical effective techniques for the treatment of anxiety disorders could be enhanced with the incorporation of 360 degrees videos providing relevant contextual triggers in which those techniques can be trained both in the clinical practice and as homework assignments. As an illustrative example, in social anxiety disorder (SAD), apart from the exposure to the stimulus that triggers the intense anxiety response, it may be of great importance the development of social skills by means of elaborating personalized tasks in which the person can train how to interact and to cope with challenging situations.

In addition to the behavioral facet, a range of cognitive and emotional aspects can also be trained by means of immersive videos. As identified in a large body of evidence, anxiety disorders are particularly characterized by a dysfunctional implementation of emotion regulation strategies. For example, a recent review of studies has shown that social anxiety disorder patients are more prone to inefficiently implement cognitive reappraisal or tend to suppress positive rewards for potential negative evaluations [97].

Indeed, some studies have shown the usefulness of 360 degrees videos to reduce the anxiety in people suffering from fear of speaking [98], social anxiety disorder [27], or fear of heights [99]. However, these are only the first studies that have served as preliminary evidence of the efficacy and feasibility of implementing this low-cost technology. Very recently, the first randomized control trial has been conducted using this approach [81], showing the efficacy in both a self-led and a therapist-led group. Future studies are expected for the coming years leading subsequently to a massive application of VR in the clinical practices.

The Potential of VR in a Transformative Context of Psychopathology and Therapeutic Interventions

So far, a wealth of developments of VR has been introduced. However, VR has an enormous potential apart from the already described value as an exposure tool. This is not only true for the interventional dimension but also for the understanding and assessment of clinical conditions, including as anxiety disorders. The principal reason that supports the fact that VR can help expand the frontiers of psychology in general and clinical psychology in particular relies on the possibility of exploring usually understudied aspects of mental processes.

Despite the controversies around the ontological status of mental processes, cognitive sciences consider information processing as a key element. However, clinical psychology has approached this process primarily from a reductive perspective. Instead, information processing not only should be reduced to the representational nature of thoughts. Classical purely representational cognitive models have been fiercely criticized for confusing individuals' thought reports with their actual thoughts [100]. Instead, embodied, situated, or extended cognition may greatly help understanding and thus intervening the whole psychopathological spectrum, and anxiety disorders are not the exception to the rule. Hence, a broad information processing should be adopted, and in that sense VR presents an undoubted potential, as a research and intervention tool.

Why Virtual Reality May Help Exploring the Hidden Reality of Mental Processes?

Virtual reality can be defined as a multisensory computer-generated interface. Usually, the VR-simulated worlds enable the user individual to feel immerse in an undistinguishable way respective to non-virtual or "real" environments. Hence, VR has challenged the limits of how perception and reality are understood. VR constitutes a strong tool to represent alternative realities, creating the illusion of being in places where actually we are not. Indeed, from a classical cognitive perspective, our supposedly real world is the intrapersonal brain-centered activity. However, latest advancements in cognitive sciences have refuted this stance [101, 102]. Embodied, situated, embedded, and enactive perspectives of cognition have added thought-provoking reasons to consider that cognitions are not only determined by internal representational contents but by the body of the individual, the contextual environment, and the actions that are deployed. Not only for anxiety disorders but also for psychopathology in general, incorporating a broad information-processing framework is essential, permitting us to move from exposure to more complex facets as recently reviewed by Riva, Wiederhold, and Mantovani [103]. In the next sections, we will present what has already been achieved and what we expect that will happen next in the field of VR and how this can be applied to the understanding, assessment, and intervention of anxiety disorders.

Embodiment and Virtual Reality

The application of VR in these 25 years has been successful in the treatment of anxiety and stress-related disorders. Although there is room for improvement, in particular the dissemination in the real world, there is no doubt that VRET has shown to be a relevant tool for the delivery of exposure. However, as recently suggested by Riva and colleagues [103], VR has the possibility to go beyond the representation and simulation of the external world. Concretely, VR can be used to represent, simulate, and ultimately modify the internal representation of the body and in particular the potential modification of those representations in order to fool the "body matrix" [104].

Specifically, VR is incorporating this internal dimension by means of the integration of bodily self-consciousness (BSC) and multisensory integration [105] with haptic, bio- and neurofeedback, and brain/body stimulation technologies. This is permitting to materialize in rigorous scientific ways, a long-standing philosophical principle that resides in modifying the body to alter the mind. Riva and colleagues have denominated this new emerging field as *embodied medicine* [105].

Among the functions of body representation, the encoding and integration of visual, visceral, somatosensory, auditory, vestibular, and motor signals is one of the most important ones. BSC is considered to emerge from those functions and to be controlled by the "body matrix," which is defined as a complex network of multisensory and homeostatic brain areas that has the purpose of protecting the organism when a threat alters the body and the space around it [105]. The multisensory integration works in a predictive way through the continuous prediction of expected sensory inputs, which in turn are used to coordinate its content, creating coherent mental representations. According to the Bayesian principle, which defines the mind as a probabilistic system [106], the experience of the body an individual has is the result of a predictive process in which the different unimodal properties from several sensory systems are integrated. Specifically, *exteroception* (e.g., vision and touch), *proprioception* (the sense of the position of the body/body segments), *vestibular input* (the sense of motion and position of the body), and *interoception* (the afferent signalling of internal bodily signals) conform the palette of sensory systems [104, 106].

New cross-modal associations between bodily stimuli that were unrelated or not systematically related before can be generated in VR environments, permitting a transformative experience in the individual. In that vein, it is essential that a new network of representations can be generated in order to let the internal system to update the predictive model of the body matrix. By doing this, the ultimate goal is to correct the prediction errors that disrupt the homeostatic balance of the organisms. That is, VR can fool the perception of the self-representation in order to help to re-elaborate the dysfunctional associations.

VR can provide the person with new self-representations, by means of modifying the appearance of one's avatar. As an illustrative example, body-swapping illusion permits to induce an illusory experience of owning a virtual body. The illusion can be achieved through the observation of a virtual body from a first-person perspective, which in turn can be synchronized or asynchronized with the real body of the observer.

When a virtual body is spatially coincident with one's real body, one sees through the eyes of this body and experiences various degrees of synchronous multisensory correlation—such as visuomotor [105]. Although the avatar's appearance is not a prerequisite to create an embodiment illusion, the sensorimotor correspondences elicit stronger senses of body agency and ownership [107]. Indeed, supported in modeling learning, users can also be influenced by means of observing the behaviors of *doppelgängers*, that is, virtual representations with which the user feels represented but does not have agency over it.

In the field of anxiety disorders, Aymerich-Franch, Kizilcec, and Bailenson [108] suggested that the use of *doppelgängers* may allow to enhance the vicarious learning, as the avatar is one's own body and the learning model at the same time. The authors analyzed whether embodying a dissimilar virtual self would decrease anxiety in a public speaking situation. They found that anxious individuals preferred dissimilar virtual self-representations and embodying a dissimilar avatar helped them reduce their anxiety.

Likewise, Falconer and colleagues [109] showed that through the increase of self-compassion in a VR task, it is possible to improve self-criticism, which in turn can be applied to clinical populations improving anxiety and depressive symptomatology [107].

A third example in VR research and embodiment for anxiety disorders is the work carried out by Guterstam, Abdulkarim, and Ehrsson [110] who revealed that the experience of having an invisible body may be helpful in social anxiety while standing in front of an audience. The experimental tasks consisted of the possibility of having an invisible body, which turned to affect both bodily self-perception and social cognition.

As described, VR represents a powerful instrument both for the study of the emotions and for the possibility to develop advanced instruments able to address the clinical complexity of anxiety disorders. From this perspective, VR solutions can be enhanced through an embodied cognition paradigm where a multisensory integration framework can be used to modulate the virtual experience. These enhanced solutions can improve embodiment and body ownership [113], but, at the same time, an "embodied VR experience" can also be used to reduce symptom severity with techniques far more advanced and effective than simple exposure.

This innovative perspective may offer new forms of treatment with direct application in the management and regulation of symptoms and other dimensions in anxiety disorders [111, 112], forging a new medical perspective based upon the concept of a regenerative virtual medicine. Riva et al. [9] introduced the concept of "sonoception" (www.sonoception.com), a novel noninvasive multisensory stimulation paradigm based on wearable acoustic and vibrotactile transducers able to stimulate both mechanoreceptors in different parts of the body—the stomach, the heart, and the muscles—and the otolith organs of the vestibular system.

As a matter of fact, technological devices able to stimulate and manipulate the perception of bodily sensations can provide new innovative VR solutions with several clinical applications beyond exposure therapy. Such technological devices have already been developed [114] and used for clinical purposes [115, 116] modulating the parasympathetic response and improving well-being. This approach stimulates the afferent ways responsible to transmit the overall physiological state of the organism and can be used to develop new VR biofeedback systems able to improve both mind and body.

Virtual Reality Biofeedback: An Embodied Medicine Technology

Biofeedback constitutes an effective and noninvasive procedure, whose basic operating principle is the conscious registration of normally unconscious body procedures (e.g., brain activity, electrocardiogram, electromyography, or skin conductance) that are represented by a visual, haptic, or audio signal [117, 118].

There is a large body of evidence showing the strong relation of emotion regulation with physiological processes such as HRV [119], and in turn emotion regulation plays an instrumental role in the appearance and development of anxiety and stress-related disorders [120, 121]. Accordingly, HRV biofeedback has shown to be effective for stress and anxiety [122]. Besides, the neural activity, in particular the activity of the amygdala which constitutes a key area for emotion activity and regulation, has shown to be successfully regulated through neurofeedback procedures [123].

Although biofeedback has shown to be effective for a vast array of medical and psychological clinical conditions, one of the long-standing limitations has been how to best the targeted physiological processes. In that vein, virtual reality (VR) permits to represent the physiological process through virtual stimuli that are connected to biosensors, strengthening the engagement of users and potentially augmenting also the effectiveness of the interventions. Indeed, this has already been shown to be effective in a range of studies for healthy and clinical populations [124–126]. Another recent example with neurofeedback procedures was carried out by Lorenzetti and colleagues [127] who implemented a real-time functional magnetic resonance imaging protocol to enhance emotional states in healthy subjects.

In our laboratory (Applied Technology for Neuro-Psychology Laboratory), two systems of HRV biofeedback were recently developed and are currently under experimental testing. One of the virtual environments is connected to an ECG sensor (Zephyr™), and thus certain stimuli of the virtual environment change according to the physiological activity. In particular, the software was developed to calculate specific measures of HRV (like RMSDD) that index the vagal activity. This system is of great complexity and also of elevated cost as it is not only developed in Unity3D but fundamentally is to displayed in a cave automatic virtual environment (CAVE). The CAVE is a room-sized cube in which four stereoscopic projectors, three rear-projection screens, and one downward projection screen are combined, permitting the 3D visualization of the virtual environments. The system is displayed in a 1:1 scale ratio, thus enhancing the feeling of being immersed in the virtual scene. The hardware setup is shown in Fig. 21.1.

The other development our group is carrying out is a mobile-based VR biofeedback, whose main advantage is constituted by the possibility of providing an application outside the laboratory or the clinical settings. Besides, its cost is much more reduced in comparison with the CAVE system, both in terms of software and

Fig. 21.1 CAVE system picture

Fig. 21.2 Mobile-based VR biofeedback

hardware (it has been developed for Oculus Go, Gear VR, and Cardboard). In this case, besides, as long as a ubiquitous sensor was needed, we opted to connect the virtual environment to the CorSense sensor. Despite having a lower sampling rate compared to other sensors like Zephyr, its usability and reduced cost led us to decide for this sensor.

Given the results of the recent meta-analysis on attrition on VR in which the incorporation of between-session was associated with better retention [128], counting with a mobile tool for VR could be of great help. Likewise, in the current advent of self-led interventions for VR, these interventions can be of great significance (Fig. 21.2).

Currently, the two systems are being tested in order to determine their usability, feasibility, and acceptance in both healthy and clinical populations. If they prove to be useful, a set of trials will be conducted as the next step.

Major Challenges

There are significant considerations that should be described regarding the main challenges for the present and near future of VR for anxiety disorders. First and foremost, there are ethical concerns regarding the extent to which VR will be open for the massive public. On the one hand, this is important to have solid and evidence-based guidelines in order to guarantee the access of patients to the best possible treatments. Besides, it is of utmost importance to have clear regulations regarding the data acquisition and the way that data is protected. This is particularly sensitive given that nowadays most VR devices are connected to the web and therefore even nonverbal data may be recorded and involuntarily delivered by the users. As described by Bailenson [128], it is urgent that the privacy issue can be solved in order to make further progresses in all VR fields. Besides, it is of utmost relevance to establish when people do need human support [129] in a context in which self-led and automated interventions are emerging.

A clear challenge is to improve the developments of VR as an assessment tool. More than 20 years ago, when VR was just emerging as a technological development for clinical psychology, one of its described potentialities was the possibility of improving clinical assessment [130]. The main argument is that clinical psychology has primarily relied on self-report measures or behavioral tasks in laboratory or consultation settings. In that sense, VR is able to recreate relevant places for the person and thus facilitate a contextualized assessment [131].

Nevertheless, scant research has been carried out in this domain, with some exceptions such as the work by Dechant and colleagues (2017) who showed that the combination of VR and eye tracking can be useful as a diagnostic tool for social anxiety. The great majority of studies of assessment and VR have been developed for the validation of new virtual environments or as a research tool to conduct psychopathological studies regarding anxiety [132].

Likewise, the integration of VR with other technologies is a pending task, which could enormously expand the field. For example, more gamified VR applications should be developed, harnessing the possibilities of engagement and potentially increasing efficacy that such type of environment may provide. Besides, big data and deep learning should be incorporated in order to foster the personalized interventions depending on the specific characteristics and preferences of the users. For therapeutic purposes this is of particular relevance. Finally, as aforementioned, it is of utmost importance to disseminate the existing developments as a way of providing therapists with feasible, cost-effective, and appealing ways of delivering a key therapeutic technique like exposure as well as incorporating other principles like embodied interventions or physiological interventions.

Concluding Remarks

Despite the existence of a long-standing tradition in VR for anxiety disorders, the great majority of the developments have been conducted for the augmentation or replacement of in vivo exposure so far. In a context marked by a significant improvement and democratization of VR, we believe that it constitutes a tool of enormous help for anxiety disorders. One the one hand for the massive delivery of a key technique that should be widely disseminated in order to help millions of people who do not receive any treatment and, on the other hand, because VR constitutes a system that operates like the brain and thus it may construct parallel realities with a full immersive sense, which therapeutically can be revolutionary, in particular for the understanding, assessment, and intervention of embodied facets.

However, in any case, we do not have to forget that VR, like any other technology, is only a tool, which can be used for better or for worse. In that sense, our core interest as clinical psychologists, either researchers or clinicians, is to understand the clinical phenomena and not be led by the usually mistaken idea of thinking that cutting-edge technologies will necessarily entail solutions for the problems we have to face with. Thus, it is of utmost importance to know how to implement the benefits of the developments in order to help the people in need. That is why one of the key challenges is to have always in mind that we need to respond the key question set by Gordon Paul more than 50 years ago: what works for whom. In other words, we will need to elucidate which people can benefit from VR, at which moments of the psychopathological evolution.

John Norcross and James Prochaska conduct a Delphi study every 10 years, trying to predict which are going to be the most significant improvements in the field of psychotherapy. In the last two articles, published in 2002 and 2013, respectively, VR was listed among the developments that were expected to flourish the most. However, in the span time between 2002 and 2013, this was not the case, and VR has not experienced a wide dissemination as a clinical tool [133, 134]. Currently, the new decade is about to finish, and VR has not sprouted out yet. However, there are relevant reasons to believe the situation may change now. Although there are still obstacles to consider, the main difficulty has been the disproportionate costs of VR software development and hardware purchase. In that sense, the penetration of VR is undoubted, and both in commercial and research domains, an exponential growth of off-the-shelf VR is being experienced. Besides, we have initial data suggesting that VR applications available in app stores are massively used but also without consistency after the initial download [135]. This emphasizes that despite VR applications being widely and easily implemented, the therapist support and indications are key to know when the intervention can be helpful and particularly to motivate users to persist using the application beyond the initial uptake.

References

1. Carl E, Stein AT, Levihn-Coon A, Pogue JR, Rothbaum B, Emmelkamp P, Asmundson GJG, Carlbring P, Powers MB. Virtual reality exposure therapy for anxiety and related disorders: a meta-analysis of randomized controlled trials. J Anxiety Disord. 2019;61:27–36. https://doi.org/10.1016/j.janxdis.2018.08.003.
2. Opriş D, Pintea S, García-Palacios A, Botella C, Szamosközi Ş, David D. Virtual reality exposure therapy in anxiety disorders: a quantitative meta-analysis. Depress Anxiety. 2012;29:85–93. https://doi.org/10.1002/da.20910.
3. Parsons TD, Rizzo AA. Affective outcomes of virtual reality exposure therapy for anxiety and specific phobias: a meta-analysis. J Behav Ther Exp Psychiatry. 2008;39:250–61. https://doi.org/10.1016/j.jbtep.2007.07.007.
4. Powers MB, Emmelkamp PMG. Virtual reality exposure therapy for anxiety disorders: a meta-analysis. J Anxiety Disord. 2008;22:561–9. https://doi.org/10.1016/j.janxdis.2007.04.006.
5. Botella C, Fernández-Álvarez J, Guillén V, García-Palacios A, Baños R. Recent progress in virtual reality exposure therapy for phobias: a systematic review. Curr Psychiatry Rep. 2017;19 https://doi.org/10.1007/s11920-017-0788-4.
6. Botella C, Serrano B, Baños RM, Garcia-Palacios A. Virtual reality exposure-based therapy for the treatment of post-traumatic stress disorder: a review of its efficacy, the adequacy of the treatment protocol, and its acceptability. Neuropsychiatr Dis Treat. 2015;11:2533–45. https://doi.org/10.2147/NDT.S89542.

7. Chesham RK, Malouff JM, Schutte NS. Meta-analysis of the efficacy of virtual reality exposure therapy for social anxiety. Behav Chang. 2018;35:152–66. https://doi.org/10.1017/bec.2018.15.
8. Morina N, Ijntema H, Meyerbröker K, Emmelkamp PMG. Can virtual reality exposure therapy gains be generalized to real-life? A meta-analysis of studies applying behavioral assessments. Behav Res Ther. 2015;74:18–24. https://doi.org/10.1016/j.brat.2015.08.010.
9. Benbow AA, Anderson PL. A meta-analytic examination of attrition in virtual reality exposure therapy for anxiety disorders. J Anxiety Disord. 2019;61:18–27. https://doi.org/10.1016/j.janxdis.2018.06.006.
10. Fernández-Álvarez J, Rozental A, Carlbring P, Colombo D, Riva G, Anderson PL, Baños RM, Benbow AA, Bouchard S, Bretón-López JM, Cárdenas G, Difede JA, Emmelkamp P, García-Palacios A, Guillén V, Hoffman H, Kampann I, Moldovan R, Mühlberger A, North M, Pauli P, Peñate Castro W, Quero S, Tortella-Feliu M, Wyka K, Botella C. Deterioration rates in virtual reality therapy: an individual patient data level meta-analysis. J Anxiety Disord. 2019;61:3–17. https://doi.org/10.1016/j.janxdis.2018.06.005.
11. Garcia-Palacios A, Botella C, Hoffman H, Fabregat S. Comparing acceptance and refusal rates of virtual reality exposure vs. in vivo exposure by patients with specific phobias. Cyberpsychol Behav. 2007;10:722–4. https://doi.org/10.1089/cpb.2007.9962.
12. Schumacher S, Betzler F, Miller R, Kirschbaum C, Ströhle A. Habituation of stress in psychotherapists performing subsequent in vivo exposures – a case series. 2017;27:218–24.
13. Botella C, Baños RR, García-Palacios A, Quero S. Virtual reality and other realities. In: The science of cognitive behavioral therapy. London: Academic; 2017.
14. Page S, Coxon M. Virtual reality exposure therapy for anxiety disorders: small samples and no controls? Front Psychol. 2016;7:1–4. https://doi.org/10.3389/fpsyg.2016.00326.
15. Carl E, Stein AT, Levihn-Coon A, Pogue JR, Rothbaum B, Emmelkamp P, Asmundson GJG, Carlbring P, Powers MB. Virtual reality exposure therapy for anxiety and related disorders: a meta-analysis of randomized controlled trials. J Anxiety Disord. 2018; https://doi.org/10.1016/j.janxdis.2018.08.003.
16. Cardoş RAI, David OA, David DO. Virtual reality exposure therapy in flight anxiety: a quantitative meta-analysis. Comput Human Behav. 2017;72 https://doi.org/10.1016/j.chb.2017.03.007.
17. Gonçalves R, Pedrozo AL, Coutinho ESF, Figueira I, Ventura P. Efficacy of virtual reality exposure therapy in the treatment of PTSD: a systematic review. PLoS One. 2012;7:1–7. https://doi.org/10.1371/journal.pone.0048469.
18. Lindner P, Miloff A, Hamilton W, Reuterskiöld L, Andersson G, Powers MB, Carlbring P. Creating state of the art, next-generation Virtual Reality exposure therapies for anxiety disorders using consumer hardware platforms: design considerations and future directions. Cogn Behav Ther. 2017; https://doi.org/10.1080/16506073.2017.1280843.
19. Levy F, Leboucher P, Rautureau G, Komano O, Millet B, Jouvent R. Fear of falling: efficacy of virtual reality associated with serious games in elderly people. Neuropsychiatr Dis Treat. 2016;12:877–81. https://doi.org/10.2147/NDT.S97809.
20. Zinzow HM, Brooks JO, Rosopa PJ, Jeffirs S, Jenkins C, Seeanner J, Mckeeman A, Hodges LF. Science direct virtual reality and cognitive-behavioral therapy for driving anxiety and aggression in veterans: a pilot study. Cogn Behav Pract. 2018;25:296–309. https://doi.org/10.1016/j.cbpra.2017.09.002.
21. Bissonnette J, Dubé F, Provencher MD, Moreno Sala MT. Evolution of music performance anxiety and quality of performance during virtual reality exposure training. Virtual Real. 2016;20:71–81. https://doi.org/10.1007/s10055-016-0283-y.
22. Gujjar KR, van Wijk A, Sharma R, de Jongh A. Virtual reality exposure therapy for the treatment of dental phobia: a controlled feasibility study. Behav Cogn Psychother. 2018;46(3):367–73.
23. Gujjar KR, van Wijk A, Kumar R, de Jongh A. Efficacy of virtual reality exposure therapy for the treatment of dental phobia in adults: a randomized controlled trial. J Anxiety Disord. 2019;62:100–8.

24. Lindner P, Milo A, Fagernäs S, Andersen J, Sigeman M, Andersson G, Furmark T, Carlbring P. Therapist-led and self-led one-session virtual reality exposure therapy for public speaking anxiety with consumer hardware and software: a randomized controlled trial. J Anxiety Disord. 2019;61:45–54. https://doi.org/10.1016/j.janxdis.2018.07.003.
25. Donker T, Cornelisz I, van Klaveren C, van Straten A, Carlbring P, Cuijpers P, van Gelder J-L. Effectiveness of self-guided app-based virtual reality cognitive behavior therapy for acrophobia. JAMA Psychiatry. 2019; https://doi.org/10.1001/jamapsychiatry.2019.0219.
26. Freeman D, Haselton P, Freeman J, Spanlang B, Kishore S, Albery E, Denne M, Brown P, Slater M, Nickless A. Automated psychological therapy using immersive virtual reality for treatment of fear of heights: a single-blind, parallel-group, randomised controlled trial. Lancet Psychiatry. 2018;5:625–32. https://doi.org/10.1016/S2215-0366(18)30226-8.
27. Kim HE, Hong YJ, Kim MK, Jung YH, Kyeong S, Kim JJ. Effectiveness of self-training using the mobile-based virtual reality program in patients with social anxiety disorder. Comput Human Behav. 2017;73:614–9. https://doi.org/10.1016/j.chb.2017.04.017.
28. Šalkevičius J, Miškinytė A, Navickas L. Cloud based virtual reality exposure therapy service for public speaking anxiety. Information. 2019;62:1–15. https://doi.org/10.3390/info10020062.
29. Baños RM, Botella C, Guillen V, García-Palacios A, Quero S, Bretón-López J, Alcañiz M. An adaptive display to treat stress-related disorders: EMMA's world. Br J Guid Couns. 2009;37:347–56. https://doi.org/10.1080/03069880902957064.
30. Loucks L, Yasinski C, Norrholm SD, Maples-Keller J, Post L, Zwiebach L, Fiorillo D, Goodlin M, Jovanovic T, Rizzo AA, Rothbaum BO. You can do that?!: feasibility of virtual reality exposure therapy in the treatment of PTSD due to military sexual trauma. J Anxiety Disord. 2019;61:55–63. https://doi.org/10.1016/j.janxdis.2018.06.004.
31. Botella C, Pérez-Ara MÁ, Bretón-López J, Quero S, García-Palacios A, Baños RM. In vivo versus augmented reality exposure in the treatment of small animal phobia: a randomized controlled trial. PLoS One. 2016;11:1–22. https://doi.org/10.1371/journal.pone.0148237.
32. Suso-Ribera C, Fernández-Álvarez J, García-Palacios A, Hoffman HG, Bretón-López J, Baños RM, Quero S, Botella C. Virtual reality, augmented reality, and *In Vivo* exposure therapy: a preliminary comparison of treatment efficacy in small animal phobia. Cyberpsychol Behav Soc Netw. 2019;22:31–8. https://doi.org/10.1089/cyber.2017.0672.
33. Botella C, Breton-López J, Quero S, Baños RM, García-Palacios A, Zaragoza I, Alcaniz M. Treating cockroach phobia using a serious game on a mobile phone and augmented reality exposure: a single case study. Comput Human Behav. 2011;27:217–27. https://doi.org/10.1016/j.chb.2010.07.043.
34. Meyerbröker K, Emmelkamp PMG. Therapeutic processes in virtual reality exposure therapy: the role of cognitions and the therapeutic alliance. J CyberTherapy Rehabil. 2008;1
35. Price M, Anderson P, Henrich CC, Rothbaum BO. Greater expectations: using hierarchical linear modeling to examine expectancy for treatment outcome as a predictor of treatment response. Behav Ther. 2008;39:398–405. https://doi.org/10.1016/j.beth.2007.12.002.
36. Price M, Anderson PL. Outcome expectancy as a predictor of treatment response in cognitive behavioral therapy for public speaking fears within social anxiety disorder. Psychotherapy. 2012;49:173–9. https://doi.org/10.1037/a0024734.
37. Price M, Maples JL, Jovanovic T, Norrholm SD, Heekin M, Rothbaum BO. An investigation of outcome expectancies as a predictor of treatment response for combat veterans with PTSD: comparison of clinician, self-report, and biological measures. Depress Anxiety. 2015;32:392–9. https://doi.org/10.1002/da.22354.
38. Norrholm SD, Jovanovic T, Gerardi M, Breazeale KG, Price M, Davis M, Duncan E, Ressler KJ, Bradley B, Rizzo A, Tuerk PW, Rothbaum BO. Baseline psychophysiological and cortisol reactivity as a predictor of PTSD treatment outcome in virtual reality exposure therapy. Behav Res Ther. 2016;82:28–37. https://doi.org/10.1016/j.brat.2016.05.002.
39. Garcia-Palacios A, Botella C, Hoffman H, Fabregat S. Comparing acceptance and refusal rates of virtual reality exposure vs. in vivo exposure by patients with specific phobias. CyberPsychol Behav. 2007;10:722–4. https://doi.org/10.1089/cpb.2007.9962.

40. Garcia-Palacios A, Hoffman HG, Kwong See S, Tsai A, Botella C. Redefining therapeutic success with virtual reality exposure therapy. CyberPsychol Behav. 2002;4:341–8. https://doi.org/10.1089/109493101300210231.
41. Lindhiem O, Bennett CB, Trentacosta CJ, McLear C. Client preferences affect treatment satisfaction, completion, and clinical outcome: a meta-analysis. Clin Psychol Rev. 2014;34:506–17. https://doi.org/10.1016/j.cpr.2014.06.002.
42. Schwartzman D, Segal R, Drapeau M. Perceptions of virtual reality among therapists who do not apply this technology in clinical practice. Psychol Serv. 2012;9:310–5. https://doi.org/10.1037/a0026801.
43. Segal R, Bhatia M, Drapeau M. Therapists' perception of benefits and costs of using virtual reality treatments. Cyberpsychology Behav Soc Netw. 2010;14:29–34. https://doi.org/10.1089/cyber.2009.0398.
44. Kramer TL, Pyne JM, Kimbrell TA, Savary PE, Smith JL, Jegley SM. Clinician perceptions of virtual reality to assess and treat returning veterans. Psychiatr Serv. 2014;61:1153–6. https://doi.org/10.1176/ps.2010.61.11.1153.
45. Lindner P, Miloff A, Zetterlund E, Reuterskiöld L, Andersson G, Carlbring P. Attitudes toward and familiarity with virtual reality therapy among practicing cognitive behavior therapists: a cross-sectional survey study in the era of consumer VR platforms. Front Psychol. 2019;10:1–10. https://doi.org/10.3389/fpsyg.2019.00176.
46. Wrzesien M, Bretón-López J, Botella C, Burkhardt J-M, Alcañiz M, Pérez-Ara MÁ, del Amo AR. How technology influences the therapeutic process: evaluation of the patient-therapist relationship in augmented reality exposure therapy and in vivo exposure therapy. Behav Cogn Psychother. 2013;41:505–9. https://doi.org/10.1017/S1352465813000088.
47. Wrzesien M, Burkhardt J-M, Botella C, Alcañiz M. Evaluation of the quality of collaboration between the client and the therapist in phobia treatments. Interact Comput. 2012;24:461–71. https://doi.org/10.1016/j.intcom.2012.09.001.
48. Ngai I, Tully EC, Anderson PL. The course of the working alliance during virtual reality and exposure group therapy for social anxiety disorder. Behav Cogn Psychother. 2015;43:167–81. https://doi.org/10.1017/S135246581300088X.
49. Anderson PLPL, Price MM, Edwards SMSM, Obasaju MAMA, Schmertz SKSK, Zimand E, Calamaras MRMR. Virtual reality exposure therapy for social anxiety disorder: a randomized controlled trial. J Consult Clin Psychol. 2013;81:751–60. https://doi.org/10.1037/a0033559.
50. Bouchard S, Dumoulin S, Robillard G, Guitard T, Klinger E, Forget H, Loranger C, Roucaut FX. Virtual reality compared with in vivo exposure in the treatment of social anxiety disorder: a three-arm randomised controlled trial. Br J Psychiatry. 2017;210:276–83. https://doi.org/10.1192/bjp.bp.116.184234.
51. Moldovan R, David D. One session treatment of cognitive and behavioral therapy and virtual reality for social and specific phobias. Preliminary results from a randomized clinical trial. J Evidence-Based Psychother. 2014;14:67–83.
52. Levy F, Leboucher P, Rautureau G, Komano O, Millet B, Jouvent R. Fear of falling: efficacy of virtual reality associated with serious games in elderly people. Neuropsychiatr Dis Treat. 2016;12:877–81. https://doi.org/10.2147/NDT.S97809.
53. Wrzesien M, Burkhardt JM, Botella C, Alcañiz M. Towards a virtual reality- and augmented reality-mediated therapeutic process model: a theoretical revision of clinical issues and HCI issues. Theor Issues Ergon Sci. 2015;16:124–53. https://doi.org/10.1080/1463922X.2014.903307.
54. Miloff A, Lindner P, Hamilton W, Reuterskiöld L, Andersson G, Carlbring P. Single-session gamified virtual reality exposure therapy for spider phobia vs. traditional exposure therapy: study protocol for a randomized controlled non-inferiority trial. Trials. 2016;17:60. https://doi.org/10.1186/s13063-016-1171-1.
55. Price MM, Mehta NN, Tone EBEB, Anderson PLPL. Does engagement with exposure yield better outcomes? Components of presence as a predictor of treatment response for virtual reality exposure therapy for social phobia. J Anxiety Disord. 2011;25:763–70. https://doi.org/10.1016/j.janxdis.2011.03.004.

56. Reger GM, Smolenski D, Norr A, Katz A, Buck B, Rothbaum BO. Does virtual reality increase emotional engagement during exposure for PTSD? Subjective distress during prolonged and virtual reality exposure therapy. J Anxiety Disord. 2019;61:75–81. https://doi.org/10.1016/j.janxdis.2018.06.001.
57. Slater M, Wilbur S. A framework for immersive virtual environments (FIVE): speculations on the role of presence in virtual environments. Presence Teleop Virt Environ. 1997;6:603–16.
58. Regenbrecht HT, Schubert TW, Friedmann F. Measuring the sense of presence and its relations to fear of heights in virtual environments. Int J Hum Comput Interact. 1998;10:233–49. https://doi.org/10.1207/s15327590ijhc1003_2.
59. Schubert T, Friedmann F, Regenbrecht H. The experience of presence: factor analytic insights. Presence Teleop Virt Environ. 2001;10:266–81. https://doi.org/10.1162/105474601300343603.
60. Riva G, Botella C, Baños R, Mantovani F, García-Palacios A, Quero S, Serino S, Triberti S, Repetto C, Dakanalis A, Villani D, Gaggioli A. Presence-inducing media for mental health applications. In: Immersed in media. Cham: Springer; 2015. p. 283–332. https://doi.org/10.1007/978-3-319-10190-3_12.
61. Gromer D, Madeira O, Gast P, Nehfischer M, Jost M, Müller M, Mühlberger A, Pauli P. Height simulation in a virtual reality CAVE system: validity of fear responses and effects of an immersion manipulation. Front Hum Neurosci. 2018;12:1–10. https://doi.org/10.3389/fnhum.2018.00372.
62. Ling Y, Nefs HT, Morina N, Heynderickx I, Brinkman WP. A meta-analysis on the relationship between self-reported presence and anxiety in virtual reality exposure therapy for anxiety disorders. PLoS ONE. 2014;9:e96144. https://doi.org/10.1371/journal.pone.0096144.
63. Gromer D, Reinke M, Christner I, Pauli P. Causal interactive links between presence and fear in virtual reality height exposure. Front Psychol. 2019;10:1–11. https://doi.org/10.3389/fpsyg.2019.00141.
64. Peperkorn HM, Diemer J, Mühlberger A. Temporal dynamics in the relation between presence and fear in virtual reality. Comput Human Behav. 2015;48:542–7. https://doi.org/10.1016/j.chb.2015.02.028.
65. Cummings JJ, Bailenson JN. How immersive is enough? A meta-analysis of the effect of immersive technology on user presence. Media Psychol. 2016;19:272. https://doi.org/10.1080/15213269.2015.1015740.
66. Krijn M, Emmelkamp PM, Biemond R, de Wilde de Ligny C, Schuemie MJ, van der Mast CAP. Treatment of acrophobia in virtual reality: the role of immersion and presence. Behav Res Ther. 2004;42:229–39. https://doi.org/10.1016/S0005-7967(03)00139-6.
67. Price M, Anderson P. The role of presence in virtual reality exposure therapy. J Anxiety Disord. 2007;21:742–51. https://doi.org/10.1016/j.janxdis.2006.11.002.
68. Schuemie, M.J., Schuemie, M.J., Bruynzeel, M., Drost, L., Brinckman, M., De Haan, G., Emmelkamp, P.M.G., van der Mast, C.A.P.G.: Treatment of Acrophobia in Virtual Reality: a Pilot Study. In: Broeckx, F., Pauwels, L. Schuemie AL. Conf. Proc. Euromedia 2000. May 8–10. 271–275 (2000).
69. Hofmann SG, Hayes SC. The future of intervention science: process-based therapy. Clin Psychol Sci. 2018; https://doi.org/10.1177/2167702618772296.
70. Norr AM, Smolenski DJ, Katz AC, Rizzo AA, Rothbaum BO, Difede JA, Koenen-Woods P, Reger MA, Reger GM. Virtual reality exposure versus prolonged exposure for PTSD: which treatment for whom? Depress Anxiety. 2018; https://doi.org/10.1002/da.22751.
71. Peskin M, Wyka K, Cukor J, Olden M, Altemus M, Lee FS, Difede JA. The relationship between posttraumatic and depressive symptoms during virtual reality exposure therapy with a cognitive enhancer. J Anxiety Disord. 2019; https://doi.org/10.1016/j.janxdis.2018.03.001.
72. Maples-Keller JL, Jovanovic T, Dunlop BW, Rauch S, Yasinski C, Michopoulos V, Coghlan C, Norrholm S, Rizzo AS, Ressler K, Rothbaum BO. When translational neuroscience fails in the clinic: dexamethasone prior to virtual reality exposure therapy increases drop-out rates. J Anxiety Disord. 2019;61:89–97. https://doi.org/10.1016/j.janxdis.2018.10.006.

73. Tardif N, Therrie C, Bouchard S. Re-examining psychological mechanisms underlying virtual reality-based exposure for spider phobia. Cyberpsychology Behav Soc Netw. 2019;22:39–45. https://doi.org/10.1089/cyber.2017.0711.
74. Côté S, Bouchard S. Cognitive mechanisms underlying virtual reality exposure. CyberPsychology Behav. 2009;12:121–9. https://doi.org/10.1089/cpb.2008.0008.
75. Shiban Y, Peperkorn H, Alpers GW, Pauli P, Mühlberger A. Influence of perceptual cues and conceptual information on the activation and reduction of claustrophobic fear. J Behav Ther Exp Psychiatry. 2016;51:19–26. https://doi.org/10.1016/j.jbtep.2015.11.002.
76. Shiban Y, Pauli P, Mühlberger A. Effect of multiple context exposure on renewal in spider phobia. Behav Res Ther. 2013;51:68–74. https://doi.org/10.1016/j.brat.2012.10.007.
77. Shiban Y, Schelhorn I, Pauli P, Mühlberger A. Effect of combined multiple contexts and multiple stimuli exposure in spider phobia: a randomized clinical trial in virtual reality. Behav Res Ther. 2015;71:45–53. https://doi.org/10.1016/j.brat.2015.05.014.
78. Shiban Y, Brütting J, Pauli P, Mühlberger A. Fear reactivation prior to exposure therapy: does it facilitate the effects of VR exposure in a randomized clinical sample? J Behav Ther Exp Psychiatry. 2015;46:133–40. https://doi.org/10.1016/j.jbtep.2014.09.009.
79. Shiban Y, Pauli P, Mühlberger A. Effect of multiple context exposure on renewal in spider phobia. Behav Res Ther. 2013;51:68–74. https://doi.org/10.1016/j.brat.2012.10.007.
80. Shiban YY, Peperkorn HH, Alpers GWGW, Pauli PP, Mühlberger AA. Influence of perceptual cues and conceptual information on the activation and reduction of claustrophobic fear. J Behav Ther Exp Psychiatry. 2016;51:19–26. https://doi.org/10.1016/j.jbtep.2015.11.002.
81. Lindner P, Miloff A, Fagernäs S, Andersen J, Sigeman M, Andersson G, Furmark T, Carlbring P. Therapist-led and self-led one-session virtual reality exposure therapy for public speaking anxiety with consumer hardware and software: a randomized controlled trial. J Anxiety Disord. 2019;61:45–54. https://doi.org/10.1016/j.janxdis.2018.07.003.
82. Miloff A, Lindner P, Dafgård P, Deak S, Garke M, Hamilton W, Heinsoo J, Kristoffersson G, Rafi J, Sindemark K, Sjölund J, Zenger M, Reuterskiöld L, Andersson G, Carlbring P. Automated virtual reality exposure therapy for spider phobia vs. in-vivo one-session treatment: a randomized non-inferiority trial. Behav Res Ther. 2019;118:130–40. https://doi.org/10.1016/j.brat.2019.04.004.
83. Mataix-Cols D, Fernández de la Cruz L, Monzani B, Rosenfield D, Andersson E, Pérez-Vigil A, Frumento P, de Kleine RA, Difede J, Dunlop BW, Farrell LJ, Geller D, Gerardi M, Guastella AJ, Hofmann SG, Hendriks G-J, Kushner MG, Lee FS, Lenze EJ, Levinson CA, McConnell H, Otto MW, Plag J, Pollack MH, Ressler KJ, Rodebaugh TL, Rothbaum BO, Scheeringa MS, Siewert-Siegmund A, Smits JAJ, Storch EA, Ströhle A, Tart CD, Tolin DF, van Minnen A, Waters AM, Weems CF, Wilhelm S, Wyka K, Davis M, Rück C. D-cycloserine augmentation of exposure-based cognitive behavior therapy for anxiety, obsessive-compulsive, and posttraumatic stress disorders. JAMA Psychiatry. 2017;22:1–10. https://doi.org/10.1001/jamapsychiatry.2016.3955.
84. Rothbaum BO, Price M, Jovanovic T, Norrholm SD, Gerardi M, Dunlop B, Davis M, Bradley B, Duncan EJ, Rizzo A, Ressler KJ. A randomized, double-blind evaluation of D-cycloserine or alprazolam combined with virtual reality exposure therapy for posttraumatic stress disorder in Iraq and Afghanistan War veterans. Am J Psychiatry. 2014;171:640–8.
85. de Quervain DJ-F, Bentz D, Michael T, Bolt OC, Wiederhold BK, Margraf J, Wilhelm FH. Glucocorticoids enhance extinction-based psychotherapy. Proc Natl Acad Sci. 2011;108:6621–5. https://doi.org/10.1073/pnas.1018214108.
86. Meyerbroeker K, Morina N, Kerkhof GA, Emmelkamp PMG. Virtual reality exposure therapy does not provide any additional value in agoraphobic patients: a randomized controlled trial. Psychother Psychosom. 2013;82:170–6. https://doi.org/10.1159/000342715.
87. Quero S, Pérez-Ara MÁ, Bretón-López J, García-Palacios A, Baños RM, Botella C. Acceptability of virtual reality interoceptive exposure for the treatment of panic disorder with agoraphobia. Br J Guid Counc. 2014;42:123–37. https://doi.org/10.1080/03069885.2013.852159.

88. Yuen EK, Herbert JD, Forman EM, Goetter EM, Comer R, Bradley J-C. Treatment of social anxiety disorder using online virtual environments in second life. Behav Ther. 2013;44:51–61. https://doi.org/10.1016/j.beth.2012.06.001.
89. Wong Sarver N, Beidel DC, Spitalnick JS. The feasibility and acceptability of virtual environments in the treatment of childhood social anxiety disorder. J Clin Child Adolesc Psychol. 2014;43:63–73. https://doi.org/10.1080/15374416.2013.843461.
90. Guillén V, Baños RM, Botella C. Users' opinion about a virtual reality system as an adjunct to psychological treatment for stress-related disorders: a quantitative and qualitative mixed-methods study. Front Psychol. 2018;9:1–14. https://doi.org/10.3389/fpsyg.2018.01038.
91. Wiederhold BK, Wiederhold MD. Three-year follow-up for virtual reality exposure for fear of flying. CyberPsychol Behav. 2003;6:441–5. https://doi.org/10.1089/109493103322278844.
92. Anderson PL, Edwards SM, Goodnight JR, Anderson PL. Virtual reality and exposure group therapy for social anxiety disorder: results from a 4–6 year follow-up. Cogn Ther Res. 2017;41:230–6. https://doi.org/10.1007/s10608-016-9820-y.
93. Benbow A, Anderson P. Long-term improvements in probability and cost biases following brief cognitive behavioral therapy for social anxiety disorder. Cognit Ther Res. 2019;43:412–8. https://doi.org/10.1007/s10608-018-9947-0.
94. Miloff A, Lindner P, Hamilton W, Reuterskiöld L, Andersson G, Carlbring P. Single-session gamified virtual reality exposure therapy for spider phobia vs. traditional exposure therapy: study protocol for a randomized controlled non-inferiority trial. Trials. 2016;17:60. https://doi.org/10.1186/s13063-016-1171-1.
95. Bun P, Gorski F, Grajewski D, Wichniarek R, Zawadzki P. Low – cost devices used in virtual reality exposure therapy. Procedia Comput Sci. 2016;104:445–51. https://doi.org/10.1016/j.procs.2017.01.158.
96. Jerdan SW, Grindle M, Van Woerden HC, Kamel Boulos MN. Head-mounted virtual reality and mental health: critical review of current research. J Med Internet Res. 2018;20:1–16. https://doi.org/10.2196/games.9226.
97. Kashdan TB, Breen WE. Social anxiety and positive emotions: a prospective examination of a self-regulatory model with tendencies to suppress or express emotions as a moderating variable. Behav Ther. 2008;39:1–12.
98. Stupar-Rutenfrans S, Ketelaars LEH, van Gisbergen MS. Beat the fear of public speaking: mobile 360° video virtual reality exposure training in home environment reduces public speaking anxiety. Cyberpsychol Behav Soc Netw. 2017;20:624–33. https://doi.org/10.1089/cyber.2017.0174.
99. Hong Y-J, Kim HE, Jung YH, Kyeong S, Kim J-J. Usefulness of the mobile virtual reality self-training for overcoming a fear of heights. Cyberpsychol Behav Soc Netw. 2017;20:753–61. https://doi.org/10.1089/cyber.2017.0085.
100. Leder G. Know Thyself? Questioning the theoretical foundations of cognitive behavioral therapy. Rev Philos Psychol. 2017;8:391–410. https://doi.org/10.1007/s13164-016-0308-1.
101. Bailey JO, Bailenson JN, Casasanto D. When does virtual embodiment change our minds? Presence Teleop Virt Environ. 2016;25:222–33. https://doi.org/10.1162/PRES_a_00263.
102. Fuchs T, Koch SC. Embodied affectivity: on moving and being moved. Front Psychol. 2014;5:508. https://doi.org/10.3389/fpsyg.2014.00508.
103. Riva G, Wiederhold BK, Mantovani F. Neuroscience of virtual reality: from virtual exposure to embodied medicine. Cyberpsychol Behav Soc Netw. 2019;22:82–96. https://doi.org/10.1089/cyber.2017.29099.gri.
104. Riva G. The neuroscience of body memory: from the self through the space to the others. Cortex. 2018;104:241–60. https://doi.org/10.1016/j.cortex.2017.07.013.
105. Riva G, Serino S, Di Lernia D, Pavone EF, Dakanalis A. Embodied medicine: mens sana in corpore virtuale sano. Front Hum Neurosci. 2017;11:1–9. https://doi.org/10.3389/fnhum.2017.00381.
106. Falconer CJ, Slater M, Rovira A, King JA, Gilbert P, Antley A, Brewin CR. Embodying compassion: a virtual reality paradigm for overcoming excessive self-criticism. PLoS One. 2014;9 https://doi.org/10.1371/journal.pone.0111933.

107. Bailey J, Bailenson J, Casasanto D. When does virtual embodiment change our minds? Presence Teleop Virt Environ. 2016;25:222. https://doi.org/10.1162/PRES.
108. Aymerich-Franch L, Kizilcec RF, Bailenson JN. The relationship between virtual self similarity and social anxiety. Front Hum Neurosci. 2014;8:944. https://doi.org/10.3389/fnhum.2014.00944.
109. Falconer CJ, Rovira A, King JA, Gilbert P, Antley A, Fearon P, Ralph N, Slater M, Brewin CR. Embodying self-compassion within virtual reality and its effects on patients with depression. Br J Psychol Open. 2016;2:74–80. https://doi.org/10.1192/bjpo.bp.115.002147.
110. Guterstam A, Abdulkarim Z, Ehrsson HH. Illusory ownership of an invisible body reduces autonomic and subjective social anxiety responses. Sci Rep. 2015;5:1–8. https://doi.org/10.1038/srep09831.
111. Crucianelli L, Metcalf NK, Fotopoulou A, Jenkinson PM. Bodily pleasure matters: velocity of touch modulates body ownership during the rubber hand illusion. Front Psychol. 2013;4:1–7. https://doi.org/10.3389/fpsyg.2013.00703.
112. Di Lernia D, Cipresso P, Pedroli E, Riva G. Toward an embodied medicine: a portable device with programmable interoceptive stimulation for heart rate variability enhancement. Sensors (Switzerland). 2018;18 https://doi.org/10.3390/s18082469.
113. Di Lernia, D., Riva, G., Cipresso, P.: iStim. A new portable device for interoceptive stimulation. In: Lecture Notes of the Institute for Computer Sciences, Social-Informatics and Telecommunications Engineering, LNICST (2018). https://doi.org/10.1007/978-3-030-01093-5_6.
114. Di Lernia D, Serino S, Polli N, Cacciatore C, Persani L, Riva G. Interoceptive axes dissociation in anorexia nervosa: a single case study with follow up post-recovery assessment. Front Psychol. 2019;9:1–8. https://doi.org/10.3389/fpsyg.2018.02488.
115. Di Lernia D, Serino S, Pezzulo G, Pedroli E, Cipresso P, Riva G. Feel the time. Time perception as a function of interoceptive processing. Front Hum Neurosci. 2018;12:1–17. https://doi.org/10.3389/fnhum.2018.00074.
116. Di Lernia D, Serino S, Riva G. Pain in the body. Altered interoception in chronic pain conditions: a systematic review. Neurosci Biobehav Rev. 2016;71:328–41. https://doi.org/10.1016/j.neubiorev.2016.09.015.
117. Castelnuovo G, Giusti EM, Manzoni GM, Saviola D, Gabrielli S, Lacerenza M, Pietrabissa G, Cattivelli R, Maria Spatola CA, Rossi A, Varallo G, Novelli M, Villa V, Luzzati F, Cottini A, Lai C, Volpato E, Cavalera C, Pagnini F, Tesio V, Castelli L, Tavola M, Torta R, Arreghini M, Zanini L, Brunani A, Seitanidis I, Ventura G, Capodaglio P, D'Aniello GE, Scarpina F, Brioschi A, Bigoni M, Priano L, Mauro A, Riva G, Di Lernia D, Repetto C, Regalia C, Molinari E, Notaro P, Paolucci S, Sandrini G, Simpson S, Wiederhold BK, Gaudio S, Jackson JB, Tamburin S, Benedetti F, Agostini M, Alfonsi E, Aloisi AM, Alvisi E, Aprile I, Armando M, Avenali M, Azicnuda E, Barale F, Bartolo M, Bergamaschi R, Berlangieri M, Berlincioni V, Berliocchi L, Berra E, Berto G, Bonadiman S, Bonazza S, Bressi F, Brugnera A, Brunelli S, Buzzi MG, Cacciatori C, Calvo A, Cantarella C, Caraceni A, Carone R, Carraro E, Casale R, Castellazzi P, Castino A, Cerbo R, Chiò A, Ciotti C, Cisari C, Coraci D, Dalla Toffola E, Defazio G, De Icco R, Del Carro U, Dell'Isola A, De Tanti A, D'Ippolito M, Fazzi E, Ferrari A, Ferrari S, Ferraro F, Formaglio F, Formisano R, Franzoni S, Gajofatto F, Gandolfi M, Gardella B, Geppetti P, Giammò A, Gimigliano R, Greco E, Ieraci V, Invernizzi M, Jacopetti M, La Cesa S, Lobba D, Magrinelli F, Mandrini S, Manera U, Marchettini P, Marchioni E, Mariotto S, Martinuzzi A, Masciullo M, Mezzarobba S, Miotti D, Modenese A, Molinari M, Monaco S, Morone G, Nappi R, Negrini S, Pace A, Padua L, Pagliano E, Palmerini V, Pazzaglia C, Pecchioli C, Picelli A, Porro CA, Porru D, Romano M, Roncari L, Rosa R, Saccavini M, Sacerdote P, Schenone A, Schweiger V, Scivoletto G, Smania N, Solaro C, Spallone V, Springhetti I, Tassorelli C, Tinazzi M, Togni R, Torre M, Traballesi M, Tramontano M, Truini A, Tugnoli V, Turolla A, Vallies G, Verzini E, Vottero M, Zerbinati P. What is the role of the placebo effect for pain relief in neurorehabilitation? Clinical implications from the Italian consensus conference on pain in neurorehabilitation. Front Neurol. 2018;9:310. https://doi.org/10.3389/fneur.2018.00310.

118. Gaume A, Vialatte A, Mora-Sánchez A, Ramdani C, Vialatte FB. A psychoengineering paradigm for the neurocognitive mechanisms of biofeedback and neurofeedback. Neurosci Biobehav Rev. 2016;68:891–910. https://doi.org/10.1016/j.neubiorev.2016.06.012.
119. Appelhans BM, Luecken LJ. Heart rate variability as an index of regulated emotional responding. Rev Gen Psychol. 2006;10:229–40. https://doi.org/10.1037/1089-2680.10.3.229.
120. Balzarotti S, Biassoni F, Colombo B, Ciceri MR. Cardiac vagal control as a marker of emotion regulation in healthy adults: a review. Biol Psychol. 2017;130:54–66. https://doi.org/10.1016/j.biopsycho.2017.10.008.
121. Aldao A, Nolen-Hoeksema S, Schweizer S. Emotion-regulation strategies across psychopathology: a meta-analytic review. Clin Psychol Rev. 2010;30:217–37. https://doi.org/10.1016/j.cpr.2009.11.004.
122. Goessl VC, Curtiss JE, Hofmann SG. The effect of heart rate variability biofeedback training on stress and anxiety: a meta-analysis. Psychol Med. 2017;47:2578–86. https://doi.org/10.1017/S0033291717001003.
123. Johnston SJ, Boehm SG, Healy D, Goebel R, Linden DEJ. Neurofeedback: a promising tool for the self-regulation of emotion networks. Neuroimage. 2010; https://doi.org/10.1016/j.neuroimage.2009.07.056.
124. Zotev V, Phillips R, Young KD, Drevets WC, Bodurka J. Prefrontal control of the amygdala during real-time fMRI neurofeedback training of emotion regulation. PLoS One. 2013; https://doi.org/10.1371/journal.pone.0079184.
125. Nicholson AA, Rabellino D, Densmore M, Frewen PA, Paret C, Kluetsch R, Schmahl C, Théberge J, Neufeld RWJ, McKinnon MC, Reiss J, Jetly R, Lanius RA. The neurobiology of emotion regulation in posttraumatic stress disorder: amygdala downregulation via real-time fMRI neurofeedback. Hum Brain Mapp. 2017;38:541–60. https://doi.org/10.1002/hbm.23402.
126. Repetto C, Gaggioli A, Pallavicini F, Cipresso P, Raspelli S, Riva G. Virtual reality and mobile phones in the treatment of generalized anxiety disorders: a phase-2 clinical trial. Pers Ubiquitous Comput. 2013;17:253–60. https://doi.org/10.1007/s00779-011-0467-0.
127. Lorenzetti V, de Melo BRP, Basilio R, Suo C, Yucel M, Tierra-Criollo CJ, Moll J. Emotion regulation using virtual environments and real-time fMRI neurofeedback. Front Neurol. 2018;9:390. https://doi.org/10.3389/FNEUR.2018.00390.
128. Bailenson J. Protecting nonverbal data tracked in virtual reality. JAMA Pediatr. 2018;172:905. https://doi.org/10.1001/jamapediatrics.2018.1909.
129. Schueller SM, Tomasino KN, Mohr DC. Integrating human support into behavioral intervention technologies: the efficiency model of support. Psychol Sci Pract. 2017;24(1):27–45. https://doi.org/10.1111/cpsp.12173.
130. Riva G. Virtual reality as assessment tool in psychology. Stud Health Technol Inform. 1998;44:71–9. https://doi.org/10.3233/978-1-60750-888-5-71.
131. Chicchi Giglioli IA, Pallavicini F, Pedroli E, Serino S, Riva G. Augmented reality: a brand new challenge for the assessment and treatment of psychological disorders. Comput Math Methods Med. 2015;19 https://doi.org/10.1155/2015/862942.
132. Freeman D, Reeve S, Robinson A, Ehlers A, Clark D, Spanlang B, Slater M. Virtual reality in the assessment, understanding, and treatment of mental health disorders. Psychol Med. 2017;47:2393. https://doi.org/10.1017/S003329171700040X.
133. Norcross JC, Pfund RA, Prochaska JO. Psychotherapy in 2022: a Delphi poll on its future. Prof Psychol. 2013;44:363–70. https://doi.org/10.1037/a0034633.
134. Norcross JC, Hedges M, Prochaska JO. The face of 2010: a Delphi poll on the future of psychotherapy. Prof Psychol Res Pract. 2002; https://doi.org/10.1037/0735-7028.33.3.316.
135. Lindner P, Miloff A, Hamilton W, Carlbring P, Pauli P. The potential of consumer-targeted virtual reality relaxation applications: descriptive usage, uptake and application performance statistics for a first-generation application. Front Psychol. 2019;10:132. https://doi.org/10.3389/fpsyg.2019.00132.

Current Research on Complementary and Alternative Medicine (CAM) in the Treatment of Anxiety Disorders: An Evidence-Based Review

22

Vladimir Trkulja and Hrvoje Barić

Introduction

Assessing *complementary and alternative medicine* (CAM) in treatment of anxiety disorders is linked to several dilemmas. One is related to terminology and definition of CAM. While the words *complementary* or *alternative* are self-understandable, the term CAM is typically used to address "therapies that lie outside the spectrum of traditional, science-based clinical medicine and surgery" [1]. Within the paradigm of evidence-based medicine (EBM), decisions about therapeutic interventions should be based on critical assessment of evidence, where *evidence* (*scientific evidence*) means empirical observations supporting a certain claim/theory and *quality evidence* means empirical observations that leave little uncertainty (because consistent, direct, unbiased, precise) about the direction and size of the effect of an intervention [2]. However, the distinction between CAM and *conventional* is not only about scientific background. Definition by the National Center for Complementary and Alternative Medicine (NCCAM; since 2014 National Center for Complementary and Integrative Health; NCCIH), USA [3, p. 19], recognizes that the boundaries between CAM and the "dominant (health) system" are not always sharp or fixed and that CAM "encompasses resources, health systems, modalities, and practices and their accompanying theories and beliefs, other than those intrinsic to the dominant health system of a particular society or culture in a given historical period. CAM includes such resources perceived by their users as associated with positive health outcomes" [3]. Hence, reasons for defining a treatment as CAM are not only scientific but also political, social, and conceptual [4]. Simply said, a CAM method would be any "nonmainstream practice" [5] aimed at achieving a beneficial health

V. Trkulja (✉)
Department of Pharmacology, Zagreb University School of Medicine, Zagreb, Croatia
e-mail: vtrkulja@mef.hr

H. Barić
Department of Neurosurgery, University Hospital Center Zagreb, Zagreb, Croatia

Y.-K. Kim (ed.), *Anxiety Disorders*, Advances in Experimental Medicine and Biology 1191, https://doi.org/10.1007/978-981-32-9705-0_22

outcome. Within the present context, *mainstream* pertains to the usual allopathic Western practice whose knowledge base derives from science. The EBM approach enables translation of seemingly incommensurable concepts into the (scientific) language of proof and evidence – the discovery of artemisinin, an antimalaric, is a prominent example [6]. Moreover, it classifies treatments as either supported by evidence or not; all other distinctions (e.g., conventional vs. CAM) are only of secondary relevance [7]. Consequently, one could doubt whether, for example, virtual reality or well-being therapy, each addressed in other chapters of this book, should be (due to this fact) considered mainstream or CAM, or whether otherwise mainstream psychiatric interventions being evaluated for further indication in anxiety disorders should be considered as such, or as CAM (on the account of thus far scarce evidence). For a practical purpose of defining a framework of this chapter, we considered all interventions *not* listed as recommended treatments for anxiety disorders in respective professional guidelines (e.g., [8–15]) as CAM, except those with a recognized use (*mainstream*) in other psychiatric/neurological or medical conditions, or in treatment of anxiety disorders but with a new mode of implementation, e.g., cognitive behavioral therapy (CBT) via the Internet or smartphones. Surveys [16, 17] have identified more than 100 different CAM practices. We address them using the classification proposed by NCCAM (NCCIH) [3], i.e., as (a) biologically based therapies (e.g., "folk medicine," diet-based therapies, natural products); (b) mind-body interventions (e.g., meditation, relaxation, hypnosis, Tai Chi, Qi gong); (c) manipulative and body-based methods (e.g., chiropractic, osteopathy, massage, movement therapies); (d) alternative medical systems (e.g., Ayurveda, homeopathy, traditional healers); (e) energy therapies (e.g., light therapy, magnetic fields); and others.

This book explicitly addresses anxiety disorders in line with the DSM 5 criteria (excludes obsessive-compulsive disorder [OCD] and post-traumatic stress disorder [PTSD]). This chapter includes reference to relevant publications on the topic where anxiety disorders were diagnosed in line with the accepted *mainstream* criteria that were in place at a given time period (e.g., previous editions of DSM). To be in line with the general concept of the entire book, the potential of CAM specifically in OCD or PTSD is not addressed, but studies comprising mixed patient populations with different anxiety disorders, including OCD or PTSD are considered. Level of anxiety is increased in different situational contexts, physiological (e.g., pregnancy) and medical conditions (e.g., malignancy, cardiovascular disorders). Even a brief search of Medline discloses a number of publications on the use of CAM to "relieve anxiety" in such settings – a recent systematic review [18] addressed randomized controlled trials (RCTs) of CAM in pregnant women reaching certain cut-off scores on anxiety rating scales, but none of the participants was actually diagnosed with *anxiety disorder due to another medical condition* [18]; another one identified 19 RCTs of transcendental meditation in trait anxiety [19] in students, prison inmates, veterans, prison staff, patients with hypertension, etc., but again, participants were not actually diagnosed with any *anxiety disorder*. This chapter focuses on evidence pertaining to the use of CAM for treatment of specific *anxiety disorders* and not for relief of anxiety in otherwise healthy subjects or in medically ill patients. While this might be objected having in mind the concept of complementary and integrative health [20] that implies the use of CAM to improve the outcomes of *mainstream* treatments in any diseased condition or to contribute to disease prevention and/or improvement of well-being (i.e., CAM as a part of a holistic approach) [20], we find the settings were anxiety is the only or the primary disorder to be more appropriate for detecting the potential of CAM procedures for its treatment as their main effect (not confounded or obscured by possible effects on the other primary underlying condition). In 2013, Jonas et al. [21] provided a comprehensive overview of the "evolution of CAM in the USA," evolution of terminology related to CAM and its conceptual perception, and a "switch" in stakeholders' reasoning resulting in a "shift from questions such as 'What is CAM?' or even 'Does this CAM work?' to questions such as 'How can we use this CAM practice to improve clinical outcomes, reduce overall health-related expenditures or increase worker productivity and quality of life'" [21]. We take a "step back" and primarily address the question "Does this CAM work?" for treatment of anxiety disorders looking for evidence in the EBM sense.

The NHIS surveys show a clear increasing trend in the use of CAM, and anxiety and depression appear to be among 5–6 conditions for which CAM methods are most frequently used by adults and children (3–4% of participants reported using CAM for anxiety or depression) [22]. In reverse, a large survey in 25 countries reported a 1-year prevalence of CAM use among patients with an anxiety disorder diagnosed in line with DSM 4 to be 3.9%; moreover, in those with high disease severity, the prevalence was 7.2% for CAM use, 1.4% for only CAM use, and 5.9% for CAM combined with other care [23]. History of psychiatry is rich with examples of treatments that, at certain times, were considered scientifically *unsound* (by the actual criteria), but were subsequently accepted as *mainstream* only to be eventually discarded [24]. In the meantime, understanding of the nature (biology) of psychiatric disorders has enormously evolved (although still being far from complete) and so is the understanding of the mechanisms through which the *mainstream* (or intended to be) treatments exert their effects, this fact giving them the image of *scientifically based*. This kind of knowledge is missing for most of the CAM treatments in most of the situations [20, 24]; however, the EBM methodology does not necessarily require understanding of the underlying mechanisms. Here we overview the body of evidence on efficacy/safety of CAM treatments for anxiety disorders using the established EBM principles that are standardly employed to evaluate the *mainstream* interventions. Considering the ever-growing number of publications on CAM (now counted in hundreds of thousands, [21, 25]), many of which pertain to anxiety disorders, a selection had to be made so this overview is far from exhaustive. In this process, we reasoned as follows:

(a) RCTs have the highest potential to accurately (unbiasedly) identify treatment effects. As a starting point, we searched PubMed/Medline and Cochrane databases to identify systematic reviews of RCTs pertaining to treatment of anxiety disorders (diagnosed as explained) published between 2000 and January 15, 2019, and selected those reporting on the use of CAM (defined as explained).

Where indicated by the systematic reviews, we further searched individual RCTs. We provided numerical data when we deemed it informative to support claims about treatment (non)effects. We considered it unfeasible to list other numerical data (interested readers are directed to the cited literature), in part due to their practical irrelevance and in part due to the fact that the area is dynamically changing and new data might have arisen in the meantime.

(b) We attempted to evaluate quality of evidence, i.e., the level of (un)certainty about the effect (or lack of it) suggested for individual treatments (how likely it is that the provided estimate hits the true "population effect"). While there are widely used tools to evaluate quality (risk of bias) of RCTs (e.g., Cochrane collaboration tool, RoB 2.0; https://sites.google.com/site/riskofbiastool/welcome/rob-2-0-tool) or of systematic reviews (e.g., updated AMSTAR tool; https://amstar.ca/) and overall quality of evidence (e.g., GRADE; https://cebgrade.mcmaster.ca/), we did not systematically employ any of them, but implemented key criteria contained in these tools, as outlined in Table 22.1.

(c) Consequently, although this approach is sometimes used in evaluation of the *mainstream* treatments (e.g., [30]), we disregarded evaluations of "post- vs. pre-" effect sizes since they cannot distinguish the treatment effect from chance, natural course of the disease, or regression to the mean.

Biologically Based Therapies

NCAM/NCCIH explicitly excludes vitamins and minerals from this subgroup of CAM, and the term refers typically to various herbal preparations [3, 5]. Several systematic reviews addressing CAM, however, included also a reference to "nutraceuticals" or "diet" (e.g., [31–33]). Some of the herbal preparations typically addressed as CAM have a long-standing medicinal use in the Western world related to *anxiety*, and some of them have a specific regulatory status. The European Medicines Agency (EMA) and its Committee on Herbal Medicinal Products (HMPC) recognize a regulatory pathway of *traditional use registration* for herbal drugs – herbal drug/medicinal plant and preparations thereof need to be in human use for more than 30 years, data (generally, only bibliographic) are needed to support safety and *plausible* efficacy, and products are intended to be used without medical supervision (i.e., for self-medication or over-the-counter use) [34].

Lavender (*Lavandula angustifolia*) Preparations

Background HMPC has adopted two community herbal monographs [35, 36] pertaining to *Lavandula angustifolia* Miller and *traditional use* of preparations for "relief of mild symptoms of mental stress and exhaustion and to aid sleep": lavender flower (*L. angustifolia, flos*) comminuted substance to be used for preparation of tea or as a tincture and essential oil obtained by steam distillation from the flowering tops (*L. angustifolia, aetheroleum*), to be used orally or as a bath additive. There

Table 22.1 Elements considered and reasoning in evaluation of quality of RCTs and meta-analysis of RCTs and of (un)certainty about utility of CAM for treatment of anxiety disorders

Individual RCTs	
Patient sample	*Sampling procedure*. Effects observed in single-center samples are at a higher risk of being by chance (sampling variation) than the effects in multicenter samples. Even with the same inclusion-exclusion criteria, single-center samples are less likely to fully represent the population. *Size*. Small samples (vs. large) are less likely to fully represent the population; observed effects are more likely to be inflated and/or by chance and fragile
Control treatments	*Representing "idea of treatment."* Placebo or adequate sham interventions are likely to transfer the "same idea of treatment" or "expectation" as the tested intervention. No-treatment control may inflate the treatment effect (e.g., due to negative expectations). Waiting-list control may introduce selection bias (trial may include only patients "susceptible" to CAM effect) combined with inflation of treatment effect (negative expectation in the "waiting-list" group). *Active treatment*. Superiority to otherwise known effective treatments supports efficacy, but size of the effect cannot be estimated. "Lack of difference" or formal non-inferiority to otherwise known effective treatments in trials lacking assay sensitivity does not unambiguously support efficacy. Superiority to treatments not otherwise confirmed as not harmful does not unambiguously support efficacy. In order to gain regulatory approval, *mainstream* pharmacological treatments are required to be tested in RCTs including placebo (with or without an active treatment control) (see, e.g., [26])
Treatment assignment	Randomization and concealment of the randomization sequence are expected to ascertain balance between groups regarding potential confounders/effect modifiers and to prevent selection bias (arising from knowledge or anticipation of the subsequent treatment). Some findings might indicate that this process has failed, e.g., excessive similarity between groups that is not likely to be by chance, a marked imbalance in group sizes (compared to the intended ratio), excessive baseline differences between groups not likely to be by chance, or imbalance in likely confounders/effect modifiers or baseline values of the outcome measures. However, such signs are difficult to assess in small trials (how to estimate "deviation" from chance?) and small but relevant imbalances might be undetected. Hence, one has to rely upon what was reported by the authors. Although this might sound prejudiced, we consider that sponsored trials (e.g., professional societies, institutes, industry) are less likely to be flawed in this respect simply due to the fact that they are more likely to be more scrutinized during the planning and conductance phase
Trial/treatment duration	Regulatory requirements for *mainstream* pharmacological treatments consider 8–12-week trials to be "short-term" trials and include also "long-term" trials (6–12 months) (e.g., [26]). The shortest recommended (*mainstream* pharmacological or psychological) treatment in anxiety disorders is 6 months (e.g., [13]). Efficacy may become apparent only after longer periods of treatment and/or may vary in time [13]. Exclusively short-term trials (e.g., <12 weeks) may demonstrate short-term efficacy, but do not unambiguously support a claim related to successful management of anxiety disorders in line with professional recommendations. Exclusively short-term trials may not unambiguously exclude possible efficacy

(continued)

Table 22.1 (continued)

Individual RCTs	
Outcome measures	*Mainstream* treatments are expected to provide benefits in terms of reduction of severity of anxious symptoms (adequate symptom scales) and a benefit in terms of proportions of subjects experiencing (a) clinically relevant response, (b) remission, or (c) relapse [13, 26]
Detection of outcomes (beyond blinding)	Validated scales with known psychometric properties are likely to accurately quantify anxiety. For detection (ascertainment) of outcomes in trials of *mainstream* treatments, clinician-implemented tools are recommended (e.g., [26]), like Hamilton anxiety rating scale, panic disorder severity scale, panic and agoraphobia scale, Liebowitz social anxiety scale, or brief social phobia scale, and investigators are expected to be trained in their implementation. Inexperienced assessors/assessors being trained through evaluation of the study subjects may introduce bias. Unequally experienced/trained assessor may introduce bias
Blinding of patients	Considering the nature of the condition (anxiety disorder) and the fact that its quantification includes subjective patient reporting, when patients are aware of their assigned treatment, effects could easily be overestimated (e.g., unequal "expectation" vs. no-treatment/waiting list, recognizable sham procedure) but also underestimated (e.g., vs. established *mainstream* treatment). Reporting of adverse events may also be biased
Blinding of investigators/ assessors	Investigators/carers aware of the assigned patient treatments may introduce performance bias. Considering the judgment component in outcome assessment (clinician-implemented rating scales), assessors aware of the assigned patient treatment may introduce detection bias
Treatments that cannot be adequately blinded	While for biologically based therapies (e.g., herbal preparations, nutraceuticals etc.) provision of adequate placebos is feasible allowing for blinding of both patients and carers/outcome assessors, adequate sham procedures (and, consequently, blinding) might be more difficult to achieve for various other CAMs. This, however, does not mean that this is impossible, i.e., that it is not possible to achieve a fair level of reduction of risk of various biases potentially arising from the patient/carer/outcome assessor's awareness about the assigned treatments. For example, some sham procedures might be reasonable enough as to allow for the blinding of patients (but not carers/assessors). Separating the roles of carers/treatment providers/outcome assessors would then allow that carers/assessors are also blinded. Sometimes, it might be uncertain whether the sham procedure conveys only the "idea of treatment" or whether it actually might be effective, which would generate "control treatment bias" and underestimate or completely blur the effect of treatment. Factorial or incomplete factorial designs could be useful in such situations, including *mainstream* pharmacological treatments and respective placebos, resulting in multiarm trials

(continued)

Table 22.1 (continued)

Individual RCTs	
Attrition and related issues	Efficacy analysis based exclusively on patients who complete the trial "as planned" (i.e., *per* protocol, PP) tends to inflate the treatment effect. Hence, tests of inequality should be based on all randomized (intent-to-treat, ITT) patients or, what appears more reasonable in the case of RCTs in anxiety disorders, all randomized patients who are evaluated at baseline and have received at least one "dose" of the assigned treatment. PP is more conservative in the case of non-inferiority trials, but in anxiety disorders, assay sensitivity of such trials if placebo/sham is not included is questionable, and a test of non-inferiority should follow after a test of inequality (expected superiority) vs. placebo/sham. Considering the nature of the disease and expected several weeks needed to observe a treatment effect, the outcome measures are based on "difference in symptoms after a period of treatment vs. baseline," thus requiring some form of data imputation for patients who drop out of the trial (for any reason) to preserve the ITT principle. There is no ideal imputation method. Fixed data imputation (e.g., LOCF) however is commonly used in this type of trials. Imbalance in the rate of attrition or in reasons for attrition may introduce bias. Even if the reasons are comparable for the evaluated and control treatment, when attrition is high (i.e., a considerable amount of data needs to be imputed) – The estimates of treatment-control differences are of questionable accuracy
Assessment of safety	*Ascertainment of adverse events*. Adequate methods for ascertainment of adverse events in a trial are essential in order to make any claims about their (non)occurrence. All patients receiving at least "one dose" of a treatment should be considered. *Reporting*. Reports not explicitly stating a lack of a particular adverse event are uninformative
Data analysis	Baseline values of a certain measure have a great impact on the value of this measure taken at a certain later time point. Regardless of an apparent balance between groups in respect to baseline severity of symptoms, outcome measures based on "change from baseline" (actual score, change in score vs. baseline, or proportion of responders or remitters) need to be adjusted for the baseline value. Randomization is expected to achieve a balance in confounders/moderators. However, even when there are no obvious imbalances between groups in characteristics that may affect the outcome, such characteristics should be accounted for in data analysis, particularly in smaller trials [27]
Treatment effect estimates across trials (pooled estimates, meta-analysis)	
High risk of bias trials	To get an accurate estimate of a true effect, estimation should be based on unbiased observations. High risk of bias trials are likely largely inaccurate; hence pooling data from such trials makes little sense – It may only increase "precision in inaccuracy" and thus be misleading. Even if a pooled effect is similar to a pooled effect across low risk of bias trials, it still makes no sense since we do not know what the estimates would be in such trials (patient samples) had they been better protected from various biases

(continued)

Table 22.1 (continued)

Individual RCTs	
Clinical homogeneity	Effects pooled across clinically inhomogeneous trials (e.g., different patient characteristics; different doses or, e.g., acupuncture points or type of exercise; different duration) make little sense even if results are "statistically homogeneous" (particularly when the number of trials is limited) as they provide different information. Showing similar effects in different clinical settings (e.g., different patient characteristics, trial duration) is a different thing and is more informative about robustness of the effect than showing "lack of statistical heterogeneity" across a limited set of trials – With a limited number of trials, formal tests, or measures of heterogeneity/ inconsistency can be underpowered/imprecise, hence not reliable
Statistical heterogeneity/ inconsistency	Whether or not to generate pooled estimates – And how to interpret them – Across trials with heterogeneous treatment effects is a delicate issue. By using the *fixed-effect* approach, pooled estimates are likely to be inaccurate. Moreover, the idea is conceptually wrong: Pooled effect tends to present something that might not exist – a single true effect. The *random-effects* method(s) will account for the between-study variance, but would not "resolve it". The produced estimate is difficult to interpret (particularly if reported without prediction intervals) – It is the estimate of the mean of the effects: If different from zero, this does not necessarily mean that the population of effects does not also contain "effects not different from zero" or some other larger effects; if not different from zero, it does not mean that there are no "different from zero" effects. In fact, if meta-regression is not possible (and it requires a substantial number of trials) that would identify effect moderators, i.e., subsets of trials each estimating one of the effects from the "population of effects" (e.g., in younger and in older; after 6 or 24 weeks of treatment, etc.) and each subset with a reasonably low heterogeneity – The random-effects estimate may not make much sense. One such scenario refers to random-effects estimates across large and small trials. Sets of smaller trials inherently show higher between-study variance (τ^2) than the comparable sets of larger trials [28]. At the same time, estimate of τ^2 in a set of small trials with imprecise estimates (high within-study variance) may commonly be 0, although there is a substantial dispersion of effects [28]. Since in standard random-effects procedures one same estimate of τ^2 is used in weighting of all the trials, this could skew the estimates and the prediction intervals. The proportion of true heterogeneity between studies within the total heterogeneity (true/true+random) – I^2 – Depends on trial size (within-study variance): In sets of large studies (low within-study variance; random component) actually small differences (low τ^2) between study effects may result in high I^2 (since the contribution of the random part is low), while in sets of small trials (high within-study variance), I^2 could be low, despite a considerable dispersion of effects; consequently, the actual dispersion of effects "behind" one value of I^2 can be low-moderate or high [29]. Main points: (i) fixed-effect meta-analysis is rarely appropriate (e.g., studies replicated in design and clinical features with similar, consistent effect estimates); (ii) random-effects meta-analysis requires careful interpretation; (iii) choice of the method should not depend on the estimated heterogeneity/ inconsistency; and (iv) assessment of heterogeneity across a limited number of trials/small trials is problematic

(continued)

Table 22.1 (continued)

Individual RCTs	
Overall assessment of efficacy and of utility in treatment of anxiety	
Clinical relevance	Treatment of anxiety disorders implies (a) reduction of anxiety, i.e., achieving a relevant clinical improvement/remission and, in responders, (b) maintenance of remission based on the use of preferred instruments. In practical clinical terms, evidence of efficacy based on increased probability of remission/clinically relevant response or reduced risk of relapse is more meaningful than the one based on a difference in symptom severity scores. Duration of recommended *mainstream* treatments (across different anxiety disorders) is at least 6 months – a conclusion about *full utility* of any CAM treatment for anxiety disorders can only be based on the ability to achieve these goals. Other conclusions may pertain to parts of the process: e.g., efficacy in the initial short-term (e.g., 8–12-week) treatment
Overall body of data	One low risk of bias trial with adequate outcome measures and duration may result in a high level of certainty about the treatment effect; however, strength of recommendations (and regulatory requirements for pharmacological treatments) about *mainstream* treatments largely relies upon replication of the results – This is related to the nature of the disorder, and independent repeated trials are generally needed in order to exclude the chance
Risk of bias	Biased trials leave uncertainty (greater with more severe bias) about the actual treatment effect (or its nonexistence). Severely biased trials, e.g., trials using a waiting-list control (selection, performance, detection) even if large and indicating a large effect, leave a high level of uncertainty about whether a true effect exists at all
Inconsistency	In a single trial, inconsistency between treatment effects based on change in symptoms scores and treatment effects based on probability of remission/response questions the existence of a true effect. Inconsistency of effects across several trials generally generates variable levels of uncertainty about the existence of an effect and/or its size. However, when large, low risk of bias trials show effects that are all in the same direction, even if the indices of heterogeneity/inconsistency are high, this does not compromise the conclusion of efficacy
Imprecision	In general, imprecision per se raises (variable degree of) uncertainty about the treatment effect, but interpretation of imprecise estimates depends also on their location as well as the level of bias in the trials: Imprecise estimates from low risk of bias trials that consistently show an effect result in a reasonably high certainty that the true effect indeed exists, but in a high uncertainty about its size; if consistently showing lack of an effect, then we can be reasonably certain that there is no (relevant) effect (although, a smaller effect cannot be excluded). Imprecise estimates from high risk of bias trials result in a very high uncertainty about the existence/nonexistence of an effect

appear to be no RCTs evaluating Lavender tea, tincture, or baths in the treatment of anxiety disorders [37, 38]. On the other hand, an industrially sponsored oral product containing an active substance defined as "lavender essential oil obtained by steam distillation" (Silexan®, WS® 1265) [39] is likely the most comprehensively evaluated herbal preparation for treatment of anxiety disorders. The product (80 mg soft gelatin capsules) has been approved (*traditional use*) in Germany [39], Sweden [40], and several other EU and non-EU countries for "relief of mild anxiety" (e.g., Sweden) or "treatment of restlessness accompanying anxious mood" (Germany) (80 mg capsule once daily). The development of the product is outlined in an "umbrella" review [39] and is addressed in several other targeted reviews/meta-analyses [41–44]. Silexan was evaluated in RCTs in patients with generalized anxiety disorder (GAD) (DSM 5 300.02), anxiety disorder not otherwise specified (DSM 4 300.00; ICD-10 F41.9), restlessness and agitation (ICD-10 R45.1), or mixed anxiety and depressive disorder (ICD-10 F41.2) (Table 22.2). The latter three conditions are referred to as "subsyndromal anxiety disorder" – patients meet some but not all criteria for diagnosis of GAD [43].

Efficacy *GAD*. In one [45] large, multicentric low risk of bias 10-week double-blind RCT, Silexan 80 mg qd was superior to placebo and tended to be superior to paroxetine 20 mg qd in respect to Hamilton anxiety rating scale (HAM-A) score reduction at the end of treatment, proportion of responders (≥50% score reduction), but not in proportion of remitters, while Silexan 160 mg qd was superior based on all outcomes (Table 22.2). No withdrawal difficulties were observed during a subsequent week after abrupt Silexan withdrawal/paroxetine taper [50]. Another 10-week identically designed/conducted multicentric unpublished double-blind trial (Table 22.2) was indicated in the reviews [42, 43]. In this trial Silexan 80 mg qd was not superior to placebo (Table 22.2) [43]. The third [46] double-blind, multicentric 6-week trial compared Silexan 80 mg qd to lorazepam 0.5 mg qd. Although non-inferiority was claimed for Silexan [46], it suffered several limitations (Table 22.2) leaving high uncertainty about efficacy of Silexan. No withdrawal symptoms were observed in either arm over a 2-week taper [46]. *Subsyndromal anxiety disorder*. Three identically designed (Silexan 80 mg qd) multicentric, double-blind 10-week placebo-controlled low risk of bias RCTs [47–49] were conducted in these conditions (Table 22.2). Data were also analyzed jointly in a meta-analysis [44]. Silexan was formally superior to placebo in all three trials regarding HAM-A reduction, in 2/3 regarding response rates (numerically better in the third trial) and in 1/3 regarding remission rates (numerically better in other two trials) (Table 22.2; [44]). Differences vs. placebo were rather large in the first trial [47] and smaller and similar in the other two [48, 49] (Table 22.2); hence, quite some heterogeneity was observed [44]. However, all effects were in the same direction (Table 22.2, [44]).

Safety Adverse events (AEs) monitoring and reporting was adequate (regulatory trials). In none of the individual trials or joint analyses was incidence of "any AE" higher with Silexan (10–160 mg/day) than with placebo. The only AE observed more commonly with Silexan was eructation [39, 41–50]. Targeted studies in healthy subjects demonstrated no effect of Silexan (160 mg qd over 11 or 28 days) on the activity of CYP1A2, CYP2C9, CYP2C19, or CYP3A4 (relevant in metabolism of *mainstream* treatments) and no pharmacokinetic interactions with oral con-

Table 22.2 Randomized controlled trials of oral Silexan (standardized lavender oil extract in soft gelatin capsules) in anxiety disorders (identified from Ref. [39, 41–44]). Data are counts (percentages), mean ± SD or differences with 95% confidence intervals[a]

	General design and quality	ITT data	HAM-A reduction	Response (≥50% reduced)	Remission (HAM-A < 10)
Generalized anxiety disorder (DSM 5, 300.02)					
Kasper 2014 [45]	Multicentric, DB, 10-week; Outpatients 18–65 years, HAM-A ≥ 18; low risk of bias all items, adequate data analysis	Placebo $n = 135$ Paroxetine 20 mg qd $n = 132$ Silexan 80 mg qd $n = 135$ Silexan 160 mg qd $n = 121$	Placebo: 9.5 ± 9.0 Paroxetine 11.3 ± 8.0 Silexan 80-placebo: 2.8 (0.9–4.8) Silexan 160-placebo: 4.0 (1.9–6.1)	Placebo: 51 (37.8) Paroxetine: 57 (43.2) Silexan 80-placebo RD: 14.1 (2.2–25.6) Silexan 160-placebo RD: 22.6 (10.3–34.1)	Placebo: 40 (29.6) Paroxetine: 45 (34.1) Silexan 80-placebo RD: 3.7 (−7.4 to 28.2) Silexan 160-placebo RD: 16.6 (4.8–28.2)
Unpublished, indicated in 43 and 44	Multicentric, DB, 10-week; Outpatients 18–65 years, HAM-A ≥ 18; low risk of bias all items, adequate data analysis	Placebo $n = 102$ Silexan 80 mg qd $n = 103$ (also 10 mg or 40 mg, but clearly not effective)	Placebo: 11.4 ± 8.0 Silexan 80: 11.6 ± 8.1 (Δ = 0.2; −1.7 to 2.3)	No data available, but apparently [43] no difference vs. placebo	No data available, but apparently [43] no difference vs. placebo
Woelk 2010 [46]	Multicentric, DB, 6-week; Outpatients 18–65 years, HAM-A ≥ 18; no attrition, but: No assay sensitivity; no adjustments for baseline HAM-A; inconsistency in score change and response/remission treatment differences, imprecise estimates	Lorazepam 0.5 mg qd $n = 37$ Silexan 80 mg qd $n = 40$	Lorazepam: 11.6 ± 6.6 Silexan 80: 11.3 ± 6.6 (Δ = −0.3; −3.3 to 2.7)	Lorazepam: 15 (40.5) Silexan 80: 21 (52.5) (RD = 12.0;- 10.4 to 33.1)	Lorazepam: 10 (27.0) Silexan 80: 16 (40.0) (RD = 13.0; −8.4 to 33.1)

(continued)

Table 22.2 (continued)

	General design and quality	ITT data	HAM-A reduction	Response (≥50% reduced)	Remission (HAM-A < 10)
"Subsyndromal" anxiety disorder(s)					
Kasper 2010 [47], DSM IV 300.00/ICD-10 F41.9	Multicentric, DB, 10-week; Outpatients 18–65 years, HAM-A ≥ 18; low risk of bias, but biased detection of response/remission (PSQI interference)	Placebo n = 108 Silexan 80 mg qd n = 104	Placebo: 9.5 ± 9.1 Silexan 80: 16.0 ± 8.3 (Δ = 6.5; 4.1–8.9)	HAM-A or PSQI ≥50% red. Placebo: 53 (49.1) Silexan 80: 80 (76.9) (RD = 27.8; 15.0–39.8)	HAM-A < 10 or PSQI <6 Placebo: 46 (42.6) Silexan 80: 63 (60.6) (RD = 18.0; 4.5 to 30.8)
Kasper 2015 [48], ICD-10 R45.1	Multicentric, DB, 10-week; Outpatients 18–65 years, HAM-A ≥ 18; low risk of bias all items, adequate data analysis	Placebo n = 84 Silexan 80 mg qd n = 86	Placebo: 9.3 Silexan-placebo 2.7 (4.5 to 30.8)	Placebo: 28 (33.3) Silexan 80: 42 (48.8) (RD = 15.5; 0.7–29.7)	Placebo: 19 (22.6) Silexan 80: 27 (31.4) (RD = 8.8; −4.7 to 22.0)
Kasper 2016 [49], ICD-10 F41.2	Multicentric, DB, 10-week; Outpatients 18–65 years, HAM-A ≥ 18; low risk of bias all items, adequate data analysis	Placebo n = 156 Silexan 80 mg qd n = 159	HAM-A Placebo: 8.4 ± 8.9 Silexan-placebo 2.5 (0.5–4.5) MADRS Placebo: 6.1 ± 7.6 Silexan-placebo 3.3 (1.6–5.4)	HAM-A Placebo: 54 (34.6) Silexan 80: 66 (41.5) (RD = 6.9; −3.8 to 17.5) MADRS Placebo: 50 (32.1) Silexan 80: 64 (40.3) (RD = 8.2; −2.4 to 18.6)	HAM-A Placebo: 45 (28.8) Silexan 80: 55 (34.6) (RD = 5.7; −4.6 to 15.9) MADRS ≤10 Placebo: 53 (34.0) Silexan: 74 (46.5) (RD = 12.5; 4.5 to 30.8)

[a]Where adjusted differences were reported, symptom score reductions for placebo or active controls are shown as mean ± SD, and for Silexan adjusted mean differences vs. placebo are shown; otherwise, score reductions (mean ± SD) and proportions for Silexan and control are shown, and differences (mean difference, Δ, or risk difference, RD) are calculated from summary data for the purpose of this overview

HAM-A Hamilton anxiety rating scale, *MADRS* Montgomery-Asberg Depression Rating Scale, *PSQI* Pittsburgh Sleep Quality Index, *qd* once daily

traception [39]. Studies are underway to evaluate the impact of Silexan on driving ability (vs. placebo and lorazepam) in healthy subjects (https://www.clinicaltrialsregister.eu/ctr-search/trial/2015-001101-14/DE).

Conclusion Safety and tolerability of this particular product over 2–3 months of use are well documented. One additional large multicentric low risk of bias trial would be needed to demonstrate its efficacy in initial short-term treatment of GAD. There is no data on long(er)-term treatment, attainment/maintenance of response/remission; hence utility of this product as a monotreatment for GAD is unknown. There is moderate evidence of efficacy of Silexan 80 mg qd for short-term treatment in patients with "subsyndromal anxiety disorder," but the size of the effect is uncertain. There is no data on long(er)-term treatment.

Kava Kava (*Piper methysticum*) Preparations

Background The use of Kava Kava (preparations of Kava rhizomes; *Piper methysticum* G. Forst., rhizome) in treatment of anxiety/anxious disorders is likely the most controversial topic in the setting of herbal (CAM) preparations for anxiety and has been addressed and debated in numerous reviews/meta-analyses (e.g., [31–33, 51–60]). In part, the topic is intertwined with the topic of recreational/ceremonial and traditional medicinal (for various ailments, including anxiety) consumption of Kava beverage (maceration of grounded dried peeled rhizome in water and coconut milk) originating in Polynesia, Melanesia, and Micronesia, with a tradition of over 2000 years [61]. In 2016, the Food and Agriculture Organization (FAO) of the United Nations World Health Organization comprehensively evaluated the use of this beverage to conclude that excessive consumption might have unwanted acute and transitory but also potentially long-term health effects and that thus it should be limited in extent [61]. In part, it was prompted by observations related to industrially manufactured food/dietary supplements (dried tableted extracts or liquid extracts frequently obtained using organic solvents) in the Western countries which, at some point, were associated with a signal of hepatotoxicity resulting in withdrawal/ban of these products throughout most of the EU (except the homeopathic ones). In Germany, approval was revoked by the medicines agency, but based on a court decision in one of the German federal states, ethanolic extracts were later reintroduced to market [62]. In 2017 HMPC provided assessment report on *Piper methysticum* stating that there *was not sufficient* evidence of "*plausible* efficacy and documented safety" that would justify a community herbal monograph on *Piper methysticum* preparations as either *well-established* use or *traditional use* herbal products [62] for treatment of anxiety/stress-related disorders: an exhaustive review was provided on phytopharmaceutical, manufacturing, nonclinical pharmacological/toxicological aspects and clinical data (overall more than 1700 publications). The following key points were emphasized [62]: (a) a variety of preparations (different extraction methods; isolated or synthetic components) were used in analytical, nonclinical, and clinical studies; (b) nonclinical toxicology/safety pharmacology studies yielded variable results, indicating hepatotoxicity or increased incidence of hepatoblastoma for some specific preparations, likely not applicable to others, and a lack of adequate reproductive and developmental toxicity studies; and (c) clinical studies varied in duration, dosage (based on kavalactones), type of extract, patient population, and outcome measures. Table 22.3 summarizes RCTs of Kava preparations "in anxiety" identified by HMPC to illustrate their clinical heterogeneity.

Efficacy Two meta-analyses are commonly used to substantiate efficacy of Kava extracts – one by Cochrane Collaboration [53] and one that included only studies with the extract WS1490 (likely based on acetonic or ethanolic extraction) [54]. The first one [53] included seven RCTs vs. placebo (trials No 1, 6, 11–14 in Table 22.3 and one additional 8-week trial of WS1490 70 mg tid, $n = 20$, vs. placebo, $n = 20$, in "climacteric syndrome, HAM-A >18") [53, 62]. Based on 197 patients on Kava and 183 on placebo, a pooled estimate (weighted mean difference, WMD) of difference (Kava-placebo) in HAM-A reduction at the end of treatment was 3.85 (95%CI 0.05 to 7.66) by random-effects method. However, $\tau^2 = 19.89$ and $I^2 = 78\%$ [53] call for caution in interpretation. For example, (a) without the "climacteric syndrome" trial (six trials), WMD is 2.36 (95%CI −0.79–5.51); $z = 1.47$, $p = 0.142$, $\tau^2 = 10.59$, $I^2 = 69\%$ (no effect); (b) with all seven trials but using the Hartung-Knapp-Sidik-Jonkman (HKSJ) correction recommended with small number of trials/small trials, the 95%CI for WMD extends from −1.81 to 9.47, $t = 1.660$, $p = 0.147$ (no effect), and without the "climacteric syndrome" trial, it extends from −1.79 to 6.50, $t = 1.467$, $p = 0.202$ (no effect); and (c) based on all seven trials, 95% prediction interval for Kava-placebo difference extends from −8.59 to 16.3 (from Kava substantially worse to substantially better), and the range of effects (WMD ± 2τ) is −5.02 to 12.72. The published pooled effect, hence, should be carefully interpreted. The second meta-analysis was based on individual patient data [54] and included the same trials as the previous one [53], except for the trial No 1 in Table 22.3 (overall, 180 on WS1490 and 165 on placebo), to conclude superiority in terms of HAM-A reduction (reported WMD 5.94, 95%CI −0.86 to 12.8; $p = 0.074$). A few points: (a) recalculation on study summary data for the purpose of this chapter indicated WMD = 4.96, 95%CI 1.14 to 8.77, $z = 2.547$, $p = 0.011$, $\tau^2 = 16.09$, and $I^2 = 74\%$; (b) with HKSJ correction, 95%CI −1.05 to 10.97, $t = 2.122$, $p = 0.087$ (no effect); (c) without "climacteric" trial and with HKSJ correction (5 RCTs, 160 WA1490, 145 placebo), WMD = 3.41, 95%CI −0.73 to 7.55, $\tau^2 = 6.25$, $I^2 = 56.1\%$, $t = 2.287$, $p = 0.084$ (no effect); (d) with all six trials, range of effects from −3.1 to 12.98 and 95%PI from −7.14 to 17.3; (e) the largest trial with WS1490 (No 5 in Table 22.3) was not included as it used Anxiety Status Inventory: reduction in scores were somewhat higher with WS1490 ($n = 71$) than with placebo ($n = 70$) but not significantly [63]. The meta-analysis [54] reported also a strong effect of WS1490 based on proportion of responders (reduction in HAM-A ≥ 50%) – pooled OR = 3.30, 95%CI 2.09–5.22. Recalculation based on summary data, without the "climacteric syndrome trial" and including the largest trial (63; response = reduction in ASI >5 score points), with HKSJ correction gives Mantel-Haenszel RR = 1.47, 95%CI 1.10–1.97, $t = 3.379$, $p = 0.02$, $\tau^2 = 0.017$, $I^2 = 22.4\%$, and 95%PI extending from 0.90 to 2.37 indicating, still, some uncertainty. We previously [51] addressed RCTs of Kava vs. placebo to point out meaninglessness of pooled estimates due to clinical and statistical heterogeneity (even after adjustment for baseline

Table 22.3 Randomized controlled trials of Kava extracts in "anxious disorders" identified in the HMPC assessment report [62]. Study data taken from Ref. [62] and individual study publications

Trial N°, design	Indication	Extract, kavalactones dose	Treatments
1, DB, 4 weeks, HAM-A primary	DSM 4 GAD HAM-A ≥ 16	Ethanolic, 70 mg bid 1 week and then 140 mg bid	Kava $n = 17$ Placebo $n = 18$
2, DB, 4 weeks HAM-A primary	DSM 4 GAD HAM-A 12–20	Ethanolic, 70 mg bid 1 week and then 140 mg bid	Kava $n = 6$ Placebo $n = 7$
3, DB, 8 weeks HAM-A primary	DSM 4 GAD HAM-A ≥ 18	Ethanolic, 70 mg bid 1 week and then 140 mg bid	Kava $n = 5$ Venlaf. XR 225 mg qd $n = 6$ Placebo $n = 5$
4, DB, 8 weeks HAM-A primary	GAD ICD-10 HAM-A ≥ 19	Ethanolic (Li 150), 120 mg qd	Kava $n = 43$ Opipramol 50 mg bid $n = 43$ Buspirone 5 mg bid $n = 43$
5, DB, 4 weeks Anxiety status inventory primary	DSM 3-R; 300.02, 22, 23, 24, 29 HAM-A ≥ 18	Acetonic (WS1490) 35 mg tid	Kava $n = 71$ Placebo $n = 70$
6, DB, 4 weeks Sleep question. Sf-B primary	DSM 3-R; 300.02, 22, 23, 24, 29 HAM-A > 15	Acetonic (WS1490), 140 mg qd (evening)	Kava $n = 34$ Placebo $n = 23$
7, DB, 6 weeks, HAM-A primary	DSM 4 GAD No cut-off score	Aqueous, 60 mg bid	Kava $n = 27$ Placebo $n = 31$
8, DB, cross-over, single-dose State trait anxiety inventory primary	DSM 4 GAD HAM-A 14–25	Aqueous, 180 mg once	Kava Oxazepam 30 mg once Placebo once 22 each treatment
9, DB, cross-over, 1 week HAM-A primary	Anxious subjects Beck anxiety inventory score > 10	Aqueous, 100 + 100 + 50 mg/day	Kava Placebo 37 both treatments
10, DB, 6 weeks HAM-A primary	Anxiety, agitation and tension, nonpsychotic; no cut-off score	Not specified, WS1490 70 mg to 100 mg tid	Kava $n = 57$ Oxazepam 5 mg tid $n = 59$ Bromazep. 3 mg tid $n = 56$

(continued)

Table 22.3 (continued)

Trial N°, design	Indication	Extract, kavalactones dose	Treatments
11, DB, 5 weeks HAM-A primary	DSM 3-R; 300.02, 22, 23, 24, 29 Previously on BZD	Not specified, WS1490 70 mg tid	Kava $n = 20$ Placebo $n = 20$
12, DB, 25 weeks HAM-A primary	DSM 3-R; 300.02, 22, 23, 24, 29 HAM-A ≥ 19	Not specified, WS1490 70 mg tid	Kava $n = 52$ Placebo $n = 48$
13, DB, 4 weeks HAM-A primary	Anxiety of nonmental origin HAM-A ≥ 19	Not specified, WS1490 70 mg tid	Kava $n = 29$ Placebo $n = 29$
14, DB, 4-weeks HAM-A primary	DSM 3-R; 300.02, 22, 23, 24, 29; HAM-A ≥ 19	Not specified, WS1490 70 mg tid	Kava $n = 25$ Placebo $n = 25$

DB double blind, *GAD* generalized anxiety disorder, *HAM-A* Hamilton Anxiety rating scale
bid twice daily, *tid* three times a day, *qd* once daily

symptom scores and trial duration) resulting in high uncertainty about the existence of an effect (but also nonexistence of a mild effect) and pointing out also that the trial with active comparators (No 4 in Table 22.3) lacked assay sensitivity, while active controls were likely sub-dosed. Joint analysis of three small trials in DSM 4 GAD patients (trials 1–3 in Table 22.3) (same investigators) indicated no difference between Kava and placebo in terms of HAM-A reduction over 4–8 weeks and proportion of responders (numerically higher with placebo: 14/30, 47% vs. 9/28, 32%) [64]. One recent meta-analysis [60] included three RCTs of Kava vs. placebo in exclusively GAD patients (trials 1, 7, and 9, period 1, in Table 22.3) to conclude "promising evidence…suggesting Kava….to be an effective treatment in GAD." This conclusion was based on a total of 63 patients on Kava and 67 on placebo over 1–6 weeks of treatment, with standardized mean difference in HAM-A reduction of 0.59 (95%CI −0.57 to 1.75) (tends to favor Kava), with $\tau^2 = 0.93$ and $I^2 = 89\%$ [60] – an apparently nonconservative interpretation of data.

Safety In all trials and observational studies [62], Kava extracts were well tolerated with no indication of withdrawal difficulties or any relevant adverse effects [62]. The *hepatotoxicity* signal, however, is real and should not be neglected [65]. Some observations suggest that the risk might be associated with products based on organic solvents, doses, and duration of use that by far exceed those seen in studies related to anxiety or also with individual subject characteristics [61, 62, 65].

Conclusion The issue of the risk of serious liver damage associated with Kava products needs to be resolved by mechanistic and epidemiological studies.

Evaluation of efficacy is perplexed by heterogeneity of products, doses, dosing regiments, patient characteristics, and trial durations. A biological rationale exists, but the currently existing RCTs should be viewed as a (cumulative) proof-of-concept material suggesting (albeit with some uncertainty) that Kava Kava extracts (as a therapeutic principle, not a specific product) might have a mild-moderate effect in terms of reducing the level of anxiety in the initial short-term treatment in patients with anxiety disorders. A judgment on utility of this "therapeutic principle" in treatment of anxiety disorders is impossible – since no data of this kind exists. In 2015, a protocol of a two-site (in Australia) double-blind placebo-controlled 16-week RCT was published [66]: it was to include non-treated DSM 5 GAD patients (HAM-A ≥ 18), randomized to 2×120 mg/day of kavalactones (tableted aqueous dry extract of a high-quality noble *Borogu* cultivar) ($n = 105$) or placebo ($n = 105$). According to the trial site (https://clinicaltrials.gov/ct2/show/NCT02219880, accessed January 29, 2019), recruitment has been completed (total $N = 178$), but no results are yet available.

Other Herbal Products

Systematic reviews published over the past 10 years (e.g., [32, 33, 51, 58, 59, 67–70]) have identified a range of herbal preparations evaluated mostly in one or two RCTs in anxiety disorders. Some of the addressed medicinal plants [e.g., passion flower (*Passiflora incarnata* L., *herba*), Valerian (essential oil, *Valeriana officinalis* L., *aetheroleum*; or root, *Valeriana officinalis* L., *radix*), Hawthorn leaf and flower (*Crataegus* spp., *folium* cum *flore*), and California poppy (*Eschscholzia californica* Cham., *herba*)] have community herbal monographs justifying *traditional use* for "relief of mild symptoms of mental stress" and/or "to aid sleep" within the EU. Data addressed here were identified in the mentioned systematic reviews.

Crataegus + Eschscholzia In one 12-week multicentric low risk of bias (assessed previously, 26) RCT in DSM 3 R GAD patients, 2×2 tablets a day of 75 mg *Crataegus* extract+20 mg of *Eschscholzia* extract+75 mg elemental magnesium ($n = 130$) resulted in a significantly greater HAM-A reduction (difference = −1.7; 95%CI −1.8 to −1.6) and greater proportion of responders (RR = 1.41; 95%CI 1.04–1.93) than placebo ($n = 134$). One additional similar trial would be needed to confirm ability of this product to reduce anxiety in initial short-term treatment of GAD.

Table 22.4 summarizes body of evidence on several other herbal preparations classified, based on criteria in Table 22.1 (size, duration, bias, design issues, precision), as those for which data could be considered as an "early proof-of-concept" justifying further research (chamomile, *Ginkgo biloba*, and a specific mix of herbal extracts), those for which data are inconclusive (do not demonstrate but do not exclude a possible effect) (*Passiflora*, *Galphimia*, *Echium amoenum*, cannabidiol, and a specific Chinese herbal mix), and those for which data strongly suggest no effect (St. John's Wort extract, a specific Korean herbal mix).

Table 22.4 Summarized body of evidence regarding different herbal preparations in anxiety disorders with evaluation based on criteria and rationale depicted in Table 22.1

Preparation	Indication	Body of data (key trial properties)	Outcomes	Limitations (evaluated in)
Indicative of effect (early proof-of-concept) –justifies further research				
Chamomile extract, tableted (*Matricaria recutita*) (apigenin) (see [51])	DSM 4 GAD	1 DB 8-week vs. PBO initial treatment, 5 × 2.6 mg/day (total $N = 57$) 1 DB 25-week vs. PBO maintenance, 3 × 6 mg/day (total $N = 93$)	Numerically greater HAM-A reduction and response rate vs. PBO Numerically lower risk of relapse vs. PBO	Low risk of bias all items, but small, with imprecise estimates [51]
Ginkgo biloba leaves extract, tableted (22–27% ginkgolides, 5–7% bilobalide) [71]	DSM-III-R GAD or 309.24	1 DB 4-week vs. PBO, $n = 37$, 30 GAD 3 × 80 mg/day $n = 36$, 25 GAD 3 × 160 mg/day $n = 34$, 27 GAD	Significantly greater HAM-A reduction overall, in GAD and in 309.24 subset	Low risk of bias all items, but short, small with relatively imprecise estimates (this chapter)
Tableted extract mix: *Passiflora incarnata, Valeriana officinalis, Crataegus oxycanta, Ballota foetida, Paullinia cupana* and *Cola nitida* [72]	DSM-III-R 309.24	1 DB 4-week vs. PBO, $n = 91$ 3 × 140 mg/day, $n = 91$	Significantly greater HAM-A reduction, response rate (HAM-A < 14) and remission (HAM-A < 10)	Selection and detection bias likely (randomization? Recruitment and scoring by general practitioners), short (this chapter)
Inconclusive – Do not demonstrate but also do not unambiguously exclude a possible effect				
Passiflora incarnata, two extracts, in Iran (see [51]) and Japan (see [67]), dose not clear	DSM 4 GAD or "anxiety"	1 DB 4-week vs. oxazepam ($N = 36$) 1 DB 4-week vs. mexazolam ($N = 150$) (both 1:1 randomization)	HAM-A reduction like oxazepam "Improvement" like with mexazolam	Both low risk of bias [51, 67], but no assay sensitivity, short, rather imprecise estimates
Valeriana officinalis extract, tableted (valepotriates) (see [51, 68])	DSM-III-R GAD	1 DB 4-week vs. PBO $n = 12$ 3 × 50 mg/day valepotriates $n = 12$ 3 × 2.5 mg/day diazepam $n = 12$	Similar HAM-A reduction to placebo and diazepam	Low risk of bias [51, 68], but no assay sensitivity, small, short

(continued)

Table 22.4 (continued)

Preparation	Indication	Body of data (key trial properties)	Outcomes	Limitations (evaluated in)
Galphimia glauca extract (galphimin B) (see [51])	DSM 4 GAD	1 DB 4-week 2 × 350 μg/day $n = 72$ vs. lorazepam 2 × 1 mg/day $n = 80$ 1 DB 12-week 2 × 350 μg/day $n = 94$ vs. lorazepam 2 × 0.5–1 mg/day $n = 97$	End-treatment HAM-A similar to lorazepam in both trials; pooled analysis [51] indicates no difference with or without baseline adjustment	Unclear selection and detection bias, huge attrition (around 40%) and reporting bias, no assay sensitivity [51]
Echium amoneum extract (see [51]) with fluoxetine	DSM 4-TR GAD	1 DB 8-week trial vs. PBO $n = 18$ 3 × 750 mg/day extract $n = 19$	Lower HAM-A at the end with extract +fluoxetine vs. PBO + fluoxetine	Unclear detection, high reporting risk of bias; small, imprecise [51]
Chinese crude powder mix 14 herbs with CBT (see [51])	DSM 4 GAD	1 DB 24-week vs. paroxetine $n = 104$ 2 × 10 g/day $n = 93$	Similar HAM-A scores and response rates CBT + mix vs. CBT + paroxetine	High risk performance bias and attrition, no assay sensitivity [51]
Cannabidiol oral [73]	DSM 4 SAD	Single dose 600 mg $n = 12$ vs. PBO $n = 12$; evaluation over 2 hours	Cannabidiol reduced anxiety, discomfort, during speech performance	Low risk of bias, but small, imprecise estimates, "surrogate" outcome (this chapter)
Strongly suggest no effect				
St. John's Wort (*Hypericum perforatum*) extract Li160 [74]	Social phobia, generalized	1 DB 12-week vs. PBO $n = 20$ Flexible 2 × 300 to 2 × 900 mg/day $n = 20$	Numerically greater reduction in Liebowitz social anxiety scale with PBO	Low risk of bias all items (this chapter)
Gamisoyo-san – Korean mix of 10 herbs, two types of extraction (see [51])	DSM 4 GAD	1 DB 8-weeks vs. PBO $n = 49$ 3 × 7.7 g/day type A $n = 49$ 3 × 7.7 g/day type B $n = 49$	Closely similar HAM-A reduction and response rates with both treatments and PBO	Low risk of bias all items [51]

CBT cognitive behavioral therapy, *DB* double blind, *GAD* generalized anxiety disorder, *HAM-A* Hamilton Anxiety rating scale, *PBO* placebo, *SAD* social anxiety disorder

Nutrients/Nutraceuticals

Systematic reviews found no RCTs evaluating lysine or lysine-arginine [32, 33], *N*-acetylcysteine, tryptophan, folate, or dehydroepiandrosterone [31] supplementation in anxiety disorders. Two reviews [33, 75] identified two small (each $N = 21$) 4-week treatment, 1-week wash-out double-blind cross-over RCTs of inositol (12 g/day) including patients with DSM 3-R or DSM 4 *panic disorder with or without agoraphobia* [76, 77]: in the first trial comparison was vs. placebo and in the second trial vs. fluvoxamine (50–150 mg/day). The first one [76] reported greater reduction of the number of panic attacks vs. placebo; the second one [77] reported no difference vs. fluvoxamine. Both used inappropriate way of data analysis (number of attacks as a continuous variable, instead as Poisson event rates; analysis not accounting for period effect); the second trial lacked assay sensitivity. Data could be viewed as an early "proof-of-concept" (but with uncertainty).

Manipulative and Body-Based Therapies

Acupuncture

Data is methodologically extremely variable [78]. Recent systematic reviews identified 13 [79] and 6 RCTs [80] of various acupuncture procedures, none in patients diagnosed with anxiety disorders. One miniature (acupuncture $n = 7$, sham $n = 6$) 4-week double-blind low risk of bias RCT in ICD-10 GAD patients (see 26) indicated a possible anxiolytic effect. A recent 5-week RCT in children with GAD indicated considerably greater HAM-A reduction in those treated with a specific acupuncture procedure ($n = 10$) than in the waiting-list controls ($n = 10$) [81]. This small single-center trial is burdened with risk of various biases inherent to waiting-list controlled trials (see Table 22.1). One non-randomized 6-week, open-label Chinese trial [82] indicated closely similar CGI score reduction with specific Jin-3-needling therapy, Western pharmacotherapy, and combined treatment in GAD patients ($N = 86$). Due to low quality and lack of assay sensitivity, the trial is uninformative. The "body of evidence" about acupuncture for anxiety disorders practically does not exist.

Exercise

Several reviews addressed aerobic exercise for "psychiatric symptoms/disorders" (e.g., [83–86]) mixing RCTs in psychiatric patients (various diagnoses) and medically ill or healthy subjects, all recognizing clinical heterogeneity (indications, exercise protocols, duration) and mostly high risk of bias in these trials. Two provided pooled estimates [83, 86] which, based on criteria of clinical heterogeneity, statistical heterogeneity and high risk of bias (Table 22.1) are uninformative. Nine RCTs identified in these reviews actually conducted in patients with anxiety disorders are summarized in Table 22.5: (a) in addition to heterogeneity of exercise protocols, all trials were small resulting in imprecise estimates/(under)powered statistical tests; (b) in panic disorder, one trial [88] indicated a greater symptom improvement vs. pill placebo (but less than with clomipramine), while another, by the same group [90], with a 2x2 factorial design indicated similar changes with exercise (plus paroxetine or pill placebo) vs. relaxation (plus paroxetine or pill placebo) and was due to the choice of reference treatment actually inconclusive about exercise (non) effect; (c) all other trials were inconclusive about exercise (non)effect due to risk of bias (performance, ascertainment, attrition) and/or lack of assay sensitivity (choice of control treatments). Overall, there is some indication (but with considerable uncertainty) that aerobic exercise as a concept might be anxiolytic in patients with anxiety disorders.

Other Procedures

A Cochrane review in 2007 [96] found no controlled trials of *therapeutic touch* in patients with anxiety disorders. A more recent one [97] identified two small 4-week RCTs (total $N = 40$) of *repetitive transcranial magnetic stimulation* (rTMS) vs. sham procedure in panic disorder patients indicating no differences in symptom reduction – trials do not indicate, but also do not exclude a possible rTMS effect. We previously [51] identified individual RCTs in GAD patients indicating (criteria in Table 22.1) an effect of a specific *balneotherapy* procedure (greater HAM-A reduction in an open-label 8-week trial vs. paroxetine 20 mg qd, $N = 237$) and a possible effect of *Swedish massage*, of a specific *flotation in water* protocol, and of *Chinese bloodletting*.

Mind-Body Therapies

A very recent scoping review of systematic reviews of CAM [98] concluded "moderate quality of evidence" (in the GRADE sense) supporting efficacy of CAM in anxiety: meditation, yoga, and mindfulness-based stress reduction (MBSR). However, different reviews included different CAMs under the same names, e.g., mediation, or included yoga, reiki, and similar interventions under "meditative therapies" [99]. Moreover, very heterogeneous patients were included in these trials. We therefore reviewed a number of systematic reviews to identified data pertaining specifically to trials in patients with anxiety disorders.

Yoga

Across reviews addressing yoga [99–103], only five RCTs were in patients diagnosed with anxiety disorders: (a) three small high risk of bias trials from India; one 8-week vs. no treatment in "anxiety disorder" ($N = 45$) indicating no difference

Table 22.5 Summary of RCTs of aerobic exercise in patients with anxiety disorders

Ref	Patients	Design	Treatments	Outcome	Limitations	Interpretation
[87]	21 "anxiety", 32 other nonpsychotic, DSM 3	Open, 8 weeks	Walking $n = 25$ Jogging $n = 27$	Similar for "anxiety change"	High RoB, no assay sensitivity	Inconclusive about exercise effect
[88]	DSM 3 panic disorder	Partly DB 10 weeks	Exercise $n = 16$ TCA $n = 15$ Pill PBO $n = 15$	HAM-A reduction: Exercise & TCA > PBO; PAS reduction: TCA > exercise >PBO	Overall low RoB, but no blinding of exercise; small, imprecise	Indicates effect
[89]	DSM 4 GAD, social phobia, panic disorder	Open, 8 weeks	CBT + exercise $n = 21$ CBT + education $n = 20$	No clear numerical data, authors "suggest" effect	High RoB (see [51]), attrition (40%), unclear data analysis, small	Inconclusive about exercise effect
[90]	DSM 4 panic disorder w/wo agoraphobia	Partly DB, 10 weeks	Exercise+ Parox $n = 21$ Exercise+ pill PBO $n = 20$ Relaxation+ Parox $n = 17$ relaxation+pill PBO $n = 17$	PAS and HAM-A reduction: Parox >PBO, exercise similar to relaxation	Overall low RoB, questionable assay sensitivity, 20% attrition, small, imprecise	Inconclusive about exercise effect
[91]	ICD 10 panic disorder, GAD or mild depression	Open, 32 weeks	Exercise $n = 27$ Usual therapy $n = 21$	HAM-A and HAM-D reduction similar in two groups	Only partly randomized, high RoB, 40% attrition, small, imprecise	Inconclusive about exercise effect
[92]	DSM 4 GAD	Open, 6 weeks, waiting-list control	Aerobic $n = 10$ Resistance $n = 10$ Wait list $n = 10$	PSWQ values consistently similar between groups	High RoB inherent to wait-list control (see [51]), small, imprecise	Inconclusive about exercise effect

[93]	DSM 4 generalized social phobia	Open, 8 weeks	Exercise $n = 25$ Mindful. Stress relief $n = 31$	LSAS-SR similar reduction both groups	High RoB – Open, no assay sensitivity	Inconclusive about exercise effect
[94]	DSM 4 panic disorder w/wo agoraphobia	Open, 8 weeks	Exercise $n = 17$ CBT = 19	Mobility inventory, ACQ, BSQ reduction CBT > exercise	High RoB – Open, no assay sensitivity, small, imprecise	Inconclusive about exercise effect
[95]	DSM 4 panic disorder w/wo agoraphobia	Assessor blind, 9 weeks	CBT 1 week + 8 week exercise $n = 27$ CBT 1 week + 8 week stretching $n = 31$	HAM-A reduction similar, exercise tends better	Overall low RoB, but not fully patient blind, small, imprecise	Inconclusive about exercise effect

ACQ Agoraphobic Cognitions Questionnaire, *BSQ* Body Sensations Questionnaire, *CBT* cognitive behavioral therapy, *DB* double-blind, *HAM-A (HAM-D)* Hamilton anxiety (Depression) rating scale, *LSAS-SR* Leibowitz Social Anxiety Scale-Self report, *PAS* panic and agoraphobia scale, *PBO* placebo, *PSQW* Penn State Worry Questionnaire, *RoB* risk of bias, *TCA* tricyclic antidepressant

regarding symptom severity; and three 3–4-week trials vs. naturopathy (massage, acupressure, breathing), relaxation, or breathing and relaxation, respectively, in GAD, DSM 3 anxiety, or "psychoneurosis," respectively (79 subjects total). One pooled estimate suggested lower anxiety scores with yoga (SMD −0.86; −1.56 to −0.15), but with $I^2 = 50\%$ [103]. An appropriate estimate would include HKSJ correction for 95%CIs −2.39 to 0.67, $t = -2.419$, $p = 0.137$ (no effect); (b) one small Canadian high risk of bias 3-week trial vs. progressive relaxation in "snake anxiety" ($N = 40$), indicating no difference; and (c) one small Brazilian 8-week trial in DSM 4 panic disorder vs. yoga + CBT indicating no difference. Based on criteria in Table 22.1, there is no relevant data on (non)effect of yoga in anxiety disorders.

Morita Therapy

A Cochrane review [104] found four small Chinese high risk of bias 4–6-week RCTs: one in GAD ($N = 31$ men), one in "anxiety" ($N = 86$ women) (neither providing any numerical data), and two in social phobia, comparing Morita ($n = 19$ or $n = 24$) to pharmacological treatment ("unspecified" $n = 20$ or alprazolam $n = 12$, respectively). The pooled fixed-effect RR for "response" was reported = 1.85, 95%CI 1.27–2.69 favoring Morita. An appropriate estimate would be random-effects with HKSJ correction = 1.77, 95%CI 0.14–21.8, $t = 2.892$, $p = 0.212$ (no effect). Based on criteria in Table 22.1, there is no relevant data on (non)effect of Morita in anxiety disorders.

Tai Chi, Baduanjin Mindfulness Exercise, Chinese Cognitive Psychotherapy, and Reiki

A recent review [105] identified one high risk of bias Chinese 6-week trial in elderly with CCMD 3 "anxiety" treated with paroxetine+*Tai Chi* ($n = 16$) vs. paroxetine ($n = 16$), indicating greater HAM-A reduction with combined treatment. Another review [106] identified one 12-week high risk of bias Chinese RCT in GAD of *Baduanjin mindfulness exercise* + pharmacological therapy vs. pharmacological therapy indicating greater improvement of symptoms. We previously [51] identified an open high risk of bias 24-week Chinese trial in CCMD 2-R GAD showing better symptoms scores with *Chinese psychotherapy* alone or combined with benzodiazepines vs. benzodiazepines alone (total $N = 131$). With limitations due to small samples and high risk of bias, data indicate "add-on" effect of these procedures to pharmacological treatment. A recent Cochrane review [107] identified no RCTs of *reiki* in anxiety disorders.

Religious and Spiritual Interventions

Across two reviews [51, 108], two small high risk of bias 12-week RCTs in DSM 4 GAD were identified comparing a multifaith spiritual intervention to CBT (total

$N = 22$) indicating no difference or to supportive psychotherapy (total $N = 23$) indicating lower HAM-A scores and more responders at the end of treatment. With limitations due to high risk of bias and small samples, data indicate possible anxiolytic effect.

Relaxation

One review [109] identified only one 14-week RCT of *applied relaxation* (AR) in DSM 4 social phobia patients [110]. In this otherwise high-quality trial, comparison of relaxation ($n = 21$) to wait-list controls ($n = 20$) showing a markedly greater LSAS score reduction was partly compromised by the limitations inherent to wait-list controlled trials. In comparison to cognitive therapy ($n = 21$), reduction of symptoms was significantly less with AR. Another review [111] included 16 RCTs of "good or acceptable quality" [111]: 9 with AR (in GAD or panic disorder) and 7 with *mindfulness-based stress reduction* to conclude an overall "superiority of relaxation" in terms of high effect size (Hedges' g = around 0.62) [111]. However (based on data from the review, 111): (a) five trials evaluated AR in GAD, one ($N = 50$) vs. wait-list showing no difference in symptoms after 12 weeks and four vs. CBT over 6–12 weeks ($N = 172$): random-effects $g = 0.662$, 95%CI (HKSJ correction) −0.1 to 1.42, $t = 2.777$, $p = 0.069$, $I^2 = 58\%$ (no effect). Hence, data do not demonstrate (due to lack of assay sensitivity), but also do not exclude an effect of AR; (b) four trials evaluated AR in panic disorder; in one 12-week trial, symptom reduction was greater with AR ($n = 16$) than in wait-list controls ($n = 16$), similar as with imipramine ($n = 16$), but considerably less than with CBT ($n = 16$) [112]; in another 12-week trial, symptom reduction was similar for AR ($n = 19$) and CBT ($n = 19$); and in two 12-week trials (total $N = 48$) vs. progressive relaxation or cognitive therapy, symptom reductions were in favor of AR, but not achieving statistical significance, $g = 0.60$, 95%CI −1.64 to 2.84 (with HKSJ correction). Therefore, there is indication about the effect of AR, but also quite some uncertainty.

Mindfulness-Based Interventions (MBIs) and Meditation

A considerable number of conceptually and methodologically variable systematic reviews/meta-analyses have been published over the past 10 years addressing RCTs of MBIs and/or meditation for anxiety disorders/symptoms. One large review [113], for example, included trials with a "clinical diagnosis" [113] referring to mindfulness-based cognitive therapy (MBCT) or stress reduction (MBSR) and mantra-based programs (e.g., transcendental or mantra mediation), but excluded acceptance and commitment therapy (ACT), dialectical behavior therapy (DBT), relaxation, spiritual treatments, and trials with wait-list controls or "no treatment" controls, and considered all MBIs jointly and trials in all anxiety disorders and all treatment durations jointly. For MBIs vs. "nonspecific controls" (any non-EBM-based treatment), based on 8 trials with 647 patients, pooled estimate for anxiety symptom scores reduction was (95%CI) from 0% to 44% greater reduction with

MBIs ("moderate strength of evidence for improvement"); based on 11 RCTs with 691 patients vs. "any EBM-based treatment," the estimate was from 39% less to 8% more symptom reduction (tends to favor EBM-based treatment) [113]. For "mantra" (clinically standardized mantra meditation) (3 trials, 247 patients) vs. "nonspecific controls," the difference was "0," and data vs. EBM treatments were too few (both judged as insufficient for conclusions about (non)effects without considerable uncertainty) [113]. Another recent large review [114] excluded ACT, DBT but also mantra, and electronically delivered treatments and included trials in patients with clinical diagnoses or with anxiety symptom scores above certain levels and with any type of non-MBI controls. All anxiety trials (diagnosis, scores, duration) and all MBIs were considered jointly [114]: (a) based on eight trials (472 patients) vs. no treatment, MBIs provided markedly greater symptom reduction ($d = 0.89$; 0.62 to 1.17), but with much inconsistency ($I^2 = 81\%$); (b) based on five trials (374 patients) vs. non-EBM-based treatments, there appeared no difference ($d = 0.15$; −0.16 to 0.46) and no inconsistency; (c) based on five trials (362 patients) vs. EBM-based treatments, difference tended to favor EBM ($d = -0.18$; −0.41 to 0.06) with mild inconsistency ($I^2 = 38\%$). Another review focused specifically on MBCT in patients with a clinical diagnosis and found only one small trial ($N = 26$) in social phobia indicating a better outcome with CBT vs. MBCT [115], while yet another focused on transcendental meditation [116] identified only one small trial ($N = 31$) in "anxiety" finding no difference vs. muscle biofeedback or relaxation. A review focused on mindfulness and acceptance-based interventions for anxiety disorders [117] disregarded randomization and analyzed pre-post differences, which we consider inappropriate. Three recent reviews [118–120] included (among others or exclusively) RCTs of mindfulness-based interventions in patients with clinical anxiety disorder diagnoses and enabled identification of RCTs by treatment, diagnosis, and comparisons, one [118] exclusively regarding electronically delivered MBIs (Internet, smartphones, recorded material) and two [119, 120] explicitly excluding such trials. All three reviews provided meta-analytical pooled estimates across a number of trials with considerable clinical (and statistical) heterogeneity. Using these three reviews, we identified all individual included trials: Table 22.6 outlines their characteristics sorted by the mode of delivery (electronic or "in person"). Overall, data suggest the following:

(a) Major reviews/meta-analyses (across diagnoses and different treatment comparisons) [113, 114] support the view that, as a concept, MBIs may be useful in treatment of anxiety disorders.
(b) There is, however, a reasonable level of uncertainty about this effect:
    (i) No-treatment control/wait-list control groups may have inflated the effect.
    (ii) There is some inconsistency in comparisons vs. "non-EBM-based treatments" – an indication of moderate superiority [113] but also of no relevant difference [114].
    (iii) A consistent indication of inferiority vs. "EBM-based" treatments, which is particularly burdened with uncertainty since the number of RCTs is small and estimates are imprecise.

Table 22.6 Summary of RCTs of mindfulness interventions (MBI) identified from systematic reviews/meta-analyses specifically in patients with clinical diagnoses of anxiety disorders. Depicted are diagnosis, duration, type of treatment, type of control, and effects (treatment-control differences, T-C) for symptoms at the end of treatment as reported for individual studies in respective meta-analyses [118–120], as Hedges' *g* or standardized mean difference, SMD. "Pooled" estimates are generated for this chapter – where it seemed feasible to provide them based on clinical homogeneity of studies (diagnoses, tested treatments, control treatments – depicted in bold letters)

RCT	Dg	Duration	Treatment, n	Control, n	T-C (95% CI)	Interpretation
Trials in which intervention was delivered in an electronic format (Internet, smartphones, video material etc.*)*						
1	*GAD*	9 weeks	*Mindful. Exercise*, 52	*Wait-list*, 51	−1.21 (−1.67, −0.75)	MBI better
2	*GAD*	8 weeks	*MBSR*, 50	*Wait-list*, 50	−0.91 (−1.32, −0.50)	MBI better
Pooled (standard DL)			102	101	−1.04 (−1.35, −0.74)	MBI better
Pooled (HKSJ corrected)			102	101	−1.04 (−2.90, 0.81)	MBI tends to better
3	*Anxiety*	8 weeks	*Mindful. Exercise*, 45	*Discussion*, 46	−0.76 (−1.20, −0.32)	MBI better
4	*Anxiety*	10 weeks	*AFPP*, 22	*Emp. Listen.* 21	−0.04 (−0.63, 0.55)	No difference
5	*SAD, PD*	10 weeks	*ACT*, 50	*Wait-list*, 51	−0.54 (−0.94, −0.15)	ACT better
Trials in which intervention was delivered in person (group)						
1	*Anxiety*	12 weeks	*MBSR*, 45	*CBT*, 60	0.38 (−0.09, 0.86)	CBT tends to better
2	*SAD*	8 weeks	*MBSR*, 26	*CBT*, 27	0.77 (0.21, 1.32)	CBT better
3	*SAD*	8 weeks	*MBCT*, 14	*CBT*, 11	"CBT num. Better"[a]	CBT tends to better
4	*GAD*	8 weeks	*MBSR*, 16	*Wait-list*, 15	−5.29 (−6.87, −3.72)	MBSR better

(continued)

Table 22.6 (continued)

RCT	Dg	Duration	Treatment, n	Control, n	T-C (95% CI)	Interpretation
5	*PD, SAD GAD*	8 weeks	*MBSR*, 39	*Wait-list*, 37	−0.70 (−1.17, −0.24)	MBSR better
6	*GAD*	8 weeks	*MBSR*, 48	*Education*, 41	−0.37 (−0.79, 0.05)	MBSR tends better
7	*SAD*	16 weeks	*MBSR*, 24	*Exercise*, 18	−0.25 (−0.86, 0.37)	No difference
8	*Hypoh.*	8 weeks	*MBCT*, 36	*Unrestricted services*, 38	−0.34 (−0.79, 0.12)	MBCT tends better

ACT acceptance-commitment therapy, *AFPP* affect-focused psychodynamic psychotherapy, *CBT* cognitive behavioral therapy, *Emp. listen.* empathic listening, *GAD* generalized anxiety disorder, *Hypoh.* hypochondriasis, *MBCT* MB cognitive therapy, *MBSR* MB stress reduction, *PD* panic disorder, *SAD* social anxiety disorder

[a]The trial provided no numerical data in a usual format but indicated numerically better outcomes with CBT

(c) In reality, the total number of individual RCTs is small with a limited number of patients, particularly when broken down by mode of delivery by disorder by type of intervention by type of the control (Table 22.6).

(d) Not going into details about potential quality issues, data in Table 22.6 show that there are only five RCTs regarding "electronic delivery," quite clinically heterogeneous (diagnoses, treatments, controls):

   (i) Only the two RCTs in GAD are similar enough as to justify a pooled estimate. In both trials MBI was superior to wait-list, but the pooled estimate (with HKSJ correction) extends from benefit to no benefit (Table 22.6) leaving still some uncertainty; the remaining three RCTs are too heterogeneous clinically and by results to justify data pooling.

(e) There are eight "live" RCTs that are, actually, too heterogeneous clinically to justify pooled estimates (Table 22.6). Three comparisons vs. CBT (in "anxiety" and social anxiety disorder) suggest that MBI might be inferior to CBT (Table 22.6), two trials vs. wait-list suggest efficacy (GAD or combined different disorders) (Table 22.6), while three trials vs. "other, non-EBM-based treatments" do not (Table 22.6).

Overall, therefore, while there might not be uncertainty about whether the concept generally "works" in anxiety disorder, there is quite some uncertainty about (non)existence and size of its effect in different anxiety disorders.

Alternative Medical Systems

Ayurveda

One dedicated systematic review [121] identified five RCTs of a herb *Ashwagandha* (*Withania somnifera*), but none in patients diagnosed with anxiety disorders. We previously identified [51] one overall low risk of bias 11-week RCT in GAD (N = 102) indicating no difference in HAM-A reduction or response to an oral Ayurvedic preparation *Sarasvata choorna* vs. placebo and one small (4 weeks, total N = 65) inconclusive (no assay sensitivity) trial of oral *Manasamitra Vatakam* vs. benzodiazepines in GAD patients.

Homeopathy

We previously identified [51] one overall low risk of bias 10-week double-blind placebo-controlled trial of homeopathy in GAD showing no difference in HAM-A reduction or response rate. Two dedicated reviews [122, 123] found no further published RCT of homeopathy in anxiety disorders.

Conclusion

A broad range of CAM modalities have been investigated in RCTs in anxiety disorders but to a considerably variable extent and with a considerable variability in trial quality and consistency of results. Some treatments considered as CAM (at this moment) do show a potential of possible utility in treatment of anxiety disorders, but so far none has been evaluated to the extent comparable to that of the *mainstream* treatments. Considering that the only reasonable "classification" of treatments is to those supported by evidence and those that (at the moment) are not, no lesser criteria should be in place for CAM than for *mainstream* treatments. There is a need for methodologically clear-cut and stringent larger clinical trials – meta-analytical estimates based on numerous small, heterogeneous trials are not likely to resolve the issue, and the currently existing meta-analytical estimates in some cases appear to be overtly nonconservative.

References

1. Goroll AH. Moving towards evidence-based complementary care. JAMA Intern Med. 2014;174:368–9.
2. Canadian Health Services Research Foundation. Annual report 2005. Ottawa: Canadian Health Services Research Foundation; 2005. p. 9.
3. Institute of Medicine of the National Academies. Complementary and alternative medicine in the United States. Washington, DC: The National Academic Press; 2005.

4. Jonas WB. Policy, the public, and priorities in alternative medicine research. Ann Am Acad Pol Soc Sci. 2002;583:29–43.
5. National Center for Complementary and Integrative Health. https://nccih.nih.gov/health/integrative-health#hed4. Accessed 3 Jan 2019.
6. Miller L, Su X. Artemisinin: discovery from the Chinese herbal garden. Cell. 2011;146:855–8.
7. Fontanarosa P. Alternative medicine meets science. JAMA. 1998;280:1618.
8. Bandelow B, Zohar J, Hollander E, Kasper S, Moller HJ. World Federation of Societies of Biological Psychiatry (WFSBP) guidelines for the pharmacological treatment of anxiety, obsessive-compulsive and post-traumatic stress disorders – first revision. W J Biol Psychiatry. 2008;9:248–312.
9. Bandelow B, Sher L, Bunevicius R, Hollander E, Kasper S, Zohar J, Moller JH. Guidelines for the pharmacological treatment of anxiety disorders, obsessive-compulsive disorder and posttraumatic stress disorder in primary care. J Psychiatry Clin Pract. 2012;16:77–84.
10. McIntosh A, Cohen A, Turnbull N, Esmonde L, Dennis P, Eatock J. Clinical guidelines and evidence review for panic disorder and generalized anxiety disorder. London: University of Scheffield/National Collaborating Center for Primary Care; 2004.
11. National Collaborating Center for Mental Health. Generalized anxiety disorder. The Nice guideline on management in primary, secondary and community care. National Clinical Guideline number 113. The British Psychological Society and The Royal College of Psychiatrists; 2011.
12. National Collaborating Center for Mental Health. Social anxiety disorder. The Nice guideline on recognition, assessment and treatment. National Clinical Guideline number 159. The British Psychological Society and The Royal College of Psychiatrists; 2013.
13. Baldwin DA, Anderson IM, Nutt DJ, Allgulander C, Bandelow B, den Boer AJ, et al. Evidence-based pharmacological treatment of anxiety disorders, post-traumatic stress disorder and obsessive-compulsive disorder: a revision of the 2005 guideline from the British Association for Psychopharmacology. J Psychopharmacol. 2014;28:403–39.
14. Lim L, Chan HN, Chew PH, Chua SM, Ho C, Kwek SKD, et al. Ministry of health clinical practice guidelines: anxiety disorders. Singap Med J. 2015;56:310–6.
15. Katzman MA, Bleau P, Blier P, Chokka P, Kjernisted K, Van Ameringen M, et al. Canadian clinical practice guidelines for the management of anxiety, posttraumatic stress and obsessive-compulsive disorders. BMC Psychiatry. 2014;14(suppl 1):S1.
16. Barnes PM, Bloom B, Nahin RL. Complementary and alternative medicine use among adults and children: United States, 2007. National health statistics reports; no 12. Hyattsville: National Center for Health Statistics; 2008.
17. von Ammon K, Cardini F, Daig U, Draman S, Frei-Erb M, Hgyl G et al. Health Technology Assessment (HTA) and map of CAM provision in the EU. A pan-European research network for Complementary and Alternative Medicine (CAM): Final Report of CAMbrella work package 5 (FP7 project). Available at: https://services.phaidra.univie.ac.at/api/object/o:300096/diss/Content/get. Accessed 5 Jan 2019.
18. Smith CA, Shewamene Z, Galbally M, Schmied V, Dahlen H. The effect of complementary medicines and therapies on maternal anxiety and depression in pregnancy: a systematic review and meta-analysis. J Affect Disord. 2019;245:428–39.
19. Orme-Johnson DW, Barnes VA. Effects of the transcendental meditation technique on trait anxiety: a meta-analysis of randomized controlled trials. J Altern Complement Med. 2014;20:330–41.
20. National Center for Complementary and Integrative Health. 2016 Strategic plan: exploring the science of complementary and integrative health. NCCIH 2016. Available at: https://nccih.nih.gov/about/strategic-plans/2016. Accessed 4 Jan 2019.
21. Jonas WB, Eisenberg D, Hufford D, Crawford C. The evolution of complementary and alternative medicine (CAM) in the USA over the last 20 years. Forsch Komplementmed. 2013;20:65–72.
22. Clarke TC, Black LI, Stussman BJ, Barnes PM, Nahin RL. Trends in the use of complementary healthy approaches among adults: United States, 2002–2012. National health statistics reports; no 79. Hyattsville: National Center for Health Statistics; 2015.

23. de Jonge P, Wardenaar KJ, Hoenders HR, Evans-Lacko S, Kovess-Masfety V, Aguilar-Gaxiola S, et al. Complementary and alternative medicine contacts by persons with mental disorders in 25 countries: results from the world mental health surveys. Epidemiol Psych Sci. 2018;27:552–67.
24. Schulz P, Hede V. Alternative and complementary approaches in psychiatry: beliefs versus evidence. Dialogues Clin Neurosci. 2018;20:207–14.
25. Triester-Goltzman Y, Peleg R. Trends in publications on complementary and alternative medicine in the medical literatura. J Complement Integr Med. 2015;12:111–5.
26. European Medicines Agency. Scientific guidelines. Clinical efficacy and safety: nervous system. Available at: https://www.ema.europa.eu/en/human-regulatory/research-development/scientific-guidelines/clinical-efficacy-safety/clinical-efficacy-safety-nervous-system. Accessed 10 Jan 2019.
27. Center for statistical methodology, London School of Hygiene&Tropical medicine. Available at https://csm.lshtm.ac.uk/centre-themes/analysis-of-clinical-trials/. Accessed 5 Feb 2019.
28. IntHout J, Ioannidis JP, Borm GF, Goeman JJ. Small studies are more heterogeneous than large ones: a meta-meta-analysis. J Clin Epidemiol. 2015;68:860–9.
29. Borenstein M, Hedges LV, Higgins JPT, Rothstein HR. Identifying and quantifying heterogeneity. In: Introduction to meta-analysis. Chichester: Wiley; 2009. p. 107–26.
30. Bandelow B, Reitt M, Rover C, Michaelis S, Gorlich Y, Wedekind D. Efficacy of treatments for anxiety disorders: a meta-analysis. Int Clin Psychopharmacol. 2015;30:183–92.
31. Ravindran AV, da Silva T. Complementary and alternative therapies as add-on to pharmacotherapy for mood and anxiety disorders: a systematic review. J Affect Disord. 2013;150:707–19.
32. Lakhan ES, Vieira KF. Nutritional and herbal supplements for anxiety and anxiety-related disorders: systematic review. Nutr J. 2010;9:42. http://www.nutritionj.com/content/9/1/42
33. Sarris J, Moylan S, Camfield DA, Pase MP, Mischoulon D, Berk M, Jacka FN, Schweitzer I. Complementary medicine, exercise, meditation, diet and lifestyle modification for anxiety disorders: a review of current evidence. Evid Based Complement Alternat Med. 2012. https://doi.org/10.1155/2012/809653.
34. Herbal medicinal products. Available at: https://www.ema.europa.eu/en/human-regulatory/herbal-medicinal-products. Accessed 15 Jan 2019.
35. Community herbal monograph *Lavandula angustifolia*, Miller, *flos*. Available at: https://www.ema.europa.eu/documents/herbal-monograph/final-community-herbal-monograph-lavandula-angustifolia-p-mill-flos_en.pdf. Accessed 25 Dec 2018.
36. Community herbal monograph *Lavandula angustifolia*, Miller, *aetheroleum*. Available at: https://www.ema.europa.eu/documents/herbal-monograph/final-community-herbal-monograph-lavandula-angustifolia-miller-aetheroleum_en.pdf. Accessed 25 Dec 2018.
37. Assessment report on *Lavandula angustifolia* Miller, *Aetheroleum* and *Lavandula angustifolia* Miller, *flos*. EMA/HMPC/143183/2010. Available at: https://www.ema.europa.eu/documents/herbal-report/final-assessment-report-lavandula-angustifolia-miller-aetheroleum-lavandula-angustifolia-miller-flos_en.pdf. Accessed 25 Dec 2018.
38. Perry R, Terry R, Watson LK, Ernst E. Is lavender an anxiolytic drug? A systematic review of randomized clinical trials. Phytomedicine. 2012;19:825–35.
39. Kasper S, Muller WE, Volz HP, Moller HJ, Koch E, Dienel A. Silexan in anxiety disorders: clinical data and pharmacological background. World J Biol Psych. 2018;19:412–20.
40. Lasea capsule, soft SmPC. Available at: https://docetp.mpa.se/LMF/Lasea%20capsule%20soft%20SmPC_09001be68141f58a.pdf. Accessed 25 Dec 2018.
41. Kasper S. An orally administered lavandula oil preparation (Silexan) for anxiety disorder and related conditions: an evidence based review. Int J Psychiatry Clin Pract. 2013;17(suppl 1):15–22.
42. Kasper S. Phytopharmaceutical treatment of anxiety, depression and dementia in the elderly: evidence from randomized controlled clinical trials. Wien Med Wochenschr. 2015;165:217–28.

43. Kasper S, Moller HJ, Volz HP, Schlafke S, Dienel A. Silexan in generalized anxiety disorder: investigation of the therapeutic dosage range in a pooled data set. Int Clin Psychopharmacol. 2017;32:195–204.
44. Moller HJ, Volz HP, Dienerl A, Schlafke S, Kasper S. Efficacy of Silexan in subtreshold anxiety: meta-analysis of randomized, placebo-controlled trials. Eur Arch Psychiatry Clin Neurosci. 2019;269:183–193.
45. Kasper S, Gastpar M, Muller WE, Volz HP, Moller JH, Schlafke S, Dienel A. Lavender oil preparation Silexan is effective in generalized anxiety disorder – a randomized, double-blind comparison to placebo and paroxetine. Int J Neuropsychopharmacol. 2014;17:859–69.
46. Woelk H, Schlafke S. A multicenter, double-blind randomized study of Lavender oil preparation Silexan in comparison to lorazepam for generalized anxiety disorder. Phytomedicine. 2010;17:94–9.
47. Kasper S, Gastpar M, Muller WE, Volz HP, Moller HJ, Dienel A, Schlafke S. Silexan, an orally administered Lavandula oil preparation, is effective in the treatment of "subsyndromal" anxiety disorder: a randomized double-blind placebo controlled trials. Int Clin Psychopharmacol. 2010;25:277–87.
48. Kasper S, Anghelescu I, Dienel A. Efficacy of orally administered Silexan in patients with anxiety-related restlessness and disturbed sleep – a randomized placebo-controlled trial. Eur Neuropsychopharmacol. 2015;25:1960–7.
49. Kasper S, Volz HP, Dienel A, Schlafke S. Efficacy of silexan in mixed anxiety-depression – a randomized placebo-controlled trial. Eur Neuropsychopharmacol. 2016;26:331–40.
50. Gastpar M, Muller WE, Volz HP, Moller HJ, Schlafke S, Dienel S, Kasper S. Silexan does not cause withdrawal symptoms even when abruptly discontinued. Int J Psychiatry Clin Pract. 2017;21:177–80.
51. Barić H, Đorđervić V, Cerovečki I, Trkulja V. Complementary and alternative treatments for generalized anxiety disorder: systematic review and meta-analysis of randomized controlled trials. Adv Ther. 2018;35:261–88.
52. Stevinson C, Huntley A, Ernst E. A systematic review of the safety of Kava extract in the treatment of anxiety. Drug Saf. 2002;25:251–61.
53. Pittler MH, Ernst E. Kava extracts versus placebo for treating anxiety. Cochrane Datab Syst Rev. 2003. Art.No.CD03383. https://doi.org/10.1002/14651858.CD003383.
54. Witte S, Loew D, Gaus W. Meta-analysis of the efficacy of the acetonic Kava-Kava extract WS1490 in patients with non-psychotic anxiety disorders. Phytother Res. 2005;19:183–8.
55. Sarris J, Kavanagh DJ. Kava and St. John's Wort: current evidence for us in mood and anxiety disorders. J Altern Complement Med. 2009;15:827–36.
56. Sarris J, LaPorte E, Schweitzer I. Kava: a comprehensive review of efficacy, safety and psychopharmacology. Aust NZ J Psychiatry. 2011;45:27–35.
57. Sarris J, Stough C, Teschke R, Wahid ZT, Bousman CA, Murray G, Savage KM, Mouatt P, Ng C, Schweitzer I. Kava for the treatment of generalized anxiety disorder RCT: analysis of adverse reactions, liver function, addiction and sexual effects. Phytother Res. 2013;27:1723–8.
58. Sarris J. Herbal medicines in the treatment of psychiatric disorders: 10-year updated review. Phytother Res. 2018;32:1147–62.
59. Sarris J, McIntyre E, Camfield DA. Plant-based medicines for anxiety disorders, part 2: a review of clinical studies with supporting preclinical evidence. CNS Drugs. 2013;27:301. https://doi.org/10.1007/s40263-013-0059-9.
60. Ooi SL, Henderson P, Pak SC. Kava for generalized anxiety disorder A review of current evidence. J Complement Altern Med. 2018. https://doi.org/10.1089/acm.2018.0001.
61. Kava: a review of the safety of traditional and recreational beverage consumption. Technical report. Food and Agriculture Organization of the United Nations World Health Organization. 2016. Available at: http://www.fao.org/3/a-i5770e.pdf. Accessed 26 Jan 2019.
62. Assessment report on *Piper methysticum* G. Forst., rhizome. EMA/HMPC/450859/2016. Available at: https://www.ema.europa.eu/documents/herbal-report/final-assessment-report-piper-methysticum-g-forst-rhizoma_en.pdf. Accessed 26 Jan 2019.

63. Gastpar M, Klimm HD. Treatment of anxiety, tension and restlessness states with Kava special extract WS1490 in general practice: a randomized placebo-controlled double-blind multicenter trial. Phytomedicine. 2003;10:631–9.
64. Connor KM, Payne V, Davidson JRT. Kava in generalized anxiety disorder: three placebo-controlled trials. Int Clin Psychopharmacol. 2006;21:249–53.
65. Ernst E. A re-evaluation of kava (*Piper methysticum*). Br J Clin Pharmacol. 2007;64:415–7.
66. Savage KM, Stough CK, Byrne GJ, Scholey A, Bousman C, Murphy J, Macdonald P, et al. Kava for the treatment of generalized anxiety disorder (K-GAD): study protocol for a randomized controlled trial. Trials. 2015;16:493. https://doi.org/10.1186/s13063-015-0986-5.
67. Miyasaka LS, Atallah AN, Soares B. Passiflora for anxiety disorders. Cochrane Datab Syst Rev. 2007; Art. No.: CD004518. https://doi.org/10.1002/14651858.CD004518.pub2.
68. Miyasaka LS, Atallah AN, Soares B. Valerian for anxiety disorders. Cochrane Datab Syst Rev. 2006; Art. No.: CD004515. https://doi.org/10.1002/14651858.CD004515.pub2.
69. Whiting PF, Wolff RF, Deshpande S, Di Nisio M, Duffy S, Hernandez AV, Keurentjes JC, et al. Cannabinoids for medical use. A systematic review and meta-analysis. JAMA. 2015;313:2456–73.
70. Hoch E, Niemann D, von Keller R, Schneider M, Friemel CM, Preuss UW, Hasan A, Pogarell O. How effective and safe is medical cannabis as a treatment of mental disorders? A systematic review. Eur Arch Psychiatry Clin Neurosci. 2019. https://doi.org/10.1007/s00406-019-00984-4.
71. Woelk H, Arnoldt KH, Kieser M, Hoerr R. Ginkgo biloba special extract EGb 761 in generalized anxiety disorder and adjustment disorder with anxious mood: a randomized double-blind placebo-controlled trial. J Psychiatry Res. 2007;41:472–80.
72. Bourin M, Gougerol T, Guitton B, Broutin E. A combination of plant extracts in the treatment of outpatients with adjustment disorder with anxious mood: controlled study versus placebo. Fundam Clin Pharmacol. 1997;11:127–32.
73. Bergamaschi MM, Costa Queiroz RH, Nisihara Chagas MH, Gomes de Oliveira DCG, De Martinis BS, Kapczinski F, Quevedo J, Roesler R, et al. Cannabidiol reduces the anxiety induced by simulated public speaking in treatment-naïve social phobia patients. Neuropsychopharmacology. 2011;36:1219–26.
74. Kobak KA, Taylor LH, Warner G, Futterer R. St. John's wort versus placebo in social phobia. Results from a placebo-controlled pilot study. J Clin Psychopharmacol. 2005;25:51–8.
75. Mukai T, Kishi T, Matsuda Y, Iwata N. A meta-analysis of inositol for depression and anxiety disorders. Hum Psychopharmacol. 2014;29:55–63.
76. Benjamin J, Levine J, Fux M, Aviv A, Levy D, Belmaker RH. Double-blind placebo-controlled crossover trial of inositol treatment for panic disorder. Am J Psychiatry. 1995;152:1084–6.
77. Palatnik A, Frolov K, Fux M, Benjamin J. Double-blind controlled cross-over trial of inositol versus fluvoxamine for the treatment of panic disorder. J Clin Psychopharmacol. 2001;21:335–9.
78. Errington-Evans N. Acupuncture for anxiety. CNS Neurosci Ther. 2012;18:277–84.
79. Amorim D, Amado J, Brito I, Fiuza SM, Amorim N, Costeira C, Machado J. Acupuncture and electroacupuncture for anxiety disorders: a systematic review of the clinical research. Complement Ther Clin Pract. 2018;31:31.37.
80. Goyata SLT, Valcanti Avelino CC, Marques dos Santos AC, de Souza Junior DI, Lopes Gurgel MDS, de Souza Terra F. Effects from acupuncture in treating anxiety: integrative review. Rev Bras Enferm. 2016;69:564–71.
81. Leung B, Takeda W, Holec V. Pilot study of acupuncture to treat anxiety in children and adolescents. J Paediatr Child Health. 2018;54:881–8.
82. Yuan Q, Li J, Liu B, Wu Z, Jin R. Effect of Jin-3-needling therapy on plasma corticosteroid, adrenocorticotrophic hormone, platelet 5-HT levels in patients with generalized anxiety disorder. Chin J Integr Med. 2007;13:264–8.
83. Bartley CA, Hay M, Bloch MH. Meta-analysis: aerobic exercise for the treatment of anxiety disorders. Prog Neuro-Psychopharmacol Biol Psychiatry. 2013;45:34–9.
84. Jayakody K, Gundasa S, Hosker C. Exercise for anxiety disorders: systematic reviews. Br J Sports Med. 2014;48:187–96.

85. Stonerock GL, Hoffman BM, Smith PJ, Blumenthal JA. Exercise as treatment for anxiety: systematic review and analysis. Ann Behav Med. 2015;49:542–56.
86. Stubbs B, Vancampfort D, Rosenbaum S, Firth J, Cosco T, Veronese N, Salum GA, Schuch FB. An examination of the anxiolytic effects of exercise for people with anxiety and stress-related disorders: a meta-analysis. Psychiatry Res. 2017;249:102–8.
87. Sexton A, Maere A, Dahl NH. Exercise intensity in reduction of neurotic symptoms. Acta Psychiatr Scand. 1989;80:231–5.
88. Broocks A, Bandelow B, Pekrun G, George A, Meyer T, Martmann U, Hillmer-Vogel U, Ruther E. Comparison of aerobic exercise clomipramine and placebo in the treatment of panic disorder. Am J Psychiatry. 1989;155:603–9.
89. Merom D, Phongsavan P, Wagner R, CHey T, Marnane C, Steel Z, Silove D, Bauman A. Promoting walking as an adjunct intervention to group cognitive behavioral therapy for anxiety disorders – a pilot group randomized trial. J Anxiety Disord. 2008;22:959–68.
90. Wedekind D, Broocks A, Weiss N, Engel K, Neubert K, Bandelow B. A randomized controlled trial of aerobic exercise in combination with paroxetine in the treatment of panic disorder. World J Biol Psychiatry. 2010;11:904–13.
91. Oeland AM, Laessoe U, Olesen AV, Munk JP. Impact of exercise on patients with depression and anxiety. Nord J Psychiatry. 2010;64:210–7.
92. Herring M, Jacob M, Suveg C, Dishman R, O'Connor P. Feasibility of exercise training for the short-term treatment of generalized anxiety disorder: a randomized controlled trial. Psychother Psychosom. 2012;81:21–8.
93. Jazaieri H, Goldin PR, Werner K, Ziv M, Gross JJ. A randomized trial of MBSR versus aerobic exercise for social anxiety disorder. J Clin Psychol. 2012;68:715–31.
94. Hovland A, Nordhus IH, Martinsen EW, Torsheim T, Pallesen S. Comparing physical exercise in groups to group cognitive behavior therapy for the treatment of panic disorder in a randomized controlled trial. Behav Cogn Psychother. 2013;41:408–32.
95. Gaudlitz K, Plag J, Dimeo F, Strohle A. Aerobic exercise training facilitates the effectiveness of cognitive behavioral therapy in panic disorder. Depress Anxiety. 2015;32:221–8.
96. Robinson J, Biley FC, Dolk H. Therapeutic touch for anxiety disorders. Cochrane Datab Syst Rev. 2007; Art. No.: CD006240. https://doi.org/10.1002/14651858.CD006240.pub3.
97. Li H, Wang J, Li C, Xiao Z. Repetitive transcranial magnetic stimulation (rTMS) for panic disorder in adults. Cochrane Datab Syst Rev. 2014; Art. No.: CD009083. https://doi.org/10.1002/14651858.CD009083.pub2.
98. Lorenc A, Feder G, MacPherson H, Little P, Mercer SW, Sharp D. Scoping review of systematic reviews of complementary medicine for musculoskeletal and mental health conditions. BMJ Open. 2018;8:e020222. https://doi.org/10.1136/bmjopen-2017-020222.
99. Chen K, Berger CC, Manheimer E, Forde D, Magidson J, Dachman L, Lejuez CW. Meditative therapies for reducing anxiety: a systematic review and meta-analysis of randomized controlled trials. Depress Anxiety. 2012;29:545–62.
100. Kirkwood G, Rampes H, Tuffrey V, Richardson J, Pilkington K. Yoga for anxiety: a systematic review of the research evidence. Br J Sports Med. 2005;39:884–91.
101. Vollbehr NK, Bartles-Velthuis AA, Hauta MH, Castelein S, Steenhuis LA, Hoenders HJR, Ostafin BD. Hatha yoga for acute, chronic and/or treatment-resistant mood and anxiety disorders: a systematic review and meta-analysis. PLoS One. 2018;13:e0204925. https://doi.org/10.1371/journal.pone.0204925.
102. Hofmann GS, Andreoli G, Carpenter JK, Curtiss J. Effect of hatha yoga on anxiety: a meta-analysis. J Evid Based Med. 2016;9:116–24.
103. Cramer H, Lauche R, Anheyer D, Piklington K, de Manincor M, Dobos G, Ward L. Yoga for anxiety: a systematic review and meta-analysis of randomized controlled trials. Depress Anxiety. 2018. https://doi.org/10.1002/da.22762.
104. Wu H, Yu D, He Y, Wang J, Xiao Z, Li C. Morita therapy for anxiety disorders in adults. Cochrane Datab Syst Rev. 2015; Art. No.: CD008619. https://doi.org/10.1002/14651858.CD008619.pub2.

105. Sharma M, Haider T. Tai Chi as an alternative and complimentary therapy for anxiety: a systematic review. J Evid Based Complement Altern Med. 2015;20:143–53.
106. Zou L, Yeung A, Quan X, Hui SSC, Hu X, Chan JSM, Wang C, Boyden SD, Sun L, Wang H. Mindfulness-based Baduanjin exercise for depression and anxiety in people with physical or mental illnesses: a systematic review and meta-analysis. Int J Enrvion Res Pub Health. 2018:321. https://doi.org/10.3390/ijerph15020321.
107. Joyce J, Herbinson GP. Reiki for depression and anxiety. Cochrane Datab Syst Rev. 2015; Art. No.: CD006833. https://doi.org/10.1002/14651858.CD006833.pub2.
108. Goncalves JPB, Lucchetti G, Menezes PR, Vallada H. Religious and spiritual interventions in mental health care: a systematic review and meta-analysis of randomized controlled clinical trials. Psychol Med. 2015;45:2937–49.
109. Manzoni GM, Pagnini F, Casterlnuovo G, Molinari E. Relaxation training for anxiety: a ten-years systematic review with meta-analysis. BMC Psychiatry. 2008;8:41. https://doi.org/10.1186/1471-244X-8-41.
110. Clarke D, Ehlers A, Hackmann A, McManus F, Fennell M, Grey N, Waddington L, Wild J. Cognitive therapy versus exposure and applied relaxation in social phobia: a randomized controlled trial. J Consult Clin Psychol. 2006;74:568–78.
111. Kim H, Kim EJ. Effects of relaxation therapy on anxiety disorders: a systematic review and meta-analysis. Arch Psychiatr Nurs. 2017. https://doi.org/10.1016/j.apnu.2017.11.015.
112. Clark DM, Salkovskis PM, Hackmann A, Hiddleton H, Anastasiades P, Gelder M. A comparison of cognitive therapy, applied relaxation and imipramine in the treatment of panic disorder. Br J Psychiatry. 1994;164:759–69.
113. Goyal M, Singh S, Sibinga EMS, Gould NF, Rowlan-Seymour A, Sharma R, Berger Z, Sleicher D, et al. Mediation programs for psychological stress and well-being. A systematic review and meta-analysis. JAMA Intern Med. 2014;174:357–68.
114. Berger SB, Tucker RP, Greene PA, Davidson RJ, Wampold BE, Kearney DJ, Simpson TL. Mindfulness-based interventions for psychiatric disorders: a systematic review and meta-analysis. Clin Psychol Rev. 2017. https://doi.org/10.1016/j.cpr.2017.10.011.
115. Chiesa A, Serretti A. Mindfulness based cognitive therapy for psychiatric disorders: a systematic review and meta-analysis. Psychiatry Res. 2011;187:442–53.
116. Krisanaprakornkit T, Sriraj W, Piyavhatkul N, Laopaiboon M. Meditation therapy for anxiety disorders. Cochrane Datab Syst Rev 2006; Art. No.: CD004998. https://doi.org/10.1002/14651858.CD004998.pub2.
117. Vollestad J, Birkeland Nielsen M, Hostmark NG. Mindfulness- and acceptance-based interventions for anxiety disorders: a systematic review and meta-analysis. Br J Clin Psychol. 2012;51:239–60.
118. Sevilla-Llewellyn-Jones J, Santesteban-Echarri O, Pryor I, McGorry P, Alvarez-Jimenez M. Web-based mindfulness interventions for mental health treatment: systematic review and meta-analysis. JMIR Ment Health. 2018;5:e10278. https://doi.org/10.2196/10278.
119. Strauss C, Cavanagh K, Oliver A, Pettman D. Mindfulness-based interventions for people diagnosed with a current episode of an anxiety or depressive disorder: a meta-analysis of randomized controlled trials. PLoS One. 2014;9:e96110. https://doi.org/10.1371/journal.pone.0096110.
120. Hedman-Lagerlof M, Hedman-Lagerlof E, Ost LG. The empirical support for mindfulness-based interventions for common psychiatric disorders: a systematic review and meta-analysis. Psychol Med. 2018;48:2116–29.
121. Pratte MA, Nanavati KB, Young V, Morley CP. An alternative treatment for anxiety: a systematic review of human trial results reported for the Ayurvedic herb Ashwagandha (Withania somnifera). J Altern Complement Med. 2014;20:901–8.
122. Pilkington K, Kirkwood G, Rampes H, Fisher P, Richardson J. Homeopathy for anxiety and anxiety disorders: a systematic review of the research. Homeopathy. 2006;95:151–62.
123. Davidson JRT, Crawford C, Ives JA, Jonas WB. Homeopathic treatments in psychiatry: a systematic review of randomized placebo-controlled studies. J Clin Psychiatry. 2011;72:795–805.

Contemporary Psychodynamic Approaches to Treating Anxiety: Theory, Research, and Practice

23

Seth R. Pitman and Daniel P. C. Knauss

Psychodynamic theory is founded on the idea that human behavior is influenced by forces and experiences that lie outside of conscious awareness. Despite what we may or may not recognize about our lived experience, an essential connection exists between unconscious processes and everyday psychological functioning. By extension, psychodynamic theory presumes that unconscious conflicts are pathognomonic of anxiety disorders and anxiety symptoms generally. The term "psychodynamic" refers not only to that which occurs within one's mind but also what happens between people and within families, groups, and systems. A comprehensive psychodynamic treatment of anxiety attempts to take into account these multiple domains of experience and functioning (including biological and genetic considerations) and applies a specific therapeutic approach to working with patients based in part on clinical techniques first developed by Sigmund Freud more than 100 years ago.

In his earliest theory of anxiety, Freud (1895) considered it to be the result of a buildup of libido that, due to repression, is not discharged. He later reformulated his understanding based on his development of a structural model of psychic functioning. In this "signal theory," anxiety is regarded as a safeguard to protect the ego's functioning from threatening impulses generated internally by the id. Anxiety *signals* the press of a threatening (often sexual or aggressive) impulse, which in turn activates the ego's defense mechanisms in the service of rendering it unconscious through repression. Freud also argued that repression can be accomplished by the process of distortion, in which the dangerous impulse becomes disguised or transformed into a more readily acceptable idea, neutralizing the threat.

Whereas in the earlier theory Freud believed anxiety to be the result of repression itself, in the later theory, anxiety was thought to be the result of a conflict between id and superego, or internal, unacceptable wish and the standards and morals

S. R. Pitman (✉) · D. P. C. Knauss
Austen Riggs Center, Stockbridge, MA, USA
e-mail: Seth.Pitman@austenriggs.net

Y.-K. Kim (ed.), *Anxiety Disorders*, Advances in Experimental Medicine and Biology 1191, https://doi.org/10.1007/978-981-32-9705-0_23

imputed onto the individual by society and his or her environment. In this structure, the ego (or self) is placed precariously between conflicting messages to both satisfy an internal urge (generated by the id) and also act in accordance with external expectations (represented psychically by the super ego). This theoretical position is easily understandable when we consider the many instances in which a wish or longing can be at odds with environmental conditions that discourage one from actualizing it. The well-known phrase "forbidden love" may come to mind. In fact, Freud believed anxiety to be so ubiquitous that he suggested it was an essential precondition for membership into civilized society.

Freud also outlined a framework of psychosexual development, each stage of which is linked to anxiety about a particular developmental conflict. Anxiety at the most mature level related to guilt feelings and fear of retribution from the father in the Oedipal phase, which Freud categorized under superego anxiety. Other sources of anxiety, from more to less mature, included castration anxiety, fear of loss of love, and fear of loss of the object itself (separation anxiety). Subsequent psychoanalytic thinkers elaborated on Freud's theory by postulating even earlier forms including persecutory anxiety and disintegration anxiety, frequently found in the psychoses. Despite the hierarchical organization of these developmental types clinical experience suggests that patients are susceptible to experiencing many different kinds of anxiety, and at times several kinds at once, depending on the nature of the disturbance and factors affecting the patient intrapsychically and interpersonally [19].

The model of psychological functioning postulated by Freud was essentially intrapsychic, giving primacy to the mind's functioning in relation to itself. Successive generations of psychoanalytic and psychodynamic thinkers have increasingly recognized the role of interpersonal, relational, familial, and group experiences in the development of psychological symptoms. Developments in attachment theory and infant research have broadened an interest in the role of early relational patterns with caregivers in clinical work with both children and adults [13, 15]. As such, postmodern influences have inspired clinicians to consider the role of intersubjectivity and the reciprocal influence of the personhood of the patient and therapist in the clinical encounter, including transference-countertransference dynamics. What has evolved over time is a heterogenous group of theories and approaches to treatment with historical roots in Freudian analysis. Given the breadth and diversity in the field, this chapter will focus primarily on recent psychodynamic approaches to anxiety that have empirical support and can be feasibly delivered within the context of the current managed care landscape and in community settings. We have also chosen this focus in order to highlight pragmatic approaches to common clinical presentations, in line with emerging trends in the psychotherapy literature regarding transdiagnostic conceptualizations and treatments of anxiety. These approaches require minimal advanced training and can be employed by both seasoned clinicians and therapists in training, making them widely disseminable. Although we do not cover long-term psychodynamic therapy or psychoanalysis proper (for which empirical evidence also exists), those approaches have their own merit and provide the foundation from which current approaches have been developed.

Clinical Vignette

Mr. X, a 30-year-old man with a history of anxiety and depression and previous diagnosis of generalized anxiety disorder, presented to treatment with a psychodynamic therapist in the context of becoming engaged to his girlfriend of several years. Whereas his anxiety had previously been effectively managed with once-weekly supportive psychotherapy and an SSRI, he found that he was having more and more difficulty managing his anxiety symptoms, which were beginning to interfere at work and in his relationship with his fiancé. He had also begun to rely on alcohol more heavily than in the past in order to quell his worry and help him sleep.

In initial therapy sessions, Mr. X wondered whether he should resume use of an antidepressant or try a benzodiazepine, out of concern that his struggles would derail his engagement and jeopardize his wedding. Instead, the therapist suggested he bring his anxiety into treatment, so that they may understand why it had emerged with such force at this time. In order to accomplish this task, the therapist suggested he verbalize his thoughts and feelings freely as they arose in sessions, with particular attention paid to the memories, fantasies, fears, and wishes that appeared alongside his symptoms. In response, the therapist would listen for and comment on any unarticulated connections between these experiences, in an effort to help find meaning in Mr. X's symptoms.

Although Mr. X had some insight into why he was anxious, he could not account for why his worries had become so debilitating. He loved his fiancé and was excited to spend his life with her, making the issue confusing. As initial sessions unfolded, he spoke about the sense of duty he felt toward giving her exactly the kind of wedding she wanted and also his worry about something going wrong on the big day. The therapist noticed that as Mr. X talked about his sense of duty to provide "the perfect wedding," he displayed noticeable contempt and anger. The therapist shared this impression with Mr. X, who quickly retorted, "Well she can be quiet demanding. In fact it's downright infuriating!" In subsequent sessions, Mr. X spoke about his experience of feeling pressured by his fiancé to live up to impossible expectations, but unable to speak about his anger.

As Mr. X developed a capacity to speak about the range of feelings he had toward his fiancé, his anxiety about the wedding itself began to diminish. However, he still struggled to share his anger with his fiancé and remained anxious about the damaging effect he imagined his negative emotions would have on their relationship. As he associated to his experiences, Mr. X reflected on his memory of his parent's marriage when he was a child. He shared that his mother appeared to him heavily dependent on his father for attention and care. His father, however, was avoidant and unavailable, and his mother was loath to make her desires known to him, becoming instead conspicuously withdrawn and sullen. This dynamic led to a tense home where feelings often went unspoken. Mr. X often found himself attempting to cheer his mother up, inventing various ways to please her and later stewing angrily at the amount of time he felt obligated to care for her.

These memories led to further elaborations and a more thorough understanding of Mr. X's anxiety. In his relationship to his fiancé, he found himself in a similar position to his mother: wanting care and connection but uncertain about how to make his desires known, as he expected them to be painfully rejected. He also found himself repeating aspects of his own childhood relationship with his mother, as he attempted to accommodate his fiancé's needs and give her exactly what he thought she wanted while feeling frustrated and angered by the effort he put into pleasing her.

This dynamic appeared to be repeated within the therapeutic relationship, particularly in moments when Mr. X seemed eager to please his therapist by offering "exciting" or "interesting" associations, and then seeming sullen and irritated when his emotional intensity wasn't matched. Over time and with his therapist's supportive encouragement, Mr. X began to speak more freely about the way he longed for his therapist's care and how he was reluctant to disclose these feelings out of a fear of being met with rejection or lack of interest. The recreation of these dynamics in the therapeutic relationship provided Mr. X's therapist with an experience-near understanding of his patient's position within his childhood family. In turn they became able to work through a problematic relational paradigm in the here-and-now while experimenting with new ways of being in a close relationship that included risking and sharing his desires and feelings. This therapeutic process gradually allowed Mr. X to be more aware of the immediate and historical context for his anxiety and equip himself with a more flexible and adaptive manner of managing his anxiety with his wife-to-be.

Evidence Base for Psychodynamic Treatment of Anxiety

Historically, the research literature has appeared to suggest psychodynamic therapy (PDT) to be less effective than cognitive behavioral therapy (CBT) and other therapies for multiple classes of disorders, including anxiety disorders. At the same time, a compelling body of evidence supports the notion that all *bone fide* therapies – that is, treatments consisting of active therapeutic elements derived from a coherent theory of psychopathology – are uniformly effective [41]. Based on substantial evidence on the importance of therapist and patient factors on outcome [2, 7, 24], the recommendations set forth by APA's Presidential Task Force on Evidence-Based Practice [1] support the notion that an effective approach rests not only on specific techniques but also on a variety of factors known to have a positive impact on all therapies, most notably the therapeutic alliance. Today the best evidence suggests that PDT therapists practicing in a flexible, responsive manner to patient needs as they arise in treatment – and not simply in fidelity to rigid, historical, theoretical understandings – have the best chance of effecting positive change in their patients.

Unlike CBT, PDT practitioners have typically eschewed disorder-specific treatments in favor of a more universal, or "generic," approach. Rather than develop individualized treatment manuals for particular manifestations of psychological distress, a PDT framework remains more or less consistent and applies a similar set of techniques across a wide range of disorders. This predilection is one reason why in the past PDT has appeared not to fare particularly well in comparison with other treatments for anxiety and other disorders. Randomized control trials (RCTs) examining the efficacy of PDT for specific mental disorders in relation to other therapies often use different, poorly operationalized, conceptions and methods of PDT interchangeably, leading many PDT treatments to be deemed only "possibly efficacious" when judged by the criteria of empirically supported treatments (ESTs) [25].

In terms of comparison trials, although PDT has been shown to be less efficacious than CBT in studies in which a disorder-specific cognitive behavioral manual was used [12], two recent and well-conducted meta-analyses challenge the notion that PDT is inferior. The first examined the controlled effects of PDT for anxiety disorders among 14 RCTs [21] and found PDT to be significantly more effective than control conditions. (The authors reported an effect size of $g = 0.64$, with g representing a more conservative estimate of effect size for small samples than Cohen's *d*.). These results were detected among primary and secondary outcome measures, and they were maintained at follow-up at 1 year ($g = -0.11$) and after a year ($g = -0.26$). The second examined 39 RCTs across a range of disorders [28] and found PDT to be efficacious for social anxiety disorder (SAD) and possibly efficacious for generalized anxiety disorder (GAD). Further, the combination of PDT with psychopharmacology was also deemed efficacious for panic disorder (PD) and SAD. Importantly, this meta-analysis included both manual-guided psychodynamic treatments, as well as short-term, long-term, and group formats of PDT, all of which included clear descriptions of the theoretical background and technical recommendations such as indications, interventions, and timing.

Additionally, a recent Cochrane network meta-analysis [37] compared the efficacy of eight different forms of psychotherapy for panic disorder. Despite the generally low quality of evidence for outcomes evidenced across the entire network, and inclusion of only two studies examining PDT for panic disorder, PDT demonstrated higher tolerability among short-term treatments than CBT (odds ratio [OR] 0.52, 95% confidence interval [CI] 0.15–1.8) and comparably high rates of remission at 6 month follow-up. The authors concluded that although PDT appears to be the best tolerated of the psychotherapies, further research needs to be conducted in order to bolster the evidence and better explore the relative efficacy of PDT and CBT.

Naturalistic and quasi-experimental studies have also provided evidence for the benefits of PDT for anxiety disorders, with some mixed results. The reader is directed to Slavin-Mulford and Hilsenroth [38] for a thorough review of this research literature. Overall, PDT has been shown to have marginally smaller effects compared to cognitive, behavioral, and CBT treatments, although many of the studies comparing these modalities found large effects for both treatments (e.g., [23, 27]). PDT studies have consistently demonstrated large effects, and following PDT patients meet criteria for anxiety diagnoses at a reduced rate, have fewer symptoms of anxiety and co-occurring depression, and report lower levels of global distress [11, 32, 39]. In an updated review of outcome studies and meta-analyses of effectiveness studies of PDT for a range of major mental disorders, Fonagy [14] concluded that treatment approaches generated from PDT principles appeared to benefit patients presenting with symptoms related to SAD, GAD, and PD.

Taken together, PDT appears to be no better or worse than other active treatments for anxiety disorders, based on results at both the conclusion of treatment and at follow-up. Despite this evidence, the few studies examining specific mechanisms of change in PDT have not yielded consistent results [8, 10, 40]. This inconsistency has led researchers to call for future trials of PDT for both specific anxiety disorders and examining the anxiety spectrum as a whole [38], as well as examining processes of change common among effective therapies and specific to PDTs [21].

A brief word should be said about how PDT appears to fare in comparison to psychopharmacological interventions for anxiety. Based on their review of the few existing studies comparing PDT to medication, Slavin-Mulford and Hilsenroth [38] assert PDT to be as efficacious as pharmacological interventions, and cite evidence that it may in fact enhance the effects of medication trials [42]. Impressively, they point to one well-designed study that showed PDT to be more efficacious for anxious patients with co-occurring personality pathology, based on primary anxiety outcome measures and improvements in social and occupational functioning [14]. A psychodynamic view of prescribing suggests that attending to the meaning of medications, and what they represent for both the patient and the prescriber, is an important aspect of the therapeutic engagement that has direct implications for the success of the intervention. More on this topic can be found elsewhere [33, 36].

Existing Psychodynamic Approaches to Anxiety

Psychodynamic theorists beginning with Freud have discussed unconscious conflicts and their defenses in relation to a wide range of anxiety disorders, but as mentioned, few manual-guided treatments exist. Two examples illustrate how PDT conceptualizes the etiology, maintenance, and treatment approach of anxiety symptoms. Panic-focused psychodynamic psychotherapy (PFPP; [9]) posits that panic attacks arise from unconscious conflicts related to dependency, attachment, and feelings of anger. In this conceptualization, panic potentially serves as a disguised means of expressing a need for care, an aggressive and coercive attempt to attract the attention of caregivers, or a form of self-punishment brought on by guilt around dependency strivings. Treatment consists of exploring and understanding the emotional significance of panic, delineating its meaning in relation to previous difficulties with caregivers, and working through these conflicts by recognizing how panic emerges across multiple, related settings. Leichsenring et al. [26] developed an integrative treatment for SAD that directs therapists to explore the affective aspects of shame and the unrealistic demands socially phobic patients make on themselves to perform in social situations. These techniques are utilized in combination with self-guided symptom exposure in the middle phase of treatment. Exposure is framed psychodynamically by explicitly relating it back to the therapeutic relationship and to underlying, repetitively experienced, relational themes.

More commonly, therapists practicing contemporary forms of PDT in clinical settings employ a collection of techniques from within a larger, and not always consistent, theoretical umbrella. Although there exists no definitive technical definition of PDT, Blagys and Hilsenroth [5] identified several key features of short-term psychodynamic psychotherapy (STPP), most notably (a) a focus on affect and the expression of emotion and (b) the identification of patterns in actions, thoughts, feelings, experiences, and relationships. These patterns are usefully conceptualized and explored using the Core Conflictual Relationship Theme (CCRT) format, as developed by Luborsky [29]. The CCRT focuses on what a patient wants from relationships (the wish; W), the response he/she expects from others (response from other; RO), and the patient's subsequent affective and behavioral response (response of self; RS). With regard to anxiety symptoms, the CCRT provides a cogent method to draw out the relational struggles that cause distress, and à la Freud, the affective and intrapsychic conflicts that can result and produce symptoms of anxiety.

Another universally applicable PDT concept that is fruitfully applied to anxiety symptoms is Luborsky's [29] model of supportive-expressive (SE) therapy. This model represents a continuum of interventions thought to work synergistically in order to bring about symptomatic improvement, and can be variously employed based on a variety of patient factors. Expressive interventions, such as transference and defense interpretation, are intended to promote insight and self-understanding, which can help the patient tolerate the uncomfortable affects and fantasies that mobilize problematic defenses and produce anxiety. Conversely, supportive techniques are used to help enhance patient's adaptive capacities and most importantly can foster to the development of a secure therapeutic alliance. Enlisting the patient's collaboration in becoming curious about the meaning and developmental origin of his or her anxiety rests on the therapist's capacity to build of a "holding" environment, so that difficult affective and relational experiences can be faced and explored collaboratively. Given the aversive nature of anxiety, many people are initially more motivated to extinguish the experience (often through avoidance) rather that tolerate its presence, which is often required along the pathway to understanding and meaning-making. Luborsky [29] and Crits-Christoph et al. [11] developed a number of principles and therapeutic actions to foster the development of the alliance in PDT, including socializing the patient to the tasks of therapy. Luborsky [29] also usefully suggested that a greater number of supportive interventions may be needed for patients with greater psychopathology or when in times of crisis.

Psychodynamic Techniques with Empirical Support

At present, few studies have examined the impact of specific techniques in psychodynamic therapy for anxiety disorders. A recent study examining the use of interpretations in PFPP [22] found that panic-focused interpretations improved panic symptoms in the middle phase of treatment (B = 1.79 [95% CI: 0.61, 2.97], $SE = 0.59$, $t = 3.04$, $p = 0.004$, adjusted $p = 0.016$, $sr = 0.37$), while interpretations about conflicts that did not include a panic-focus did not ($B = -0.47$ [95 CI: −1.54, 0.60], $SE = 0.54$, t [59] $= -0.88$, $p = 0.382$, adjusted $p = 0.437$, $sr = -0.09$). In light of these findings, the authors argued for the importance of taking a symptom-focused approach in STPP for anxiety, although they could not confirm the presence of a causal relationship between the intensity of panic interpretations and symptom change. Within a naturalistic/effectiveness model of STPP for anxiety, Slavin-Mulford et al. [39] conducted one of the first studies to examine treatment fidelity, credibility, and satisfaction in an outpatient community population. Examining the relationship between the use of specific psychodynamic interventions (PI) in the third session and reduction in anxiety symptoms at posttreatment ($r = .46$, $p = .04$; [39]), the authors identified several individual PI techniques that were found to be meaningfully related to outcome. A follow-up study using the same sample of anxiety disorder patients [34] found a significant relationship between psychodynamic technique and posttreatment change in anxiety symptoms at the ninth session ($r = .49$, $p = .03$) and at mean levels of psychodynamic technique across both third and ninth sessions ($r = .53$, $p = .02$). Based on these findings, the authors suggested that there is a positive relationship between techniques intended to increase understanding of cyclical relational-affective patterns early in treatment and subsequent improvement in anxiety symptoms.

A second follow-up study [35] extended this work by applying it to a transdiagnostic sample of patients, in order to clarify whether the same, or different, PI interventions would be found effective for patients with subclinical levels of anxiety as they were with patients suffering from severe anxiety disorders. Given the implications for emerging trends in the psychotherapy literature regarding transdiagnostic approaches to psychological problems, this study will be discussed below.

Providing Alternate Understandings of Symptoms

In Luborsky's supportive-expressive continuum, providing an alternative understanding of anxiety symptoms represents an expressive intervention aimed at deepening patient's exploration of his or her anxiety and gaining an appreciation for its meaning that may not be readily apparent. Another way to describe this intervention is "interpretation," which throughout the history of PDT has been a cornerstone of its approach. By maintaining a consistent focus on deepening patients' exploration of their internal experience and attending carefully to the treatment frame, therapists can learn to recognize the relational and emotional circumstances in which anxiety arises. Oftentimes, this may occur in moments of therapeutic rupture or even expectable separation, such as weekends or the therapist taking a vacation. Therapists can then use the knowledge gained through their own experience with the patient's anxiety in order to generalize to other relationships, including their earliest experiences with caregivers, thereby gaining a subjective appreciation for the etiology and function of these symptoms.

One example of how the CCRT is used in treatment for anxiety disorders can be found in Leichsenring et al.'s SE-based treatment manual for social phobia. In this protocol, the therapist is instructed to utilize the CCRT as a way to provide an alternative understanding of symptoms at the outset of therapy and as they recur throughout the course of treatment. Based on a formulation by Gabbard [18], a CCRT for a patient with SAD may be described as wishing to be affirmed by others (W), expecting that others will humiliate him or her (RO), and feeling ashamed and afraid of being together with others, therefore deciding to avoid exposing him- or herself (RS = symptoms of social phobia). During the termination phase, the CCRT can also be used to address the resurgence of phobic symptoms in anticipation of the loss of the therapist and the patient's fear that the wish to be cared for and accepted will not be fulfilled.

Recalling the clinical vignette presented in the beginning of this chapter, a CCRT for a patient with GAD might be articulated as follows:

> *You want others to take care of you (W), but you are afraid that they'll fail you, either because they aren't emotionally capable of being there or are preoccupied with something else, such as preparing for an important event like a wedding (RO). This leads you to withdraw, to feel lonely and sad, and to worry a great deal about whether you are able to meet your obligations. You become angry or critical of others' inadequacies and may even become critical of yourself (RS).*

The process of outlining the CCRT aims to help the patient achieve insight into the underlying meaning of their anxiety symptoms and to begin to develop more adaptive means of coping with the preoccupying symptoms of worry and anxious distress.

Focus on Affect and the Expression of Emotions

As described above, gaining a new understanding of anxiety symptoms as they relate to the patient's self and relationships entails an affective component, namely, by identifying and exploring the warded off emotional experiences that the patient finds too painful or overwhelming to bear. The defensive function of worrying is an animating principle of Affect Phobia Therapy (APT; [31]), which views inhibitory emotional responses such as anxiety, guilt, and shame as obstructing the experience of painful but genuine emotions. Using the triangle of conflict and triangle of person models developed by Malan [30], APT helps patients develop new understandings of their symptoms in relation to present and past relational-affective events.

As previously described, Leichsenring et al.'s [26] manual-guided treatment for SAD incorporates self-guided symptom exposure in the middle of treatment. Although customarily exposure activities have been more associated with CBT than PDT [6], it has been suggested that exposure and exploration of uncomfortable feelings are active in PDT, though the context is different from that provided in a traditional CB format. Helping the patient become aware and tolerant of their disavowed emotions in the here-and-now with the therapist has been conceptualized in terms of lowering experiential avoidance, which is thought to be a core process in the development and maintenance of anxiety [20].

Thus, whereas Freud suggested that treatment of anxiety should consist of bringing the threat into consciousness in order to help the patient understand that it no longer represents a present danger, current dynamic approaches emphasize the affective and relational aspects of symptoms with the intention of reducing related emotions such as shame, grief, rage, and humiliation. Emotional insight is achieved in part by the therapists' making connections between past and current relational-affective events, including those that arise within the therapeutic relationship. The emphasis on developing more adaptive patterns of interpersonal relatedness is a feature of most contemporary, brief forms of PDT, while longer forms including psychoanalysis may require greater patient autonomy in learning from and applying insights gained in therapy to real-world experience.

Anxiety and Reflective Function

The ability to learn, make use of new insights, and improve relationships requires a range of cognitive, affective, and interpersonal reflective capacities. Fonagy et al. [17] have proposed the concept of *mentalizing*, operationally defined as *reflective function*, to describe the process by which an individual makes sense of behavior (their own and others') as meaningfully predicated on underlying mental states such as thoughts, feelings, wishes, or intentions. This theory, grounded in psychodynamic theory, attachment research, and cognitive-affective neuroscience, proposes that the capacity to mentalize is developed within the context of our earliest attachment relationships. Given average, expectable conditions, the capacity to mentalize emerges over the course of development. Conversely, early childhood trauma or disruption in our attachment relationships may disrupt the development of the mentalizing capacity, leaving one vulnerable to further difficulties later in life.

Regardless of one's developmental trajectory and baseline ability to mentalize, the capacity is not static. Mentalizing theory asserts that activation of the attachment system derails the ability to reflect, leaving the individual to revert to mental operations that antedate the development of the mentalizing capacity [3]. Thus, the capacity is fluid and inhibited by anxiety aroused in close interpersonal relationships. While interpersonal disruptions and subsequent attachment anxiety are more prevalent in some groups than others, individuals with Borderline Personality Disorder being a prime example, such ruptures are ubiquitous to the human experience [4].

Mentalization-informed approaches to treatment are uniquely interested in anxiety aroused in interpersonal contexts. Because anxiety diminishes one's capacity to flexibly reflect, learn, and develop new insights, it must be intentionally addressed in treatment to restore robust mentalizing and maximize the potential for therapeutic gain. In mentalization-informed approaches, interventions made by the therapist are offered based on a careful assessment of the level of anxious arousal in the patient at a given moment. The more anxious the patient, the less likely they will be able to make use of interventions that would require robust mentalizing. For example, a therapist may be misguided in offering an expressive, interpretive intervention in these moments. Instead, supportive interventions may restore mentalizing capacities by reestablishing an empathic working alliance (akin to a secure base in attachment language), from which arousal might be diminished and the ability to achieve insight restored.

Future Directions: Unified and Transdiagnostic Approaches to Anxiety Symptoms

There is currently a growing trend toward transdiagnostic treatments based on the still relatively unsatisfactory outcome and remission rates for PDT for anxiety disorders. One promising example is the recently developed *Unified Psychodynamic Protocol for Anxiety Disorders* (UPP-ANXIETY) [25]. UPP-ANXIETY integrates empirically supported psychodynamic treatment mechanisms, including many of those mentioned in this chapter, across a wide range of anxiety and related disorders. In line with the unified, transdiagnostic approach, Pitman et al. [35] recently conducted the first study to systematically investigate whether anxiety symptoms respond to PDT techniques regardless of diagnosis, if the symptoms are at clinical or subclinical levels, or if they occurred in the presence of personality disorder pathology. The authors found significant decreases in anxiety symptoms across no anxiety disorder ($p = 0.004$) and total (anxiety disorder and no anxiety disorder) ($p = 0.0001$) samples, with improvements representing small to medium effects ($d = 0.48$ and 0.56, respectively). Results also showed several PDT techniques to be meaningfully related to anxiety reduction at outcome for the total sample, including (1) a focus on exploration of uncomfortable feelings ($r = 0.25$, $p = 0.03$); (2) linking current feelings and perceptions to experiences of the past ($r = 0.24$, $p = 0.04$); (3) focusing attention on similarities among the patient's relationships repeated over time, settings, or people ($r = 0.24$, $p = 0.04$); (4) identifying recurrent patterns in actions, feelings, or experiences ($r = 0.26$, $p = 0.02$); and (5) focusing on wishes, dreams, and early childhood memories ($r = 0.25$, $p = 0.03$). One technique, providing alternative understanding not previously recognized by the patient (i.e., interpretation), was meaningfully related to outcome for both the total sample and sample with no anxiety disorder (anxiety disorder sample, $r = 0.58$, $p = 0.01$; sample with no anxiety disorder, $r = 0.33$, $p = 0.02$; total sample, $r = 0.34$, $p = 0.003$). The authors suggested that these findings indicate there may be several PDT interventions that help with anxiety symptoms generally, and one intervention in particular that is helpful when anxiety is the specific focus of PDT.

Importantly, PDT techniques were found to be equally effective in reducing anxiety regardless of whether they presented with personality disorder pathology ($b = 0.414$, SE = 0.307, $t = 1.349$, $p = 0.185$). The fact that the PDT techniques identified as being effective for anxiety symptoms did not appear to be influenced or mitigated by Axis II diagnosis may provide further support for this type of broad, transdiagnostic, PDT approach. The authors suggested that practitioners utilizing PDT transdiagnostic protocols should feel confident in applying broad inclusion criteria when selecting eligible patients for this approach, provided that improvement on a targeted disorder or symptom (e.g., anxiety) is the primary goal of treatment.

In line with the Presidential Task Force on Evidence-Based Practice (2006), we believe it is crucial to consider the role of patient variables in affecting treatment outcome. Although presence of Axis II diagnosis alone did not moderate the relationship between PDT technique use and anxiety symptom improvement, this linear relationship was not maintained when the authors examined the presence of Axis I and II diagnoses together ($b = -0.548$, SE = 0.254, $t = -2.157$, p = 0.037). These findings suggest that practitioners applying a PDT transdiagnostic protocol should be aware that patients with co-occurring Axis I and II disorders may require a somewhat different configuration of PDT techniques in order for their anxiety symptoms to improve, in line with the flexibility advocated by Luborsky [29] and akin to the approach outlined in the UPP-ANXIETY [25] protocol.

Concluding Remarks

Psychodynamic therapists are fundamentally interested in the meaning and context for anxiety, actively attempting to foster curiosity in the patient about what the anxiety might signal. As such, anxiety might be thought of like the check engine light in a car. The light begs further investigation and careful assessment of the whole system. Finding a way to turn the light off without understanding the underlying problems it signals would be a partial solution at best. Rather than finding a way to quickly shut off the light, psychodynamic therapists are curious about the unseen disruptions that led to the signal in the first place. From this perspective, anxiety emerges when conflict reaches a fever pitch, often outside conscious awareness, eluding a coherent narrative in which it can be contextualized. The simmering conflict and related thoughts, feelings, and wishes that aren't expressed consciously then find their way to awareness in the form of anxious disturbance.

For therapists treating patients who report experiencing anxiety symptoms, current evidence suggests one effective, specific, approach to employ, especially early in treatment: the therapist should attempt to help the patient establish an understanding of his or her anxiety symptoms linked to the intrapsychic and interpersonal context in which symptoms occur. This effort includes helping the patient make connections between his or her current feelings and perceptions to past experiences, as well as across different times, settings, and relationships. Within this framework, the therapist can clarify for the patient the recurrent ways in which he or she internally experiences the actions, feelings, and occurrences that exacerbate anxiety symptoms.

A psychodynamic therapist accomplishes this task by focusing on helping the patient understand what it is he or she hopes to get out of their important relationships (i.e., the wish). From there, the therapist can help identify what the patient expects will be the response to that wish, and the characteristic way in which the patient in turn reacts to that perceived expectation. While these important clarifying issues should be ascertained early on, the therapist should continue to monitor these interpersonal patterns throughout the course of treatment, as they occur outside of therapy and also within the therapeutic relationship. In this manner, the therapist can help provide the patient with insight into the circumstances and processes that maintain his or her anxiety symptoms, give the space and perspective necessary to see how these patterns manifest in important relationships, and provide a safe and supportive environment in which to help the patient build up a tolerance for the difficult emotions that underlay the anxiety. Therapists at all levels of training and experience who utilize this model of treatment will likely find that providing this context for helping patients understand anxiety symptoms, and supportively engaging in the patient's relational world, is a beneficial and effective approach across many apparently different anxiety clinical presentations.

References

1. American Psychological Association (APA) Presidential Task Force on Evidence-Based Practice. Evidence-based practice in psychology. Am Psychol. 2006;61:271–85.
2. Baldwin SA, Imel ZE. Therapist effects: findings and methods. In: Lambert MJ, editor. Bergin and Garfield's handbook of psychotherapy and behavior change. 6th ed. New York: Wiley; 2013. p. 258–97.
3. Bateman AW, Fonagy P. Handbook of mentalizing in mental health practice. Arlington, VA: American Psychiatric Publishing; 2011.
4. Bateman A, Fonagy P. Mentalization based treatment for personality disorders: A practical guide. New York: Oxford University Press; 2016.
5. Blagys M, Hilsenroth M. Distinctive features of short-term psychodynamic–interpersonal psychotherapy: a review of the comparative psychotherapy process literature. Clin Psychol Sci Pract. 2000;7:167–88.
6. Blagys M, Hilsenroth M. Distinctive features of short-term cognitive–behavioral psychotherapy: a review of the comparative psychotherapy process literature. Clin Psychol Rev. 2002;22:671–706.
7. Bohart AC, Wade AG. The client in psychotherapy. In: Lambert MJ, editor. Bergin and Garfield's handbook of psychotherapy and behavior change. 6th ed. Hoboken, NJ: Wiley; 2013. p. 219–57.
8. Borkovec TD, Costello E. Efficacy of applied relaxation and cognitive-behavioral therapy in the treatment of generalized anxiety disorder. J Consult Clin Psychol. 1993;61:611–9.
9. Busch FN, Milrod BL, Singer MB, Aronson AC. Manual of panic focused psychodynamic psychotherapy–extended range (Vol. 36). Routledge; 2012.
10. Connolly Gibbons MA, Crits-Christoph P, Barber JP, Wiltsey Stirman S, Gallop R, Goldstein LA, Temes CM, Ring-Kurtz S. Unique and common mechanisms of change across cognitive and dynamic psychotherapies. J Consult Clin Psychol. 2009;77:801–13.
11. Crits-Christoph P, Connolly MB, Azarian K, Crits-Christoph K, Shappell S. An open trial of brief supportive-expressive psychotherapy in the treatment of generalized anxiety disorder. Psychotherapy. 1996;33:418–29.
12. Durham RC, Murphy T, Allan T, Richard K, Treliving LR, Fenton GW. Cognitive therapy, analytic psychotherapy and anxiety management training for generalised anxiety disorder. Br J Psychiatry. 1994;165:315–23.
13. Eagle MN. Attachment and psychoanalysis: theory, research and clinical implications. New York: Guilford Press; 2013.
14. Ferrero A, Pierò A, Fassina S, Massola T, Lanteri A, Daga GA, Fassino S. A 12-month comparison of brief psychodynamic psychotherapy and pharmacotherapy treatment in subjects with generalised anxiety disorders in a community setting. Eur Psychiatry. 2007;22(8):530–9.
15. Fonagy P. Attachment theory and psychoanalysis. New York: Other Press; 2001.
16. Fonagy P. The effectiveness of psychodynamic psychotherapies: an update. World Psychiatry. 2015;14(2):137–50.
17. Fonagy P, Gergely G, Jurist E, Target M. Affect regulation, mentalization and the development of the self. New York: Other Press; 2005.
18. Gabbard GO. Psychodynamics of panic disorder and social phobia. Bull Menn Clin. 1992;56(suppl 2A):A3–13.
19. Gabbard GO. Psychodynamic psychiatry in clinical practice. Washington, DC: American Psychiatric Press; 2000.
20. Kashdan TB, Barrios V, Forsyth JP, Steger MF. Experiential avoidance as a generalized psychological vulnerability: comparisons with coping and emotion regulation strategies. Behav Res Ther. 2006;44(9):1301–20.
21. Keefe JR, McCarthy KS, Dinger U, Zilcha-Mano S, Barber JP. A meta-analytic review of psychodynamic therapies for anxiety disorders. Clin Psychol Rev. 2014;34(4):309–23.
22. Keefe JR, Solomonov N, Derubeis RJ, Phillips AC, Chambless DL, Busch FN, Barber JP, Milrod BL. Focus is key: Panic-focused interpretations are associated with symptomatic improvement in panic-focused psychodynamic psychotherapy. Psychother Res. 2018;1–12.

23. Klein DF, Zitrin CM, Woerner MG, Ross DC. Treatment of phobias. II. Behavior therapy and supportive psychotherapy: are there any specific ingredients. Arch Gen Psychiatry. 1983;40:139–45.
24. Lambert MJ. The efficacy and effectiveness of psychotherapy. In: Lambert MJ, editor. Bergin and Garfield's handbook of psychotherapy and behavior change. 6th ed. New York: Wiley; 2013. p. 169–218.
25. Leichsenring F, Salzer S. A unified protocol for the transdiagnostic psychodynamic treatment of anxiety disorders: An evidence-based approach. Psychotherapy. 2014;51(2):224.
26. Leichsenring F, Beutel M, Leibing E. Psychodynamic psychotherapy for social phobia: A treatment manual based on supportive-expressive therapy. Bull Menn Clin. 2007;71(1):56.
27. Leichsenring F, Hoyer J, Beutel M, Herpertz S, Hiller W, Irle E, et al. The Social Phobia Psychotherapy Research Network (SOPHO-NET) – the first multi-center randomized controlled trial of psychotherapy for social phobia: Rationale, methods and patient characteristics. Psychother Psychosom. 2009;78:35–41.
28. Leichsenring F, Leweke F, Klein S, Steinert C. The empirical status of psychodynamic psychotherapy – an update: Bambi's alive and kicking. Psychother Psychosom. 2015;84(3):129–48.
29. Luborsky L. Principles of psychoanalytic psychotherapy. Manual for supportive-expressive treatment. New York: Basic Books; 1984.
30. Malan DH. The frontier of brief psychotherapy: an example of the convergence of research and clinical practice. New York: Plenum Medical Book Co; 1976.
31. McCullough L, Kuhn N, Andrews S, Kaplan A, Wolf J, Hurley C. Treating affect phobia: a manual for short term dynamic psychotherapy. New York: Guilford Press; 2003.
32. Milrod B, Busch F, Leon AC, Aronson A, Roiphe J, Rudden M, et al. A pilot open trial of brief psychodynamic psychotherapy for panic disorder. J Psychother Pract Res. 2001;10(4):239.
33. Mintz D, Belnap B. A view from Riggs: treatment resistance and patient authority—III. What is psychodynamic psychopharmacology? An approach to pharmacologic treatment resistance. J Am Acad Psychoanal Dyn Psychiatry. 2006;34(4):581–601.
34. Pitman SR, Slavin-Mulford J, Hilsenroth M. Psychodynamic techniques related to outcome for anxiety disorder patients at different points in treatment. J Nerv Ment Dis. 2014;202(5):391–6.
35. Pitman SR, Hilsenroth MJ, Weinberger J, Conway F, Owen J. Psychotherapy technique related to changes in anxiety symptoms with a transdiagnostic sample. J Nerv Ment Dis. 2017;205(6):427–35.
36. Plakun EM. Psychoanalytic treatment of anxiety disorders, obsessive-compulsive and trauma-related disorders. In: Kaplan & Sadock's comprehensive textbook of psychiatry, 10th edn. , Philadelphia: Walters Kluwer; 2017. p. 2666–2672.
37. Pompoli A, Furukawa TA, Imai H, Tajika A, Efthimiou O, Salanti G. Psychological therapies for panic disorder with or without agoraphobia in adults: a network meta-analysis. The Cochrane Library; 2016.
38. Slavin-Mulford J, Hilsenroth MJ. Evidence-based psychodynamic treatments for anxiety disorders: a review. In: Psychodynamic psychotherapy research. New York: Humana Press; 2012. p. 117–37.
39. Slavin-Mulford J, Hilsenroth M, Weinberger J, Gold J. Therapeutic interventions related to outcome in psychodynamic psychotherapy for anxiety disorder patients. J Nerv Ment Dis. 2011;199(4):214–21.
40. Stangier U, Von Consbruch K, Schramm E, Heidenreich T. Common factors of cognitive therapy and interpersonal psychotherapy in the treatment of social phobia. Anx Str Cop Int J. 2010;23:289–301.
41. Wampold BE, Mondin GW, Moody M, Stich F, Benson K, Ahn HN. A meta-analysis of outcome studies comparing bona fide psychotherapies: empirically, "all must have prizes.". Psychol Bull. 1997;122(3):203.
42. Wiborg IM, Dahl AA. Does brief dynamic psychotherapy reduce the relapse rate of panic disorder? Arch Gen Psychiatry. 1996;53(8):689–94.

24 Well-Being Therapy in Anxiety Disorders

Fiammetta Cosci

Introduction

Psychiatry and clinical psychology have been criticized for having had a strong "disease bias," which is related to negativity bias [1]. Indeed, when something negative has more impact on thoughts, emotions, and social interactions than something positive of equal relevance, we have a negativity bias [1]. In this framework, it has been proposed to identify and understand as much as possible those negativity biases not to let them command scientific and clinical activities [1]. As it has been pointed out [2], in the majority of treatments proposed for the most common psychiatric disorders, great attention has been devoted to stress and its consequences, and this has triggered an automatic and inexorable reinforcement of negative aspects of life.

One possible explanation of a so great attention of researcher on a negative way of seeing the world can be methodological. Indeed, humans and animals are more similar in their response to exposure to negative events than in their response to exposure to positive events [1], while the processing of reward shows important interindividual and interspecies differences as well as the predilections for positive things change considerably during the life-span [1]. Thus, since positive things are mutable, it is not easy to find an adequate method to assess predilections for them. In this vein, assessment and experimental methods allowing the comparisons within individuals, between individuals, and across species are more easily applied to research which study the stress system rather than to research which study the reward system.

It is also evident from the literature that the short-term negative events have more power in the capacity to development of a sense of meaning than the short-term

F. Cosci (✉)
Department of Health Sciences, University of Florence, Florence, Italy
e-mail: fiammetta.cosci@unifi.it

Y.-K. Kim (ed.), *Anxiety Disorders*, Advances in Experimental Medicine and Biology 1191, https://doi.org/10.1007/978-981-32-9705-0_24

positive aspects of life. This is extremely important, being the production of meaning a relevant part of clinical and scientific activity [1].

In everyday life we can observe the greater strength of negative events if compared to good events; some examples are major life events, such as trauma, but also events which might occur in the framework of close relationships, social network interactions, and interpersonal relationships, and also in learning processes. Experimental studies confirmed that negative events activate the meaning system more effectively than positive events [3].

In everyday life, negative events have stronger and more lasting consequences than positive events, although they present comparable relevance. Intimate relationships are more deeply troubled by destructive actions than by constructive ones, by conflict than by accord, and by negative communication than by positive communication. In the same vein, fighting or sour relationships showed greater negative effects than friendly and amicable relationships. Negative moods and emotions seem to mark the cognitive processes more strongly than positive mood and emotions. Indeed, the largest efforts of people are usually aimed at escaping from negative mood rather than at entering in or empowering positive mood [3].

In this vein, words describing bad emotions are more numerous than words describing good emotions. This opposes the greater frequency of good emotions that people may live and suggests that bad emotions have more power than good emotions.

According to the literature on patterns of learning, bad things are more rapidly and successfully learned than good things as well as negative events have effects much more lasting and important than positive or neutral events; this is confirmed also by the fact that there are no words and concepts to describe the opposite of trauma. For instance, bad parental attitudes can be stronger than genetic influences, but good parental attitudes are not. Studies focusing on social support found that people presenting negative or conflictual behaviors with their friends and peers have stronger negative effects than people acting via positive and sympathetic behaviors. In addition, bad things receive more attention and mark more strongly cognitive processing than good things; negative stereotypes and bad reputation are easier to be acquired and harder to abandon than positive stereotypes and good reputation; bad feedbacks have stronger effects than good feedbacks; poor health has a more considerable impact on well-being than good health; and health per se is more damaged by pessimism than by optimism [3]. Generally speaking, people appear to be more strongly motivated to avoid the bad rather than to receive the good.

Negativity bias may also be related to the sense of lack of social connectedness. For instance, talking about negative events leads to more interpersonal confidentiality than talking about own future projects [4]. There are also other social factors to be considered: people who talk about the bad are perceived as more capable and thoughtful than those talking about the good. Experimental studies showed that slight reminders of death increased the preference for a charismatic leader and decreased the preference for a relationship-oriented leader [5].

Fortunately, exceptions which indicate greater strength of good things on bad things can still be found in the literature. Examples are when positive information is found to be more diagnostic than negative information [6] or in anticipation of future events; thus the dominance of optimism on pessimism about the future reflects a higher power of good future events on bad future events [7]. Boucher and Osgood [8] noted that there is a common human tendency to use evaluatively positive words more frequently than evaluatively negative words; similarly, Matlin and Stang [9] highlighted that negative words consist more often of a positive root that becomes negative by a prefix (e.g., unfavorable) rather than the other way around. In addition, when people are instructed to attribute traits to a target person, they tend to use more positive traits than negative traits, and the proportion is 62% for positive versus 38% for negative traits [10, 11].

There are also some studies suggesting that positive and negative events have different effects on memory; people tend to remember positive events more than negative events [12].

Finally, there are individual differences in the level in which people are oriented toward good and bad things. Healthy subjects look for positive information and shrink negative information more than depressed subjects in several domains which include optimism regarding the future [13] and recall bias for positive information [14].

Because of the important influence of individual and cultural aspects on positive health, measuring well-being is an indispensable challenge, and the application of a therapy which might enhance well-being is fundamental.

An extensively speculative literature has addressed the meaning of positive psychological functioning. This literature includes Maslow's view of self-actualization [15], Rogers' conceptualization of the fully functioning person [16], Jung's notion of individuation [17], and Allport's definition of maturity [18]. Another domain of theory for defining psychological well-being comes from life-span developmental perspectives which emphasize the differing challenges faced in the different ages of life [19–21]. Jahoda's positive criteria of mental health, generated to replace previous definitions of well-being which were mainly focused on the absence of the illness, offer a great description of what good psychological health means [22]. More recently, Paul Dolan suggested that well-being is composed of pleasure and purpose [23]. This conceptualization reflects the fact that we usually do not want to just feel good but to feel that we are doing good. Thus, we need to balance and combine these two aspects to enhance well-being; more precisely, we should balance our psychic forces in order to acquire a unifying outlook on life which might guide our actions and feelings with the aim of shaping our future accordingly [22].

Well-Being Dimensions

Well-being therapy (WBT) is a short-term psychotherapeutic intervention which is based on the model developed in 1958 by Marie Jahoda [22] (Fig. 24.1). In her model, Marie Jahoda outlined six criteria for describing positive mental health:

(1) Autonomy that is regulation of behavior from within

Fig. 24.1 Jahoda's model of psychological well-being [22]

(2) Environmental mastery that is the capacity to manage the environment in a way which allows to get benefits from it
(3) Satisfactory interactions with others and the milieu, which refer to interpersonal and intimate relationships
(4) The individual's style and degree of growth, which includes development and self-actualization
(5) The attitudes of an individual toward his own self, which means also self-perception and self-acceptance
(6) The individual's balance and integration of psychic forces, which refer to the concept of psychological flexibility

This latter dimension represents a unifying outlook on life that should guide individual actions and feelings for shaping future accordingly. It also includes resistance to stress, which has been defined over the time as resilience and tolerance to anxiety and frustration. This sixth dimension is not purely a generic and clinically futile advice of avoiding excesses and extremes; it is how the individual adjusts his own psychological dimensions of well-being to face changing needs [24].

Ryff and Singer [25] further elaborated the first five dimensions of positive functioning proposed by Marie Jahoda (Table 24.1) and proposed their conceptualization as follows:

- Self-acceptance: this is a central aspect which implies self-actualization, optimal functioning, and maturity. Adult developmental stage theories have strongly stressed the importance of acceptance of one's self and one's past life. Thus, having positive attitudes toward oneself emerges as a central characteristic of positive psychological functioning [25].
- Positive relations with others: many theories supported the importance of warm and trusting interpersonal intimate relationships, and the ability to love is

Table 24.1 The six dimensions of psychological well-being according to Ryff's model

Dimensions	High scores	Low scores
Autonomy	The subject is autonomous and independent; he can resist social pressures; he regulates behavior from within; he evaluates self by personal references	The subject is worried about the opinions, the expectations, and the point of view of others; he is influenced by judgment of other people to make relevant decisions; he conforms to social pressure
Self-acceptance	The subject has a positive attitude toward the self and knows and accepts the various aspects of self including good and bad qualities; he has positive feelings about the past	The subject is dissatisfied with self and is disappointed with the past; he is uncomfortable with specific own personal qualities; he would like to be different
Environmental mastery	The subject has mastery and competence in managing the environment; he is able to manage a complex pattern of activities; he uses effectively surrounding opportunities; he can choose or build context patterns which are appropriate for his personal needs and values	The subject has difficulties in managing everyday environment; he feels to be unable to change or ameliorate the surrounding environment; he is not aware of surrounding opportunities and has no sense of control over the external activities
Purpose in life	The subject has purposes in life and knows the direction of his life; he sees and perceives the meaning of life; he has beliefs that support life purposes and has aims and objectives for living	The subject has a poor sense of meaning in life; has few purposes or aims, no sense of direction of his life; does not feel and see goals in past life; and has few or no beliefs that give life meaning
Positive relations with others	The subject has intimate, positive, trusting relationships with other people; is worried about well-being of people close to him; and can feel deep empathy, love, and intimacy toward others; he is able to have relationships with others	The subject has limited close relationships with others; he finds it difficult to be open and close to others; he is alone, isolated, not willing to sustain close relationships with others
Personal growth	The subject has a feeling of continued growing; he sees self as expanding and continuously changing; he is open to novelties; he sees improvements over time and changes in a way that reflects self-knowledge and effectiveness	The subject has a sense of personal immobility and no or poor sense of improvement over time; he is bored and not interested in life; he had the perception not to be able to develop new projects, activities, and behaviors

Adapted from [26]

considered a central component of mental health. This means having strong feelings of empathy and love for the human real and being capable of deeper affect, closer friendship, and more complete connection with others. Life-span developmental theories also emphasized the achievement of close unions with others, the so-called intimacy, and the guidance and direction of others, the so-called generativity [25].
- Autonomy: this dimension refers to individual qualities such as self-determination, independence, and regulation of behavior from within. It means autonomous functioning, resistance to enculturation, and having an internal locus of evaluation, which means evaluating oneself by personal standards. Autonomy also includes emancipation from convention, in which the person no longer clings to the collective fears, beliefs, and laws of the mass. This may give the person a sense of freedom from the norms governing everyday life [25].
- Environmental mastery: this is the individual's ability to choose or create environments suitable to own psychic conditions; it also means having the ability to manipulate and control complex environments. This conceptualization emphasizes people's ability to progress in the world and change it creatively through mental and physical engagement. Successful aging also stresses the extent to which the individual takes advantages of environmental opportunities [25].
- Purpose in life: this dimension is characterized by having the feeling that there is a purpose, an aim, and a meaning in life which includes a sense of directedness, intentionality, and a clear comprehension of life's goals [25].
- Personal growth: it requires that individuals continue to develop their potential to grow and expand as persons; therefore, a central position is devoted to the need to actualize one's self and realize one's potential. In this vein, openness to experience becomes a key characteristic of the fully functioning persons. Such individuals are continuously developing, rather than achieving a fixed state, which also means continuous growing and facing new challenges [25].

Ryff also introduced a method for the assessment of these five dimensions, the so-called psychological well-being scales [26].

Although initially WBT was simply aimed at increasing psychological well-being, its goal was subsequently refined in the achievement of a state of euthymia, which corresponds to Jahoda's sixth dimension [22]. A specific conceptualization of euthymia has been recently proposed by Fava and Bech [24] who identified the following peculiar characteristics: (a) absence of mood disturbances that can be diagnosed according to the current nosography; if the subject has had at least one prior mood disorder episode, it should be in full remission. If sadness, anxiety, and irritable mood are present, they should be short-lived and related to specific situations and do not significantly affect everyday life functioning. (b) The subject feels cheerful, calm, active, and interested in things, and his sleep is refreshing or restorative. (c) The subject has a balance and integration of his own psychic forces (the so-called psychological flexibility), which means having a unifying outlook on life which guides actions and feelings to shape the future accordingly. The subjects also show resistance to stress (the so-called resilience and tolerance toward anxiety or frustration). In this framework, Fava and Bech [24] proposed a 10-item self-administered scale assessing euthymia according to the above definition; the euthymia scale has been recently validated by our group [27].

Well-Being Therapy

The well-being therapy [2, 28, 29] is a short-term psychotherapeutic intervention which is based on self-observation, via the use of a structured diary, patient-therapist interaction, and homework. WBT is based on the model of psychological well-being developed in 1958 by Marie Jahoda [22] and later refined by Ryff and Singer [25] who identified five dimensions of positive functioning. Ryff and Singer [25] noted that over the time, mental health research has been too influenced by the concept of psychological dysfunction and that health has been erroneously equated to the absence of illness rather than to the presence of wellness. Ryff and Singer [25] thus suggested that the absence of well-being might create the conditions of vulnerability to possible future adversities and that the route to recovery should lay not exclusively in alleviating the negative but in creating or empowering the positive. More recently, WBT has been further refined via the definition of the concept of state of euthymia which mirrors to Jahoda's sixth dimension of balance among psychic forces [22].

WBT Structure and Features

Well-being therapy may be used as a only therapeutic strategy or as a second-line treatment strategy. In case it is used as a only therapeutic strategy, the number of sessions ranges from 8 to 20, and the duration of each session ranges from 45 min to 60 min. In case WBT is used as a second-line treatment, it should be used in sequential combination with other psychotherapeutic strategies, mainly cognitive behavioral therapy (CBT) since the sequential combination of CBT/WBT has been largely studied so far [2]. When WBT is used as a second-line treatment, the number of sessions may be reduced to 4–6 [2].

The initial phase of WBT is based on patient's self-observation of situations in which he feels to experience well-being. As soon as the patient is able to recognize adequately these instances of well-being, he is encouraged to identify thoughts, beliefs, and behaviors which might lead to premature interruption of well-being. This is the intermediate phase. The final phase involves cognitive restructuring of dysfunctional dimensions of psychological well-being; thus the clinician invites the patient to criticize the content of thoughts and beliefs which might lead to premature interruption of well-being as well as to criticize the behaviors which might lead to premature interruption of well-being, observing such thoughts, believes, and behaviors from a different point of view. In all the phases, the patient is also encouraged to look for optimal experiences, increase the probability to live optimal experiences in the daily life, and meet the challenge that optimal experiences may entail [2]. The main phases of WBT are summarized in Fig. 24.2.

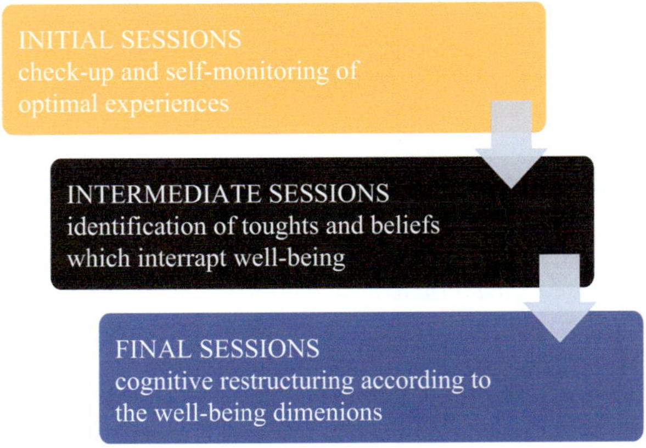

Fig. 24.2 Well-being therapy main phases [28]

WBT was originally conceived as an individual therapy [30], but it may be delivered also in a group format [31]. This modality may increase sharing optimal experiences and personal meanings of psychological well-being. Based on Kauffman and Silberman's [32] observation, positive psychology interventions were found to be suitable also for couple therapy and able to improve the outcomes; thus, it is also hypothesized, although not yet tested, that WBT may increase the effectiveness of couple and family interventions.

WBT can be differentiated from positive interventions [33] on the basis of the following features: monitoring of psychological well-being in a diary; identification of low tolerance to well-being by seeking automatic thoughts; behavioral exposure; cognitive restructuring using specific psychological well-being models; and individualized and balanced focus. These features are here described in detail:

1. Monitoring of psychological well-being in a diary: patients are encouraged to identify episodes in which they feel to experience well-being and describe in the diary the situational context in which they occur. In practice, they are asked to report in the structured diary the circumstances surrounding their episodes of well-being and rate the level of psychological well-being experienced in such circumstance based on a 0–100 scale, where 0 means absence of well-being and 100 means the most intense well-being ever experienced. The patient is informed that well-being can be a feeling which follows the break from distress, anxiety, anguish, worry, sadness, hopelessness, or similar; well-being might also be related to a specific experience here named "optimal" experience. Optimal experiences are characterized by clear aims, immediate feedback, high challenges matched with adequate personal skills, a combination of action and awareness, concentration on the task that is taking place, perceived control of the situation, loss of self-consciousness, altered sense of time, and intrinsic motivation [34].
2. Once the patient is able to recognize the instances of well-being, he is encouraged to identify those situations in which there is low tolerance to psychological well-being. Thus, the therapist encourages the patient to seek automatic thoughts and beliefs which lead to premature interruption of well-being. These thoughts

are automatic and must be managed as it is performed in cognitive therapy [35]; the focus of self-observation is, however, based on well-being instead of distress.

3. The therapist also supports and encourages activities which might trigger or empower well-being as well as optimal experiences. Thus, behavioral exposure [35] (for instance, assigning the task to undertake specific pleasurable activities) is proposed as a reinforcement. Meeting the challenge that optimal experiences may require is emphasized since it is through this challenge that growth and improvement of self can be achieved.
4. The therapist verifies whether specific impairments or excessive levels in well-being dimensions, according to Jahoda-Ryff's conceptual framework, occur [22, 26]. This is possible by means of the monitoring of the course of episodes of well-being. At this point, the patient should be able to adequately identify moments in which he feels to experience well-being; should be aware of those interfering thoughts, beliefs, or behaviors which interrupt the feelings of well-being; should be able to use cognitive behavioral techniques to address these interruptions; and finally should have learned strategies to pursue optimal experiences.
5. Contrary to what positive interventions promote [33], patients are not encouraged to seek the highest possible level of psychological well-being, which is illustrated in the sixth dimension model elaborated by Jahoda and Ryff [22, 26]. On the contrary, patients are encouraged to obtain a balanced functioning, which can be subsumed under the rubric of state of euthymia [24]. This optimal-balanced well-being can be different from patient to patient, according to factors such as personality traits, social roles, and cultural and social contexts [24].

WBT Framework

Unlike many other psychotherapeutic interventions, WBT was conceived as a therapeutic tool which should be incorporated in a therapeutic plan rather than as an only treatment for mental disorders. Thus, well-being therapy should be applied as a second- or third-line intervention and not as the first-line treatment of an acute psychiatric disorder. Second- and third-line interventions are nowadays precious for clinicians since most of the patients have complex and chronic disorders and always more often need second- and third-line interventions [36]. In this vein, WBT has been tested in several controlled trials but mostly as an additional treatment ingredient of CBT.

WBT was also conceived to be used in patients whose clinical case is formulated via the clinical reasoning and via the use of macroanalysis [37]; it was not conceived to be used in every patient meeting certain diagnostic criteria [2].

Macroanalysis consists of establishing a relationship among individual's co-occurring syndromes and problematic areas. Thus, macroanalysis should not be applied only to diagnostic entities, as in the conceptualization of comorbidity which is proposed by the current nosography (for instance, the *Diagnostic and Statistical*

Manual of Mental Disorders or the International Classification of Diseases), but it should be applied to any clinical manifestations and any ailments that are referred as problematic by the patient and are judged clinically relevant by the clinician since they affect the patient's life, although they do not reach the diagnostic threshold or are not mentioned in the current nosographic systems. Macroanalysis is based on the assumption that, in most of the cases, there are functional relationships among different patient's problematic areas and that the targets of treatment may vary according to the problematic area taken into account as well as during the course of the disturbances.

There are four steps within the process of macroanalysis:

1. Detecting the complexity of the clinical case: macroanalysis assumes a detailed and systematic collection of the different ailments, clues, or problems affecting the patient's life independently from the fact they are included or not included in a classificatory system. This collection of data is run by the clinician from the first meeting with the patient until the end of the treatment since new information can always come as the clinician-patient relationship develops and becomes stronger [37].
2. Establishing a relationship among the problematic areas which affect the patient's life: macroanalysis starts from the assumption that, in most of the cases, there are functional relationships among problematic areas. Such functional relationships are usually based on the primary/secondary onset as well as on the predominance of a more severe clinical picture on another [37]. The relationships established allow the clinician to hierarchically organize the problematic areas.
3. Choosing the first clinical target and the related treatment: according to the relationships established among the patient's problematic areas and according to their hierarchical organization, the first target of treatment is chosen. Of course, further treatments may be needed, and they may vary during the course of the disturbances. Usually, some treatments are postponed after the outcome of the first-line treatment is measured. The hierarchical organization that is chosen may depend on a variety of factors such as the urgency of symptoms (i.e., severity of distress or functional impairments), the availability of treatment tools, and the patient's preferences and priorities. [37–39]
4. Choosing further targets and related treatments: after the first-line treatment is completed, the outcome of the first-line treatment is measured, and a new macroanalysis is performed. On the basis of this new macroanalysis, different targets of treatments might be chosen by the clinician as well as related different types of treatments [37].

While macroanalysis involves an assessment of the relationship between co-occurring clinically relevant problematic areas, microanalysis is a detailed analysis of symptoms for functional assessment [37]. It involves consideration of the onset of complaints, their course, circumstances that worsen or improve symptoms, short-term and long-term impact of symptoms on quality of life, and work and social adjustment [37]. Microanalysis may also include specific tests and rating scales which must be integrated into the rest of the assessment and not viewed in isolation [37]. The euthymia scale as well as the PWB scales might be very helpful when psychological well-being and WBT are considered.

Staging is another key tool to formulate the case via the clinical reasoning. It allows to reason in terms of longitudinal development of a disorder and in terms of treatment response [40]. Indeed, staging allows an assessment of the longitudinal development of mental disorders defining where a person is currently along the continuum of the course of illness which includes prodromes (i.e., early signs and symptoms that differ from the acute clinical phase), the acute phase of the disease, residual symptoms (i.e., persistent signs and symptoms despite apparent remission or recovery), and in some cases, chronicity (i.e., signs and symptoms become chronic in an attenuated or in a severe level of severity). More specifically, stage 1 is the prodromal phase, defined as the time interval between the onset of prodromal symptoms and the onset of the characteristic manifestations of the fully developed acute illness [40]. After the acute phase (stage 2), it might be difficult to assess whether partial or full remission has occurred; attenuated symptoms, the so-called residual symptoms, might be observed (stage 3); they are due to partial persistence of the disorder or to the worsening of a pre-existing abnormal personality trait. Stage 4 represents chronicity of the psychiatric disorder [40].

Staging differs from the conventional diagnostic practice which only defines the extent of progression of a disorder at a particular point in time.

Staging models have been proposed for mental disorders such as schizophrenia, unipolar depression, bipolar disorder, panic disorder, substance use disorders, and eating disorders [40].

Staging can be used to reason in terms of longitudinal development of a disorder and in terms of treatment response also thanks to the model provided by Detre and Jarecki [41], the so-called rollback phenomenon. The rollback phenomenon establishes a relationship between residual and prodromal symptoms since as the illness remits, it progressively recapitulates in the reverse order. This means that certain prodromal symptoms may be overshadowed by the acute manifestations of the mental disorder and persist as residual symptoms after the acute phase of the illness fades away. In most of the cases, such residual symptoms can also progress to become prodromes of the following relapse. If this is the case, prodromal symptoms of relapse tend to mirror those of the initial episode [30]. According to the rollback phenomenon, there is also a temporal relationship between the time of the development of a disorder, that is, the time in which prodromal symptoms become more and more severe and finally reach the level of severity of the acute manifestations, and the duration of the phase of recovery, that is, the time in which symptoms are no more acute but persist at an attenuate level of severity. Indeed, these two times tend to be similar in duration.

Table 24.2 reports the staging model proposed for panic disorder. This model as a peculiarity: it does not only include the clinical features attributed to panic disorder according to the current nosography; on the contrary it includes clinical features which can be observed in what the current nosography identifies as different

Table 24.2 Staging of panic disorder

Stage 1	Prodromal phase: subclinical symptoms of agoraphobia and/or social phobia and/or generalized anxiety disorder and/or hypochondriasis
Stage 2	Acute manifestations of agoraphobia and/or social phobia and/or generalized anxiety disorder and/or hypochondriasis
Stage 3	Panic disorder with worsening of anxiety and hypochondriacal symptoms demoralization and/or major depression may occur
Stage 4	Chronic panic disorder and agoraphobia and/or social phobia and/or generalized anxiety disorder and/or hypochondriasis (in attenuated or persistent form). Increased liability to major depression

Adapted from [40, 44]

diagnostic entities. Thus, this staging model is named staging of panic disorder, but, in essence, it is the staging of anxiety disorders.

Stage 1 is characterized by subclinical manifestations of agoraphobia, social phobia, generalized anxiety disorder, and/or hypochondriasis. If we use the current nosography terminology, we might say that at this stage, we have subthreshold manifestations of agoraphobia, social phobia, generalized anxiety disorder, and/or hypochondriasis.

Stage 1 was proposed on the basis of the literature findings which suggested that:

(a) Truly spontaneous attacks, not preceded by anxiety-provoking cognitions, are uncommon
(b) Patients meeting the diagnostic criteria for panic disorder tend to suffer from the whole range of anxiety disorders
(c) A unique relationship between panic disorder and agoraphobia cannot be established
(d) Other diagnoses, particularly social phobia and generalized anxiety disorder, frequently predate the onset of panic and agoraphobia [42]

Stage 2 is characterized by the acute manifestations of agoraphobia, social phobia, generalized anxiety disorder, and/or hypochondriasis. Thus, the onset of the acute manifestations of panic disorder is not identified when panic attacks occur, as suggested by the current nosography, but when anxiety disorders (agoraphobia or social phobia or generalized anxiety disorder) or hypochondriasis develop and are florid. The stage 2 thus allows to formulate the diagnosis according to criteria which differ from the current nosography; the direct implication is higher diagnostic sharpness and lower risks of using undifferentiated treatment, which may leave substantial areas of non-response. A further direct clinical implication is that undifferentiated treatments are substituted with stage-guided therapeutic interventions [42].

Stage 3 is characterized by panic attacks with worsening of anxiety and hypochondriacal symptoms; demoralization and/or major depression may also occur. This means that panic attacks are not the acute manifestation of panic disorder but the expression of a development toward chronicity. Already in 2008, Fava et al. [42] suggested to consider panic disorder in a longitudinal way which implies that there is a development of anxiety disorders and hypochondriasis instead of a specific disease. In addition, stage 3 allows to overcome the concept of comorbidity between panic disorder and major depression, which is proposed by the current nosography, since major depression, as well as demoralization, is an integral part of stage 3 of the illness. Similarly, also the concept of comorbidity between panic disorder and hypochondriasis is superseded, not via the deletion of the diagnosis of hypochondriasis (as proposed by the *Diagnostic and Statistical Manual of Mental Disorders*) but via the inclusion of hypochondriasis as a clinical characteristic of panic at different stages of its development.

Stage 4 is characterized by chronic panic disorder and agoraphobia and/or social phobia, generalized anxiety disorder, and hypochondriasis. The latter ones can be in an attenuated or in a persistent form. The patient also has increased liability to develop major depression. Also stage 4 allows to supersede the concept of comorbidity which, in this framework, becomes useless from a clinical point of view.

This model, widely supported in the literature [40], has the advantage to provide a monitoring of the patient's progress [43]. Because of the chronic nature of panic disorder, the emphasis of treatment has been more and more shifted to long-term outcome, and disappearance of all symptoms may become the final target of therapy since they constitute a substantial risk of relapse [44]. This has been clearly highlighted in the long-term outcome of patients with panic disorder treated with behavioral interventions [45]. Another interesting proposal could be the stage-oriented therapy, that is, a specific therapy planned to treat a specific stage of the illness to produce additional benefits.

The Role of WBT in the Treatment of Anxiety Disorders

Generalized Anxiety Disorder

A seminal study was performed in patients with a diagnosis of generalized anxiety disorder [46]. Twenty patients were randomly assigned to eight sessions of CBT or to the sequential administration of CBT (four sessions) and WBT (four sessions). Patients who received the two types of treatments presented a significant reduction of the level of anxiety; however, the CBT/WBT sequential combination produced significant advantages over CBT in terms of symptom reduction and in terms of psychological well-being improvement, as measured by the Clinical Interview for Depression [47] and the PWB scales [26], respectively. Thus, the findings suggested that the sequential combination of CBT and WBT was able to increase the level of recovery, and these results disentangled the feasibility and clinical advantages of adding WBT to the cognitive behavioral treatment of generalized anxiety disorder. Of course, the decrease of anxiety, as assessed via the Clinical Interview for Depression, could be also due to the fact that self-monitoring the episodes of well-being may have empowered the identification of automatic thoughts much more than what happens with the customary monitoring of episodes of distress in cognitive therapy. Therefore, the cognitive restructuring might have been more effective.

In addition, the level of psychological well-being, as assessed via the PWB scales, also increased in those patients who were treated with CBT alone; however, the degree of such an increase was lower than the degree of the increase observed in those who received the CBT/WBT combination, suggesting a specific mechanism for WBT. Both aspects support a potential role for WBT in increasing the level of recovery.

Panic Disorder

Another seminal study showed, for the first time, the potential positive effects of WBT in panic disorder and agoraphobia patients. Six subjects with a diagnosis of agoraphobia and panic attacks underwent a controlled trial with a crossover design which included three different interventions (i.e., exposure alone, cognitive therapy with exposure, exposure associated with imipramine) [48]. At the end of the trial, they were still suffering from panic attacks; thus well-being therapy was offered. WBT was delivered associated with the prolongation of exposure in vivo homework. Three out of the six patients accepted the WBT intervention [30], and, interestingly, two of the three patients achieved a panic-free status after it.

It is of course extremely difficult to draw conclusions from this trial since the sample under study was very small and only half of the patients who were still suffering from panic attacks and were offered the well-being therapy were examined. A placebo effect, thus a nonspecific effect, could be claimed although it seems quite unlikely to see a placebo effect in patients who had unsuccessfully undergone three consecutive trials. In addition, it should be noted that the controlled trial disclosed a significant effect of the time factor [48]; thus the results might have been due to the prolongation of the exposure in vivo homework. However, it is also possible that WBT helped the two patients undergoing exposure because it increased their compliance to exposure homework; this, indeed, appeared to be the case according to the therapist's ratings [48].

Cosci [49] described the case of a patient with panic disorder, agoraphobia, and a major depressive episode who failed to respond to paroxetine and CBT and successfully responded to WBT. This patient was unable to identify automatic thoughts by monitoring distress with cognitive therapy, whereas she was able to do so while monitoring well-being with WBT. Interestingly, after WBT she was able to complete also the cognitive therapy [49].

These results would not, of course, justify the use of WBT in panic disorder patients. However, a clinical justification for such a use might be based on the comparison between patients with panic disorder and agoraphobia and healthy controls, matched for sociodemographic variables, which showed lower levels of psychological well-being in remitted panic disorder patients treated with WBT than in healthy subjects [48].

There are also some suggestions that WBT can play a role in drug treatment discontinuation in anxiety disorders such as panic [50]. Finally, future research may disclose whether WBT has a place in treatment resistance in anxiety disorders.

Drug Tapering and Discontinuation

In the last decade, a progressive change in the pharmacotherapy of anxiety disorders has occurred. Selective serotonin reuptake inhibitors (SSRIs), originally approved for the treatment of depressive disorders, have been approved also for the treatment of anxiety disorders, despite the lack of sufficient evidences. The major implication was that benzodiazepines were progressively replaced by antidepressants, mainly SSRIs, venlafaxine, and duloxetine, which are now proposed as the first-line treatment of anxiety disorders. Despite the shift of guidelines from benzodiazepines to antidepressants, benzodiazepines have still unquestionable benefits and are widely prescribed by clinicians. Apparently, a campaign against benzodiazepines has been turned on and has been mainly based on the risk of abuse of these drugs. In reality, benzodiazepines abuse is low, especially for certain benzodiazepines, and usually occurs in co-occurrence of other substance use disorders [51]. Notwithstanding this, the literature has gradually been inflated by articles on negative properties of benzodiazepines. And even though many of these publications were not based on good science or were biased, on one hand they activated suspicion on benzodiazepines; on the other hand they suggested difficulties in using them, invariably denying their benefits [51, 52].

SSRIs, venlafaxine, and duloxetine have very frequent and specific steps [53]. They include withdrawal reactions that have been disguised as discontinuation syndromes with the aim to avoid any clue of a potential for dependence from SSRIs, venlafaxine, and duloxetine that may affect marketing. These withdrawal syndromes may occur during drug reduction or discontinuation and are characterized by a broad range of somatic symptoms (e.g., headache, dizziness, fatigue, diminished appetite, sleep disturbances, flu-like symptoms) and psychological manifestations clinically relevant (e.g., agitation, anxiety, dysphoria, and confusion) [50, 54]. These reactions are particularly pronounced with paroxetine and venlafaxine, they may persist for months or even years after medication discontinuation, and they might lead to what have been identified as persistent post-withdrawal disorders [55].

Three types of withdrawal syndromes which might occur after antidepressant reduction or discontinuation have been described: new withdrawal symptom, rebound, and persistent post-withdrawal disorders [55].

New withdrawal symptoms are withdrawal symptoms that are not part of the patient's original illness; they are associated with antidepressant decrease, discontinuation, or switch [55]. They can occur up to 6 weeks after drug discontinuation or dose reduction, depending on elimination half-life of the drug. New withdrawal symptoms are transient, reversible with complete recovery, relatively short-lasting (from a few hours to a few days); usually they last less than 6 weeks and rarely persist for weeks and months. Some common new withdrawal symptoms are nausea, headaches, tremor, sleep disturbances, decreased concentration, anxiety, irritability, agitation, aggression, depression, and dysphoria [55].

Rebound symptoms are a rapid return of patient's original symptoms at greater intensity than before treatment. They can occur up to 6 weeks after drug discontinuation or dose reduction, depending on elimination half-life of the drug. Rebound symptoms are usually transient, reversible with complete remission, relatively short-lasting since they usually last less than 6 weeks, and rarely persist for weeks and months. [55]

Persistent post-withdrawal disorders occur after several years of continuous antidepressant use (1–4 years for SSRIs or for serotonin-noradrenaline reuptake inhibitors – SNRIs) [50, 56]. SSRI/SNRI dose is usually not considered relevant, but the length of treatment is. Persistent post-withdrawal disorder entitles return of the original illness after drug discontinuation or dose reduction at a greater intensity and/or with additional features of the illness and/or emergence of new psychiatric disorders. Unlike the other withdrawal syndromes, persistent post-withdrawal disorders usually persist at least 6 weeks (but they may last months and even years) and are severe [55].

Fava and colleagues explored the prevalence of withdrawal symptoms occurring after gradual tapering of SSRIs in patients with panic disorder and agoraphobia [57]. The conditions could be judged as optimal since all patients had fully remission upon behavioral treatment and were psychologically prepared to drug reduction and discontinuation. Nine of the 20 patients (45%) experienced a withdrawal syndrome that faded away within a month in all but 3. These three patients all received paroxetine and displayed alternation of worsened mood, fatigue, and emotional instability with trouble sleeping, irritability, and hyperactivity [57]. Thereafter, Belaise and co-workers analyzed online self-reporting from a variety of websites visited by patients who had discontinued SSRI antidepressants and were reporting spontaneously significant withdrawal symptoms and post-withdrawal psychopathology that they attributed to discontinuation of their SSRI [56]. Paroxetine ($n = 3$), sertraline ($n = 2$), citalopram ($n = 2$), fluoxetine ($n = 1$), fluvoxamine ($n = 1$), and escitalopram ($n = 3$) were self-reported by patients as the cause of withdrawal symptoms when discontinued. In addition, 58% of patients (7 out of 12) reported persistent post-withdrawal symptoms: 3 of 3 paroxetine patients, 2 of 2 citalopram, 1 of 1 fluvoxamine, 1 of 3 escitalopram, and none of both sertraline and fluoxetine patients.

A sequential combination of CBT/WBT that consists of 6–16 weekly sessions was proposed [50] to treat persistent post-withdrawal disorders induced by antidepressant reduction or discontinuation. The protocol consists of:

(1) Explanatory therapy, which includes providing accurate information on withdrawal, repeated reassurance, and teaching the physiological principles underlying withdrawal phenomena.
(2) Monitoring of emergent symptoms in a diary according to the cognitive behavioral model, followed by cognitive restructuring consisting of alternative interpretations of patient thoughts about his symptoms.
(3) Homework exposure for avoidance patterns. For example, the patient, who thinks that without the antidepressant, he cannot go out with friends, is encouraged to go out facing social situation of increasing level of difficulty.

(4) Lifestyle modifications. They include avoidance of alcohol, increased physical exercise, limited caffeine consumption, and no change in cigarette smoking habits.
(5) Techniques of decreasing abnormal reactivity to the social environment, consisting in learning ways to cope with stressful situations related to the level of arousal increased by drug withdrawal.
(6) Teaching well-being therapy to clarify that there is life after antidepressants.

More recently, Fava and Belaise [58] illustrated more extensively which are the specific circumstances to discontinue antidepressants (i.e., medical reasons, pregnancy and breast feeding, paradoxical effects, lack of loss of efficacy of antidepressants, unclear reasons of initial prescription, clinical improvement, patient preference), the pharmacological strategies to apply, and the psychotherapeutic management. This latter includes three modules.

The first module is based on explanatory therapy and includes 4–6 sessions. The second module is based on reassessment and cognitive behavioral treatment and counts 6–10 sessions. The third module is based on WBT and includes 6–8 sessions. Indeed, many patients are convinced that they cannot make it without antidepressants, because they are weak and vulnerable [2]. The difficulties in discontinuing the antidepressant may lead to impairments in various dimensions of psychological well-being: environmental mastery ("I cannot work without antidepressants"); autonomy ("I am ill. I cannot do things I did in the past"); purpose in life ("There is no life after antidepressants"); positive relations with others ("Without antidepressants I do not want to go out with friends"); and personal growth ("Things will get worse and worse"). Use of the well-being diary, associated with cognitive restructuring and exposure homework, may provide a different perspective to patients [58].

A decreased risk of depression or anxiety occurrence has been shown after WBT in mood and anxiety disorder patients, and this may lead to a lower likelihood to use antidepressant in the future [2, 28].

Side Effects During Long-Term Antidepressant Treatment

A negative aspect of long-term antidepressant treatments is concerned with side effects, particularly those observed with SSRIs [53], such as high rates of sexual dysfunction, bleeding (in particular gastrointestinal), weight gain (after initial weight loss), risk of fracture and osteoporosis, and hyponatremia [53]. In this framework, the sequential treatment represents a great opportunity for antidepressant drug tapering and discontinuation. Indeed, it is possible to offer, for instance, a psychological intervention while the patient is reducing the psychotropic medication; this means that the clinician has the opportunity to monitor the patient in one of the most delicate aspects of treatment.

Despite the use of a sequential approach aiming at discontinuing antidepressants, there are patients who take medications in the long-term. Still nowadays, most of the patients with a diagnosis of anxiety disorder never have a real chance to get appropriate psychotherapy for their disease. In addition, if these patients try to stop their psychotropic medication, they are at risk of relapse, and, if they were under SSRIs or SNRIs, they are at risk of having developed dependence; [54, 59] thus withdrawal symptoms mixed to residual anxious symptomatology may occur. These patients are, thus, often suggested by clinicians to come back to the drugs.

Even when patients never stop antidepressants, they may encounter problems. Loss of efficacy of antidepressant in the long-term treatment is a problem, for instance, in depression that increases with the duration of treatment (from 23% with 1 year to 34% within 2 years and to 45% within 3 years) [60].

Conclusions

The applications of WBT to the treatment of anxiety disorders that we have here discussed are largely speculative and in need of further investigations. Up to now, WBT appears to be uniquely applicable to the treatment of generalized anxiety disorder and panic disorder patients who did not benefit from CBT. However, an increasing attention to psychological well-being might reveal that this is a clinical dimension in need of being assessed, studied, and, when appropriate, enhanced.

Although customary clinical taxonomy and common clinical evaluations do not include psychological well-being, a clinical interview assessing psychological well-being, if incorporated in standard, symptom-oriented psychiatric examinations [61], might give additional information and define major prognostic and therapeutic differences among patients who otherwise seem to be deceptively similar since they share the same diagnosis.

Well-being therapy, being based on self-observation of psychological well-being associated with specific homework, has a different perspective from interventions that are labelled as distress-oriented, such as CBT. Another important feature of WBT is the assumption that imbalances in well-being are object of clinical interest and may vary from one illness to another and from patient to patient [62]. Thus, pursuing euthymia [24] can only be possible by means of a personalized approach that characterizes the treatment protocol and requires a comprehensive initial evaluation [63].

References

1. Hasler G. Well-being: an important concept for psychotherapy and psychiatric neuroscience. Psychother Psychosom. 2016;85(5):255–61. https://doi.org/10.1159/000447268.
2. Fava GA. Well-being therapy. treatment manual and clinical applications. Basel: Karger; 2016a.
3. Baumeister RF, Bratslavsky E, Finkenauer C, Vohs KD. Bad is stronger than good. Rev Gen Psychol. 2001;5:323–70.

4. Aron A, Melinat E, Aron EN, Vallone RD, Bator RJ. The experimental generation of interpersonal closeness: a procedure and some preliminary findings. Pers Soc Psychol Bull. 1997;23:363–77.
5. Cohen F, Solomon S, Maxfield M, Pyszczynski T, Greenberg J. Fatal attraction: the effects of mortality salience on evaluations of charismatic, task-oriented, and relationship-oriented leaders. Psychol Sci. 2004;15:846–51.
6. Skowronski JJ, Carlston DE. Caught in the act: when impressions based on highly diagnostic behaviours are resistant to contradiction. Eur J Soc Psychol. 1992;22:435–52.
7. Weinstein ND. Unrealistic optimism about future life events. J Personal Soc Psychol. 1980;39:806–20.
8. Boucher J, Osgood CE. The Pollyanna hypothesis. J Ver Learn Ver Behav. 1969;8:1–8.
9. Matlin MW, Stang DJ. The pollyanna principle. Cambridge: Schenkman; 1978.
10. Benjafield J. On the relation between the Pollyanna and golden section hypotheses. Br J Soc Psychol. 1984;23:83–4.
11. Tuohy AP, Stradling SG. Maximum salience vs. golden section proportions in judgemental asymmetry. Br J Psychol. 1987;78:457–64.
12. Brewer WF. Memory for randomly sampled autobiographical events. In: Neisser U, Winograd E, editors. Remembering reconsidered: ecological and traditional approaches to the study of memory. Cambridge: Cambridge University Press; 1988. p. 21–90.
13. Taylor SE, Brown JD. Illusion and well-being: a social psychological perspective on mental health. Psychol Bull. 1988;103:193–210.
14. Matt J, Vasquez C, Campbell WK. Mood-congruent recall of affectively toned stimuli: a meta-analytic review. Clin Psychol Rev. 1992;12:227–55.
15. Maslow A. Toward a psychology of being, ed. 2. New York: Van Nostrand; 1968.
16. Rogers CR. On becoming a person. Boston: Houghton Mifflin; 1961.
17. Jung CG. Modern man in search of a soul. New York: Harcourt, Brace, & World; 1933.
18. Allport GW. Pattern and growth in personality. New York: Holt, Rinehart, & Winston; 1961.
19. Erikson E. Identity and life cycle. Psychol Issues. 1959;1:18–164.
20. Buhler C, Massarik F. The course of human life. New York: Springer; 1968.
21. Neugarten BL. Personality change in late life: a developmental perspective. In: Eisdorfer C, Lawton MP, editors. The psychology of adult development and aging. Washington: American psychological Association; 1973. p. 311–35.
22. Jahoda M. Current concepts of positive mental health. New York: Basic Books; 1958.
23. Dolan P. Happiness by design. Finding pleasure and purpose in everyday life. London: Allen Lane; 2014.
24. Fava GA, Bech P. The concept of euthymia. Psychother Psychosom. 2016;85:1–5.
25. Ryff CD, Singer B. Psychological well-being: meaning, measurement and implication for psychotherapy research. Psychother Psychosom. 1996;65:14–23.
26. Ryff CD. Psychological well-being revisited. Psychother Psychosom. 2014;83:10–28.
27. Carrozzino D, Svicher A, Patierno C, Cosci F. The Euthymia Scale: a clinimetric analysis. In press.
28. Fava GA. Well-being therapy: current indications and emerging perspectives. Psychother Psychosom. 2016b;85(3):136–45.
29. Fava GA, Cosci F, Guidi J, Tomba E. Well-being therapy in depression: new insights into the role of psychological well-being in the clinical process. Depress Anxiety. 2017;34(9):801–8. https://doi.org/10.1002/da.22629.
30. Fava GA. Well-being therapy: conceptual and technical issues. Psychother Psychosom. 1999;68:171–9.
31. Moeenizadeh M, Salagame KK. The impact of well-being therapy on symptoms of depression. Int J Psychol Stud. 2010;2:223–30.
32. Kauffman C, Silberman J. Finding and fostering the positive in relationships: positive interventions in couples therapy. J Clin Psychol. 2009;65:520–31.
33. Quoidbach J, Mikolajczak M, Gross JJ. Positive interventions: an emotion regulation perspective. Psychol Bull. 2015;141:655–93.

34. Csikszentmihalyi M, Csikszentmihalyi I. Optimal experiences. In: Psychological studies of flow of consciousness. New York: Cambridge University Press; 1988.
35. Beck AT, Rush AJ, Shaw BF, Emery G. Cognitive therapy of depression. New York: The Guilford Press; 1979.
36. Fava GA, Rafanelli C, Tomba E. The clinical process in psychiatry. J Clin Psychiatry. 2012;73:177–84.
37. Emmelkamp PMG, Bouman TK, Scholing A. Anxiety Disorders. Chichester: Wiley; 1993.
38. Fava GA, Ruini C, Rafanelli C. Psychometric theory is an obstacle to the progress of clinical research. Psychother Psychosom. 2004;73(3):145–8.
39. Stangier U, Hilling C, Heidenreich T, Risch AK, Barocka A, Schlösser R, Kronfeld K, Ruckes C, Berger H, Röschke J, Weck F, Volk S, Hambrecht M, Serfling R, Erkwoh R, Stirn A, Sobanski T, Hautzinger M. Maintenance cognitive-behavioral therapy and manualized psychoeducation in the treatment of recurrent depression: a multicenter prospective randomized controlled study. Am J Psychiatry. 2013;170(6):624–32.
40. Cosci F, Fava GA. Staging of mental disorders: systematic review. Psychother Psychosom. 2013;82(1):20–34.
41. Detre TP, Jarecki H. Modern psychiatric treatment. Philadelphia: Lippincott; 1971.
42. Fava GA, Rafanelli C, Tossani E, Grandi S. Agoraphobia is a disease: a tribute to Sir Martin Roth. Psychother Psychosom. 2008;77:133–8.
43. Shear MK, Clark D, Feske U. The road to recovery in panic disorder: response, remission, and relapse. J Clin Psychiatry. 1998;59:4–8.
44. Cosci F. The psychological development of panic disorder: implications for neurobiology and treatment. Rev Bras Psiquiatr. 2012;34(Suppl 1):S9–19.
45. Fava GA, Rafanelli C, Grandi S, Conti S, Ruini C, Mangelli L, Belluardo P. Long-term outcome of panic disorder with agoraphobia treated by exposure. Psychol Med. 2001;31:891–8.
46. Fava GA, Ruini C, Rafanelli C, Finos L, Salmaso L, Mangelli L, Sirigatti S. Well-being therapy of generalized anxiety disorder. Psychother Psychosom. 2005;74(1):26–30.
47. Guidi J, Fava GA, Bech P, Paykel E. The clinical interview for depression: a comprehensive review of studies and clinimetric properties. Psychother Psychosom. 2011;80(1):10–27. https://doi.org/10.1159/000317532.
48. Fava GA, Savron G, Zielezny M, Grandi S, Rafanelli C, Conti S. Overcoming resistance to exposure in panic disorder with agoraphobia. Acta Psychiatr Scand. 1997;95:306–12.
49. Cosci F. Well-being therapy in a patient with panic disorder who failed to respond to paroxetine and cognitive behavior therapy. Psychother Psychosom. 2015;84:318–9.
50. Belaise C, Gatti A, Chouinard V-A, Chouinard G. Persistent postwithdrawal disorders induced by paroxetine, a selective serotonin reuptake inhibitor, and treated with specific cognitive behavioral therapy. Psychother Psychosom. 2014;83:247–8.
51. Balon R, Chouinard G, Cosci F, Dubovsky SL, Fava GA, Freire RC, Greenblatt DJ, Krystal JH, Nardi AE, Rickels K, Roth T, Salzman C, Shader R, Silberman EK, Sonino N, Starcevic V, Weintraub SJ. International task force on benzodiazepines. Psychother Psychosom. 2018;87(4):193–4. https://doi.org/10.1159/000489538.
52. Nardi AE, Cosci F, Balon R, Weintraub SJ, Freire RC, Krystal JH, Roth T, Silberman EK, Sonino N, Fava GA, Starcevic V, Dubovsky SL, Salzman C, Rickels K, Greenblatt DJ, Shader RI, Chouinard G. International task force on benzodiazepines. The prescription of benzodiazepines for panic disorder: time for an evidence-based educational approach. J Clin Psychopharmacol. 2018;38(4):283–5. https://doi.org/10.1097/JCP.0000000000000908.
53. Fava GA. Rational use of antidepressant drugs. Psychother Psychosom. 2014;83:197–204.
54. Fava GA, Gatti A, Belaise C, Guidi J, Offidani E. Withdrawal symptoms after selective serotonin reuptake inhibitor discontinuation: a systematic review. Psychother Psychosom. 2015;84:72–81.
55. Chouinard G, Chouinard VA. new classification of selective serotonin reuptake inhibitor withdrawal. Psychother Psychosom. 2015;84(2):63–71.

56. Belaise C, Gatti A, Chouinard VA, Chouinard G. Patient online report of selective serotonin reuptake inhibitor-induced persistent postwithdrawal anxiety and mood disorders. Psychother Psychosom. 2012;81:386–8.
57. Fava GA, Bernardi M, Tomba E, Rafanelli C. Effects of gradual discontinuation of selective serotonin reuptake inhibitors in panic disorder with agoraphobia. Int J Neuropsychopharmacol. 2007;10:835–8.
58. Fava GA, Belaise C. Discontinuing antidepressant drugs: lesson from a failed trial and extensive clinical experience. Psychother Psychosom. 2018;87(5):257–67. https://doi.org/10.1159/000492693.
59. Fava GA, Benasi G, Lucente M, Offidani E, Cosci F, Guidi J. Withdrawal symptoms after serotonin-noradrenaline reuptake inhibitor discontinuation: systematic review. Psychother Psychosom. 2018;87(4):195–203. https://doi.org/10.1159/000491524.
60. Meulenbeek P, Christenhusz L, Bohlmeijer E. Well-being therapy in the Netherlands. Psychother Psychosom. 2015;84:316–7.
61. Fava GA, Tomba E. Increasing psychological well-being and resilience by psychotherapeutic methods. J Pers. 2009;77:1902–34.
62. Rafanelli C, Park SK, Ruini C, Ottolini F, Cazzaro M, Fava GA. Rating well-being and distress. Stress Med. 2000;72:268–75.
63. Tomba E, Fava GA. Treatment selection in depression: the role of clinical judgment. Psychiatr Clin North Am. 2012;35:87–98.

Part V

Anxiety and Precision Psychiatry

Personalized Clinical Approaches to Anxiety Disorders

25

Giampaolo Perna, Alessandra Alciati, Erika Sangiorgio, Daniela Caldirola, and Charles B. Nemeroff

Personalized Medicine in Anxiety Disorders

Anxiety disorders (ADs) are common mental diseases, whose lifetime prevalence is estimated at 33.7% according to data collected by the US National Comorbidity Survey Replication (NCS-R) [1]. The current edition of the *Diagnostic and Statistical Manual of Mental Disorders* (DSM-5) [2] codified as ADs panic disorder (PD), agoraphobia, generalized anxiety disorder (GAD), social anxiety disorder (SAD), specific phobia, separation anxiety disorder and selective mutism. In the past, the last two disorders were classified as childhood disorders but are now considered to be appropriate also in adulthood.

ADs affect personal, social and occupational functioning, produce serious disability and lead to a relevant impairment in quality of life. The World Health Organization [3] ranked ADs as the sixth contributor of global disability, on the basis of years of life spent with disability, in both developed and developing countries. Moreover, ADs are crucial in the development and prognosis of medical illness. The meta-analysis of Emdin et al. [4] found that patients suffering from ADs had a relevant risk for several cardiovascular conditions, including stroke, coronary heart disease and heart failure, and for cardiovascular mortality. In the last decade,

G. Perna (✉) · E. Sangiorgio · D. Caldirola
Department of Biomedical Sciences, Humanitas University, Milan, Italy

Department of Clinical Neurosciences, Hermanas Hospitalarias, Villa San Benedetto Menni Hospital, Como, Italy

A. Alciati
Department of Clinical Neurosciences, Hermanas Hospitalarias, Villa San Benedetto Menni Hospital, Como, Italy

Humanitas Clinical and Research Center, Milan, Italy

C. B. Nemeroff
Institute for Early Life Adversity Research, University of Texas Dell Medical School in Austin, Austin, TX, USA

Y.-K. Kim (ed.), *Anxiety Disorders*, Advances in Experimental Medicine and Biology 1191, https://doi.org/10.1007/978-981-32-9705-0_25

there has been an increased research interest into ADs, both because of a greater recognition of their burden and because they are among the most prevalent untreated psychiatric disorders [5]. A large European study showed that only 20.6% of participants with ADs asked for professional help; of those who approach to healthcare resources, 23.2% were not treated at all, 19.6% received only psychological treatment, 30.8% were treated only with psychotropic medications, and 26.5% received both pharmacotherapy and psychotherapy [6].

International guidelines indicated that psychotropic and psychotherapeutic treatments are effective for ADs. Medications include selective serotonin reuptake inhibitors (SSRIs), serotonin-noradrenaline reuptake inhibitors (SNRIs), pregabalin, tricyclic antidepressants (TCA) and benzodiazepines. Among psychotherapies, cognitive-behavioural therapy (CBT) is the most investigated, but a number of trials have also focused on interpersonal therapy (IPT) and psychodynamic therapy (PDTh) [7].

Unfortunately, 30–70% of patients failed to respond to the first pharmacologic intervention [8]. This is likely because the criteria that guide the choice of the best drug for the individual patient did not derive from the integration of the biological and environmental factors but rather from clinician's personal experience and a balance of potential side effects and costs.

The development of preventive and/or early treatment measures should be a high priority because of the high prevalence of ADs, their serious personal and socioeconomic impact as well their role as a complicating factor in somatic disease. However, universal preventive measures, applied to unselected target groups regardless of the susceptibility of the individual, would be both very expensive and demanding.

Personalized medicine, an approach to prevention and treatment of medical and psychiatric diseases that advocates tailoring healthcare to each individual by integrating genetics and epigenetics, brain imaging, clinical aspects, and environmental factors, can be an answer to these needs. In fact, the aim of personalized medicine is to predict more accurately illness susceptibility and to tailor the most effective treatment for each individual [9].

According to the two main components of personalized medicine [10], the present chapter summarizes evidence for the possible predictors of vulnerability to disease and response to treatment in the three most prominent and/or debilitating ADs [1], i.e. PD, GAD and SAD.

Panic Disorder

Patients with PD suffer from a chronic and disabling condition, whose lifetime prevalence has been estimated at 3–4% in the general population [1]. PD is associated with medical/mental comorbidity, significant social and occupational impairment and poor quality of life [11, 12]. In the DSM-5 [2], a diagnosis of PD is defined as the fulfilment of both recurrent unexpected panic attacks (PAs) (criterion A) and the existence of one or more of the following persistent PA-related conditions for at least 1 month: concern, worry and behavioural changes, such as agoraphobic avoidance (criterion B). PAs are characterized by sudden and severe fear/discomfort episodes, accompanied by intense and unpleasant somatic symptoms, such as chest pain, palpitations and dyspnoea [2].

Several medications, as well as CBT, are effective for PD. Medications include, as first-line agents, SSRIs and serotonin-norepinephrine reuptake inhibitor venlafaxine and, as second-line agents, TCAs and benzodiazepines. Among psychotherapies, CBT is recommended as second-line agents [11, 13–15]. Despite these treatment options, in clinical trials, 17–64% of participants with PD fail to adequately respond to pharmacological treatments and continued to have PAs and/or avoidance behaviours; in clinical settings, about 20–40% of patients did not achieve complete remission with the recommended medications or CBT [10].

Because drugs based on novel mechanisms are far from being implemented in clinical settings [15], examining markers of vulnerability and predictors of response to current pharmacological/psychotherapeutic treatments appears to be the most feasible strategy in the next future for applying a personalized approach to PD and increasing the rate of successful outcomes.

Endophenotypes/Biomarkers

Prediction of Disease Vulnerability

Hypersensitivity to CO_2

Hypersensitivity to carbon dioxide (CO_2) has received decades of research investigations as a putative endophenotype of PD. According to the early hypotheses, interaction between hyperventilation and fear was considered the causal mechanism of PAs. "Bad breathing habits" caused a burst of hyperventilation, during which more CO_2 is eliminated than produced, causing respiratory alkalosis, which elicited panic symptoms and, thus, fear and anxiety [16, 17]. This model has been exceeded by subsequent studies, involving CO_2 challenge test with enriched air mixtures, which proposed hypersensitivity to CO_2 as a reliable panicogen challenge.

Hypersensitivity to CO_2 meets all the criteria to be considered an endophenotype of PD, which are association with the disorder, trait stability and heritability [18].

Several studies have shown that inhalation of 35% CO_2 can elicit a clear anxiety reaction only in patients with PD, who experience a sudden rise of both subjective anxiety and panic symptoms, compared with healthy subjects [19] and patients with other anxiety disorders or psychiatric diseases, such as GAD and mood disorders [19, 20].

This effect has been found to be independent from the clinical features of the disorder (i.e. frequency of PAs, agoraphobia and anticipatory anxiety and duration of illness), suggesting that hypersensitivity to CO_2 is a trait marker of PD [19].

Family studies have suggested that CO_2 hypersensitivity can represent a heritable hallmark of PD. Thus, the first-degree relatives of probands with PD had greater reactions to 35% CO_2 inhalation compared with healthy controls [21]. Moreover, in a twin study, monozygotic twins showed higher concordance for 35% CO_2-induced

PAs than dizygotic twins (55.6% versus 12.5%) [22]. The main source of familial risk for PD appears due to genetic factors, as supported by a population-based twin study of shared genetic influence for 35% CO_2 hypersensitivity and uncued PAs [23]. Some studies have investigated whether CO_2 hypersensitivity may be related to the allelic variation of the serotonin transporter gene (5-HTTLPR), in fulfilment of current theories on the role of serotonin in panic and respiration. Schmidt et al. [24] found that polymorphisms (ll or l′l′ genotype) in the 5-HTTLPR predicted greater fearful responses to a 35% CO_2 challenge in healthy volunteers. In contrast, these findings were not replicated by Perna et al. [25] in patients with PD. However, the authors suggested that their findings may be explained by a possible influence of other polymorphisms in the gene coding for 5-HTT or in the genes coding for other structures involved in serotonergic modulation. A more recent study [26] indicated that these contradictory results may be due to the dose-response relationship between CO_2 dosage and CO_2-induced PAs. Indeed, the investigators found that variances in 5-HTTLPR genotype (LL, SL and SS) may moderate panic response to different CO_2 mixtures (0%, 9%, 17.5% and 35%) in healthy subjects, with dose-dependent effects. Subjects with the SL and SS genotype reported less fear than LL subjects, with those with the SS genotype showing lower fear scores, particularly in response to the 17.5% CO_2 dose.

Neurochemical Markers

Neurotransmitter Systems

Plasma and serum 5-HT levels were reported to be lower in patients with PD than in controls, while platelet 5-HT concentrations were unchanged or lower [27]. Studies on platelet 5-HT uptake and platelet 5-HT reuptake site binding reported inconsistent results. Cerebrospinal fluid (CSF) concentrations of the 5-HT metabolite 5-hydroxyindoleacetic acid (5-HIAA) were similar in patients with PD and healthy subjects [27].

No significant changes were found in CSF levels of homovanillic acid (HVA), the main metabolite of dopamine, in patients with PD compared with healthy subjects [28].

Plasma noradrenaline has not consistently been shown to be enhanced, nor has the CSF levels of 3-methoxy-4-hydroxyphenyglycol (MHPG), the main metabolite of central noradrenaline [27]. In the absence of baseline alterations in noradrenaline physiology, dynamic noradrenergic dysfunction has been carried out using noradrenergic challenge paradigm. Yohimbine is a selective α2-adrenergic antagonist, whose primary activity is at the presynaptic α2-autoreceptor; its effect is to increase central and peripheral noradrenergic outflow. With yohimbine challenge, patients with PD consistently reported a raise in the rate of PAs and greater increases in Visual Analogue Scale (VAS) anxiety, plasmatic cortisol concentrations and systolic blood pressure (BP) than controls [29].

Finally, because platelet and central nervous system (CNS) α2-adrenoreceptors are similar in terms of pharmacology, binding to platelet α2-adrenoreceptors with the agonist clonidine and/or the antagonist yohimbine has been studied as a proxy measure of CNS receptor function. The main used parameters were the binding maximum (Bmax), which measures the number of receptors per given mass of membrane, and the dissociation coefficient (Kd), which measures the affinity for the receptor, with higher Kd indicative of lower affinity. Results of platelet α-2-adrenoreceptors studies on PD are not consistent [30].

Other Blood or Peripheral Measures

Some growing evidence suggested the role of blood cells and peripheral measures as possible biomarkers for the vulnerability to PD.

Platelet activation and changes in reactivity indicators have been linked to chronic stress, cardiovascular conditions, nutritional deficiencies and various other medical conditions. According to Bhad [31], measurement of serotonin levels, platelet proinflammatory and immune-modulatory secretory compounds (e.g. platelet factor-4, P-selectin and β-thromboglobulin, monoamine oxidase activity and platelet activity indicators) may be useful for development of biomarkers for diagnosis of PD and/or various related conditions. Indeed, some studies found that patients with PD have higher platelet distribution width (PDW), red cell distribution width (RDW) and mean platelet volume compared with healthy subjects [32]. Furthermore, some findings indicated that it is possible to predict development of psychiatric comorbidities in patients with PD using biological serum markers, such as tetranectin and creatine kinase MB, and differentiate comorbidities associated with PD measuring serum ghrelin levels and lipid profile [31].

As regards immunological markers, PD patients showed a decreased percentage and number of circulating CD19+ B lymphocytes relative to control individuals, but no differences emerged in number and percentage of other lymphocyte populations, mitogen responses and NK cell activity. These immune measure changes in PD contrast with those found in major depressive disorder (MDD), suggesting some specificity of immune changes in psychiatric diseases [33]. Levels of the proinflammatory cytokines IL-1β and IL-6 were reported to be higher in PD patients compared to control subjects, while conflicting results were found regarding IL-2, IL-12 and INF-γ [34].

Finally, in line with the hypothesis that PAs might result from a disruption in stress response regulation by the hypersensitivity of the hypothalamic-pituitary-adrenal (HPA) axis, a body of studies have addressed the role of HPA axis activation in PD, using cortisol secretion as the marker of HPA functioning, in basal conditions, during PAs, or psychological challenge test. However, results of these studies are not consistent regarding baseline cortisol and adrenocorticotropic hormone (ACTH) levels, cortisol and ACTH response to spontaneous and provoked PAs, exposure to feared situations, dexamethasone suppression or corticotropin-releasing hormone (CRH) challenge [27]. Thus, further examinations are needed to establish if cortisol secretion could be a biomarker of PD [32].

Physiological Markers

Several studies have suggested aberrant respiratory and cardiovascular patterns as specific biomarker for PD.

Although patients with PD do not display any full-blown respiratory diseases, their respiratory system does appear to be more sensitive than that of individuals without PD.

In a meta-analysis of baseline respiratory and hematologic parameters related to respiration, Grassi et al. [35] showed that patients with PD are characterized by chronic hyperventilation (e.g. higher baseline mean minute ventilation, lower end-tidal partial pressure of CO_2 and reduced HCO_3- and PO_4-blood concentrations), higher indices of respiratory variability/irregularity and higher rate of sighs and apneas, compared with healthy subjects. In a following meta-analytic study, the same research group found that respiratory abnormalities are specific to patients with PD, compared with those with GAD or SAD [36]. As recently reported by Perna and Caldirola [37], these patients are also characterized by a higher-than-expected prevalence of obstructive respiratory diseases in childhood, irregular breathing pattern in their sleep, increased respiratory variability during mild physical activity and impaired diaphragmatic breathing with reduced vital capacity; furthermore, they have behavioural and respiratory hypersensitivity to hypercapnic challenges, with altered respiratory patterns during the challenge and recovery phases, and to other respiratory stimuli, such as hypoxic challenge and hyperventilation.

Some evidence suggested that respiratory abnormalities may be useful in predicting the risk of PAs/PD in both symptomatic and asymptomatic subjects. Meuret et al. [38] found a specific pattern of autonomic and respiratory instability in the 47 min preceding the onset of PAs, which may predict occurrence of PAs in affected patients. Furthermore, Perna et al. [39] showed higher respiratory rate and higher irregularities in respiratory measures in healthy children of patients with PD than healthy children of individuals without psychiatric diagnosis, suggesting that such respiratory abnormalities may help in the identification of children at risk for PD.

Similarly, although most patients with PD do not report any full-blown cardiovascular diseases, several studies indicated a series of subtle abnormalities in their cardiovascular system. As extensively reported by Perna and Caldirola [37], these patients exhibit imbalance in their autonomic regulation, reduction in heart rate variability (HRV), increase in time variability of ventricular repolarization, higher regional heterogeneity of ventricular repolarization and atrial depolarization and higher variability of ECG-R wave amplitude after beta-adrenergic stimulation with isoproterenol. Moreover, PD patients who are normotensive present an impaired circadian BP pattern with an inappropriate depletion in night-time BP (non-dipper pattern) as well as an unstable heart rate (HR) during sleep. In addition, at least in some patients with PD, PAs may be induced by paroxysmal supraventricular tachycardia. Finally, these patients present enhanced arterial stiffness and several factors that negatively influence endothelial function, such as heightened homocysteine levels and platelet aggregation or volume and reduced levels of nitric oxide.

Despite the absence of longitudinal evidence, preliminary reports suggest that cardiovascular abnormalities may be useful in predicting the risk of PAs/PD in both symptomatic and asymptomatic subjects. In their study Meuret et al. [38] showed that PA onset was characterized by a specific increase of HR, in addition to the respiratory changes mentioned above, suggesting that PAs may be predicted with specific cardiac measures in affected patients. Finally, Srinivasan et al. [40] demonstrated a heritable effect of certain measures of HRV, given that children of parents with PD exhibit lower cardiac vagal function, relative to children of normal controls. However, some authors argued that cardiac abnormalities may be not specific of PD. For instance, Chalmers et al. [41] reported similar findings comparing patients with different ADs [i.e. PD, GAD, SAD, post-traumatic stress disorder (PTSD)] with healthy controls, suggesting that reduced HRV is a non-specific marker across ADs.

Neuroimaging

Structural and Functional MRI Studies

An early hypothesis proposed a central role of amygdala and limbic regions (i.e. the "fear network") in the pathophysiology of PD, conceiving PAs as a result of dysfunctions in this network [42]. Even though some structural abnormalities (i.e. smaller volume) in amygdala and its subnuclei were found in patients with PD [43], other evidence did not support this view, suggesting that amygdala may not be crucial for PAs. Individuals with neurological damage of amygdala had both spontaneous and laboratory-induced PAs (i.e. with 35% CO2 inhalation or isoproterenol infusion), indicating that amygdala is not required to the occurrence of PAs [43–45]. Consistently, several neuro-structural findings showed volumetric difference in patients with PD compared with healthy controls in different brain areas, such as brainstem, left insula, right dorsal and rostral anterior cingulate cortex (ACC), temporal and frontal lobes, right posterior-medial orbitofrontal cortex (OFC) and right basal ganglia [27, 46].

In line with these findings, functional magnetic resonance imaging (fMRI) studies provided support to the idea that brain areas different from the amygdala may play a role in the occurrence of PAs. fMRI utilizes intensity changes in a magnetic resonance signal to track haemodynamic changes in the brain (i.e. blood oxygenation level dependent or BOLD) as proxy measures for neuronal activity. This technique allows identification of brain regions that are activated during a given behaviour/task and the synchrony of brain function across two or more regions. Through this technique, brainstem and insular regions were found to be crucial in the occurrence of PAs [46, 47]. Goossens et al. [47], investigating brain activation in response to a hypercapnic inhalation, found increased activation in the brainstem in response to the challenge in patients with PD compared with controls. Moreover, subjective feelings of breathing discomfort were positively correlated with brain activation in the anterior insula in all groups. In their review on neuroimaging studies, Perna et al. [46] specifically considered the potential involvement of the brainstem in the pathophysiology of PD. The authors concluded that unexpected PAs may be related to phylogenetically older brain structures, such as the brainstem areas, which process basic functions related to the organism's internal milieu, while amygdala and limbic system may be mainly involved in anticipatory anxiety and phobic behaviours.

Other neuroimaging studies investigated neural activation in response to stimulation with panic-specific pictures. The findings showed that patients with PD had a greater activation of neural structures associated with interoception and somatosensory processing compared to controls, which comprise the insular cortices, along with the left inferior frontal gyrus, the dorsomedial prefrontal cortex, the left hippocampal formation and the left caudate [48].

In a recent review on the state of neuroanatomy in PD, Sobanski and Wagner [49] concluded that neuro-functional findings mainly indicated abnormal activations in an extended network, which included brainstem, insula, anterior and midcingulate cortex and lateral, as well as medial, parts of the prefrontal cortex. Furthermore, the authors reported that abnormal activations in the amygdala were not as consistently found, suggesting that the activation of amygdala in functional studies appeared to depend on stimuli and experimental paradigms, as well as on limitations of neuroimaging techniques.

Prediction of Treatment Outcome

Hypersensitivity to CO_2

The available literature mainly focused on the effects of treatments on hypersensitivity to CO_2. Decrease of reactivity to 35% CO_2 was shown in all groups of patients treated with SSRIs (paroxetine, sertraline, citalopram or fluvoxamine), TCAs (imipramine or clomipramine) or CBT (CBT with/without respiratory training) compared with the placebo/untreated groups [50–55]. Some differences in effects of psychotropic therapies on hypersensitivity to 35% CO_2 were shown. Although fluvoxamine and clomipramine resulted in a similar decrease of 35% CO_2 reactivity, paroxetine resulted in a greater decrease than reboxetine, across and/or after 1 week of treatment [52, 54]. Furthermore, in patients with hypersensitivity to CO_2, paroxetine, sertraline and clomipramine showed greater antipanic properties than fluvoxamine and imipramine after 30 days of treatment [53]. All of these findings suggested a relevant role of the serotonergic system in the modulation of the CO_2 hyperreactivity.

One interesting study [53] investigated whether the decrease in CO_2 reactivity, in the short-term, could be a significant predictor for good clinical treatment outcome. The authors found that the decrease in CO_2 reactivity, after the first week, was a significant predictor for improvement following 1 month of treatment with paroxetine, sertraline, fluvoxamine, imipramine or clomipramine.

Neurochemical Markers, Blood or Other Peripheral Measures

Because of their critical role in PD, pretreatment indices of noradrenergic systems have been evaluated as putative predictors of treatment response. Studies on the predictive role of plasma MHPG concentration yielded inconclusive results. Conversely, Lee et al. [56] demonstrated that the acute patients with PD who responded to 12-week paroxetine treatment had a lower pretreatment lymphocyte β-adrenoceptor affinity in comparison with the non-responders. This result is consistent with a previous study that reported that pretreatment lymphocyte β-adrenoceptor density was lower in responders to adinazolam [57]. Finally, high pretreatment α2-adrenoreceptor density predicted low severity of anxiety after treatment with imipramine [58]. Because results are scant and mixed, it is not possible to reach a conclusion about the role of noradrenergic markers in the prediction of treatment response in patients with PD.

Although the role of other blood or peripheral measures in predicting treatment outcomes was not directly examined, preliminary data suggested that serum phosphate change is a promising biomarker in PD. Hypophosphatemia has been proposed as a metabolic mechanism for adapting to the respiratory alkalosis induced by the hyperventilation associated with PD [59]. In this context, Perez-Costillas et al. [60] found normalization of hypophosphatemia after a successful 12-week treatment with clomipramine or alprazolam; Beria et al. [61] showed increased phosphate levels in responders to a 12-week group CBT.

Finally, the only study on brain-derived neurotrophic factor (BDNF) found that poor responders to CBT had lower serum BDNF levels compared with patients exhibiting a good response [62].

Physiological Markers

Several studies have indicated that antipanic medications, particularly SSRIs, modulate various brain areas, neural pathways and peripheral sites directly implicated in respiratory and cardiovascular functions [37]. There are a body of studies indicating that SSRIs, particularly paroxetine, citalopram and sertraline, may improve respiratory irregularity and dyspnoea, both in chronic respiratory diseases and in PD [37]. SSRIs or CBT may have several positive effects also on the cardiovascular system of patients with PD, leading to (a) an increase of HRV and decrease in the cardiac sympathetic activity; (b) inhibition of platelet aggregation, collagen, and thrombin, exerted anti-inflammatory effects; (c) improvement of endothelial function, which may be protective against cardiovascular diseases; and (d) reduction of the symptoms of paroxysmal hypertension [37, 63].

Although few studies are available, some evidence indicated that baseline respiratory and cardiovascular patterns may predict responses to pharmacotherapy and/or psychotherapy. Although respiratory parameters were not assessed, Nardi et al. [64] found that the respiratory subtype of PD predicted a faster response to clonazepam, compared with the non-respiratory subtype, in a group of patients treated with clonazepam for a 3-year period. In addition, Tolin al [65]. suggested that low end-tidal CO_2 ($ETCO_2$) at baseline was a negative prognostic marker for CBT in PD, because it predicted dropout in non-responders affected by ADs [i.e. PD, GAD, SAD or specific phobia (SP)] or other diagnosis [i.e. obsessive-compulsive disorder (DOC) or PTSD], prior to complete treatment. In contrast, Slaap et al. [66] found that abnormalities in autonomic nervous system (ANS), as indicated by reduced HRV, predicted non-response to a 12-week treatment with mirtazapine. Furthermore, Wendt et al. [67] showed that patients with PD with reduced pretreatment HRV were more likely to show residual symptoms after completion of exposure-focused therapy.

Neuroimaging

Patients with PD remitted after escitalopram treatment showed an increased grey matter volume in the left superior frontal gyrus [68] and an increased white matter microstructural integrity in the right uncinate fasciculus and the left fronto-occipital fasciculus and a significant decrease in the right precentral gyrus [69], but none of these studies has been designed to evaluate the predictive role of neural structural measures on response to treatment. As reported by Shin et al. [70] in their review on neuroimaging predictors of treatment response in ADs, no studies have previously assessed whether pretreatment structural or functional neuroimaging measures can predict outcomes of pharmacotherapy in PD.

A more recent systematic review that focused on outcome of CBT reported that a beneficial response to CBT was predicted by elevated pretreatment neural activation in the ACC, hippocampus, insula, dorsolateral prefrontal cortex (dlPFC), amygdala and inferior frontal gyrus (iFG) [71].

Genetics

To elucidate genetic association, two main approaches have been proposed. The hypothesis-driven studies have focused on candidate genes potentially implicated in PD pathogenesis and the non-hypothesis-driven genome-wide association studies (GWAS) that utilize the statistical power emerging from thousands of samples without an a priori selection of risk genes.

Prediction of Disease Vulnerability

Studies on the genetic basis of PD focused on genes involved in traditional neurotransmitters, in particular monoamines. Within the serotonin system, it has been shown that high-expression alleles of the monoamine oxidase A (MAOA) promoter region (MAOA-uVNTR) increase the risk for PD, especially in women [72], and that HTR1A, HTR2A and the tryptophan hydroxylase 2 (TPH2) genes may contribute to the pathogenesis of PD [73].

The most consistently reported gene of the dopaminergic/noradrenergic system found to be implicated in vulnerability, the gene encoding catechol-O-methyltransferase (COMT) (an enzyme involved in 5-HT/other monoamines catabolism). The COMT Val158Met polymorphism, which is related to an increase in COMT enzymatic activity and decreased cortical dopamine quantitatively, has been shown to be associated with PD in females only and in an ethnic-specific manner [74].

Several polymorphisms located within the 5’ end of the norepinephrine transporter (NET) have been shown to be significantly associated with PD [56, 75], though discordant findings have appeared [76].

A meta-analytic support for the bradykinin receptor B2 and NPY receptor Y5 gene loci in Japanese patients with PD emerged from the few GWAS conducted in PD [77].

Prediction of Treatment Outcome

Pharmacogenetic studies have centred on serotonergic candidate genes and in particular on the functional polymorphisms of the 5-HTTLPR in PD. In a Caucasian population, homozygotes (l/l) and heterozygotes (l/s) females for the long variant (l) exhibited a greater reduction of PAs after 12 weeks of treatment with paroxetine when compared to female homozygotes (s/s) for the short variant (s), but no difference in the improvement of anticipatory anxiety/phobic avoidance [78]. Conversely, in Japanese patients, after 2 weeks of paroxetine treatment, the homozygotes (s/s) had a greater decrease of panic-phobic symptoms than heterozygotes (l/s), without gender differences [79]. In a separate Japanese sample, heterozygotes (l/s) reported a higher tendency to interrupt therapy for adverse side effects than homozygotes (s/s), within the first 2 weeks of treatment with paroxetine [80]. In two other Asian samples, no effects of 5-HTTLPR polymorphisms were found on panic-phobic symptoms or clinical global improvement after 4 or 10 weeks of treatment with paroxetine, respectively [81, 82]. In these last two studies, no effects of tryptophan hydroxylase polymorphisms were found [82], while the 1019C/C genotype in the regulatory region of the 5-HT1A receptor gene was associated with a better clinical outcome than with C/G or G/G genotype [81]. Similar results were reached in a Caucasian sample, where the G/G genotype was associated with a minimal decrease of PAs after 6 weeks of paroxetine or sertraline treatment and no relationship with 5-HTTLPR polymorphisms was found [83].

Few studies have examined the association of polymorphisms of COMT with treatment outcomes. Functional polymorphisms of COMT were not associated with response after 12 weeks of treatment with paroxetine in Caucasian patients with PD [84]; Korean patients with the L/L COMT genotype had poorer clinical global improvement after the same treatment [85].

Epigenetics

Prediction of Disease Vulnerability

A preliminary epigenome-wide association study showed DNA methylation abnormalities in patients with PD, with association with several CpG sites across the genome [86]. Some studies investigated methylation of the genes corticotropin-releasing hormone receptor 1 (CRHR1) [87], glutamate decarboxylases (GAD67/65; GAD1/GAD2) and MAOA [88], which are crucially involved in the regulation of the stress response and/or in the pathogenesis of ADs, including PD. Findings showed hypomethylation of these candidate genes, with a significant influence of sex and positive/negative life events for GAD1 and MAOA, suggesting that methylation of these genes may represent an epigenetic signature of the risk of PD.

Prediction of Treatment Outcome

Only one study looked into MAOA methylation changes during the course of a 6-week exposure-based CBT in PD [89]. The authors noted lower MAOA methylation in PD patients than in healthy subjects, which was associated with PD severity.

Furthermore, MAOA methylation increased up to the level of healthy controls along with CBT response, while non-responders had a further decline in methylation. The authors suggested that the reversibility of MAOA hypomethylation is a potential epigenetic marker of response to CBT.

Summary

Almost all of the candidate biomarkers described above (i.e. neurotransmitters markers, physiological markers and structural and functional MRI patterns) distinguish PD patients and healthy subjects.

Among the biomarkers/endophenotypes reviewed, hypersensitivity to CO_2 remains the most promising, because of the clear evidence in predicting both vulnerability to PD and outcomes following pharmacological/psychotherapeutic treatments.

Although studies suggest the presence of noradrenergic and serotoninergic dysfunction in PD, it must be considered that the evidence is based on studies involving a range of relatively crude methodologies, which mostly applied indirect and peripheral measures. For instance, there is consistent evidence of an increased reactivity to challenges of the noradrenergic system in PD. However, challenge studies in PD are subject to confounding, because the physiological and psychological response to the challenge may come from the amplification of bodily sensations, typical of patients with PD, rather than a noradrenergic dysfunction.

In neuroimaging studies, the central role of amygdala, in the pathophysiology of PD, is not consistently reported and its changes seem to depend on the stimuli employed and experimental paradigms. Current state of neuro-structural and functional research in PD has underlined the role of multiple cortical and subcortical areas, comprising brainstem, insula, hypothalamus, ACC and lateral and medial parts of the PFC [49]. In particular, brainstem and insular regions seem to have a central role in the occurrence of PAs [46, 47].

There are more than 20 identified candidate genes for susceptibility to PD involving different neurotransmitter and neuropeptides. Unfortunately, in most of the studies, the small sample size renders it difficult to generalize, and the majority of findings of identified genes have not often been replicated in larger cohorts and in different populations.

Genetic variations thus far do not predict an individual's responses to antidepressants in PD patients. Moreover, the effectiveness of genetic prediction is affected by the paucity of the studies, the small sample sizes, the use of unsuitable measures of outcome and the impact of ethnic differences.

Generalized Anxiety Disorder

GAD can be considered a chronic condition. The core feature of GAD is excessive and uncontrollable worry occurring alongside somatic and cognitive symptoms, such as restlessness, fatigue, muscle tension, difficulty concentrating, irritability and sleep disturbances. It affects up to 6.2% of the general population and is the most prevalent AD described in primary care patients, where its prevalence is estimated at 8% [90]. The impairment of functioning and quality of life is similar to that experienced by patients with MDD, arthritis and diabetes [91].

Endophenotypes/Biomarkers

Prediction of Disease Vulnerability

Neurochemical Biomarkers

In patients with GAD, analysis of serotonin (5-HT)-related biomarkers has found decreased 5-HT reuptake site binding platelets compared to controls [92], but not in lymphocytes [93]. Moreover, concentrations of 5-HT and 5-hydroxyindoleacetic acid (5-HIAA) in platelet-rich and platelet-poor plasma, as well as in lymphocytes, were within the normal range [93].

Peripheral markers of HPA axis activity, such as cortisol [94] and dehydroepiandrosterone sulphate (DHEAS) and cortisol/DHEAS ratios, urinary free cortisol and plasma cortisol levels after a stressful test, did not show significant alterations compared to normal controls [95]. The analysis of hair cortisol concentrations, which represented the long-term cortisol secretion independent of acute HPA axis responses, demonstrated up to 50–60% lower hair cortisol concentrations in patients with GAD compared with healthy subjects [96, 97]. This result is in agreement with the view that chronic anxiety – the core hallmark of GAD – may result in downregulation of HPA axis activity.

Although there is some evidence for a link between neurotrophic factors and mood disorders, evidence is not consistent in patients with GAD; plasma levels of BDNF are similar to [98], higher [99] or lower [100] than normal controls.

Among inflammatory and immunological factors, studies have found increased level of C-reactive protein (CRP) [101, 102], plasma interleukin (IL)-10 and α-melanocyte-stimulating hormone (α-MSH), but no significant changes in IL-2A.

Cell-mediated immune functions, measured through the lymphocyte proliferative response to phytohaemagglutinin, IL-2 production and natural killer cell activity, were lower in patients affected by GAD than in healthy subjects [103].

Neuroimaging

SPECT Studies

Dopamine transporters (DAT) availability in the striatum and fronto-cortical GABAA receptors were significantly lower in patients with GAD in comparison with healthy subjects. Serotonin transporter (5-HTT) binding is unchanged in comparison to normal controls [27].

Structural MRI Studies

Larger right amygdala [104] and dorsomedial prefrontal cortical volumes [105] have been reported in women with GAD.

Significantly greater grey matter volumes have been frequently found in the amygdala [106] and less frequently in the right putamen (with a positive correlation with childhood maltreatment) [107], basal ganglia and superior temporal pole [108].

White matter volumes were lower in the dlPFC of medication-free GAD subjects [108] and in the dlPFC, anterior limb of the internal capsule (ALIC) and midbrain in patients with GAD who had working memory dysfunction [108].

Functional MRI Studies

The BOLD imaging studies in GAD have focused on brain responses to facial expressions of negative emotions. Fear faces strongly activate limbic circuitry, including the amygdala and insula. The amygdala is well-known to mediate fear responses to an external stimulus and emotion perception [109]; the insula is involved in interoception, a process that provides information on changes in internal body signals.

Findings regarding amygdala function in GAD are mixed. In children and adolescents with GAD, greater amygdala activation has been demonstrated in response to masked angry [110] or fearful faces [111]. However, in the latter study, amygdala hyperactivity was detected only when GAD adolescents were attending to their own fear in response to a fearful face, but not when attending to the emotional facial expression. In adults with GAD, some studies have observed hyperactive amygdala responses to processing fearful faces [112], whereas others have found lower amygdala activation [113] or no differences [114].

To better understand amygdala function in GAD, it is important to disentangle stimulus anticipation and stimulus response processes. Adults with GAD showed greater anticipatory activity in the bilateral dorsal amygdala preceding both aversive and neutral pictures than did healthy controls. These findings suggested that patients with GAD may have an increased and indiscriminate anticipatory response to both pathology-specific and non-specific cues [114].

Some studies have investigated amygdala interactions with other brain structures, through the degree of synchrony of BOLD responses in two regions in interaction during one task condition relative to another. In patients with GAD, elevated amygdala-insula connectivity, during processing of fearful and/or angry versus happy facial expressions, supports the role of these regions in emotion processing and anxiety. However, studies demonstrated a variable pattern of prefrontal and limbic connectivity during the processing of facial affect, such as increased connectivity among the amygdala, dorsomedial PFC (dmPFC) and dorsal ACC (dACC) or a deficient amygdala-ventral ACC (vACC) connectivity [115].

Imaging Genetics

Imaging genetics approaches seek to elucidate the relationship of genetic polymorphisms with physiological correlates of cerebral activity or connectivity.

Among the very few imaging genetic studies that have investigated GAD directly, Oathes, Hilt and Nitschke [116] reported that individuals who were S/L(G) carriers of the 5-HTTLPR-rs25531 showed less activity than their L(A)/L(A) counterparts in both the amygdala and anterior insula, in a paradigm designed to elicit activations in these neural areas during the anticipation of, and response to, unpleasant images.

Prediction of Treatment Outcome

Neuroimaging

Few fMRI studies have studied prediction of treatment outcome in GAD.

Higher pretreatment reactivity to fearful faces in the rostral ACC (rACC) and lesser reactivity in the amygdala predicted a favourable response to 8-week venlafaxine treatment, in absence of any difference between patients and controls in neuronal activation within these regions before treatment [114]. Patients with GAD with increased activity of the pregenual ACC, in anticipation of both aversive and neutral images, showed greater reductions in anxiety and worry symptoms after 8 weeks of venlafaxine treatment [117]. These two studies suggest that ACC-amygdala responsiveness could be considered a possible predictor of antidepressant therapeutic response in patients with GAD. Finally, greater activation in cortico-limbic circuitry predicted better CBT response in mixed samples of patients with PD and GAD [118].

Genetics

Prediction of Disease Vulnerability

Given that GAD has a moderate genetic risk (heritability of approximately 30%), association studies have been pursued to identify chromosomal risk loci and susceptibility genes for GAD.

Candidate gene studies have paid attention on serotonergic/catecholaminergic systems and neurotrophic signalling. The less active serotonin transporter (SLC6A4) polymorphic region (5-HTTLPR) s/s (with "s" indicating the short allele) genotype was found significantly more prevalent in Chinese patients with GAD compared with healthy subjects [12].

Genes for MAOA, SLC6A4 and, with the partial mediation of comorbidity with major depression, 5-hydroxytryptamine receptor 1A (5-HTR1A) have all been considered to be potentially implicated in the pathogenesis of GAD.

The less active Val66Met polymorphism of the functional BDNF gene has been associated with GAD susceptibility, along with an increase in serum BDNF levels in a Brazilian [119], but not Chinese Han, population [120, 121].

GWAS approach has targeted mainly GAD-related dimensional traits rather than GAD directly, on the basis of twin studies which have reported high genetic correlations between GAD and GAD-related dimensional traits [122], in particular neuroticism [12].

Prediction of Treatment Outcome

A large part of data on genetic treatment predictors in patients with GAD comes from studies derived from an open-label prospective trial with venlafaxine XR. These investigations have shown that carriers of the G-allele single-nucleotide polymorphism rs7997012 of the serotonin receptor 2A gene (HTR2A) had a superior response to venlafaxine XR [123], while carriers of the Met variant of the

COMT gene [124] and of SNP rs2856966 of the pituitary adenylate cyclase-activating polypeptide gene (PACAP) responded only slightly better [125]. Functional genetic polymorphisms in the dopamine active transporter 1 (DAT1) and dopamine receptor D2 (DRD2) did not impact response to venlafaxine XR in GAD [126].

Narasimhan et al. [127] did not report any relationship between the BDNF Val66Met polymorphism and the therapeutic effect of 6-week treatment with venlafaxine XR. BDNF Val66Met polymorphism was also not related to the efficacy of escitalopram (SSRI) and venlafaxine XR (SNRI) after 8 weeks of treatment in GAD [120].

Summary

Despite the official status as a stand-alone diagnosis, the never-ending debate of whether GAD is truly a separate disorder or a possible expression of other disorder such as major depression may have contributed to the dearth of studies of possible markers of susceptibility or treatment response.

The studies on demographic and clinical characteristics, biological markers and genetics as possible predictors of treatment response have not yet yielded any robust findings.

Neuroimaging studies have shown that the pattern of limbic/prefrontal activation and connectivity is highly variable in patients with GAD, different from other ADs, in which this pattern is more consistent. Understanding the sources of this variability will be essential in future investigation. These results preclude, at the moment, the use of neuroimaging data in predicting GAD vulnerability and the response to pharmacological or psychotherapeutic interventions. One possible exception is the pattern of ACC-amygdala responsiveness as a predictor of response to antidepressants in GAD, though confirmatory studies are needed.

Thus far, genetic studies have not led to the discovery of GAD vulnerability genes. Due to the high phenotypic (and probably also etiological) heterogeneity of GAD categorical diagnosis, focusing on dimensional measures of anxiety-related endophenotypes, such as excessive worry, fear of uncertainty and neuroticism, might enhance detection of genetic risk variants.

Social Anxiety Disorder

Patients with SAD are marked by a pervasive and often disabling fear, anxiety and/or avoidance of social interactions and/or situations in that they are likely of being scrutinized. The cognitive ideation is of being negatively judged by others, by being embarrassed, humiliated or rejected, or being offended by others [2]. Its lifetime prevalence, reaching 13% in adults, is one of the highest among mental health disorders [1]. SAD has a significant social and economic burden, because patients commonly have a reduced probability of marrying, a lower economic status and a higher probability of employment termination [128]. SAD has serious economic implications for society, because of the loss of occupation productivity and costs of healthcare utilization [97, 129].

Clinical recommendation for treatment of SAD [130] suggests individual CBT as the first-step intervention for the treatment of SAD, because of its large effect sizes [131]. The available pharmacological treatments (i.e. SSRIs) are recommended when patients decline CBT and express a preference for a psychotropic medication. Second-stage treatments, for patients who do not respond and/or do not tolerate side effects, mainly involve combination of SSRIs with individual CBT, offering alternative SSRIs or switching to SNRIs (i.e. venlafaxine) [130].

In accordance with a recent meta-analysis [132], about 60% of patients undergoing CBT as monotherapy do not achieve remission at post-treatment or at follow-up. Unsatisfactory response rates were also found after initial pharmacological treatment with SSRIs [133, 134], with about 37–47% of patients failing to respond, and after proposing second-stage strategies, with more than 50% of patients remaining refractory [135].

Endophenotypes/Biomarkers

Up to now, available data do not allow to identify clear neurobiological and psychophysiological endophenotypes/biomarkers in SAD. However, there are several promising candidates that deserve further examination.

Prediction of Disease Vulnerability

Behavioural Inhibition

BI is a particular temperamental trait defined by a particular pattern of emotional and behavioural responses to unfamiliar people or unusual situations. Fearfulness, shyness, hypersensitivity to novel sensorial stimuli, proximity to attachment figures and caution or avoidance in facing unfamiliar people or situations reflect the core features of inhibited children [136, 137]. Such patterns can be identified early in childhood (i.e. 2 years of age or earlier) and overlap with several characteristics of SAD [138]. Furthermore, such patterns are associated with specific physiological responses. Inhibited children, compared with non-inhibited children, present with signs of greater physiological arousal at rest, as reflected by higher cortisol levels, and in reaction to novel situations, as indicated by higher HR and HR acceleration, poorer HR habituation, pupillary dilation and laryngeal muscle tension [138].

BI has been consistently reported as a possible endophenotype for the development of SAD. Prevalence of lifetime social anxiety (i.e. DSM-IV social phobia or DSM-III-R avoidant disorder) has been found to be elevated twofold in inhibited children versus non-inhibited children, without association with other ADs [139]. Moreover, a meta-analysis of longitudinal studies [136] reported that a BI classification in toddlers or early childhood augmented the risk of developing SAD during childhood or adolescence, independently from study differences in temperament assessment, control group, familiar risk and age at anxiety diagnosis.

Blood or Other Peripheral Measures

A growing body of studies suggests that endocrine systems may be involved in the aetiology of SAD [140–142]. According to van Honk et al. [142], hormonal alterations, often established early in life, by the interplay between biological and environmental factors, may predispose to socially fearful, avoidant and submissive behaviour.

The developmental plasticity of the social brain, and thus of social behaviour in adulthood, critically depends on peptide or steroid hormones, such as the oxytocin (OXT) or testosterone (T) hormones. In rodents, OXT and T have shown prosocial and anxiolytic properties (i.e. facilitation of relational approach and maintenance of closeness, increase of social exploration, decrease of social anxiety and avoidance) [143]. Furthermore, there are some intriguing data showing that OXT and T levels, measured in plasma or saliva, are lowered in SAD [144–146].

Other neuroendocrine studies described possible imbalance of the HPA axis in SAD. In line with animal models of submissiveness, enhanced cortisol stress responses have been found in SAD in comparison with healthy or PTSD subjects, with high cortisol responses being related to elevated social avoidance behaviours on a social approach-avoidance (AA) task in patients with SAD [140].

Neuroimaging

Although longitudinal studies offer a window on developmental changes and, thus, can uniquely inform on the aetiology of SAD, most of what we know about the mechanisms involved is solely based on cross-sectional studies [147].

In a recent critical overview, Cremers and Roelofs [140] reported that patients with SAD are characterized by (1) hyperresponsive emotion network, which includes the amygdala, the bed nucleus of the stria terminalis (BNST), several brainstem nuclei, the insula and the fusiform gyrus, both in response to social threatening stimuli and at rest; (2) a diminished cognitive control and emotional regulation network, which involves PFC regions, during socially stressful situations; (3) an excessively active and extensively connected default-mode network (DMN), which traditionally encompasses the posterior cingulate and medial prefrontal regions, both at rest and during self-referential processes; and (4) an active motivational system, which comprises a set of regions in the brainstem, striatum and medial PFC (mPFC) integral to mesolimbic dopaminergic activity, especially in obtaining nonsocial rewards and avoiding social punishment.

Some fMRI studies evaluated the neural signatures of risk of SAD testing the role of BI in functional activation and connectivity in the brain regions mainly involved in SAD. The investigators [148] found that inhibited subjects had more activation of a prefrontal network when anticipating viewing fear faces, relative to uninhibited subjects (similar activity was identified in the amygdala). Furthermore, a greater negative connectivity between the rostral ACC and the bilateral amygdala was found in inhibited subjects, while a greater ACC activation and functional connectivity between the ACC and amygdala was found in those with fewer social anxiety symptoms and better emotion regulation skills. Another study [149] examined if defects in brain reward systems, which was hypothesized to be the brain basis of BI, were evident early in inhibited children who were susceptible to develop SAD. Greater caudate activation and stronger striatal connectivity was found in high, compared to low, BI children, with a relationship between caudate activation and presence of social anxiety symptoms. The authors concluded that elevated striatal responsivity reliably manifests among highly BI children as early as age 10, reflecting hypersensitivity to reward or excessive tendency to avoid errors.

Electroencephalography

A recent family study [150] investigated electrocortical endophenotypes of SAD during a social judgment task in which they believed to be social accepted or rejected. The feedback-related electroencephalogram (EEG) brain potentials (N1, FRN, P3) and theta power were tested as candidate endophenotypes, based on the criteria of familiar co-segregation of the disorder and heritability. Findings showed that increases in N1 in response to expected rejection feedback and in P3 in response to acceptance feedback might be candidates as electrocortical endophenotypes, although the heritability analyses were not significant after multiple testing correction. The authors suggested that increased N1 probably represents hypervigilance to socially threatening stimuli, and increased P3 might indicate that positive feedback is more important for, and/or less expected by, patients with SAD. Finally, increased feedback-related negativity and theta power in response to unexpected rejection feedback, compared to the other conditions, co-segregated with SAD, but without heritability of these EEG measures.

A study [151] on autonomic arousal [HR, HRV, electrodermal activity (EDA) and blushing] in children of parents with or without lifetime SAD found that children at risk for SAD exhibited extreme blushing in socially challenging situations when compared to those of parents without SAD. Furthermore, children of parents with more social anxiety symptoms had heightened EDA throughout the tasks. At last, a reduction in HRV and increases in blushing and EDA were associated with more severe child social anxiety. A subsequent investigation by the same research group [152] similarly found enhanced EDA in children of parents with more severe SAD throughout the task phases. Furthermore, increased HR and reduced HRV during the stranger approach and elevated EDA throughout the task procedures were associated with later development social anxiety in childhood (i.e. from 2.5 to 4.5 years old).

Prediction of Treatment Outcome

Blood or Other Peripheral Measures

Although there are no studies that have directly examined blood or other peripheral measures as predictors of treatment outcome, some preliminary studies, on the potential therapeutic use of OXT and T in patients affected by SAD or in those with severe levels of social anxiety, deserve attention.

Intranasal OXT administration has been reported to decrease and, thus, normalize attention bias for emotional faces in highly social-anxious subjects [153]. Furthermore, OXT has been proposed as a complementary intervention to exposure therapy for SAD. Patients administered intranasal oxytocin exhibited improvement of positive evaluations of appearance and speech performance as exposure sessions progressed, though these effects did not generalize to improve general outcome from exposure therapy (they may be either short term or situation specific) [154]. Similarly, T administration can reduce gaze avoidance [155] and attentional bias for social threat (i.e. angry faces) in females with SAD [156].

Neuroimaging

Several studies on neuroimaging predictors of treatment response, which largely comprises antidepressants and CBT, found the role of amygdala and other putative brain structures, particularly linked to emotional regulation, as candidate sites of action for therapeutic improvement in patients with SAD.

In one of the pioneer investigations in generalized SAD, Kilts et al. [157] explore neurocortical effects of 8 weeks of nefazodone. Social anxiety-related brain activations were compared before and after nefazodone. Findings demonstrated greater activity in the precentral gyrus, insula, midbrain/hypothalamus and middle frontal and anterior cingulate gyrus prior to treatment and greater activity in the left middle occipital and bilateral lingual gyri, postcentral gyrus, gyrus rectus and hippocampus following treatment. An analysis on treatment-related changes in symptom severity resulted in differential neural responses associated with symptom remission vs partial response.

Exaggerated amygdala activation in patients with SAD, which was positively associated with symptom severity, appeared to decrease after successful outcomes of citalopram or CBT [70]. A recent review [158] highlighted that activity of amygdala and other regions, including frontal structures (e.g. ACC, mPFC) and areas implicated in visual processes (e.g. occipital regions, superior temporal gyrus), predicted SSRIs/CBT outcomes in SAD and was altered following treatments. In particular, pretreatment functional connectivity between amygdala and frontal areas was shown to predict benefits; treatment, regardless of modality, produced reductions in amygdala reaction and altered functional connectivity in amygdala pathways. These findings were expanded by a recent meta-analysis focused on CBT outcomes in SAD, which confirmed the predictive role of amygdala and frontal regions (i.e. dlPFC and ACC) and indicated insula activity as a putative predictor of treatment response [71].

Electroencephalography

Few studies focused on electroencephalographic endophenotypes of treatment outcome.

In the investigation of Slaap et al. (1996) [159], a disturbed autonomic nervous system was found in non-responders to drug treatment (i.e. brofaromine or fluvoxamine), who exhibited higher HR and BP levels compared with responders. However, these results were not replicated by Stein et al. [160] in their follow-up study. Two other preliminary studies [161, 162] investigated EEG activity and coupling between EEG delta and beta oscillations in adults with a primary diagnosis of SAD following CBT. Findings showed that patients shifted significantly from greater relative right to greater relative left resting neural brain activity throughout the treatment. Greater left frontal EEG activity at pretreatment predicted substantial reduction in social anxiety from pre- to post-treatment and lower post-treatment social anxiety [162]. Furthermore, responders also showed a depletion in delta-beta coupling at rest and speech anticipation, suggesting a normalization of the EEG cross-frequency profiles with treatment [161]. Finally, in a more recent investigation [163], it was assessed whether vMMN amplitude, i.e. an event-related potential sensitive to violations of learned statistical contingencies, predicts attention bias modification (ABM) therapy outcome. Findings supported vMMN as a putative brain marker for contingency learning in ABM, because vMMN amplitude predicted clinical improvement after ABM treatment.

Genetics

The pathophysiology of SAD and the mechanisms of action of pharmacological/psychotherapeutic therapies remain mostly unknown. However, family, twin and, to a lesser extent, adoption studies contribute to the hypothesis that genetic factors may be involved both in susceptibility and response to treatment in SAD.

Prediction of Disease Vulnerability

There is strong evidence that familiar aggregation of SAD has a crucial role in determining vulnerability for this disorder [164], though the familiar risk differs across ages. The heritability rate for SAD/social anxiety symptoms measured in children is approximately 54%, with the remaining variance attributable to environmental influences. In contrast, a lesser estimate of genetic influences was found in adults (27%), with the most variance attributable to non-shared environmental influences (69%) [165].

Although there is a high genetic risk for susceptibility to SAD [166], genetic research on candidate genes is too meagre to draw conclusions. A strong focus has been posited on genetic polymorphisms associated with components of the serotoninergic (5-HT) and dopaminergic pathways; however, to date, evidence on their role is scarce [138, 167–169].

Prediction of Treatment Outcome

Research on genetic predictors of response to treatment is scarce.

Variations in the 5-HTTLPR and in a regulator of G-protein signalling 2 (RGS2) genes have been implicated to the outcomes after 10–12 weeks of drug treatment with SSRIs in patients with SAD [170, 171]. However, no genetic polymorphisms (5-HTTLPR, COMTval158met, BDNFval66met and TPH2 G-703T] predicted or moderated outcomes in a study on patients with SAD treated with CBT [172, 173].

Epigenetics

Prediction of Disease Vulnerability

Applying a multilevel epigenetic approach, Ziegler et al. [174] have examined whether OXT receptor (OXTR) gene methylation may play a role in categorical, dimensional and intermediate neuroendocrinological phenotypes of social anxiety. The investigators found that a significant decrease of OXTR methylation (particularly at CpG Chr3: 8 809 437) was linked to (1) the categorical phenotype of SAD, (2) enhanced Social Phobia Scale (SPS) and Social Interaction Anxiety Scale (SIAS) scores, (3) enhanced cortisol response to the TSST and (4) enhanced amygdala responsiveness during social anxiety-related verbal stimuli. The authors suggested that OXTR methylation patterns might serve as peripheral surrogates of oxytocin tone and aid in establishing accessible biomarkers of SAD susceptibility. Another intriguing study [175] showed differences in epigenetic regulation of the OXTR gene methylation during detection of, and orientation to, social information in non-clinical individuals. A functional epigenetic modification of the OXTR methylation was related to increased response within – and reduced functional coupling between – brain regions responsible for guiding attention to social information during selective social attention tasks. The authors suggest that OXTR methylation, reflecting a decrease in the extent to which social information automatically captures attention, may be implicated in regulation of social behaviour and social anxiety traits.

Growing evidence argues an interactional model, in which intrinsic factors, such as genetic predisposition, interplay with environmental factors, such as childhood trauma (CT), to increase susceptibility to SAD [138]. Higher rates of CT experiences, particularly emotional abuse/neglect, have been reported in patients affected by SAD, in comparison with healthy subjects [176] and patients with other ADs (i.e. PD) [177]. Reinelt et al. [178], examining a specific diathesis-stress model of SAD, found that the l/l genotype of 5-HTTLPR, emotional abuse/neglect, female sex and younger age were all linked to increased risk for SAD.

Summary

Although not prolific, the current literature on endophenotypes/biomarkers, genetics and epigenetics in SAD offer hope toward early prediction of vulnerability of SAD and improvement of outcomes resulting from the available pharmacological/psychotherapeutic interventions. First, among the clinical endophenotypes, BI has been consistently recognized as playing a prominent role in development of SAD. Second, studies on neurobiological endophenotypes found that particular neural activity in regions primarily devoted to emotion control and/or regulation (e.g. amygdala, brainstem, insula, ACC and PFC) may be indicative of disease vulnerability or predictive of negative treatment outcomes. Third, some preliminary electroencephalographic findings show that particular EEG activity and autonomic arousal (e.g. HR, HRV, EDA) may reflect the risk of developing SAD or failing to respond to pharmacological/CBT treatments. Furthermore, research on blood or other peripheral measures, even if at an early stage, give intriguing inputs that dysregulations in hormonal systems, particularly related to OXT, T and the HPA axis, may be implicated in vulnerability and treatment response of SAD. Finally, although genetic and epigenetic studies can only give some preliminary cues, it makes sense that some genetic polymorphisms (e.g. 5-HTTLPR) and epigenetic mechanisms (e.g. OXTR methylation and CT) should receive further investigation in prediction of vulnerability and treatment outcome in SAD.

Conclusions and Future Directions

Despite significant efforts expended into finding possible indicators of susceptibility for ADs, no result is robust enough to be considered acceptable to incorporate in the clinical practice. Why is this the case?

First, given that ADs develop from complex interaction between biological and environmental factors, it is implausible that a single biomarker might capture the multifaceted pathogenesis. A clearly multimodal approach based on a combination of different biomarkers will likely be implemented. At a more general level, more fruitful recent paradigms shift from a reductionist perspective toward a multiple biological system-based approaches. These approaches investigate complex phenotypes, such as psychiatric disorders, by combining mathematical models, molecular information from in vitro/in vivo/in silico studies, and multi-omics data.

Second, the diagnostic taxonomy of mental disorders, as introduced by DSM and ICD, is only partially useful for personalized psychiatry, as it does not take into account all experimental findings regarding neurobiological and behavioural systems involved in mental disorders. To address this limitation, the National Institute of Mental Health initiated the Research Domain Criteria (RDoC) project. This ongoing project is aimed to establish a classification system for mental disorders based on neurobiological findings and behaviours, integrating biomarkers, neuropsychological assessment and brain imaging data. Currently, no results translatable to clinical practice are available.

Third, although the use of neuroimaging techniques, such as volumetric and functional MRI, may hold promise for advancing personalized psychiatry, several methodological limitations such as small study populations, comorbidities, variability in neuroimaging protocols, medication confounders and poor study design account for inconsistent and non-reproducible findings. The ENIGMA (Enhancing Neuroimaging Genetics through Meta-Analysis) network is now managing to overcome some of these limitations through an imaging genomics consortium.

Finally, although GWAS has provided a powerful approach to address the main limitations of studies on putative genes, the susceptibility genes that have been found are not of diagnostic or predictive value in ADs, though polygenic risk factor scores may represent a significant advance. In order to obtain reliable results, large biological samples, collected on the basis of international cooperation of genetic researchers and analysed with the new methods of joint analysis, are needed.

In recent years, several statistical methods have been introduced to detect association using multiple phenotypes simultaneously. Traditionally, GWAS have been performed on individual phenotypes, but, to study complex diseases, several phenotypes related to each disorder should be taken into account and jointly analysed for increasing the chance of detecting variants with weak genetic effects.

As with research in other mental disorders, such as depression, there are still no neurobiological, genetic or epigenetic predictors of sufficient clinical utility to inform the selection of a specific pharmacological or psychotherapeutic treatment, or a compound, for an individual patient. Similarly, clinical characteristics of ADs, such as clinical subtypes and anxiety-relevant endophenotypes, have a marginal role in predicting disease vulnerability or treatment response. The Coordinated Anxiety Learning and Management (CALM) [179] study has offered a unique possibility to compare socio-demographic and clinical predictors between ADs undergoing similar baseline assessment and treatment in a primary care setting. The study has demonstrated that, across multiple ADs, low socioeconomic condition was related to worse outcomes and positive treatment expectancy/belief in self-efficacy with treatment response. SAD showed different predictors of treatment outcome compared with the other ADs, primarily including measures of perceived social support.

Current biomarker research progression is shifting toward the proteins that are the true effectors of physiological functions.Proteomics is the study of any protein expressed by a given cell, tissue or organism at a particular time under known and determined conditions and includes the modification of proteins under the influence of certain internal and environmental factors. Proteomic biomarker research is an area with enormous potential for easy accessibility of the source (blood, urine and saliva) of putative proteomic biomarker candidate, though the relationship of such findings to CNS processes remains to be determined.

Conflicts of Interest Dr. Charles B. Nemeroff's disclosures are as follows:

Dr. Nemeroff has received funding from NIH and the Stanley Medical Research Institute; he has served as a consultant to Bracket (Clintara), Gerson Lehrman Group Healthcare and Biomedical Council, Fortress Biotech, Sunovion Pharmaceuticals, Janssen Research and Development, Magstim, Navitor Pharmaceuticals, Intra-Cellular Therapies, Takeda, Taisho Pharmaceutical and Xhale; he has served on the boards of directors for the American Foundation for Suicide Prevention, Gratitude America and the Anxiety Disorders Association of America; he is a stockholder in AbbVie, Antares, Bracket Intermediate Holding Corp., Celgene, OPKO Health, Seattle Genetics and Xhale; he serves on the scientific advisory boards of the American Foundation for Suicide Prevention, the Anxiety Disorders Association of America, the Brain and Behavior Research Foundation, Bracket (Clintara), the Laureate Institute for Brain Research, RiverMend Health, Skyland Trail and Xhale; he reports income sources or equity of $10,000 or more from American Psychiatric Publishing, Bracket (Clintara), CME Outfitters, Takeda and Xhale; he has patents on a method and devices for transdermal delivery of lithium (US 6,375,990B1) and a method of assessing antidepressant drug therapy via transport inhibition of monoamine neurotransmitters by ex vivo assay (US 7,148,027B2).

Dr. Giampaolo Perna's disclosures are as follows:

Giampaolo Perna has received funding from Cariplo Foundation. He has served in the scientific advisor board of Medibio Ltd and has served as consultant for Lundbeck and Pfizer.

References

1. Kessler RC, Petukhova M, Sampson NA, Zaslavsky AM, Wittchen HU. Twelve-month and lifetime prevalence and lifetime morbid risk of anxiety and mood disorders in the United States. Int J Methods Psychiatr Res. 2012;21(3):169–84.
2. American Psychiatric Association. Diagnostic and statistical manual of mental disorders. 5th ed. Arlington, VA: American Psychiatric Association; 2013.
3. World Health Organization (WHO). Depression and other common mental disorders: global health estimates 2017. Available from: https://apps.who.int/iris/bitstream/handle/10665/254610/WHO-MSD-MER-2017.2-eng.pdf.
4. Emdin CA, Odutayo A, Wong CX, Tran J, Hsiao AJ, Hunn BH. Meta-analysis of anxiety as a risk factor for cardiovascular disease. Am J Cardiol. 2016;118(4):511–9.
5. Bandelow B, Michaelis S. Epidemiology of anxiety disorders in the 21st century. Dialogues Clin Neurosci. 2015;17(3):327–35.
6. Alonso J, Lepine JP, Committee ESMS. Overview of key data from the European Study of the Epidemiology of Mental Disorders (ESEMeD). J Clin Psychiatry. 2007;68(Suppl 2):3–9.
7. Bandelow B, Michaelis S, Wedekind D. Treatment of anxiety disorders. Dialogues Clin Neurosci. 2017;19(2):93–107.
8. Koen N, Stein DJ. Pharmacotherapy of anxiety disorders: a critical review. Dialogues Clin Neurosci. 2011;13(4):423–37.
9. Prendes-Alvarez S, Nemeroff CB. Personalized medicine: prediction of disease vulnerability in mood disorders. Neurosci Lett. 2018;669:10–3.
10. Perna G, Nemeroff CB. Personalized medicine in psychiatry: back to the future. Pers Med Psychiatry. 2017;1–2(1)
11. Batelaan NM, Van Balkom AJ, Stein DJ. Evidence-based pharmacotherapy of panic disorder: an update. Int J Neuropsychopharmacol. 2012;15(3):403–15.
12. Smith DJ, Escott-Price V, Davies G, Bailey ME, Colodro-Conde L, Ward J, et al. Genome-wide analysis of over 106 000 individuals identifies 9 neuroticism-associated loci. Mol Psychiatry. 2016;21(11):1644.
13. American Psychiatric Association. Practice guidelines for the treatment of patients with panic disorder. 2009.
14. Baldwin DS, Anderson IM, Nutt DJ, Allgulander C, Bandelow B, den Boer JA, et al. Evidence-based pharmacological treatment of anxiety disorders, post-traumatic stress disorder and obsessive-compulsive disorder: a revision of the 2005 guidelines from the British Association for Psychopharmacology. J Psychopharmacol. 2014;28(5):403–39.
15. Perna G, Schruers K, Alciati A, Caldirola D. Novel investigational therapeutics for panic disorder. Expert Opin Investig Drugs. 2015;24(4):491–505.
16. Perna G, Caldirola D, Bellodi L. Panic disorder: from respiration to the homeostatic brain. Acta Neuropsychiatr. 2004;16(2):57–67.
17. Roberson-Nay R, Klein DF, Klein RG, Mannuzza S, Moulton JL 3rd, Guardino M, et al. Carbon dioxide hypersensitivity in separation-anxious offspring of parents with panic disorder. Biol Psychiatry. 2010;67(12):1171–7.
18. Gottesman II, Gould TD. The endophenotype concept in psychiatry: etymology and strategic intentions. Am J Psychiatry. 2003;160(4):636–45.
19. Perna G, Barbini B, Cocchi S, Bertani A, Gasperini M. 35% CO2 challenge in panic and mood disorders. J Affect Disord. 1995;33(3):189–94.
20. Perna G, Bussi R, Allevi L, Bellodi L. Sensitivity to 35% carbon dioxide in patients with generalized anxiety disorder. J Clin Psychiatry. 1999;60(6):379–84.
21. Perna G, Cocchi S, Bertani A, Arancio C, Bellodi L. Sensitivity to 35% CO2 in healthy first-degree relatives of patients with panic disorder. Am J Psychiatry. 1995;152(4):623–5.
22. Bellodi L, Perna G, Caldirola D, Arancio C, Bertani A, Di Bella D. CO2-induced panic attacks: a twin study. Am J Psychiatry. 1998;155(9):1184–8.

23. Battaglia M, Pesenti-Gritti P, Spatola CA, Ogliari A, Tambs K. A twin study of the common vulnerability between heightened sensitivity to hypercapnia and panic disorder. Am J Med Genet B Neuropsychiatr Genet. 2008;147B(5):586–93.
24. Schmidt NB, Storey J, Greenberg BD, Santiago HT, Li Q, Murphy DL. Evaluating gene × psychological risk factor effects in the pathogenesis of anxiety: a new model approach. J Abnorm Psychol. 2000;(2):308–20.
25. Perna G, di Bella D, Favaron E, Cucchi M, Liperi L, Bellodi L. Lack of relationship between CO2 reactivity and serotonin transporter gene regulatory region polymorphism in panic disorder. Am J Med Genet B Neuropsychiatr Genet. 2004;129B(1):41–3.
26. Schruers K, Esquivel G, van Duinen M, Wichers M, Kenis G, Colasanti A, et al. Genetic moderation of CO2-induced fear by 5-HTTLPR genotype. J Psychopharmacol. 2011;25(1):37–42.
27. Bandelow B, Baldwin D, Abelli M, Bolea-Alamanac B, Bourin M, Chamberlain SR, et al. Biological markers for anxiety disorders, OCD and PTSD: a consensus statement. Part II: Neurochemistry, neurophysiology and neurocognition. World J Biol Psychiatry. 2017;18(3):162–214.
28. Eriksson E, Westberg P, Alling C, Thuresson K, Modigh K. Cerebrospinal fluid levels of monoamine metabolites in panic disorder. Psychiatry Res. 1991;36(3):243–51.
29. Charney DS, Woods SW, Heninger GR. Noradrenergic function in generalized anxiety disorder: effects of yohimbine in healthy subjects and patients with generalized anxiety disorder. Psychiatry Res. 1989;27(2):173–82.
30. Kalk NJ, Nutt DJ, Lingford-Hughes AR. The role of central noradrenergic dysregulation in anxiety disorders: evidence from clinical studies. J Psychopharmacol. 2011;25(1):3–16.
31. Bhad R. Red blood cell and platelet indices: a potential biomarker for panic disorder. J Neurosci Rural Pract. 2017;8(2):164.
32. Cosci F, Mansueto G. Biological and clinical markers in panic disorder. Psychiatry Investig. 2019;16(1):27–36.
33. Schleifer SJ, Keller SE, Bartlett JA. Panic disorder and immunity: few effects on circulating lymphocytes, mitogen response, and NK cell activity. Brain Behav Immun. 2002;16(6):698–705.
34. Quagliato LA, Nardi AE. Cytokine alterations in panic disorder: a systematic review. J Affect Disord. 2018;228:91–6.
35. Grassi M, Caldirola D, Vanni G, Guerriero G, Piccinni M, Valchera A, et al. Baseline respiratory parameters in panic disorder: a meta-analysis. J Affect Disord. 2013;146(2):158–73.
36. Grassi M, Caldirola D, Di Chiaro NV, Riva A, Dacco S, Pompili M, et al. Are respiratory abnormalities specific for panic disorder? A meta-analysis. Neuropsychobiology. 2014;70(1):52–60.
37. Perna G, Caldirola D. Is panic disorder a disorder of physical fitness? A heuristic proposal. F1000Research. 2018;7:294.
38. Meuret AE, Rosenfield D, Wilhelm FH, Zhou E, Conrad A, Ritz T, et al. Do unexpected panic attacks occur spontaneously? Biol Psychiatry. 2011;70(10):985–91.
39. Perna G, Ieva A, Caldirola D, Bertani A, Bellodi L. Respiration in children at risk for panic disorder. Arch Gen Psychiatry. 2002;59(2):185–6.
40. Srinivasan K, Ashok MV, Vaz M, Yeragani VK. Decreased chaos of heart rate time series in children of patients with panic disorder. Depress Anxiety. 2002;15(4):159–67.
41. Chalmers JA, Quintana DS, Abbott MJ, Kemp AH. Anxiety disorders are associated with reduced heart rate variability: a meta-analysis. Front Psych. 2014;5:80.
42. Gorman JM, Kent JM, Sullivan GM, Coplan JD. Neuroanatomical hypothesis of panic disorder, revised. Am J Psychiatry. 2000;157(4):493–505.
43. Asami T, Nakamura R, Takaishi M, Yoshida H, Yoshimi A, Whitford TJ, et al. Smaller volumes in the lateral and basal nuclei of the amygdala in patients with panic disorder. PLoS One. 2018;13(11):e0207163.
44. Feinstein JS, Buzza C, Hurlemann R, Follmer RL, Dahdaleh NS, Coryell WH, et al. Fear and panic in humans with bilateral amygdala damage. Nat Neurosci. 2013;16(3):270–2.

45. Wiest G, Lehner-Baumgartner E, Baumgartner C. Panic attacks in an individual with bilateral selective lesions of the amygdala. Arch Neurol. 2006;63(12):1798–801.
46. Perna G, Guerriero G, Brambilla P, Caldirola D. Panic and the brainstem: clues from neuro-imaging studies. CNS Neurol Disord Drug Targets. 2014;13(6):1049–56.
47. Goossens L, Leibold N, Peeters R, Esquivel G, Knuts I, Backes W, et al. Brainstem response to hypercapnia: a symptom provocation study into the pathophysiology of panic disorder. J Psychopharmacol. 2014;28(5):449–56.
48. Engel KR, Obst K, Bandelow B, Dechent P, Gruber O, Zerr I, et al. Functional MRI activation in response to panic-specific, non-panic aversive, and neutral pictures in patients with panic disorder and healthy controls. Eur Arch Psychiatry Clin Neurosci. 2016;266(6):557–66.
49. Sobanski T, Wagner G. Functional neuroanatomy in panic disorder: status quo of the research. World J Psychiatry. 2017;7(1):12–33.
50. Bertani A, Caldirola D, Bussi R, Bellodi L, Perna G. The 35% CO2 hyperreactivity and clinical symptomatology in patients with panic disorder after 1 week of treatment with citalopram: an open study. J Clin Psychopharmacol. 2001;21(3):262–7.
51. Bertani A, Perna G, Arancio C, Caldirola D, Bellodi L. Pharmacologic effect of imipramine, paroxetine, and sertraline on 35% carbon dioxide hypersensitivity in panic patients: a double-blind, random, placebo-controlled study. J Clin Psychopharmacol. 1997;17(2):97–101.
52. Perna G, Bertani A, Caldirola D, Di Pasquale D, Migliarese G, Bellodi L. Modulation of hyperreactivity to 35% CO2 after one week of treatment with paroxetine and reboxetine: a double-blind, randomized study. J Clin Psychopharmacol. 2004;24(3):277–82.
53. Perna G, Bertani A, Caldirola D, Gabriele A, Cocchi S, Bellodi L. Antipanic drug modulation of 35% CO2 hyperreactivity and short-term treatment outcome. J Clin Psychopharmacol. 2002;22(3):300–8.
54. Perna G, Bertani A, Gabriele A, Politi E, Bellodi L. Modification of 35% carbon dioxide hypersensitivity across one week of treatment with clomipramine and fluvoxamine: a double-blind, randomized, placebo-controlled study. J Clin Psychopharmacol. 1997;17(3):173–8.
55. Schmidt NB, Trakowski JH, Staab JP. Extinction of panicogenic effects of a 35% CO2 challenge in patients with panic disorder. J Abnorm Psychol. 1997;106(4):630–8.
56. Lee IS, Kim KJ, Kang EH, Yu BH. beta-adrenoceptor affinity as a biological predictor of treatment response to paroxetine in patients with acute panic disorder. J Affect Disord. 2008;110(1-2):156–60.
57. Maddock RJ, Carter CS, Magliozzi JR, Gietzen DW. Evidence that decreased function of lymphocyte beta adrenoreceptors reflects regulatory and adaptive processes in panic disorder with agoraphobia. Am J Psychiatry. 1993;150(8):1219–25.
58. Gurguis GN, Vo SP, Griffith JM, Rush AJ. Platelet alpha2A-adrenoceptor function in major depression: Gi coupling, effects of imipramine and relationship to treatment outcome. Psychiatry Res. 1999;89(2):73–95.
59. Roestel C, Hoeping W, Deckert J. Hypophosphatemia in panic disorder. Am J Psychiatry. 2004;161(8):1499–500.
60. Perez-Costillas L, Montes MR, Martinez-Ortega JM, Carretero MD, Gutierrez-Rojas L, Gurpegui M. Phosphate levels as a possible state marker in panic disorder: preliminary study of a feasible laboratory measure for routine clinical practice. J Psychiatr Res. 2013;47(10):1357–62.
61. Beria P, Viana ACW, Behenck A, Heldt E, Manfro GG, Dreher CB. Respiratory subtype of panic disorder: can serum phosphate levels be a possible outcome to group cognitive-behavior therapy? J Affect Disord. 2018;235:474–9.
62. Kobayashi K, Shimizu E, Hashimoto K, Mitsumori M, Koike K, Okamura N, et al. Serum brain-derived neurotrophic factor (BDNF) levels in patients with panic disorder: as a biological predictor of response to group cognitive behavioral therapy. Prog Neuro-Psychopharmacol Biol Psychiatry. 2005;29(5):658–63.
63. Garakani A, Martinez JM, Aaronson CJ, Voustianiouk A, Kaufmann H, Gorman JM. Effect of medication and psychotherapy on heart rate variability in panic disorder. Depress Anxiety. 2009;26(3):251–8.

64. Nardi AE, Valenca AM, Nascimento I, Lopes FL, Mezzasalma MA, Freire RC, et al. A three-year follow-up study of patients with the respiratory subtype of panic disorder after treatment with clonazepam. Psychiatry Res. 2005;137(1-2):61–70.
65. Tolin DF, Billingsley AL, Hallion LS, Diefenbach GJ. Low pre-treatment end-tidal CO2 predicts dropout from cognitive-behavioral therapy for anxiety and related disorders. Behav Res Ther. 2017;90:32–40.
66. Slaap BR, Boshuisen ML, van Roon AM, den Boer JA. Heart rate variability as predictor of nonresponse to mirtazapine in panic disorder: a preliminary study. Int Clin Psychopharmacol. 2002;17(2):69–74.
67. Wendt J, Hamm AO, Pane-Farre CA, Thayer JF, Gerlach A, Gloster AT, et al. Pretreatment cardiac vagal tone predicts dropout from and residual symptoms after exposure therapy in patients with panic disorder and agoraphobia. Psychother Psychosom. 2018;87(3):187–9.
68. Lai CH, Wu YT. Changes in gray matter volume of remitted first-episode, drug-naive, panic disorder patients after 6-week antidepressant therapy. J Psychiatr Res. 2013;47(1):122–7.
69. Lai CH, Wu YT, Yu PL, Yuan W. Improvements in white matter micro-structural integrity of right uncinate fasciculus and left fronto-occipital fasciculus of remitted first-episode medication-naive panic disorder patients. J Affect Disord. 2013;150(2):330–6.
70. Shin LM, Davis FC, Vanelzakker MB, Dahlgren MK, Dubois SJ. Neuroimaging predictors of treatment response in anxiety disorders. Biol Mood Anxiety Disord. 2013;3(1):15.
71. Santos VA, Carvalho DD, Van Ameringen M, Nardi AE, Freire RC. Neuroimaging findings as predictors of treatment outcome of psychotherapy in anxiety disorders. Prog Neuropsychopharmacol Biol Psychiatry. 2018;
72. Reif A, Richter J, Straube B, Hofler M, Lueken U, Gloster AT, et al. MAOA and mechanisms of panic disorder revisited: from bench to molecular psychotherapy. Mol Psychiatry. 2014;19(1):122–8.
73. Maron E, Toru I, Must A, Tasa G, Toover E, Vasar V, et al. Association study of tryptophan hydroxylase 2 gene polymorphisms in panic disorder. Neurosci Lett. 2007;411(3):180–4.
74. Domschke K, Deckert J, O'Donovan MC, Glatt SJ. Meta-analysis of COMT val158met in panic disorder: ethnic heterogeneity and gender specificity. Am J Med Genet B Neuropsychiatr Genet. 2007;144B(5):667–73.
75. Buttenschon HN, Kristensen AS, Buch HN, Andersen JH, Bonde JP, Grynderup M, et al. The norepinephrine transporter gene is a candidate gene for panic disorder. J Neural Transm. 2011;118(6):969–76.
76. Sand PG, Mori T, Godau C, Stober G, Flachenecker P, Franke P, et al. Norepinephrine transporter gene (NET) variants in patients with panic disorder. Neurosci Lett. 2002;333(1):41–4.
77. Otowa T, Kawamura Y, Nishida N, Sugaya N, Koike A, Yoshida E, et al. Meta-analysis of genome-wide association studies for panic disorder in the Japanese population. Transl Psychiatry. 2012;2:e186.
78. Perna G, Favaron E, Di Bella D, Bussi R, Bellodi L. Antipanic efficacy of paroxetine and polymorphism within the promoter of the serotonin transporter gene. Neuropsychopharmacology. 2005;30(12):2230–5.
79. Saeki Y, Watanabe T, Ueda M, Saito A, Akiyama K, Inoue Y, et al. Genetic and pharmacokinetic factors affecting the initial pharmacotherapeutic effect of paroxetine in Japanese patients with panic disorder. Eur J Clin Pharmacol. 2009;65(7):685–91.
80. Aoki A, Ishiguro S, Watanabe T, Ueda M, Hayashi Y, Akiyama K, et al. Factors affecting discontinuation of initial treatment with paroxetine in panic disorder and major depressive disorder. Neuropsychiatr Dis Treat. 2014;10:1793–8.
81. Ishiguro S, Watanabe T, Ueda M, Saeki Y, Hayashi Y, Akiyama K, et al. Determinants of pharmacodynamic trajectory of the therapeutic response to paroxetine in Japanese patients with panic disorder. Eur J Clin Pharmacol. 2011;67(12):1213–21.
82. Kim W, Choi YH, Yoon KS, Cho DY, Pae CU, Woo JM. Tryptophan hydroxylase and serotonin transporter gene polymorphism does not affect the diagnosis, clinical features and treatment outcome of panic disorder in the Korean population. Prog Neuro-Psychopharmacol Biol Psychiatry. 2006;30(8):1413–8.

83. Yevtushenko OO, Oros MM, Reynolds GP. Early response to selective serotonin reuptake inhibitors in panic disorder is associated with a functional 5-HT1A receptor gene polymorphism. J Affect Disord. 2010;123(1-3):308–11.
84. Favaron E, Biffi S, Grassi M, Bellodi L, Perna G. Response to paroxetine and catecol-o-metil transferasi (comt) polymorphisms in panic disorder. 11th National Congress of the Italian Society of Psychopathology. Rome, Italy; 2006.
85. Woo JM, Yoon KS, Choi YH, Oh KS, Lee YS, Yu BH. The association between panic disorder and the L/L genotype of catechol-O-methyltransferase. J Psychiatr Res. 2004;38(4):365–70.
86. Shimada-Sugimoto M, Otowa T, Miyagawa T, Umekage T, Kawamura Y, Bundo M, et al. Epigenome-wide association study of DNA methylation in panic disorder. Clin Epigenetics. 2017;9:6.
87. Schartner C, Ziegler C, Schiele MA, Kollert L, Weber H, Zwanzger P, et al. CRHR1 promoter hypomethylation: an epigenetic readout of panic disorder? Eur Neuropsychopharmacol. 2017;27(4):360–71.
88. Domschke K, Tidow N, Schrempf M, Schwarte K, Klauke B, Reif A, et al. Epigenetic signature of panic disorder: a role of glutamate decarboxylase 1 (GAD1) DNA hypomethylation? Prog Neuro-Psychopharmacol Biol Psychiatry. 2013;46:189–96.
89. Ziegler C, Richter J, Mahr M, Gajewska A, Schiele MA, Gehrmann A, et al. MAOA gene hypomethylation in panic disorder-reversibility of an epigenetic risk pattern by psychotherapy. Transl Psychiatry. 2016;6:e773.
90. Kroenke K, Spitzer RL, Williams JB, Monahan PO, Lowe B. Anxiety disorders in primary care: prevalence, impairment, comorbidity, and detection. Ann Intern Med. 2007;146(5):317–25.
91. Hoffman DL, Dukes EM, Wittchen HU. Human and economic burden of generalized anxiety disorder. Depress Anxiety. 2008;25(1):72–90.
92. Iny LJ, Pecknold J, Suranyi-Cadotte BE, Bernier B, Luthe L, Nair NP, et al. Studies of a neurochemical link between depression, anxiety, and stress from [3H]imipramine and [3H] paroxetine binding on human platelets. Biol Psychiatry. 1994;36(5):281–91.
93. Hernandez E, Lastra S, Urbina M, Carreira I, Lima L. Serotonin, 5-hydroxyindoleacetic acid and serotonin transporter in blood peripheral lymphocytes of patients with generalized anxiety disorder. Int Immunopharmacol. 2002;2(7):893–900.
94. Gerra G, Zaimovic A, Zambelli U, Timpano M, Reali N, Bernasconi S, et al. Neuroendocrine responses to psychological stress in adolescents with anxiety disorder. Neuropsychobiology. 2000;42(2):82–92.
95. Phillips AC, Batty GD, Gale CR, Lord JM, Arlt W, Carroll D. Major depressive disorder, generalised anxiety disorder, and their comorbidity: associations with cortisol in the Vietnam Experience Study. Psychoneuroendocrinology. 2011;36(5):682–90.
96. Steudte S, Stalder T, Dettenborn L, Klumbies E, Foley P, Beesdo-Baum K, et al. Decreased hair cortisol concentrations in generalised anxiety disorder. Psychiatry Res. 2011;186(2-3):310–4.
97. Stuhldreher N, Leibing E, Leichsenring F, Beutel ME, Herpertz S, Hoyer J, et al. The costs of social anxiety disorder: the role of symptom severity and comorbidities. J Affect Disord. 2014;165:87–94.
98. Molendijk ML, Bus BA, Spinhoven P, Penninx BW, Prickaerts J, Oude Voshaar RC, et al. Gender specific associations of serum levels of brain-derived neurotrophic factor in anxiety. World J Biol Psychiatry. 2012;13(7):535–43.
99. Pallanti S, Tofani T, Zanardelli M, Di Cesare ML, Ghelardini C. BDNF and Artemin are increased in drug-naive non-depressed GAD patients: preliminary data. Int J Psychiatry Clin Pract. 2014;18(4):255–60.
100. Carlino D, Francavilla R, Baj G, Kulak K, d'Adamo P, Ulivi S, et al. Brain-derived neurotrophic factor serum levels in genetically isolated populations: gender-specific association with anxiety disorder subtypes but not with anxiety levels or Val66Met polymorphism. PeerJ. 2015;3:e1252.
101. Bankier B, Barajas J, Martinez-Rumayor A, Januzzi JL. Association between C-reactive protein and generalized anxiety disorder in stable coronary heart disease patients. Eur Heart J. 2008;29(18):2212–7.

102. Copeland WE, Shanahan L, Worthman C, Angold A, Costello EJ. Generalized anxiety and C-reactive protein levels: a prospective, longitudinal analysis. Psychol Med. 2012;42(12):2641–50.
103. Koh KB, Lee BK. Reduced lymphocyte proliferation and interleukin-2 production in anxiety disorders. Psychosom Med. 1998;60(4):479–83.
104. Makovac E, Meeten F, Watson DR, Garfinkel SN, Critchley HD, Ottaviani C. Neurostructural abnormalities associated with axes of emotion dysregulation in generalized anxiety. NeuroImage Clin. 2016;10:172–81.
105. Schienle A, Ebner F, Schafer A. Localized gray matter volume abnormalities in generalized anxiety disorder. Eur Arch Psychiatry Clin Neurosci. 2011;261(4):303–7.
106. Etkin A, Prater KE, Schatzberg AF, Menon V, Greicius MD. Disrupted amygdalar subregion functional connectivity and evidence of a compensatory network in generalized anxiety disorder. Arch Gen Psychiatry. 2009;66(12):1361–72.
107. Liao M, Yang F, Zhang Y, He Z, Song M, Jiang T, et al. Childhood maltreatment is associated with larger left thalamic gray matter volume in adolescents with generalized anxiety disorder. PLoS One. 2013;8(8):e71898.
108. Hilbert K, Pine DS, Muehlhan M, Lueken U, Steudte-Schmiedgen S, Beesdo-Baum K. Gray and white matter volume abnormalities in generalized anxiety disorder by categorical and dimensional characterization. Psychiatry Res. 2015;234(3):314–20.
109. Moon CM, Jeong GW. Abnormalities in gray and white matter volumes associated with explicit memory dysfunction in patients with generalized anxiety disorder. Acta Radiol. 2017;58(3):353–61.
110. Monk CS, Telzer EH, Mogg K, Bradley BP, Mai X, Louro HM, et al. Amygdala and ventrolateral prefrontal cortex activation to masked angry faces in children and adolescents with generalized anxiety disorder. Arch Gen Psychiatry. 2008;65(5):568–76.
111. McClure EB, Monk CS, Nelson EE, Parrish JM, Adler A, Blair RJ, et al. Abnormal attention modulation of fear circuit function in pediatric generalized anxiety disorder. Arch Gen Psychiatry. 2007;64(1):97–106.
112. Fonzo GA, Ramsawh HJ, Flagan TM, Sullivan SG, Letamendi A, Simmons AN, et al. Common and disorder-specific neural responses to emotional faces in generalised anxiety, social anxiety and panic disorders. Br J Psychiatry. 2015;206(3):206–15.
113. Blair K, Shaywitz J, Smith BW, Rhodes R, Geraci M, Jones M, et al. Response to emotional expressions in generalized social phobia and generalized anxiety disorder: evidence for separate disorders. Am J Psychiatry. 2008;165(9):1193–202.
114. Whalen PJ, Johnstone T, Somerville LH, Nitschke JB, Polis S, Alexander AL, et al. A functional magnetic resonance imaging predictor of treatment response to venlafaxine in generalized anxiety disorder. Biol Psychiatry. 2008;63(9):858–63.
115. Fonzo GA, Etkin A. Affective neuroimaging in generalized anxiety disorder: an integrated review. Dialogues Clin Neurosci. 2017;19(2):169–79.
116. Oathes DJ, Hilt LM, Nitschke JB. Affective neural responses modulated by serotonin transporter genotype in clinical anxiety and depression. PloS One. 2015;10(2):e0115820.
117. Nitschke JB, Sarinopoulos I, Oathes DJ, Johnstone T, Whalen PJ, Davidson RJ, et al. Anticipatory activation in the amygdala and anterior cingulate in generalized anxiety disorder and prediction of treatment response. Am J Psychiatry. 2009;166(3):302–10.
118. Ball TM, Stein MB, Ramsawh HJ, Campbell-Sills L, Paulus MP. Single-subject anxiety treatment outcome prediction using functional neuroimaging. Neuropsychopharmacology. 2014;39(5):1254–61.
119. Moreira FP, Fabiao JD, Bittencourt G, Wiener CD, Jansen K, Oses JP, et al. The met allele of BDNF Val66Met polymorphism is associated with increased BDNF levels in generalized anxiety disorder. Psychiatr Genet. 2015;25(5):201–7.
120. Lin M, Zhu J, Yuan Y, Ren L, Qian M, Shen Z, et al. Association analysis of the brain-derived neurotrophic factor gene Val66Met polymorphism and gender with efficacy of antidepressants in the Chinese han population with generalized anxiety disorder. Genet Test Mol Biomarkers. 2018;22(3):199–206.

121. Wang Y, Zhang H, Li Y, Wang Z, Fan Q, Yu S, et al. BDNF Val66Met polymorphism and plasma levels in Chinese Han population with obsessive-compulsive disorder and generalized anxiety disorder. J Affect Disord. 2015;186:7–12.
122. Hettema JM, Prescott CA, Kendler KS. Genetic and environmental sources of covariation between generalized anxiety disorder and neuroticism. Am J Psychiatry. 2004;161(9):1581–7.
123. Lohoff FW, Aquino TD, Narasimhan S, Multani PK, Etemad B, Rickels K. Serotonin receptor 2A (HTR2A) gene polymorphism predicts treatment response to venlafaxine XR in generalized anxiety disorder. Pharmacogenomics J. 2013;13(1):21–6.
124. Narasimhan S, Aquino TD, Multani PK, Rickels K, Lohoff FW. Variation in the catechol-O-methyltransferase (COMT) gene and treatment response to venlafaxine XR in generalized anxiety disorder. Psychiatry Res. 2012;198(1):112–5.
125. Cooper AJ, Narasimhan S, Rickels K, Lohoff FW. Genetic polymorphisms in the PACAP and PAC1 receptor genes and treatment response to venlafaxine XR in generalized anxiety disorder. Psychiatry Res. 2013;210(3):1299–300.
126. Saung WT, Narasimhan S, Lohoff FW. Lack of influence of DAT1 and DRD2 gene variants on antidepressant response in generalized anxiety disorder. Hum Psychopharmacol. 2014;29(4):316–21.
127. Narasimhan S, Aquino TD, Hodge R, Rickels K, Lohoff FW. Association analysis between the Val66Met polymorphism in the brain-derived neurotrophic factor (BDNF) gene and treatment response to venlafaxine XR in generalized anxiety disorder. Neurosci Lett. 2011;503(3):200–2.
128. Nardi AE. Social anxiety disorder has social and economic burden. BMJ. 2003;327(7426):1287.
129. Dams J, Konig HH, Bleibler F, Hoyer J, Wiltink J, Beutel ME, et al. Excess costs of social anxiety disorder in Germany. J Affect Disord. 2017;213:23–9.
130. National Institute for Health and Care Excellence. Social anxiety disorder: recognition, assessment and treatment 2013. Available from: nice.org.uk/guidance/cg159.
131. Mayo-Wilson E, Dias S, Mavranezouli I, Kew K, Clark DM, Ades AE, et al. Psychological and pharmacological interventions for social anxiety disorder in adults: a systematic review and network meta-analysis. Lancet Psychiatry. 2014;1(5):368–76.
132. Springer KS, Levy HC, Tolin DF. Remission in CBT for adult anxiety disorders: A meta-analysis. Clin Psychol Rev. 2018;61:1–8.
133. Liebowitz MR, Gelenberg AJ, Munjack D. Venlafaxine extended release vs placebo and paroxetine in social anxiety disorder. Arch Gen Psychiatry. 2005;62(2):190–8.
134. Van Ameringen MA, Lane RM, Walker JR, Bowen RC, Chokka PR, Goldner EM, et al. Sertraline treatment of generalized social phobia: a 20-week, double-blind, placebo-controlled study. Am J Psychiatry. 2001;158(2):275–81.
135. Pollack MH, Van Ameringen M, Simon NM, Worthington JW, Hoge EA, Keshaviah A, et al. A double-blind randomized controlled trial of augmentation and switch strategies for refractory social anxiety disorder. Am J Psychiatry. 2014;171(1):44–53.
136. Clauss JA, Blackford JU. Behavioral inhibition and risk for developing social anxiety disorder: a meta-analytic study. J Am Acad Child Adolesc Psychiatry. 2012;51(10):1066–75.. e1
137. Kagan J, Reznick JS, Snidman N. Biological bases of childhood shyness. Science. 1988;240(4849):167–71.
138. Spence SH, Rapee RM. The etiology of social anxiety disorder: An evidence-based model. Behav Res Ther. 2016;86:50–67.
139. Hirshfeld-Becker DR, Biederman J, Henin A, Faraone SV, Davis S, Harrington K, et al. Behavioral inhibition in preschool children at risk is a specific predictor of middle childhood social anxiety: a five-year follow-up. J Dev Behav Pediatr. 2007;28(3):225–33.
140. Cremers HR, Roelofs K. Social anxiety disorder: a critical overview of neurocognitive research. Wiley Interdiscip Rev Cogn Sci. 2016;7(4):218–32.
141. Marazziti D, Abelli M, Baroni S, Carpita B, Ramacciotti CE, Dell'Osso L. Neurobiological correlates of social anxiety disorder: an update. CNS Spectr. 2015;20(2):100–11.
142. van Honk J, Bos PA, Terburg D, Heany S, Stein DJ. Neuroendocrine models of social anxiety disorder. Dialogues Clin Neurosci. 2015;17(3):287–93.

143. Jones C, Barrera I, Brothers S, Ring R, Wahlestedt C. Oxytocin and social functioning. Dialogues Clin Neurosci. 2017;19(2):193–201.
144. Giltay EJ, Enter D, Zitman FG, Penninx BW, van Pelt J, Spinhoven P, et al. Salivary testosterone: associations with depression, anxiety disorders, and antidepressant use in a large cohort study. J Psychosom Res. 2012;72(3):205–13.
145. Hoge EA, Lawson EA, Metcalf CA, Keshaviah A, Zak PJ, Pollack MH, et al. Plasma oxytocin immunoreactive products and response to trust in patients with social anxiety disorder. Depress Anxiety. 2012;29(11):924–30.
146. Hoge EA, Pollack MH, Kaufman RE, Zak PJ, Simon NM. Oxytocin levels in social anxiety disorder. CNS Neurosci Ther. 2008;14(3):165–70.
147. Haller SPW, Mills KL, Hartwright CE, David AS, Cohen KK. When change is the only constant: The promise of longitudinal neuroimaging in understanding social anxiety disorder. Dev Cogn Neurosci. 2018;33:73–82.
148. Clauss JA, Avery SN, VanDerKlok RM, Rogers BP, Cowan RL, Benningfield MM, et al. Neurocircuitry underlying risk and resilience to social anxiety disorder. Depress Anxiety. 2014;31(10):822–33.
149. Lahat A, Benson BE, Pine DS, Fox NA, Ernst M. Neural responses to reward in childhood: relations to early behavioral inhibition and social anxiety. Soc Cogn Affect Neurosci. 2018;13(3):281–9.
150. Harrewijn A, van der Molen MJW, van Vliet IM, Tissier RLM, Westenberg PM. Behavioral and EEG responses to social evaluation: A two-generation family study on social anxiety. Neuroimage Clin. 2018;17:549–62.
151. Nikolic M, de Vente W, Colonnesi C, Bogels SM. Autonomic arousal in children of parents with and without social anxiety disorder: a high-risk study. J Child Psychol Psychiatry. 2016;57(9):1047–55.
152. Nikolic M, Aktar E, Bogels S, Colonnesi C, de Vente W. Bumping heart and sweaty palms: physiological hyperarousal as a risk factor for child social anxiety. J Child Psychol Psychiatry. 2018;59(2):119–28.
153. Clark-Elford R, Nathan PJ, Auyeung B, Mogg K, Bradley BP, Sule A, et al. Effects of oxytocin on attention to emotional faces in healthy volunteers and highly socially anxious males. Int J Neuropsychopharmacol. 2014;18(2)
154. Guastella AJ, Howard AL, Dadds MR, Mitchell P, Carson DS. A randomized controlled trial of intranasal oxytocin as an adjunct to exposure therapy for social anxiety disorder. Psychoneuroendocrinology. 2009;34(6):917–23.
155. Enter D, Terburg D, Harrewijn A, Spinhoven P, Roelofs K. Single dose testosterone administration alleviates gaze avoidance in women with Social Anxiety Disorder. Psychoneuroendocrinology. 2016;63:26–33.
156. van Peer JM, Enter D, van Steenbergen H, Spinhoven P, Roelofs K. Exogenous testosterone affects early threat processing in socially anxious and healthy women. Biol Psychol. 2017;129:82–9.
157. Kilts CD, Kelsey JE, Knight B, Ely TD, Bowman FD, Gross RE, et al. The neural correlates of social anxiety disorder and response to pharmacotherapy. Neuropsychopharmacology. 2006;31(10):2243–53.
158. Klumpp H, Fitzgerald JM. Neuroimaging predictors and mechanisms of treatment response in social anxiety disorder: an overview of the Amygdala. Curr Psychiatry Rep. 2018;20(10):89.
159. Slaap BR, van Vliet IM, Westenberg HG, Den Boer JA. Responders and non-responders to drug treatment in social phobia: differences at baseline and prediction of response. J Affect Disord. 1996;39(1):13–9.
160. Stein DJ, Stein MB, Pitts CD, Kumar R, Hunter B. Predictors of response to pharmacotherapy in social anxiety disorder: an analysis of 3 placebo-controlled paroxetine trials. J Clin Psychiatry. 2002;63(2):152–5.
161. Miskovic V, Moscovitch DA, Santesso DL, McCabe RE, Antony MM, Schmidt LA. Changes in EEG cross-frequency coupling during cognitive behavioral therapy for social anxiety disorder. Psychol Sci. 2011;22(4):507–16.

162. Moscovitch DA, Santesso DL, Miskovic V, McCabe RE, Antony MM, Schmidt LA. Frontal EEG asymmetry and symptom response to cognitive behavioral therapy in patients with social anxiety disorder. Biol Psychol. 2011;87(3):379–85.
163. Arad G, Abend R, Pine DS, Bar-Haim Y. A neuromarker of clinical outcome in attention bias modification therapy for social anxiety disorder. Depress Anxiety. 2018;
164. Merikangas KR, Lieb R, Wittchen HU, Avenevoli S. Family and high-risk studies of social anxiety disorder. Acta Psychiatr Scand Suppl. 2003;417:28–37.
165. Scaini S, Belotti R, Ogliari A. Genetic and environmental contributions to social anxiety across different ages: a meta-analytic approach to twin data. J Anxiety Disord. 2014;28(7):650–6.
166. Elizabeth J, Gullone E, Tonge B, Watson SD. Social anxiety disorder in children and youth: a research update on aetiological factors. Couns Psychol Q. 2006;19(2):151–63.
167. Bergman O, Ahs F, Furmark T, Appel L, Linnman C, Faria V, et al. Association between amygdala reactivity and a dopamine transporter gene polymorphism. Transl Psychiatry. 2014;4:e420.
168. Domschke K, Stevens S, Beck B, Baffa A, Hohoff C, Deckert J, et al. Blushing propensity in social anxiety disorder: influence of serotonin transporter gene variation. J Neural Transm. 2009;116(6):663–6.
169. Furmark T, Marteinsdottir I, Frick A, Heurling K, Tillfors M, Appel L, et al. Serotonin synthesis rate and the tryptophan hydroxylase-2: G-703T polymorphism in social anxiety disorder. J Psychopharmacol. 2016;30(10):1028–35.
170. Stein MB, Keshaviah A, Haddad SA, Van Ameringen M, Simon NM, Pollack MH, et al. Influence of RGS2 on sertraline treatment for social anxiety disorder. Neuropsychopharmacology. 2014;39(6):1340–6.
171. Stein MB, Seedat S, Gelernter J. Serotonin transporter gene promoter polymorphism predicts SSRI response in generalized social anxiety disorder. Psychopharmacology. 2006;187(1):68–72.
172. Andersson E, Ruck C, Lavebratt C, Hedman E, Schalling M, Lindefors N, et al. Genetic polymorphisms in monoamine systems and outcome of cognitive behavior therapy for social anxiety disorder. PLoS One. 2013;8(11):e79015.
173. Hedman E, Andersson E, Ljotsson B, Andersson G, Andersson E, Schalling M, et al. Clinical and genetic outcome determinants of Internet- and group-based cognitive behavior therapy for social anxiety disorder. Acta Psychiatr Scand. 2012;126(2):126–36.
174. Ziegler C, Dannlowski U, Brauer D, Stevens S, Laeger I, Wittmann H, et al. Oxytocin receptor gene methylation: converging multilevel evidence for a role in social anxiety. Neuropsychopharmacology. 2015;40(6):1528–38.
175. Puglia MH, Connelly JJ, Morris JP. Epigenetic regulation of the oxytocin receptor is associated with neural response during selective social attention. Transl Psychiatry. 2018;8(1):116.
176. Kuo JR, Goldin PR, Werner K, Heimberg RG, Gross JJ. Childhood trauma and current psychological functioning in adults with social anxiety disorder. J Anxiety Disord. 2011;25(4):467–73.
177. Lochner C, Seedat S, Allgulander C, Kidd M, Stein D, Gerdner A. Childhood trauma in adults with social anxiety disorder and panic disorder: a cross-national study. Afr J Psychiatry. 2010;13(5):376–81.
178. Reinelt E, Stopsack M, Aldinger M, John U, Grabe HJ, Barnow S. Testing the diathesis-stress model: 5-HTTLPR, childhood emotional maltreatment, and vulnerability to social anxiety disorder. Am J Med Genet B Neuropsychiatr Genet. 2013;162B(3):253–61.
179. Jakubovski E, Bloch MH. Anxiety Disorder-Specific Predictors of Treatment Outcome in the Coordinated Anxiety Learning and Management (CALM) Trial. Psychiatry Q. 2016;87(3):445–64.

The Role of Hormonal and Reproductive Status in the Treatment of Anxiety Disorders in Women

26

Samantha Tang and Bronwyn Margaret Graham

Introduction

A well-established phenomenon is that anxiety and trauma and stressor-related disorders disproportionately affect women [1]. For instance, women are over twice as likely as men to be diagnosed with post-traumatic stress disorder (PTSD) in their lifetime (male/female prevalence ratio 1:2.69). Lifetime prevalence rates of other anxiety disorders, such as generalised anxiety disorder (male/female prevalence ratio 1:1.83), panic disorder (male/female prevalence ratio 1:1.70) and specific phobia (male/female prevalence ratio 1:1.96), are also significantly higher among women compared to men [2]. Additionally, anxiety disorders in women are associated with greater symptom severity and chronicity, higher rates of comorbidity and a higher burden of disease compared to men [2, 3].

Notably, sex differences in the prevalence of anxiety disorders appear to emerge after puberty (but see Anderson and colleagues [4] for evidence of sex differences emerging during childhood) and cease postmenopause [5–8]. Fluctuations in the prevalence of anxiety disorders in women also occur at other points in their reproductive life span. For instance, while pregnancy was previously thought to be a protective factor against the development of anxiety disorders, recent statistics indicate that the prevalence of anxiety disorders among pregnant women may be similar to, if not greater than that of, nonpregnant women [9]. Moreover, rates of generalised anxiety disorder and obsessive-compulsive disorder are significantly higher among postpartum women compared to the general population [10]. This is of concern not only due to the impact of having an anxiety disorder on the mother but also due to its impact on the development and well-being of the child [11]. Fluctuations in the prevalence of anxiety disorders across different stages of the female

S. Tang (✉) · B. M. Graham
School of Psychology, UNSW Sydney, Sydney, NSW, Australia
e-mail: samantha.tang@unsw.edu.au; bgraham@psy.unsw.edu.au

Y.-K. Kim (ed.), *Anxiety Disorders*, Advances in Experimental Medicine and Biology 1191, https://doi.org/10.1007/978-981-32-9705-0_26

reproductive life span indicate that sex hormones may be involved in the aetiology and maintenance of such disorders.

Given the high prevalence of anxiety disorders in women, it is important to ensure that current treatments are optimised for women, taking into account potential impacts of sex and sex hormones on treatment outcomes. Currently, cognitive behavioural therapy (CBT) is the gold-standard treatment for anxiety disorders, with numerous clinical studies and randomised controlled trials demonstrating the short-term efficacy and effectiveness of this treatment [12, 13]. However, despite the overall success of CBT, it is notable that a significant minority of patients either fail to respond to treatment or experience relapse following treatment [14]. This has prompted increased research into the neurobiological mechanisms underlying the treatment of anxiety, with the view that such research may provide insight into how treatment outcomes can be optimised [15]. To date, however, the majority of animal studies examining the neurobiological mechanisms underlying the treatment of anxiety have only included male subjects or have failed to take into account the hormonal status of female subjects [16]. A more comprehensive understanding of the hormonal and neurobiological mechanisms underlying the treatment of anxiety in females may be helpful in developing more effective treatments for this population and will therefore serve as the focus of the current chapter.

Laboratory Models of Anxiety and Its Treatment

Our current understanding of the mechanisms underlying the development and treatment of anxiety in females is primarily based on laboratory models, including Pavlovian fear conditioning and extinction. During Pavlovian fear conditioning, a neutral stimulus (conditioned stimulus, CS) is paired with an aversive outcome (unconditioned stimulus, US). In cued fear conditioning, the CS is a discrete cue (e.g. tone), whereas in contextual fear conditioning, the CS refers to a specific context. After a number of paired CS-US presentations, the CS alone elicits a conditioned fear response (CR) as subjects learn that the CS predicts the US. Freezing, defined as the absence of all movement aside from that which is required for respiration [17], is a commonly measured fear response in rodents. In humans, fear responses are often measured using differential skin conductance responses (SCR), whereby the subject shows higher SCR responses to a neutral stimulus (CS+) that has been paired with a US compared to an unpaired stimulus (CS−). The fear responses produced by fear conditioning mimic the symptoms exhibited by individuals with an anxiety disorder [18].

CS-elicited fear responses that occur as a result of fear conditioning can be diminished by repeatedly presenting subjects with the CS in the absence of the US. This process is known as Pavlovian fear extinction. During extinction training, subjects learn that the CS no longer predicts the US, thus causing fear responses to subside over training [19]. Within-session extinction refers to the decrement in conditioned responding that takes place over the course of extinction training. As with conditioning memories, extinction memories must undergo consolidation in order to be retained. This time-dependent process involves molecular and neurological changes that facilitate the long-term stability of the memory [20]. Subjects' memory of extinction can be tested by measuring their fear responses to the CS at a later time (extinction recall). High levels of conditioned responding during recall reflect poor consolidation or poor retrieval of the extinction memory.

Fear extinction is a useful model for understanding both the underlying pathology and the treatment of anxiety disorders. Impairments in extinction acquisition and extinction recall are thought to partly contribute to the excessive fear that is characteristic of anxiety disorders. Indeed, such impairments have been demonstrated in laboratory studies of a number of anxiety and trauma and stressor-related disorders, including PTSD, social anxiety disorder and panic disorder [21–24]. Impaired fear extinction may represent a number of dysregulated processes among those with an anxiety disorder, such as poor emotion regulation capacity [25] or poor safety learning (i.e. the persistence of fear responses in the presence of cues signalling safety; [26]). Understanding the neurobiological mechanisms underlying fear extinction can therefore provide insight into dysfunctional processes that may contribute to the development and maintenance of anxiety disorders.

Fear extinction models not only the underlying pathology of anxiety disorders but also their treatment. Specifically, fear extinction is a laboratory model for exposure therapy, a key component of CBT that involves repeatedly presenting patients with anxiety-provoking stimuli in the absence of feared outcomes [27]. Over repeated exposures, patients exhibit reduced anxiety as they learn that the feared stimuli are, in fact, safe. Since fear extinction serves as a laboratory model for exposure therapy, an understanding of the processes underlying the acquisition and recall of fear extinction may allow for improved exposure therapy outcomes. For instance, an understanding of the neurobiology of fear extinction may aid the development/identification of pharmacological adjuncts that enhance the learning that takes place during exposure [28]. Importantly, it is also necessary to consider whether proposed pharmacological adjuncts to exposure therapy are equally effective in both sexes, particularly given that the neurobiology of fear extinction may be sex-specific, as will be examined below.

Hormonal Contributions to Fear Extinction in Females

As noted earlier, the prevalence of anxiety disorders among women fluctuates across the reproductive life span, indicating that sex hormones may be involved in the aetiology and maintenance of such disorders. The two major sex hormones in women are oestrogens and progesterone. Estradiol (the primary form of oestrogen) and progesterone co-fluctuate during a woman's menstrual cycle, with low levels of these hormones seen during the early follicular phase and high levels of these hormones seen during the mid-luteal phase [29]. In rodents, estradiol and progesterone levels are low during the metestrus phase, and high during the proestrus phase of the estrous cycle [30]. In the past decade, there has been a growing literature indicating that fluctuations in levels of estradiol and progesterone across the estrous/menstrual cycle influence fear extinction in both female rodents and women, as will be described below.

The current literature examining the role of sex hormones in fear extinction has consistently demonstrated that estradiol facilitates the consolidation of extinction memories in females. While female rats extinguished during the metestrus (low estradiol) phase of their estrous cycle show impaired extinction recall, this impairment is not evident in female rats extinguished during the proestrus (high estradiol) phase of their cycle [16, 31–35]. Systemic administration of estradiol or selective oestrogen receptor agonists pre- or post-extinction training improve extinction recall [16, 35–40], while systemic administration of an oestrogen receptor antagonist pre-extinction training impairs extinction recall [16], suggesting that estradiol not only enhances but is also necessary for fear extinction.

The enhancing effect of both endogenous and exogenous estradiol on fear extinction has also been demonstrated in healthy and clinical samples of naturally cycling women [33, 38, 40–45]. For instance, it has been shown that both spider phobic and healthy women extinguished during periods of low estradiol demonstrate impaired extinction recall, while phobic and healthy women extinguished periods of high estradiol do not demonstrate such impairments [41]. Moreover, Glover and colleagues [43] demonstrated that women with PTSD who undergo extinction during periods of low estradiol exhibit significantly higher levels of fear-potentiated startle during extinction training compared to women undergoing extinction during periods of high estradiol. However, it is of note that Pineles and colleagues [46] found that women with PTSD exhibited poor extinction recall when extinguished during the mid-luteal phase (typically associated with high estradiol levels) compared to the early follicular phase (typically associated with low estradiol levels). Opposite findings were made in trauma-exposed women without PTSD. However, when absolute hormone levels were analysed in place of menstrual phase, results indicated that fear extinction was most impaired at times of low estradiol and high progesterone among women with PTSD and times of low estradiol and low progesterone among healthy women.

As discussed earlier, estradiol and progesterone co-fluctuate across a female's estrous or menstrual cycle. However, the increase in progesterone levels observed during proestrus in rats and the mid-luteal phase in women occurs shortly after the increase in estradiol levels. When progesterone levels reach their peak, levels of estradiol rapidly decline as a female begins to move into metestrus or the early follicular phase [47, 48]. Studies examining the involvement of progesterone in fear extinction in females have yielded inconsistent findings. A number of studies have shown no involvement of progesterone in fear extinction in female rats and women [36, 38, 42, 49], while other studies have demonstrated that progesterone may facilitate fear extinction. For instance, Milad and colleagues [16] demonstrated that impairments in extinction consolidation associated with extinguishing female rats during metestrus (low estradiol/progesterone phase) can be overcome with pre-extinction administrations of progesterone. They also found that pre-extinction administration of a progesterone receptor antagonist significantly decrease the high levels of extinction recall that are normally observed when a female is extinguished during proestrus [16]. Given that increases in estradiol precede increases in progesterone during proestrus, it is possible that progesterone regulates fear extinction in females through its interaction with estradiol. Indeed, Graham and Daher [34] demonstrated that progesterone initially potentiates, before rapidly reversing the enhancing effects of estradiol on fear extinction in ovariectomised rats, thus suggesting that both sex hormones are critical in modulating fear extinction in females.

Further support for the notion that estradiol and progesterone regulate fear extinction in females comes from studies examining the effect of hormonal contraceptive pills on fear extinction. Combined oral contraceptive pills, which are used by at least 100 million women worldwide [50], inhibit ovulation by reducing the production of ovarian estradiol and progesterone [51]. Women using this form of contraception therefore experience reduced levels of these hormones [52]. The chronically low levels of oestrogens resulting from the use of hormonal contraceptives have been shown to result in impaired extinction recall [40, 49] and an impaired ability to discriminate between the CS+ and CS− after extinction training [53] among women. Graham and Milad [40] further demonstrated that impairments in extinction recall associated with the use of hormonal contraceptives could be alleviated by administering selective oestrogen receptor agonists or terminating the use of hormonal contraceptives prior to extinction training in rats. Together, the findings reviewed above suggest that low levels of estradiol and progesterone resulting from the use of hormonal contraceptives may impair fear extinction in female rats and women.

Given that fear extinction forms the basis of exposure therapy, one may predict that exposure therapy outcomes in women are also dependent on estradiol and/or progesterone levels. Indeed, there is preliminary evidence in support of this prediction in women with spider phobia [54]. In this study, spider phobic women underwent a single session of exposure therapy. Women who underwent exposure while using hormonal contraceptives exhibited higher behavioural avoidance of spiders (as measured using a task in which participants were required to approach a live spider) compared to naturally cycling women at both post-treatment and follow-up, despite not demonstrating any differences in avoidance pre-treatment. Avoidance of phobic stimuli is a defining symptom of anxiety disorders (American Psychiatric [55]). Graham and colleagues [54] also found that women using contraceptive pills exhibited slower improvement in symptoms across the exposure session and participated in fewer self-initiated exposure tasks post-treatment. Moreover, correlational analyses across the whole sample demonstrated that estradiol levels were positively correlated with improvement in symptoms across the exposure session and negatively correlated with self-reported spider fear and behavioural avoidance at post-treatment. Together, such findings indicate that low estradiol levels, particularly from the use of hormonal contraceptive pills, may be associated with reduced responsiveness to exposure therapy.

In summary, both preclinical and clinical studies have demonstrated that the sex hormones estradiol and progesterone may be involved in fear extinction in females. Estradiol, in particular, appears to be necessary for fear extinction in this population.

Low levels of estradiol either from cyclic fluctuations or the use of contraceptive pills are likely to impair fear extinction and, potentially, exposure therapy outcomes.

Potential Mechanisms by Which Estradiol Facilitates Fear Extinction

As reviewed above, there is a substantial body of evidence to suggest that estradiol enhances fear extinction in females. It is possible that estradiol enhances fear extinction by regulating activity in the neural fear circuit. When female rats are administered estradiol immediately pre- or post-extinction training, there is increased activity in the infralimbic (IL) subregion of the prefrontal cortex (PFC) at extinction recall [37, 38]. Similarly, women extinguished during the high estradiol phase of their menstrual cycle show higher activation of the ventromedial PFC (the functional homologue of the IL in humans) and the hippocampus during extinction recall compared to those extinguished during the low estradiol phase of their cycle [38, 56]. Increased activation of the IL and hippocampus are associated with lower fear responses during extinction recall in adult male rats [57]. Additionally, high endogenous estradiol has been shown to be associated with reduced activity in the lateral nuclei of the amygdala [58], while administration of estradiol prior to extinction training has been shown to reduce activity in the amygdala during extinction recall [38]. The amygdala is a key structure involved in the acquisition, storage and expression of fear conditioning memories [59]. Together, these findings indicate that estradiol may enhance fear extinction by increasing activity in brain regions involved in fear inhibition and reducing activity in brain regions involved in fear expression.

One way in which estradiol may alter activity in the brain regions involved in fear extinction is through its effect on N-methyl-D-aspartate (NMDA) receptors. In adult male rats, studies have consistently demonstrated that NMDA receptors are critically involved in the acquisition and consolidation of extinction memories [28, 60]. The pharmacological blockade of NMDA receptors prior to or immediately after extinction training impairs extinction recall [61–65], while administration of an NMDA receptor agonist immediately post-extinction enhances extinction recall [66, 67]. Recently, we have shown that the pharmacological blockade of NMDA receptors prior to extinction training also impairs extinction recall in female rats [68], indicating that the activation of these receptors may be required for fear extinction in females, as it is in male rats.

The dependence of fear extinction on both estradiol [16] and NMDA receptors [68] in females suggests that estradiol might enhance the consolidation of extinction through its interactions with NMDA receptors. When administered onto hippocampal brain slices, estradiol increases NMDA receptor expression, binding and phosphorylation and the amplitude of NMDA receptor-mediated excitatory postsynaptic potentials [69–73]. Estradiol also increases the magnitude of long-term potentiation (LTP) via increasing NMDA receptor transmission in the hippocampus [74]. LTP is a process that involves the strengthening of synapses following instances of learning and other forms of stimulation and is thought to be one of the ways in which memories are formed in the brain [75]. NMDA receptors have been heavily implicated in LTP. The molecular cascade that follows the activation of these receptors results in gene transcription and structural and functional changes that allow for the strengthening of synapses [76]. Based on the literature reviewed above, it is possible that estradiol enhances the consolidation of fear extinction in females by upregulating NMDA receptor activity and subsequent NMDA receptor-mediated LTP. Consistent with this notion is research demonstrating that the blockade of NMDA receptor activity inhibits estradiol-induced improvements in other memory tasks, such as the object recognition task [77, 78].

Estradiol may enhance fear extinction not only through its effects on cell activity but also through its effects on dendritic spine density. Dendritic spines are the microscopic projections on a neuron's dendrites that receive synaptic input from other neurons and are thought to be the storage site for synaptic plasticity [79]. There is evidence to suggest that both endogenous and exogenous estradiol increase dendritic spine density in various parts of the neural circuitry. For instance, hippocampal dendritic spine density is high during proestrus and low during metestrus [80–82]. Moreover, ovariectomy is associated with a decrease in hippocampal dendritic spine density – an effect that is reversed following oestrogen replacement [83–86]. Estradiol-induced spinogenesis has also been found in the PFC and amygdala [87–90]. Since dendritic spine density in the frontal association cortex during extinction is positively correlated with levels of extinction recall [91], it is possible that estradiol facilitates extinction consolidation through its effect on these microscopic projections. Graham and Daher [34] provided indirect evidence in support of this premise by examining the effect of different oestrogen and progesterone replacement regimens on extinction consolidation in ovariectomised female rats. The oestrogen replacement regimens they used produced high or low dendritic spine densities at different time points, as demonstrated by Woolley and McEwen [92]. Extinction consolidation was enhanced when rats were extinguished during periods of high dendritic spine density and impaired when rats were extinguished during periods of low dendritic spine density. Estradiol-induced increases in dendritic spine density are dependent on NMDA receptor activation [93]. It therefore follows that the enhancements in fear extinction that are typically observed when ovariectomised rats are given an oestrogen replacement regimen that produces high dendritic spine densities are diminished when an NMDA receptor antagonist is administered alongside this regimen [39]. Such findings provide further support for the notion that estradiol may facilitate fear extinction in females through its effect on NMDA receptors. Continued focus on the mechanisms by which estradiol facilitates extinction may help to develop novel means of improving exposure therapy for women via hormonal manipulations.

The Impact of Reproductive Experience

While it is promising that the last decade has seen a growing interest in the mechanisms underlying fear extinction in females, it is of note that the existing literature is limited by the fact that most animal studies examining fear extinction only use reproductively inexperienced subjects. Moreover, most human studies have used young women, aged 18–30, whose reproductive status is unknown. This problematic given that 84% of women will become mothers by the age of 44 [94]. 8–12% of these women will experience an anxiety disorder during the postpartum period [95], and an unknown percentage of these women will experience an anxiety disorder after the postpartum period. As will be outlined below, pregnancy and motherhood are times of vast hormonal, neural and behavioural change. Many of these changes persist long after weaning and may alter the mechanisms by which reproductively experienced females regulate fear.

Hormonal Changes Associated with Reproductive Experience

Pregnancy and motherhood are times of profound hormonal change. The most distinguishing hormonal feature of pregnancy relates to changes in levels of estradiol and progesterone. Levels of both of these hormones increase continually throughout pregnancy in both female rats and women [96]. Specifically, there is a 100-fold increase in estradiol production and a 20-fold increase in progesterone production across the course of pregnancy [96]. Importantly, a female will not be exposed to such an extended surge in levels of these hormones at any other time in her life. Levels of both estradiol and progesterone decrease dramatically following parturition, returning to pre-pregnancy levels by the fifth postpartum day in women [97]. There is also research demonstrating that there are long-term changes to endogenous estradiol levels post-weaning. For example, rodent studies have shown that reproductive experience significantly reduces levels of circulating estradiol during proestrus [33, 98]. Similar findings have also been made in women [99].

Pregnancy and motherhood are also associated with changes in levels of other hormones, including oxytocin and glucocorticoids. Levels of both of these hormones increase over the course of pregnancy, before drastically decreasing following parturition [96, 100, 101]. However, there are brief increases in oxytocin levels in response to suckling during the postpartum period [96]. There is a substantial body of evidence showing that both oxytocin and glucocorticoids are involved in fear extinction. For instance, pre-extinction administrations of oxytocin facilitate extinction learning in male rodents [102] and extinction consolidation in humans [103]. Moreover, pre-extinction administrations of glucocorticoid receptor agonists improve extinction recall in male rodents, whereas glucocorticoid receptor antagonists block extinction recall [104]. The administration of glucocorticoids prior to exposure therapy in humans has also been found to reduce fear responses at post-treatment and follow-up [105].

Given that the hormonal milieu of pregnancy and motherhood is largely characterised by the presence of hormones known to play a role in facilitating fear extinction, it is possible that such experiences will alter the characteristics of fear extinction. For instance, changes in levels of estradiol and progesterone both during and after pregnancy may alter the involvement of these hormones in fear extinction in mothers. Evidence in support of this premise will be reviewed later in this chapter.

Neurological Changes Associated with Reproductive Experience

The hormonal changes accompanying pregnancy and motherhood result in vast morphological changes in the female brain. Such changes are apparent not only in brain structures involved in maternal behaviour but also in areas associated with learning, memory and anxiety [106]. For instance, Lemaire and colleagues [107] found that both NMDA and non-NMDA receptor-mediated LTP in the CA1 region of the hippocampus are enhanced in primiparous (one prior reproductive experience) females compared to nulliparous (virgin) females. Moreover, Tomizawa and colleagues [108] found evidence to suggest that motherhood reduces the threshold for LTP induction in the hippocampus.

In addition to inducing functional changes in the hippocampus, pregnancy and motherhood are also associated with morphological changes in this brain region. Late-pregnant females, lactating mothers and multiparous (more than one reproductive experience) females exhibit higher dendritic spine density in the CA1 region of the hippocampus compared to nulliparous females tested during both the high and low estradiol phases of their estrous cycle [109, 110]. Such increases in dendritic spine density are also evident in ovariectomised rats treated with a pregnancy-mimicking hormone replacement regimen consisting of estradiol and progesterone [110], demonstrating that these hormones play a key role in motherhood-induced neurological changes. Increases in spine density associated with reproductive experience are also observable in other parts of the neural fear circuit, including the medial PFC and the amygdala [111, 112].

Reproductive experience has also been shown to alter the expression and activity of oestrogen receptors. For instance, Byrnes and colleagues [113] demonstrated that the expression of oestrogen receptors differs between nulliparous and primiparous females across various brain regions, including the medial preoptic area and the medial amygdala. They also found that primiparous rats require a higher dose of estradiol to elicit an increase in prolactin levels compared to nulliparous rats, suggesting that oestrogen receptor sensitivity may decrease following reproductive experience [98]. Further supporting this idea is the finding that while a low dose of estradiol biases nulliparous females towards using response memory over place memory, no such effect is detected in primiparous females [114]. However, it has also been shown that 17α-estradiol, 17β-estradiol and estrone each trigger significant increases in hippocampal cell proliferation in multiparous rats, but not in nulliparous rats [115]. Moreover, Byrnes and colleagues [116] demonstrated that the administration of an oestrogen receptor-alpha agonist reduces anxiety-like behaviour on the elevated plus maze in primiparous females, but not nulliparous females. It is therefore possible that reproductive experience alters the function of oestrogen receptors, rather than their sensitivity.

Behavioural Changes Associated with Reproductive Experience

The hormonal and neurological changes accompanying pregnancy and motherhood are reflected in the behaviour of parous females [117]. This is particularly true of tasks involving the hippocampus and the amygdala – brain structures that, as discussed earlier, undergo anatomical and functional modifications as a result of reproductive experience. For instance, mothers have been found to show enhanced spatial working and reference memory, outperforming virgins in such tasks as the dry land maze, Morris water maze and radial arm maze [107, 118–120]. Importantly, these enhancements persist long after weaning [107, 118]. Pregnancy and motherhood also augment females' performance in the object placement and object recognition tasks, both of which involve the hippocampus [83, 121]. Multiparous rats spend significantly more time with an object placed in a novel location, and more time with a novel object, than nulliparous rats [122, 123].

Parous females also differ from nulliparous females in their behavioural response to stress and anxiety-provoking situations. Compared to nulliparous females, parous females show decreased fearfulness, evident by the greater amount of time they spend in the open arms of an elevated plus maze and their reduced anxiety-like behaviour in an open field [107, 118, 124–127]. Additionally, the effect of stress on learning that is observed in nulliparous females is absent in parous females [128]. When nulliparous rats are exposed to an acute stressor, they show impairments in subsequent associative eyeblink conditioning. Parous females, however, show a resistance to the effect of stress on learning, even long after they have given birth [129]. This effect is not evident in parous females who were separated from their pups at birth, but is evident in nulliparous females that fostered pups, suggesting that mothering alone plays an essential role in facilitating enhanced learning. Attenuated stress responses among parous females may be attributable to functional changes in the brain following pregnancy and motherhood. Specifically, it has been demonstrated that primiparous (one prior reproductive experience) and multiparous females show reduced activity in the basolateral amygdala and CA3 region of the hippocampus after being exposed to an anxiety-provoking situation, as compared to nulliparous females [125].

The Impact of Reproductive Experience on Fear Conditioning and Extinction

The vast and persistent hormonal, neurological and behavioural changes associated with pregnancy and motherhood raise the possibility that fear conditioning and extinction may differ between mothers and non-mothers. Currently, the limited studies examining the effect of reproductive experience on fear conditioning show mixed results. A study by Rima and colleagues [130] found that parous females show lower levels of conditioned fear compared to nulliparous (virgin) female rats. Such results may suggest that reproductive experience increases a female's resilience to developing an anxiety disorder. However, Milligan-Saville and Graham [33] demonstrated no differences in fear conditioning between nulliparous and primiparous females, evident by similar levels of freezing during fear conditioning, and at the beginning of extinction training between these groups. Contrastingly, Tang and Graham [68] found that primiparous females showed enhanced fear conditioning relative to nulliparous females. Given these mixed findings, further investigations into the impact of reproductive experience on fear conditioning are necessary.

Changes associated with pregnancy and mothering have also been shown to affect the features of fear extinction in females. Specifically, it has been shown that while estrous phase affects fear extinction in nulliparous female rats (i.e. these rats show impaired extinction recall when extinguished during metestrus), it has no such effect in primiparous female rats. Primiparous females extinguished during metestrus show comparable levels of extinction recall to those extinguished during proestrus [33, 131]. Similar findings have been observed in women, whereby there is a positive correlation between estradiol levels and extinction recall in non-mothers, but no such correlation in mothers [33]. These findings indicate that fear extinction may be estrous-dependent in females without prior reproductive experience, but not in females with prior reproductive experience.

In addition to differences in the hormonal features of fear extinction, nulliparous and primiparous females differ in their behavioural features of fear extinction. Nulliparous female rats extinguished during proestrus showed fear relapse when presented with the CS outside of the extinction context (renewal) and when exposed to an unsignalled presentation of the US (reinstatement). Primiparous females, however, show neither renewal nor reinstatement, regardless of the phase in which they underwent extinction training [33]. However, in contrast to the findings of Milligan-Saville and Graham [33], Tang and Graham [68] found that, when extinguished during proestrus, primiparous females show poorer extinction recall compared to nulliparous females. This inconsistency may be due to a variety of factors, such as individual differences in the quality and characteristics of the pregnancy and motherhood experience. Indeed, there is evidence to show that stress during the postpartum period may have persistent effects on fear extinction in female rats. Specifically, Graham [131] compared extinction recall between primiparous rats whose pups had been removed for 3 h daily from postpartum 2 to 14 (i.e. maternal separation; MS) and primiparous rats whose pups were standard reared (SR). When tested approximately 3 months post-weaning, MS rats were shown to exhibit reduced conditioned fear at the beginning of extinction training and poor extinction recall relative to SR rats. Such findings indicate that postpartum stress may lead to long-term impairments in fear extinction.

Given that the hormonal and behavioural features of fear extinction are altered as a consequence of reproductive experience, it is unsurprising that the neural features of fear extinction also differ between mothers and non-mothers. As noted earlier, the activation of NMDA receptors is required for the consolidation of fear extinction in nulliparous female rats, as it is in male rats [68]. However, it has been demonstrated that blocking the activation of NMDA receptors prior to extinction training does not impair extinction recall in primiparous females, indicating that NMDA receptors are not required for extinction in primiparous rats [68]. Given that fear extinction does not appear to be modulated by estradiol (no impact of estrous cycle [33, 131]), or NMDA receptors [68] in primiparous females, it is possible that primiparous females rely on a different molecular signalling process for fear extinction compared to nulliparous females. However, further research is required to explore this possibility. The potential clinical implications of alterations in the hormonal and neurobiological mechanisms underlying fear extinction following pregnancy and motherhood will be discussed next.

Potential Clinical Implications

In light of the research findings reviewed in this chapter, one may speculate that exposure therapy outcomes in women are partly influenced by estradiol and progesterone levels. For instance, given that low levels of estradiol are associated with impaired fear extinction, it is possible that women undergoing exposure therapy while using hormonal contraceptives or during a low estradiol phase of their menstrual cycle are less able to consolidate the learning that takes place during exposures that occur both in-session and between treatment sessions. Low estradiol levels may also impair the learning that takes place when women are incidentally exposed to anxiety-provoking stimuli in the absence of a negative outcome in their day-to-day lives, thereby preventing the natural extinction of fear. Therefore, it is possible that timing exposure therapy with respect to a woman's menstrual cycle, or addressing the impact of hormonal contraceptives on estradiol levels (e.g. by ceasing their use for the duration of treatment or administration of estradiol prior to exposure sessions), may be beneficial in enhancing exposure therapy outcomes. However, these potential clinical implications are purely speculative at this stage and require further research.

The literature discussed in this chapter also highlighted that the hormonal, behavioural and molecular features of fear extinction appear to be altered by reproductive experience; it is therefore possible that reproductive status is important to consider when determining how to enhance exposure therapy outcomes. For instance, given that fear extinction appears to be independent of cyclic fluctuations in estradiol in mothers, it may not be of benefit to time exposure therapy to periods of high estradiol in this population. Moreover, changes in the signalling processes underlying fear extinction resulting from reproductive experience may indicate that the effectiveness of pharmacological adjuncts to psychological therapy could differ between mothers and non-mothers. For instance, D-cycloserine (DCS; a partial NMDAR agonist) has recently been proposed as a pharmacological adjunct to exposure therapy given that it enhances extinction recall in adult male rats and humans. Indeed, the administration of DCS prior to CBT has been shown to improve outcomes in the treatment of various anxiety disorders, such as PTSD and specific phobia [28]. However, given that primiparous females may not rely on NMDARs for fear extinction, it is possible that DCS may not be as effective in augmenting exposure therapy outcomes in mothers. Similarly, while estradiol may be an effective pharmacological adjunct to exposure therapy in non-mothers, it may not be effective in mothers given that fear extinction appears to be independent of endogenous fluctuations of estradiol in this population. Again, it is of note that these potential clinical implications are purely speculative at this stage. Further investigation of the hormonal and neurobiological mechanisms underlying fear extinction in both non-mothers and mothers is necessary to determine how to best enhance treatment outcomes in these populations.

Conclusion

In summary, it appears that the extinction-based mechanisms on which exposure therapy relies are modulated by estradiol and progesterone in female rodents and women. There is converging evidence to suggest that high levels of these hormones facilitate fear extinction, while low levels of these hormones impair fear extinction. Estradiol and progesterone appear to interact with various signalling processes, including those involving NMDA receptors, within the neural fear circuit to facilitate the consolidation of fear extinction. Notably, however, the involvement of estradiol and progesterone, and the signalling processes with which they interact, may be altered by long-term changes stemming from pregnancy and motherhood. Namely, it appears that fear extinction may be independent of cyclic fluctuations in estradiol and progesterone, and the activation of NMDA receptors, in reproductively experienced females. Given that fear extinction models both the underlying pathology and treatment of anxiety disorders, such findings indicate that both hormonal and reproductive status may alter the way in which anxiety disorders are maintained and treated in women. Continuing to build upon our understanding of the hormonal and neurobiological mechanisms underlying fear extinction in females may facilitate the development of more tailored and effective treatments for anxiety disorders among women.

References

1. Bandelow B, Michaelis S. Epidemiology of anxiety disorders in the 21st century. Dialogues Clin Neurosci. 2015;17(3):327–35.
2. McLean CP, Asnaani A, Litz BT, Hofmann SG. Gender differences in anxiety disorders: prevalence, course of illness, comorbidity and burden of illness. J Psychiatr Res. 2011;45(8):1027–35.
3. Bekker MH, van Mens-Verhulst J. Anxiety disorders: sex differences in prevalence, degree, and background, but gender-neutral treatment. Gend Med. 2007;4:S178–S93.
4. Anderson JC, Williams S, McGee R, Silva PA. DSM-III disorders in preadolescent children: prevalence in a large sample from the general population. Arch Gen Psychiatry. 1987;44(1):69–76.

5. Faravelli C, Alessandra Scarpato M, Castellini G, Lo Sauro C. Gender differences in depression and anxiety: the role of age. Psychiatry Res. 2013;210(3):1301–3.
6. Paus T, Keshavan M, Giedd JN. Why do many psychiatric disorders emerge during adolescence? Nat Rev Neurosci. 2008;9(12):947–57.
7. Lewinsohn PM, Gotlib IH, Lewinsohn M, Seeley JR, Allen NB. Gender differences in anxiety disorders and anxiety symptoms in adolescents. J Abnorm Psychol. 1998;107(1):109–17.
8. Patton G, Hibbert M, Carlin J, Shao Q, Rosier M, Caust J, et al. Menarche and the onset of depression and anxiety in Victoria, Australia. J Epidemiol Community Health. 1996;50(6):661–6.
9. Goodman JH, Chenausky KL, Freeman MP. Anxiety disorders during pregnancy: a systematic review. J Clin Psychiatry. 2014;75(10):e1153–84.
10. Ross LE, McLean LM. Anxiety disorders during pregnancy and the postpartum period: a systematic review. J Clin Psychiatry. 2006;67(8):1285–98.
11. Vythilingum B. Anxiety disorders in pregnancy. Curr Psychiatry Rep. 2008;10(4):331–5.
12. Hofmann SG, Smits JA. Cognitive-behavioral therapy for adult anxiety disorders: a meta-analysis of randomized placebo-controlled trials. J Clin Psychiatry. 2008;69(4):621–32.
13. Stewart RE, Chambless DL. Cognitive–behavioral therapy for adult anxiety disorders in clinical practice: a meta-analysis of effectiveness studies. J Consult Clin Psychol. 2009;77(4):595–606.
14. Loerinc AG, Meuret AE, Twohig MP, Rosenfield D, Bluett EJ, Craske MG. Response rates for CBT for anxiety disorders: need for standardized criteria. Clin Psychol Rev. 2015;42:72–82.
15. McNally RJ. Mechanisms of exposure therapy: how neuroscience can improve psychological treatments for anxiety disorders. Clin Psychol Rev. 2007;27(6):750–9.
16. Milad MR, Igoe SA, Lebron-Milad K, Novales JE. Estrous cycle phase and gonadal hormones influence conditioned fear extinction. Neuroscience. 2009;164(3):887–95.
17. Fanselow MS. Conditional and unconditional components of post-shock freezing. Pavlov J Biol Sci. 1980;15(4):177–82.
18. Lonsdorf TB, Menz MM, Andreatta M, Fullana MA, Golkar A, Haaker J, et al. Don't fear 'fear conditioning': methodological considerations for the design and analysis of studies on human fear acquisition, extinction, and return of fear. Neurosci Biobehav Rev. 2017;77:247–85.
19. Myers KM, Davis M. Behavioral and neural analysis of extinction. Neuron. 2002;36:567–84.
20. Quirk GJ, Mueller D. Neural mechanisms of extinction learning and retrieval. Neuropsychopharmacology. 2008;33(1):56–72.
21. Duits P, Cath DC, Lissek S, Hox JJ, Hamm AO, Engelhard IM, et al. Updated meta-analysis of classical fear conditioning in the anxiety disorders. Depress Anxiety. 2015;32(4):239–53.
22. Michael T, Blechert J, Vriends N, Margraf J, Wilhelm FH. Fear conditioning in panic disorder: enhanced resistance to extinction. J Abnorm Psychol. 2007;116(3):612–7.
23. VanElzakker MB, Dahlgren MK, Davis FC, Dubois S, Shin LM. From Pavlov to PTSD: the extinction of conditioned fear in rodents, humans, and anxiety disorders. Neurobiol Learn Mem. 2014;113:3–18.
24. Hermann C, Ziegler S, Birbaumer N, Flor H. Psychophysiological and subjective indicators of aversive pavlovian conditioning in generalized social phobia. Biol Psychiatry. 2002;52(4):328–37.
25. Hartley CA, Phelps EA. Changing fear: the neurocircuitry of emotion regulation. Neuropsychopharmacology. 2010;35(1):136–46.
26. Jovanovic T, Kazama A, Bachevalier J, Davis M. Impaired safety signal learning may be a biomarker of PTSD. Neuropharmacology. 2012;62(2):695–704.
27. Graham BM, Milad MR. The study of fear extinction: implications for anxiety disorders. Am J Psychiatry. 2011;168(12):1255–65.
28. Singewald N, Schmuckermair C, Whittle N, Holmes A, Ressler K. Pharmacology of cognitive enhancers for exposure-based therapy of fear, anxiety and trauma-related disorders. Pharmacol Ther. 2015;149:150–90.
29. Farage MA, Osborn TW, MacLean AB. Cognitive, sensory, and emotional changes associated with the menstrual cycle: a review. Arch Gynecol Obstet. 2008;278(4):299–307.

30. Goldman JM, Murr AS, Cooper RL. The rodent estrous cycle: characterization of vaginal cytology and its utility in toxicological studies. Birth Defects Res B Dev Reprod Toxicol. 2007;80(2):84–97.
31. Rey CD, Lipps J, Shansky RM. Dopamine d1 receptor activation rescues extinction impairments in low-estrogen female rats and induces cortical layer-specific activation changes in prefrontal–amygdala circuits. Neuropsychopharmacology. 2014;39(5):1282–9.
32. Gruene TM, Roberts E, Thomas V, Ronzio A, Shansky RM. Sex-specific neuroanatomical correlates of fear expression in prefrontal-amygdala circuits. Biol Psychiatry. 2015;78(3):186–93.
33. Milligan-Saville JS, Graham BM. Mothers do it differently: reproductive experience alters fear extinction in female rats and women. Transl Psychiatry. 2016;6(10):e928.
34. Graham BM, Daher M. Estradiol and progesterone have opposing roles in the regulation of fear extinction in female rats. Neuropsychopharmacology. 2016;41(3):774–80.
35. Graham BM, Scott E. Effects of systemic estradiol on fear extinction in female rats are dependent on interactions between dose, estrous phase, and endogenous estradiol levels. Horm Behav. 2018;97:67–74.
36. Chang YJ, Yang CH, Liang YC, Yeh CM, Huang CC, Hsu KS. Estrogen modulates sexually dimorphic contextual fear extinction in rats through estrogen receptor β. Hippocampus. 2009;19(11):1142–50.
37. Maeng LY, Cover KK, Taha MB, Landau AJ, Milad MR, Lebron-Milad K. Estradiol shifts interactions between the infralimbic cortex and central amygdala to enhance fear extinction memory in female rats. J Neurosci Res. 2017;95(1–2):163–75.
38. Zeidan MA, Igoe SA, Linnman C, Vitalo A, Levine JB, Klibanski A, et al. Estradiol modulates medial prefrontal cortex and amygdala activity during fear extinction in women and female rats. Biol Psychiatry. 2011;70(10):920–7.
39. Graham BM, Scott E. Estradiol-induced enhancement of fear extinction in female rats: the role of NMDA receptor activation. Prog Neuropsychopharmacol Biol Psychiatry. 2018;86:1–9.
40. Graham BM, Milad MR. Blockade of estrogen by hormonal contraceptives impairs fear extinction in female rats and women. Biol Psychiatry. 2013;73(4):371–8.
41. Li S, Graham BM. Estradiol is associated with altered cognitive and physiological responses during fear conditioning and extinction in healthy and spider phobic women. Behav Neurosci. 2016;130(6):614–23.
42. Milad MR, Zeidan MA, Contero A, Pitman RK, Klibanski A, Rauch SL, et al. The influence of gonadal hormones on conditioned fear extinction in healthy humans. Neuroscience. 2010;168(3):652–8.
43. Glover EM, Jovanovic T, Mercer KB, Kerley K, Bradley B, Ressler KJ, et al. Estrogen levels are associated with extinction deficits in women with posttraumatic stress disorder. Biol Psychiatry. 2012;72(1):19–24.
44. Wegerer M, Kerschbaum H, Blechert J, Wilhelm FH. Low levels of estradiol are associated with elevated conditioned responding during fear extinction and with intrusive memories in daily life. Neurobiol Learn Mem. 2014;116:145–54.
45. Antov MI, Stockhorst U. Stress exposure prior to fear acquisition interacts with estradiol status to alter recall of fear extinction in humans. Psychoneuroendocrinology. 2014;49:106–18.
46. Pineles SL, Nillni YI, King MW, Patton SC, Bauer MR, Mostoufi SM, et al. Extinction retention and the menstrual cycle: different associations for women with posttraumatic stress disorder. J Abnorm Psychol. 2016;125(3):349–55.
47. Butcher R, Collins W, Fugo N. Plasma concentration of LH, FSH, prolactin, progesterone and estradiol-17β throughout the 4-day estrous cycle of the rat. Endocrinology. 1974;94(6):1704–8.
48. Reed BG, Carr BR. The normal menstrual cycle and the control of ovulation. In: de Groot LJ, Chrousos G, Dungan K, Feingold KR, Grossman A, Hershman JM, et al. Endotext [Internet]. South Darmouth, MA: MDText. com, Inc.; 2000. [cited 2019 Mar 5]. Available from: https://www.ncbi.nlm.nih.gov/books/NBK279054/

49. White EC, Graham BM. Estradiol levels in women predict skin conductance response but not valence and expectancy ratings in conditioned fear extinction. Neurobiol Learn Mem. 2016;134(Pt B):339–48.
50. Petitti DB. Combination estrogen–progestin oral contraceptives. N Engl J Med. 2003;349(15):1443–50.
51. Fleischman DS, Navarrete CD, Fessler DMT. Oral contraceptives suppress ovarian hormone production. Psychol Sci. 2010;21(5):750–2.
52. Mishell DR, Thorneycroft IH, Nakamura RM, Nagata Y, Stone SC. Serum estradiol in women ingesting combination oral contraceptive steroids. Am J Obstet Gynecol. 1972;114(7):923–8.
53. Lonsdorf TB, Haaker J, Schümann D, Sommer T, Bayer J, Brassen S, et al. Sex differences in conditioned stimulus discrimination during context-dependent fear learning and its retrieval in humans: the role of biological sex, contraceptives and menstrual cycle phases. J Psychiatry Neurosci. 2015;40(6):368–75.
54. Graham BM, Li SH, Black MJ, Öst L-G. The association between estradiol levels, hormonal contraceptive use, and responsiveness to one-session-treatment for spider phobia in women. Psychoneuroendocrinology. 2018;90:134–40.
55. American Psychiatric Association. Diagnostic and statistical manual of mental disorders. 5th ed. Arlington, VA: American Psychiatric Association; 2013.
56. Hwang MJ, Zsido RG, Song H, Pace-Schott EF, Miller KK, Lebron-Milad K, et al. Contribution of estradiol levels and hormonal contraceptives to sex differences within the fear network during fear conditioning and extinction. BMC Psychiatry. 2015;15:295.
57. Milad MR, Wright CI, Orr SP, Pitman RK, Quirk GJ, Rauch SL. Recall of fear extinction in humans activates the ventromedial prefrontal cortex and hippocampus in concert. Biol Psychiatry. 2007;62(5):446–54.
58. Blume SR, Freedberg M, Vantrease JE, Chan R, Padival M, Record MJ, et al. Sex-and estrus-dependent differences in rat basolateral amygdala. J Neurosci. 2017;37(44):10567–86.
59. Tovote P, Fadok JP, Lüthi A. Neuronal circuits for fear and anxiety. Nat Rev Neurosci. 2015;16(6):317–31.
60. Myers KM, Davis M. Mechanisms of fear extinction. Mol Psychiatry. 2007;12(2):120–50.
61. Sotres-Bayon F, Bush DE, LeDoux JE. Acquisition of fear extinction requires activation of NR2B-containing NMDA receptors in the lateral amygdala. Neuropsychopharmacology. 2007;32(9):1929–40.
62. Baker JD, Azorlosa JL. The NMDA antagonist MK-801 blocks the extinction of Pavlovian fear conditioning. Behav Neurosci. 1996;110(3):618–20.
63. Burgos-Robles A, Vidal-Gonzalez I, Santini E, Quirk GJ. Consolidation of fear extinction requires NMDA receptor-dependent bursting in the ventromedial prefrontal cortex. Neuron. 2007;53(6):871–80.
64. Liu JL, Li M, Dang XR, Wang ZH, Rao ZR, Wu SX, et al. A NMDA receptor antagonist, MK-801 impairs consolidating extinction of auditory conditioned fear responses in a Pavlovian model. PLoS One. 2009;4(10):e7548.
65. Kim JH, Richardson R. Extinction in preweanling rats does not involve NMDA receptors. Neurobiol Learn Mem. 2010;94(2):176–82.
66. Ledgerwood L, Richardson R, Cranney J. Effects of D-cycloserine on extinction of conditioned freezing. Behav Neurosci. 2003;117(2):341–9.
67. Walker DL, Ressler KJ, Lu K-T, Davis M. Facilitation of conditioned fear extinction by systemic administration or intra-amygdala infusions of D-cycloserine as assessed with fear-potentiated startle in rats. J Neurosci. 2002;22(6):2343–51.
68. Tang S, Graham BM. Reproductive experience alters the involvement of N-methyl-D-aspartate receptors in fear extinction, but not fear conditioning, in female Sprague Dawley rats. Psychopharmacology (Berl). 2019;236(1):251–64.
69. Bi R, Foy MR, Vouimba R-M, Thompson RF, Baudry M. Cyclic changes in estradiol regulate synaptic plasticity through the MAP kinase pathway. Proc Natl Acad Sci U S A. 2001;98(23):13391–5.

70. Foy M, Xu J, Xie X, Brinton R, Thompson R, Berger T. 17β-estradiol enhances NMDA receptor-mediated EPSPs and long-term potentiation. J Neurophysiol. 1999;81(2):925–9.
71. Galvin C, Ninan I. Regulation of the mouse medial prefrontal cortical synapses by endogenous estradiol. Neuropsychopharmacology. 2014;39(9):2086–94.
72. Weiland NG. Estradiol selectively regulates agonist binding sites on the N-methyl-D-aspartate receptor complex in the CA1 region of the hippocampus. Endocrinology. 1992;131(2):662–8.
73. Woolley CS, Weiland NG, McEwen BS, Schwartzkroin PA. Estradiol increases the sensitivity of hippocampal CA1 pyramidal cells to NMDA receptor-mediated synaptic input: correlation with dendritic spine density. J Neurosci. 1997;17(5):1848–59.
74. Smith CC, McMahon LL. Estrogen-induced increase in the magnitude of long-term potentiation occurs only when the ratio of NMDA transmission to AMPA transmission is increased. J Neurosci. 2005;25(34):7780–91.
75. Whitlock JR, Heynen AJ, Shuler MG, Bear MF. Learning induces long-term potentiation in the hippocampus. Science. 2006;313(5790):1093–7.
76. Lüscher C, Malenka RC. NMDA receptor-dependent long-term potentiation and long-term depression (LTP/LTD). Cold Spring Harb Perspect Biol. 2012;4(6):a005710.
77. Lewis MC, Kerr KM, Orr PT, Frick KM. Estradiol-induced enhancement of object memory consolidation involves NMDA receptors and protein kinase A in the dorsal hippocampus of female C57BL/6 mice. Behav Neurosci. 2008;122(3):716–21.
78. Vedder LC, Smith CC, Flannigan AE, McMahon LL. Estradiol-induced increase in novel object recognition requires hippocampal NR2B-containing NMDA receptors. Hippocampus. 2013;23(1):108–15.
79. Segal M. Dendritic spines and long-term plasticity. Nat Rev Neurosci. 2005;6(4):277–84.
80. Woolley CS, McEwen BS. Estradiol mediates fluctuation in hippocampal synapse density during the estrous cycle in the adult rat. J Neurosci. 1992;12(7):2549–54.
81. Woolley CS, Gould E, Frankfurt M, McEwen BS. Naturally occurring fluctuation in dendritic spine density on adult hippocampal pyramidal neurons. J Neurosci. 1990;10(12):4035–9.
82. Kato A, Hojo Y, Higo S, Komatsuzaki Y, Murakami G, Yoshino H, et al. Female hippocampal estrogens have a significant correlation with cyclic fluctuation of hippocampal spines. Front Neural Circuits. 2013;7:149.
83. Wallace M, Luine VN, Arellanos A, Frankfurt M. Ovariectomized rats show decreased recognition memory and spine density in the hippocampus and prefrontal cortex. Brain Res. 2006;1126(1):176–82.
84. Gould E, Woolley CS, Frankfurt M, McEwen BS. Gonadal steroids regulate dendritic spine density in hippocampal pyramidal cells in adulthood. J Neurosci. 1990;10(4):1286–91.
85. Smith CC, Vedder LC, Nelson AR, Bredemann TM, McMahon LL. Duration of estrogen deprivation, not chronological age, prevents estrogen's ability to enhance hippocampal synaptic physiology. Proc Natl Acad Sci U S A. 2010;107(45):19543–8.
86. Velázquez-Zamora DA, González-Tapia D, González-Ramírez MM, Flores-Soto ME, Vázquez-Valls E, Cervantes M, et al. Plastic changes in dendritic spines of hippocampal CA1 pyramidal neurons from ovariectomized rats after estradiol treatment. Brain Res. 2012;1470:1–10.
87. de Castilhos J, Forti CD, Achaval M, Rasia-Filho AA. Dendritic spine density of posterodorsal medial amygdala neurons can be affected by gonadectomy and sex steroid manipulations in adult rats: a Golgi study. Brain Res. 2008;1240:73–81.
88. Hao J, Rapp PR, Leffler AE, Leffler SR, Janssen WG, Lou W, et al. Estrogen alters spine number and morphology in prefrontal cortex of aged female rhesus monkeys. J Neurosci. 2006;26(9):2571–8.
89. Inagaki T, Frankfurt M, Luine V. Estrogen-induced memory enhancements are blocked by acute bisphenol A in adult female rats: role of dendritic spines. Endocrinology. 2012;153(7):3357–67.
90. Khan MM, Dhandapani KM, Zhang QG, Brann DW. Estrogen regulation of spine density and excitatory synapses in rat prefrontal and somatosensory cerebral cortex. Steroids. 2013;78(6):614–23.

91. Lai CSW, Franke TF, Gan W-B. Opposite effects of fear conditioning and extinction on dendritic spine remodelling. Nature. 2012;483(7387):87–91.
92. Woolley CS, McEwen BS. Roles of estradiol and progesterone in regulation of hippocampal dendritic spine density during the estrous cycle in the rat. J Comp Neurol. 1993;336(2):293–306.
93. Woolley CS, McEwen BS. Estradiol regulates hippocampal dendritic spine density via an N-methyl-D-aspartate receptor-dependent mechanism. J Neurosci. 1994;14(12):7680–7.
94. Monte LM, Ellis R. Fertility of women in the United States, 2012. Washington, DC: U.S. Census Bureau; 2014.
95. Pawluski JL, Lonstein JS, Fleming AS. The neurobiology of postpartum anxiety and depression. Trends Neurosci. 2017;40(2):106–20.
96. Brett M, Baxendale S. Motherhood and memory: a review. Psychoneuroendocrinology. 2001;26(4):339–62.
97. Hendrick V, Altshuler LL, Suri R. Hormonal changes in the postpartum and implications for postpartum depression. Psychosomatics. 1998;39(2):93–101.
98. Bridges RS, Byrnes EM. Reproductive experience reduces circulating 17beta-estradiol and prolactin levels during proestrus and alters estrogen sensitivity in female rats. Endocrinology. 2006;147(5):2575–82.
99. Bernstein L, Pike MC, Ross RK, Judd HL, Brown JB, Henderson BE. Estrogen and sex hormone-binding globulin levels in nulliparous and parous women. J Natl Cancer Inst. 1985;74(4):741–5.
100. Prevost M, Zelkowitz P, Tulandi T, Hayton B, Feeley N, Carter CS, et al. Oxytocin in pregnancy and the postpartum: relations to labor and its management. Front Public Health. 2014;2:1.
101. Duthie L, Reynolds RM. Changes in the maternal hypothalamic-pituitary-adrenal axis in pregnancy and postpartum: influences on maternal and fetal outcomes. Neuroendocrinology. 2013;98(2):106–15.
102. Toth I, Neumann ID, Slattery DA. Central administration of oxytocin receptor ligands affects cued fear extinction in rats and mice in a timepoint-dependent manner. Psychopharmacology (Berl). 2012;223(2):149–58.
103. Acheson D, Feifel D, de Wilde S, Mckinney R, Lohr J, Risbrough V. The effect of intranasal oxytocin treatment on conditioned fear extinction and recall in a healthy human sample. Psychopharmacology (Berl). 2013;229(1):199–208.
104. Yang Y-L, Chao P-K, Lu K-T. Systemic and intra-amygdala administration of glucocorticoid agonist and antagonist modulate extinction of conditioned fear. Neuropsychopharmacology. 2006;31(5):912–24.
105. Dominique J-F, Bentz D, Michael T, Bolt OC, Wiederhold BK, Margraf J, et al. Glucocorticoids enhance extinction-based psychotherapy. Proc Natl Acad Sci U S A. 2011;108(16):6621–5.
106. Kinsley CH, Lambert KG. Reproduction-induced neuroplasticity: natural behavioural and neuronal alterations associated with the production and care of offspring. J Neuroendocrinol. 2008;20(4):515–25.
107. Lemaire V, Billard JM, Dutar P, George O, Piazza PV, Epelbaum J, et al. Motherhood-induced memory improvement persists across lifespan in rats but is abolished by a gestational stress. Eur J Neurosci. 2006;23(12):3368–74.
108. Tomizawa K, Iga N, Lu YF, Moriwaki A, Matsushita M, Li ST, et al. Oxytocin improves long-lasting spatial memory during motherhood through MAP kinase cascade. Nat Neurosci. 2003;6(4):384–90.
109. Pawluski JL, Galea LAM. Hippocampal morphology is differentially affected by reproductive experience in the mother. J Neurobiol. 2006;66(1):71–81.
110. Kinsley CH, Trainer R, Stafisso-Sandoz G, Quadros P, Marcus LK, Hearon C, et al. Motherhood and the hormones of pregnancy modify concentrations of hippocampal neuronal dendritic spines. Horm Behav. 2006;49(2):131–42.
111. Leuner B, Gould E. Dendritic growth in medial prefrontal cortex and cognitive flexibility are enhanced during the postpartum period. J Neurosci. 2010;30(40):13499–503.

112. Rasia-Filho AA, Fabian C, Rigoti KM, Achaval M. Influence of sex, estrous cycle and motherhood on dendritic spine density in the rat medial amygdala revealed by the Golgi method. Neuroscience. 2004;126(4):839–47.
113. Byrnes EM, Babb JA, Bridges RS. Differential expression of oestrogen receptor alpha following reproductive experience in young and middle-aged female rats. J Neuroendocrinol. 2009;21(6):550–7.
114. Hussain D, Hoehne A, Woodside B, Brake WG. Reproductive experience modifies the effects of estradiol on learning and memory bias in female rats. Horm Behav. 2013;63(3):418–23.
115. Barha CK, Galea LAM. Motherhood alters the cellular response to estrogens in the hippocampus later in life. Neurobiol Aging. 2011;32(11):2091–5.
116. Byrnes EM, Casey K, Bridges RS. Reproductive experience modifies the effects of estrogen receptor alpha activity on anxiety-like behavior and corticotropin releasing hormone mRNA expression. Horm Behav. 2012;61(1):44–9.
117. Macbeth AH, Luine VN. Changes in anxiety and cognition due to reproductive experience: a review of data from rodent and human mothers. Neurosci Biobehav Rev. 2010;34(3):452–67.
118. Love G, Torrey N, McNamara I, Morgan M, Banks M, Hester NW, et al. Maternal experience produces long-lasting behavioral modifications in the rat. Behav Neurosci. 2005;119(4):1084–96.
119. Gatewood JD, Morgan MD, Eaton M, McNamara IM, Stevens LF, Macbeth AH, et al. Motherhood mitigates aging-related decrements in learning and memory and positively affects brain aging in the rat. Brain Res Bull. 2005;66(2):91–8.
120. Kinsley CH, Madonia L, Gifford GW, Tureski K, Griffin GR, Lowry C, et al. Motherhood improves learning and memory. Nature. 1999;402(6758):137–8.
121. Broadbent NJ, Gaskin S, Squire LR, Clark RE. Object recognition memory and the rodent hippocampus. Learn Mem. 2010;17(1):5–11.
122. Paris JJ, Frye CA. Estrous cycle, pregnancy, and parity enhance performance of rats in object recognition or object placement tasks. Reproduction. 2008;136(1):105–15.
123. Macbeth AH, Scharfman HE, MacLusky NJ, Gautreaux C, Luine VN. Effects of multiparity on recognition memory, monoaminergic neurotransmitters, and brain-derived neurotrophic factor (BDNF). Horm Behav. 2008;54(1):7–17.
124. Byrnes EM, Bridges RS. Reproductive experience alters anxiety-like behavior in the female rat. Horm Behav. 2006;50(1):70–6.
125. Wartella J, Amory E, Macbeth A, McNamara I, Stevens L, Lambert KG, et al. Single or multiple reproductive experiences attenuate neurobehavioral stress and fear responses in the female rat. Physiol Behav. 2003;79(3):373–81.
126. Pereira M, Uriarte N, Agrati D, Zuluaga M, Ferreira A. Motivational aspects of maternal anxiolysis in lactating rats. Psychopharmacology (Berl). 2005;180(2):241–8.
127. Scanlan VF, Byrnes EM, Bridges RS. Reproductive experience and activation of maternal memory. Behav Neurosci. 2006;120(3):676–86.
128. Leuner B, Shors TJ. Learning during motherhood: a resistance to stress. Horm Behav. 2006;50(1):38–51.
129. Maeng LY, Shors TJ. Once a mother, always a mother: maternal experience protects females from the negative effects of stress on learning. Behav Neurosci. 2012;126(1):137–41.
130. Rima BN, Bardi M, Friedenberg JM, Christon LM, Karelina KE, Lambert KG, et al. Reproductive experience and the response of female Sprague–Dawley rats to fear and stress. Comp Med. 2009;59(5):437–43.
131. Graham BM. Postnatal stress is associated with impaired fear conditioning and extinction, and heightened hippocampal fibroblast growth factor 2, in mother rats. Horm Behav. 2018;105:110–4.

27 Risk Factors and Prevention Strategies for Anxiety Disorders in Childhood and Adolescence

Maria Demma Cabral and Dilip R. Patel

Introduction

Most youth have appropriate adaptive anxiety response to any given situation or environment. In those who have maladaptive responses, an anxiety disorder may ensue. Anxiety in young children causes impairment, interferes with family functioning and parent-child and peer relationships, and negatively impacts quality of life resulting in having low self-esteem and poor academic performance [1–3]. Continuous research in this field identifies risk factors and key steps to early detection and prevention. There are several evidence-based treatments for anxiety disorders available, yet affected individuals are less likely to seek treatment, lack response to treatment, or experience a new anxiety episode over time. Studying risk factors for anxiety disorders in childhood and adolescence allow earlier identification of at-risk children for provision of prevention strategies.

Prevalence

Of the psychiatric illnesses, anxiety is one of the most common diagnosed mental illnesses in children and adolescents with known persistence into adulthood [2–4]. From the 2016 National Survey of Children's Health (NSCH), among children 3–17 years old, 7.1% currently had anxiety symptoms, compared to 3.2% having depression, both most common in the adolescent age group 12–17 years [5]. From the National Comorbidity Study-Adolescent Supplement (NCS-A) in the United States, lifetime prevalence of anxiety disorders was 31.9% in adolescents aged 13–18 years, much higher compared to 19.1% for behavior disorders, 14.3% for

M. D. Cabral (✉) · D. R. Patel
Division of Adolescent Medicine, Department of Pediatric and Adolescent Medicine,
Western Michigan University Homer Stryker MD School of Medicine, Kalamazoo, MI, USA
e-mail: mariademma.cabral@med.wmich.edu

Y.-K. Kim (ed.), *Anxiety Disorders*, Advances in Experimental Medicine and Biology 1191, https://doi.org/10.1007/978-981-32-9705-0_27

mood disorders, and 11.4% for substance use disorders (SUD), with 2 in 5 adolescents with at least 1 mental disorder [6]. Actual rates could be higher than reported due to subclinical anxiety. Prevalence for specific subtypes of anxiety disorders from the NCS-A was 2.2% for generalized anxiety disorder (GAD), 2.3% for panic disorder (PD), 2.4% for agoraphobia, 5% for posttraumatic stress disorder (PTSD), 7.6% for separation anxiety disorder (SAD), 9.1% for social phobia, and 19.3% for specific phobia. PTSD, PD, social phobia, and GAD prevalence increased with age. Of the total sample of adolescents, severe anxiety disorders were seen in 8.3% majority of which are those individuals with agoraphobia and PDs [6].

Risk and Protective Factors

Though there are children without known predisposing factors, development of anxiety disorders is a complex interplay of genetics, temperament in childhood, sociodemographic traits, environmental triggers, and physiologic factors. According to the National Institute of Mental Health, risk factors can be generalized for all subtypes of anxiety disorders [7]. These include associated introversion traits or behavioral inhibition in childhood, having a family history of anxiety or other psychopathological conditions, and presence of negative life experiences. From a national sample, Blanco et al. also identified other common risk factors: disturbed family environment, sexual abuse in childhood, low self-esteem, and lower academic achievement [8].

Developmental Stages

Kingston et al. looked at the association between specific prenatal, postnatal, and early life predictors by examining the demographic, obstetrical, psychosocial, medical, behavioral, and infant factors on development of anxiety in childhood by age 5 [9]. The authors found that at 5 min of life having an APGAR score of 7 or less was a significant predictor of childhood anxiety symptoms likely from neural circuitry disruption [9]. Psychological distress in the mother during the first 5 years of life was another significant predictor, though mechanisms may be poorly understood. Protective factors for a child born to a mother <20 years old, having more than one child, and prematurity were found to be protective [9].

Essau et al. examined the incidence, recurrence, and comorbidity rates of anxiety disorders using four developmental periods: childhood (5 to <13 years), adolescence (13 to <18 years), emerging adulthood (18 to <24 years), and adulthood (24–30 years) [4]. Incidence rates were found to be significantly higher in childhood due to separation anxiety and adulthood due to PD. Increased recurrence rates noted in individuals ages 13 and up. Total incidence in adulthood was identified to be significantly increased likely due to having child and adolescent anxiety increased likelihood of having anxiety during the emerging adulthood period [4].

Social anxiety has been especially common during in early adolescence and can result in poor academic performance and social skills. Appearance of social anxiety symptoms was found to be dependent on gender and puberty. Girls with advance physical changes were found to have higher levels of anxiety, likely due to body dissatisfaction, anticipated unwanted weight gain, and beginning sexuality [10].

Comorbidities

An estimated 20–25% of youth in the United States will have a mental health diagnosis with associated severe impairment with lifetime comorbidity present in over 80% of cases [6]. Comorbidity of current and lifetime anxiety and mood disorders have been significantly high overtime [11], while comorbidity with illicit drug and alcohol use was around 50% [12]. From the NSCH, 1 in 3 children age 3–17 years with anxiety had comorbid depression, while 3 in 4 children with depression had comorbid anxiety [5]. Other important comorbid conditions include attention-deficit/hyperactivity disorder (ADHD), behavioral problems, learning disorders, and certain medical conditions [3]. Assessment of anxiety disorders necessitates identifying underlying physical health conditions that can possibly cause or worsen them, such as thyroid and heart diseases and SUD.

Anxiety disorders have both homotypic and heterotypic continuity, described as likelihood of having the disorder by the same disorder and by another disorder, respectively [13]. The separation anxiety hypothesis was thought to explain the proximity of having PD and SAD, since both have mutual somatic symptoms. SAD is one of the most common subtypes seen in school-age children and found to be predictive of increased risk of adult PD [4, 13]. What has been shown is that having anxiety previously is the best predictor of anxiety persistence [2]. There is strong heterotypic continuity from depression to anxiety and vice versa that is most pronounced in the female gender [14]. Essau et al. demonstrated this finding and as well stated that degree of comorbidity between anxiety disorders, major depressive disorder, and SUD was statistically significant across all four developmental stages with exception of the adolescence stage for SUD [4].

Sociodemographic Risk Factors

Results from a nationally representative survey showed that the median age of onset for anxiety disorders was 11 years old, though from the NCS-A found earlier onset in childhood around age 6 plateauing off after 12 years of age [6, 15]. Though from both studies, anxiety disorders appear much earlier compared to mood disorders, behavior disorders, and SUD. There is a wide range of distributions for different subtypes. Specific phobia and SAD had earlier ages of onset compared to social phobia and other anxiety disorders [15], though one recent meta-analysis found 21.3 years as the mean age of onset for all anxiety disorders, with SAD, specific phobia, and social phobia having earlier onset at age 15 years, compared to a later onset between 21 and 35 years old for agoraphobia, obsessive-compulsive disorder (OCD), PTSD, PD, and GAD [16]. Anxiety disorders that started in adolescence had a more negative progression compared to those with childhood onset resulting in poor adjustments at work and relationships, low satisfaction and coping skills, and increased stress as an adult [4]. What has been consistent is that onset in childhood has been associated with more severe psychopathology, psychiatric comorbidity, avoidance behaviors, and increased suicide rates [14–17].

Women were found to have two- to threefold increased risk for anxiety compared to men, though Lijster et al. found no gender difference from their meta-analysis [4, 8, 15, 16]. From a nationally representative sample of adolescents in the United States as published in the NCS-A, all subtypes of anxiety disorder were more common in females with the highest difference seen PTSD [6]. Being female was associated with higher prevalence, persistence, and incidence rates of anxiety, especially following a traumatic event and having PD [2, 8, 18, 19]. Analyzing course of development, initial overall anxiety symptoms were higher in girls, followed by slight decrease in GAD, PD, and SAD, while boys had a more stable course on all subtypes [20]. Essau et al. found that being female was not predictive of anxiety recurrence during the developmental periods of adolescence and emerging adulthood [4]. Maeng and Milad found that stress, fear, and sex hormones related to amplified stress sensitivity and extinction memory impairment are correlated with hypoestrogenism in females [19]. A two-hit hypothesis explaining one's susceptibility to anxiety was identified by Catuzzi and Beck. The authors found evidence that the female sex and behavioral inhibition temperament are cumulative resulting in more development of anxiety disorders [21].

Results from NCSH showed anxiety problems prevalent among non-Hispanic White children, while results from the NCS-A found non-Hispanic Black adolescents to have increased rates compared to peers [5, 6]. In an adult sample population, White race was found to have increased anxiety and MDD [8].

Childhood Temperament

Temperament in children is variable and can be characterized in two as presented by Mian et al. [1]. Behavioral inhibition is one and a significant risk factor for developing later anxiety. It is described as the unfailing tendency to exhibit fear, withdrawal, or caution in new or uncomfortable situations. The other risk factor that is more correlated with later depression is negative emotionality described as irritability, challenge of ease, negative mood, and undesirable emotional responses. The authors in their longitudinal, prospective study found certain anxiety symptoms and temperament in childhood to be strongly predictive of anxiety outcomes in kindergarten and second grade students, also mediating impact of maternal and family risk factors [1].

Anxiety sensitivity, on the other hand, defined as fear of experiencing symptoms of anxiety emerging in middle childhood and said to be cognitive bias, predicted similar future anxiety symptoms, commonly seen in PDs. This risk factor exhibits homotypic continuity as mentioned earlier. Waszczuk et al. found that anxiety sensitivity at age 8 significantly predicts symptoms in 2 years. Such is a strong broad risk factor with specific predilection for occurrence of somatic/panic and separation anxiety symptoms, less so for having generalized anxiety and social phobia [13].

Family History and Parental Characteristics

Anxiety tends to aggregate within families. Having a parent with psychiatric disorders significantly increases the child's risk for similar psychopathologies compared to those whose parents have no history [22]. Maternal depression, anxiety, and negative child attachment were associated with anxiety-related symptoms that mediated temperament in early childhood [1–3, 23]. Much of available studies on parental anxiety highlight the maternal role, whereas the paternal role was thought to boost the child's autonomy, exploration, and adaptability through spontaneous and competitive play by providing sets of skills to overcome stressful situations. Yet, arguably paternal overinvolvement was said to be predictive of development of anxiety in young children. One study identified an indirect link of paternal anxiety to future anxiety in the child as mediated by maternal restriction of independence, but not by paternal parenting behaviors [24].

Considerable amount of literature focuses on parenting behaviors and its impact on childhood anxiety. Parental overcontrol involves restriction of autonomy and independence and overprotection, with the latter found to increase levels of anxiety [2]. One study looked at three major domains of factors (parental, child, and parent-child relationship) and if certain negative life events made these domains worse, and in turn resulted in worse anxiety symptoms [25]. The authors found that levels of parenting stress, parental anxious nurturing, and parent-child dysfunctional interaction mediated the association between stressful life events and severity of anxiety symptoms [25]. An interesting concept was presented by Gouze et al. that anxiety in younger children drives parental behavior, and when older, it would be the parental behavior mediating anxiety symptoms [26]. In their study, they examined reciprocal relations between parental regard for autonomy, hostility, and support in relation to childhood anxiety development at a critical transitional school period and found that parenting and anxiety interact mutually over time. One finding was parental support at age 5 lowered anxiety at age 6–7, while high anxiety at age 5 lowered parental support at age 6–7 [26].

Parental insecure relationships and parental problems at any time during one's life period and low income levels were found to have high correlation with anxiety disorders [2], though from the NCS-A, parental poverty level and urbanicity were not associated with any mental illness diagnosis [6]. Higher rates of anxiety disorder were seen in adolescents with divorced or separated parents and whose parents did not receive a college degree had higher rates [6].

Non-traumatic and Traumatic Stressful Life Events and Experiences

Stressful life events considered non-traumatic or traumatic lead to increased preponderance for anxiety disorders. There is neural evidence of early stress and childhood emotional maltreatment leading to adult anxiety [27]. Examples of non-traumatic life events include parental divorce or separation, arrival of a stepparent, loss or death of a family member, moving to a new home or school or acculturation, and teasing. Anxiety was increased in those with divorced parents [6]. One study looked at divorce occurring at different times in a child's life and found that its occurrence during elementary school was correlated with increased adverse effects on internalizing and externalizing problems, noting anxiety or fear of abandonment and guilt in the affected child [28].

On the other hand, traumatic life events include victimization, interpersonal violence, discrimination, and child abuse. Peer victimization, such as rejection and ostracism, in adolescence has been found to correlate with short- and long-term anxiety and depression symptoms due to persistent psychological stress [29]. Exposure to violence in the community was associated with childhood anxiety symptoms and temperament [1]. A systematic review looked at the role of child sexual abuse in the development of anxiety disorders. The authors found that sexual abuse in childhood is indeed a very important risk factor, especially those with PTSD, and said to be unrelated to how severe the abuse was. Males have the same risk profile as females for anxiety-related symptoms due to abuse [30].

Sleep

Sleep is especially important in brain development specifically memory and learning. One review showed preliminary evidence highlighting increased sleep and anxiety difficulties arising during late childhood to early adolescence, relating to the sleep changes occurring in this developmental transition [31]. Common reason for poor sleep is media exposure and its digital imprints on the child's emotional regulation. Hoge et al. summarized the growing areas of interest with the use of technology devices relating to anxiety and depression [32]. They have described current states of fear and anxiety to both traditional and substituted media, concomitant social anxiety, and lacking social interactions and being disconnected, such with victimization from being cyberbullied [32].

Intolerance of Uncertainty

Intolerance of uncertainty, as the name implies, is considered a core feature of worry and has been studied more in adults. Not much has been described in the pediatric population despite children's awareness of such uncertainty and ability to respond despite being young. Since such neuronal processes persist through middle childhood and adolescence, one meta-analysis did find similar results of strong correlation between intolerance of uncertainty and anxiety and worry in young individuals [33].

Special Populations

Anxiety in youth with chronic disease had been correlated with negative health and psychosocial outcomes and unfortunately found to be either underrecognized or undertreated. This has been true for children with special health care needs, and adolescents with congenital heart disease [34], asthma, mixed connective tissue disorders and diabetes mellitus, with higher rates in those using insulin pump [35]. One study looked at anxiety symptoms among teens who survived childhood cancer. Anxiety was far more common than depression, and the authors found the subtype PTSD as the most common form of anxiety with rates higher than the general population but lower than those who experienced other significant traumas [36]. Social anxiety in this study group was likely due to illness outcome, poor body image, and long interval posttreatment.

Lo and Cheng found Latinos, more than Blacks and Asians, to have higher prevalence of anxiety and mood disorders [37]. They also found that protective factors in minority individuals include being male, good financial status, being marriage, being more educated, and experiencing few discriminations [37]. Sexual minority youth are known to express more anxiety than their heterosexual counterparts. One prospective birth cohort study from the United Kingdom identified this at-risk youth to have higher risk of anxiety disorders at age 17.5 with contributing factors such as bullying during middle adolescence and poor self-esteem [38].

Prevention of Anxiety Disorders

Prevention of anxiety symptoms and disorders should be considered within the broader context of promotion of mental health and prevention of mental disorders in general. A dynamic interplay of multiple risk and protective factors contributes to the development, severity, progression, remission, resolution, and prevention of mental disorders across the life span of an individual. As described by Mrazek and Haggerty (1994), the prevention of mental disorders aims at "reducing incidence, prevalence, recurrence of mental disorders, the time spent with symptoms, or the risk condition for mental illness, preventing or delaying recurrences and also decreasing the impact of illness in the affected person, their families and the society" [39, 40].

The assessment of risk and protective factors is the first step in the development and application of the most appropriate preventive intervention. Effective preventive intervention focuses on eliminating or reducing the impact of risk factors and facilitating and strengthening the positive impact of protective factors on the symptoms of anxiety or anxiety disorder. Both risk and protective factors are specific characteristics, experiences, or events that occur at individual, family, community, or state or policy level. Biological factors play a specific role at individual level. Risk factors are associated with a higher likelihood of subsequent development of a disorder and generally precede the development of symptoms or disorder, whereas protective factors are associated with a lower likelihood of development of a disorder or its symptoms and generally mitigate the negative influence of risk factors [40]. In a broader sense, risk factors are associated with negative outcomes, while protective factors are associated with positive outcomes. It is also useful to recognize that while some risk and protective factors are constant and non-modifiable, others are more fluid and variable over time. For example, an individual's inherent genetic and biological characteristics remain constant over time, whereas an individual's socioeconomic status, peer group, or adverse childhood experiences may vary over time [39–41].

While the risk and protective factors at in individual level may be less variable, the increased likelihood of variability beyond individual level necessitates prioritization of particular risk and protective factors to be addressed based on the degree of their impact. When designing and evaluating a preventive intervention program, certain general characteristics of risk and protective factors should be taken into account – multiple contexts, correlational and cumulative characteristics, outcomes characteristics, and long-term and life span influence [39–41].

Multiple Contexts

Risk and protective factors exist in multiple contexts [39, 40]. Specific biological and psychological characteristics of an individual contribute to either a higher or a lesser likelihood for him or her for developing a mental disorder such as an anxiety disorder. Each person has specific relationship within the community in which he or she resides and to the broader society. Because of such relationships, an individual's biological and psychological characteristics exist in multiple contexts, rather than in isolation. Multiple risk and protective factors operate within each such context and may influence each other. Design and evaluation of preventive interventions need consideration of addressing risk and protective factors in multiple contexts at the same time.

For example, within the context of individual and family relationships, risk factors include alcohol use by parents, parents with mental illness, abuse and neglect of child, or lack of suboptimal adult supervision of children. On the other hand, protective factors in this context include appropriate and optimal parental involvement, good parental mental and physical health, and absence of substance use. Within the context of the community, examples of risk factors include neighborhood poverty and violence, whereas protective factors could include the availability of faith-based resources and after-school activities [39, 40]. At the society level, examples of risk factors can include norms and laws favorable to substance use, as well as racism and a lack of economic opportunity, whereas protective factors in this context would include hate crime laws or policies limiting the availability of alcohol [39, 40].

Correlational and Cumulative Characteristics

A given risk factor tends to be positively correlated with one or more additional risk factors while negatively correlated with one or more protective factors [39, 40]. An individual with some risk factors is more likely to have additional risk factors while at the same time less likely to have protective factors. Both risk and protective factors have been shown to have a long-term cumulative impact on the likelihood of development of an anxiety or other mental health disorders. Consideration of the impact of the correlation and cumulative characteristics of risk and protective factors provides the fundamental basis for the critical importance of early screening, identification and preventive intervention programs during prenatal period, and infancy and early childhood [39–41].

Outcomes Characteristics

One or more risk factors can increase the likelihood of development of one or more mental health disorder, for example, adverse life experiences have been associated with the development of depression, anxiety, or substance abuse [40]. Similarly, one or more protective factors can reduce the likelihood of development of one or more mental health disorder, for example, healthy parenting practices and positive childhood experiences have been shown to reduce the likelihood of or protect against development of multiple negative outcomes later in life. Preventive interventions that address multiple risk and protective factors are more desirable and tend to influence outcomes in multiple areas.

Long-Term and Life Span Influence

Multiples studies provide evidence that both risk and protective factors have lifelong influence on an individual's mental and physical health and long-term outcomes [39, 40]. Adverse early childhood experiences have been shown to have long-term negative mental health consequences [42]. The influence of risk or protective factors often transcend between different contexts; for example, effective parenting has been shown to mediate the effects of multiple risk factors, including poverty, divorce, parental bereavement, and parental mental illness.

Types of Prevention Interventions

Not all people or populations are at the same risk of developing mental health problems. Prevention interventions are most effective when they are matched to their target population's level of risk. Three types of preventive interventions have been described [39–41]:

1. *Universal prevention* is defined as those interventions that are targeted at the general public or to a whole population group that has not been identified on the basis of increased risk.
2. *Selective prevention* targets individuals or subgroups of the population whose risk of developing a mental disorder is significantly higher than average, as evidenced by biological, psychological, or social risk factors.
3. *Indicated prevention* targets high-risk people who are identified as having minimal but detectable signs or symptoms foreshadowing mental disorder or biological markers indicating predisposition for mental disorder but who do not meet diagnostic criteria for disorder at that time.

For each of the three types of preventive interventions described, which are widely applied in the development of prevention programs, multiple factors need specific consideration. Universal prevention takes a broad approach and is applied to larger groups or populations, such as communities, workplaces, or school systems. Universal prevention programs are generally more expensive and require broad collaboration and participation at community and state level. Because the universal intervention is applied to persons who do not have a given mental health condition or may not be at risk for developing a disorder, it is challenging to convince a high percentage of the population to participate. A high level of participation increases sample representativeness and reduces cost of program administration. For both selective and targeted preventive interventions, the relative cost is less than universal programs. For both the selective and indicated prevention programs, a methodological consideration is protecting confidentiality while at the same time encouraging participation.

In addition to the type of preventive intervention, the timing of intervention should be considered. The risk for developing anxiety disorder varies at different times during childhood and adolescence based on exposure to specific life events and environmental factors. Also, the appropriate level of intervention – individual, family, community, or multiple – should be a consideration in the development of preventive strategy.

Cognitive Behavioral Therapy-Based Approaches

Multiple studies provide evidence to support variable effectiveness of cognitive behavioral therapy-based prevention interventions for symptoms of anxiety or anxiety disorder. Evidence is insufficient to support other prevention approaches. Several investigators have conducted comprehensive systematic reviews and meta-analytic reviews of published studies of CBT-based prevention interventions for anxiety disorders.

Topper et al. (2017) reported findings of their randomized controlled trial that evaluated the efficacy of a preventive intervention for anxiety disorders and depression by targeting excessive levels of repetitive negative thinking (RNT; worry and rumination) in adolescents and young adults [43]. A total of 257 study participants

(83.7% female) reported to have repetitive negative thinking, worry, and rumination were randomly allocated to a 6-week cognitive behavioral training delivered in a group, via the Internet, or to a waitlist control condition. Self-control measures were collected at pre-intervention, post-intervention, and 3-month and 12-month follow-up. Both modes of preventive intervention were found to significantly reduce RNT (d = 0.53–0.89) and symptom levels of anxiety and depression (d = 0.36–0.72). Effects were maintained until 12-month follow-up. The interventions resulted in a significantly lower 12-month prevalence rate of depression (group intervention, 15.3%; Internet intervention, 14.7%) and generalized anxiety disorder (group intervention, 18.0%; Internet intervention, 16.0%), compared to waitlist (32.4% and 42.2%, respectively). Mediation analyses demonstrated that reductions in RNT mediated the effect of the interventions on the prevalence of depression and generalized anxiety disorder. Topper et al.'s (2017) study provides evidence for the efficacy of cognitive behavioral intervention in preventing repetitive negative thinking and rumination in adolescents and young adults.

Moreno-Peral et al. (2017) conducted a systematic review, meta-analysis, and meta-regression study to evaluate the effectiveness of preventive psychological and/or educational interventions for anxiety disorders in varied populations [44]. The review included 29 studies representing 10,430 patients from 11 countries on 4 continents. Based on an extensive analysis of the data, the authors reported that psychological and/or educational interventions had a small but statistically significant benefit for anxiety prevention in all populations evaluated.

Rasing et al. (2017) conducted a meta-analytic review of the effects of school-based and community-based prevention programs that are based on cognitive behavioral therapy with the primary goal of preventing depression, anxiety, or both in high-risk adolescents [45]. Thirty-six studies met the inclusion criteria, 23 on depression, and 13 on anxiety. For anxiety prevention aimed at high-risk adolescents, no short-term or 12-month follow-up effect was found. Three to six months after the prevention intervention, symptoms of anxiety were significantly decreased. The analysis showed that CBT-based prevention programs for anxiety appear to be effective after 3–6 months, but this effect disappears after 12 months.

Fisak et al. (2011) conducted a meta-analytic review of the effectiveness of child and adolescent anxiety prevention programs [46]. The analysis included 35 studies that met inclusion criteria. Of these, 30 studies included a comparison group and 5 studies included pretest-posttest data only. Four of the studies provided follow-up data to previously published studies. Mean weighted effect sizes were calculated, and studies were encoded for potential moderator variables. A statistically significant effect size of.18 was obtained at post-intervention. However, the effect sizes obtained at follow-up yielded mixed results. Significant moderators of program effectiveness were provider type (professional versus lay provider) and the use of FRIENDS program [47]. Program duration, participant age, gender, and program type (universal versus targeted) were not found to moderate program effectiveness.

A substantial number of studies identified by Fisak et al. (2011) for inclusion in their analysis utilized Barrett's FRIENDS program [46, 47]. FRIENDS program is a well-established and effective program for the prevention of general anxiety symptoms. Studies utilizing the FRIENDS program were found to be more effective than programs not utilizing FRIENDS. Particular strengths of the FRIENDS program include the following: the program is manualized and well-structured, and it can be easily integrated into school curriculums. Overall, based on their analysis, Fisak et al. (2011) found that almost all of the prevention program for anxiety were either behavioral or cognitive behavioral and Barrett's manualized FRIENDS program [46, 47].

Zalta (2011), in a meta-analysis, assessed efficacy of cognitive behavioral interventions in preventing anxiety symptoms [48]. A systematic review identified 15 independent pretest-posttest randomized or quasi-randomized efficacy trials for analysis. At posttest, intervention groups demonstrated significantly greater symptom reduction compared to control groups resulting in weighted mean effect sizes of 0.25 for general anxiety after removal of outliers. These effects appeared to diminish over 6- and 12-month follow-up. Exploratory moderator analyses indicated that individually administered media interventions were more effective than human-administered group interventions at preventing general anxiety and depression symptoms. Deady et al. (2017) reported findings of a systematic review and meta-analysis of eHealth interventions for the prevention of depression and anxiety in the general population [49]. Ten trials were included in the systematic review and meta-analysis. All studies utilized cognitive behavioral techniques. For anxiety outcome studies, at posttreatment, the overall mean difference between the intervention and control groups was 0.31 (95% confidence interval: 0.10, 0.52; $p = 0.004$), indicating a small but positive effect of the eHealth interventions. The effect sizes for universal and indicated interventions were similar. However, there was inadequate evidence to suggest that such interventions have an effect on long-term disorder incidence rates.

Clarke et al. (2015) conducted a systematic review to provide a narrative synthesis of the evidence on the effectiveness of online mental health promotion and prevention interventions for youth aged 12–25 years [50]. Their analysis included 28 studies conducted between 2000 and 2015. The results from the online prevention interventions indicate the significant positive effect of computerized cognitive behavioral therapy on adolescents' and emerging adults' anxiety and depression symptoms. The rates of non-completion were moderate to high across a number of studies. Implementation findings provide some evidence that participant face-to-face and/or web-based support was an important feature in terms of program completion and outcomes. Additional research examining factors affecting exposure, adherence, and outcomes is required. The quality of evidence across the studies varied significantly, thus highlighting the need for more rigorous, higher-quality evaluations conducted with more diverse samples of youth. Although future research is warranted, this study highlights the potential of online mental health promotion and prevention interventions in promoting youth well-being and reducing mental health problems.

Conclusion

With anxiety being prevalent in childhood and adolescence, identification of different risk and protective factors (Tables 27.1 and 27.2) has been crucial in understanding treatment approaches [8, 9, 25, 30, 40, 41, 42, 51, 52]. Anxiety disorders have a multifactorial etiology from genetics to environmental triggers, with considerable individual and familial factors that either mediate or moderate the symptoms. Because persistent anxiety symptoms and anxiety disorders can have long-term and lifelong adverse impact and outcomes, it is important to develop and implement effective prevention programs. Prevention interventions should be guided by assessment of risk and protective factors for anxiety disorders. Effective preventive

Table 27.1 Risk factors for the development of anxiety disorders in childhood and adolescence

Risk factors	
Individual	Academic underachievement Anxiety sensitivity Avoidant coping style Behavior problems Child abuse and neglect Childhood temperament – behavioral inhibition Chronic illness Exposure to violence Heritability of anxiety disorder Learning disabilities Low perceived self-control Other mental health disorders – such as depression, attention deficit hyperactivity disorder Poverty Sleep problems Stressful life events and loss Subclinical anxiety symptoms
Familial	Divorce or separation Familial psychopathologies Insecure attachment Lower parental education Lower socioeconomic status Parental avoidant problem-solving strategies Parental chronic illness or disability Parental mental illness Parental social inhibition, negative affectivity Parenting overprotection Perinatal complications Prenatal exposure to substances
Community	Access to illicit substances Isolation and estrangement Lack of resources: Education, transportation, housing Neighborhood violence Poor social support network Social injustice and discrimination Unemployment

Table 27.2 Protective factors for the development of anxiety disorders in childhood and adolescence

Protective factors	
Individual	Ability to overcome adversity Adaptability Adequate sleep Conflict management skills Self-esteem Self-sufficiency Stress coping skills present
Family	Affirmative parent-child relationship Cohesive family unit Higher parental education Parental employment and higher socioeconomic status Parental security Positive parenting Support from family
Community	Community networks Empowered social relationships Integrated ethnic minority groups Positive environment in the school system Social awareness and involvement Social responsibility Support from friends and the community

intervention should focus on either eliminating or reducing the impact of risk factors and facilitating and strengthening the positive impact of protective factors on the symptoms of anxiety or anxiety disorder. The risk for developing anxiety disorders varies among individuals, and therefore preventive interventions should be appropriately matched with specific type of intervention programs – universal, selective, or indicated. Numerous studies have provided evidence to support variable effectiveness of cognitive behavioral therapy-based prevention interventions for symptoms of anxiety or anxiety disorder; however, evidence is insufficient to support other prevention approaches.

References

1. Mian ND, Wainwright L, Briggs-Gowan M, Carter AS. An ecological risk model for early childhood anxiety: the importance of early child symptoms and temperament. J Abnorm Child Psychol. 2011;39(4):501–12.
2. Voltas N, Hernández-martínez C, Arija V, Canals J. The natural course of anxiety symptoms in early adolescence: factors related to persistence. Anxiety Stress Coping. 2017;30(6):671–86.
3. Bagnell AL. Anxiety and separation disorders. Pediatr Rev. 2011;32(10):440–5.
4. Essau CA, Lewinsohn PM, Lim JX. Moon-ho RH, avenue H. Incidence, recurrence and comorbidity of anxiety disorders in four major developmental stages. J Affect Disord. 2018;228:248–53.

5. Ghandour RM, Sherman LJ, Vladutiu CJ, Ali MM, Lynch SE, Bitsko RH, et al. Prevalence and treatment of depression, anxiety, and conduct problems in US children. J Pediatr [Internet] 2018;206:256–267.e3. Available from: https://doi.org/10.1016/j.jpeds.2018.09.021.
6. Merikangas K, He J, Burstein M, Swanson S, Avenevoli S, Cui L, et al. Lifetime prevalence of mental disorders in US adolescents: results from the National Comorbidity Study-Adolescent Supplement (NCS-A). J Am Acad Child Adolesc Psychiatry. 2011;49(10):980–9.
7. The National Institute of Mental Health. Anxiety disorders [Internet]. 2018 [cited 2019 Feb 21]. Available from: https://www.nimh.nih.gov/health/topics/anxiety-disorders/index.shtml#part_145335
8. Blanco C, Rubio J, Wall M, Wang S, Jiu C, Kendler K. Risk factors for anxiety disorders: common and specific effets in a national sample. Depress Anxiety. 2014;31(9):756–64.
9. Kingston D, Heaman M, Brownell M, Ekuma O. Predictors of childhood anxiety : a population-based cohort study. PLoS One. 2015;10(7):e0129339.
10. Deardorff J, Ph D, Hayward C, Wilson KA, Hammer LD, Agras S. Puberty and gender interact to predict social anxiety symptoms in early adolescence. J Adolesc Health. 2007;41(1):102–4.
11. Brown T, Campbell L, Lehman C, Grisham J, Mancill R. Current and lifetime comorbidity of the DSM-IV anxiety and mood disorders in a large clinical sample. J Abnorm Psychol. 2001;110(Nov):1–2.
12. Man H, Lai X, Cleary M, Sitharthan T, Hunt GE. Prevalence of comorbid substance use, anxiety and mood disorders in epidemiological surveys, 1990–2014 : a systematic review and meta-analysis. Drug Alcohol Depend. 2015;154:1–13.
13. Waszczuk MA, Zavos HMS, Eley TC. Genetic and environmental influences on relationship between anxiety sensitivity and anxiety subscales in children. J Anxiety Disord [Internet]. 2013;27(5):475–84. Available from: https://doi.org/10.1016/j.janxdis.2013.05.008
14. Costello E, Mustillo S, Erkanli A, Keeler G, Angold A. Prevalence and development of psychiatric disorders in childhood and adolescence. Arch Gen Psychiatry. 2003;60:837–44.
15. Kessler RC, Berglund P, Demler O. Lifetime prevalence and age-of-onset distributions of DSM-IV disorders in the national comorbidity survey replication. Arch Gen Psychiatry. 2005;62(6):593–602.
16. De Lijster JM, Dierckx B, Utens EMWJ, Verhulst FC, Zieldorff C, Dieleman GC, et al. The age of onset of anxiety disorders. Can J Psychiatr. 2017;62(4):237–46.
17. Ramsawh H, Weisberg R, Dyck I, Stout R, Keller M. Age of onset, clinical characteristics, and 15-year course of anxiety disorders in a prospective, longitudinal, observational study. J Affect Disord. 2011;132(1–2):260–4.
18. Zahn-waxler C, Shirtcliff E, Marceau K. Disorders of childhood and adolescence: gender and psychopathology. Annu Rev Clin Psychol. 2008;4:275–303.
19. Maeng LY, Milad MR. Sex differences in anxiety disorders: interactions between fear, stress,and gonadal hormones. Horm Behav. 2015;76:106–17.
20. Ohannessian CM, Milan S, Vannucci A. Gender differences in anxiety trajectories from middle to late adolescence. J Youth Adolesc. 2017;46(4):826–39.
21. Catuzzi JE, Beck KD. Anxiety vulnerability in women: a two-hit hypothesis. Exp Neurol. 2014;259:75–80.
22. Ginsburg G, Drake K, Tein J, Teetsel R, Riddle M. Preventing the onset of anxiety disorders in offspring of anxious parents: a randomized controlled trial of a family-based intervention. Am J Psychiatry. 2015;172(12):1207–14.
23. Hopkins J, Lavigne JV, Gouze KR, Lebailly SA, Bryant FB. Multi-domain models of risk factors for depression and anxiety symptoms in preschoolers : evidence for common and specific factors. J Abnorm Child Psychol. 2013;41:705–22.
24. Gibler RC, Kalomiris AE, Kiel EJ. Paternal anxiety in relation to toddler anxiety: the mediating role of maternal behavior. Child Psychiatry Hum Dev. 2018;49(4):512–22.
25. Platt R, Williams S, Ginsburg G. Stressful life events and child anxiety: examining parent and child mediators. Child Psychiatry Hum Dev. 2016;47(1):23–34.

26. Gouze KR, Hopkins J, Bryant FB, Lavigne JV. Parenting and anxiety: bi-directional relations in young children. J Abnorm Child Psychol. 2017;45:1169–80.
27. Fonzo G, Ramsawh H, Flagan T, Simmons A, Sullivan S, Allard C, et al. Early life stress and the anxious brain: evidence for a neural mechanism linking childhood emotional maltreatment to anxiety in adulthood. Psychol Med. 2016;46(5):1037–54.
28. Lansford J, Malone P, Castellino D, Dodge K, Pettit G, Bates J. Trajectories of internalizing, externalizing, and grades for children who have and have not experienced their parents' divorce or separation. J Fam Psychol. 2009;20(2):292–301.
29. Stapinski LA, Araya R, Heron J, Montgomery AA, Stallard P. Peer victimization during adolescence: concurrent and prospective impact on symptoms of depression and anxiety. Anxiety Stress Coping [Internet]. 2015;28(1):105–20. Available from: https://doi.org/10.1080/10615806.2014.962023
30. Maniglio R. Child sexual abuse in the etiology of anxiety disorders: a systematic review of reviews. Trauma Violence Abuse. 2013;14(2):96–112.
31. McMakin D, Alfano C. Sleep and anxiety in late childhood and early adolescence. Curr Opin Psychiatry. 2015;28(6):483–9.
32. Hoge E, Bickham D, Cantor J. Digital media, anxiety, and depression in children. Pediatrics. 2017;140(Supplement 2):S76–80.
33. Osmanağaoğlu N, Creswell C, Dodd HF. Intolerance of uncertainty, anxiety, and worry in children and adolescents: a meta-analysis. J Affect Disord. 2018;225:80–90.
34. Wang Q, Hay M, Clarke D, Menahem S. The prevalence and predictors of anxiety and depression in adolescents with heart disease. J Pediatr. 2012;161:943–6.
35. Knight A, Weiss P, Morales K. Identifying differences in risk factors for depression and anxiety in pediatric chronic disease: a matched cross-sectional study of youth with lupus/mixed connective tissue disease and their diabetic peers. J Pediatr. 2015;167(6):1397–403.
36. Mcdonnell GA, Salley CG, Barnett M, DeRosa A, Werk RS, Hourani A, et al. Anxiety among adolescent survivors of pediatric cancer. J Adolesc Health. 2017;61(4):409–23.
37. Lo CC, Cheng TC. Social status, discrimination, and minority individuals' mental health: a secondary analysis of US National Surveys. J Racial Ethn Health Disparities. 2018;5(3):485–94.
38. Jones A, Robinson E, Oginni O, Rahman Q, Rimes KA. Anxiety disorders, gender nonconformity, bullying and self-esteem in sexual minority adolescents: prospective birth cohort study. J Child Psychol Psychiatry. 2017;58(11):1201–9.
39. Mzarek P, Haggerty R. Reducing risks for mental disorders: frontiers for preventive intervention research. Washington: National Academy Press; 1994.
40. National Research Council and Institute of Medicine. Preventing mental, emotional, and behavioral disorders among young people: Progress and possibilities. Committee on the prevention of mental disorders and substance abuse among children, youth, and young adults: research advances and promising interventions. In: O'Connell ME, Boat T, Warner KE, editors. Board on children, youth, and F D of B and SS and E, editor. Washington, DC: The National Academies Press; 2009.
41. World Health Organization. Prevention of mental disorders: effective interventions and policy options. Geneva: World Health Organization; 2004.
42. Shonkoff J, Garner A. American academy of pediatrics. The lifelong effects of early childhood adversity and toxic stress. Pediatrics. 2012;129(1):e232–46.
43. Topper M, Emmelkamp PMG, Watkins E, Ehring T. Prevention of anxiety disorders and depression by targeting excessive worry and rumination in adolescents and young adults : a randomized controlled trial. Behav Res Ther. 2017;90:123–36.
44. Moreno-Peral P, Conejo-Ceron S, Rubio-Valera M, et al. Effectiveness of psychological and/or educational interventions in the prevention of anxiety: a systematic review, meta-analysis, and meta-regression. JAMA Psychiat. 2017;74(10):1021–9.
45. Rasing SPA, Creemers DHM, Janssens JMAM, Noel X. Depression and anxiety prevention based on cognitive behavioral therapy for at-risk adolescents: a meta-analytic review. Front Psychol. 2017;8(1066):1–17.

46. Fisak B, Richard D, Mann A. The prevention of child and adolescent anxiety: a meta-analytic review. Prev Sci. 2011;12(3):255–68.
47. Barrett P, Farrell L, Ollendick T, Dadds M. Long-term outcomes of an Australian universal prevention trial of anxiety and depression symptoms in children and youth: an evaluation of the friends program. J Clin Child Adolesc Psychol. 2006;35:403–11.
48. Zalta AK. A meta-analysis of anxiety symptom prevention with cognitive-behavioral interventions. J Anxiety Disord [Internet]. 2011;25(5):749–60. Available from: https://doi.org/10.1016/j.janxdis.2011.02.007
49. Deady M, Choi I, Calvo R, Glozier N, Christensen H, Harvey S. EHealth interventions for the prevention of depression and anxiety in the general population: a systematic review and meta-analysis. BMC Psychiatry. 2017;17:310.
50. Clarke A, Kuosmanen T, Barry M. A systematic review of online youth mental health promotion and prevention interventions. J Youth Adolesc. 2015;44:90–113.
51. Donovan CL, Spence SH. Prevention of childhood anxiety disorders. Clin Psychol Rev. 2000;20(4):509–31.
52. American Psychiatric Association. Diagnostic and statistical manual of mental disorders. Anxiety disorders. 5th ed. Washington DC: American Psychiatric Press; 2013. p. 189–290.

Anxiety Disorders in the Elderly

28

Carmen Andreescu and Soyoung Lee

Introduction

Anxiety disorders are the most prevalent psychiatric disorders in the United States and worldwide. In the United States, it has a lifetime prevalence of 31.1%, with 19.1% of adults being affected in the last year. In older age group, anxiety disorders remain highly prevalent causing a wide range of functional impairment and distress. Recent data has linked anxiety and its disorders in late life to increased morbidity and mortality, especially related to a higher cardiovascular burden and an increased cognitive decline. However, late-life anxiety disorders are underdiagnosed and undertreated in clinical practice, and research in this area has lagged behind.

In this chapter, we will review epidemiology, including prevalence, age of onset, and course of late-life anxiety disorders, comorbidities, and risk factors as well as the neurobiological basis of anxiety disorders in late life. We will emphasize new perspectives on late-life anxiety, including its role in risk for dementia and several cardiovascular diseases. Then, we will discuss assessment and treatment of late-life anxiety disorders.

Epidemiology

Prevalence

Prevalence reports of anxiety disorders in late life vary between 1.2% and 15% in community samples and between 1% and 28% in medical settings [1]. The prevalence of clinically relevant anxiety symptoms was between 15% and 52% in

C. Andreescu (✉) · S. Lee
Department of Psychiatry, Western Psychiatric Institute and Clinic University of Pittsburgh, Pittsburgh, USA
e-mail: andrcx@upmc.edu; Lees23@upmc.edu

Y.-K. Kim (ed.), *Anxiety Disorders*, Advances in Experimental Medicine and Biology 1191, https://doi.org/10.1007/978-981-32-9705-0_28

community samples and 15% and 56% in medical settings [1]. This wide variety stems most likely from methodological differences such as discrepancies in sampling procedures, differences in assessment tools, exclusion of individuals with more than one anxiety disorders, different age cutoffs [2], but also phenomenological differences in anxiety symptoms reported by older adults [3]. The prevalence of anxiety disorders also varies across different cultures or different cultural groups [4].

It is generally agreed that the prevalence of anxiety disorders appears to decrease in late life, except for GAD which maintains similar prevalence rates in young and old [5–7]. However, the prevalence of anxiety disorders in late life is considered underestimated due to the challenges of assessing and diagnosing anxiety in the elderly, which we will discuss in detail later in this chapter.

The high prevalence of late life anxiety is emerging as a significant public health burden since higher severity of anxiety symptoms in late life is associated with greater disability and poorer health-related quality of life after controlling for medical burden and depressive symptoms [8]. According to the 2013 State of Health report in the United States, anxiety was one of the four diseases with the largest number of years lived with disability (YLDs) in 2010 following low back pain, other musculoskeletal disorders, and major depressive disorder [9].

Age of Onset and Course

While the majority of anxiety disorders develop sometime between childhood and young adulthood and are carried over in midlife [2], generalized anxiety disorder (GAD) and agoraphobia have a relatively later onset [10]. Almost half of older patients with GAD report the onset of their disorder after age 50. Compared with older adults with late-onset GAD, older adults with an early-onset GAD appear to have a more severe course, characterized by more profound and impairing pathological worry [11, 12]. In contrast, late-onset GAD was distinguished from early-onset GAD by a more frequent association with medical comorbidities (e.g., hypertension), greater disability, and a poorer health-related quality of life [8, 12]. Several age-related stressors are specific risk factors for late onset of GAD, including chronic illness and disability, caregiver status, social isolation, institutionalization, and bereavement [13].

Major longitudinal studies of late-life anxiety disorders demonstrated the high risk of relapse, with recurrence and chronicity rate up to 39–52% after 3–6 years [14–16]. For late-life GAD, the main source of relapse and chronicity appears to be the persistence of pathologic worry even when other associated symptoms respond to treatment [17, 18]. Although severe worry represents the cardinal criterion of GAD, only 20% of older adults with severe worry qualify for a GAD diagnosis [19]. Comparison between tertiary center trials and community samples shows that the anxiety disorders are under-recognized and undertreated especially in the nonspecialist setting [20] and only 36.3% of the patients with late-life GAD receive treatment [21].

Psychiatric Comorbidity

Anxiety and depressive disorders are frequently comorbid. They share several symptoms, such as irritability, restlessness, decreased concentration, sleep changes, and fatigue. The comorbidity persists in the older age. 13–23% of older adults with anxiety disorders also met criteria for major depressive disorder [22, 23], and 26.1% met criteria for any depressive disorder [24]. Likewise, 47.5% of patients with a major depressive disorder also had a current anxiety disorder diagnosis [24], and 27.5% of depressed elderly would meet diagnostic criteria for GAD [25].

Late-life GAD comorbid with depressive symptoms is associated with greater risk for chronicity [15], lower social functioning, and higher severity of GAD symptoms [26]. Likewise, the presence of GAD symptoms among depressed older adults has been associated with greater suicidal risk, greater somatic symptoms, and increased risk of cognitive impairment [25, 27, 28].

Comorbidity of depression and anxiety negatively affects response to treatment [29]. Older adults with depression and concurrent anxiety symptoms required 50% more time to respond to antidepressants and were more likely to discontinue treatment [29, 30]. Anxiety symptoms often persist after remission of depression and increase risk for depressive relapse [29, 31, 32].

Late-Life Anxiety and Cognitive Decline

It is well-established in cross-sectional studies and observational research that anxiety in the elderly is frequently comorbid with changes in neuropsychological functioning. Older adults with clinically significant anxiety have poorer cognitive function [33], and those with late-life GAD show deficits in multiple cognitive domains including language, processing speed, immediate and delayed memory, and executive function [34]. Several studies have shown that anxiety symptoms or disorders in the elderly may increase the risk of cognitive decline [28, 35–37].

The association between dementia and anxiety symptoms rendered ambiguous results with some studies reporting no significant difference in the prevalence of anxiety symptoms in the elderly with dementia compared to non-demented subjects [38] while others reporting a positive association [39, 40]. The wide range of reports may result from the diagnostic difficulties in the patients with dementia as well as from the methodological differences between the studies. The symptoms that are seen in both anxiety disorders and dementia, such as restlessness, fatigue, and poor concentration, make it difficult to distinguish anxiety in the context of dementia [2]. Additionally, AD subjects may have difficulties relaying information about themselves, making researchers to rely on caregiver reports which are not well suited to capture internal symptoms, such as worry and rumination [41].

The causal relationship between anxiety and cognitive decline may be bidirectional; chronic anxiety may contribute to cognitive decline through various stress-related pathways that increase allostatic load [17, 42–46], but also worsening cognition may trigger increased anxiety and worry [47–49]. Thus, higher amyloid burden was recently associated with increasing anxiety and depressive symptoms over time in cognitively normal older individuals, suggesting that emerging symptoms (e.g., anxiety) may represent an early manifestation of preclinical AD [50].

More evidence is becoming available to support the association between anxiety and neurodegeneration. Several groundbreaking recent studies have reported on the association between anxiety/worry and amyloid burden, including a moderating effect of anxiety on the association between beta-amyloid burden and cognitive decline [50–53]. Higher worry severity in older adults has been positively associated with cortical atrophy – more precisely with decreased cortical thickness and decreased mean diffusivity (a marker of grey matter integrity) in the prefrontal cortex [54]. Overall, the recent studies support a link between chronic anxiety/severe worry and cognitive decline. However, the pathways through which anxiety/severe worry alters cognitive decline remain complex and may include chronic stress, increased cerebrovascular burden, chronic inflammation, and unopposed glutaminergic excitotoxicity in vulnerable brain regions (e.g., hippocampus and prefrontal cortex) [42, 55–57].

Medical Comorbidity

Anxiety disorders are a well-known public health burden as they are associated with greater disability and poorer health-related quality of life [8]. Subjects with anxiety disorders have overall increased mortality both from natural and unnatural causes [58]. Certain medical conditions have a higher association with anxiety symptoms, such as gastrointestinal problems, hypo- or hyperthyroidism, diabetes, cardiovascular disease, and respiratory disorders [59–62].

More recently, the link between cardiovascular disease and anxiety has been the focus of several studies. Anxiety is a risk factor for mortality after coronary bypass surgery [63], and panic disorders have been found to increase risk for cardiovascular morbidity and mortality [64], after controlling for other risk factors including major depression. According to a recent meta-analysis, anxiety was associated with a significantly elevated risk of cardiovascular mortality [relative risk 1.41], coronary heart disease [relative risk 1.41], stroke [relative risk 1.71], and heart failure [relative risk 1.35]) [56]. In a prospective longitudinal study, higher anxiety symptom levels were associated with increased risk for incident stroke independent of other risk factors, including depression [65]. Also, GAD was significantly associated with major cardiovascular events at 5 years in patients following coronary artery bypass surgery [63].

All these studies emphasize a highly relevant but relatively poorly recognized comorbidity between anxiety and cardiovascular disease risk. This association is even more salient for older anxious subjects. The recognized role and treatment priority received by depression associated with cardiovascular disease should be extended to anxiety disorders [66].

Substance Use and Late-Life Anxiety

Another category of disorders that are highly comorbid with anxiety is substance use disorder. According to the National Comorbidity Survey Replication in 2005, one in five substance use disorders also meets the criteria for an anxiety disorder. Substance use disorders in older adults are an emerging public health concern in the United States and in the world as population ages [67]. However, there is a scarcity of research in this area. There have been increases in treatment admissions for illicit and prescription drug abuse in older adults in the United States [68, 69]. Among prescription drug misuse, opioid and benzodiazepine misuse is a particular concern given risk of fatal overdoses [69, 70]. Older patients are more likely to receive opioid and benzodiazepine prescriptions, and they are more likely to have side effects such as increased risk of falls and cognitive impairment [71–74]. More recently, interest in the medicinal use of cannabis has been increased. Use of cannabis/cannabinoids is particularly relevant for older individuals due to the increased frequency of symptoms such as chronic pain, insomnia, and mood symptoms [75]. However, so far there is very little evidence of its efficacy in older adults, and the potential of side effects and drug-drug interactions should not be underestimated.

The comorbidity between anxiety and substance use disorders (SUD) complicates treatment and worsens prognosis of SUD. Although the mechanisms underlying the relationship between anxiety and substance use disorders are unclear, there are several hypotheses: self-medication hypothesis (substances used to self-treat anxiety), substance-induced anxiety (anxiety caused by other intoxication or withdrawal), and common factors theory (shared personality/neurobiological vulnerabilities between anxiety and substance use disorders) [76, 77].

Assessment

The standard epidemiological instruments in mental health are often not well suited for older adults. Because of diagnostic challenges in late-life anxiety, diagnostic criteria and screening tools have lower sensitivity and cannot efficiently detect and observe anxiety disorders or relevant anxiety symptoms in older adults [78]. Both older adults and clinicians may mistakenly view anxiety/fear/avoidance as a part of normal aging process. Older adults are less accurate in identifying anxiety symptoms compared to younger adults as they tend to minimize symptoms and to use different languages to describe anxiety to attribute symptoms to physical illness or they may simply have difficulties remembering or identifying symptoms [79]. Older people may experience anxiety differently in terms of autonomic responses (e.g., panic response) as central and peripheral physiology change with aging [80]. Functional impairment due to anxiety symptoms may require to be tailored to the functional limitations and overall abilities of older people. Lastly, some aging-specific anxiety conditions may not be detected by the standard anxiety screening questions. These atypical anxiety syndromes often encountered in late-life include fear of falling [81], hoarding syndrome [82], PTSD in the elderly, as well as less well-defined but frequent somatic symptoms such as dizziness and shakiness, anxiety/agitation in the context of dementia, COPD, heart disease [61], Parkinson's disease, irritable bowel syndrome, and vestibular symptoms [49].

While the majority of the most commonly used anxiety measures lack sufficient evidence to validate their use in older adults, Beck Anxiety Inventory, Penn State Worry Questionnaire, Generalized Anxiety Disorder Severity Scale, and Geriatric Mental Status Examination have been validated in assessing anxiety in older adults [78, 83]. Some anxiety measures that were specifically designed to assess late-life anxiety, such as the Geriatric Anxiety Scale and Geriatric Anxiety Inventory, are also reliable to use in older adults [84].

Treatment

The treatment response rates in late-life anxiety disorders are generally lower than in younger adults. Course of disease is more likely to be chronic, with frequent relapses and uncommon remission [17].The treatment of late-life anxiety disorders raises several specific challenges. Thus, older adults are more reluctant to seek help from mental health professionals [85], and they are more likely to drop out of treatment due to perceived stigma related to mental health [86]. Compared with younger adults, older adults also seem to have different treatment preference as a group. In one study, older adults were more reluctant about participating in group therapy but were willing to attend psychoeducational classes [87]. In another study, psychotherapy was selected as the preferred treatment by the majority of older adults who answered a survey about anxiety treatment [88]. Adequate treatment of late-life anxiety disorders is particularly important given the morbidity and mortality risks associated with untreated anxiety [85].

Pharmacotherapy

According to a meta-analysis of 32 studies, pharmacotherapy is more effective than psychotherapy for late-life anxiety [89]. Several pharmacological trials showed favorable results for active pharmacological treatments compared with control conditions [90]. Pharmacotherapy for older individuals requires a comprehensive assessment, as elderly patients have several pharmacokinetics and dynamics changes related to reduced glomerular filtration and hepatic metabolization, lower cardiac output, and decreased activity of target receptors [91]. These physiologic changes increase the risk of medication-related side effects, such as anticholinergic (urinary retention, constipation, delirium, cognitive difficulties), antiadrenergic (orthostatic hypotension), and antihistaminergic effects (drowsiness, dizziness, confusion) [92–94].

Benzodiazepines remain the most pharmacological option for anxiety in late life [95–97]. Some studies have found benzodiazepines efficacious in reducing anxiety symptoms in late life [98, 99]. However, the mass scale use of benzodiazepines in late-life anxiety remains problematic due to well-known risks such as falls, cognitive impairment [73, 74], and the risk for misuse [68–70].

SSRIs and SNRIs are considered the first-line treatment for late-life anxiety disorders. Three randomized controlled trials showed that SSRIs are efficacious in treating late-life anxiety disorders, predominantly GAD [100–102]). Venlafaxine XR and duloxetine have been found to be equally efficacious in older adults, with side effect profile similar to younger adults [103, 104]. Importantly, SSRIs and SNRIs are relatively well tolerated. However, in the elderly populations, there are several specific potential risks that need to be monitored such as the increasing risk for falls in the very old [105–107], gastrointestinal bleeding [108], bone loss [109], and hyponatremia [110]. For higher doses of SNRIs (especially venlafaxine), it is recommended a closer monitoring of blood pressure [111].

Few studies have focused on second-line pharmacological strategies. Pregabalin was found efficacious in older adults with GAD, although it is difficult to assess the clinical impact of pregabalin (i.e., pregabalin was associated with a two-point greater reduction on the Hamilton Anxiety Scale than placebo) [112]. A recent study found that buspirone was as effective and well tolerated as sertraline for the treatment of late-life GAD [113].

Off-label use of antipsychotic drugs in the treatment of late-life anxiety is often practiced in the community. The only antipsychotic with some evidence of efficacy is quetiapine in generalized anxiety disorder, with doses between 50 and 150 mg being as efficacious as 20 mg of paroxetine or 10 mg of escitalopram [114]. Quetiapine has demonstrated efficacy and tolerability both as monotherapy and as adjunctive to SSRIs [115]. In a large study (N = 450), quetiapine XR monotherapy (50–300 mg/day) was efficacious in treating GAD in the elderly (number needed to treat = 8) [116]. The study reported multiple side effects such as drowsiness, dry mouth, dizziness, headache, and nausea [116]. One small study of risperidone showed its efficacy as an augmentation strategy [117]. Importantly, clinicians need to balance the potential benefits of second-generation antipsychotics in the elderly with substantial metabolic side effects, including weight gain, hyperglycemia, and increased cholesterol [118–120] as well as the increased risk of sudden death and cardiovascular events [121].

Mirtazapine (Remeron) is a popular choice for the treatment of anxiety in late-life mainly because of its effects on sleep and appetite, but the evidence for its efficacy is limited and inconsistent [122].

Other antidepressant drugs, such as tricyclic antidepressants (TCA) and irreversible monoamine oxidase inhibitors (MAOI) may be efficacious, but they should be considered only for cases resistant to other treatment options due to their side effect profile and safety concerns.

Two studies in late-life panic disorder examined the efficacy of several SSRIs. Escitalopram was found superior over citalopram in time to response [123], and sertraline showed favorable results [124]. Few studies involving mixed-age populations have suggested that citalopram [125] and mirtazapine [126] and prazosin [127] are effective in late life PTSD.

Psychotherapy

Cognitive behavioral therapy (CBT) for late-life GAD has been studied the most extensively. The results remain mixed, with some trials and meta-analyses showing a moderate efficacy of CBT in late-life GAD [128, 129], while others have failed to prove the superiority of CBT over supportive therapy or waiting lists [130–132].

One study has indicated that for older patients, relaxation therapy may be the most efficacious module of CBT [131]. Although there is evidence that older adults with GAD can learn new cognitive skills and use them effectively [132], the cognitive reappraisal may be more challenging in a population with diminishing cognitive abilities [133–135]. Importantly, naturalistic follow-up studies of CBT in late-life GAD indicated that many patients maintained CBT gains for up to 1 year after treatment was completed [129]. Similarly, another study showed that compared with supportive therapy, CBT remained more beneficial at 1-year follow-up than immediately following treatment in patients who received at least 3 months of pharmacotherapy prior to receiving CBT. Their results indicate that CBT in older adults continues to bring long-term benefits [136].

Some modifications can be made to fit the needs of older subjects, such as between-session reminder phone calls, weekly review of concepts, in-home assignments, and inclusion of relatively easily instructed component and more simplified approach [137]. Also, more group-specific adjustments can be implemented to facilitate the impact of psychotherapeutic interventions. Incorporating religions and/or spirituality for older African American subjects was proven to be effective [137]. Telephone-delivered CBT for rural populations was superior to telephone-delivered nondirective supportive therapy in reducing worry, additional GAD symptoms, and depressive symptoms in older adults with GAD [138]. Older adults were more likely to complete Internet-delivered CBT than younger adults, although younger adults had more robust response [139].

The comparison between psychotherapy and another active control condition (e.g., discussion group) was not significant. There was no difference between different types of psychotherapy [90]. Also, although psychotherapy is not proven to be more effective than pharmacotherapy, it might be clinically relevant to question whether combining CBT and medication would be more effective. Augmenting SSRI treatment with CBT improved worry symptoms and reduced relapse rates in older adults with GAD [140].

Different innovative psychotherapeutic approaches are actively being examined recently. For example, modular CBT uses a personalized protocol that takes into account the specific symptoms of the patients and allows flexibility to tailor the treatment components to meet the specific need of each patient [141]. Mindfulness-based stress reduction (MBSR) is proven to reduce anxiety symptoms in older adults [142]. MBSR for anxious elderly patients with cognitive dysfunction has also shown promising preliminary results [18]. Acceptance and Commitment therapy has also showed improvement in worry severity in elderly [143]. These promising results suggest the need of further investigation and validation on larger samples.

Conclusion and Key Points

In conclusion, anxiety in late-life remains highly prevalent, causing a wide range of functional impairment and distress. Recent studies support a link between chronic anxiety/severe worry and cognitive decline as well as between anxiety and cardiovascular disease. These new data reflect a larger impact of anxiety and its disorders beyond the classic comorbidity with mood and substance use disorders. These links are singularly relevant for an increasingly older population burdened by cardiovascular disease and higher rates of Alzheimer's disease. Clinically, anxiety symptoms may be more difficult to elicit in older adults who are less accurate in identifying anxiety symptoms and tend to minimize symptoms and to attribute symptoms to physical illness. SSRIs remain the first-line treatment for anxiety in late life, but benzodiazepines continue to be the most commonly used pharmacological treatment for late-life anxiety. Although SSRIs are proven more effective than psychotherapy in late-life anxiety, many elderly anxious subjects prefer psychotherapeutic interventions. These interventions appear to work best when tailored for the needs, expectations, and cultural background of older anxious subjects.

References

1. Bryant C, Jackson H, Ames D. The prevalence of anxiety in older adults: methodological issues and a review of the literature. J Affect Disord. 2008;109(3):233–50.
2. Wolitzky-Taylor KB, Castriotta N, Lenze EJ, Stanley MA, Craske MG. Anxiety disorders in older adults: a comprehensive review. Depress Anxiety. 2010;27(2):190–211.
3. Andreescu C, Varon D. New research on anxiety disorders in the elderly and an update on evidence-based treatments. Curr Psychiatry Rep. 2015;17(7):595.
4. Lewis-Fernández R, Hinton DE, Laria AJ, Patterson EH, Hofmann SG, Craske MG, et al. Culture and the anxiety disorders: recommendations for DSM-V. Depress Anxiety. 2010;27(2):212–29.
5. Kessler RC, Berglund P, Demler O, Jin R, Merikangas KR, Walters EE. Lifetime prevalence and age-of-onset distributions of DSM-IV disorders in the national comorbidity survey replication. Arch Gen Psychiatry. 2005;62(6):593–602.
6. Flint AJ. Epidemiology and comorbidity of anxiety disorders in later life: implications for treatment. Clin Neurosci. 1997;4(1):31–6.
7. Beekman AT, Bremmer MA, Deeg DJ, van Balkom AJ, Smit JH, de Beurs E, et al. Anxiety disorders in later life: a report from the longitudinal aging study Amsterdam. Int J Geriatr Psychiatry. 1998;13(10):717–26.
8. Porensky EK, Dew MA, Karp JF, Skidmore E, Rollman BL, Shear MK, et al. The burden of late-life generalized anxiety disorder: effects on disability, health-related quality of life, and healthcare utilization. Am J Geriatr Psychiatry. 2009;17(6):473–82.
9. Murray CJ, Atkinson C, Bhalla K, Birbeck G, Burstein R, Chou D, et al. The state of US health, 1990–2010: burden of diseases, injuries, and risk factors. JAMA. 2013;310(6):591–608.
10. Goncalves DC, Byrne GJ. Sooner or later: age at onset of generalized anxiety disorder in older adults. Depress Anxiety. 2012;29(1):39–46.
11. Le Roux H, Gatz M, Wetherell JL. Age at onset of generalized anxiety disorder in older adults. Am J Geriatr Psychiatry. 2005;13(1):23–30.
12. Chou KL. Age at onset of generalized anxiety disorder in older adults. Am J Geriatr Psychiatry. 2009;17(6):455–64.

13. Baldwin DS, Anderson IM, Nutt DJ, Bandelow B, Bond A, Davidson JR, et al. Evidence-based guidelines for the pharmacological treatment of anxiety disorders: recommendations from the British Association for Psychopharmacology. J Psychopharmacol. 2005;19(6):567–96.
14. Schoevers RA, Deeg DJ, van Tilburg W, Beekman AT. Depression and generalized anxiety disorder: co-occurrence and longitudinal patterns in elderly patients. Am J Geriatr Psychiatry. 2005;13(1):31–9.
15. Schoevers RA, Beekman AT, Deeg DJ, Jonker C, van Tilburg W. Comorbidity and risk-patterns of depression, generalised anxiety disorder and mixed anxiety-depression in later life: results from the AMSTEL study. Int J Geriatr Psychiatry. 2003;18(11):994–1001.
16. Schuurmans J, Comijs HC, Beekman AT, de Beurs E, Deeg DJ, Emmelkamp PM, et al. The outcome of anxiety disorders in older people at 6-year follow-up: results from the longitudinal aging study Amsterdam. Acta Psychiatr Scand. 2005;111(6):420–8.
17. Lenze EJ, Wetherell JL. A lifespan view of anxiety disorders. Dialogues Clin Neurosci. 2011;13(4):381–99.
18. Lenze EJ, Hickman S, Hershey T, Wendleton L, Ly K, Dixon D, et al. Mindfulness-based stress reduction for older adults with worry symptoms and co-occurring cognitive dysfunction. Int J Geriatr Psychiatry. 2014;29(10):991–1000.
19. Golden J, Conroy RM, Bruce I, Denihan A, Greene E, Kirby M, et al. The spectrum of worry in the community-dwelling elderly. Aging Ment Health. 2011;15(8):985–94.
20. Sami MB, Nilforooshan R. The natural course of anxiety disorders in the elderly: a systematic review of longitudinal trials. Int Psychogeriatr. 2015;27(7):1061–9.
21. Zhang X, Norton J, Carriere I, Ritchie K, Chaudieu I, Ancelin ML. Generalized anxiety in community-dwelling elderly: prevalence and clinical characteristics. J Affect Disord. 2015;172:24–9.
22. van Balkom AJ, Beekman AT, de Beurs E, Deeg DJ, van Dyck R, van Tilburg W. Comorbidity of the anxiety disorders in a community-based older population in the Netherlands. Acta Psychiatr Scand. 2000;101(1):37–45.
23. Cairney J, Corna LM, Veldhuizen S, Herrmann N, Streiner DL. Comorbid depression and anxiety in later life: patterns of association, subjective Well-being, and impairment. Am J Geriatr Psychiatry. 2008;16(3):201–8.
24. Beekman AT, de Beurs E, van Balkom AJ, Deeg DJ, van Dyck R, van Tilburg W. Anxiety and depression in later life: co-occurrence and communality of risk factors. Am J Psychiatry. 2000;157(1):89–95.
25. Lenze EJ, Mulsant BH, Shear MK, Schulberg HC, Dew MA, Begley AE, et al. Comorbid anxiety disorders in depressed elderly patients. Am J Psychiatry. 2000;157(5):722–8.
26. Hopko DR, Bourland SL, Stanley MA, Beck JG, Novy DM, Averill PM, et al. Generalized anxiety disorder in older adults: examining the relation between clinician severity ratings and patient self-report measures. Depress Anxiety. 2000;12(4):217–25.
27. Jeste ND, Hays JC, Steffens DC. Clinical correlates of anxious depression among elderly patients with depression. J Affect Disord. 2006;90(1):37–41.
28. DeLuca AK, Lenze EJ, Mulsant BH, Butters MA, Karp JF, Dew MA, et al. Comorbid anxiety disorder in late life depression: association with memory decline over four years. Int J Geriatr Psychiatry. 2005;20(9):848–54.
29. Flint AJ, Rifat SL. Anxious depression in elderly patients. Response to antidepressant treatment. Am J Geriatr Psychiatry. 1997;5(2):107–15.
30. BH M. In: BA W, editor. Psychological treatment of older adults: an introductory text. New York: Plenum Press; 1996. p. 91–101.
31. Andreescu C, Lenze EJ, Dew MA, Begley AE, Mulsant BH, Dombrovski AY, et al. Effect of comorbid anxiety on treatment response and relapse risk in late-life depression: controlled study. Br J Psychiatry. 2007;190:344–9.
32. Dombrovski AY, Lenze EJ, Dew MA, Mulsant BH, Pollock BG, Houck PR, et al. Maintenance treatment for old-age depression preserves health-related quality of life: a randomized, controlled trial of paroxetine and interpersonal psychotherapy. J Am Geriatr Soc. 2007;55(9):1325–32.

33. Schultz SK, Moser DJ, Bishop JR, Ellingrod VL. Phobic anxiety in late-life in relationship to cognition and 5HTTLPR polymorphism. Psychiatr Genet. 2005;15. England:305–6.
34. Butters MA, Bhalla RK, Andreescu C, Wetherell JL, Mantella R, Begley AE, et al. Changes in neuropsychological functioning following treatment for late-life generalised anxiety disorder. Br J Psychiatry. 2011;199(3):211–8.
35. Sinoff G, Werner P. Anxiety disorder and accompanying subjective memory loss in the elderly as a predictor of future cognitive decline. Int J Geriatr Psychiatry. 2003;18(10):951–9.
36. Andreescu C, Teverovsky E, Fu B, Hughes TF, Chang CC, Ganguli M. Old worries and new anxieties: behavioral symptoms and mild cognitive impairment in a population study. Am J Geriatr Psychiatry. 2014;22(3):274–84.
37. Palmer K, Berger AK, Monastero R, Winblad B, Backman L, Fratiglioni L. Predictors of progression from mild cognitive impairment to Alzheimer disease. Neurology. 2007;68(19):1596–602.
38. Forsell Y, Winblad B. Anxiety disorders in non-demented and demented elderly patients: prevalence and correlates. J Neurol Neurosurg Psychiatry. 1997;62(3):294–5.
39. Hwang TJ, Masterman DL, Ortiz F, Fairbanks LA, Cummings JL. Mild cognitive impairment is associated with characteristic neuropsychiatric symptoms. Alzheimer Dis Assoc Disord. 2004;18(1):17–21.
40. Geda YE, Smith GE, Knopman DS, Boeve BF, Tangalos EG, Ivnik RJ, et al. De novo genesis of neuropsychiatric symptoms in mild cognitive impairment (MCI). Int Psychogeriatr. 2004;16(1):51–60.
41. Seignourel PJ, Kunik ME, Snow L, Wilson N, Stanley M. Anxiety in dementia: a critical review. Clin Psychol Rev. 2008;28(7):1071–82.
42. Schulkin J, McEwen BS, Gold PW. Allostasis, amygdala, and anticipatory angst. Neurosci Biobehav Rev. 1994;18(3):385–96.
43. Mantella RC, Butters MA, Amico JA, Mazumdar S, Rollman BL, Begley AE, et al. Salivary cortisol is associated with diagnosis and severity of late-life generalized anxiety disorder. Psychoneuroendocrinology. 2008;33(6):773–81.
44. Narita K, Murata T, Hamada T, Kosaka H, Sudo S, Mizukami K, et al. Associations between trait anxiety, insulin resistance, and atherosclerosis in the elderly: a pilot cross-sectional study. Psychoneuroendocrinology. 2008;33(3):305–12.
45. Simon NM, Smoller JW, McNamara KL, Maser RS, Zalta AK, Pollack MH, et al. Telomere shortening and mood disorders: preliminary support for a chronic stress model of accelerated aging. Biol Psychiatry. 2006;60(5):432–5.
46. Sibille E, Su J, Leman S, Le Guisquet AM, Ibarguen-Vargas Y, Joeyen-Waldorf J, et al. Lack of serotonin1B receptor expression leads to age-related motor dysfunction, early onset of brain molecular aging and reduced longevity. Mol Psychiatry. 2007;12(11):1042–56.. 975
47. Ballard C, Neill D, O'Brien J, McKeith IG, Ince P, Perry R. Anxiety, depression and psychosis in vascular dementia: prevalence and associations. J Affect Disord. 2000;59(2):97–106.
48. Lyketsos CG, Lopez O, Jones B, Fitzpatrick AL, Breitner J, DeKosky S. Prevalence of neuropsychiatric symptoms in dementia and mild cognitive impairment: results from the cardiovascular health study. JAMA. 2002;288(12):1475–83.
49. Starkstein SE, Jorge R, Petracca G, Robinson RG. The construct of generalized anxiety disorder in Alzheimer disease. Am J Geriatr Psychiatry. 2007;15(1):42–9.
50. Donovan NJ, Locascio JJ, Marshall GA, Gatchel J, Hanseeuw BJ, Rentz DM, et al. Longitudinal Association of Amyloid Beta and Anxious-Depressive Symptoms in Cognitively Normal Older Adults. Am J Psychiatry. 2018: appiajp201717040442.
51. Pietrzak RH, Lim YY, Neumeister A, Ames D, Ellis KA, Harrington K, et al. Amyloid-beta, anxiety, and cognitive decline in preclinical Alzheimer disease: a multicenter, prospective cohort study. JAMA Psychiat. 2015;72(3):284–91.
52. Pietrzak RH, Scott JC, Neumeister A, Lim YY, Ames D, Ellis KA, et al. Anxiety symptoms, cerebral amyloid burden and memory decline in healthy older adults without dementia: 3-year prospective cohort study. Br J Psychiatry. 2014;204:400–1.

53. Krell-Roesch J, Lowe VJ, Neureiter J, Pink A, Roberts RO, Mielke MM, et al. Depressive and anxiety symptoms and cortical amyloid deposition among cognitively normal elderly persons: the Mayo Clinic study of aging. Int Psychogeriatr. 2018;30(2):245–51.
54. Andreescu C, Tudorascu D, Sheu LK, Rangarajan A, Butters MA, Walker S, et al. Brain structural changes in late-life generalized anxiety disorder. Psychiatry Res. 2017;268:15–21.
55. Bisht K, Sharma K, Tremblay ME. Chronic stress as a risk factor for Alzheimer's disease: roles of microglia-mediated synaptic remodeling, inflammation, and oxidative stress. Neurobiol Stress. 2018;9:9–21.
56. Emdin CA, Odutayo A, Wong CX, Tran J, Hsiao AJ, Hunn BH. Meta-analysis of anxiety as a risk factor for cardiovascular disease. Am J Cardiol. 2016;118(4):511–9.
57. Delvecchio G, Stanley JA, Altamura AC, Brambilla P. Metabolic alterations in generalised anxiety disorder: a review of proton magnetic resonance spectroscopic studies. Epidemiol Psychiatr Sci. 2017;26(6):587–95.
58. Meier SM, Mattheisen M, Mors O, Mortensen PB, Laursen TM, Penninx BW. Increased mortality among people with anxiety disorders: total population study. Br J Psychiatry. 2016;209(3):216–21.
59. Byrne G, Rosenfeld G, Leung Y, Qian H, Raudzus J, Nunez C, et al. Prevalence of anxiety and depression in patients with inflammatory bowel disease. Can J Gastroenterol Hepatol. 2017;2017:6496727.
60. Lustman PJ, Griffith LS, Clouse RE, Cryer PE. Psychiatric illness in diabetes mellitus. Relationship to symptoms and glucose control. J Nerv Ment Dis. 1986;174(12):736–42.
61. Todaro JF, Shen BJ, Raffa SD, Tilkemeier PL, Niaura R. Prevalence of anxiety disorders in men and women with established coronary heart disease. J Cardiopulm Rehabil Prev. 2007;27(2):86–91.
62. Ittermann T, Volzke H, Baumeister SE, Appel K, Grabe HJ. Diagnosed thyroid disorders are associated with depression and anxiety. Soc Psychiatry Psychiatr Epidemiol. 2015;50(9):1417–25.
63. Tully PJ, Baker RA, Knight JL. Anxiety and depression as risk factors for mortality after coronary artery bypass surgery. J Psychosom Res. 2008;64(3):285–90.
64. Machado S, Sancassiani F, Paes F, Rocha N, Murillo-Rodriguez E, Nardi AE. Panic disorder and cardiovascular diseases: an overview. Int Rev Psychiatry. 2017;29(5):436–44.
65. Lambiase MJ, Kubzansky LD, Thurston RC. Prospective study of anxiety and incident stroke. Stroke. 2014;45(2):438–43.
66. Tully PJ, Harrison NJ, Cheung P, Cosh S. Anxiety and cardiovascular disease risk: a review. Curr Cardiol Rep. 2016;18(12):120.
67. Simoni-Wastila L, Yang HK. Psychoactive drug abuse in older adults. Am J Geriatr Pharmacother. 2006;4(4):380–94.
68. Wu LT, Blazer DG. Substance use disorders and psychiatric comorbidity in mid and later life: a review. Int J Epidemiol. 2014;43(2):304–17.
69. Wu LT, Blazer DG. Illicit and nonmedical drug use among older adults: a review. J Aging Health. 2011;23(3):481–504.
70. Schulden JD, Lopez MF, Compton WM. Clinical implications of drug abuse epidemiology. Psychiatr Clin North Am. 2012;35(2):411–23.
71. Maree RD, Marcum ZA, Saghafi E, Weiner DK, Karp JF. A systematic review of opioid and benzodiazepine misuse in older adults. Am J Geriatr Psychiatry. 2016;24(11):949–63.
72. Airagnes G, Pelissolo A, Lavallee M, Flament M, Limosin F. Benzodiazepine misuse in the elderly: risk factors, consequences, and management. Curr Psychiatry Rep. 2016;18(10):89.
73. Markota M, Rummans TA, Bostwick JM, Lapid MI. Benzodiazepine use in older adults: dangers, management, and alternative therapies. Mayo Clin Proc. 2016;91(11):1632–9.
74. Bogunovic OJ, Greenfield SF. Practical geriatrics: use of benzodiazepines among elderly patients. Psychiatr Serv. 2004;55(3):233–5.
75. Minerbi A, Hauser W, Fitzcharles MA. Medical Cannabis for older patients. Drugs Aging. 2019;36(1):39–51.

76. Smith JP, Randall CL. Anxiety and alcohol use disorders: comorbidity and treatment considerations. Alcohol Res. 2012;34(4):414–31.
77. Pasche S. Exploring the comorbidity of anxiety and substance use disorders. Curr Psychiatry Rep. 2012;14(3):176–81.
78. Therrien Z, Hunsley J. Assessment of anxiety in older adults: a systematic review of commonly used measures. Aging Ment Health. 2012;16(1):1–16.
79. Wetherell JL, Petkus AJ, McChesney K, Stein MB, Judd PH, Rockwell E, et al. Older adults are less accurate than younger adults at identifying symptoms of anxiety and depression. J Nerv Ment Dis. 2009;197(8):623–6.
80. Flint A, Bradwejn J, Vaccarino F, Gutkowska J, Palmour R, Koszycki D. Aging and panicogenic response to cholecystokinin tetrapeptide: an examination of the cholecystokinin system. Neuropsychopharmacology. 2002;27(4):663–71.
81. Gagnon N, Flint AJ, Naglie G, Devins GM. Affective correlates of fear of falling in elderly persons. Am J Geriatr Psychiatry. 2005;13(1):7–14.
82. Roane DM, Landers A, Sherratt J, Wilson GS. Hoarding in the elderly: a critical review of the recent literature. Int Psychogeriatr. 2017;29(7):1077–84.
83. Andreescu C, Belnap BH, Rollman BL, Houck P, Ciliberti C, Mazumdar S, et al. Generalized anxiety disorder severity scale validation in older adults. Am J Geriatr Psychiatry. 2008;16(10):813–8.
84. Balsamo M, Cataldi F, Carlucci L, Fairfield B. Assessment of anxiety in older adults: a review of self-report measures. Clin Interv Aging. 2018;13:573–93.
85. de Beurs E, Beekman AT, van Balkom AJ, Deeg DJ, van Dyck R, van Tilburg W. Consequences of anxiety in older persons: its effect on disability, Well-being and use of health services. Psychol Med. 1999;29(3):583–93.
86. Sirey JA, Bruce ML, Alexopoulos GS, Perlick DA, Friedman SJ, Meyers BS. Stigma as a barrier to recovery: perceived stigma and patient-rated severity of illness as predictors of antidepressant drug adherence. Psychiatr Serv. 2001;52(12):1615–20.
87. Arean PA, Alvidrez J, Barrera A, Robinson GS, Hicks S. Would older medical patients use psychological services? Gerontologist. 2002;42(3):392–8.
88. Mohlman J. A community based survey of older adults' preferences for treatment of anxiety. Psychol Aging. 2012;27(4):1182–90.
89. Pinquart M, Duberstein PR. Treatment of anxiety disorders in older adults: a meta-analytic comparison of behavioral and pharmacological interventions. Am J Geriatr Psychiatry. 2007;15(8):639–51.
90. Goncalves DC, Byrne GJ. Interventions for generalized anxiety disorder in older adults: systematic review and meta-analysis. J Anxiety Disord. 2012;26(1):1–11.
91. Pollock B, Forsyth C, Bies R. The critical role of clinical pharmacology in geriatric psychopharmacology. Clin Pharmacol Ther. 2009;85(1):89–93.
92. B.H. M. In: Thakur MB, Blazer DG, Steffens, DC, editor. Clinical manual of geriatric psychiatry. Washington DC: American Psychiatric Publishing, Inc; 2014.
93. Zubenko GS, Sunderland T. Geriatric psychopharmacology: why does age matter? Harv Rev Psychiatry. 2000;7(6):311–33.
94. Brooks JO, Hoblyn JC. Neurocognitive costs and benefits of psychotropic medications in older adults. J Geriatr Psychiatry Neurol. 2007;20(4):199–214.
95. Benitez CI, Smith K, Vasile RG, Rende R, Edelen MO, Keller MB. Use of benzodiazepines and selective serotonin reuptake inhibitors in middle-aged and older adults with anxiety disorders: a longitudinal and prospective study. Am J Geriatr Psychiatry. 2008;16(1):5–13.
96. Olfson M, Marcus SC, Wan GJ, Geissler EC. National trends in the outpatient treatment of anxiety disorders. J Clin Psychiatry. 2004;65(9):1166–73.
97. Olfson M, King M, Schoenbaum M. Benzodiazepine use in the United States. JAMA Psychiat. 2015;72(2):136–42.
98. Bresolin N, Monza G, Scarpini E, Scarlato G, Straneo G, Martinazzoli A, et al. Treatment of anxiety with ketazolam in elderly patients. Clin Ther. 1988;10(5):536–42.

99. Koepke HH, Gold RL, Linden ME, Lion JR, Rickels K. Multicenter controlled study of oxazepam in anxious elderly outpatients. Psychosomatics. 1982;23(6):641–5.
100. Lenze EJ, Mulsant BH, Shear MK, Dew MA, Miller MD, Pollock BG, et al. Efficacy and tolerability of citalopram in the treatment of late-life anxiety disorders: results from an 8-week randomized, placebo-controlled trial. Am J Psychiatry. 2005;162(1):146–50.
101. Lenze EJ, Rollman BL, Shear MK, Dew MA, Pollock BG, Ciliberti C, et al. Escitalopram for older adults with generalized anxiety disorder: a randomized controlled trial. JAMA. 2009;301(3):295–303.
102. Schuurmans J, Comijs H, Emmelkamp PM, Gundy CM, Weijnen I, van den Hout M, et al. A randomized, controlled trial of the effectiveness of cognitive-behavioral therapy and sertraline versus a waitlist control group for anxiety disorders in older adults. Am J Geriatr Psychiatry. 2006;14(3):255–63.
103. Katz IR, Reynolds CF 3rd, Alexopoulos GS, Hackett D, Venlafaxine ER. As a treatment for generalized anxiety disorder in older adults: pooled analysis of five randomized placebo-controlled clinical trials. J Am Geriatr Soc. 2002;50(1):18–25.
104. Davidson J, Allgulander C, Pollack MH, Hartford J, Erickson JS, Russell JM, et al. Efficacy and tolerability of duloxetine in elderly patients with generalized anxiety disorder: a pooled analysis of four randomized, double-blind, placebo-controlled studies. Hum Psychopharmacol. 2008;23(6):519–26.
105. Laghrissi-Thode F, Pollock BG, Miller MC, Mulsant BH, Altieri L, Finkel MS. Double-blind comparison of paroxetine and nortriptyline on the postural stability of late-life depressed patients. Psychopharmacol Bull. 1995;31(4):659–63.
106. Gebara MA, Lipsey KL, Karp JF, Nash MC, Iaboni A, Lenze EJ. Cause or effect? Selective serotonin reuptake inhibitors and falls in older adults: a systematic review. Am J Geriatr Psychiatry. 2015;23(10):1016–28.
107. Kerse N, Flicker L, Pfaff JJ, Draper B, Lautenschlager NT, Sim M, et al. Falls, depression and antidepressants in later life: a large primary care appraisal. PLoS One. 2008;3(6):e2423.
108. Yuan Y, Tsoi K, Hunt RH. Selective serotonin reuptake inhibitors and risk of upper GI bleeding: confusion or confounding? Am J Med. 2006;119(9):719–27.
109. Diem SJ, Blackwell TL, Stone KL, Yaffe K, Haney EM, Bliziotes MM, et al. Use of antidepressants and rates of hip bone loss in older women: the study of osteoporotic fractures. Arch Intern Med. 2007;167(12):1240–5.
110. Fabian TJ, Amico JA, Kroboth PD, Mulsant BH, Corey SE, Begley AE, et al. Paroxetine-induced hyponatremia in older adults: a 12-week prospective study. Arch Intern Med. 2004;164(3):327–32.
111. Baldwin DS, Waldman S, Allgulander C. Evidence-based pharmacological treatment of generalized anxiety disorder. Int J Neuropsychopharmacol. 2011;14(5):697–710.
112. Montgomery S, Chatamra K, Pauer L, Whalen E, Baldinetti F. Efficacy and safety of pregabalin in elderly people with generalised anxiety disorder. Br J Psychiatry. 2008;193(5):389–94.
113. Mokhber N, Azarpazhooh MR, Khajehdaluee M, Velayati A, Hopwood M. Randomized, single-blind, trial of sertraline and buspirone for treatment of elderly patients with generalized anxiety disorder. Psychiatry Clin Neurosci. 2010;64(2):128–33.
114. Merideth C, Cutler AJ, She F, Eriksson H. Efficacy and tolerability of extended release quetiapine fumarate monotherapy in the acute treatment of generalized anxiety disorder: a randomized, placebo controlled and active-controlled study. Int Clin Psychopharmacol. 2012;27(1):40–54.
115. Kreys TJ, Phan SV. A literature review of quetiapine for generalized anxiety disorder. Pharmacotherapy. 2015;35(2):175–88.
116. Mezhebovsky I, Magi K, She F, Datto C, Eriksson H. Double-blind, randomized study of extended release quetiapine fumarate (quetiapine XR) monotherapy in older patients with generalized anxiety disorder. Int J Geriatr Psychiatry. 2013;28(6):615–25.
117. Morinigo A, Blanco M, Labrador J, Martin J. Noval D. Risperidone for resistant anxiety in elderly persons. Am J Geriatr Psychiatry. 2005;13. England:81–2.

118. Finkel S, Kozma C, Long S, Greenspan A, Mahmoud R, Baser O, et al. Risperidone treatment in elderly patients with dementia: relative risk of cerebrovascular events versus other antipsychotics. Int Psychogeriatr. 2005;17(4):617–29.
119. Bullock R, Saharan A. Atypical antipsychotics: experience and use in the elderly. Int J Clin Pract. 2002;56(7):515–25.
120. Gareri P, De Fazio P, Manfredi VG, De Sarro G. Use and safety of antipsychotics in behavioral disorders in elderly people with dementia. J Clin Psychopharmacol. 2014;34(1):109–23.
121. Ray WA, Chung CP, Murray KT, Hall K, Stein CM. Atypical antipsychotic drugs and the risk of sudden cardiac death. N Engl J Med. 2009;360(3):225–35.
122. Schutters SI, Van Megen HJ, Van Veen JF, Denys DA, Westenberg HG. Mirtazapine in generalized social anxiety disorder: a randomized, double-blind, placebo-controlled study. Int Clin Psychopharmacol. 2010;25(5):302–4.
123. Rampello L, Alvano A, Raffaele R, Malaguarnera M, Vecchio I. New possibilities of treatment for panic attacks in elderly patients: escitalopram versus citalopram. J Clin Psychopharmacol. 2006;26(1):67–70.
124. Sheikh JI, Lauderdale SA, Cassidy EL. Efficacy of sertraline for panic disorder in older adults: a preliminary open-label trial. Am J Geriatr Psychiatry. 2004;12. England:230.
125. English BA, Jewell M, Jewell G, Ambrose S, Davis LL. Treatment of chronic posttraumatic stress disorder in combat veterans with citalopram: an open trial. J Clin Psychopharmacol. 2006;26(1):84–8.
126. Chung MY, Min KH, Jun YJ, Kim SS, Kim WC, Jun EM. Efficacy and tolerability of mirtazapine and sertraline in Korean veterans with posttraumatic stress disorder: a randomized open label trial. Hum Psychopharmacol. 2004;19(7):489–94.
127. Raskind MA, Peskind ER, Hoff DJ, Hart KL, Holmes HA, Warren D, et al. A parallel group placebo controlled study of prazosin for trauma nightmares and sleep disturbance in combat veterans with post-traumatic stress disorder. Biol Psychiatry. 2007;61(8):928–34.
128. Hofmann SG, Smits JA. Cognitive-behavioral therapy for adult anxiety disorders: a meta-analysis of randomized placebo-controlled trials. J Clin Psychiatry. 2008;69(4):621–32.
129. Stanley MA, Beck JG, Novy DM, Averill PM, Swann AC, Diefenbach GJ, et al. Cognitive-behavioral treatment of late-life generalized anxiety disorder. J Consult Clin Psychol. 2003;71(2):309–19.
130. Wetherell JL, Gatz M, Craske MG. Treatment of generalized anxiety disorder in older adults. J Consult Clin Psychol. 2003;71(1):31–40.
131. Thorp SR, Ayers CR, Nuevo R, Stoddard JA, Sorrell JT, Wetherell JL. Meta-analysis comparing different behavioral treatments for late-life anxiety. Am J Geriatr Psychiatry. 2009;17(2):105–15.
132. Wetherell JLHD, Diefenbach GJ, et al. Cognitive-behavioral therapy for late-life generalized anxiety disorder: who gets better? Behav Ther. 2005;36:147–56.
133. Mohlman J, Gorman JM. The role of executive functioning in CBT: a pilot study with anxious older adults. Behav Res Ther. 2005;43(4):447–65.
134. Mohlman J. Executive skills in older adults with GAD: relations with clinical variables and CBT outcome. J Anxiety Disord. 2013;27(1):131–9.
135. Mohlman J, Price RB, Vietri J. Attentional bias in older adults: effects of generalized anxiety disorder and cognitive behavior therapy. J Anxiety Disord. 2013;27(6):585–91.
136. Barrowclough C, King P, Colville J, Russell E, Burns A, Tarrier N. A randomized trial of the effectiveness of cognitive-behavioral therapy and supportive counseling for anxiety symptoms in older adults. J Consult Clin Psychol. 2001;69(5):756–62.
137. Stanley MA, Wilson N, Shrestha S, Amspoker AB, Armento M, Cummings JP, et al. Calmer life: a culturally tailored intervention for anxiety in underserved older adults. Am J Geriatr Psychiatry. 2016;24(8):648–58.
138. Brenes GA, Danhauer SC, Lyles MF, Anderson A, Miller ME. Effects of telephone-delivered cognitive-behavioral therapy and nondirective supportive therapy on sleep, health-related quality of life, and disability. Am J Geriatr Psychiatry. 2016;24(10):846–54.

139. Hobbs MJ, Mahoney AEJ, Andrews G. Integrating iCBT for generalized anxiety disorder into routine clinical care: treatment effects across the adult lifespan. J Anxiety Disord. 2017;51:47–54.
140. Wetherell JL, Stoddard JA, White KS, Kornblith S, Nguyen H, Andreescu C, et al. Augmenting antidepressant medication with modular CBT for geriatric generalized anxiety disorder: a pilot study. Int J Geriatr Psychiatry. 2011;26(8):869–75.
141. Wetherell JL, Ayers CR, Sorrell JT, Thorp SR, Nuevo R, Belding W, et al. Modular psychotherapy for anxiety in older primary care patients. Am J Geriatr Psychiatry. 2009;17(6):483–92.
142. Hazlett-Stevens H, Singer J, Chong A. Mindfulness-based stress reduction and mindfulness-based cognitive therapy with older adults: a qualitative review of randomized controlled outcome research. Clin Gerontol. 2018;
143. Wetherell JL, Afari N, Ayers CR, Stoddard JA, Ruberg J, Sorrell JT, et al. Acceptance and commitment therapy for generalized anxiety disorder in older adults: a preliminary report. Behav Ther. 2011;42(1):127–34.

MIX
Papier aus verantwortungsvollen Quellen
Paper from responsible sources
FSC® C105338

If you have any concerns about our products,
you can contact us on
ProductSafety@springernature.com

In case Publisher is established outside the EU,
the EU authorized representative is:
Springer Nature Customer Service Center GmbH
Europaplatz 3, 69115 Heidelberg, Germany

Printed by Libri Plureos GmbH
in Hamburg, Germany